T0269105

CAMBRIDGE LIBRARY COLLECTION

Books of enduring scholarly value

Zoology

Until the nineteenth century, the investigation of natural phenomena, plants and animals was considered either the preserve of elite scholars or a pastime for the leisured upper classes. As increasing academic rigour and systematisation was brought to the study of 'natural history', its sub-disciplines were adopted into university curricula, and learned societies (such as the London Zoological Society, founded in 1826) were established to support research in these areas. These developments are reflected in the books reissued in this series, which describe the anatomy and characteristics of animals ranging from invertebrates to polar bears, fish to birds, in habitats from Arctic North America to the tropical forests of Malaysia. By the middle of the nineteenth century, this work and developments in research on fossils had resulted in the formulation of the theory of evolution.

Manual of British Vertebrate Animals

Widely respected by contemporaries for his work in natural history, Leonard Jenyns (1800–93) combined research with his duties as an Anglican clergyman. He published and lectured extensively on zoology and botany. Having recommended Darwin for the *Beagle* voyage, he later produced a paper, 'On the Variation of Species', which Darwin personally requested to see. This 1835 work catalogues five classes of vertebrates: mammals, birds, reptiles, amphibians and fish. Native, introduced and extinct species of the British Isles are included, with binomial and common names given, along with the dimensions and a description. An improvement on previous works which had overly relied on secondary sources, Jenyns's manual also provides information on location, diet and propagation. The catalogue testifies to the diligent work being carried out in natural history in the era prior to Darwin's revolution. Jenyns's *Observations in Natural History* (1846) and *Observations in Meteorology* (1858) are also reissued in this series.

Cambridge University Press has long been a pioneer in the reissuing of out-of-print titles from its own backlist, producing digital reprints of books that are still sought after by scholars and students but could not be reprinted economically using traditional technology. The Cambridge Library Collection extends this activity to a wider range of books which are still of importance to researchers and professionals, either for the source material they contain, or as landmarks in the history of their academic discipline.

Drawing from the world-renowned collections in the Cambridge University Library and other partner libraries, and guided by the advice of experts in each subject area, Cambridge University Press is using state-of-the-art scanning machines in its own Printing House to capture the content of each book selected for inclusion. The files are processed to give a consistently clear, crisp image, and the books finished to the high quality standard for which the Press is recognised around the world. The latest print-on-demand technology ensures that the books will remain available indefinitely, and that orders for single or multiple copies can quickly be supplied.

The Cambridge Library Collection brings back to life books of enduring scholarly value (including out-of-copyright works originally issued by other publishers) across a wide range of disciplines in the humanities and social sciences and in science and technology.

Manual of British Vertebrate Animals

Or, Descriptions of All the Animals Belonging to the Classes Mammalia, Aves, Reptilia, Amphibia, and Pisces Which Have Been Hitherto Observed in the British Islands

LEONARD JENYNS

CAMBRIDGE
UNIVERSITY PRESS

CAMBRIDGE
UNIVERSITY PRESS

University Printing House, Cambridge, CB2 8BS, United Kingdom

Cambridge University Press is part of the University of Cambridge.

It furthers the University's mission by disseminating knowledge in the pursuit of education, learning and research at the highest international levels of excellence.

www.cambridge.org
Information on this title: www.cambridge.org/9781108070034

This edition first published 1835
This digitally printed version 2014

ISBN 978-1-108-07003-4 Paperback

MANUAL

OF

BRITISH VERTEBRATE ANIMALS:

OR

DESCRIPTIONS

OF

ALL THE ANIMALS BELONGING TO THE CLASSES,

MAMMALIA, AVES, REPTILIA, AMPHIBIA, AND PISCES,

WHICH HAVE BEEN

HITHERTO OBSERVED IN THE BRITISH ISLANDS:

INCLUDING THE

DOMESTICATED, NATURALIZED, AND EXTIRPATED SPECIES:

THE WHOLE SYSTEMATICALLY ARRANGED.

BY THE

REV. LEONARD JENYNS, M.A.

FELLOW OF THE LINNEAN, ZOOLOGICAL, AND ENTOMOLOGICAL SOCIETIES
OF LONDON; AND OF THE CAMBRIDGE PHILOSOPHICAL SOCIETY.

CAMBRIDGE:

PRINTED AT THE PITT PRESS, BY JOHN SMITH,
PRINTER TO THE UNIVERSITY.

SOLD BY J. & J. J. DEIGHTON; AND T. STEVENSON, CAMBRIDGE;
AND LONGMAN & CO., LONDON.

———

M.DCCC.XXXV.

PREFACE.

THE present Work contains descriptions of all the Vertebrate Animals, including the domesticated, naturalized, and extirpated species, which have been hitherto observed in the British Islands. The object of its author is to present Naturalists with a Manual in this department of our Fauna, adapted to the existing state of our knowledge, and such as shall be calculated to meet the wants of the Science in that advanced stage to which it has attained since the publication of former works of this nature. In furtherance of this end two points appeared necessary to be attended to. One was to ascertain, as far as practicable, the additions which had been made of late years to our lists of British Animals, to inquire into the respective claims of those which had been admitted into these lists previously, and carefully to distinguish between such species as have unquestionably occurred within the limits of our own Islands, or in the adjoining seas, and others of which reasonable doubts might be entertained in re-ference to this matter. The other important point, as it appeared to the author, was to take care that the descrip-tions should as far as possible be obtained from the animals themselves, and nothing inserted upon the credit of other writers which was capable of being verified by personal examination. The day is for ever gone by in which mere compilations will be thought to be of any service to the science of Zoology. So far from advancing its progress, it may be said unhesitatingly that they tend only to retard it. It is through such channels that errors already of long standing become more widely circulated, at the same time that new ones, to a greater or less extent, are infallibly introduced. The author who is too indolent to examine and describe for himself, has often spared himself the trouble of even investigating the nature of the materials which he has obtained from others. The consequence is, that he has perhaps blended together the descriptions of two or more perfectly distinct species, or out of one made several, or, led away by the identity of mere names, has transferred to our native animals the descriptions of exotic, nearly allied species, with which he has confounded them.

After the above enunciation of the two leading points which have been borne in mind whilst preparing this work for publication, the author hopes that it will not be thought uncalled for by those for whose use and guidance it is principally intended. The latest previous work upon the same subject and conducted upon the same general plan is the " History of British Animals" by Dr. Fleming. This was completed in 1827 and published the year following, since which period a great variety of species have been added to the Fauna of this country, more particularly in the Class of Fish, though many occur in the Classes of Quadrupeds and Birds. As these additions have been already indicated in the " Systematic Catalogue" lately published by the author of the present work, it is unnecessary to dwell upon them individually. It may, however, afford interest to present a comparative view of the aggregate numbers of species in each Class, as they appear in the " British Animals" of Dr. Fleming and in the Manual now offered to the public.

The total number of Mammalia noticed by that Naturalist, excluding the domesticated and naturalized species, as well as all those not met with at the present day, is fifty-three, from which deducting three as of rather doubtful character, there remain fifty. The number of Birds, excluding in like manner the domesticated and naturalized kinds, but including Stragglers, which Dr. Fleming has only briefly alluded to and not allowed to form part of the regular numbering, amounts to two hundred and eighty-one, from which deducting seventeen as either decidedly not British or of doubtful character, there remain two hundred and sixty-four. The number of Reptiles and Amphibious Animals, which are united in one Class, is fourteen, from which deducting one as a doubtful native, and two others as very doubtful species, there remain eleven. The number of indigenous Fish amounts to one hundred and seventy, from which deducting eleven not distinct from others, one not British, and three of doubtful character, there remain one hundred and fifty-five, which added to about seven noticed only as stragglers*, gives one hundred and sixty-two. Hence it appears that the total number of truly British *Vertebrata*, restricting the expression in the manner above indicated, described or mentioned by Dr. Fleming, amounts to only four hundred and eighty-seven. In the present work, excluding in a similar way the domesticated, naturalized, and extirpated species, as well as all those of doubtful character, the number of described Mammalia amounts to sixty-one; that of Birds to two hundred and ninety-seven; that of Reptiles and Amphibious Animals together to thirteen;

---

* Dr. Fleming has alluded briefly to several other species which have been included by some authors in the British Fauna, but the claims of these to be considered as true natives being extremely doubtful, they are not here taken into the account.

that of Fish to two hundred and ten. These added together give five hundred and eighty-one as the total number, exceeding that above stated by ninety-four. But independently of these additions which have been made of late years to our native Animals, Dr. Fleming's work professing to be in a great measure only a compilation, it was thought that one which partook more of an original character, would not prove unacceptable at the present day. It must not be forgotten that two other works connected with the Vertebrate Division of the British Fauna have appeared since the one just spoken of. I allude to the excellent " Illustrations of British Ornithology " by Mr. Selby, published in 1833, in which the number of Birds is raised to two hundred and eighty*, and the valuable " British Fishes " of Mr. Yarrell now in course of publication. But not to dwell upon the circumstance, that these works are upon totally different plans from that of the present one, the author conceives that a single volume comprising all the Vertebrate Animals which have been observed hitherto in these Islands will form a useful addition to the library of the British Naturalist, as well as prove a convenient travelling companion.

A few remarks may now be made upon the particular plan which is adopted in the present work. And first, it must be stated why a deviation has been made from Dr. Fleming's method of separating from the other species, and partially excluding from the British Fauna, those animals which have occurred as Stragglers only in a few rare instances. The reason is founded in the impossibility of drawing any marked line where such separation shall com-

---

* The exact number of species given by Mr. Selby is two hundred and eighty-seven, but six, if not seven, of these cannot be considered otherwise than as doubtful natives.

mence. However just it may be in general terms to speak of the animals of any country under the three heads of Residents, Periodical Visitants, and Irregular Visitants or Stragglers, it must be obvious to every one that between the last two there exists so little boundary, that there are many species of which it is difficult to say whether they belong to the former, or to the latter of these divisions *. Every intermediate degree of frequency, as regards the visits of different species of Birds to this country, may be found between what constitutes an annual migration and a solitary appearance. There are some which occur most years in greater or less numbers, though not with that unerring punctuality which attends the movements of our more regular migrants; there are others which only shew themselves at uncertain and more distant intervals; there are others, again, which have not been observed in more than a few instances; and of these last a series might easily be made out in which such instances became continually fewer, till the number was reduced ultimately to one. Now under such circumstances it is clearly impossible to draw the line which Dr. Fleming has attempted to point out. For this reason, in the present work, the species above alluded to are noticed exactly in the same manner as those which are permanently resident in this country, or which visit it at fixed intervals. In so doing there is no wish to raise them to a higher rank in the Fauna than they really deserve. They are still considered only as Stragglers to a greater or less extent. But if it be the object of a Fauna to present naturalists with an account

---

* Dr. Fleming himself must have experienced some difficulty in determining what species were to be "degraded to the rank of Stragglers," judging from the circumstance, that some which he has only briefly noticed as of this character, have in fact occurred in this country oftener than, many quite as often as, others described at length in the body of his work.

of such animals as are to be met with in their own country, it must surely include all those which have been known to occur in it hitherto. Neither is the reason obvious why the characters, or even a detailed description, of such species, should be suppressed. If it be quite certain that they have appeared in one instance, it cannot be deemed improbable, clearly not impossible, that they may occur in other instances *, and should this prove to be the case, it is very desirable that the student should have the means of identifying them.

A line of quite as much importance, in the opinion of the author, as that on which Dr. Fleming has insisted, and one more easily drawn, though never yet attempted to be drawn with accuracy †, is that between species, the occurrence of which in this country at one time or another there is no ground for questioning, and others whose claims to be considered as British have not yet been made substantially good. In distinguishing between these two classes, the author has generally been guided, at least in the case of Birds, by the fact of the existence or not of British-killed specimens in any known collection, or by the comparative recency of the occurrence of any species, and the circumstances under which that occurrence has been announced. Species which are not to be found in any of our Museums, for which no authority is known, or whose claims rest on statements made many years back, at a time when specific differences were but little attended to, he has no hesitation in saying, ought to have a separate place

---

* Several cases might be mentioned of species, which only a few years back had occurred but in a single instance, and which have since been met with more than once.

† This is said with reference only to the Vertebrate Division of the British Fauna. Mr. Stephens has taken great pains to draw the line spoken of in the case of our native Insects.

allotted to them as *doubtful natives*, and not to be mixed up with others which labour under no uncertainty of this nature.

Another line which the author has attempted to draw, is that between what may be termed *good* or *genuine* species, and such as are probably not distinct from others, or which are involved in some obscurity from the circumstance of their true characters not being well understood. In this latter division he has placed such animals as the *Catodon Sibbaldi* of Fleming, the *Red Lark* of Lewin the *Lacerta cedura* and *L. anguiformis* of Sheppard, the *Comber Wrasse* of Pennant, the *Lesser Fork-beard* of Jago (*Raniceps Jago*, Flem.), and a vast many others, especially amongst the Fish, all which he would designate as *doubtful species;* but he has *not* placed in it animals, such as the *Exocœtus* and *Hippocampus*, which are perfectly good species, as well as undoubtedly British, but of which the *exact* species met with in our seas or islands remains to be determined *. These are often cases in which it was not supposed that more than *one* species existed, at the time of their being enrolled in the British Fauna.

It must now be explained in what way these and other divisions, under which the Animals of this country may be parcelled, are distinguished in the present work. For this purpose recourse has been had to different types, and to two distinct sets of numbers. Firstly, all those well-ascertained, or at least *genuine* species, which are met with at the present day, or which have undoubtedly occurred at one time or another within the record of history, have one continuous numbering in each Class respectively, and form

---

* If there are cases which appear exceptions to this rule, such as those of the *Manatus borealis* of Fleming and the *Albacore* of Couch, they are instances in which not only the *species*, but even the *genus*, is as yet unascertained.

what may be termed the body of the work *.   These are
then divided into, firstly, such as are found now existing
in a *truly wild* state; secondly, such as are *domesticated*,
or, when at large, are supposed to have been originally
introduced into these Islands; and thirdly, such as were
indigenous in former times, but are now *extirpated.*   In
the case of the *truly wild* and *now existing* species, which
of course constitute the great bulk of the Fauna, and which
may be subdivided by those who please into *Residents,*
Periodical Visitants, and *Stragglers,* the names and specific
characters are printed in Small Pica; the synonyms and
descriptions in Bourgeois.   In the case of the *domesticated,*
naturalized, and *extirpated* species, these types are exchanged for Bourgeois and Minion respectively; the first two
are, however, particularly distinguished by having an asterisk (*) prefixed to their names, the last by a dagger (†)
in like manner.   The second principal division comprises
the *doubtful natives* and *doubtful species,* which terms are
employed in the sense in which they have been already
explained.   These, although inserted in their proper places
in the system, are cut off, as it were, from all those above
mentioned, and marked by a distinct numbering enclosed
in brackets.   The types resorted to are the same as those
adopted in the case of the naturalized and extirpated species,
but these types are here set a little way in from the margin.
By such an arrangement the attention is more readily drawn
to these animals which stand in such particular need of further investigation by the naturalist.

---

* It may be here observed, that a single species has been inserted in the
body of the work as British, which perhaps is hardly entitled to be considered
in that light.   The author alludes to the *Balæna Physalus,* Linn., of the occurrence of which in our seas, he can find no certain instance on record.   He
also entertains some doubts respecting the *Sparus Aurata,* from the circumstance
of its having been so often confounded with the *S. centrodontus.*

Some other matters remain now to be spoken of. It has been already intimated that the descriptions are as far as possible original. It may be added, that in a large number of instances they are derived from recent specimens. It must be obvious, however, that in preparing a work of this nature, there will always be many species, especially among the large marine animals, which it is impossible for any individual to describe from his own observation. In such cases, then, recourse has been had to what have appeared the best authorities, more especially to such persons as have published any thing original on the species in question; and where under such circumstances the accounts of authors differ, the discrepancies are pointed out. It is believed that in almost all the above cases the name of the author, from whom any thing is borrowed, is subjoined *, who of course is responsible for the accuracy of what is stated. Desmarest and Scoresby (the last for the Cetaceous Animals) are the authorities mostly resorted to in the Class Mammalia; Temminck in that of Birds; Cuvier and Valenciennes, as well as Bloch, and occasionally Pennant and Donovan, in that of Fish. By some it may be thought that the descriptions are too long, and run needlessly into detail. But when it is considered how many species have been overlooked from their supposed identity with others; how many, some even of the most common occurrence, have been misunderstood, and referred

---

* A few instances occur in the Class *Mammalia*, in which this authority was omitted to be annexed. The cases in question are those of the *Extirpated Quadrupeds*, the *Common* and *Great Seals*, the *Walrus*, the *Red Deer*, the *Roe*, and two or three of the *Cetacea*, the descriptions of which have been borrowed, partly from Desmarest, and partly from other sources. There are also three species of *Birds* similarly circumstanced; the *Shore Lark*, the *Rock Ptarmigan*, and the *Virginian Partridge*. The description of the first is taken from Temminck and Wilson; that of the second from Sabine, and the "Fauna Boreali-Americana;" that of the third from Temminck.

to others which perhaps are not inhabitants of this country ;
and that these and similar errors have arisen not merely
from the imperfect, but, it must be added, careless descriptions which have been given of such animals, it is
hoped that the pains which have been taken to render
this portion of the work as complete and accurate as possible, will not be thought entirely thrown away. The plan
which the author has adopted in most instances, in the
case of the Fish especially, which of all our British *Vertebrata* have till lately been the least attended to, and,
in regard to the distinction of species, the worst understood, has been to describe as minutely as possible the first
species in each genus, or sub-genus if it present a marked
modification of form, and then, when there were other nearly
allied species, to restrict himself principally to the *differences*
which were observable in these last, with reference to the
first and to one another. By this method, which is adopted
from MM. Cuvier and Valenciennes, the enquirer is more
readily enabled to identify the particular species he may
have before him.   In the Class of Birds, the different
variations of plumage, arising from age and season, have
been pointed out and characterized so far as they are
known ; and this plan has been pursued even in the case
of those species which have occurred in this country as
yet only in the immature state, for the more complete
guidance of our own naturalists, in the event of their being
met with in the adult plumage.

What has been said of the descriptions applies also to
the measurements. They are to be considered as original,
excepting where the name of any author stands attached to
them.   They have been taken with much care, in by far the
larger number of instances, from recently killed specimens,
and many of them are the mean results obtained from mea-

suring a large number of individuals. In the Class of Birds, they are generally those of the *male*, unless stated to the contrary. It may be observed that in this Class the *entire length* is measured from the tip of the bill to the extremity of the tail; the length of the bill is estimated in two directions, firstly from the frontal feathers, following the curvature of the ridge, secondly from the gape or angle of the mouth; the length of the tail is measured from the extremity of the longest quill to its insertion in the coccyx.

Fish being very variable in size, and having no very well marked limit of growth, it was thought that the *absolute* dimensions of any species, would, if given, prove of little value. Hence in this Class, the author has generally restricted himself to noting the average length to which the species attains; the *relative proportions*, which are often of great importance, being introduced into the body of the description. With reference to these proportions, it may be stated, that the *entire length*, unless mentioned to the contrary, is always measured to the extremity of the caudal fin. The length of the head is measured from the end of the snout to the posterior margin of the gill-cover. The *depth* (termed the *breadth* in the case of the *Pleuronectidæ*) is reckoned vertically from the most elevated point in the line of the back, termed the *dorsal line*, to the corresponding point in the line of the abdomen, or *ventral line*, the dorsal and anal fins, unless mentioned to the contrary, not being included.

It may be added that in computing the fin-ray formula, which is similar to that adopted by MM. Cuvier and Valenciennes, all the rays are included which it was possible to distinguish, on the ground that if the short rays be omitted, which often pass insensibly into the longer ones,

it were difficult to know where to begin the reckoning.
The mention of this circumstance will serve to explain why
the number of fin-rays as stated in this work will be often
found to exceed the number given by former authors, who
appear in general to have made their computation without
much attempt at accuracy. The above remark, however,
does not apply to the caudal fin, in which, generally, though
not always, there is a tolerably well marked line of separation between what may be termed the *principal* rays, and
the *accessory* or shorter ones. In many instances, in
which this distinction is evident, these two kinds of rays
are reckoned separately.

Appended to the description of each species, are a few
general remarks illustrative of its habits; more especially
those connected with locality, food, and propagation. It
was thought that these would render the work more generally useful. Many of them are the result of the author's
own observation, though some are confessedly obtained from
other sources.

On the subject of Classification, it must be remarked
that the system of no one individual author has been rigidly
adhered to. Regard has been paid to what has been written
on this subject by the most recent writers in each department,
and all the larger groups, as well as, in most instances, their
mode of collocation, have been derived from such sources.
The arrangement of the Mammalia has been drawn up
from a combined view of the system of Cuvier, and the
systems of Gray * and MacLeay †. That of the Birds
from a similar view of the system of Vigors ‡, and the
modifications of that system as adopted by Swainson and

---

* *Ann. of Phil.* vol. xxvi. p. 337.   † *Linn. Trans.* vol. xvi. p. 1, &c.
‡ *Linn. Trans.* vol. xiv. p. 395. See also some Articles by the same author
in the first and second volumes of the " *Zoological Journal.*"

Selby; assistance has been also gained, in regard to the situation of a few genera, from the system of Lesson *. The arrangement of the Reptiles and Amphibious Animals is for the most part in accordance with that in the "Regne Animal" of Cuvier. The author has, however, followed Latreille and many modern naturalists in considering these groups as two distinct classes. The arrangement of the Fish is likewise similar to that in the *second edition* of the "Regne Animal," excepting a slight alteration in the value of the larger groups, adopted from the Prince of Musignano†. In no case must it be imagined that the order of affinities is exactly the same as the order of arrangement given, since, if it be true that all natural groups be circular, an opinion now generally prevalent, and one in which the author is strongly disposed to join, it is clearly impossible to preserve such a coincidence under circumstances which necessarily entail the appearance of a linear series.

The author has also exercised his own judgment in the adoption of certain genera and sub-genera. With respect to these last, having expressed his views elsewhere‡, it is unnecessary to repeat them in this place. He may simply state that he has endeavoured to acknowledge, and to act up to, the principle which he has there advocated; although, from the ignorance in which we are upon the subject of the real value of many groups, it cannot be doubted that numerous instances occur, in which he has failed making a correct application of it. Being aware of

---

* *Traité d'Ornithologie; ou Tableau méthodique des Ordres, Sous-ordres, Familles, Tribus, Genres, Sous-genres et Races d'Oiseaux.* Par. 1831. 8vo.

† *Saggio di una Distribuzione metodica degli Animali Vertebrati.* Rom. 1831. 8vo. The alteration above alluded to has been adopted by the author since the publication of his "Systematic Catalogue."

‡ *Lond. Mag. of Nat. Hist.* vol. vi. p. 385. and vol. vii. p. 97.

the dislike which many of our own naturalists have to the adoption of sub-genera, he begs to state that they are inserted in such a manner, that any one who chooses may place them upon the same footing with the genera, or, if he be not a friend to the subdivision of the old-established groups, take no notice of them at all.

Instead of prefixing the generic characters to the several genera respectively as they occur in order, they are presented in a synoptic form at the head of each of the Classes, by which means a better view is obtained of the relative collocation and affinities of the larger groups. Moreover, the same kind of arrangement with respect to types prevails here, as that which occurs in the other part of the work. The characters of all those genera and sub-genera which contain *truly wild*, as well as *genuine* and *now existing* species, are printed in Small Pica and Bourgeois respectively, the names of the genera standing in LARGE CAPITALS, those of the sub-genera in SMALL CAPITALS. In the case of the *domesticated, naturalized,* and *extirpated* animals, these types are exchanged respectively, as before, for Bourgeois and Minion; the first two, being, also in like manner, particularly distinguished by an asterisk, the last by a dagger. The same types are employed for the characters of those genera and sub-genera which contain only *doubtful natives* or *doubtful species,* but these may be readily distinguished from the last by the circumstance of the names standing in *ITALIC* Capitals, which are large or small, as in the former instances, according as the group in question is either a genus or sub-genus. Moreover, in this division, the genera have a distinct numbering enclosed in brackets: the sub-genera no numbering at all. It may be further observed, that the characters of the orders and families are printed in Pica, and that here likewise the type is exchanged for one of subordinate size in the case of two families (*Phasianidæ* and *Siluridæ*), one containing no true natives, the other only a doubtful native. In accordance, also, with the rule before given, the former is distinguished by an asterisk; the latter, by the name standing in *ITALIC CAPITALS*, and its being, as in the case of sub-genera similarly circumstanced, without a number.

The specific characters, which are for the most part entirely new, have been drawn up, as far as practicable, with a view to the exact differences of species; but the difficulty of doing this in the instance of the aquatic Birds, which are subject to such great changes of plumage from age and other circumstances, it must be confessed is very great, and has not in all cases been satisfactorily got over. In the Class of Fish, specific characters have seldom been added in the case of genera or sub-genera containing not more than one *British* species, or when given, bear reference solely to other, nearly allied, *European* species; the essential characters of those met with in other countries, in many groups, not having been hitherto determined with precision.

The terms employed in characterizing the larger groups, or in drawing up the descriptions of species, have for the most part been derived from the works of Illiger*, Desmarest †, and Cuvier. The " Histoire Naturelle des Poissons" of this last author has been exclusively resorted to in the Class of Fish. To these works, the reader, to whom the terms are not familiar, is accordingly referred.

* *Prodromus Systematis Mammalium et Avium.* Berol. 1811. 8vo. This work will be found to contain a very complete terminology with reference to the two classes of Mammalia and Birds.

† *Terminologie des Mammifères;* prefixed to his " Mammalogie."

On the subject of Nomenclature it may be remarked that the oldest names have been adopted in most instances, unless a different one has been sanctioned by general use. As the author is of opinion that in common parlance there is no occasion to name the sub-genus, when speaking of any particular species, he has not thought it necessary to change the names of species which are the same as those of the sub-genera to which they belong.

The synonyms will be found to embrace references to some of the principal writers, more especially those in our own country, who have treated of the different Classes, or of any subordinate group, or any particular species. These references have, in every instance, been examined personally by the author. To facilitate the inquiries of such persons, as may wish to examine for themselves on this subject, a complete list has been annexed of all those works, with their several editions, which are quoted, either for the synonyms, or for any other purpose.

It has been thought proper to annex two distinct indexes, one containing the Latin, the other the English names. No pains have been spared to render them as complete as possible. It is believed that every name and synonym, in the above two languages, occurring in the body of the work, will be found in its proper place.

It now only remains for the author to express his acknowledgments to those friends who have assisted him in this undertaking. To Mr. Yarrell in particular, he begs publicly to return his sincere thanks for the able help which he has experienced at his hands, and such as alone has enabled him to complete the work upon the plan first contemplated. This help has been especially felt upon the subject of the British Fishes. Had it not been for the very liberal manner in which that gentleman offered him

the almost unlimited use of his Manuscripts and rich collections, the author has no hesitation in saying that he
could never have extended the Manual to that department, or presumed to enter upon a field, to which he was
previously almost an entire stranger. Assistance, however,
has been not the less afforded him in the other Classes.
Mr. Yarrell's well-known practical acquaintance with our
British Birds has enabled the author to detail more at
length the changes of plumage to which some species are
liable, and to correct a few errors into which previous
writers had fallen on this subject. The same gentleman
kindly volunteered an accurate description, accompanied
by measurements, of the egg of every species, of which
his extensive collection afforded specimens; thus enhancing
the utility of the work by an addition, which, but for
this circumstance, the author would have been unable to
supply. He begs it may be distinctly understood that
this portion of it is from the pen of Mr. Yarrell.

To Mr. Gray, he desires to make an acknowledgment
of the readiness with which he has at all times allowed
him to consult the specimens in the British Museum. The
same return must be made to the officers of the Zoological Society for similar liberty to examine a few specimens
contained in their collection.

In conclusion, the author may state that he has no
intention of extending his work to the Invertebrate Division of the British Fauna. In the present advanced state
of the Science, a complete Manual of all the Animals occurring in these Islands can only be accomplished by the
united labours of many individuals. With the view, however, of continuing his researches into that portion of it
which is here treated of, he begs to solicit such observations, notices of new or rare species, and, where it may

not be inconvenient, specimens, as it may be in the power
of any of his readers to supply. Even the Vertebrate
Animals of our own country are far yet from being
thoroughly understood. Much confusion prevails in several groups, and, without doubt, many additional species
remain to be detected. The *Shrews* and *Bats* amongst
Quadrupeds; the *Cetaceous* Animals; the *Wrasses*, *Gobies*,
Blennies, the *Salmon* Tribe, the *Sharks* and *Rays*, amongst
Fish; these and other families might be pointed out, with
respect to which we want more information, and which
particularly invite the attention of the British Naturalist.
Should any one be disposed to honour him with specimens
in these or other instances, the author begs to state that
they will not be reserved for his own use exclusively, but
will, with the permission of the donor, be deposited in
the Museum of the Cambridge Philosophical Society, which
already possesses an extensive collection of British Animals,
and which the author is anxious to render as perfect as
possible in that department of the Fauna, to the advance-
ment of which the Work now offered to the Public is
directed.

SWAFFHAM BULBECK,
   Oct. 24, 1835.

THE Author takes this opportunity of expressing his grateful acknowledgments to the SYNDICS of the University Press, for their liberality in taking upon themselves the expense of printing this Work.

ERRATA.

PAGE	LINE		ERROR.	CORRECTION.
52	16		LANIIDÆ.	LANIADÆ.
53	11	from the bottom;	SYLVIIDÆ.	SYLVIADÆ.
59	12	....	CERTHIIDÆ.	CERTHIADÆ.
64	8	....	CHARADRIIDÆ.	CHARADRIADÆ.
126	7	....	vol. ii.	vol. i. part ii.
137	2	....	dele &.	
160	13		*Europæan.*	*European.*
252	14		*C. Urinator.*	*Colymbus Urinator.*
360	14	from the bottom, and elsewhere }	*Mackarel.*	*Mackerel.*
434	2		Clupea, *Linn.*	Clupea, *Cuv.*
445	7		*peritonæum.*	peritoneum.

ALPHABETICAL LIST

OF

WORKS QUOTED.

Ann. des Sci.—ANNALES des Sciences Naturelles. Paris, 1824, &c. 8vo.

Ann. du Mus.—Annales du Muséum d'Histoire Naturelle. Paris, 1802–1813. 4to. 20 vols.

Ann. of Phil.—The Annals of Philosophy. New Series. Lond. 1821–1826. 8vo. (*Vol. 6, or 22 from the commencement, quoted.*)

Berkenh. Syn.—Berkenhout (J.). Synopsis of the Natural History of Great Britain and Ireland; being a Second Edition of the Outlines, &c. Lond. 1789. 8vo. 2 vols.

Bew. Brit. Birds.—Bewick (T.). A History of British Birds. Sixth Edition. Newcastle, 1826. 8vo. 2 vols.

—— *Quad.*—Bewick (T.). A General History of Quadrupeds. The Figures engraved on Wood. Eighth Edition. Newcastle-upon-Tyne, 1824. 8vo.

Bloch, Ichth.—Bloch (M. E.). Ichthyologie; ou Histoire Naturelle des Poissons. Berlin, 1785–1795. fol. 12 parts.

Bon. Am. Orn.—Bonaparte (C. L.). American Ornithology; or the Natural History of Birds inhabiting the United States, not given by Wilson. Philadelph. 1825, &c. 4to. 4 vols.

—— *Faun. Ital.*—Bonaparte (C. L.). Iconographia della Fauna Italica. Rom. 1832, &c. fol. (*In course of publication.*)

—— *Syn.*—Bonaparte (C. L.). The Genera of North American Birds, and a Synopsis of the Species found within the Territory of the United States. New York, 1828. 8vo. (*From the Annals of the Lyceum of Natural History of New York.*)

Borl. Cornw.—Borlase (W.). The Natural History of Cornwall. Oxford, 1758. fol.

Bowd. Brit. fr. wat. Fish.—Bowdich (Mrs. T. E.). The Fresh-water Fishes of Great Britain; drawn and described. Lond. 1828, &c. 4to. (*In course of publication.*)

Briss. Orn.—Brisson. Ornithologia, sive Synopsis Methodica Avium, &c. Lug. Bat. 1763. 8vo. 2 vols.

—— *Reg. An.*—Brisson. Regnum Animale in classes ix. distributum, cum duarum primarum classium, Quadrupedum scilicet et Cetaceorum particulari divisione. Paris, 1756. 4to.

Brown, Illust.—Brown (P.). New Illustrations of Zoology. Lond. 1776. 4to.

Buff Hist. Nat.—Buffon. Histoire Naturelle générale et particulière, avec la description du Cabinet du Roi. Paris, 1750–1767. 4to. 15 vols.

Bull. des Sci. Nat.—Bulletin des Sciences Naturelles et de Géologie. Paris, 1824–1831. 8vo. (*Vol.* 4, *for the year* 1825, *quoted.*)

Camb. Phil. Trans.—Transactions of the Cambridge Philosophical Society. Cambridge, 1822, &c. 4to.

Child. Add. to Zool. Cl.—Children (J. G.). An Address delivered at the Anniversary Meeting of the Zoological Club of the Linnean Society, held Nov. 29, 1827. Lond. 1827. 8vo.

Cuv. Oss. Foss.—Cuvier (G.). Recherches sur les Ossemens Fossiles. Paris, 1821–1824. 4to. 5 vols.

—— *Reg. An.*—Cuvier, (G.) Le Règne Animal distribué d'après son organisation. Paris, 1817. 8vo. 4 vols. The same. Paris, 1829. 8vo. 5 vols. (*The second edition is the one quoted, unless stated to the contrary.*)

—— *& Val. Poiss.*—Cuvier et Valenciennes. Histoire Naturelle des Poissons. Paris, 1828–1833. 4to. 9 vols. (*In course of publication.*)

Dan. Rur. Sports.—Daniel (W. B.). Rural Sports. Lond. 1801–1802. 4to. 2 vols.

Daud. Hist. Nat. des Rept.—Daudin (F. M.). Histoire Naturelle des Reptiles. Paris, 1805. 8vo. 8 vols.

Desm. Mammal.—Desmarest (A. G.). Mammalogie; ou description des espèces de Mammifères. Paris, 1820–1822. 4to. 2 parts.

Don. Brit. Birds.—Donovan (E.). The Natural History of British Birds. Lond. 1794–1819. 8vo. 10 vols.

—— *Brit. Fish.*—Donovan (E.). The Natural History of British Fishes. Lond. 1808. 8vo. 5 vols.

—— *Brit. Quad.*—Donovan (E.). The Natural History of British Quadrupeds. Lond. 1820. 8vo. 3 vols.

Edinb. Journ. of Nat. and Geog. Sci.—The Edinburgh Journal of Natural and Geographical Science. Edinb. 1830, &c. Directed by W. Ainsworth and H. Cheek. 8vo. 3 vols.

Edinb. New Phil. Journ.—The Edinburgh New Philosophical Journal. Conducted by R. Jameson. Edinb. 1826, &c. 8vo.

Edinb. Phil. Journ.—The Edinburgh Philosophical Journal. Conducted by R. Jameson. Edinb. 1819–1826. 8vo. 14 vols.

Edw. Glean.—Edwards (G.). Gleanings of Natural History. Lond. 1758–1764. 4to. 3 parts.

——— *Nat. Hist.*—Edwards (G.). A Natural History of uncommon Birds, and of some other rare and undescribed Animals. Lond. 1743–1751. 4to. 4 parts.

Encycl. Brit.—Encyclopædia Britannica. Seventh Edition. Edited by Professor Napier. Edinb. 1834, &c. 4to. (*In course of publication.*)

Faun. Bor. Am.—Fauna Boreali-Americana; or the Zoology of the Northern parts of British America. Part second, The Birds. By W. Swainson, and J. Richardson. Lond. 1831. 4to.

——— *Franç.*—Faune Française, ou Histoire Naturelle, générale et particulière, des Animaux qui se trouvent en France, &c. Par MM. Vieillot, Desmarest, De Blainville, Audinet-Serville, St. Fargeau, et Walkenäer. Paris, 1824, &c. 8vo. (*Incomplete.*)

Flem. Brit. An.—Fleming (J.). A History of British Animals, &c. Edinb. 1828. 8vo.

——— *Phil. Zool.*—Fleming (J.). The Philosophy of Zoology; or a general view of the structure, functions, and classification of Animals. Edinb. 1822. 8vo. 2 vols.

Gmel. Linn.—Gmelin (J. F.). Caroli a Linné, Systema Naturæ, &c. Lips. 1788, &c. 8vo. vol. i., Animal Kingdom.

Gould, Europ. Birds.—Gould (J.). The Birds of Europe. Lond. 1832, &c. fol. (*In course of publication.*)

Gray, Syn. Rept.—Gray (J. E.). Synopsis Reptilium; or short descriptions of the Species of Reptiles. Part i. Cataphracta. Lond. 1831. 8vo. (*No more has been hitherto published.*)

Grew, Rar.—Grew (N.). Musæum Regalis Societatis; or a Catalogue and Description of the natural and artificial Rarities belonging to the Royal Society and preserved at Gresham College. Lond. 1681. fol.

Hast. Worcest.—Hastings (C.). Illustrations of the Natural History of Worcestershire, &c. Lond. 1834. 8vo.

Heysh. Cumb. An.—Heysham (J.). A Catalogue of Cumberland Animals. (*Prefixed to the 1st volume of Hutchinson's History of the County of Cumberland.* Carlisle, 1794. 4to.)

Horsf. Zool. Research.—Horsfield (T.). Zoological Researches in Java, and the neighbouring Islands. Lond. 1824. 4to.

Jago.—Catalogus quorundam Piscium rariorum, quos oris Cornubiæ maritimis nuper observavit, et delineavit, &c. G. Jago, &c. (*In Ray's Synopsis Piscium.* p. 162–166.)

Jard. and Selb. Orn.—Jardine (W.) and Selby (P. J.). Illustrations of
Ornithology. Edinb. 1827, &c. 4to. (*In course of publication.*)

Jen. Cat. of Brit. Vert. An.—Jenyns (L.). A Systematic Catalogue of
British Vertebrate Animals. Cambridge, 1835. 8vo.

Jesse, Glean.—Jesse (E.). Gleanings in Natural History. Second Series.
Lond. 1834. 8vo.

Journ. de Phys.—Journal de Physique, de Chimie, d'Histoire Naturelle,
et des Arts. Paris, 1773-1823. 4to. (*Vol. 71, the only one quoted.*)

Lacép Cétac.—La Cépède. Histoire Naturelle des Cétacées. Paris,
1803. 4to.

—— *Poiss.*—La Cépède. Histoire Naturelle des Poissons. Paris, 1798-
1803. 4to. 5 vols.

Lath. Ind. Orn.—Latham (J.). Index Ornithologicus; sive Systema
Ornithologiæ; &c. Lond. 1790. 4to. 2 vols.

—— *Syn.*—Latham (J.). A General Synopsis of Birds. Lond. 1781-
1785. 4to. 3 vols. in 6 parts.

—— *Supp.*—Supplement to the General Synopsis of Birds. Lond. 1787.
4to. Supplement II. Lond. 1801. 4to.

Latr. Salam. de France.—Latreille. Histoire Naturelle des Salamandres
de France, précédé d'un Tableau méthodique des autres Reptiles
indigènes. Paris, 1800. 8vo.

Leach, Zool. Misc.—Leach (W. E.). The Zoological Miscellany; being
descriptions of new, or interesting Animals. Lond. 1814-1817. 8vo.
3 vols.

Leigh, Lancash.—Leigh (C.). The Natural History of Lancashire,
Cheshire, and the Peak, in Derbyshire, &c. Oxford, 1700. fol.

Lew. Brit. Birds.—Lewin (W.) The Birds of Great Britain, with their
Eggs, accurately figured. Lond. 1789-1794. 4to. 7 vols.

Linn. Syst. Nat.—Linnæus (C.). Systema Naturæ, &c. Edit. 12. Holm.
1766-1768. 8vo. 3 vols. Vol. 1, Animal Kingdom.

Linn. Trans.—Transactions of the Linnean Society of London. Lond.
1791, &c. 4to.

L'Instit.—L'Institut; Journal des Académies et Sociétés scientifiques
de la France et de l'Etranger. Paris, 1833, &c. 4to.

Lond. & Edinb. Phil. Mag.—The London and Edinburgh Philosophical
Magazine and Journal of Science. By Sir D. Brewster, R. Taylor,
and R. Philips. Lond. 1832, &c. 8vo.

Lond. Quart. Journ. of Sci.—The Quarterly Journal of Science, Litera-
ture, and the Arts. Lond. 1816-1830. 8vo. 29 vols.

Lond. Mag. of Nat. Hist.—The Magazine of Natural History, and
Journal of Zoology, Botany, Mineralogy, Geology, and Meteorology.
Conducted by J. C. Loudon. Lond. 1829, &c. 8vo.

Low, Faun. Orc.—Low (G.). Fauna Orcadensis; or the Natural History of the Quadrupeds, Birds, Reptiles, and Fishes, of Orkney and Shetland. Edinb. 1813. 4to.

Lyell, Geol.—Lyell (C.). Principles of Geology, &c. Lond. 1830–1833. 8vo. 3 vols.

Mac Cull. West. Isles.—Mac Culloch (J.). A Description of the Western Islands of Scotland, including the Isle of Man. Lond. 1819. 2 vols. 8vo. and 4to atlas of plates.

Mém. de la Soc. de Gen.—Mémoires de la Societé de Physique, et d'Histoire Naturelle de Genève. Genev. 1821, &c. 4to. (*Vol.* 3 quoted.)

Mém. du Mus.—Mémoires du Muséum d'Histoire Naturelle. Paris, 1815–1832. 4to. 20 vols.

Merr. Pin.—Merrett (C.). Pinax Rerum Naturalium Britannicarum. Lond. 1666. 8vo.

Mont. Orn. Dict.—Montagu (G.). Ornithological Dictionary; or Alphabetical Synopsis of British Birds. Lond. 1802. 8vo. 2 vols.

—— *Supp.*—Montagu (G.). Supplement to the Ornithological Dictionary. Exeter, 1813. 8vo.

Mull. Zool. Dan.—Muller (O. F.). Zoologia Danica. Haun. 1788–1806. fol. 4 vols.

Nash, Worcest.—Nash (T.). Collections for the History of Worcestershire. Lond. 1781. fol. 2 vols.

Newcast. Nat. Hist. Trans.—Transactions of the Natural History Society of Northumberland, Durham, and Newcastle. Newcast. 1831, &c. 4to.

Nichol. Journ.—A Journal of Natural Philosophy, Chemistry, and the Arts. Lond. 1802–1813. 8vo. 36 vols. (*Vol.* 22, *for the year* 1809, quoted.)

Nilss. Prod. Ichth. Scand.—Nilsson (S.). Prodromus Ichthyologiæ Scandinavicæ. Lund. 1832. 8vo.

Nutt. Orn. of Un. St.—Nuttall (T.). A Manual of the Ornithology of the United States and of Canada. Boston, 1832–1834. 8vo. 2 vols.

Osbeck, Voy. to China.—Osbeck (P.). A Voyage to China and the East Indies. Lond. 1771. 8vo. 2 vols. (*Translated from the German by J. R. Forster.*)

Paget, Nat. Hist. of Yarm.—Paget (C. J. & J.). Sketch of the Natural History of Yarmouth and its Neighbourhood, containing Catalogues of the Species of Animals, Birds, Reptiles, Fish, Insects, and Plants, at present known. Yarmouth, 1834. 8vo.

Penn. Arct. Zool.—Pennant (T.). Arctic Zoology. Lond. 1792. 4to. 2 vols.

Penn. Brit. Zool.—Pennant (T.). British Zoology. Fourth Edition. Lond. 1776–1777. 8vo. 4 vols. The same. New Edition. Lond. 1812. 8vo. 4 vols. (*The Fourth Edition is the one quoted, unless stated to the contrary.*)

Phil. Trans.—Philosophical Transactions of the Royal Society of London. Lond. 1665, &c. 4to.

Plot, Oxfordsh.—Plot (R.). The Natural History of Oxfordshire. Oxford, 1705. fol. (*Second Edition.*)

Proceed. of Berw. Nat. Club.—Proceedings of the Berwickshire Naturalist's Club. Edinb. 1834, &c. 8vo.

———— *Zool. Soc.*—Proceedings of the Zoological Society of London. Lond. 1830, &c. 8vo.

Pult. Cat. Dors.—Pulteney (R.). Catalogue of the Birds, Shells, and some of the more rare Plants of Dorsetshire. From the new and enlarged Edition of Mr. Hutchin's History of that county. Lond. 1813. fol.

Quart. Rev.—The Quarterly Review. vol. 47. Lond. 1832. 8vo.

Ray, Syn. Av.—Ray (J.). Synopsis methodica Avium. Lond. 1713. 8vo.

———— *Syn. Pisc.*—Ray (J.). Synopsis methodica Piscium. Lond. 1713. 8vo.

———— *Syn. Quad.*—Ray, (J.). Synopsis methodica Animalium Quadrupedum et Serpentini Generis. Lond. 1693. 8vo.

Richards. App. Parr. Voy.—Richardson (J.). Zoological Appendix to Parry's Second Voyage. 1824. 4to.

Riss. Hist. Nat.—Risso (A.). Histoire Naturelle des principales productions de l'Europe Méridionale, et particulièrement de celles des environs de Nice et des Alpes maritimes. Par. 1826. 8vo. 5 vols. (*Vol. 3, quoted.*)

Rœs. Ran.—Rœsel (A. J.). Historia Naturalis Ranarum nostratium. Nurnberg. 1758. fol.

Rond. Pisc.—Rondeletius (G.). De Piscibus marinis. Lug. Bat. 1554. fol.

Ross, Voy. App.—Appendix to Ross's Voyage of Discovery. Lond. 1819. 4to.

Sab. Supp. Parr. Voy.—A Supplement to the Appendix of Captain Parry's Voyage for the discovery of a N.W. passage in the years 1819, 1820, containing an account of the subjects of Natural History. Lond. 1824. 4to. The Birds by Capt. Sabine.

Salmon.—Salmonia; or Days of Fly-fishing. By an Angler. Second Edition. Lond. 1829. 8vo.

Scoresb. Arct. Reg.—Scoresby (W.). An Account of the Arctic Regions, with a history and description of the Northern Whale-fishery. Edinb. 1820. 8vo. 2 vols.

Selb. Illust.—Selby (P. J.). Illustrations of British Ornithology. Edinb. 1833. 8vo. 2 vols. Plates to the same. Eleph. fol.

Shaw, Gen. Zool.—Shaw (G.). General Zoology. Lond. 1800–1826. 8vo. 14 vols. (*The last six volumes are by Mr. Stephens.*)

—— *Nat. Misc.*—Shaw (G.). The Naturalists Miscellany. Lond. 1790, &c. 8vo. 24 vols.

Sibb. Phalain.—Sibbald (R.). Phalainologia Nova; sive observationes de rarioribus quibusdam Balænis in Scotiæ Littus nuper ejectis. Edinb. 1773. 8vo.

—— *Scot. Illust.*—Sibbald (R.). Scotia Illustrata; sive Prodromus Historiæ Naturalis, &c. Edinb. 1684. fol.

Sloane, Jam.—Sloane (Hans). A Voyage to the Islands Madeira, Barbados, Nieves, St. Christopher's and Jamaica, with the Natural History of the last of those Islands. Lond. 1707–1725. fol.

Sow. Brit. Misc.—Sowerby (J.). The British Miscellany; or coloured figures of new, rare, or little known Animal subjects; many not before ascertained to be inhabitants of the British Isles. vol. 1. Lond. 1806. 8vo. (*The work was not continued beyond the first volume.*)

Stew. El. of Nat. Hist.—Stewart (C.). Elements of the Natural History of the Animal Kingdom. Second Edition. Edinb. 1817. 8vo. 2 vols.

Swains. Zool. Ill.—Swainson (W.). Zoological Illustrations; or original figures and descriptions of new, rare, or interesting Animals, &c. Lond. 1820–1823. 8vo. 3 vols.

Temm. Man. d'Orn.—Temminck (C. J.). Manuel d'Ornithologie, ou Tableau Systématique des Oiseaux qui se trouvent en Europe; &c. Second Edition. Parts 1 & 2. Paris, 1820. Part 3. Par. 1835. 8vo.

—— *Monog. de Mamm.*—Temminck (C. J.). Monographies de Mammalogie. Paris, 1827. 4to.

—— *Pig. et Gall.*—Temminck (C. J.). Histoire Naturelle Générale des Pigeons et des Gallinacés. Amsterdam, 1813–1815. 8vo. 3 vols.

Turt. Brit. Faun.—Turton (W.). British Fauna; containing a Compendium of the Zoology of the British Islands. Swansea, 1807. 12mo.

—— *Linn.*—Turton (W.). A General System of Nature, &c. Translated from Gmelin's last edition of the Systema Naturæ, by Sir Charles Linné. Swansea, 1800. 8vo. Vols. 1–4, Animal Kingdom.

Vieill. Gal. Ois.—Vieillot (L. P.). La Gallerie des Oiseaux. Paris, 1825. 4to. 2 vols.

Wagl. Syst. Av.—Wagler (J.). Systema Avium. Pars piima. Stuttgard, 1827. 8vo. (*Incomplete.*)

Wern. Mem.—Memoirs of the Wernerian Natural History Society.
   Edinb. 1811, &c. 8vo.

White, Selb.—White (G.). The Natural History of Selborne, to which
   are added The Naturalist's Calendar, Miscellaneous Observations,
   and Poems. Lond. Edit. 1813. 8vo. 2 vols.

Will. Hist. Pisc.—Willughby (F.). De Historia Piscium Libri Quatuor,
   &c. Totum opus recognovit, coaptavit, supplevit, Librum etiam
   primum et secundum integros adjecit, Johannes Raius. Curâ
   Cromwelli Mortimeri, M.D. Lond. 1743. fol.

Wils. Amer. Orn.—Wilson (A.). American Ornithology; or the Natural
   History of the Birds of the United States. Philadelph. 1808—1814.
   4to. 8 vols.

Yarr. Brit. Fish.—Yarrell (W.). A History of British Fishes. Illustrated
   by Wood-cuts of all the species. vol. 1. Lond. 1835. 8vo. (*In course
   of publication.*)

Zool. Journ.—The Zoological Journal. Lond. 1825, &c. 8vo.

VERTEBRATA.

Vertebrate animals are characterized by having the brain and principal trunk of the nervous system included in a bony articulated case, composing the skull and vertebral column. They have all red blood, and a muscular heart. The mouth is furnished with two jaws moving vertically. They have distinct organs of sight, hearing, smell, and taste. The limbs are never more than four in number. The sexes are in all cases separate.

They are divisible into the following five classes:

* *With warm blood: heart with two auricles and two ventricles.*

I. MAMMALIA.—Viviparous animals, suckling their young: breathing by lungs. Body generally covered with hair, and furnished in most cases with four feet.

II. AVES.—Oviparous animals; breathing by lungs. Body covered with feathers; furnished with two wings and two feet.

A

** *With cold blood: heart with one or two auricles, and one ventricle.*

III. REPTILIA.——Oviparous animals; breathing by lungs. Body covered with scales or shelly plates. Impregnation effected by sexual union: the young undergoing no metamorphosis.

IV. AMPHIBIA.——Oviparous animals; breathing by lungs and gills. Body covered with a soft naked skin: feet without claws. Eggs impregnated after exclusion: the young undergoing metamorphosis.

V. PISCES.——Oviparous animals; breathing by gills only. Body covered with scales, and furnished with fins instead of feet.

CLASS I. MAMMALIA.

§ *1. Incisor, canine, and molar teeth all present, forming a continuous series.*

ORDER I.   FERÆ.

Feet armed with strong claws: mammæ abdominal: penis sheathed.

* *Six incisors in each jaw : grinders of three sorts.*

† *Walk on the soles of the feet.*

† 1. URSUS.—Lower incisors set in the same line: grinders varying in number, the three last large and tubercular: no glandular pouch under the tail.

2. MELES.—Second incisor on each side in the lower jaw placed behind the others: grinders five above and six below on each side; the first very small, the last tubercular: body low on the legs: a glandular pouch or scent-bag under the tail.

†† *Walk on the extremities of the toes.*

3. MUSTELA.—Grinders varying in number: body very much elongated: feet short; the toes separate: ears short and rounded.

(1. MUSTELA.) Grinders five above and six below on each side: tongue smooth.

(2. PUTORIUS.) Grinders four above and five below on each side: tongue rough.

4. LUTRA.—Grinders five above and five or six below: body long: feet short, with five palmated toes: head depressed: ears very short: tongue slightly rough: tail flattened horizontally.

5. CANIS.—Grinders six above and seven below on each side; the two last in each jaw tubercular: ears moderately large: tongue smooth: claws not retractile.

(1. Canis.) Pupil circular.

(2. Vulpes.) Pupil linear: tail bushy.

6. FELIS.—Grinders four above and three below on each side; no tubercular grinder in the lower jaw: tongue armed with prickly papillæ pointing backwards: claws retractile.

** *Teeth various: feet short and fin-shaped, adapted to swimming; those behind horizontal.*

7. PHOCA.—Incisors and tusks in both jaws: grinders uniform; each of three lobes; the central lobe triangular, large and cutting, the anterior and posterior lobes small.

8. TRICHECHUS.—No incisors or tusks in the lower jaw: tusks in the upper jaw greatly produced, directed downwards: grinders cylindrical, short, and obliquely truncated.

*** *Incisors varying in number: summits of the grinders with conical points: soles of the hinder feet applied to the ground.*

9. TALPA.—Incisors six above and eight below, nearly equal: tusks large: grinders seven above and six below on each side: body very thick: fore feet short and broad, formed for digging: tail short: no external ears: eyes very minute.

10. SOREX.—Two middle incisors produced beyond the others; those above bent and notched at the base: lateral incisors or false tusks very small: grinders four above and three below on each side: body covered with hair: snout attenuated: tail long.

11. ERINACEUS.—Two middle incisors produced beyond the others, longer than the tusks; those above cylindrical, with a space between: grinders five above and four below on each side: body covered with spines: tail very short.

ORDER II.   PRIMATES.

Claws flat and small: anterior extremities with a distinct thumb: mammæ pectoral: penis free and pendulous.

VESPERTILIONIDÆ.—*Incisors varying in number; summits of the grinders armed with sharp points: anterior and posterior extremities connected by a naked expansion of the skin, adapted to flight: fingers very long.*

12. RHINOLOPHUS.—Incisors two above and four below: grinders five on each side above and below: nostrils with two foliaceous appendages, the posterior one erect: ears free; tragus wanting.

13. VESPERTILIO.—Incisors four above and six below: grinders varying in number: nostrils without foliaceous appendages: ears sometimes free, sometimes united at their base; tragus always present.

(1. VESPERTILIO.) Ears moderately large, lateral, separate: grinders from four to six, above, and from three to six below on each side.

(2. PLECOTUS.) Ears very large, much longer than the head, with their inner edges united at the base above the eyes: grinders five above and six below on each side.

(3. BARBASTELLUS.) Ears moderate, united at the base above the eyes: a flat naked space on the forehead surrounded by a membraneous edge: grinders four on each side above and below.

---

§ *II. Incisor, canine, and molar teeth not all present, or not forming a continuous series.*

ORDER III.   GLIRES.

Incisors two in each jaw, large, and strong, remote from the grinders: tusks none: toes distinct, with small conical claws.

* *Furnished with strong clavicles.*

14. SCIURUS.—Grinders simple, with tubercular summits, five above and four below on each side: upper incisors chisel-shaped; lower ones pointed, compressed laterally: toes long, armed with sharp claws: tail long and bushy.

15. MYOXUS.—Grinders simple, their summits marked with transverse ridges of enamel, four on each side above and below: hair very soft and fine: tail very long, somewhat bushy.

16. MUS.—Grinders simple, with tubercular summits, three on each side above and below: upper incisors wedge-shaped, lower ones compressed and pointed: tail nearly naked, annulated with scales.

17. ARVICOLA.—Grinders compound, with flat summits, the enamel appearing in ridges on the surface: upper

incisors chisel-shaped, lower ones pointed: tail round and hairy, shorter than the body.

† 18. CASTOR.—Grinders compound, with flat summits, four on each side above and below: toes of the hinder feet palmated: tail oval, depressed, and covered with scales.

** *Clavicles rudimentary.*

19. LEPUS.—Two subsidiary incisors in the upper jaw immediately behind the others : grinders six above and five below on each side, the summits flat, with transverse plates of enamel: inside of the cheeks hairy: ears very large: toes five before and four behind : tail short, and turned up.

* 20. CAVIA.—Grinders compound; each with one simple and one fork-shaped plate of enamel; four on each side above and below: toes four before and three behind: tail none.

ORDER IV.   UNGULATA.

Teeth various: incisors and tusks often wanting in one or both jaws: grinders of one sort: toes large, covered with hoofs.

* *No incisors in the upper jaw; eight in the lower: two middle toes separate : frontal bones furnished with horns : ruminate.*

† *Horns hollow, growing on a bony core ; persistent.*

* 21. BOS.—Horns smooth; directed laterally at first, afterwards recurved: body thick and heavy: limbs strong: tail moderately long, terminated by a tuft of hair: four inguinal mammæ.

* 22. OVIS.—Horns rough and angular; directed backwards and laterally, more or less spirally twisted: chin without a beard: limbs rather slender: tail moderately short: two inguinal mammæ.

* 23. CAPRA.—Horns rough and angular; directed upwards and backwards: chin bearded: tail short: mammæ two.

†† *Horns bony and solid: deciduous.*

24. CERVUS.—Horns branched or palmated; while growing covered with a soft velvety skin: body and limbs slender: four inguinal mammæ.

** *Six incisors in each jaw : two middle toes soldered in one: no horns: do not ruminate.*

* 25. EQUUS.

*** *Six incisors in each jaw; lower ones projecting forwards: four toes on each foot; two middle ones large and hoofed, lateral ones much smaller and not touching the ground.*

†* 26. SUS.—Tusks exserted, inclining both upwards and to one side: snout elongated, and truncated at the extremity: body covered with stiff bristles.

ORDER V. CETACEA.

Teeth various; sometimes entirely wanting: body fish-shaped: anterior extremities in the form of fins; posterior united, forming a flat horizontal tail.

* *Grinders with flat summits : nostrils at the extremity of the snout, not acting as blow-holes: whiskers very distinct: mammæ pectoral.*

(1.) *MANATUS.*—Grinders eight or nine on each side above and below; with two transverse ridges on the summits: no incisors or tusks in the adult state: vestiges of nails on the margins of the pectoral fins.

(2.) *STELLERUS.*—A single grinder on each side above and below: summit flat with projecting plates of enamel: fins without vestiges of nails.

** *Teeth, when present, conical: nostrils opening on the crown of the head, acting as blow-holes: whiskers none: mammæ near the anus.*

† *Head small: blow-holes united.*

27. DELPHINUS.—Both jaws with numerous teeth, all simple and equal.

(1. DELPHINUS.) Snout produced into a beak, broad at its base, rounded at its extremity, and separated from the forehead by a kind of furrow: a dorsal fin.

(2. PHOCŒNA.) Snout short and blunt; not beaked: a dorsal fin.

(3. DELPHINAPTERA.) Head blunt; snout not produced: no dorsal fin.

28. MONODON.—Head blunt: no true teeth; upper jaw with one, rarely two, very long straight tusks projecting forwards in a line with the body: no dorsal fin.

29. HYPEROODON.—Snout produced: teeth generally but two in number, placed in front of the lower jaw: palate studded with tubercles: a dorsal fin.

†† *Head very large; half or one third of the entire length.*

30. PHYSETER.—Head enormously large, truncated in front: upper jaw without whalebone or visible teeth; teeth in the lower jaw numerous: blow-holes united.

(1. CATODON.) No dorsal fin.

(2. PHYSETER.) An elevated dorsal fin.

31. BALÆNA.—Head very large: palate furnished with whalebone: no teeth: blow-holes separate.

(1. BALÆNA.) No dorsal fin.

(2. BALÆNOPTERA.) A dorsal fin.

ORDER I. FERÆ.

---

† GEN. 1. URSUS, *Linn.*

† 1. U. *Arctos*, Linn. (*Brown Bear.*)—Blackish brown: forehead convex above the eyes: snout suddenly tapering: soles of the hinder feet moderately long.

U. Arctos, *Desmar. Mammal.* p. 163. Common Bear, *Shaw, Gen. Zool.* vol. i. p. 450. pl. 102.

DIMENS. Length of the head and body three feet seven inches; of the head one foot; of the fore foot seven inches seven lines; of the hinder foot eight inches ten lines.

DESCRIPT. Body entirely covered with thick shaggy hair, varying in colour from chestnut brown to black: soles of the fore feet with their anterior half naked; those behind naked throughout: ears short and rounded: eyes small: tail very short.

Formerly an inhabitant of Great Britain; but extirpated many centuries back. Infested Scotland (according to Pennant) so late as the year 1057. Still common on many parts of the European continent.

GEN. 2. MELES, *Cuv.*

2. M. *Taxus*, Flem. (*Badger.*)—Gray above, black beneath; a black band on each side of the head extending from the nose over the eyes to behind the ears.

M. Taxus, *Flem. Brit. An.* p. 9. M. vulgaris, *Desm. Mammal.* p. 173. Badger, *Penn. Brit. Zool.* vol. i. p. 85. pl. 8. no. 13. *Shaw, Gen. Zool.* vol. i. p. 467. pl. 106.

DIMENS. Length of the head and body two feet six inches; of the head six inches eight lines; of the ears one inch four lines; of the tail seven inches.

DESCRIPT. Body thick and clumsy: hair rigid and very long; gray on the upper parts, black on the throat, breast, belly, and legs: head above white, with a longitudinal black spot on each side, which takes its origin between the extremity of the nose and the eye, and terminates behind the ear: toes five on each foot: claws long and bent: eyes very small: ears short and rounded, almost concealed in the hair: a transverse glandular pouch between the tail and the anus, secreting a fœtid substance.

Found in several parts of the kingdom, but not of very general occurrence. Burrows in the ground, concealing itself during the day, and coming abroad at night. Feeds indiscriminately on animal and vegetable substances.

GEN. 3.   MUSTELA, *Linn.*

(1. Mustela, *Cuv.*)

3.   M.   *Foina*, Linn.   (*Common   Martin.*)—Brown :
throat and breast white.

> M. Foina, *Desm. Mammal.* p. 182.   Martes Fagorum, *Flem. Brit. An.*
> p. 14.   Martin, *Penn. Brit. Zool.* vol. i. p. 92. pl. 6. no. 15.

Dimens.   Length of the head and body eighteen inches; of the head
four inches three lines; of the tail nine inches six lines.

Descript.   Hair of two sorts ; the shorter very fine and soft, and of a
pale ash-colour ; the longer somewhat rigid, less abundant than the last,
ash-coloured at the roots, dusky brown towards the extremity, with a
tinge of chestnut red : legs and tail dusky : under parts somewhat paler
than the upper : a white spot upon the throat extending itself over the
under surface of the neck and anterior portion of the breast.

More generally diffused than the next species, and said to prefer the
vicinity of houses.   Preys on poultry, game, rats, moles, &c.   Breeds
frequently in hollow trees, producing from three to seven young at a
time.

4.   M.   *Martes*, Linn. (*Pine Martin.*)—Brown : throat
and breast yellow.

> M. Martes, *Desm. Mammal.* p. 181.   Martes Abietum, *Flem. Brit.*
> *An.* p. 14.   Pine Martin, *Penn. Brit. Zool.* vol. i. p. 94.   *Don. Brit.*
> *Quad.* pl. 13.

Dimens.   Length of the head and body one foot seven inches; of the
head four inches; of the tail ten inches.

Descript.   Differs from the last species in having the throat, under
surface of the neck, and anterior portion of the breast, yellow : the head
also shorter, the hair rather darker, and the legs a little longer.

Inhabits the fir woods of Scotland : occurs also sparingly in the West
of England.   Builds its nest on the tops of trees, and produces from two
to three young.   Frequents wild situations.   Preys on the smaller quadrupeds and birds.

(2. Putorius, *Cuv.*)

5.   M.   *Putorius*, Linn.   (*Polecat.*)—Dusky brown with
a tinge of yellow ; beneath paler : some white spots about
the ears and mouth.

> M. Putorius, *Desm. Mammal.* p. 177.   *Flem. Brit. An.* p. 14.   Fitchet,
> *Penn. Brit. Zool.* vol. i. p. 89. pl. 6. no. 14.   Polecat, *Shaw, Gen.*
> *Zool.* vol. i. p. 415. pl. 98.

Dimens.   Length of the head and body one foot six inches; of the
head two inches ten lines ; of the ears six lines : of the tail five inches
six lines.

DESCRIPT. Tail proportionably shorter than in the last sub-genus: space round the mouth, and edge of the ears, white: hair on the body of two kinds; the longer sort somewhat harsh, shining, and of a dusky brown colour; the shorter more woolly, and of a yellowish or tawny white; from the mixing of these two the general tint appears brown with a slight cast of tawny yellow: the legs and tail are of a uniform dusky brown.

A common inhabitant of woods and plantations in all parts of the country. Preys on game, poultry, eggs, and the smaller quadrupeds: is particularly fond of blood. Produces in the Spring from five to six young.

* 6. **M. *Furo*, Linn. (*Ferret*.)**—Yellowish white with the eyes red.

M. Furo, *Desm. Mammal.* p. 178. Ferret, *Shaw, Gen. Zool.* vol. I. p. 418. pl. 98.

DIMENS. Length of the head and body one foot two inches; of the head two inches six lines; of the ears six lines; of the tail five inches six lines.

DESCRIPT. In general somewhat smaller than the polecat, and of a more slender shape; the snout is also proportionably longer: colour of the fur bright yellow, here and there tinged with white; sometimes a mixture of white, black, and tawny, with the tail entirely black.

Originally a native of Africa, from whence imported into Europe. Known in this country only in a domesticated state, where it is much used in the destruction of rabbits and rats. Supposed by some to be a mere variety of the polecat; but independently of the differences above mentioned, the circumstance of its being a native of a warmer climate seems to militate against that idea. Habits similar to those of that species.

7. **M. *vulgaris*, Gmel. (*Weasel*.)**—Reddish brown above; beneath white: tail of the same colour with the body.

M. vulgaris, *Desm. Mammal.* p. 179. *Flem. Brit. An.* p. 13. Common Weasel, *Penn. Brit. Zool.* vol. I. p. 95. pl. 7. no. 17. *Shaw, Gen. Zool.* vol. I. p. 420. pl. 98.

DIMENS. Length of the head and body (*male*) eight inches three lines, (*female*) seven inches; of the head (*male*) one inch nine lines, (*female*) one inch six lines; of the ears (*male*) four lines, (*female*) three lines; of the tail (*male*) two inches four lines, (*female*) two inches.

DESCRIPT. Upper part of the head, neck and body, shoulders, external and anterior portions of the fore legs, and hind legs wholly, reddish brown tinged with yellowish; the under parts, from the extremity of the lower jaw to the vent, white: a brown spot below each corner of the mouth.

Var. β. White, with a few black hairs at the extremity of the tail.

Common everywhere in the vicinity of barns and outhouses. Devours young birds, eggs, rats, mice, moles, &c. Breeds twice or thrice in the year, and produces from four to six young at a birth. The white variety is rare.—*Obs.* The female of this species is constantly much smaller than the male, and is probably the animal alluded to in White's *Nat. Hist. of Selb.* (vol. I. p. 73.) under the name of *Cane.*

8. M. *Erminea*, Linn. (*Stoat or Ermine.*)—Reddish brown above, beneath white; or wholly white: extremity of the tail always black.

> M. Erminea, *Desm. Mammal.* p. 180.   *Flem. Brit. An.* p. 13.   Stoat, *Penn. Brit. Zool.* vol. i. p. 89. pl. 17. no. 18.   *Shaw, Gen. Zool.* vol. i. p. 426. pl. 99.

Dimens. Length of the head and body ten inches; of the head one inch eleven lines; of the ears six lines; of the tail five inches.

Descript. Larger than the last species, to which it is closely allied.— (*Summer dress.*) Upper part of the head, neck, and back, as well as a considerable portion of the tail, reddish brown; under parts white tinged with yellow; tail terminating in a tuft of black hair.—(*Winter dress.*) Wholly white, or white with a slight tinge of yellow, the extremity of the tail excepted, which remains black.—*Obs.* In spring and autumn these two liveries are found intermixed.

Equally common with the weasel. Habits similar.

GEN. 4.   LUTRA, *Cuv.*

9. L. *vulgaris*, Desm. (*Otter.*)—Deep brown: sides of the head, throat and breast, cinereous.

> L. vulgaris, *Desm. Mammal.* p. 188.   *Flem. Brit. An.* p. 16.   Otter, *Penn. Brit. Zool.* vol. i. p. 92. pl. 8. no. 19.   *Shaw, Gen. Zool.* vol. i. p. 437. pl. 100.

Dimens. Length of the head and body two feet three inches; of the head five inches; of the ears eight lines; of the tail one foot four inches and a half: girth one foot four inches.

Descript. Head broad and flattened : muzzle obtuse: upper lip very thick and muscular, projecting over the lower: mouth rather small: whiskers strong, nearly three inches in length: eyes small, situate one inch behind the nostrils: ears short and rounded, almost hid in the fur: hair on the body of two kinds; the finer sort grayish white; the longer and coarser grayish white at the roots, deep brown at the extremity, the latter colour alone appearing externally: sides of the head, throat, under surface of the neck, and breast, cinereous: hair on the feet short, brown with a reddish tinge: tail dusky brown.

Inhabits the banks of rivers, lakes, and marshes. Swims and dives with great facility, and is destructive to fish, on which it preys. Breeds in March: goes with young nine weeks, and produces from four to five at a birth.

GEN. 5.   CANIS, *Linn.*

(1. Canis, *Flem.*)

* 10. C. *familiaris*, Linn. (*Dog.*)—Tail recurved.

> C. familiaris, *Desm. Mammal.* p. 190. *Flem. Brit. An.* p. 10. Dog, *Penn. Brit. Zool.* vol. i. p. 59. *Shaw, Gen. Zool.* vol. i. p. 273. pl. 75.

Descript. Known only in the domesticated state. Varieties extremely numerous, offering every gradation of size, form, colour, and quality of the hair.

Primitive stock supposed to have been allied to the *Shepherd's Dog*, a race chiefly characterized by straight ears, long hair, and a bushy tail.

Feeds principally on animal substances, and is particularly attached to carrion. Female goes with young sixty-three days, and produces from six to twelve at a time. The puppies are born blind, and continue so for the first ten days.

† 11. C. *Lupus*, Linn. (*Wolf.*)—Tail straight : eyes oblique.

C. Lupus, *Desm. Mammal.* p. 197. Wolf, *Penn. Brit. Zool.* vol. i. p. 75. pl. 5. *Shaw, Gen. Zool.* vol. i. p. 290. pl. 75.

DIMENS. Length of the head and body three feet nine inches ; of the head ten inches six lines ; of the ears four inches nine lines ; of the tail one foot five inches.
DESCRIPT. Larger than the dog ; the limbs stronger, and the body more muscular : muzzle sharp and pointed : eyes set obliquely : ears erect : tail straight and pendent ; somewhat long and bushy : fur tawny or yellowish gray, with a black streak on the fore legs in the adult state.
Formerly abundant throughout Great Britain, but long since extirpated. Continued in Scotland till the year 1680, and in Ireland so late as 1710. Still found in most parts of the continent.

(2. VULPES, *Flem.*)

12. C. *Vulpes*, Linn. (*Fox.*)—Tawny brown above ; white beneath : ears externally black : tail thick and bushy, tipped with white.

C. Vulpes, *Desm. Mammal.* p. 201. Vulpes vulgaris, *Flem. Brit. An.* p. 13. Fox, *Penn. Brit. Zool.* vol. i. p. 71.

DIMENS. Length of the head and body two feet three inches ; of the head six inches ; of the ears four inches ; of the tail one foot four inches : girth one foot one inch.
DESCRIPT. Muzzle sharp : head rather large, and somewhat flattened on the forehead : tusks relatively longer than in the dog, and more slender : ears erect and pointed : eyes oblique : tail very thick and bushy : fur thick and long, of a tawny or reddish brown colour : lips, lower jaw, forepart of the neck, abdomen, and inside of the thighs, white : back of the ears blackish brown ; a streak of the same colour passing from the corner of each eye to the nose : extreme tip of the tail white. Varies occasionally in size and colour, as well as in the quality of the hair.
Frequents woods in the vicinity of farmyards and villages. Prowls about during the night, preying on poultry, small birds, rabbits, &c. Brings forth its young under ground, and produces from three to six at a birth.

GEN. 6. FELIS, *Linn.*

13. F. *Catus*, Linn. (*Wild Cat.*)—Yellowish gray, with longitudinal and transverse bars of black : tail annulated with black ; of equal thickness throughout.

F. Catus, *Desm. Mammal.* p. 232. *Flem. Brit. An.* p. 15. Wild Cat, *Penn. Brit. Zool.* vol. i. p. 80.

DIMENS. Length of the head and body (*male*) twenty-seven inches nine lines, (*female*) twenty-one inches three lines ; of the tail (*male*) thirteen inches six lines, (*female*) twelve inches four lines.

DESCRIPT. Hair long and bushy; gray or brownish gray above, more or less tinged with yellow; beneath paler: a longitudinal black line down the middle of the back, with transverse bars of the same colour branching out laterally in parallel waves and extending over the flanks, shoulders, and thighs; several narrow parallel black lines on the top of the head between the ears: lips black: feet yellowish; the soles black: tail annulated with black; the tip entirely of that colour; of equal thickness throughout its whole length, and as it were truncated at the extremity.

Confined entirely to the northern parts of the island. Resides in woods; and climbs trees with great facility. Preys on birds and small quadrupeds. Goes with young fifty-six days, and produces from four to five at a birth.

* 14. F. *maniculata*, Rüpp. (*Domestic Cat.*)—Tail relatively longer than in the last species, and gradually tapering to a point.

F. maniculata, *Temm. Monog. de Mammal.* tom. I. p. 128. Domestic Cat, *Penn. Brit. Zool.* vol. I. p. 82.

DESCRIPT. Invariably smaller than the wild cat: the head larger in proportion to the body; the tail longer, more slender, and gradually terminating in a point: colours extremely variable.

I have followed Rüppel and Temminck in referring the domestic Cat to this species, which in a wild state inhabits the North of Africa, where it was discovered by the former of the two naturalists above mentioned. He supposes it to have been first reclaimed by the Egyptians.

GEN. 7. PHOCA, *Linn.*

15. P. *vitulina*, Linn. (*Common Seal.*)—Fur yellowish gray more or less marked and spotted with brown; whiskers with the bristles undulated.

P. vitulina, *Desm. Mammal.* p. 244. *Flem. Brit. An.* p. 17. Common Seal, *Penn. Brit. Zool.* vol. I. p. 137. pl. 12. *Shaw, Gen. Zool.* vol. I. p. 250. pl. 70.

DIMENS. Entire length from five to six feet, sometimes more.

DESCRIPT. Body elongated and somewhat conical, tapering gradually from the breast to the tail: neck very short: head large and round: muzzle broad, flat, and as it were truncated: upper lip furnished with long whiskers, the bristles of which have alternate contractions and dilatations so as to appear of unequal thickness: incisors six above and four below, the lower ones separated in pairs by an intermediate space; grinders all of the false kind, on each side five above and below: no external ears: feet very short; the toes, which are five in number, enveloped in a membrane; claws very strong, larger on the hind than on the fore feet: hair fine and close-set, abundant on the young: colour variable; the general tint yellowish gray with different shades of brown, usually deeper on the head and back than on the sides and abdomen: tail shorter than the hind feet.

Common on many parts of the coast, but prefers rocky shores. Swims and dives readily. Preys on fish, which it devours under the water. Breeds about Midsummer, and produces two young.—*Obs.* The *Pied Seal* of Pennant is probably only a variety of this species.

16. P. *barbata*, Mull. (*Great Seal*.)—Fur blackish: middle toes on the forefeet longer than the lateral ones.

P. barbata, *Desm. Mammal.* p. 246. *Flem. Brit. An.* p. 18. Great Seal, *Penn. Brit. Zool.* vol. I. p. 136. *Don. Brit. Quad.* pl. 11.

DIMENS. Entire length from ten to twelve feet.
DESCRIPT. Teeth as in the last species: head somewhat elongated: muzzle broad: upper lip rounded, fleshy, divided into two lobes by a deep furrow; each lobe furnished with numerous strong white bristles, semi-transparent, and curled at the ends: eyes large; irides dark hazel: auricular apertures larger than in the last species: fore feet rather long, with the middle toe more developed than the lateral ones; claws of these black, horny, and curved; those of the hind feet long and straight: body elongated; when young, clothed with a long woolly fur, which after the space of fourteen or fifteen days is cast and superseded by a new covering of close short hair*: general colour dusky gray, in very old individuals black.
Inhabits the coast of Scotland, and (according to Selby) the Farn Islands, but is much less common than the last species. Breeds in November.

GEN. 8. TRICHECHUS, *Linn.*

17. T. *Rosmarus*, Linn. (*Walrus*.)

T. Rosmarus, *Desm. Mammal.* p. 253. *Flem. Brit. An.* p. 18. Arctic Walrus, *Shaw, Gen. Zool.* vol. I. p. 234. pl. 68. *Nat. Miscell.* vol. VIII. pl. 276.

DIMENS. Entire length from eleven to fifteen feet.
DESCRIPT. Head small, rounded, obtuse: lips very thick and swollen; upper lip divided into two large rounded lobes, over which are scattered numerous semitransparent bristles, somewhat flattened towards their roots, and slightly pointed at their extremities: eyes small and brilliant: auricular apertures situate very much behind: mouth small, armed with two enormous tusks bent downwards, and attaining in some individuals the length of two feet: neck short: body very thick and heavy: tail relatively longer than in the last genus: skin of a dusky hue, with a very few short, scattered, reddish hairs: hind feet very broad.
Very rare in the British seas. A solitary individual shot on the east coast of Harris in December 1817: a second killed in Orkney in June 1825. Habits resembling those of the Seals.

* See Selby's observations on this species in *Zool. Journ.* vol. II. p. 465.

GEN. 9. TALPA, *Linn.*

18. T. *Europæa*, Linn. (*Mole.*)

T. Europæa, *Desm. Mammal.* p. 160.   *Flem. Brit. An.* p. 9.   Mole,
  Penn. Brit. Zool. vol. I. p. 128.   *Shaw, Gen. Zool.* vol. I. p. 515.
  pl. 117.

DIMENS. Length of the head and body five inches three lines; of the
head one inch seven lines; of the tail one inch two lines.

DESCRIPT. Body thick, oblong, almost cylindrical: snout sharp and
slender: eyes extremely small; entirely concealed in the fur: no external
ears: feet extremely short; the anterior pair larger, very robust, and
inclining sideways: claws strong, as long as the toes themselves. Fur
very soft and silky, shining, black or deep ash-colour, according to
the direction in which it is viewed, sometimes white, or yellowish
white.

Common in England and Scotland, but said to be unknown in Ireland.
Habits subterraneous: constructs galleries beneath the surface of the
soil, which it throws up at intervals in hillocks. Feeds on worms and
insects, but principally on the former. Breeds twice in the year, producing from four to six at each birth.

GEN. 10. SOREX, *Linn.*

19. S. *Araneus*, Linn. (*Common Shrew.*)—Reddish brown above; paler beneath: tail shorter than the body, somewhat square, not ciliated on its under surface.

S. Araneus, *Desm. Mammal.* p. 149.   *Flem. Brit. An.* p. 8.   Fetid
  Shrew, *Penn. Brit. Zool.* vol. I. p. 125.   Common Shrew, *Shaw,*
  Gen. Zool. vol. I. p. 527. pl. 118.

DIMENS. Length of the head and body two inches five lines; of the
head one inch; of the ears two lines; of the tail one inch nine lines.

DESCRIPT. Somewhat variable in size and colours. Upper parts generally dusky brown, more or less deep, with a tinge of red; under parts
grayish white with a tinge of yellow: in some specimens a triangular
whitish patch upon the throat: ears small, hardly showing themselves
above the fur, furnished internally with two lobes or duplicatures of the
skin placed one above the other and fringed with hair: incisors deep
ferruginous brown*: tail varying in length, always shorter than the
body, roundish approaching to square, rather stout, of equal thickness
throughout and blunt at its extremity, uniformly clothed with short
dusky hairs, but having no fringe along its under surface: feet much
smaller than in the other species of this genus, the toes scarcely ciliated.

A very common species inhabiting gardens and hedge rows. Feeds
on insects, and also on vegetable substances. Possesses a strong musky
odour.

---

* In the *Ann. du Mus.* (tom. XVII. p. 176.) M. Geoffroy St Hilaire describes this species as
having the incisors *entirely white.* This circumstance, together with one or two others, induces
me to suspect that the *S. Araneus* of the continental authors may be distinct from ours.

20. S. *fodiens*, Gmel. (*Water Shrew.*)—Blackish brown
above ; grayish white beneath : tail two-thirds the length of
the body : feet and tail strongly ciliated with white hairs.

> S. fodiens, *Flem. Brit. An.* p. 8.   S. Daubentonii, *Desm. Mammal.*
> p. 150.   Water Shrew, *Penn. Brit. Zool.* vol. i. p. 126. pl. 11. no. 33.
> *Don. Brit. Quad.* pl. 6.

Dimens. Length of the head and body three inches two lines ; of the
head one inch ; of the ears two lines ; of the tail two inches.

Descript. Upper parts deep brown, approaching to black ; under
parts pale ash gray, in some individuals silvery white ; in general the
two colours separated by a well-defined line ; a triangular dusky spot on
the anus : snout long, somewhat depressed, emarginated at the extremity :
whiskers long : eyes small, almost concealed : ears very short, not projecting above the fur, furnished internally with three valves or lobes, on
one of which is a tuft of white hair, giving the effect of a white spot upon
the auricle : intermediate incisor teeth with the tips ferruginous : tail
more slender than in the last species, quadrangular throughout the greater
part of its extent, the extreme tip being flattened ; dusky, with a cilium
of white hairs along its under surface : feet dusky ; the toes fringed
with white hairs. *Weight* three drachms.—*Obs.* English authors do
not quite agree in their descriptions of this animal, from which circumstance it would seem that it is either subject to much variation, or that
there is some other indigenous species with which it has been confounded.

Not uncommon in many parts of the country, inhabiting marshy districts. Swims and dives with great facility. Preys on insects : is also
said to attack frogs. Produces in the Spring from six to eight young.

21. S. *remifer*, Geoff. ? (*Oared Shrew.*)—Black above ;
scarcely paler beneath ; throat brownish ash : tail quadrangular at the base, flattened at the extremity, and, as well as
the feet, ciliated.

> S. remifer, *Geoff. Ann. du Mus.* tom. xvii. p. 182. pl. 2. f. 1 ?   *Desm.*
> *Mammal.* p. 152 ?   *Yarr. in Loud. Mag. of Nat. Hist.* vol. v. p. 598.
> S. ciliatus, *Sow. Brit. Misc.* pl. 49.

Dimens. Length of the head and body three inches two lines ; of the
head one inch ; of the ears two lines ; of the tail two inches one line ; of
the fore foot five lines ; of the hind foot eight lines and a half.

Descript. Body more thick and bulky for its length than in the last
species : snout broad, and rather obtuse : feet and tail ciliated ; the latter
distinctly quadrangular for two-thirds of its length, compressed towards
the tip, where the hairs of the cilium become longer. Colour darker, and
more uniform than that of the *S. fodiens ;* all the upper parts, sides of
the abdomen, and region of the pubes, black ; throat, breast, and middle
of the abdomen, dusky ash, with a faint tinge of yellowish ; sometimes,
but not always, a white spot on the ears ; incisors ferruginous at their
extremities ; feet and tail dusky gray, the latter with the cilium of hairs
underneath of the same colour as above. *Weight* three drachms fifty-
one grains.

This species, which has been taken in Norfolk, Cambridgeshire, and
Surrey, as well as in Scotland, appears to be identical with a foreign one
in the British Museum, ticketed *S. remifer :* nevertheless I feel some

doubts as to its being the species originally described by Geoffroy under that name. It however approaches more nearly to it than to any other I am acquainted with. My specimen was taken in a corn-field at some distance from any water. The others have occurred in ditches. The ciliated feet and tail evidently mark it to be of aquatic habits.

GEN. 11.   ERINACEUS, *Linn.*

22.   E. *Europæus*, Linn. (*Hedgehog.*) — Ears short: spines moderately long.

E. Europæus, *Desm. Mammal.* p. 147. *Flem. Brit. An.* p. 7. Common Urchin, *Penn. Brit. Zool.* vol. i. p. 133. Hedgehog, *Shaw, Gen. Zool.* vol. i. p. 542. pl. 121.

Dimens. Length of the head and body nine inches six lines; of the head two inches eleven lines; of the ears one inch one line; of the tail nine lines.

Descript. Body oblong, convex above, very low on the legs: head produced into a snout which is truncated at the extremity: nostrils narrow: ears short, broad, and rounded: neck thick and short: upper part of the body clothed with round sharp-pointed spines, barely an inch in length, which cross and interlace one another in all directions; the colour of them is whitish with a black ring a little higher up than the middle: snout, forehead, sides of the head, under part of the neck, breast, and legs, covered with coarse stiff hairs of a yellowish white colour.

Resides in hedges, thickets, &c. Habits nocturnal. Omnivorous; devouring roots, insects, worms, flesh, and even snakes. Becomes torpid during the Winter. Produces in the early part of the Summer from two to four young. Spines soft at birth, and all inclining backwards; become hard and sharp in twenty-four hours. In the adult state, has the power of rolling itself into a ball to avoid danger.

ORDER II.   PRIMATES.

GEN. 12.   RHINOLOPHUS, *Geoff.*

23.   R. *Ferrum-equinum*, Gmel. (*Greater Horse-shoe Bat.*)—Posterior foliaceous appendage lanceolate, expanding laterally at the base: ears notched on their external margins.

R. Ferrum-equinum, *Flem. Brit. An.* p. 5. R. unihastatus, *Desm. Mammal.* p. 125. Horse-shoe Bat, *Penn. Brit. Zool.* vol. i. p. 147. pl. 14.

Dimens. Length of the head and body two inches five lines; of the head eleven lines and a half; of the tail one inch two lines and a half; of the ears nine lines; breadth of the ears six lines; length of the thumb two lines and a half: extent of wing thirteen inches.

Descript. Upper incisors very small, separated from each other by a space; lower incisors each with three lobes: ears nearly as long as the head, somewhat triangular, broad at the base, terminating upwards in an

acute point; the external margin notched at the base, from which point
it becomes inflexed and rises into an elevated round lobe that guards the
orifice of the ear and appears to act the part of the tragus, which is wanting: nostrils placed at the bottom of a cavity close to each other, surrounded by a naked membrane in the form of a horse-shoe arising from
the upper lip; anterior foliaceous appendage rising vertically immediately
behind the nostrils, of a somewhat pyramidal form, sinuous at the margins and at the apex, which last is obliquely truncated; the posterior one
situate on the forehead, placed transversely with respect to the first and
standing more erect, lanceolate, expanding laterally at the base, in front
of which are two small cup-shaped cavities formed by a duplicature of the
skin. Colour of the fur reddish ash, inclining to gray beneath: membranes dusky: ears within and without slightly hairy.

A local species, inhabiting caves and buildings. Found in Bristol and
Rochester Cathedrals, Dartford Powder Mills, and in some other parts of
the country.

24. R. *Hipposideros*, Bechst. (*Lesser Horse-shoe Bat.*)
—Posterior foliaceous appendage lanceolate, without any
lateral expansions: ears deeply notched on their external
margins.

R. Hipposideros, *Leach, Zool. Misc.* vol. III. p. 2. pl. 121. *Flem.
Brit. An.* p. 5. R. bihastatus, *Desm. Mammal.* p. 125. Vespertilio minutus, *Mont. in Linn. Trans.* vol. IX. p. 163. pl. 8. f. 5.

Dimens. Length of the head and body one inch four lines; of the
head eight lines; of the tail nine lines; of the ears five lines; breadth of
the ears four lines and a half; length of the thumb two lines: extent of
wing eight inches four lines.

Descript. Principally distinguished from the last by its very inferior
size. General appearance of the ears, nose, and foliaceous appendages
similar; the anterior appendage is however less obliquely truncated at
the apex, and the posterior one narrower at the base, and without the
lateral expansions: the ears likewise are more deeply notched and the
external margin altogether more sinuous. Fur soft and rather long; pale
rufous brown above, grayish ash beneath with a tinge of yellow.

Found in Wiltshire, Dorsetshire, Devonshire, Somersetshire, and also
in Wales. Is often met with in company with the last species, but is
considered as less common.

GEN. 13. VESPERTILIO, *Linn.*

(1. Vespertilio, *Geoff.*)

25. V. *murinus*, Desm.—Ears oval, as long as the
head; tragus falciform, half the length of the auricle:
fur reddish brown above; dirty white beneath.

V. murinus, *Desm. Mammal.* p. 134. La Chauve-souris, *Buff. Hist.
Nat.* tom. VIII. pl. 16.

Dimens. Length of the head and body three inches five lines; of the
head eleven lines; of the tail one inch eight lines; of the ears eleven
lines and a half; of the tragus five lines; of the thumb five lines: extent
of wing fifteen inches.

Descript. Face almost naked: forehead very hairy: eyes rather large, with a few dusky hairs immediately above them: ears inclining backwards, as long as the head, oval, naked, grayish ash-colour externally, somewhat yellowish within: tragus falciform, about half the length of the auricle. Fur above pale reddish brown; beneath dirty white, inclining to yellowish: flying and interfemoral membranes brownish.

This must not be confounded with the *V. murinus* of English authors, from which it may be readily distinguished by its very superior size. It is a common species on the Continent, but apparently rare in Britain. The only indigenous specimens known were taken in the gardens of the British Museum, and are in that collection. Said to inhabit churches and buildings: never found in trees.

26. V. *Bechsteinii*, Leisl.—Ears oval, a little longer than the head; tragus falciform, bending outwards at the extremity, not half the length of the auricle: fur reddish gray above; white beneath.

V. Bechsteinii, *Desm. Mammal.* p. 135.

Dimens. Length of the head and body two inches one line; of the head nine lines; of the tail one inch three lines; of the ears ten lines; of the tragus four lines; of the thumb four lines: extent of wing eleven inches.

Descript. Allied to the last species, but distinguished by its smaller size, relatively larger ears, and remarkably slender thumb. Face almost naked: muzzle long, and conical: ears oval, somewhat longer than the head, rounded at the extremity; tragus lanceolate, pointed, a little bent outwards towards the extremity. Fur reddish gray on the upper parts, whitish on the under.

A rare species, of which specimens are in the British Museum from the New Forest. Resides constantly amongst trees, and is never found in buildings.

27. V. *Nattereri*, Kuhl.—Ears oblong-oval, about as long as the head; tragus lanceolate, nearly two-thirds the length of the auricle: fur reddish gray above, white beneath: interfemoral ample, the margin crenate on each side towards the tip, and fringed with short bristly hairs.

V. Nattereri, *Desm. Mammal.* p. 135.

Dimens. Length of the head and body one inch eleven lines; of the head eight lines and a half; of the tail one inch seven lines; of the ears eight lines and a half; of the tragus five lines; breadth of the ears three lines and three quarters; of the tragus at the base one line; length of the fore-arm one inch six lines; of the thumb two lines and three quarters: extent of wing ten inches eight lines.

Descript. Head rather small; snout attenuated: nose a line in breadth at the extremity, slightly emarginated between the nostrils, convex above: all the face, excepting immediately above the nose, hairy, but the hair very thinly scattered about the eyes and chin, with a few bristly hairs longer than the others intermixed: gape extending as far as the posterior angle of the eye: a row of longish hairs on the upper lip forming a moustache : a large sebaceous gland above the eye advancing towards the nose, of a yellowish white colour; another gland on the upper lip immediately above the canine tooth, of the same colour with the rest of the face, causing a protuberant swelling at that spot, and furnished with long bristles: ears oblong-oval, rather broad, about the length of the head, the inner margin bending outwards, the external margin nearly straight, with a shallow notch about midway, nearly naked, excepting towards the base on the outside; tragus longer than in the last species, lanceolate, curving slightly outwards towards the tip, naked: flying and interfemoral membranes naked; the latter ample, furnished with a long spur*, between which and the end of the tail the margin has a crenate or puckered appearance, and is set with short bristly hairs; tip of the tail somewhat blunt and compressed, scarcely exserted for more than half a line: hind claws very strong, furnished with long hairs: thumb much smaller than in the last species. Fur long and silky, particularly about the upper part of the head and neck, of a light rufous brown, approaching to reddish gray above, the tips of the hairs being of this colour, the roots dusky brown; beneath silvery gray at the tips, black towards the roots: ears yellowish gray, the yellow tinge being most obvious within towards the base; tragus almost entirely yellowish: flying and interfemoral membranes dusky, but the latter paler than the former. In the *female*, the colour of the fur on the upper parts has a more reddish tinge than in the *male*.

This species occurs in hollow trees in the neighbourhood of buildings at Swaffham Prior in Cambridgeshire. Mr Yarrell has received it from Colchester and Norwich. There are also specimens in the British Museum taken near London. It appears however to be very locally distributed.

28. V. *Serotinus*, Gmel. (*Serotine*.)—Ears oval-triangular, shorter than the head; tragus semicordate : muzzle short, and obtuse: fur chestnut brown above; pale beneath.

V. Serotinus, *Desm. Mammal.* p. 137.   La Sérotine, *Buff. Hist. Nat.* tom. VIII. pl. 18. f. 2.

DIMENS. Length of the head and body two inches seven lines; of the head ten lines; of the tail one inch ten lines; of the ears eight lines; of the tragus three lines ; of the thumb three lines: extent of wing twelve inches six lines.

DESCRIPT. Face almost naked: muzzle remarkably short, broad, and obtuse; extremity of the nose one line and a half across: ears oval, approaching triangular, shorter than the head, with the inner margin bending outwards in an arcuate form, externally hairy on their basal half, naked above; tragus semicordate, somewhat elongated, pointed at the extremity. Fur in the *male* of a deep chestnut brown on the upper parts, passing beneath into a yellowish gray ; that of the *female* said to be much brighter.

Apparently a rare species in Great Britain : has hitherto only occurred in the neighbourhood of London. Is said to frequent trees, and also houses occasionally. Habits somewhat solitary.

---

* By the *spur* I mean a long tendinous process from the heel of the foot which runs along the margin of the interfemoral membrane and serves to stretch it. It in fact represents the *os calcis*. It will be found of very different length in different species, varying from three to seven lines or more.

29. V. *Noctula*, Gmel. (*Noctule.*)—Ears oval-triangular, shorter than the head; tragus small, arcuate, terminating in a broad round head: fur short, of a uniform reddish brown above and below.

V. Noctula, *Desm. Mammal.* p. 136.   *Flem. Brit. An.* p. 6.   La Noctule, *Buff. Hist. Nat.* tom. viii. pl. 18. f. 1.   Great Bat, *Penn. Brit. Zool.* vol. i. p. 146. pl. 13. no. 38.

Dimens. Length of the head and body two inches eleven lines; of the head ten lines; of the tail one inch eight lines; of the ears seven lines and a half; of the tragus two lines and a half; breadth of the ears six lines; of the tragus one line and a half; length of the fore-arm two inches; of the thumb two lines and a half: extent of wing fourteen inches.

Descript. Head very broad; muzzle short and thick in the adult state, though less so than in the last species, somewhat elongated when young; nostrils tumid at the edges, slightly bilobated; forehead very hairy; rest of the face almost naked: ears shorter than the head, somewhat triangular, rounded at the extremity; the posterior margin folded back, with a projecting ridge internally, and a small protuberance at the base, which extends round nearly to the corners of the mouth; tragus very small, somewhat arcuate with the bend directed inwards, terminating above in a broad round head. Fur rather short, but soft and thick, of a uniform reddish brown colour above and below: membranes dusky, with a ridge of hair along the bones of the arm. Tail shorter than the fore-arm, protruding from the interfemoral membrane to the extent of a line and a quarter.

Common in many parts of the country, but only to be seen on wing during the summer months. Flight high and rapid. Habits gregarious. Retires early in the Autumn into hollow trees, and beneath the roofs of large buildings. Has a strong disagreeable smell.

30. V. *Leisleri*, Kuhl.—Ears oval-triangular, shorter than the head; tragus terminating in a round head: fur long, bright chestnut above, darker beneath: under surface of the flying membrane with a broad band of hair along the fore-arm.

V. Leisleri, *Desm. Mammal.* p. 138.

Dimens. Length of the head and body two inches two lines; of the head seven lines and a half; of the tail one inch eight lines; of the ears five lines; of the tragus two lines and a half; breadth of the ears four lines; of the tragus one line and a quarter; length of the fore-arm one inch six lines and a half; of the thumb one line and three quarters: extent of wing eleven inches.

Descript. Muzzle rather more elongated than in the *Noctule;* nose depressed, naked, as is also the region of the eyes: ears oval-triangular, shorter than the head, broad, the outer basal margin advancing forwards to nearly the corners of the mouth; tragus terminating in a round head, bulging out about the middle of the external margin, slightly bending inwards at the extremity, considered relatively, somewhat larger and longer than in the last species. Fur on the upper parts bright chestnut, this

colour occupying the tips of the hairs, the roots of which are deep brown;
under parts darker than the upper, the hair being grayish brown at the
tips, dusky at bottom: flying membranes dusky; the portion contiguous
to the body very hairy; a band of scattered hair also extends along and
beneath the fore arm, about five lines in breadth: upper half of the interfemoral membrane both above and below likewise very hairy. Thumb
short: bones of the arm and fore-arm much slenderer than in the
Noctule.

The only indigenous specimen which I have seen of this species is in
the British Museum. It is not known where it was taken. According
to Desmarest it resides constantly in hollow trees in large companies, and
is attached to the neighbourhood of stagnant waters.

31. V. *discolor*, Natt.——Ears shorter than the head, oval, bending outwards, with a projecting lobe on the inner margin; tragus of equal breadth throughout: fur above reddish brown, with the tips of the hairs white; beneath dirty white.

V. discolor, *Desm. Mammal.* p. 139.

DIMENS. Length of the head and body two inches four lines; of the
head nine lines; of the tail one inch five lines; of the ears six lines and a
half; of the tragus two lines and a quarter; of the thumb three lines:
extent of wing ten inches six lines.

DESCRIPT. Forehead broad and hairy; muzzle long, and very broad;
nose thick and obtuse, measuring one line and two-thirds across the
extremity: eyes very small: ears shorter than the head, rounded, oval,
bending outwards and reaching almost to the corners of the mouth, with
a projecting lobe near the base of the internal margin, clothed externally
on their lower half with thick woolly hair; tragus short, and nearly of
equal breadth throughout. Fur on the back reddish brown, with the
extreme tips of the hairs white, presenting a marbled appearance; underneath dirty white, with a large patch of a somewhat darker tint covering
the breast and abdomen; throat pure white. Tail protruding from the
interfemoral membrane to the extent of three lines.

A single individual of this species, taken at Plymouth by Dr Leach,
is now in the British Museum. Said to be peculiar to houses, and
never to be found in trees.

32. V. *Pipistrellus*, Gmel. (*Pipistrelle.*)——Ears shorter than the head, oval-triangular, deeply notched on their external margins; tragus nearly straight, terminating in a blunt rounded head: fur reddish brown above; somewhat paler beneath.

V. Pipistrellus, *Desm. Mammal.* p. 139. *Jen. in Linn. Trans.* vol.
XVI. p. 163.   Le Pipistrelle, *Buff. Hist. Nat.* tom. VIII. pl. 19.
f. 1.   Common Bat, *Penn. Brit. Zool.* vol. I. p. 148.

DIMENS. Length of the head and body one inch seven lines; of the
head six lines; of the tail one inch two lines; of the ears four lines; of
the tragus two lines; breadth of the ears three lines; of the tragus three-

quarters of a line; length of the fore-arm one inch two lines; of the thumb one line and three quarters: extent of wing eight inches four lines.

Descript. Resembling the *Noctule* in many of its characters but much smaller: head depressed in front, convex behind; muzzle short in the adult state, somewhat elongated when young; nose obtuse at the extremity, and slightly emarginated between the nostrils; a protuberant swelling on each side of the face above the upper lip formed by a congeries of sebaceous glands: eyes very small; above each an elevated wart furnished with a few black hairs: ears broad, oval-triangular, rather more than half as long as the head, with their external margins deeply notched about midway down: tragus half the length of the auricle, nearly straight, oblong, with a blunt rounded head: tail as long as the fore-arm. Fur rather long and silky, yellowish red on the forehead and at the base of the ears, on the rest of the upper parts reddish brown, with the lower half of each hair dusky; on the under parts wholly dusky, except the extreme tips of the hairs which are of the same colour as above, but paler. Young specimens generally brownish gray, sometimes black, without any tinge of red. Nose, lips, ears, and membranes dusky.

The most common species in this country, although for a long time confounded with the *V. murinus* described above. Congregates in large numbers in the crevices of old walls, decayed door-frames, &c. Resists cold more than the other species, continuing on wing till near the end of the year. Is first seen about the beginning of March.

33. V. *pygmæus*, Leach. (*Pygmy Bat.*)—Forehead with an impressed longitudinal furrow: ears a little shorter than the head, broad at the base, obtuse and rounded at the extremity; tragus linear: fur above brown, darkest on the head and dorsal line; paler beneath.

V. pygmæus, *Leach, in Zool. Journ.* vol. i. p. 560. pl. 22.

Dimens. Length of the head and body one inch two lines and a half; of the head five lines; of the tail nine lines; of the ears four lines: extent of wing five inches four lines.

Descript. Differs from the last species in its smaller dimensions, as well as in its relative proportions. Head high; the forehead marked with a longitudinal furrow; muzzle short and obtuse, nearly of equal breadth throughout; nostrils small, opening laterally: ears somewhat shorter than the head, broad at the base, obtuse and rounded at the extremity: anterior margin nearly straight, posterior slightly concave and convolute: tragus about half the length of the auricle, regularly linear, simple and rounded at the extremity: tail as long as the body exclusively of the head; the tip naked, protruding one line beyond the interfemoral membrane. Fur short and delicate, dark brown on the upper parts, the colour being deepest on the head and on the highest part of the back along the spine; inclining to gray underneath: flying membrane dark brown.—Leach.

A new species discovered by Dr Leach at Spitchweek, near the Forest of Dartmoor, where it is said to be extremely common. Has not hitherto occurred elsewhere.

34. V. *emarginatus*, Geoff.—Ears oblong, nearly as long as the head, deeply notched on their external margins ; tragus half the length of the auricle, subulate, bending slightly inwards towards the tip: fur reddish gray above, ash-colour beneath.

V. emarginatus, *Geoff. in Ann. du Mus.* tom. VIII. p. 198. pl. 46. *Desm. Mammal.* p. 140.

DIMENS. Length of the head and body two inches; of the head eight lines; of the tail one inch five lines; of the ears six lines; of the tragus three lines; breadth of the ears three lines and three quarters; of the tragus at the base three quarters of a line; length of the fore-arm one inch four lines; of the thumb two lines and three quarters: extent of wing nine inches six lines.

DESCRIPT. Bearing some relation to the *Pipistrelle* in its physiognomy, but distinguished from that species by its somewhat superior size. Muzzle rather obtuse; head flattish; face hairy, but the hair on the nose and chin thinly scattered and longer than on the other parts; a moustache of soft longish hair on the upper lip, above which is a congeries of glands extending from the eyes to the nostrils: ears moderate, a little shorter than the head, oblong approaching triangular, rounded at the extremity, the inner margin bending outwards, the outer one sinuate and rather deeply notched about half way down; tragus half the length of the auricle, linear approaching to subulate, nearly straight but having a slight bend inwards: thumb longer and stouter than in the *Pipistrelle :* hind feet remarkably large and strong, much more so than in either the *Pipistrelle* or the next species, the toes set with bristly hairs: interfemoral moderately ample; the spur rather short: tail a little longer than the fore-arm, exserted for about one line. Fur long; the hair on the upper parts with the basal half dusky, the tips reddish brown, this last colour alone appearing externally; beneath black at the roots, ash-gray at the tips: ears and flying membrane dusky; under surface of the interfemoral whitish.

Stated to have been found near Dover by M. A. Brongniart, and in Fifeshire by Dr Fleming. The description given above is from a specimen in my possession taken at Milton Park in Northamptonshire. It may possibly be distinct from the true *V. emarginatus* of the continental authors, but it appears to approach nearer to that species than to any other I am acquainted with. Mr Yarrell has also a specimen which was taken at Islington.

35. V. *mystacinus*, Leisl.—Ears oblong, shorter than the head, bending outwards and notched on their external margins ; tragus half the length of the auricle, lanceolate, straight: a moustache of fine close hairs on the upper lip : fur blackish chestnut above, dusky ash beneath.

V. mystacinus, *Desm. Mammal.* p. 140.

DIMENS. Length of the head and body one inch eight lines; of the head seven lines and a half; of the tail one inch five lines; of the ears five lines and a half; of the tragus three lines; breadth of the ears three lines and a half; of the tragus at the base one line; length of the fore-arm one inch three lines; of the thumb two lines and a half: extent of wing eight inches six lines.

DESCRIPT. Head small and flattish; muzzle short; nose swollen, with a shallow cleft in the middle; face much more hairy than in the last species, with a few scattered hairs on the nose and chin, longer than the rest, intermixed; a row of fine, soft, close-set hairs on the upper lip forming a conspicuous moustache; a similar row crossing the forehead: ears shorter than the head, moderately broad, oblong, rounded at the extremities, curving outwards and rather deeply notched on their external margins; tragus half or rather more than half the length of the auricle, lanceolate, perfectly straight, narrowing regularly from the base upwards to the tip which terminates in a sharp point: thumb moderate: hind feet much smaller than in the last species: interfemoral with the spur of about the same length: tail longer than the fore-arm, exserted for about one line. Fur long, thick and woolly; hair dusky, approaching to black, throughout the greater part of its length, the extreme tips being reddish brown on the upper parts and ash-gray beneath: ears, flying and interfemoral membranes dusky; this last sometimes transversely marked on its under surface with numerous white ciliated lines.

This species occurs, though rarely, in houses in Cambridgeshire. I have also received specimens from Milton Park in Northamptonshire. Mr Yarrell has others which were taken at Colchester in the caverns under the castle. Said to frequent the neighbourhood of water, and to retire into hollow trees and houses.

(2. PLECOTUS, *Geoff.*)

36. V. *auritus*, Linn. (*Greater Long-eared Bat.*)— Ears more than double the length of the head ; tragus oval-lanceolate : fur brownish gray on the upper parts; paler beneath.

V. auritus, *Desm. Mammal.* p. 144. Long-eared Bat, *Penn. Brit. Zool.* vol. I. p. 147. pl. 13. no. 40. *Shaw, Gen. Zool.* vol. I. p. 123. pl. 40.

DIMENS. Length of the head and body one inch ten lines; of the head eight lines; of the tail one inch eight lines; of the ears one inch five lines; of the tragus seven lines; breadth of the ears nine lines; of the tragus two lines and a half; length of the fore-arm one inch five lines; of the thumb two lines and three quarters: extent of wing ten inches two lines.

DESCRIPT. Head and face flattened; muzzle somewhat swollen about the nose; nostrils with their anterior and inner edges tumid, elongated posteriorly into a sort of *cul de sac:* eyes small: ears extremely large, more than twice the length of the head, oblong-oval, thin, and semi-transparent; the inner margin presenting a broad longitudinal fold, which doubles back nearly at right angles to the rest of the auricle and is ciliated with hair along its external and internal edges; near the base of this fold is a small projecting lobe, also ciliated; tragus long, oval-lanceolate, with the outer margin somewhat sinuous, the inner one straight: ears united over the head to the height of one line and a half; extending round at the base to the corners of the mouth: flying and interfemoral membranes broad and ample: tail longer than the fore-arm, the tip obtuse, protruding to the extent of three quarters of a line: forehead, and anterior surface of the connecting membrane of the ears, hairy; posterior surface of the same membrane naked. Fur long and silky, brownish gray on the upper parts, paler beneath; the hair every where dusky at the roots.

A common and generally diffused species, resorting principally to the roofs of houses and churches. Flight swifter than that of the Pipistrelle. In the living animal the ears are generally curled, the bend being directed outwards. when at rest they are sometimes wholly concealed beneath the fore-arm, the tragus alone remaining erect.

37. V. *brevimanus*, Jenyns. (*Lesser Long-eared Bat.*) —Ears not double the length of the head; tragus ovallanceolate: fur reddish brown on the upper parts; yellowish white beneath.

V. brevimanus, *Linn. Trans.* vol. xvi. p. 55. pl. l. f. 2.

Dimens. Length of the head and body one inch six lines; of the head seven lines; of the tail one inch two lines; of the ears one inch; of the tragus five lines and a half; breadth of the ears five lines; of the tragus two lines; length of the fore-arm one inch two lines; of the thumb three lines: extent of wing six inches six lines.

Descript. Similar to the last in general appearance, but much smaller, and differing in its relative proportions: ears shorter with respect to the head, and rather narrower at the extremity; tragus relatively longer: bones of the hands much shorter: thumb somewhat longer: tail the length of the fore-arm; the tip more exserted from the interfemoral membrane, and terminating in a fine point. Fur on the upper parts reddish brown, presenting a marked contrast with that on the under which is yellowish white; the hair everywhere of the same colour throughout its whole length, and not dusky at the roots as in the last species: ears and membranes dusky with a tinge of red.

A new species first described by myself in the Linnæan Transactions as above quoted. Only one individual has as yet occurred, which is a female. It was found adhering to the bark of a pollard-willow, in Gruntey Fen in the Isle of Ely.

(3. Barbastellus, *Gray.*)

38. V. *Barbastellus*, Gmel. (*Barbastelle.*)—Ears shorter than the head, broad, triangular, notched on their external margins; tragus semicordate: fur black, with the tips of the hairs grayish white.

V. Barbastellus, *Desm. Mammal.* p. 145.   *Sow. Brit. Misc.* pl. 5. *Mont. Linn. Trans.* vol. ix. p. 171.

Dimens. Length of the head and body two inches; of the head seven lines and a half; of the tail one inch nine lines; of the ears five lines; of the tragus three lines and a half; breadth of the ears five lines; of the tragus one line and a half; length of the fore-arm one inch four lines and a half; of the thumb two lines and a half: extent of wing ten inches three lines.

Descript. Muzzle short and obtuse, somewhat swollen at the extremity; a naked space above the nose extending up to the ears, sunk and hollowed out in front, with the nostrils placed in the cavity; on each side of the face above the upper lip a large protuberant swelling covered with black hair : eyes very small, placed within the auricle and almost concealed : ears united over the head, oval-triangular, as broad as long, externally hairy, and notched on their outer margins ; tragus semicordate, very protuberant near the base of the external margin, above which it suddenly bends inward and terminates in a point; inner margin straight. Fur very long and silky, particularly on the top of the head behind the ears, blackish brown, with the tips of the hairs silvery gray on the upper parts, and ash-coloured beneath ; flying and interfemoral membranes dusky.

A rare species. Has occurred in Devonshire, Kent, Northamptonshire, and Cambridgeshire. Habits more diurnal than usual with this family. Flight slow and near the ground. Resorts to buildings for retirement.

ORDER III. GLIRES.

GEN. 14.   SCIURUS, *Linn.*

39.   S. *vulgaris*, Linn. (*Common Squirrel.*) — Fur brownish red above ; white beneath : ears with a pencil of long hairs at the extremity.

S. vulgaris, *Desm. Mammal.* p. 330.   *Flem. Brit. An.* p. 20.
Common Squirrel, *Penn. Brit. Zool.* vol. i. p. 107.

DIMENS. Length of the head and body nine inches; of the head two inches one line; of the ears nine lines and a half; of the tail (to the end of the bone) six inches six lines, (to the end of the hair) eight inches six lines.

DESCRIPT. Head thick, with the cheeks rather flattened ; nose prominent; the upper lip projecting considerably beyond the lower; the first grinder in the upper jaw extremely small, consisting of a single tubercle, and disappearing altogether at a certain age; eyes black, large, and round ; ears erect, moderately large, ornamented at the extremity with a tuft or pencil of long hairs: neck short: legs strong and muscular, much longer behind than before; fore feet with four long, deeply divided toes, and a claw in the place of a thumb; hind feet with five toes ; claws sharp and strong: tail long and bushy, with the hair spreading out laterally. Colour of the upper parts reddish, or bright chestnut brown, the red tint being deepest on the sides of the head and neck, the shoulders, and the external surface of the legs; lower parts, including the under portion of the neck, breast, abdomen, and inside of the legs, white : tail the colour of the back.

Common in extensive woods, residing and building in trees. Climbs and leaps with great agility. Feeds on buds, acorns, nuts, and other fruits. When at rest, often sits erect, using its fore feet as hands, with its tail turned back over the body. Produces in the Spring from three to four young.

GEN. 15.   MYOXUS, *Gmel.*

40. M. *avellanarius*, Desm. (*Dormouse.*)—Fur tawny red above; white beneath: tail bushy, the length of the body.

> M. avellanarius, *Desm. Mammal.* p. 295.   *Flem. Brit. An.* p. 22.
> Dormouse, *Penn. Brit. Zool.* vol. i. p. 110.   *Shaw, Gen. Zool.*
> vol. ii. p. 167. pl. 154.

DIMENS. Length of the head and body two inches ten lines; of the head eleven lines and a half; .of the ears four lines; of the tail two inches eight lines.

DESCRIPT. Head broad; nose blunt; eyes black, large, and prominent; ears broad, oval, rounded at the extremity: body plump: hind legs not disproportionably longer than the fore; toes four in front with the rudiment of a thumb, five behind: tail very long, somewhat flattened horizontally, thickly clothed with hair on every side. Colour of the upper parts bright tawny red; under parts white, with here and there a yellow tinge: tail red throughout.

Not uncommon in some parts of the country, frequenting woods and thick hedges. Less active than the squirrel: food similar. Builds in the hollows of trees, and produces from three to four young. Becomes torpid in the Winter.

GEN. 16.   MUS, *Linn.*

41. M. *sylvaticus*, Linn. (*Field Mouse.*)—Fur yellowish brown above; whitish beneath, with a ferruginous spot on the breast: ears more than half the length of the head.

> M. sylvaticus, *Desm. Mammal.* p. 301.   *Flem. Brit. An.* p. 19.   Field
> Mouse, *Penn. Brit. Zool.* vol. i. p. 120.   Wood Mouse, *Shaw,*
> *Gen. Zool.* vol. ii. p. 58. pl. 132.

DIMENS. Length of the head and body three inches nine lines; of the head one inch one line; of the ears seven lines; of the tail three inches seven lines.

DESCRIPT. Forehead somewhat convex; nose blunt; eyes black, large and prominent; ears oblong-oval, with the anterior margin doubled in at the base, and a projecting lobe opposite to it, this last arising within the auricle near the base of the posterior margin; whiskers very long, measuring one inch two lines. Upper part of the head and body, cheeks, sides, and external portion of the legs, of a tawny gray or yellowish brown colour, each hair being dusky ash from the root upwards throughout two-thirds of its length, then tawny yellow, and lastly (more particularly some longer ones than the others) black at the extremity; under part of the head and body, and inside of the legs, whitish, with here and there a tinge of dusky ash, the hair on these parts being dusky from the root upwards the same as above, but the extremities white; the line of separation between the above colours is tolerably well defined, and the red tinge is most prevalent just at that part; there is also a faint tawny spot upon the breast: tail a little shorter than the body, more slender and tapering than in the *M. Musculus;* dusky on the upper surface, whitish beneath: toes furnished above with long white hairs extending beyond the claws.

Common in gardens; and is occasionally, though very rarely, found in houses. A larger variety, measuring four inches and a half in length, exclusively of the tail, which is four inches, is sometimes met with in woods. Resides in holes under ground, where it amasses seeds, roots, &c. on which it feeds. Produces nine or ten young at a litter.

42.   M. *messorius*, Shaw. (*Harvest Mouse.*) — Fur tawny brown on the upper parts; beneath white: ears one-third the length of the head.

M. messorius, *Desm. Mammal.* p. 302.   *Flem. Brit. An.* p. 19.
Harvest Mouse, *Shaw, Gen. Zool.* vol. ii. p. 62. *Fig. on Frontisp.*

DIMENS. Length of the head and body two inches six lines; of the head ten lines; of the ears three lines; of the tail two inches six lines.

DESCRIPT. Much smaller than the last species, and of a more elongated form: head narrower in proportion: eyes and ears smaller, the latter barely one-third the length of the head, with the projecting lobe at the base of the posterior margin relatively larger: tail varying in length, equalling the body, or somewhat shorter. Fur above of a bright tawny red colour, with a tinge of brown, the basal half of each hair being dusky, and the upper portion red; beneath almost pure white, the hair being of one tint throughout its whole length; a distinct line of separation between the above colours running parallel with the sides.

First found in Hampshire by White. Since met with in Wiltshire, Devonshire, Cambridgeshire, and other parts of the country. Frequents cornfields during the harvest. Nest round, composed of dry straws and grasses, and suspended at a small height from the ground. Produces seven or eight at a litter.

43.   M. *Musculus*, Linn. (*House Mouse.*)—Fur dusky gray above with a tinge of yellow; beneath cinereous: ears about half the length of the head.

M. Musculus, *Desm. Mammal.* p. 301.   *Flem. Brit. An.* p. 19.
Common Mouse, *Penn. Brit. Zool.* vol. i. p. 122. pl. 11. no. 30.
Shaw, Gen. Zool. vol. ii. p. 56. pl. 131.

DIMENS. Length of the head and body three inches two lines; of the head eleven lines; of the ears five lines; of the tail two inches eleven lines.

DESCRIPT. Head of a somewhat triangular form; muzzle pointed, sharper than in the *M. sylvaticus:* eyes and ears, especially the former, smaller than in that species; the latter oval, rounded at the extremity, and narrower in proportion to their length: whiskers shorter: tail stouter: feet and toes, particularly those behind, shorter. All the upper parts of the body and sides, dusky gray with a faint tinge of yellow, this last colour occupying a small portion of each hair near the summit, the basal half and extreme tip being dusky; under parts cinereous, paler than those above, likewise tinged with yellow: ears, feet, and tail, clothed with a very short fine hair: the last dusky, of one colour above and below.—*Obs.* The colours vary a little in different specimens, and some-times individuals occur wholly white; or white with dark spots.

Found in houses and stacks. Those which occur in the latter situation are remarkable for the brightness of their colour, which has a more decided yellow tinge than in house specimens. They also sometimes attain a larger size, measuring nearly four inches in length. Feeds on various animal and vegetable substances. Breeds several times in the year, and is extremely prolific.

44. M. *Rattus*, Linn. (*Black Rat.*) — Fur grayish-black above; ash-colour beneath : tail a little longer than the body.

M. Rattus, *Desm. Mammal.* p. 300. *Flem. Brit. An.* p. 20. Black Rat, *Penn. Brit. Zool.* vol. i. p. 113. *Shaw, Gen. Zool.* vol. ii. p. 52. pl. 130.

Dimens. Length of the head and body seven inches four lines; of the head one inch ten lines; of the ears eleven lines and a half; of the tail seven inches eleven lines.

Descript. Head long; muzzle sharp-pointed; lower jaw very short; eyes large and prominent; ears oval, broad, and naked: whiskers long: fore feet with four toes, and a claw in the place of a thumb; hind feet with five: tail longer than the body, almost entirely naked, and covered with small scales disposed in rings. Colour of the upper parts deep iron-gray, or grayish black; those beneath dull ash-colour: feet and tail dusky.

Frequents houses, and is truly indigenous; but is now much less general than the next species. Food various. Breeds frequently in the year, producing from six to eleven at a time.

* 45. **M. *decumanus*, Pall. (*Brown Rat.*)**—Fur above grayish brown, tinged with yellow; beneath whitish: tail scarcely so long as the body.

M. decumanus, *Desm. Mammal.* p. 299. *Flem. Brit. An.* p. 20. Norway Rat, *Penn. Brit. Zool.* vol. i. p. 115. *Shaw, Gen. Zool.* vol. ii. p. 51. pl. 130.

Dimens. Length of the head and body eleven inches; of the head two inches four lines and a half; of the ears eight lines and a half; of the tail eight inches four lines.

Descript. Larger than the last species; thicker and of a stronger make : muzzle not quite so sharp at the extremity: eyes large and prominent: ears as broad as long, rounded at the extremity, and almost naked : tail naked and scaly, with about one hundred and eighty rings, generally not quite so long as the body, sometimes equalling it in length. Hair on the upper parts dusky ash at the roots, and reddish yellow at the extremity, with longer hairs intermixed of a deep brown colour throughout their whole length; general resulting tint grayish brown with a cast of tawny yellow; under parts dirty white, inclining here and there to ash-colour.

Introduced originally by shipping from Asia: unknown in England before the year 1730. Abundant now throughout the country, inhabiting houses, barns, granaries, drains, and other situations. Habits similar to those of the last species Takes occasionally to the water, and swims readily.

GEN. 17. ARVICOLA, *Lacép.*

46. A. *amphibia*, Desm. (*Water Campagnol.*)—Fur above dusky gray tinged with yellow; beneath paler: tail more than half the length of the body.

> A. amphibia, *Desm. Mammal.* p. 280. A. aquatica, *Flem. Brit. An.* p. 23. Water Rat, *Penn. Brit. Zool.* vol. I. p. 118. *Shaw, Gen. Zool.* vol. II. p. 73. pl. 129.

DIMENS. Length of the head and body eight inches four lines; of the head one inch ten lines; of the ears five lines and a half; of the tail four inches nine lines.

DESCRIPT. Head thick and blunt; eyes small; ears short, scarcely projecting beyond the fur; incisors of a deep yellow colour, very strong, and measuring in the lower jaw half an inch in length: fore feet with four toes and the rudiment of a thumb; hind feet with five; these last connected for a little way at their base, but not regularly webbed: tail more than half as long as the body, covered with short scattered hairs. Fur very thick, grayish black above, or reddish brown with scattered black hairs; beneath iron gray.

Var. β. A. ater, *Macgillivray in Wern. Mem.* vol. VI. p. 429. Fur of a deep black colour above and below.

Frequents the banks of rivers and ditches, in which it burrows. Swims and dives readily. Feeds principally on roots and tender plants. Produces in June from six to eight young. The black variety is not uncommon in the fens of Cambridgeshire: it differs in no respect but that of colour.

47. A. *agrestis*, Flem. (*Field Campagnol.*)—Fur reddish brown above; pale gray beneath: tail one-third the length of the body.

> A. agrestis, *Flem. Brit. An.* p. 23. A. vulgaris, *Desm. Mammal.* p. 282. Short-tailed Mouse, *Penn. Brit. Zool.* vol. I. p. 123. Meadow Mouse, *Shaw, Gen. Zool.* vol. II. p. 81. pl. 136.

DIMENS. Length of the head and body four inches one line; of the head one inch two lines; of the ears five lines; of the tail one inch three lines and a half.—*Obs.* These dimensions are rather larger than usual.

DESCRIPT. Resembles the last species, but is much smaller. Head large; muzzle blunt at the extremity; ears rounded, moderately projecting, relatively longer than in the *A. amphibia;* incisors white: tail not more than one-third the length of the body, sometimes less. Fur soft and silky, reddish brown mixed with dusky on the upper parts, grayish ash beneath.

Common in meadows, preferring moist situations. Food seeds and roots. Is very prolific; and in certain seasons multiplies to a great extent. Nest constructed of dried grass.

C

48. A. *riparia*, Yarr. (*Bank Campagnol.*)—Fur bright
chestnut-red above, ash-colour beneath : tail half the length
of the body; the hairs at the tip a little elongated.

> A. riparia, *Yarrell in Proceed. of Zool. Soc.* (1832) p. 109.   *Id. in
> Loud. Mag. of Nat. Hist.* vol. v. p. 598.

DIMENS. Length of the head and body three inches four lines; of the
head one inch half a line; of the ears five lines; of the tail one inch
eight lines.

DESCRIPT. Distinguished from the *A. agrestis* by its brighter and
more rufous colour; tail longer, with the hairs at the tip extending beyond the bone; ears rather larger, and more prominent: under parts
cinereous, with a faint yellow tinge along the mesial line of the abdomen
and on each side of it: tail blackish above, white beneath, the two
colours separated by a well-defined line.

First discovered by Mr Yarrell at Birchanger in Essex. Has since
occurred in Hertfordshire, Middlesex, Berkshire, and Cambridgeshire,
but is not so plentiful as the *A. agrestis*. Frequents hedge-bottoms and
ditch-banks, also occasionally stacks of corn. Is said to construct its
nest of wool.—*Obs.* Independently of the above external differences between this and the last species, there are others connected with their
anatomy, which will be found detailed at length in *Loud. Mag.* l. c.

† GEN. 18. CASTOR, *Linn.*

† 49. C. *Fiber*, Linn. (*Beaver.*)—Fur of a deep chestnut-brown;
glossy above, dull beneath.

> C. Fiber, *Desm. Mammal.* p. 277.   Common Beaver, *Shaw, Gen. Zool.* vol. II.
> p. 30. pl. 128.

DIMENS. Length of the head and body two feet six lines; of the head five
inches; of the tail one foot; breadth of the tail four inches two lines.—DESM.

DESCRIPT. Head short and thick, somewhat flattened at the top; muzzle obtuse:
eyes small and black: ears short, and rounded at the extremity: neck short: body
thick, very convex on the back: tail depressed, broad, and of an oval form ; the
surface naked and scaly. Fur of a deep chestnut-brown colour, smooth and glossy
on the upper parts, more dull underneath.

Formerly an inhabitant of Wales and Scotland. Observed in the former country
by Giraldus de Barri in the year 1188. Not exactly known when the species
was extirpated. Has occurred in a fossil state in Cambridgeshire.

GEN. 19. LEPUS, *Linn.*

50. L. *timidus*, Linn. (*Common Hare.*)—Tawny
gray, shaded with brown : ears longer than the head;
black at the tips : tail black above, white beneath.

> L. timidus, *Desm. Mammal.* p. 347.   *Flem. Brit. An.* p. 21.   Hare,
> *Penn. Brit. Zool.* vol. I. p. 98.   *Shaw, Gen. Zool.* vol. II. p. 197.
> pl. 162.

DIMENS. Length of the head and body twenty-one inches nine lines;
of the head four inches; of the ears four inches ten lines; of the tail
three inches six lines.

DESCRIPT. Head thick and large: inside of the cheeks hairy; eyes
placed laterally, large and prominent; ears longer than the head: limbs
slender, much longer behind than before: soles of the feet hairy. Fur
composed of a fine down with longer hairs intermixed, of a tawny gray or
rusty brown colour, the red tint prevailing in certain parts more than in
others; each individual hair gray at the roots, black in the middle, and
tawny at the tip; abdomen, inside of the thighs, and a transverse patch
beneath the lower jaw, white; ears externally cinereous towards the base
of the outer margin, above that colour a black spot reaching to the extremity of the auricle: tail black above, white underneath.

Var. β. Irish Hare. *Yarr. in Proceed. of Zool. Soc.* (1833) p. 88.
Head shorter and more rounded than in the *Common Hare;* ears shorter,
not equalling the head in length; limbs less lengthened; fur composed
of only one sort of hair, the long dark hairs, observable in the English
Hare, being wanting.

Frequents chiefly open fields. Feeds entirely on vegetables, coming
abroad in the evening for that purpose. Breeds frequently in the year.
Goes with young thirty days, and produces from one to four, rarely five,
at a time. Young born with their eyes open, and the body clothed with
fur. When full grown, has been known to weigh thirteen pounds one
ounce and a half.

Var. β, which is the only Hare found in Ireland, might almost deserve
to be considered as a distinct species. From the shortness and inferior
quality of the hair, its fur is useless in trade.

51. **L. *Cuniculus*, Linn. (*Rabbet*.)**—Brownish gray,
mixed with tawny: ears scarcely longer than the head:
tail brown above, white beneath.

L. Cuniculus, *Desm. Mammal.* p. 348. *Flem. Brit. An.* p. 21.
Rabbet, *Penn. Brit. Zool.* vol. I. p. 104. pl. 10. no. 22. *Shaw,
Gen. Zool.* vol. II. p. 204. pl. 162.

DIMENS. Length of the head and body sixteen inches; of the head
three inches four lines; of the ears three inches six lines; of the tail
three inches.

DESCRIPT. Approaching the last species in general appearance, but
with the ears and hind legs proportionably shorter: fur of a less ferruginous colour; the general tint brownish gray, or dusky brown, with the
nape reddish; throat and abdomen white; ears gray, without the black
spot at the extremity; tail dusky brown above, white beneath. In the
domesticated state the colours vary extremely.

Resides in holes under ground. Like the Hare, comes abroad in the
evening to feed. Still more prolific than that species, breeding repeatedly in the year, and producing from four to eight at a litter. Young
blind and naked at birth.

52. **L. *albus*, Briss. (*Alpine Hare*.)**—Dusky gray
in summer, with a tinge of tawny; white in winter: ears
shorter than the head, always black at the extremity.

L. albus, *Briss. Reg. An.* p. 139. L. variabilis, *Flem. Brit. An.*
p. 22. Alpine Hare, *Penn. Brit. Zool.* vol. I. p. 102. pl. 10.
no. 21. Varying Hare, *Shaw, Gen. Zool.* vol. II. p. 201.

DIMENS. Length of the head and body twenty-one inches six lines;
of the head four inches six lines; of the ears three inches three lines

of the tail two inches six lines; of the fore leg, from the olecranon to the
end of the toe-nails, seven inches three lines; of the hind foot, from heel
to toe, six inches.

DESCRIPT. In general somewhat smaller than the *Common Hare*, with
shorter ears and more slender legs: tail shorter. Fur in summer grayish
tawny mixed with black, here and there inclining to ash-colour; in
winter wholly white, with the exception of the tips of the ears, which
remain always black.

Inhabits the Scotch mountains: keeps near the summits, and never
descends into the valleys, or mixes with the common species. Change of
dress effected in the months of September and April; caused by an
actual change of colour in the hair itself, without the shedding of the
fur.—*Obs.* The *L. variabilis* of Pallas is probably a distinct species.

* GEN. 20. CAVIA, *Gmel.*

* 53. C. *Cobaya*, Gmel. (*Guinea Pig.*)—Reddish yellow, variegated
with black and white.

C. Cobaya, *Desm. Mammal.* p.356. Variegated Cavy, *Shaw, Gen. Zool.* vol. II.
p. 17. pl. 126.

DIMENS. Length of the head and body twelve inches; of the head three inches
one line; of the ears ten lines and a half.

DESCRIPT. Body thick and short; the neck scarcely to be distinguished from it:
ears broader than long, straight, naked, transparent, partly concealed by the hair:
eyes round, large, and prominent. Hair sleek and smooth; varying in colour in
different individuals; entirely white; or variegated with black, white, and tawny
yellow.

A native of Brazil; but domesticated in Europe. Is very prolific in confinement,
breeding repeatedly in the year, and producing from four to twelve at a litter.
Feeds on different vegetable substances.

ORDER IV. UNGULATA.

---

* GEN. 21. BOS, *Linn.*

* 54. B. *Taurus*, Linn. (*Common Ox.*)—Forehead flat, longer than
broad: horns taking their origin from the extremities of the occipital
ridge.

B. Taurus, *Desm. Mammal.* p. 499. *Flem. Brit. An.* p.24. Ox, *Penn. Brit.
Zool.* vol. I. p. 18. pl. 2. *Shaw, Gen. Zool.* vol. II. pl. 208.

DESCRIPT. Varying extremely in the domesticated state in size and colour, as
well as in the form and direction of the horns, which last are sometimes wholly
wanting. A wild breed, (*Bewick, Quad.* p. 38.) formerly met with in Scotland,
but now extinct, said to have been characterized by their white colour, with the
muzzle and ears black.

Is capable of breeding at the age of two years. Goes nine months with young,
and produces one, rarely two, at a time. The two central incisors are shed in the
tenth month; the adjoining ones in the sixteenth; by the end of the third year the
change is wholly completed.

* GEN. 22.   OVIS, *Linn.*

* 55.   O. *Aries*, Linn.   (*Sheep.*)—Horns compressed, and lunated.

O. Aries, *Desm. Mammal.* p. 488.   *Flem. Brit. An.* p. 25.   Sheep, *Penn.*
   Brit. Zool. vol. i. p. 27.   *Shaw, Gen. Zool.* vol. ii. p. 385.

Descript.   Varying much in size and colour: with or without horns : wool
coarse or fine, long or short, according to the breed, sometimes approaching the
quality of hair: tail short, or reaching below the knees.
   Probably not originally indigenous.   Primitive stock supposed to be the Argali*
(*Ovis Ammon*, Gmel.), which is found in a wild state in the mountainous parts of
Asia.   Period of gestation about five months.   Produces one or two at a birth,
rarely more.   The two middle incisors fall at the end of the first year, and are
replaced by others ; the next in succession at about the age of two and a half :
by the end of the third year, or soon after, all have been renewed, and the individual is said to be *full-mouthed*.

* GEN. 23.   CAPRA, *Linn.*

* 56.   C. *Hircus*, Linn.   (*Goat.*)—Horns edged in front, rounded
posteriorly.

C. Hircus, *Desm. Mammal.* p. 482.   *Flem. Brit. An.* p. 25.   Domestic Goat,
   Penn. Brit. Zool. vol. i. p. 35. pl. 3.   *Shaw, Gen. Zool.* vol. ii. p. 369.
   pl. 199.

Descript.   Subject to less variation than the *Sheep.*   Usual colour black, white,
or pied ; occasionally brown, approaching more or less to a tawny red : horns in
the *male*, in some instances upwards of three feet long; in the *female* much smaller,
or wanting altogether : tail about seven inches in length ; often black.
   Ranges in a state of liberty on the mountains of Wales, Scotland, and Ireland,
and is often domesticated ; but it is doubtful whether it be indigenous.   Supposed
to be derived either from the *C. Ibex*, Linn. or the *C. Ægagrus*, Gmel., the former
of which species is found wild on the Alps of Europe, the latter on the mountains
of Persia.   Goes with young five months.

GEN. 24.   CERVUS, *Linn.*

57.   C. *Elaphus*, Linn.   (*Stag, or Red Deer.*)—Horns
branched; round; diverging at the base, somewhat converging at the extremity.

C. Elaphus, *Desm. Mammal.* p. 434.   *Flem. Brit. An.* p. 26.   Stag,
   Penn. Brit. Zool. vol. i. p. 41.   *Shaw, Gen. Zool.* vol. ii. p. 276.
   pl. 177.

Dimens.   Length of the body about six feet six inches; of the horns
about two feet ; of the tail seven inches : height about three feet eight
inches.
   Descript.—Varying in size and colour : usually reddish brown in
summer, with a dusky line along the spine; in winter brownish gray;
under parts whitish: horns at first simple, afterwards branched; the
number of antlers increasing with age till they amount to ten or twelve;
three of them being always directed forwards: eyes large, with a distinct
lachrymal furrow: ears long and pointed: tail of moderate length.   Female or *Hind* smaller, and without horns.   The young or *Calf* is generally spotted with white, or as it is termed *menilled*, on the upper parts:
the first indication of horns takes place during the latter part of the first
year, when it is called a *Knobber.*

---

* *Shaw, Gen. Zool.* vol. ii. p. 379. pl. 201.

Formerly abundant throughout the kingdom, but now chiefly confined to the Highlands of Scotland. The horns are shed in March, and reappear in the course of the summer. Rutting season from Michaelmas to the end of November. Period of gestation rather more than eight months. Usually but one at a birth.

* 58. C. *Dama*, Linn. (*Buck, or Fallow Deer.*)—Horns branched; compressed; palmated at the top, diverging.

C. Dama, *Desm. Mammal.* p. 438.   *Flem. Brit. An.* p. 26.   Fallow Deer, Penn. *Brit. Zool.* vol. I. p. 41.   *Shaw, Gen. Zool.* vol. II. p. 282. pl. 178-9.

DIMENS. Length of the body about five feet ; of the tail seven inches five lines : height about two feet ten inches.

DESCRIPT. Body much smaller than in the last species ; tawny brown, with the back, flanks, shoulders, and thighs, more or less spotted with white ; a dusky line down the middle of the back ; buttocks white, bounded on each side by a descending black line ; tail longer than in the *Stag*, blackish brown above, white beneath ; abdomen and inside of the thighs whitish : horns round at bottom, with two antlers directed forwards ; their upper portion compressed, and dilated into a broad palm, with tooth-like processes along the outer margin. Female or *Doe* without horns, and likewise the *Fawns* during the first year.

Var. β. Entirely white.

Var. γ. Deep brown, approaching to black.

Abundant in parks and forests in a half-reclaimed state, but doubtful whether indigenous. The black variety said to have been introduced from Sweden by King James the First ; the others supposed to have come from Asia. Congregate in small herds. Rutting season in the Autumn. Female goes with young eight months, and produces one or two, rarely three, at a birth. The first two central incisors shed at the age of a year and a half ; the change completed by the end of the fourth year.

59. C. *Capreolus*, Linn. (*Roe-Buck.*)—Horns branched ; cylindrical ; small and erect, with furcate summits.

C. Capreolus, *Desm. Mammal.* p. 439.   *Flem. Brit. An.* p. 26. Roe, *Penn. Brit. Zool.* vol. I. p. 49. pl. 4.   *Shaw, Gen. Zool.* vol. II. p. 291.

DIMENS. Length of the body three feet nine inches; of the horns eight to nine inches ; of the tail one inch : height (in front) two feet three inches, (behind) two feet seven inches.

DESCRIPT. Smaller than either of the preceding species, but of similar form. Colour variable; the general tint yellowish gray or reddish brown, more or less deep in different individuals, sometimes dusky ; each hair being ash-coloured at the roots, then black, with the extreme tip tawny yellow ; lower part of the neck, abdomen, and inside of the thighs, grayish white ; contour of the anus pure white : ears long, furnished internally with long whitish hairs ; extremity of the nose dusky, with a white spot on each side of the upper lip ; chin white : horns very rugged, sulcated longitudinally, about as long as the head, with only two antlers, the first arising about the middle, directed forwards,—the second higher up directed backwards : no lachrymal furrow : tail very short. Summer coat much shorter and finer, and of a redder tint, than the winter one. *Doe* without horns. In the *young Buck*, the horns are plain and unbranched during the second year ; in the third year furnished with a single antler, and in the fourth year complete.

Common formerly in Wales, in the North of England, and in Scotland, but at present almost confined to the Scottish Highlands. Loses its horns at the latter end of Autumn. Rutting season during the first half of November. Doe goes with young five months and a half, and produces in April. Two at a birth, which are always male and female.

* GEN. 25. EQUUS, *Linn.*

* 60. E. *Caballus*, Linn. (*Horse.*)—Tail uniformly covered with long hair; mane long and flowing : ears of moderate size : no dorsal line or transverse band.

E. Caballus, *Desm. Mammal.* p. 416. *Flem. Brit. An.* p. 27. Horse, *Penn. Brit. Zool.* vol. i. p. 1. pl. 1. Common Horse, *Shaw, Gen. Zool.* vol. ii. p. 419. pl. 214.

DESCRIPT. Offering every variety of size and colour : generally bay, or chestnut-brown, more or less deep; black, or grayish white : head long and tapering : teeth, incis. $\frac{6}{6}$, can. $\frac{1-1}{1-1}$ (seldom present in the *mare*), mol. $\frac{6-6}{6-6}$, = 40 : ears erect and pointed, much smaller than in the next species : a naked callosity on the inside of the fore legs above the knee; another on the hind legs just under the knee.

Probably brought originally from Asia, where the species still exists in a truly wild state. A small variety occurs in the Highlands of Scotland, and in the Shetland Islands, half-reclaimed, but can scarcely be considered as indigenous. Period of gestation eleven months. Seldom more than one at a birth. Central incisors cast at the age of two years and a half, and replaced by permanent ones; the adjoining pair at three and a half; the remaining ones at four and a half; these last replaced more slowly than the others.

* 61. E. *Asinus*, Linn. (*Ass.*)—Gray, inclining to reddish ; with the dorsal line, and a transverse band across the shoulders, black : ears very large : tail terminating in a tuft of long hair.

E. Asinus, *Desm. Mammal.* p. 414. Ass, *Penn. Brit. Zool.* vol. i. p. 13. *Shaw, Gen. Zool.* vol. ii. p. 429. pl. 216.

DESCRIPT. Smaller than the *Horse*, and subject to less variation than that species. Colour generally gray, more or less dark; sometimes approaching to silvery white, obscurely spotted with stains of a reddish cast, at other times dark brown or dusky : a transverse black stripe upon the shoulders, crossing another of the same colour down the middle of the back ; these marks always more or less obvious. Head shorter and thicker than that of the horse : ears long and slouching : tail tipped with long hair: no naked callosities on the hind legs.

A native of the East. Introduced into this country towards the close of the tenth century. Goes eleven months with young, and produces one at a birth. Breeds occasionally with the Horse: the hybrid production termed a *Mule* or a *Hinny*, according as the Ass is the *male* or *female* parent.

GEN. 26. SUS, *Linn.*

† 62. S. *Scrofa*, Linn. (*Boar.*)—Tusks strong, triangular, of moderate length, directed to one side.

S. Scrofa, *Desm. Mammal.* p. 389. *Flem. Brit. An.* p. 28. Hog, *Penn. Brit. Zool.* vol. i. p. 55. Wild Boar, and Common Hog, *Shaw, Gen. Zool.* vol. ii. p. 459. pl. 221 and 222.

DIMENS. (*Wild Boar.*) Length of the head and body three feet three inches ; of the head one foot ; of the tail seven inches six lines: height about one foot nine inches.

DESCRIPT. Head elongated: neck short: body thick and muscular: legs short and strong : ears rather short : eyes small : mouth large, with the upper lip pushed up by the tusks : body covered with long stiff bristles, intermixed at the roots with a soft woolly hair ; the longest and strongest bristles on the back.   General colour dusky gray.—Desm.

*Var. β. domestica. (Domestic Hog.)—In this variety the ears are longer and more or less pendulous ; the bristles more thinly scattered, and of one sort ; the tusks comparatively short ; the tail more or less twisted ; the size and colour very variable.

Formerly abundant in a wild state throughout the country. Continued to inhabit the forests about London so late as the reign of Henry the Second, but it is not exactly known how soon after that period they were extirpated.   In the domestic state the Sow goes with young about four months, and is very prolific, producing sometimes as many as twenty at a litter. Food extremely various.

ORDER V.   CETACEA.

GEN (1.) MANATUS, Cuv.?

GEN. (2.) STELLERUS, Cuv.?

(1.)   ————? Manatus borealis, Flem. Brit. An. p. 29.   Mermaid of the Shetland Seas, Edinb. New Phil. Journ. (1829) vol. vi. p. 57.

Two instances at least are on record in which specimens of the Herbivorous Cetacea have been observed in the British seas, but there is no satisfactory evidence by which the species can be determined, or even the genus, to which they belonged*.   In the Elements of Nat. Hist. by Mr Stewart, mention is made (vol. i. p. 125.) of the carcase of one of these animals which was thrown ashore near Leith ; and in the Edinb. New Philos. Journ. as quoted above, there will be found some account of another individual, which occurred a few years ago off the Shetland Islands. See also Dr Fleming's Brit. An. p. 30.

GEN. 27. DELPHINUS, Linn.

(1. Delphinus, Cuv.)

63.   D. Delphis, Linn. (Common Dolphin.)—Jaws moderately produced ; nearly of equal length : teeth more than forty on each side above and below ; slender, slightly bent, pointed.

D. Delphis, Desm. Mammal. p. 514.   Flem. Brit. An. p. 35.   Dolphin, Penn. Brit. Zool. vol. iii. p. 65.   Shaw, Gen. Zool. vol. ii. p. 507. pl. 229.

Dimens.   Entire length from six to seven feet.

Descript.   Body slender, thickest in the middle, gradually tapering towards the head and tail : muzzle or beak, measured from the forehead, equalling in length the rest of the head ; much depressed, narrow, and

---

* On this account I have introduced, in the Synoptic Arrangement of Genera the characters of both Manatus and Stellerus Cuv., which are perhaps equally likely to occur in the British seas. At any rate they will prove useful in determining the true genus of any of these animals which may be met with hereafter.

somewhat pointed at the extremity: jaws nearly equal, the upper being slightly shorter than the lower: teeth very numerous, $\frac{42-42}{42-42}$ to $\frac{47-47}{47-47} = 168$ to 188, slender, somewhat bent inwards, sharp-pointed; placed at equal distances from each other, and locking in between each other when the jaws are closed: eyes small, almost in a line with the mouth: blow-hole situate on the top of the head a little above the eyes: pectoral fins placed very low, in form somewhat falcate: dorsal fin pointed, taking its origin from a little beyond the middle of the back; rather elevated, and when measured along the line of flexure, equalling one-sixth of the entire length of the body: tail crescent-shaped. Skin smooth; of a dusky colour above, white beneath, and grayish on the sides.

Met with occasionally on the British shores, but not of very frequent occurrence. Feeds on animal substances. Period of gestation said to be ten months. Produces one or two at a birth.

64. D. *Tursio*, Fab. (*Blunt-toothed Dolphin.*)—Jaws moderately produced; the lower longer than the upper: teeth rather more than twenty on each side above and below; straight, with obtuse summits.

D. Tursio, *Desm. Mammal.* p. 514. *Flem. Brit. An.* p. 37. D. truncatus, *Mont. in Wern. Mem.* vol. III. p. 75. pl. 3. Bottle-nose Whale, *Hunter in Phil. Trans. for* 1787. pl. 18.

DIMENS. Entire length eleven feet: girth seven feet four inches: length of the mouth fourteen inches; from the snout to the eye sixteen inches; from the same to the pectoral fin two feet; from the same to the dorsal four feet eight inches; from the same to the vent seven feet three inches; length of the dorsal twenty-three inches; height of the same ten inches.

DESCRIPT. Larger than the last species: lower jaw projecting further beyond the upper: teeth less numerous, $\frac{21-21}{21-21}$ to $\frac{23-23}{23-23} = 84$ to 92, straight, somewhat conical, but blunt and truncated at the extremity.

But little known as a British species. Hunter's specimen was caught upon the sea-coast near Berkeley, and is now in the Museum of the College of Surgeons. In the same collection is a second individual from the Thames. The one described by Montagu, which appears to have been large and aged, with the summits of the teeth more than usually truncated, was taken in the river Dart, in 1814. The measurements given above are those of a fourth individual which occurred a few years since in the river at Preston. They were obligingly sent me by Mr Gilbertson of that place.

(2. PHOCŒNA, *Cuv.*)

65. D. *Phocœna*, Linn. (*Porpesse.*)—Under jaw slightly projecting beyond the upper: teeth twenty-two to twenty-five on each side above and below; straight, compressed, and rounded at the summits.

D. Phocœna, *Desm. Mammal.* p. 516. *Flem. Brit. An.* p. 33. Porpesse, *Penn. Brit. Zool.* vol. III. p. 69. *Shaw, Gen. Zool.* vol. II. p. 504. pl. 229.

DIMENS. Entire length from four feet to five feet and a half.

DESCRIPT. Body elongated, gradually tapering towards the tail: snout short, rather obtuse at the extremity: under jaw somewhat longer than the upper: teeth numerous; from twenty-two to twenty-five in each jaw on each side, compressed, and nearly straight: eyes small, almost in a line with the mouth: blow-hole crescent-shaped, with the concavity directed forwards: pectoral fins placed low down, oval, and somewhat pointed: dorsal fin straight, triangular, rather beyond the middle of the body. Skin smooth; dusky on the back, whitish on the belly, the two colours meeting on the sides.

A constant inhabitant of the British seas, often entering the mouths of rivers. Preys on mackarel, herrings, and other fish. A specimen which occurred in the London market in May 1833 was found to contain a full-formed fœtus; it is probable, therefore, that they produce their young at about that period of the year.

66.   D. *Orca*, Fab. (*Grampus.*)—Upper jaw projecting a little beyond the lower : teeth about eleven on each side above and below; conical, bent at the summits : pectoral fins broad and oval.

D. Orca, *Flem. Brit. An.* p. 34.   D. Grampus, *Desm. Mammal.* p. 517. Grampus, *Hunter in Phil. Trans.* 1787, pl. 16.

DIMENS.   Entire length from twenty to twenty-five feet.

DESCRIPT.   Much larger than the last species; the body deeper and thicker in proportion to its length.   Snout very short and obtuse : upper jaw somewhat longer than the lower, but this last broader than the upper : teeth unequal, conical, a little bent at the summits; varying in number according to the age of the individual, generally about twenty two in each jaw: eyes almost in the same line with the mouth: dorsal fin nearly in the middle, very much elevated, and pointed at the extremity : pectorals very broad, of an oval form : tail crescent-shaped.   Skin smooth; glossy black above, white beneath, the two colours meeting on the sides but separated by a well-defined line; an oval white spot behind each eye.

Inhabits the British seas in large herds, and occasionally enters rivers. Is of a fierce and voracious disposition, preying upon the larger species of fish.—*Obs.* The *Delphinus Gladiator* and the *D. ventricosus* of Lacépede, two species constituted by that author from individuals taken in the Thames, are considered by Cuvier as not really distinct from the above.

67.   D. *melas*, Traill. (*Ca'ing Whale.*) — Top of the head very convex : teeth conical, varying in number : pectoral fins long and narrow.

D. melas, *Traill in Nichol. Journ.* vol. XXII. p. 81.   *Flem. Brit. An.* p. 34.   D. globiceps, *Cuv. Ann. du Mus.* tom. XIX. p. 14. pl. 1. f. 2, 3.   D. Deductor, *Scoresby, Arct. Reg.* vol. I. p. 496. pl. 13. f. 1.

DIMENS.   Entire length from twenty to twenty-four feet.

DESCRIPT.   Equalling the *Grampus* in size, but differing essentially from that species in the form and character of the fins; the dorsal much shorter, the pectorals longer, narrower, and more pointed : head short and round, with the forehead remarkably convex and prominent: upper jaw projecting a little beyond the lower : teeth conical, sharp, and a little bent; varying much in number in different individuals, not visible in very young specimens, and falling in advanced life; average number from twenty-two to twenty-four in all. Skin smooth, of a deep bluish black colour, with the exception of a whitish band beneath the body extending from the throat to the anus.

Common in large herds off the Orkney and Shetland Islands.

(3. DELPHINAPTERA, *Lacép.*)

68.   D. *albicans,* **Fab. (*Beluga.*)**—Head blunt : teeth about nine on each side above and below ; short, with obtuse summits.

D. leucas, *Desm. Mammal.* p. 519.   Delphinaptera albicans, *Flem. Brit. An.* p. 36.   Beluga, *Barclay and Neill in Wern. Mem.* vol. III. p. 371. pl. 17.   *Scoresby, Arct. Reg.* vol. I. p. 500. pl. 14.

DIMENS.   Entire length from twelve to eighteen feet.

DESCRIPT.   Body thick in the middle, somewhat tapering towards each extremity, but more especially towards the tail : head small, blunt, and round ; the forehead rising abruptly : eyes and mouth small : jaws equal : teeth in the adult state about nine on each side above and below, short, straight, slightly compressed, with obtuse summits ; those in the upper jaw falling in advanced life : a longitudinal ridge on the back supplying the place of the dorsal fin : pectorals short, broad, and oval.   Skin smooth : colour wholly white ; sometimes with a tinge of yellow.

A native of high Northern latitudes, where it is found in herds of thirty or forty together.   Must be considered as a very rare visitant of the British seas.   The individual described by Mr. Neill l. c. was taken in the Frith of Forth in June 1815.

GEN. 28.   MONODON, *Linn.*

69.   M. *Monoceros,* **Linn. (*Narwhal.*)**—Body sub-conical, with a dorsal and ventral ridge : head obtuse, about one-seventh of the entire length.

M. Monoceros, *Scoresby, Arct. Reg.* vol. I. p. 486. pl. 15.   *Flem. Brit. An.* p. 37.   Small-headed Narwhal, *Flem. in Wern. Mem.* vol. I. p. 131. pl. 6.

DIMENS.   Entire length, exclusive of the tusk, thirteen to sixteen feet ; circumference (at the thickest part) eight to nine feet ; length of the tusk about five feet *.

DESCRIPT.   Anterior half of the body nearly cylindrical ; posterior half conical : this latter portion furnished with a dorsal and ventral ridge, which take their origin about three feet from the extremity, and extend half way across the tail ; the edges of the tail run in like manner six or eight inches along the body, forming ridges on the sides of the rump : head about one-seventh of the entire length ; small, blunt, and round ; the forehead very prominent, rising suddenly from the snout : mouth small ; intermaxillary bones furnished each with one tooth directed forwards ; in the *female* these teeth generally remain through life concealed in the sockets, not appearing externally ; in the *male,* that on the left side is exserted, growing to the length of several feet ;

* The above measurements are from Scoresby, from whose excellent work on the Arctic Regions much assistance has been derived in drawing up the characters of this species, as well as of some others of the *Cetacea.*

it is spirally striated from right to left, nearly straight, and tapering to a round blunt point; very rarely, in this last sex both teeth are equally developed, and both exserted: blow-hole semicircular, situate directly over the eyes: pectoral fins short: no dorsal fin, but instead of it an irregular sharpish fatty ridge, two inches in height, extending two feet and a half along the back, nearly midway between the snout and the tail: tail divided by a notch into two lobes, which project laterally and are somewhat pointed. Prevailing colour white, or yellowish white, with dark gray or blackish spots of different degrees of intensity.

Has only occurred hitherto in two or three instances on the British shores. In the Northern seas is said to be gregarious; each sex herding separately. Feeds on *sepiæ* and other molluscous animals.

GEN. 29. HYPEROODON, *Lacép.*

70. H. *bidens*, Flem. (*Bottle-head.*)—Teeth two only in the fore part of the lower jaw.

H. bidens, *Flem. Brit. An.* p. 36. H. Butskopf, *Lacép. Cétac.* p. 319. Bottle-nose Whale with two teeth, *Hunter in Phil. Trans.* 1787. pl. 19. Beaked Whale, *Penn. Brit. Zool.* vol. III. p. 59. pl. 5. f. 1. Two-toothed Cachalot, *Sow. Brit. Misc.* pl. 1.

DIMENS. Entire length from twenty to twenty-five feet.

DESCRIPT. Body elongated; greatest circumference in the region of the pectoral fins: forehead high, very convex, rising suddenly from the snout; this last short and depressed, terminating in a kind of beak somewhat similar to that of the genus *Delphinus :* lower jaw rather longer and larger than the upper: teeth conical, and pointed; only two, situate in the fore part of the lower jaw; sometimes altogether wanting, or not appearing above the gums: palate studded with little horny eminences, considered by Cuvier as rudimentary vestiges of *whalebone :* eyes large, a little above the line of the lips: blow-hole crescent-shaped, with the horns of the crescent directed towards the tail: dorsal fin placed considerably beyond the middle of the body, but little elevated, lanceolate, pointed, inclining backwards: pectorals small, oval, in the same horizontal line with that of the mouth. Skin smooth and glossy, blackish leadcolour above, whitish underneath, the two colours mixing on the sides.

Occasionally met with on the British shores. Hunter's specimen was taken in the Thames above London Bridge in 1783.—*Obs.* This species has been unnecessarily split into several by many authors: it probably embraces all the following of Desmarest's " Mammalogie ;" *Delphinus Chemnitzianus, D. Hunteri, D. edentulus, D. Hyperoodon,* and *D. Sowerbyi.*

GEN. 30. PHYSETER, *Linn.*

(1. CATODON, *Lacép.*)

71. P. *macrocephalus*, Shaw. (*Blunt-headed Cachalot.*) —Teeth in the lower jaw from twenty to twenty-four on each side; mostly conical with obtuse summits.

Catodon macrocephalus and C. Trumpo, *Lacép. Cétac.* pp. 165 and 212. C. macroceph. *Flem. Brit. An.* p. 39. Blunt-headed Cachalot, *Penn. Brit. Zool.* vol. III. p. 61. pl. 6. Spermaceti Whale, *Alderson in Camb. Phil. Trans.* vol. II. p. 253. pl. 12—14.

Dimens. Entire length from fifty to sixty-three feet.

Descript. Head enormously large, forming more than one-third of the entire bulk; the body gradually tapering from the posterior part of it towards the tail: upper part of the snout very thick and swollen, as it were truncated in front, and overhanging considerably the lower jaw; this last of a narrow elongated form, fitting, when the mouth is closed, into a grooved cavity above: upper jaw without whalebone or visible teeth, although a few teeth may be found concealed within the gums on cutting through the integuments; in the lower jaw from forty to forty-nine teeth (there being occasionally an odd one), entering likewise when the mouth is shut into corresponding cavities above; the last three or four on each side smaller than the others, and somewhat hooked; the rest projecting above the gums about two inches, cylindrical or conical, with bluntish summits; those in front inclining *backwards*, those situate more behind *forwards*, the middle tooth on each side being nearly vertical: blow-hole single, near the extremity of the snout, and placed rather to the left of the median line: no dorsal fin; instead of it a callous ridge commencing gradually, and terminating behind abruptly in a sort of hook-like process: pectorals small. General colour black or dusky, somewhat paler beneath *.

Occasionally stranded on different parts of the coast, but not of frequent occurrence in the British seas. The upper part of the head in this species consists of large cavities, separated from each other by a cartilaginous substance, and filled with an oily fluid, which, in its congealed state, forms the *spermaceti* of commerce. These cavities are quite distinct from that of the cranium which is situate beneath.

(2.) *P. Catodon*, Linn.—*Catodon Sibbaldi*, Flem. Brit. An. p. 39.

> This supposed species is too imperfectly characterized, and rests on too doubtful authority, to rank as distinct. In the opinion of Cuvier, (*Oss. Foss.* tom. v. p. 335.) the herd stated by Sibbald (*Phalainolog. Nov.* p. 24.) as having occurred at Kairston in Orkney, consisting of an hundred individuals, were probably Belugas (*Delphinus albicans*), some of which had lost the teeth in the upper jaw through age. Moreover Sibbald appears only to have had his account from others, and not to have seen any of the individuals himself.

(2. Physeter, *Lacép.*)

72. P. *Tursio*, Linn. (*High-finned Cachalot.*)—Teeth very slightly bent, with flat summits.

P. Tursio, *Flem. Brit. An.* p. 38. P. Mular, *Lacép. Cétac.* p. 239. *Desm. Mammal.* p. 526. High-finned Cachalot, *Penn. Brit. Zool.* vol. iii. p. 64.

Descript. Said to be characterized by an erect dorsal fin, very high, and pointed at the extremity: blow-hole in front: teeth very slightly bent, with flat or obtuse summits: in other respects similar to the *P. macrocephalus*.

But very little is known of this species. The few particulars on record respecting it are principally taken from Sibbald, who briefly describes a Cachalot with the above characters, which came on shore on the Orkney Isles in 1687.

---

* Some parts of this description are borrowed from Alderson's account of this species, l. c.

(3.) *P. microps*, Linn. Flem. Brit. An. p. 38.

> This species, which entirely owes its existence to Sibbald's *Phalain-ologia*, rests upon very vague and uncertain authority. It is said to resemble the *P. Tursio*, excepting in having the teeth more pointed. Cuvier does not admit that there is any well-founded distinction between the two. Indeed, in his *Ossemens Fossiles*, (tom. v. p. 328.) he would seem to entertain some doubts with respect to the existence of either.

GEN. 31. BALÆNA, *Linn.*

(1. BALÆNA, *Lacép.*)

73. B. *Mysticetus*, Linn. (*Common Whale.*)—Gape of the mouth arched : upper jaw with about six hundred and fifty laminæ of whalebone.

> B. Mysticetus, *Scoresby, Arctic Reg.* vol. i. p. 449. pl. 12. *Flem. Brit. An.* p. 33. Common Whale, *Penn. Brit. Zool.* vol. iii. p. 50.

DIMENS. Entire length averaging from fifty to sixty-five feet : greatest circumference from thirty to forty feet.

DESCRIPT. One of the most bulky, but not in general the longest of the Cetaceous tribe. Body thickest in the middle, a little behind the fins, from which point it gradually tapers, in a conical form, towards the tail, and slightly towards the head : this last very large, of a somewhat triangular form ; "the under part, the arched outline of which is given by the jawbones, flat, and measuring sixteen to twenty feet in length, and ten to twelve in breadth : the lips, which are five or six feet high, and form the cavity of the mouth, are attached to the under jaw, and rise from the jaw-bones, at an angle of about eighty degrees, having the appearance, when viewed in front, of the letter U : the upper jaw, including the crown-bone or skull, bent down at the extremity, so as to shut the front and upper parts of the cavity of the mouth, and overlapped by the lips in a squamous manner at the sides * :" no teeth ; but the palate furnished with two extensive rows of whalebone, generally curved longitudinally, and giving an arched form to the roof of the mouth ; each series consists of upwards of three hundred laminæ, the interior edges of which are covered with a fringe of hair : eyes remarkably small : pectoral fins situate about two feet beyond the angle of the mouth : tail horizontal, of great breadth, and of a semilunar form ; the lateral lobes somewhat pointed, and turned a little backward. Colour black, or blackish gray, with the exception of the fore part of the under jaw and a portion of the belly, which are white.

Appears to have been formerly of not unfrequent occurrence in the British seas, but must be considered in these days as an extremely rare visitant. Sibbald mentions one which came ashore near Peterhead in 1682. A small one is stated to have been taken near Yarmouth, July 8, 1784 †. The food of this species is said to consist principally of shrimps and molluscous animals.

---

* Scoresby.
† C. and J. Paget's *Nat. Hist. of Yarmouth and its Neighbourhood.*

(2. BALÆNOPTERA, *Lacép.*)

74. B. *Physalus*, Linn. (*Fin-Fish.*)—Pectoral skin without longitudinal folds.

Balænoptera Gibbar, *Scoresb. Arct. Reg.* vol. I. p. 478. Physalis vulgaris, *Flem. Brit. An.* p. 32. Fin-Fish, *Penn. Brit. Zool.* vol. III. p. 57. Fin-backed Mysticete, *Shaw, Gen. Zool.* vol. II. p. 490. pl. 227.

DIMENS. Entire length about one hundred feet: greatest circumference from thirty to thirty-five feet.

DESCRIPT. The longest of the Cetaceous tribe. Body more slender, and less cylindrical than that of the *B. Mysticetus;* considerably compressed at the sides, and angular on the back: head smaller than in that species; the snout more pointed, with the jaws nearly equal; the whalebone shorter, the longest lamina measuring about four feet: a small horny protuberance, or dorsal fin, near the extremity of the back: pectorals long and narrow. Colour pale bluish black, or dark bluish gray.

Apparently of equal rarity in the British seas with the species last described. It is included by Pennant in his *British Zoology*, but it is not said on what authority.

75. B. *Boops*, Linn. (*Sharp-lipped Whale.*) —Pectoral skin with longitudinal folds admitting of dilatation: jaws pointed.

Balænoptera Jubartes, *Scoresb. Arct. Reg.* vol. I. p. 484. B. Boops, *Flem. Brit. An.* p. 31. Pike-headed Mysticete, *Shaw, Gen. Zool.* vol. II. p. 492. pl. 227. Fin-Whale, *Neill in Wern. Mem.* vol. I. p. 201.

DIMENS. Entire length about forty-six feet: greatest circumference about twenty feet.

DESCRIPT. Body very thick, and somewhat elevated, immediately over the pectoral fins; gradually tapering from that point towards the tail: head moderately large, becoming narrower towards the extremity of the snout, which terminates however in a somewhat broadish tip: lower jaw one-third of the entire length: palate with about three hundred laminæ of whalebone on each side, the longest measuring about eighteen inches in length: dorsal protuberance or fin placed far down the back; two feet and a half high: pectorals four or five feet long, scarcely a foot broad. Colour black above, whitish on the belly, inclining to red between the pectoral folds.

Represented as being of not unfrequent occurrence in the Scotch seas; and occasionally stranded on different parts of the English coast. There is some doubt whether the whale described and figured by Dr Johnston in the *Trans. of Newcastle Nat. Hist. Soc.* (vol. I. p. 6.) be referable to this species or not, as it possessed pectorals nine feet in length, being one-fourth of the length of the body. It was thrown on shore about two miles north of Berwick, in Sept. 1829.

76. B. *Musculus*, Linn. (*Round-lipped Whale.*)— Pectoral skin with longitudinal folds: margin of the under lip semicircular.

Balænoptera Rorqual, *Scoresb. Arct. Reg.* vol. I. p. 482. B. Musculus, *Flem. Brit. An.* p. 30. Round-lipped Whale, *Penn. Brit. Zool.* vol. III. p. 58.

DIMENS. Entire length from seventy to eighty-six feet.

DESCRIPT. Strongly resembling the last species, but said to grow to a much larger size: differs also in having the lower lip much broader than the upper, and semicircularly turned at its extremity, while the upper is somewhat sharp or pointed at the tip: gape large; longest lamina of whalebone three feet in length.

Found in the Scotch seas with the preceding. Said to feed principally on herrings.

Obs. The characters of the species in this sub-genus are not sufficiently well understood to admit of being laid down with any degree of certainty, and there can be little doubt that the species themselves have been unnecessarily multiplied. Cuvier appears to be of opinion (*Regne An.* tom. I. p. 298.) that even the *B. Physalus* may eventually turn out to be the same with the *B. Boops*, and that the supposed difference between them may be the result of hasty and imperfect examination. There is likewise but little dissimilarity between the *B. Boops* and the *B. Musculus*, and it is not as yet satisfactorily ascertained that even that little is to be depended upon as constant in all cases. With respect to the *Balæna rostrata* of Fabricius and Hunter (*Phil. Trans.* 1787. pl. 20.) which occurs not unfrequently on the British coasts, it is probably nothing more than the young state of one of the two last mentioned; in the opinion of Cuvier*, of the *B. Boops*.

---

* See his *Ossemens Fossiles* (tom. v. p. 365.) which contains a valuable dissertation on the different species of the *Cetacea*.

CLASS II. AVES.

§ I.   Feet formed for grasping.

ORDER I.   RAPTORES.

Bill strong, covered at the base with a cere, hooked towards the extremity: legs strong and muscular, short, or of moderate length: toes four in number, three directed forwards and one behind; rough underneath; armed with powerful, sharp, curved, retractile talons.

I. VULTURIDÆ.—*Head and neck more or less divested of feathers: nostrils lateral, placed in the cere, oval or elongated: feet generally naked: claws moderately curved.*

1. NEOPHRON.—Bill long, slender, straight, hooked at the extremity; lower mandible bending downward; no gonys: nostrils longitudinal, lateral, situate near the ridge: fore part of the head naked; neck feathered: feet strong, moderate, naked: wings long; third quill the longest: tail of fourteen feathers.

II. FALCONIDÆ.—*Head feathered: nostrils lateral, placed in the cere, rounded or oval: legs feathered to the toes, or naked: claws strongly curved, very sharp.*

2. AQUILA. — Bill long, very robust, convex or slightly angular above, straight at the base, much hooked

D

at the tip: cere hispid: nostrils rounded, or lunulate:
wings long.

> (1. AQUILA.) Tarsi feathered to the toes: claws unequal, grooved
> beneath: fourth quill longest.
>
> (2. HALIÆETUS.) Tarsi half-feathered; acrotarsia scutellated: claws
> unequal, grooved beneath: fourth quill longest: nostrils trans-
> verse.
>
> (3. PANDION.) Tarsi naked; acrotarsia reticulated with small
> rounded scales: outer toe reversible: claws equal, rounded
> beneath: second quill longest.

3. FALCO. — Bill short, strong, bending from the
base; upper mandible strongly toothed, lower one emarginated: nostrils round: tarsi short and strong; acrotarsia
reticulated: wings long; second quill longest; first and
second with the inner webs deeply notched near the extremity.

4. ACCIPITER.—Bill short, rather strong, bending
from the base: nostrils oval: tarsi slender; acrotarsia
scutellated: middle toe much longer than the lateral ones:
wings short, not reaching beyond two-thirds of the length
of the tail: fourth quill longest.

> (1. ASTUR.) Tarsi moderate; rather robust: acrotarsia scutellated.
>
> (2. ACCIPITER.) Tarsi long and slender: acrotarsia scutellated;
> the sutures of the scales obsolete.

5. MILVUS. — Bill moderate, rather weak, bending
slightly from the base, somewhat angular above: nostrils
oblique, oval or elliptic: tarsi short: toes and claws weak:
wings very long: tail long, and forked.

> (1. MILVUS.) Tarsi feathered a little below the knee; acrotarsia
> scutellated: fourth quill longest.
>
> (ELANUS.) Tarsi very short, half-feathered; acrotarsia reticulated: second
> quill generally the longest.

6. BUTEO.—Bill moderate, bending from the base:
nostrils oval or roundish: tarsi partly naked, or clothed
with feathers: wings long; third and fourth quills generally the longest; the four first emarginated on their inner
webs: tail even.

> (1. BUTEO.) Lore without feathers: tarsi short and strong; naked,
> with the acrotarsia scutellated; or feathered to the toes.

(2. PERNIS.) Lore closely covered with small scaly feathers: tarsi moderate, half-feathered; acrotarsia reticulated.

(3. CIRCUS.) Sides of the head furnished with a circle of feathers approaching that of the owls: tarsi elongated, feathered a little below the joint; acrotarsia scutellated: tail long; somewhat rounded.

III. STRIGIDÆ.—*Head large, feathered; base of the bill clothed with stiff bristly feathers directed forwards, concealing the cere and nostrils: eyes surrounded by a circle of radiating feathers: legs feathered to the toes: outer toe reversible.*

7. BUBO.—Bill strong, bending from the base: nostrils oval or rounded: facial disk small and incomplete: auditory aperture small, oval, without an operculum: head furnished with large conspicuous egrets: wings moderate; third quill generally the longest.

(1. BUBO.) Legs robust, feathered to the claws.

(2. SCOPS.) Tarsi feathered: toes naked.

8. OTUS.—Bill bending from the base: nostrils oval, oblique: facial disk complete, of moderate size: auditory conch large, the aperture covered by an operculum: egrets more or less conspicuous: wings long; second quill longest: legs feathered to the claws.

9. STRIX.—Bill somewhat elongated, bending at the tip only: nostrils oval: facial disk complete and full-sized: auditory aperture large, furnished with an operculum: head without egrets: wings long; second quill longest: tarsi feathered: toes hairy; middle claw serrated beneath.

10. SYRNIUM.—Bill bending from the base: nostrils round: facial disk large and complete: auditory aperture large, furnished with an operculum: head without egrets: wings short and rounded; fourth and fifth quills longest: legs feathered to the claws.

11. NOCTUA.—Bill bending from the base: nostrils oval: facial disk small, and generally very incom-

plete: auditory aperture small, oval: head without egrets:
wings moderate; third quill longest.

(1. SURNIA.)   Legs and toes thickly feathered.

(2. NOCTUA.)   Tarsi feathered: toes either feathered or hairy.

ORDER II.   INSESSORES.

Bill various: legs short, or of moderate length: feet
adapted for perching: toes four, varying in position, flat underneath; hind toe articulated on the
same plane with the fore toes: claws slender, somewhat retractile, curved and acute.

I.   DENTIROSTRES.—*Bill moderate, more or*
less weak, furnished with a notch or tooth on
each side towards the tip: rictus armed with
bristles: feet generally slender; toes three before
and one behind.

(LANIIDÆ.)

12.   LANIUS.—Bill robust, convex above, much compressed at the sides, furnished at the base with hairy feathers directed forwards; upper mandible hooked at the
extremity, and strongly notched: nostrils basal, lateral,
nearly round, partly closed by a membrane: wings with
the first quill short; the second shorter than the third and
fourth, which are longest.

(MUSCICAPIDÆ.)

13.   MUSCICAPA. — Bill moderate, somewhat triangular, depressed at the base, compressed towards the
tip, which is deflected; the upper mandible emarginated:
rictus armed with long stiff bristles: tarsus a little longer

than the middle toe: side toes of equal length: wings with the first quill very short; third and fourth longest.

(MERULIDÆ.)

14. CINCLUS.—Bill rather slender, straight, rounded, compressed at the sides; upper mandible emarginated at the tip, slightly bending over the lower one: nostrils longitudinally cleft, and partly covered by a membrane: tarsus longer than the middle toe: wings and tail short; the former with the first quill very short, the third and fourth longest.

15. TURDUS.—Bill moderate, convex above, slightly bending towards the point, which is rather compressed; upper mandible emarginated: rictus furnished with a few bristles: nostrils oval; partly covered by a naked membrane: tarsus longer than the middle toe: wings and tail moderate; the former with the first quill extremely short, almost abortive; the second somewhat shorter than the third and fourth, which are longest.

16. ORIOLUS.—Bill subconic, depressed at the base; upper mandible carinated above, slightly emarginated, inclining over the lower one: nostrils opening longitudinally in a large membrane: tarsus not longer than the middle toe: wings moderate; first quill very short; third longest

(SYLVIIDÆ.)

17. ACCENTOR. — Bill rather strong, subconic, straight, acute, broader than high at the base, very much compressed in front; upper mandible emarginated at the tip; the tomia of both mandibles inflected: nostrils basal, naked, pierced in a large membrane: tarsi strong: wings moderate; first quill very short; third and fourth longest.

18. SYLVIA.—Bill slender, depressed at the base, compressed in front; upper mandible bending over the lower one, and slightly emarginated at the tip; lower man-

dible straight : nostrils basal, pierced in a membrane, more
or less exposed : wings short, or moderate; first quill very
short, sometimes almost abortive; third, or third and
fourth, generally the longest.

(1. Erithaca.) Bill rather strong, as broad as high at the base;
gonys slightly ascending : rictus with long weak diverging bristles: tarsi slender, somewhat elongated: wings moderate; fourth
and fifth quills longest: tail divaricated; the tips of the feathers
pointed.

(2. Phœnicura.) Bill slender, as broad as high at the base: rictus
nearly smooth: tarsi moderate: wings rather long; third and
fourth quills longest: tail even; the feathers obtuse, and generally rufous.

(3. Salicaria.) Bill very slender, broader than high at the base:
rictus with long diverging bristles: tarsi moderate: wings short,
and somewhat rounded; first quill nearly abortive; third longest: tail cuneiform; the tips of the feathers rounded.

(4. Philomela.) Bill rather strong, as broad as high at the base;
gonys slightly ascending : rictus with a few short bristles:
tarsi elongated : wings moderate; third and fourth quills longest: tail with the two middle feathers longest; the tips of all
rounded.

(5. Curruca.) Bill rather strong, broader than high at the base;
gonys slightly ascending: rictus nearly smooth: tarsi moderate:
wings moderate; first quill nearly abortive; third longest: tail
slightly forked.

(6. Sylvia.) Bill very slender, broader than high at the base:
rictus with a few longish bristles: tarsi rather elongated: wings
moderate; third and fourth quills longest: tail slightly forked;
the tips of the feathers rounded.

19. MELIZOPHILUS. — Bill very slender, short,
slightly arched from the base, compressed, with the tip
finely emarginated; tomia of both mandibles inflected towards the middle : rictus with a few bristles: tarsi strong,
longer than the middle toe: wings short; first quill very
small; third, fourth, and fifth, equal and longest: tail
much elongated, cuneiform.

20. REGULUS.—Bill very slender, subulate, straight,
compressed throughout its whole length; the tomia inflected
inwards: nostrils concealed by small bristly feathers directed
forwards: tarsi rather long: wings with the first quill very
short; second much shorter than the third; fourth and fifth
longest : tail moderate; divaricated, the tips of the feathers
pointed.

21. MOTACILLA.—Bill slender, subulate, straight, carinated, angulated between the nostrils, emarginated at the tip; the tomia of both mandibles slightly compressed inwards: tarsus considerably longer than the middle toe: wings with the first quill extremely short; second and third longest; one of the scapulars as long as the quills: tail elongated, even.

22. ANTHUS. — Bill slender, straight, subulate towards the extremity, with the tomia compressed inwards about the middle; upper mandible carinated at the base, bending downwards at the tip, and slightly emarginated: tarsus generally longer than the middle toe; hind claw more or less produced: wings with the first quill abortive; second shorter than the third and fourth, which are longest: two of the scapulars as long as the quills.

23. SAXICOLA.—Bill rather strong, straight, carinated above, dilated at the base, advancing on the forehead; upper mandible emarginated, slightly bending at the tip: rictus with some stiff longish bristles: tarsi elevated; hind claw shorter than the toe: wings with the first quill not half the length of the second; third and fourth longest: coverts and scapulars short.

24. PARUS.—Bill strong, short, straight, subconical, slightly compressed, sharp-pointed, without any notch: nostrils basal, round, concealed by some short bristly feathers reflected over them: tarsi robust: hind claw strongest, and most hooked: wings with the first quill of moderate length; second shorter than the third; fourth and fifth longest.

(1. PARUS.) Bill moderate: tail slightly forked.
(2. MECISTURA.) Bill very short: tail elongated, cuneiform.

25. CALAMOPHILUS.—Bill short, subconical, very slightly compressed; the upper mandible convex above, curved at the extremity, and projecting over the lower one, which is nearly straight: nostrils covered by reflected bristles: tarsi slender: wings with the first quill very short, almost abortive; fourth and fifth longest: tail elongated, cuneiform.

(AMPELIDÆ.)

26. BOMBYCILLA.—Bill strong, short, and straight, broad at the base; upper mandible convex above, slightly bending at the tip, and emarginated: nostrils oval, covered by small hairy feathers directed forwards: tarsus shorter than the middle toe: wings long, and pointed; first and second quills longest; secondaries tipped with wax-like appendages.

II. CONIROSTRES.—*Bill moderate or elongated, strong, more or less conic, entire or slightly emarginated: feet robust; toes three before and one behind.*

(FRINGILLIDÆ.)

27. ALAUDA.—Bill short, subconic; mandibles of equal length; the upper one convex, and slightly curved: nostrils basal, oval, covered by small bristly feathers directed forwards: hind claw nearly straight, longer than the toe: wings long; first quill almost abortive; second a little shorter than the third, which is longest: wing-coverts shorter than the quills.

28. EMBERIZA.—Bill short, strong, conic, sharp-pointed; the mandibles a little distant from each other, and forming an angle at the gape, compressed in front, with the tomia bending inwards; upper mandible smaller and narrower than the lower; palate furnished with a hard bony knob.

> (1. PLECTROPHANES.) Wings long, and acuminated; first and second quills longest, and nearly equal: hind claw produced, and nearly straight.

> (2. EMBERIZA.) Wings moderate; first quill shorter than the second and third, which are longest: hind claw short and hooked.

29. FRINGILLA.—Bill thick, strong, more or less perfectly conic: nostrils basal, lateral, partly concealed by short bristly feathers directed forwards: tarsi short, or moderate: wings with the first three quills nearly equal; second and third generally a little the longest: tail more or less forked.

(1. FRINGILLA.) Bill straight, perfectly conic, the cone rather elongated, sharp-pointed; mandibles nearly equal, with the tomia of the lower one a little inflected: first quill a little shorter than the second and third, which are longest and equal.

(2. PYRGITA.) Bill strong, conic; the culmen somewhat arched, and the tip a little obtuse; lower mandible rather smaller than the upper: three first quills nearly equal, longer than the fourth.

(3. COCCOTHRAUSTES.) Bill conic, very thick and strong; the sides bulging, and the culmen much rounded; lower mandible nearly equal to the upper, with the tomia inflected: wings rather long, acuminated; first and fourth quills equal, shorter than the second and third, which are longest.

(4. CARDUELIS.) Bill straight, perfectly conic, the cone rather elongated, much compressed at the tip, and sharp-pointed; upper mandible angulated at the base, and slightly sinuated: wings moderate; first quill a little shorter than the second and third, which are nearly equal and longest.

(5. LINARIA.) Bill straight, perfectly conic, short, compressed anteriorly, sharp-pointed; commissure straight; wings long, and acuminated; first three quills equal, longer than the fourth.

30. PYRRHULA.——Bill short and thick, the sides inflated and bulging; upper mandible convex above, advancing on the forehead at the base, deflected at the tip and overhanging the lower one: nostrils concealed by hairy feathers directed forwards: wings rather short; first quill shorter than the second, which is longest; third and fourth nearly equal to the second.

31. LOXIA.——Bill moderate, strong, thick at the base, much compressed anteriorly; both mandibles equally convex, hooked at the tips, and crossing each other when at rest; the tomia bending inwards: nostrils round, concealed by bristly feathers directed forwards: wings with the first and second quills equal, shorter than the third, which is longest.

(STURNIDÆ.)

32. STURNUS.——Bill moderate, straight, subconic; depressed, especially at the tip, which is rather blunt;

upper mandible advancing on the forehead, the edges rather dilated and extending beyond those of the lower: nostrils basal, half closed by an arched membrane: tarsus longer than the middle toe: wings long; first quill very short; second longest; the rest gradually decreasing.

33. PASTOR.—Bill subconic, compressed, slightly arched, the tip somewhat emarginated: nostrils basal, oval, partly covered by a plumose membrane: feet robust; tarsus much longer than the middle toe: wings with the first quill very short, almost abortive; second and third longest.

(CORVIDÆ.)

34. FREGILUS.—Bill long, rather slender, arched from the base, a little compressed, subulate and pointed at the extremity : nostrils basal, oval, concealed by reflected bristles : feet robust; tarsus longer than the middle toe : wings long; the fourth and fifth quills longest: tail square.

35. CORVUS.—Bill strong, thick, convex above, compressed at the sides, with the tomia sharp and cutting; upper mandible slightly bending towards the tip; lower one nearly straight: nostrils basal, oval, generally concealed by setaceous feathers directed forwards: wings acuminated; first quill much shorter than the second and third; fourth longest.

(1. Corvus.) Tail moderate, rounded.

(2. Pica.) Tail long, cuneiform.

36. GARRULUS.—Bill rather short, strong, compressed, clothed at the base with feathers directed forwards; both mandibles inclining equally towards the tip, the upper one slightly emarginated : crown feathers long, and capable of erection : wings rounded; fifth and sixth quills longest: tail moderate, square, or slightly rounded.

37. NUCIFRAGA.—Bill long, and straight; upper mandible rounded, and longer than the lower one; both of them terminating in a slightly obtuse and depressed point :

nostrils basal, round, covered by reflected feathers : wings rather acuminated; first quill shorter than the second and third ; fourth longest : tail rounded.

III. SCANSORES.—*Bill various: feet short; adapted for climbing; toes four, two each way, or three in front and one behind.*

* *Bill straight, robust : feet zygodactyle.*

(PICIDÆ.)

38. PICUS.—Bill robust, long, straight, angular, compressed, cuneated at the tip: tongue long and extensile, lumbriciform, the tip barbed: nostrils concealed by reflected bristly feathers : wings with the first quill very short; fourth and fifth longest : feet robust: tail feathers stiff, and pointed at the extremity.

39. YUNX.—Bill short, straight, somewhat conical, depressed, rounded above, the tip sharp and pointed : tongue long and extensile, lumbriciform : nostrils naked, partly closed by a membrane: wings moderate ; first quill a little shorter than the second, which is longest : tail feathers soft and flexible.

** *Bill generally slender : feet anisodactyle.*

(CERTHIIDÆ.)

40. CERTHIA.—Bill moderate, slender, arcuate, compressed, angulated above, sharp-pointed : tongue short : nostrils naked, partly covered by an arched membrane: wings with the first quill very short ; fourth longest: hind toe strong, and longer than the others: tail feathers stiff and pointed at the extremity, deflected.

41. TROGLODYTES.—Bill slender, slightly compressed, a little arcuate : nostrils basal, oval, partly covered by an arched naked membrane : wings short, and rounded ; first quill very short ; second shorter than the third ; fourth and fifth equal and longest : tail short, rounded, carried erect.

42. UPUPA.—Bill long and slender, arcuate, compressed, convex above, sharp-pointed : nostrils oval, open :
crown furnished with a crest: wings rather long ; the first
quill very short; fourth and fifth longest : claws short, and
not much hooked : tail square at the extremity.

43. SITTA.—Bill moderate, rather strong, straight,
compressed ; both mandibles equally inclining to the tip
which is somewhat cuneated : nostrils oval, open, covered
by setaceous feathers directed forwards : wings rather short ;
third and fourth quills longest : hind toe strong, and lengthened : tail short, nearly even.

*** *Bill more or less curved : feet zygodactyle.*

(CUCULIDÆ.)

44. CUCULUS.—Bill as long as the head, rather
compressed, moderately curved, and a little hooked at the
tip : gape wide : nostrils round, surrounded by a naked and
prominent membrane : wings long, acuminated : first quill
short ; third longest; tarsi short, feathered a little below
the knee : outer hind toe partly reversible : tail long, more
or less cuneated.

45. COCCYZUS.—Bill strong, arched, the culmen
convex, triangular at the base, compressed at the sides and
tip : nostrils basal, longitudinally cleft in a membrane :
orbits naked : wings short, and concave : tarsi very long :
tail long, and cuneated.

IV. FISSIROSTRES.—*Bill broad at the base; gape wide : wings long : feet short, and weak.*

* *Bill strong, more or less elongated.*

(MEROPIDÆ.)

46. CORACIAS.—Bill moderate, robust, higher than
broad, compressed, straight; upper mandible bending at
the tip : nostrils basal, linear, pierced diagonally, half
closed by a plumose membrane : toes three before and one

behind, entirely divided : wings long, and acuminated ; first quill shorter than the second, which is longest.

47. MEROPS.——Bill longer than the head, triangular at the base, slightly curved, carinated above, sharp-pointed : nostrils basal, oval, open : toes three before and one behind ; the outer toe connected with the middle one as far as the second joint, the middle with the inner one as far as the first : wings long, and pointed ; first quill very short, the second longest.

(HALCYONIDÆ.)

48. ALCEDO.——Bill very long, straight, quadrangular, thick and pointed : nostrils basal, pierced obliquely, linear, almost closed by a naked membrane : toes three before and one behind ; the outer toe connected with the middle one as far as the second joint, the middle with the inner one as far as the first : wings moderate ; the second and third quills equal and longest : tail short.

** Bill short and weak : gape extremely large.

(HIRUNDINIDÆ.)

49. HIRUNDO.——Bill short, and much depressed ; upper mandible carinated, deflected at the tip ; gape extending nearly to the eyes : toes three before and one behind ; the outer toe connected with the middle one as far as the first joint : wings very long ; first quill longest : tail forked.

50. CYPSELUS.——Bill very short, triangular, depressed at the base ; the upper mandible deflected at the tip ; gape extending to the posterior angle of the eye : toes divided to their origin ; all directed forwards : wings extremely long ; first quill a little shorter than the second : tail forked.

(CAPRIMULGIDÆ.)

51. CAPRIMULGUS.——Bill very short, rather curved, broad and depressed at the base ; the upper mandible de-

flected at the tip, furnished at the basal edge with strong vibrissæ directed forwards; gape extending beyond the eyes: anterior toes connected by a membrane as far as the first joint; hind toe reversible: wings long; first quill shorter than the second, which is longest: tail rounded, or forked.

---

§ *11.* *Feet not formed for grasping.*

ORDER III. RASORES.

Bill short, or moderate, convex, often furnished with a cere at the base; upper mandible arched: wings short: legs strong and muscular, adapted for walking: hind toe sometimes wanting; when present, generally articulated high on the tarsus: claws robust, short, slightly curved, somewhat blunt.

I. COLUMBIDÆ.—*Hind toe present, articulated nearly on the same plane with the others: tarsi without spurs: tail of twelve feathers.*

52. COLUMBA.—Bill moderate, straight, compressed, deflected at the tip; upper mandible covered at the base with a soft tumid membrane, in which the nostrils are situate: tarsi short, reticulated; toes entirely divided.

(1. COLUMBA.) Wings with the first quill shorter than the second, which is longest: tail even: side toes equal.

(2. ECTOPISTES.) Wings pointed; two first quills equal, longer than the third: tail long, and cuneated: side toes unequal; the inner one longest.

* II. PHASIANIDÆ.—*Hind toe present, elevated above the others: tarsi generally armed with spurs: tail of more than twelve feathers: head more or less naked.*

*53. MELEAGRIS.—Bill short and thick, furnished at the base with an elongated, pendulous, fleshy appendage: head and neck naked;

throat with a pendulous carunculated wattle: tarsi of the male armed with spurs: tail broad, expansile, consisting of from fourteen to eighteen feathers.

*54. PAVO.—Bill naked at the base, thick, convex above, deflected at the tip: cheeks partially naked: head ornamonted with a crest: tarsi of the male spurred: tail of eighteen feathers: upper tail-coverts longer than the tail, broad and expansile, ocellated.

*55. GALLUS.—Bill smooth at the base, thick, slightly curved: nostrils covered by an arched scale: generally an erect fleshy crest on the head; throat with fleshy wattles on each side of the lower mandible: ears naked: tarsi with strong spurs: anterior toes united by a membrane as far as the first joint: tail of fourteen feathers, compressed, more or less arched, ascending.

*56. PHASIANUS.—Bill short and thick, naked at the base; upper mandible very convex, with the tip deflected: nostrils basal, lateral, covered by an arched membrane: cheeks naked, adorned with scarlet papillæ: ears covered: tarsi of the male spurred: anterior toes united by a membrane as far as the first joint: wings short; fourth and fifth quills longest: tail very long, cuneated, of eighteen feathers.

*57. NUMIDA.—Bill thick, covered at the base with a warty membrane, in which the nostrils are placed: head naked, the crown with a callous horny protuberance; beneath the cheeks pendulous carunculated wattles: tarsi without spurs: anterior toes united by a membrane as far as the first joint: tail short, bent down, of fourteen or sixteen feathers.

III. TETRAONIDÆ.—*Hind toe short, and weak; sometimes altogether wanting: tarsi generally armed with spurs: tail short.*

58. TETRAO.—Bill short and strong, convex above, bending towards the tip: nostrils partly closed by an arched scale, concealed by the frontal feathers: eyebrows naked, adorned with scarlet papillæ: tarsi feathered; without spurs: wings with the first quill much shorter than the second; third and fourth longest: tail of sixteen or eighteen feathers.

(1. TETRAO.) Anterior toes naked, with pectinated margins: hind toe longer than the nail: tail broad, and slightly rounded, or lyrate.

(2. LAGOPUS.) Anterior toes feathered, the margins not pectinated: hind toe shorter than the nail: tail nearly even.

59. PERDIX.—Bill short, strong, naked at the base, convex above, deflected towards the tip: nostrils basal, half closed by an arched naked scale: tarsi naked: feet with

four toes, naked: wings short, and concave; the fourth and
fifth quills generally longest: tail short, of from twelve to
eighteen feathers.

 (1. Perdix.) Bill short, and strong: orbits naked: tarsus in the
  male armed with a blunt tubercle: tail short, and bent down.

 *(2. Ortyx.) Bill thick, strong, higher than broad: orbits feathered: tarsi
  unarmed: wings with the first quill very short; fifth longest: tail short
  or moderate, of twelve feathers.

 (3. Coturnix.) Bill short, and slender: orbits feathered: tarsi
  unarmed: wings with the first quill longest: tail very short,
  almost concealed by the upper coverts.

IV. STRUTHIONIDÆ.—*Hind toe wanting: tarsi without spurs: wings very short, often unfit for flight.*

60. OTIS.—Bill moderate, subconic, nearly straight,
compressed; the upper mandible arched towards the tip:
nostrils oval, open, a little remote from the base: legs
long, naked above the knee; toes connected by a membrane at the base: wings moderate; third quill longest.

ORDER IV. GRALLATORES.

Bill various: legs moderate or elongated, slender, with
the lower part of the tibia generally naked, adapted
for wading: toes long, three or four in number,
more or less connected by a membrane at the base,
sometimes lobated.

I. CHARADRIIDÆ.—*Bill short or moderate, rarely elongated; robust or slender: legs moderate, or elongated: toes three, all directed forwards; rarely the rudiment of a fourth; the outer and middle toes generally united at the base by a membrane.*

61. CURSORIUS.—Bill shorter than the head, depressed at the base, somewhat arched and bent down to-

AVES. 65

wards the tip, pointed: nostrils basal, oval, covered above
by a protuberant membrane: legs long and slender: toes
three, very short; almost divided to their origin: claws
very small, the middle one serrated: wings moderate;
second quill longest.

62. ŒDICNEMUS.—Bill longer than the head,
straight, very robust, a little depressed at the base, compressed laterally; culmen elevated towards the tip; lower
mandible angulated beneath: nostrils medial, longitudinally
cleft: tarsi long and slender: toes three, united by a membrane at the base: wings moderate; second quill longest.

63. CHARADRIUS. — Bill shorter than the head,
straight, slender, compressed, somewhat tumid and enlarged towards the tip: nasal channel extending two-
thirds of its length, covered by a large membrane; nostrils basal, linear, pierced in the membrane: tarsi moderate,
slender: toes three; outer and middle ones connected by a
short membrane; inner toe free: wings moderate; first
quill longest.

64. VANELLUS.—Bill moderate, straight, compressed,
slender at the base, tumid towards the extremity; upper
mandible slightly bending at the tip: nasal channel large,
covered by a membrane; nostrils linear, pierced in the
membrane: tarsi moderate, slender: toes four; outer and
middle toes connected at the base; hind toe small, and
elevated: wings long and acuminated.

(1. Squatarola.) Nasal channel extending half the length of the
bill: tarsi reticulated: hind toe rudimentary: first quill longest.

(2. Vanellus.) Nasal channel two-thirds the length of the bill:
acrotarsia scutellated: hind toe moderately developed: fourth
and fifth quills longest.

65. STREPSILAS.—Bill moderate, strong, nearly
straight, forming a lengthened cone, with the culmen a
little depressed, and the apex subtruncated; the commissure slightly ascending: nasal channel half the length of
the bill: toes four; the anterior ones almost entirely divided; hind toe moderately developed, touching the
ground: wings long; first quill longest.

E

66. CALIDRIS. — Bill moderate, straight, slender and flexible throughout, compressed at the base, depressed and dilated towards the tip: nasal groove extending nearly the whole length; nostrils lateral, linear: tarsi moderate, slender: toes three, divided nearly to their origin: wings moderate; first quill longest.

67. HÆMATOPUS. — Bill long, robust, straight, compressed, the tip cuneated: nostrils lateral, nearly basal, oblong-linear, placed in a groove: tarsi moderate, robust: toes three; the outer and middle ones connected as far as the first joint; all the toes bordered by a narrow membrane: wings long; first quill longest.

II. GRUIDÆ.—*Bill strong, moderate or elongated : legs long: toes four in number; the outer and middle ones united at the base by a small membrane; hind toe short, elevated on the tarsus.*

68. GRUS.—Bill a little longer than the head, straight, compressed, deeply channelled at the base of the upper mandible on each side, the tip forming a lengthened cone: nostrils medial, placed in the lateral groove, closed behind by a membrane: head generally more or less naked: a considerable part of the tibia naked: wings moderate; third quill longest.

III. ARDEIDÆ.—*Bill strong, elongated: legs long: feet with four toes; anterior ones more or less united at the base; hind toe long, resting on the ground.*

69. ARDEA. — Bill as long as, or a little longer than the head, strong, straight or very slightly inclined, thick at the base, gradually tapering towards the tip; upper mandible channelled for about two-thirds of its length: nostrils nearly basal, placed in the groove, and partly closed by a membrane: orbits and lore naked:

lower part of the tibia more or less naked : middle claw
with the inner margin pectinated : wings ample ; second
and third quills longest.

(1. ARDEA.) Bill longer than the head, the upper mandible nearly
straight: occiput usually ornamented with a pendent crest: neck
slender, elongated.

(2. BOTAURUS.) Bill scarcely longer than the head, higher than
broad at the base, very much compressed; upper mandible
slightly curved: neck thick, rather short; the feathers on the
fore part loose and elongated, capable of erection; those behind
short and downy.

(3. NYCTICORAX.) Bill scarcely longer than the head, higher than
broad, much compressed; upper mandible slightly curved: occiput with three very long subulate feathers: neck thickish, rather
short.

70. CICONIA. — Bill much longer than the head,
straight, robust, cylindrical, acute, forming a lengthened
cone ; upper mandible convex above ; under mandible inclining a little upwards at the tip: nostrils nearly basal,
open, longitudinally pierced in the horny substance of the
bill: orbits naked: tarsi very long: a considerable part
of the tibia naked: claws short, depressed, not pectinated: wings moderate ; third, fourth, and fifth quills
longest.

71. PLATALEA. — Bill very long, robust, much
depressed, dilated and rounded at the extremity in the
form of a spatula : nostrils dorsal, oblong, open, bordered
by a membrane: head and face more or less naked : tibiæ
semiplumed : tarsi long: the three anterior toes united as
far as the second joint by a deeply cut membrane : wings
moderate ; second quill longest.

72. IBIS.—Bill very much elongated, slender, arcuate,
broad at the base, somewhat quadrangular, the tip depressed,
obtuse, and rounded ; upper mandible deeply channelled
throughout its whole length: nostrils nearly basal, placed
in the groove, oblong, narrow, surrounded by a membrane :
face and throat more or less naked: legs long, and slender ;
lower part of the tibia naked: wings moderate ; second and
third quills longest.

IV. SCOLOPACIDÆ.—*Bill long and slender; legs moderate, or elongated: toes four in number; anterior ones entirely divided, or united by a small basal membrane; hind toe weak and elevated, very rarely altogether wanting.*

73. NUMENIUS.—Bill much longer than the head, slender, arcuate, rounded, the tip rather obtuse; upper mandible projecting beyond the lower, channelled throughout three-fourths of its length: nostrils basal, oblong-linear, placed in the lateral groove: lore, and rest of the face feathered: legs slender; lower part of the tibia naked: anterior toes connected as far as the first joint: wings moderate; first quill longest.

74. TOTANUS.—Bill long or moderate, slender, generally straight, rarely inclining a little upwards, soft at the base, solid and sharp-pointed at the extremity; upper mandible channelled through half its length or more, slightly bending over the lower one at the tip: nostrils lateral, linear, placed in the groove: legs long, slender, with the lower part of the tibia naked: anterior toes, or the outer ones only, connected by a small membrane.

75. RECURVIROSTRA.—Bill very long, slender, weak, depressed, pointed, flexible at the tip, and considerably recurved; upper mandible channelled on each side: nostrils long, linear, placed at the base of the groove: legs long and slender; lower half of the tibia naked: anterior toes connected by a membrane as far as the second joint; hind toe very small.

76. HIMANTOPUS.—Bill very long, slender, attenuated, nearly straight, rounded, a little compressed at the tip; mandibles channelled laterally for half their length: nostrils linear, elongated: legs very long and slender: nearly the whole of the tibia naked: outer and middle toes connected by a broad membrane; hind toe wanting: wings very long; first quill longest.

77. LIMOSA. — Bill very long, slender, soft and flexible throughout, slightly recurved, depressed, the tip obtuse and somewhat dilated; upper mandible channelled for nearly its whole length: nostrils linear, pierced in the membrane of the groove: legs long and slender; lower part of the tibia naked: outer and middle toes connected by a membrane as far as the first joint: wings moderate; first quill longest.

78. SCOLOPAX. — Bill very long, straight, slender, soft, the tip obtuse and dilated; upper mandible a little longer than the lower; both mandibles channelled for nearly two-thirds of their length: nostrils lateral, basal, linear, covered by a membrane: legs moderate, slender; naked space on the tibia very small: wings with the first quill longest; second nearly equal to it.

(1. RUSTICOLA.) Tibia feathered to the joint: tarsi short: anterior toes entirely divided: hind claw shorter than the toe.

(2. SCOLOPAX.) Lower part of the tibia naked: tarsi more elongated: toes entirely divided, or united at the base by an extremely short membrane: hind claw longer than the toe.

(3. MACRORAMPHUS.) Lower part of the tibia naked: tarsi rather elongated: outer toe united to the middle one as far as the first joint: hind claw longer than the toe.

79. TRINGA. — Bill as long as, or a little longer than the head; straight, or slightly curved, soft and flexible throughout, compressed at the base, depressed and dilated at the extremity; upper mandible longer than the lower, with the nasal channel extending nearly to the tip: nostrils lateral, short, pierced in the membrane of the groove: tarsi moderate, or elongated: wings with the first quill longest.

(1. MACHETES.) Tarsi elongated: outer and middle toes united by a membrane as far as the first joint: neck (in the male) with an ornamental ruff during the breeding season.

(2. TRINGA.) Tarsi moderate: anterior toes divided to their origin: hind toe very small.

80. LOBIPES. — Bill long, slender, weak, a little depressed at the base, subulate at the tip; upper mandible slightly curved, and channelled throughout: tongue slender,

and pointed : nostrils basal, oblong-linear, prominent, surrounded by a membrane : tarsi somewhat elongated, and
compressed : anterior toes rather long, bordered by narrow
lobated membranes : wings with the first quill longest.

81. PHALAROPUS. — Bill as long as the head,
straight, strong, broad and depressed at the base, a little
compressed at the extremity ; upper mandible inclining over
the lower, with the nasal channel extending the whole
length : tongue short, broad, and rounded at the tip : nostrils basal, oval, prominent, surrounded by a membrane :
tarsi short and strong, scarcely compressed : anterior toes
rather short, united as far as the first joint ; bordered
beyond with broad and deeply scalloped membranes : wings
with the first and second quills longest.

V. RALLIDÆ.—*Bill short or moderate, rarely
elongated ; rather strong : body generally compressed : legs moderate : toes four ; anterior
ones elongated, entirely divided, or united by
a membrane at the base, sometimes pinnated.*

82. GLAREOLA.—Bill short, convex, compressed
towards the tip ; the upper mandible curved for half its
length : nostrils basal, lateral, obliquely cleft : a small part
of the tibia naked : tarsi moderate, slender : outer and
middle toes connected by a short membrane ; hind toe
small, and elevated : claws long, and pointed : wings very
long ; first quill considerably the longest.

83. RALLUS. — Bill longer than the head, rather
slender, very slightly arched, compressed at the base, somewhat cylindrical at the tip ; upper mandible channelled for
two-thirds of its length : nostrils submedial, oblong-linear,
placed in the groove, partly covered by a membrane lower :
part of the tibia naked : toes long, and slender ; anterior
ones entirely divided : wings short, rounded ; first quill
shorter than the two next, which are longest : bastard
winglet with a short spur.

84. CREX.—Bill shorter than the head, somewhat thick, straight, convex above, higher than broad at the base, attenuated and compressed at the tip; upper mandible a little deflected; gonys of the lower mandible ascending to meet it: nostrils lateral, medial, longitudinally cleft, half closed by the membrane of the nasal channel: anterior toes long, entirely divided: wings short; first quill shorter than the second and third, which are longest.

85. GALLINULA. — Bill shorter than the head, straight, robust, convex above, much compressed; upper mandible expanding at the base on the forehead, forming a naked disk: nostrils lateral, medial, oblong-linear: anterior toes very long, divided to their origin, bordered by a narrow, entire membrane: wings with the second and third quills longest.

86. FULICA.—Bill shorter than the head, straight, robust, somewhat convex above, compressed, thick at the base: a naked disk on the forehead: nostrils lateral, medial, longitudinally cleft: anterior toes very long, united at the base, bordered by a scalloped membrane: wings moderate; second and third quills longest.

ORDER V.   NATATORES.

Bill various: legs short, often placed far behind, adapted for swimming: lower part of the tibia more or less naked: tarsi compressed: toes generally four; the three anterior ones palmated, semipalmated, or fissopalmated; or all four united: hind toe sometimes wanting.

I. ANATIDÆ.—*Bill stout, straight, covered with a thin membranous skin; the tomia armed with lamellæ, or with small denticulations; more or*

*less depressed; the tip rounded and obtuse, fur-
nished with a nail: wings moderate: legs placed
in or near the equilibrium: feet four-toed, palmated; hind toe free, placed high on the tarsus.*

* (ANSERINÆ.)

87. ANSER.——Bill as long as, or shorter than the
head, higher than broad at the base, subconic; the upper
mandible deflected at the tip; the lower one flat and
straight, narrower: nostrils a little behind the middle: neck
of moderate length: wings tuberculated, or spurred: legs
in the equilibrium: tarsi slightly elongated.

>   (1. ANSER.) Bill as long as the head; the tips of the lamellæ
>   exserted, appearing like sharp teeth.
>
>   (2. BERNICLA.) Bill shorter than the head, somewhat slender; the
>   edges of the mandibles concealing the tips of the lamellæ.
>
>   (3. PLECTROPTERUS.) Bill moderate; furnished with a fleshy
>   tubercle at its base: wings armed with spurs.

88. CYGNUS.——Bill moderate, of nearly equal breadth
throughout, furnished at its base with a knob or fleshy
tumour, convex above; the upper mandible deflected at
the tip; the lower one flat: nostrils medial: neck very
long: legs a little beyond the equilibrium: tarsi short.

** (ANATINÆ.) — *Hind toe simple: tarsi somewhat round.*

89. TADORNA.——Bill shorter than the head, with an
elevated tubercle at the base, depressed or concave in the
middle, of nearly equal breadth throughout; the upper
mandible laterally grooved near the tip, which is slightly
recurved: face and lore feathered: nostrils large, oval,
a little before the base: tail even.

* 90. CAIRINA.—Bill furnished with an elevated tubercle at the
base; the edges of the mandibles sinuated: face and lore covered with a
naked warty skin: nostrils round, basal: wings tuberculated.

91. ANAS.——Bill as long as, or rather longer than the
head, simple at the base, broad and depressed generally
throughout its whole length, sometimes widening ante-

riorly: nostrils small, oval, a little before the base: tail more or less acute, sometimes elongated.

(1. ANAS.) Bill longer than the head, semicylindric at its origin, depressed and very much dilated towards the tip; tomia furnished with long pectinated lamellæ, projecting far beyond the margin: tail a little longer than the wings, of fourteen feathers, moderately acute.

(2. CHAULIODUS.) Bill the length of the head, of nearly equal breadth throughout; pectinated lamellæ distinct, appearing below the edge of the upper mandible: tail slightly lengthened; the two middle feathers pointed, and rather longer than the others.

(3. DAFILA.) Bill full as long as the head; the upper mandible of equal breadth throughout; lamellæ not projecting beyond the margin: tail of sixteen feathers, elongated, acute.

(4. BOSCHAS.) Bill as long as the head, of nearly equal breadth throughout; the denticulations of the upper mandible scarcely projecting beyond the margin: tail of sixteen feathers, moderate, rather acute.

92.   MARECA.——Bill shorter than the head, higher than broad at the base, depressed and narrowing towards the tip; lamellæ slightly projecting: nostrils small, oval, a little before the base: tail of fourteen feathers, short, acute.

* 93. DENDRONESSA.—Bill shorter than the head, as broad as high at the base, towards the tip narrow and contracted: nostrils large, pervious, submedial: head with an occipital crest: tertials ornamented: tail of sixteen feathers, rounded.

*** (FULIGULINÆ.)—*Hind toe with a lobated membrane: tarsi compressed.*

94.   SOMATERIA.——Bill swollen and elevated at the base, nearly straight, extending on the forehead, where it is divided by an angular projection of feathers, narrow and obtuse anteriorly, the nail strong and hooked; tomia set with coarse distant lamellæ: nostrils small, oval, medial: tail of fourteen feathers.

95.   OIDEMIA.——Bill short, broad, with an elevated tumour at the base, towards the tip much depressed and flattened; the nail obtuse and rounded, slightly deflected; lamellæ coarse, widely set, scarcely projecting: nostrils oval, submedial, elevated: tail short, graduated, acute.

96. FULIGULA.—Bill as long as the head, simple at the base or very slightly elevated, broad and very much depressed anteriorly, a little dilated towards the tip; upper lamellæ not projecting beyond the edges of the mandible: nostrils rather small, oblong-oval, a little before the base: wings and tail short, the latter generally of fourteen acute feathers.

97. CLANGULA.—Bill shorter than the head, elevated at the base, much more contracted than in the last genus, and becoming continually narrower towards the tip; lamellæ not projecting: nostrils roundish-oval, medial: tail of moderate length, graduated, the feathers semi-acute.

98. HARELDA.—Bill much shorter than the head, rather high at the base, slender, suddenly contracting towards the tip; nail strong and arched; upper lamellæ projecting considerably below the edge of the mandible: nostrils large, oblong-linear, a little before the base: forehead high: tail of fourteen feathers; the central pair acute, and very much elongated.

**** (MERGANINÆ.)

99. MERGUS.—Bill short or moderate, straight, slender, narrow, approaching to cylindrical; upper mandible high at the base, tapering towards the tip, which is armed with a strong hooked nail; tomia of both mandibles set with acute teeth directed backwards: nostrils oblong-oval, pervious, medial: legs short, placed far behind: tarsi compressed: hind toe lobated: tail of twelve feathers, rounded.

II. COLYMBIDÆ.—*Bill slightly compressed, smooth, not covered by a membranous skin; the tomia simple, or only slightly denticulated: wings short: legs placed at the extremity of the body: tarsi very much compressed: feet four-toed, palmated, or fissopalmated: hind toe free.*

100. PODICEPS.—Bill moderate, straight, robust, forming a lengthened cone, sharp-pointed; upper mandible

slightly inclining at the tip ; lower mandible with the gonys ascending : feet short ; anterior toes fissopalmated ; hind toe lobated : claws broad, and much depressed : wings with the three first quills nearly equal, and longest : tail wanting.

101. COLYMBUS.—Bill rather longer than the head, straight, robust, nearly cylindrical, sharp-pointed ; the tomia of both mandibles inflected ; gonys ascending : feet short ; anterior toes long, entirely palmated ; hind toe short, and lobated : claws depressed : wings with the first or second quill longest : tail very short, rounded.

III. ALCIDÆ.—*Bill very much compressed, often transversely grooved at the sides : wings short, sometimes wholly unfit for flight : legs placed at the extremity of the body : feet three-toed, entirely palmated.*

102. URIA.—Bill moderate, strong, straight, convex above, much compressed at the sides, sharp-pointed ; upper mandible slightly deflected at the tip ; gonys of the lower mandible ascending ; tomia curving inwards from the base to the middle : nostrils basal, lateral, pervious, elliptic-linear, half concealed by a plumose membrane : tarsi very much compressed : wings very short ; first quill longest : tail short, of twelve feathers.

103. MERGULUS.—Bill shorter than the head, rather thick, convex, conical, slightly curved, covered at the base with downy feathers ; both mandibles notched at the tip : nostrils basal, lateral, round, partly concealed by a plumose membrane : wings not reaching to the tail ; first and second quills equal, and longest : tail short, of twelve feathers.

104. FRATERCULA.—Bill shorter than the head, thick at the base, very much compressed anteriorly, higher than long, the culmen more elevated than the crown of the head ; both mandibles arched, transversely furrowed, and notched at the tip : nostrils basal, marginal, oblong-linear,

almost entirely closed by a large naked membrane : wings short, narrow, and acuminated ; first and second quills equal, and longest : tail short, of sixteen feathers.

105. ALCA.——Bill a little shorter than the head, rather longer than high, much compressed, cultrated ; both mandibles laterally sulcated ; upper one hooked at the extremity ; lower one with the gonys ascending, the mental angle prominent and rounded : nostrils marginal, nearly concealed by the advancing frontal feathers : wings short, and narrow : tail short, of twelve to sixteen feathers.

IV. PELECANIDÆ.—*Bill strong, sometimes compressed ; the tomia denticulated : wings long, and powerful : legs very short, in or near the equilibrium : feet with four toes, all united by one continuous membrane.*

106. PHALACROCORAX.——Bill a little longer than the head, straight, compressed, culmen rounded ; upper mandible with a longitudinal furrow on each side, the dertrum distinct and strongly hooked ; lower one with the tip truncated : nostrils concealed : face and throat naked : tibiæ feathered : middle claw with the inner margin denticulated : wings moderate ; first quill shorter than the second, which is longest : tail rounded.

107. SULA.——Bill a little longer than the head, robust, straight, large at the base, compressed and attenuated towards the tip ; culmen rounded ; upper mandible laterally sulcated, the tip slightly hooked : gape extending beyond the eyes : face naked : nostrils concealed : tibiæ naked at bottom : middle claw with the inner margin denticulated : wings long, and acuminated ; first quill longest, or equal to the second : tail graduated.

V. LARIDÆ.—*Bill without denticulations : wings very long : legs in or near the equilibrium :*

three anterior toes united; hind toe free, sometimes represented by a simple nail, or altogether wanting.

108. STERNA. — Bill as long as, or a little longer than the head, nearly straight, rather compressed, slender, attenuated towards the tip, sharp-pointed; mandibles of equal length, the upper one slightly bending at the tip, lower one with the mental angle a little projecting: nostrils longitudinal, oblong-linear, pervious: feet weak, small, four-toed; membranes deeply notched: wings long, acuminated; first quill longest: tail generally forked.

    (1. STERNA.) Wings very long: tail deeply forked.
    (2. ANOUS.) Wings moderate: tail even.

109. LARUS.—Bill moderate, strong, cultrated, much compressed; the culmen continuous, bending downwards at the tip; gonys obliquely ascending, straight; mental angle strongly projecting; tomia sharp: nostrils longitudinal, medial, oblong-linear, pervious: legs moderate: anterior toes with the membranes entire; hind toe small, high on the tarsus, rarely wanting: wings long; first or second quill longest: tail generally even.

    (1. XEMA.) Tail slightly forked.
    (2. LARUS.) Tail even.

110. LESTRIS.—Bill moderate, strong, thick, rounded above, compressed towards the tip; culmen and dertrum distinct, the latter convex and strongly hooked; gonys straight, obliquely ascending; mental angle projecting: nostrils immediately behind the dertrum, diagonal, narrow, closed behind, pervious: hind toe very small, nearly on a level with the others: wings with the first quill longest: two middle tail feathers elongated.

111. PROCELLARIA.—Bill more or less elongated, straight, subcylindrical, depressed and dilated at the base; culmen and dertrum distinct; the latter convex, much compressed, and strongly hooked; lower mandible with a longi-

tudinal furrow on each side : gonys concave, more or less ascending : nostrils at the extremity of an open truncated tube extending along the culmen : a simple claw in the place of a hind toe : wings long; first or second quill longest.

(1. PROCELLARIA.) Bill moderate, robust; lower mandible straight, somewhat truncated at the tip; mental angle projecting : nostrils opening by a single orifice, the tube extending half the length of the bill : tail slightly rounded.

(2. PUFFINUS.) Bill longer than the head, slender; lower mandible pointed, deflected at the tip, following the curvature of the upper; mental angle obsolete : nostrils in a double tube extending one-fourth of the length of the bill : tarsi moderate : tail rounded.

(3. THALASSIDROMA.) Bill shorter than the head, slender; lower mandible pointed, deflected at the tip, following the curvature of the upper : mental angle obsolete : nostrils opening by a single orifice, the tube extending half the length of the bill : tarsi slender, elevated : tail square, or slightly forked.

ORDER I.　RAPTORES.

GEN. 1.　NEOPHRON, *Sav.*

1.　N. *Percnopterus*, Sav. (*Egyptian Neophron.*)—
Plumage above and below white; quills alone black:
occipital feathers long, and filiform: tail cuneated.

N. Percnopterus, *Jard. and Selb. Orn.* pl. 33.　Cathartes Percnopterus, *Temm. Man. d'Orn.* tom. i. p. 8.　Egyptian Neophron,
Selb. Illust. vol. i. p. 4. pl. A.

DIMENS. Entire length two feet seven inches: length of the bill
(from the forehead) two inches six lines; of the tarsus three inches;
of the middle toe, claw included, three inches: breadth, wings extended,
five feet nine inches. SELBY.

DESCRIPT. (*Immature plumage.*) " Bill brownish black, or horn-
coloured: cere somewhat bulging at the base, and occupying half the
length of the bill, wine yellow: nostrils in the middle of the cere,
large and open: crown of the head, cheeks, and throat, covered with
a naked skin, of a livid flesh-coloured red, with a few straggling bristly
feathers between the bill and the eyes, and upon the margins of the
mandibles: ears round, open, and large: occiput and nape covered with
a close, thick-set white down, with small black feathers intermixed:
neck clothed with long, arched, and acuminated feathers, forming a
kind of ruff of a deep umber brown, tipped with cream yellow: back
and scapulars cream white, the latter intermixed and varied with umber
brown: lesser wing-coverts nearest the body deep umber brown, margined with a paler shade; these are succeeded by two rows of cream-
coloured sharp-pointed feathers: greater coverts umber brown, varied
with cream white: secondaries pale umber brown, their tips and margins yellowish white: quills black: tail cuneiform, umber brown at the
base, the tip yellowish white: under parts mixed with umber brown: legs
strong and fleshy, of a pale yellowish gray: tarsi covered with a rough
reticulated skin: middle toe with four entire scales upon the last pha-
lanx; the exterior and interior each with three; hinder toe short and
strong: claws blackish brown, strong, but not greatly arched." SELBY.
In the *adult state*, the entire plumage is white; the greater quills alone
excepted, which are black.

Common in Egypt, and in some parts of the Continent, but not known
as a visitant in this country till 1825, in the Autumn of which year, two
specimens were observed on the shores of the Bristol Channel near Kilve
in Somersetshire, and one killed.　From this individual, which proved to
be in immature plumage, the above description was taken by Mr Selby.
This species is said to feed principally upon carrion, but occasionally
on lizards and other reptiles; more rarely attacks living birds and the
smaller quadrupeds.　Builds in the crevices and hollows of rocks.

GEN. 2.   AQUILA, *Briss.*

(1. Aquila, *Cuv.*)

2.   A. *Chrysaëtos*, Vigors. (*Golden Eagle.*) — Tail longer than the wings; rounded at the extremity: the last phalanges of all the toes with only three scales.

Falco fulvus, *Temm. Man. d'Orn.* tom. i. p. 38.   Golden Eagle, *Mont. Orn. Dict. and Supp. Selb. Illust.* vol. i. p. 12. pls. 1, 1*, and 2. *Bew. Brit. Birds,* vol. i. p. 5.

Dimens. Entire length three feet and a half: breadth eight feet. Mont.

Descript. (*Adult.*) Top of the head, and nape of the neck, bright rust-colour; the feathers long and acuminated: rest of the plumage dull brown, approaching to dusky: inside of the thighs, and feathers on the tarsi, light brown: tail deep gray, barred and tipped with blackish brown: bill bluish at the base, black at the extremity: irides brown: cere and feet yellow. (*Young.*) Plumage throughout of a uniform reddish brown: vent and under tail-coverts whitish: inside of the thighs, and feathers on the tarsi, white: tail white for two-thirds of its length from the base; the remainder dark brown. Perfect plumage not attained till the fourth year. (*Egg.*) Dirty white, mottled all over with pale reddish brown: longitudinal diameter three inches; transverse diameter two inches five lines.

Found principally in the mountainous parts of Scotland and Ireland: rare in England, but has been killed as far south as in Sussex. Preys on lambs, fawns, &c. as well as on the larger birds. Builds on rocks and tall trees, and lays two, rarely three, eggs. *Obs.* The Ring-tail Eagle (*F. fulvus,* Linn.) is the young of this species.

(2. Haliæetus, *Sav.*)

3.   A. *Albicilla*, Briss. (*Cinereous Eagle.*) — Plumage brown; with spots of a darker tint in the immature state: tail not extending beyond the wings: all the phalanges of the toes scaled.

Falco Albicilla, *Temm. Man. d'Orn.* tom. i. p. 49.   Cinereous Eagle, *Mont. Orn. Dict. and Supp. Selb. Illust.* vol. i. p. 18. pls. 3, and 3*. *Bew. Brit. Birds,* vol. i. p. 9.

Dimens. Entire length two feet four to ten inches.

Descript. (*Adult.*) Head, and upper part of the neck, brownish ash: rest of the plumage on the body, including the wings, brown clouded with cinereous: tail white: bill whitish: cere and feet pale yellow. (*Young.*) Head and neck deep brown, with the tips of the feathers somewhat paler: upper parts in general reddish brown, each feather being pale at the root and dark along the shaft: quills dusky: under parts brown, with deeper coloured spots, and here and there a few white feathers: tail whitish at the base, irregularly spotted with brown on the outer webs, the tip also being of this colour: bill bluish black, paler at the base. (*Egg.*) Dirty white, with a few pale red marks: long. diam. three inches; trans. diam. two inches four lines.

More generally diffused than the last species. Most plentiful in Scotland, Ireland, and in the Orkney and Shetland Isles, but is occasionally met with, particularly in the immature state, in various parts of England. Feeds principally on fish, but attacks also birds and quadrupeds. Breeds generally on the most inaccessible cliffs, and lays two eggs. *Obs.* The Sea Eagle (*F. Ossifragus*, Linn.) is the young of this species.

(3. Pandion, *Sav.*)

4.   A. *Haliæetus*, Meyer. (*Osprey.*) — Wings longer than the tail.

> Falco Haliaetus, *Temm. Man. d'Orn.* tom. i. p. 47. Osprey, *Mont. Orn. Dict. and Supp. Selb. Illust.* vol. i. p. 24. pl. 4. *Bew. Brit. Birds*, vol. i. p. 13.

Dimens. Entire length one foot eleven inches: breadth five feet four inches. Flem.

Descript. (*Adult.*) Crown of the head and nape of the neck deep brown, the feathers, which are long and narrow, edged with white: a streak of blackish brown on each side of the neck, proceeding from the posterior angle of the eye, and reaching almost to the shoulders: upper parts in general deep umber-brown: under parts white, with some faint indications of a darker patch on the breast: tail-feathers transversely barred with white on their inner webs, the two middle ones excepted, which are wholly brown: bill blackish: irides yellow: cere and feet grayish blue. (*Young.*) Chiefly characterized by having the under parts more or less mottled; a large patch especially of tawny yellow on the breast, spotted with brown: feathers on the upper parts edged with reddish yellow. (*Egg.*) Reddish white; the larger end blotched with dark red brown, the smaller end spotted with the same colour: long. diam. two inches four lines; trans. diam. one inch ten lines.

Not a common species. Met with in Scotland, and in the West of England. Has occurred also in Hampshire, Hertfordshire, Norfolk, and Suffolk. Resides chiefly in the neighbourhood of large pieces of water, preying on fish. Builds on trees and rocks, and lays from three to four eggs.

GEN. 3.   FALCO, *Linn.*

5.   F. *Islandicus*, Lath. (*Jer-Falcon.*)—Plumage white, with dusky lines and spots.

> F. Islandicus, *Temm. Man. d'Orn.* tom. i. p. 17. Jer-Falcon, *Mont. Orn. Dict. and Supp. Selb. Illust.* vol. i. p. 36. pl. 14. *Bew. Brit. Birds*, vol. i. p. 15.

Dimens. Entire length (*male*) one foot ten inches, (*female*) two feet. Temm.

Descript. (*Old male.*) Ground of the plumage white: upper parts marked with narrow streaks of brown; under parts with spots of the same colour, which become larger and more numerous on the sides: bill yellowish: cere and orbits livid yellow: irides brown: feet bright yellow. (*Old female, and immature male.*) Spots on the under parts more numerous; confluent on the sides, forming transverse bars: the markings above more extended, and occupying a larger portion of the entire plumage. (*Young of the year.*) Upper parts of a uniform brownish

ash, the white appearing only in small spots at the tips of the feathers: under parts with large brown spots disposed longitudinally upon
a white ground: cere and orbits bluish: feet lead-colour, with a tinge of
yellow. (*Egg*). "Spotted, of the size of a Ptarmigan's." Flem.

Occurs in the northern parts of Scotland, particularly in the Orkney
and Shetland Isles; but is of very rare occurrence in England. Sheppard
mentions one that was shot on Bungay Common in Suffolk. Preys on
birds and the smaller quadrupeds. *Obs.* The *Spotted Falcon* of Pennant
is the young of this species.

6. F. *peregrinus*, Gmel. (*Peregrine Falcon.*)—Above
bluish ash, with darker fasciæ; beneath yellowish white,
with brown transverse bars; a broad black moustache:
wings reaching to the extremity of the tail.

> F. peregrinus, *Temm. Man. d'Orn.* tom. i. p. 22. Peregrine Falcon,
> *Mont. Orn. Dict. and Supp. Selb. Illust.* vol. i. p. 39. pls. 15 and
> 15*. *Bew. Brit. Birds*, vol. i. p. 17.

Dimens. Entire length (*male*) fifteen inches, (*female*) seventeen
inches; from the carpus to the end of the wing fourteen inches; tarsus
two inches.

Descript. (*Adult male.*) Head, upper part of the neck, and a broad
moustache descending from the corners of the mouth, blackish blue: rest
of the upper parts bluish ash, with shades of a darker tint: throat and
breast white, with a few fine longitudinal streaks; belly, vent, and thighs,
grayish white, with brown transverse bars: quills dusky, the inner webs
with a series of reddish white spots: tail alternately barred with black
and gray: bill bluish at the base, black at the tip: cere, space surrounding the eyes, irides, and feet, yellow. (*Adult female.*) Colour of the
upper plumage more dull: under parts with a faint reddish tinge.
(*Young.*) Upper parts ash-brown, the feathers with pale reddish edges;
region of the eyes, and moustaches, dusky, the latter ill-defined: throat
white; rest of the under parts whitish, with large brown longitudinal
spots of an oblong heart-shaped form: tail with irregular reddish bars;
the tip white: irides brown. (*Egg.*) Mottled all over with two shades
of red brown: long. diam. two inches one line; trans. diam. one inch
eight lines.

Not very abundant in England: more plentiful in the northern parts
of Scotland. Preys on birds. Builds on rocks, or in trees, and lays from
three to four eggs. *Obs.* The *Lanner* of Pennant is the young of this
species: the *Falco Lanarius* of Temminck is a distinct species, which
has not hitherto been found in this country.

7. F. *Subbuteo*, Linn. (*Hobby.*)—Above bluish black;
beneath reddish yellow, with longitudinal streaks of brown:
a broad black moustache: wings reaching beyond the extremity of the tail.

> F. Subbuteo, *Temm. Man. d'Orn.* tom. i. p. 25. Hobby, *Mont. Orn.
> Dict. Selb. Illust.* vol. i. p. 43. pl. 16. *Bew. Brit. Birds*, vol. i. p. 42.

Dimens. Entire length twelve to fourteen inches: breadth two feet
to two feet three inches.

Descript. (*Adult male.*) Upper parts bluish black: under parts whitish, passing into reddish yellow, with longitudinal brown streaks: throat white: a black patch or moustache proceeding from the corners of the mouth down each side of the neck: rump, thighs, and under tail-coverts, rust-red: quills, and outer tail-feathers, barred with reddish brown on their inner webs: bill bluish black: irides reddish brown: cere, eye-lids, and feet, yellow. (*Female.*) Colours generally more obscure; upper parts approaching to dusky brown; the red on the rump and thighs not so bright; the spots underneath deeper. (*Young.*) Plumage above dusky, all the feathers edged with reddish yellow; this last colour most pre-valent on the head: some white spots on the occiput: under parts yel-lowish white, passing into reddish yellow, with long streaks of brown: tip of the tail reddish: cere and feet greenish yellow. (*Egg.*) Yellowish white, speckled all over with reddish brown: long. diam. one inch eight lines; trans. diam. one inch four lines.

A summer visitant: arriving in April, and departing in October. Preys on larks, and other small birds. Builds in tall trees, and lays from three to four eggs.

8.　F. *Æsalon*, Gmel. (*Merlin.*)—Above bluish ash, beneath reddish yellow, with longitudinal dark spots: no moustache: wings reaching to two-thirds the length of the tail.

F. Æsalon, *Temm. Man. d'Orn.* tom. i. p. 27. Merlin, *Mont. Orn. Dict. Bew. Brit. Birds*, vol. i. pp. 46 and 48. *Selb. Illust.* vol. i. p. 51. pls. 18 and 18*.

Dimens. Entire length eleven to twelve inches: breadth two feet.

Descript. (*Adult male.*) Upper parts, including the tail, bluish ash, with the shafts of the feathers black: throat white: under parts reddish yellow, with dark brown, oblong-oval spots: quills with the inner webs barred with white: a broad black bar near the extremity of the tail; the tip itself white: bill bluish: irides brown: cere, orbits, and feet, yellow. (*Female.*) Crown of the head dusky brown, streaked with black; nape white, with reddish brown spots: back and scapulars grayish brown, with reddish brown spots and black shafts: under parts yellowish white, with spots larger and more numerous than in the male bird: tail brown, with six or seven yellowish white bars. (*Young.*) Upper plumage brown, the feathers edged with red: tail dusky, with several narrow bars of red-dish brown; the tip of this last colour: quills brownish black, spotted or barred with reddish brown: under parts as in the female. (*Egg.*) Mot-tled all over with two shades of dark red brown: long. diam. one inch seven lines; trans. diam. one inch three lines.

Supposed to be migratory in the South of England, which it visits during the winter months. Breeds in the counties of Northumberland, Cumberland, and Westmoreland, making its nest on the ground. Preys on small birds. *Obs.* This is a very variable species. The male in adult plumage is the *Stone-Falcon* of Latham, and other authors.

9.　F. *rufipes*, Bechst. (*Red-legged Falcon*)—General plumage dark bluish gray, without spots: cere and feet red: claws yellow: wings reaching to the extremity of the tail.

> F. rufipes, *Temm. Man. d'Orn.* tom. i. p. 33.   Orange-legged Hobby, *Lath. Syn. Supp.* ii. p. 46.   *Selb. Illust.* vol. i. p. 45. pl. B.   Red-footed Falcon, *Gould, Europ. Birds.*

DIMENS.   Entire length ten inches six lines.   TEMM.

DESCRIPT.   (*Adult male.*) Head, neck, breast, belly, and all the upper parts, dark lead-gray, without spots; thighs, vent, and under tail-coverts, deep ferruginous: cere, orbits, and feet, bright red: claws yellow, the tips brown.   (*Female.*) Head and neck of a uniform reddish colour, without streaks; rest of the upper parts dusky blue; all the feathers, except the quills, edged with bluish black; under parts deep reddish, streaked with dusky brown: thighs red: tail bluish gray, with six or seven dusky bars, the tip also of this colour: cere, orbits, and feet, orange-red.   (*Young.*) Top of the head brown, with dusky streaks; throat and ear-coverts white; eyes encircled with black; also a small black moustache extending from beneath the eye downwards: upper parts brown, the feathers with reddish brown edges; under parts yellowish white, with longitudinal brown streaks on the breast and abdomen: thighs without spots: tail with numerous alternate bars of brown and reddish white; the tip white.   (*Egg.*) Unknown.

Three specimens of this species, unknown before as a native of this country, occurred at Horning in Norfolk, in May 1830.   They consisted of an adult male, a female, and a young male in immature plumage.   A fourth specimen was killed afterwards in Holkham Park, in the same county; and a fifth in Yorkshire.   Said to feed on insects.   Nidification unknown.

10.   F. *Tinnunculus*, Linn. (*Kestril.*)—Upper plumage reddish brown, spotted with black: cere and feet yellow: claws black: wings reaching to three-fourths the length of the tail.

> F. Tinnunculus, *Temm. Man. d'Orn.* tom. i. p. 29.   Kestril, *Mont. Orn. Dict. Bew. Brit. Birds*, vol. i. pp. 39 and 41.   *Selb. Illust.* vol. i. p. 47. pls. 17 and 17*.

DIMENS.   Entire length fifteen inches: length of the bill (from the forehead) nine lines and a half, (from the gape) ten lines and a half; of the tarsus one inch eight lines; of the tail seven inches seven lines; from the carpus to the end of the wing nine inches ten lines: breadth, wings extended, two feet six inches.

DESCRIPT.   (*Male.*) Crown of the head and nape bluish gray: upper parts reddish brown, with scattered black spots of an angular form: breast, belly, and thighs, white, inclining to reddish, with oblong brown spots: rump bluish gray: tail cinereous, with one broad black bar near the extremity; the tip white: bill bluish: cere, orbits, irides, and feet, yellow.   (*Female.*) All the upper parts reddish orange, with transverse bars of dusky brown: tail reddish, with numerous black bars; one broad bar near the extremity, which last is reddish white.   (*Young.*) Upper part of the head, and nape, reddish brown with dark streaks; rest of the plumage above with numerous arrow-shaped spots: a blackish streak under each eye, arising from the corners of the mouth, and extending to the neck: tail reddish, with transverse dusky bars in both sexes, as in the adult female: irides brown: cere reenish yellow. (*Egg.*) Mottled all over with dark red brown: in some specimens blotches of red brown upon a pale reddish white ground: long diam. one inch seven lines; trans. diam. one inch three lines.

A common species; preying principally on mice and insects. Builds in tall trees, and lays from three to five eggs. Hatches towards the end of April or beginning of May.

GEN. 4. ACCIPITER, *Willugh.*

(1. ASTUR, *Vig.*)

11. A. *palumbarius*, Will. (*Goshawk.*)—A broad white streak over the eye.

Falco palumbarius, *Temm. Man. d'Orn.* tom. i. p. 55. Goshawk, *Mont. Orn. Dict. Selb. Illust.* vol. i. p. 29. pls. 12 and 12*. *Bew. Brit. Birds*, vol. i. p. 28.

DIMENS. Entire length (*female*) two feet, (*male*) one-third less. TEMM.

DESCRIPT. (*Adult male.*) Upper parts deep bluish ash-colour: a broad white streak over the eyes: under parts white, with transverse bars and fine longitudinal streaks of dark brown: tail cinereous, with four or five bars of blackish brown; the tip white: bill bluish gray: cere greenish yellow: irides and feet bright yellow. (*Adult female.*) Upper parts with a brownish tinge: spots on the throat more numerous. (*Young.*) Top of the head and upper part of the neck reddish brown, the feathers edged with white: under parts reddish white, with long lanceolate streaks of deep brown: tail brown, with four broad bars of a darker colour; the tip white: irides gray: cere and feet livid yellow. (*Egg.*) "Skim-milk white, marked with streaks and spots of reddish brown." SELBY.

Nearly confined to the mountainous parts of Scotland, where it breeds, and is said to be a great destroyer of game. Very rare in England. *Obs.* The Gentil Falcon (*Falco gentilis*, Gmel.) is the young of this species.

(2. ACCIPITER, *Vig.*)

12. A. *fringillarius*, Will. (*Sparrow-Hawk.*)—A white spot on the nape.

Falco Nisus, *Temm. Man. d'Orn.* tom. i. p. 56. Sparrow-Hawk, *Mont. Orn. Dict. Selb. Illust.* vol. i. p. 32. pls. 13 and 13*. *Bew. Brit. Birds*, vol. i. p. 44.

DIMENS. Entire length (*male*) twelve inches, (*female*) fifteen inches: length of the bill (from the forehead) eight lines, (from the gape) ten lines and a half; of the tarsus two inches five lines; of the middle toe, claw included, one inch ten lines; of the tail seven inches six lines; from the carpus to the end of the wing nine inches three lines: breadth, wings extended, (*male*) two feet, (*female*) two feet four inches.

DESCRIPT. (*Adult male.*) Upper parts bluish ash-colour: above the eye an indistinct pale line, conducting to a white spot on the nape: under parts white, with transverse waves of deep brown; on the throat a few longitudinal streaks: tail grayish ash, with broad brownish black bars: bill bluish gray, the tip black: cere greenish yellow: feet and irides bright gamboge-yellow. (*Female.*) All the upper parts blackish brown, passing into blackish gray: throat whitish, with fine longitudinal dusky streaks; rest of the under parts reddish white, undulated with transverse bars of dusky brown. (*Young.*) Upper parts brown, the feathers with pale reddish edges: some whitish spots on the shoulders: under parts yellowish white, with transverse reddish bars; or whitish, with brown bars: irides grayish: feet livid yellow. (*Egg.*) Pale bluish white, blotched and spotted with dark red brown: long. diam. one inch seven lines; trans. diam. one inch four lines.

Common throughout the country. Preys upon the smaller quadrupeds and birds, and is very destructive to partridges. Builds in trees, forming a shallow nest of slender twigs. Lays from three to six eggs. This species is remarkable for the great difference of size between the male and female.

GEN. 5.　MILVUS, *Bechst.*

(1. Milvus, *Vig.*)

13.　M. *Ictinus*, Sav. (*Kite.*)—Reddish brown above; beneath ferruginous, with dark longitudinal streaks.

Falco Milvus, *Temm. Man. d'Orn.* tom. i. p. 59.　Kite, *Mont. Orn. Dict. and Supp.　Selb. Illust.* vol. i. p. 74. pl. 5.　*Bew. Brit. Birds,* vol. i. p. 32.

Dimens. Entire length twenty-five inches: length of the bill (from the forehead) one inch eight lines, (from the gape) one inch eleven lines; of the tarsus two inches two lines; of the tail twelve inches eight lines; from the carpus to the end of the wing twenty inches: breadth, wings extended, five feet two inches.

Descript. (*Male.*) Head and neck grayish white, with fine streaks of dusky brown; the feathers on these parts long, and acuminated: rest of the upper parts reddish brown; the feathers with pale edges: under parts ferruginous, with longitudinal brown streaks: tail long and deeply forked, reddish orange, with obsolete brown bars: bill yellowish brown at the base, towards the tip dusky: cere and irides yellow. (*Female.*) Upper plumage of a deeper brown, with less of the ferruginous tinge; the edges of the feathers paler, approaching to white. (*Young of the year.*) Feathers on the head shorter, and less acuminated; bright red without streaks, tipped with white: upper parts redder than in the adult; feathers on the back and wings dusky in the centre, reddish yellow at the edges: on the lower part of the neck some large white spots. (*Egg.*) Dirty white; the larger end spotted with red brown: long. diam. two inches two lines; trans. diam. one inch nine lines.

Common in some parts of the country, but not generally diffused. Frequents wooded districts, and builds in trees. Food young game, and the smaller quadrupeds.

(Elanus, Sav.)

(1.) *M. furcatus*, Nob.　*Falco furcatus*, Linn. Syst. Nat. tom. i. p. 129.　*Nauclerus furcatus*, Vig. in Zool. Journ. vol. ii. p. 387. *Swallow-tailed Falcon*, Shaw, Nat. Misc. vol. vi. pl. 204.　*Swallow-tailed Elanus*, Selb. Illust. vol. i. p. 77.

This species, which is a native of North America, is stated by Dr Fleming (*Brit. An.* p. 52) as having occurred to the late Dr Walker in Argyleshire, in 1772. A second individual is said to have been taken alive in Yorkshire, in Sept. 1805. (*Linn. Trans.* vol. xiv. p. 583.) No British-killed specimen, however, is known to exist in any of our museums.

GEN. 6. BUTEO, *Nob.*

(1. Buteo, *Bechst.*)

14. B. *vulgaris*, Will. (*Common Buzzard*).—Tarsi naked: tail barred transversely throughout its whole length.

Falco Buteo, *Temm. Man. d'Orn.* tom. i. p. 63. Common Buzzard, *Mont. Orn. Dict. Selb. Illust.* vol. i. p. 55. pl. 6. *Bew. Brit. Birds*, vol. i. p. 22.

Dimens. Entire length twenty inches six lines: length of the bill (from the forehead) one inch four lines, (from the gape) one inch six lines; of the tarsus three inches one line; of the tail eight inches six lines; from the carpus to the end of the wing fifteen inches three lines : breadth, wings extended, three feet eleven inches.

Descript. (*Adult.*) All the upper parts, as well as the neck and breast, deep brown : throat whitish, with a few brown streaks on the shafts of the feathers; abdomen grayish white, with longitudinal spots of deep brown : tail with about twelve transverse dusky bars, the extremity slightly rounded: bill lead-colour : cere, irides, and feet, yellow. (*Young.*) Upper plumage brown, much varied with yellowish white and brownish yellow : throat pure white, with longitudinal brown spots ; breast and belly yellowish white, with large oval stains and blotches of deep brown, prevailing most upon the breast : irides pale brownish yellow. *Obs.* The plumage of this species varies considerably, even in individuals of the same age : the brown above is more or less uniform, and the markings beneath more or less regular : occasionally the brown extends nearly over the whole body, and the white appears only on the throat, and in transverse bars on the middle of the abdomen. (*Egg.*) Dirty white, spotted with pale brown : long. diam. two inches three lines ; trans. diam. one inch ten lines.

Of frequent occurrence throughout the country. Habits sluggish and inactive. Builds in old trees, and lays from two to four eggs. Feeds on rats, mice, moles, leverets, small birds, and occasionally on reptiles.

15. B. *Lagopus*, Vig. (*Rough-legged Buzzard.*) — Tarsi feathered nearly to the toes: basal portion of the tail white, the remainder brown.

Falco Lagopus, *Temm. Man. d'Orn.* tom. i. p. 65. Rough-legged Falcon, *Mont. Orn. Dict.* Rough-legged Buzzard, *Selb. Illust.* vol. i. p. 58. pl. 7. *Bew. Brit. Birds*, vol. i. p. 20.

Dimens. Entire length twenty inches six lines : from the carpus to the end of the wing sixteen inches six lines.

Descript. (*Adult.*) Head, upper part of the neck, throat, breast, and thighs, yellowish white, spotted and streaked with brown; back and wing-coverts deep brown, the feathers edged with reddish yellow : quills dusky; with one half of the inner webs white, spotted with deep brown : lower portion of the abdomen deep umber-brown : rump and under tail-coverts yellowish white : tail white at the base, the remaining part brown ; the extreme tip dirty white : bill bluish black, darkest at the tip; cere yellow ; irides brown. (*Young.*) Upper parts more or less varied with white : a considerable portion of the abdomen whitish, spotted with brown; the sides with larger spots of the same colour: tail with two or three transverse bars towards the extremity: irides yellowish brown. (*Egg.*) Pale brownish white, blotched with darker brown: long. diam. two inches two lines; trans. diam. one inch eight lines.

A rare and only occasional visitant in this country. Has been twice killed in Cambridgeshire; been met with also in Norfolk, Suffolk, Kent, the Isle of Wight, Gloucestershire, Northumberland, and other parts of England. Food and habits much resembling those of the last species.

(2. Pernis, *Cuv.*)

16.   B. *apivorus*, Ray. (*Honey Buzzard.*)—Upper plumage brown, with cinereous bars on the wings: under parts white, with triangular brown spots.

Falco apivorus, *Temm. Man. d'Orn.* tom. i. p. 67.   Honey Buzzard, *Mont. Orn. Dict. Selb. Illust.* vol. i. p. 62. pl. 8.   *Bew. Brit. Birds,* vol. i. p. 24.

Dimens.   Entire length one foot ten inches: length of the bill (from the forehead) eleven lines, (from the gape) one inch six lines; of the tarsus two inches three lines; of the tail eight inches; from the carpus to the end of the wing one foot four inches: breadth, wings extended, four feet one inch.

Descript.   (*Adult male.*) Top of the head bluish ash; upper parts of the body brown, inclining to cinereous: secondaries barred alternately with dusky brown and bluish gray: under parts whitish, with brown spots, which assume a triangular form on the breast and abdomen: tail with three transverse dusky bars: cere greenish grey: inside of the mouth, irides, and feet, yellow.   (*Female and young.*) Forehead alone bluish ash; rest of the upper parts pale reddish brown, with a large dusky spot in the centre of each feather: throat nearly white: fore part of the neck, breast and belly, yellowish red, with large brown spots; sometimes light brown, with reddish brown bars   (*Young of the year.*) Upper parts like those of the adult female, but the feathers on the head, neck, and greater wing-coverts, tipped with white: quills dusky, tipped with white: throat whitish; rest of the under parts reddish white with large brown spots, or reddish brown with a few partially white feathers intermixed, occasionally of a uniform pale brown with fine longitudinal dusky streaks: cere yellow: irides brown.   *Obs.* This species appears to vary considerably in plumage, especially in the immature state.   (*Egg.*) Mottled nearly all over with two shades of dark red brown, the white ground only here and there visible: long. diam. two inches one line; trans. diam. one inch nine lines.

Of rare occurrence in this country, especially in the adult state.   Has been met with in Hampshire, Berkshire, Norfolk, Northumberland, and several times in Cambridgeshire.   Feeds principally upon the larvæ and pupæ of wasps during the season in which they are to be obtained; at other times, preys on small quadrupeds and birds.   Is said to build in tall trees.

(3. Circus, *Bechst.*)

17. B. *rufus*, Nob. (*Marsh Harrier.*)—General plumage deep reddish brown: crown of the head yellowish white: wings a little shorter than the tail.

Falco rufus, *Temm. Man. d'Orn.* tom. I. p. 69.   Marsh Harrier,
  Selb. Illust. vol. I. p. 66. pl. 9.   Moor-Buzzard, *Mont. Orn.
  Dict.   Bew. Brit. Birds,* vol. I. p. 26.

DIMENS. Entire length one foot seven to nine inches : breadth, wings
extended, three feet eleven inches to four feet two inches.
DESCRIPT. (*Adult in the fourth year.*) Head and neck yellowish
brown, with longitudinal darker streaks: scapulars and wing-coverts
reddish brown: first five primaries dusky lead-colour; all the secondaries and tail-feathers cinereous: rest of the plumage bright ferruginous : bill black : irides reddish yellow : cere and feet yellow. (*Second
year.*) Crown of the head, throat, and nape, straw-yellow, with fine longitudinal dusky streaks: rest of the body dark umber-brown, tinged
beneath with rust-red: a few yellowish spots upon the wings: irides
brown. (*Young of the year.*) Whole plumage deep chocolate-brown,
inclining to yellowish on the throat, crown, and back of the head : quills,
wing-coverts, and tail-feathers, tipped with yellowish brown: occasionally
a few white patches on the lower part of the abdomen : irides dusky
brown. (*Egg.*) Rather pointed : white : long. diam. two inches one line ;
trans. diam. one inch six lines.
  Common in marshy districts. Preys on water-fowl, small quadrupeds,
and reptiles. Nest on the ground amongst rushes: eggs three to five in
number. *Obs.* The *Falco æruginosus* of Linnæus is the young of this
species.

18. B. *cyaneus*, Nob. (*Hen-Harrier.*)—Upper plumage (*male*) bluish gray, (*female*) reddish brown : quills dusky, without transverse bars : third and fourth primaries of equal length : wings reaching to three-fourths the length of the tail.

Falco cyaneus, *Temm. Man. d'Orn.* tom. I. p. 72.   Hen-Harrier and
  Ring-tail, *Mont. Orn. Dict. and Supp.   Bew. Brit. Birds,* vol. I.
  pp. 34 and 36.   *Selb. Illust.* vol. I. p. 68. pl. 10.

DIMENS. Entire length (*male*) one foot six or seven inches, (*female*)
one foot eight or nine inches: length of the tarsus three inches: breadth,
wings extended, three feet two to four inches.
DESCRIPT. (*Adult male.*) Head, neck, back, scapulars, and wing-
coverts, bluish gray; rump white: quills black, whitish at the base:
breast, belly, sides, under wing and tail-coverts, pure white, without
spots of any kind: upper part of the tail ash-gray, whitish at the tip:
bill bluish black: irides, cere, and feet, yellow. (*Adult female.*) Space
surrounding the eyes white: upper parts of the plumage deep brown;
feathers on the head, neck, top of the back, and wing-coverts, edged
with rust-red: under parts pale reddish yellow, with deep orange-brown
longitudinal streaks and spots: quills dusky, barred underneath with
white: tail with alternate broad bars of deep brown and pale yellowish
rust. *Obs.* The *young* of both sexes resemble the old female: *after the
first autumnal moult,* the male begins to assume the adult plumage, and
exhibits a mixture of ash-gray and reddish brown ; the transverse bars
on the tail also gradually disappear. (*Egg.*) White : long. diam. one inch
eight lines ; trans. diam. one inch four lines.
  Pretty generally distributed throughout the country, but seemingly
most partial to fenny districts. Preys on small birds and quadrupeds.
Makes its nest on the ground : lays from three to five eggs, and hatches

in the beginning or middle of June.   *Obs.* The *Ring-tail* of English authors is the female of this species.

19. B. *cineraceus,* Flem. (*Ash-coloured Harrier.*)—Upper plumage (*male*) bluish ash, (*female*) reddish brown: third primary much longer than the others: secondaries with three black transverse bars: wings reaching to the extremity of the tail.

Falco cineraceus, *Mont. in Linn. Trans.* vol. IX. p. 188. *Temm. Man. d'Orn.* tom. I. p. 76. Ash-coloured Harrier, *Selb. Illust.* vol. I. p. 70. pl. 11. Ash-coloured Falcon, *Mont. Orn. Dict. and Supp. with fig. Bew. Brit. Birds,* vol. I. p. 37.

DIMENS. Entire length seventeen inches: length of the bill (from the forehead) ten lines, (from the gape) eleven lines and a half; of the tarsus two inches three lines; of the tail eight inches two lines; from the carpus to the end of the wing thirteen inches nine lines. *Female.*

DESCRIPT. (*Adult male.*) Head, throat, breast, and all the upper parts of the body, deep bluish ash: belly, sides, and thighs, white, with longitudinal rust-coloured streaks: primaries black; secondaries ash-gray above, paler beneath, with three transverse dusky bars, one of which is visible externally: under wing-coverts barred with reddish brown: two middle tail-feathers brownish gray; the rest cinereous, their inner webs barred with reddish brown: bill bluish black: irides and feet yellow. (*Adult female.*) Crown of the head reddish brown, with dusky spots: nape yellowish red, sometimes approaching to white: above and below the eye a pale fascia: upper parts of the body deep brown, the feathers with reddish edges: lower part of the rump, and tail-coverts, white, streaked with pale orange-brown: all the under parts bright ferruginous, with the shafts of the feathers somewhat darker, appearing like fine slender streaks: tail with the two middle feathers of an uniform brown; the rest with brown and ferruginous bars. (*Young of the year.*) Upper plumage much resembling that of the adult female: under parts of an uniform rust-red colour, without any spots or streaks: irides brown. (*Egg.*) White: long. diam. one inch seven lines; trans. diam. one inch four lines.

First discovered by Montagu in Devonshire: has since been occasionally met with in the North of England, and likewise in the fens of Cambridgeshire. Nest placed on the ground; often amongst furze: the eggs, which are generally four in number, are hatched about the second week in June.

GEN. 7.   BUBO, *Isid. Geoff.*

(1. BUBO, *Cuv.*)

20. B. *maximus,* Flem. (*Eagle Owl.*)—Upper plumage variegated with black and ochre: under parts ochre-yellow, with longitudinal black spots.

Strix Bubo, *Temm. Man. d'Orn.* tom. I. p. 100. Eagle Owl, *Selb. Illust.* vol. I. p. 82. pl. 19. *Bew. Brit. Birds,* vol. I. p. 52. Great-eared Owl, *Mont. Orn. Dict.*

DIMENS. Entire length two feet. TEMM.
DESCRIPT. All the upper parts elegantly varied and spotted with black, ochre-yellow, and yellowish gray: under parts ochre-yellow, with longitudinal black spots and streaks: throat white: egrets composed of six or eight elongated feathers, coloured like the rest of the plumage: tarsal feathers reddish yellow: bill and claws horn-colour: irides bright orange. (*Female.*) Rather larger than the male, and without the white throat. (*Egg.*) White: long. diam. two inches five lines; trans. diam. one inch ten lines.

An occasional visitant of extreme rarity in this country. According to Montagu has been shot in Yorkshire and in Sussex, as well as in Scotland. More recently four specimens are said to have been observed on the northern coast of Donegal, Ireland. Inhabits mountainous and rocky situations, where it breeds. Food small quadrupeds, and reptiles.

(2. Scops, *Sav.*)

21. B. *Scops*, Nob. (*Scops-eared Owl.*) — Plumage variegated with brown, gray, black, and rufous.

Strix Scops, *Temm. Man. d'Orn.* tom. I. p. 103. Scops-eared Owl, *Selb. Illust.* vol. I. p. 92. pl. 22. *Bew. Brit. Birds,* vol. I. p. 67. Little-horned Owl, *Mont. Orn. Dict. Supp. App.*

DIMENS. Entire length seven inches and a half. FLEM.
DESCRIPT. Head, neck, and egrets, (the latter consisting of six or eight feathers,) brownish gray, speckled with black; rest of the upper parts reddish ash, with spots and zigzag streaks of black and brown: under parts ash-gray, streaked and speckled with black and reddish brown: quills barred with white: tail variegated with black, brown, and white: bill black: irides yellow: toes bluish gray. (*Egg.*) Of a short oval form: white, with a few faint dusky specks: long. diam. one inch three lines; trans. diam. one inch and half a line.

Very rare. Has been killed near York, and in one or two other parts of the country. Is probably migratory. Preys on mice and insects. *Obs.* The *Strix pulchella* of Donovan (*Brit. Birds*, vol. VII. pl. 165.) appears to be the same as this species.

GEN. 8. OTUS, *Cuv.*

22. O. *vulgaris*, Flem. (*Long-eared Owl.*)—Plumage reddish yellow, variegated with gray and dusky brown: egrets elongated, composed of eight or ten feathers.

Strix Otus, *Temm. Man. d'Orn.* tom. I. p. 102. Long-eared Owl, *Mont. Orn. Dict. Selb. Illust.* vol. I. p. 85. pl. 20. *Bew. Brit. Birds,* vol. I. p. 56.

DIMENS. Entire length fourteen inches: length of the tail six inches; of the egrets one inch three lines; from the carpus to the end of the wing eleven inches six lines.
DESCRIPT. Upper parts of a yellowish rust-colour, irregularly marked and spotted with deep brown, gray and cinereous: egrets of eight or ten black feathers, edged with yellow and white: under parts orange-yellow, with oblong and arrow-shaped streaks of dusky brown: quills and tail barred with brownish black: tarsal feathers yellow, without spots: bill

dusky: irides rich orange-yellow. (*Female.*) Throat and face white: rest of the plumage like that of the male, but much mottled in places with grayish white. (*Egg.*) Oval; smooth: white: long. diam. one inch eight lines and a half; trans. diam. one inch three lines and a half.

Found in wooded districts not very uncommonly. Partial to fir-plantations and old ivy, where it breeds early in the Spring, laying from four to five eggs: these are hatched towards the end of April. Remains with us the whole year.

23. O. *Brachyotos*, Flem. (*Short-eared Owl.*)—Upper plumage dusky brown, edged with ochre-yellow: egrets small and inconspicuous, composed of three or four feathers.

Strix Brachyotos, *Temm. Man. d'Orn.* tom. I. p. 99.   Short-eared Owl, *Mont. Orn. Dict. and Supp.   Selb. Illust.* vol. I. p. 88. pl. 21.   *Bew. Brit. Birds*, vol. I. pp. 58 and 60.

Dimens. Entire length fifteen inches: length of the bill (along the ridge) one inch three lines; of the tarsus one inch; of the tail six inches; of the egrets six lines; from the carpus to the end of the wing eleven inches eight lines: breadth, wings extended, three feet one inch.

Descript. Head small: facial circle dirty white, with dark streaks radiating outwards; immediate contour of the eyes black: egrets black, with tawny edges: head, neck, back, and wing-coverts, variegated with dusky brown and tawny yellow, the feathers being mostly edged with this last colour, with a dark spot in the centre of each: breast and belly pale orange-yellow, with brownish black streaks on the shafts of the feathers; these streaks most abundant on the breast: vent and under tail-coverts yellowish white: quills with alternate bars of dark brown and ochre-yellow: tail barred like the quills; the two outermost feathers much paler than the others; the four central ones with dusky spots in the middle of the yellow bars: feet and toes thickly coated with downy ochreous feathers: bill dusky: irides rich golden yellow: claws black. (*Egg.*) Smooth: white, with a slight blush of red: long. diam. one inch eight lines; trans. diam. one inch three lines and a half.

A migratory species, visiting England in October, and departing in April. Has been known, however, to breed in Norfolk, as well as in some parts of Scotland. Habits somewhat diurnal. Met with chiefly on moors and in open fields, often in flocks consisting of several individuals. Resorts to plantations in the evening to roost. Food principally the field-campagnol. Nest placed on the ground.

GEN. 9.   STRIX, *Linn.*

24. S. *flammea*, Linn. (*White Owl.*)—Upper plumage tawny yellow, variegated with gray and brown: underneath white.

S. flammea, *Temm. Man. d'Orn.* tom. I. p. 91.   White Owl, *Mont. Orn. Dict.   Selb. Illust.* vol. I. p. 99. pl. 24.   *Bew. Brit. Birds*, vol. I. p. 61.

Dimens. Entire length fourteen inches six lines: length of the bill (from the forehead) one inch four lines, (from the gape) one inch seven lines; of the tarsus two inches six lines; of the tail four inches eight lines; from the carpus to the end of the wing eleven inches: breadth, wings extended, two feet eleven inches six lines.

DESCRIPT. Facial circle white: all the upper parts bright tawny yellow, variegated with fine zigzag lines of brown and gray, and sprinkled with numerous white dots: under parts pure white, or with a faint tinge of reddish yellow, occasionally speckled with a few brownish points: bill and claws yellowish white: irides bluish black. (*Egg.*) Dull white: long. diam. one inch six lines; trans. diam. one inch three lines.

Common in every part of the kingdom. Frequents churches and the eaves of old buildings, where it breeds, laying from three to five eggs. Habits nocturnal: comes abroad about sunset. Feeds principally on mice, but will occasionally devour other of the smaller quadrupeds: rejects the shrew. Screams in its flight, but does not hoot.

GEN. 10. SYRNIUM, *Sav.*

25. S. *Aluco*, Nob. (*Tawny Owl.*) — Upper plumage reddish brown, variegated with black and ash-gray: scapulars and wing-coverts spotted with white: under parts white, with reddish bars, and longitudinal dusky streaks.

Strix Aluco, *Temm. Man. d'Orn.* tom. i. p. 89. Tawny Owl, *Mont. Orn. Dict. Selb. Illust.* vol. i. p. 102. pl. 25. *Bew. Brit. Birds*, vol. i. p. 63.

DIMENS. Entire length sixteen inches: length of the bill (from the forehead) one inch four lines, (from the gape) one inch six lines; of the tarsus two inches; of the tail six inches eight lines; from the carpus to the end of the wing ten inches ten lines: breadth, wings extended, three feet.

DESCRIPT. Head large: facial circle white, with a tinge of reddish brown: upper parts of the plumage ferruginous brown, variously marked and spotted with dark brown, black, and ash-gray: several large white spots upon the scapulars and wing-coverts, disposed in rows: under parts white, with transverse bars of reddish brown, and longitudinal dusky streaks on the shafts of the feathers: quills barred alternately with dusky and tawny brown: two middle tail-feathers of a uniform tawny brown; the others barred like the quills: bill yellowish white: eyes very large; irides bluish black. (*Female.*) General colour more ferruginous than in the male, with less white: the transverse bars on the wings and tail alternately red and brown. (*Egg.*) Smooth: dull white: long. diam. one inch ten lines; trans. diam. one inch six lines.

Equally common with the last species, but found only in woods. Builds in the hollows of old trees, or amongst ivy, and lays four eggs, which are hatched in the beginning of April. Preys upon various small quadrupeds and birds. Comes abroad only during the night, and has a clamorous hooting note.

GEN. 11. NOCTUA, *Sav.*

(1. SURNIA, *Selby.*)

26. N. *nyctea*, Nob. (*Snowy Owl.*) — Plumage snow-white, more or less marked with transverse brown bars.

Strix nyctea, *Temm. Man. d'Orn.* tom. I. p. 82.   Snowy Owl, *Mont.
Orn. Dict. Supp. App. Selb. Illust.* vol. I. p. 95. pl. 23.   *Bew.
Brit. Birds,* vol. I. p. 54.

DIMENS. Entire length (*female*) two feet one inch: breadth, wings
extended, five feet. *Male* of inferior size. SELBY.
DESCRIPT. Head small: plumage above and below snow-white, more
or less variegated with brown spots and transverse bars, which are largest
and most conspicuous in *immature* birds; in very old individuals (*males*
more especially) these spots wholly disappear, leaving the entire plumage
pure white: bill black, almost concealed by the bristly feathers at its
base: irides bright orange-yellow: tail rounded, rather short, extending
nearly two inches beyond the folded wings: legs and toes very thickly
clothed with hair-like feathers partly concealing the claws. (*Egg.*) Pure
white.
Found in the Orkney and Shetland Isles, where it remains the whole
year. Very rare in England. Has been killed in Northumberland, and
even as far south as in Norfolk. In the latter county it has occurred
twice. Habits said to be diurnal. Food hares, rats, mice, grouse, and
other birds. Builds in rocky situations, and lays two eggs.

(2. NOCTUA, *Selby.*)

27. N. *passerina,* Selby. (*Little Owl.*) — Tarsi feathered; toes thinly covered with a few white hairs.

Strix passerina, *Temm. Man. d'Orn.* tom. I. p. 92. Little Owl, *Mont.
Orn. Dict.* Little Night-Owl, *Selb. Illust.* vol. I. p. 107. pl. 27.

DIMENS. Entire length nine inches. TEMM.
DESCRIPT. Facial circle mostly white: upper plumage grayish brown;
head and nape rather thickly spotted with white; wing-coverts and scapulars with large irregular white spots of an oval-oblong form; middle of
the back nearly free from spots; sides of the nape much variegated with
white; quills with a regular series of yellowish white spots on their outer
webs; tail with alternate bars of brown and yellowish white: under parts
white, with oblong longitudinal spots of grayish brown; chin and under
tail-coverts free from spots: tarsi clothed with short downy white feathers; longer (according to Selby) than in the next species; toes thinly
covered with a few white hairs. (*Egg.*) Of a short oval form: white:
long. diam. one inch four lines; trans. diam. one inch one line.
A very rare and occasional visitant in this country. Has been killed
in a few instances, but from having been formerly confounded with the
next species, such instances can hardly be particularized with certainty.
Said to frequent towers and old buildings, in which situations it breeds,
laying from two to four eggs. Food, according to Temminck, mice, bats,
small birds, and insects.

28. N. *Tengmalmi,* Selby. (*Tengmalm's Owl.*) — Tarsi
and toes thickly clothed with downy feathers.

Strix Tengmalmi, *Temm. Man. d'Orn.* tom. I. p. 94. Tengmalm's
Night-Owl, *Selb. Illust.* vol. I. p. 105. pl. 26. Little Owl, *Bew.
Brit. Birds,* vol. I. p. 65.

DIMENS. Entire length eight inches four lines. TEMM.
DESCRIPT. Facial circle more developed than in the last species;
blackish round the eyes, and at the base of the bill; the rest of it white,

bordered externally with dark brown and small white spots intermixed: upper parts nearly similar; liver-brown, spotted with white, but the spots on the head and nape not so numerous; the white spots on the scapulars and wing-coverts disposed more in lines: wings and tail relatively longer; the latter with interrupted white bars: under parts white; the dark spots much as in *N. passerina*, but having a tendency to unite to form transverse bars: bill yellowish white: legs and toes thickly clothed with soft downy feathers. (*Egg.*) Elongated; pointed: white: long. diam. one inch four lines; trans. diam. eleven lines.

Of equal rarity with the last in this country. Has been killed near Morpeth in Northumberland, and probably in other instances. Said to feed on mice and insects, and to breed in the hollows of firs.

ORDER II. INSESSORES.

GEN. 12. LANIUS, *Linn.*

* *Tail long; graduated.*

29. L. *Excubitor*, Linn. (*Cinereous Shrike.*)—Head, neck, and back, cinereous; a black band beneath the eyes: under parts white.

> L. Excubitor, *Temm. Man. d'Orn.* tom. i. p. 142. Cinereous Shrike,
> *Mont. Orn. Dict. and Supp.*   *Selb. Illust.* vol. i. p. 148. pl. 43.
> f. 1.   *Bew. Brit. Birds*, vol. i. p. 71.

DIMENS. Entire length ten inches: length of the bill (from the forehead) eight lines, (from the gape) one inch; of the tarsus one inch; of the tail four inches three lines; from the carpus to the end of the wing four inches six lines: breadth, wings extended, fourteen inches.

DESCRIPT. (*Male.*) Upper plumage, including the head neck and back, ash gray; scapulars, rump, and upper tail-coverts, somewhat paler: a black band beneath the eye, commencing at the base of the upper mandible, and extending beyond the orifice of the ear: under parts pure white: wings black; primaries white at the base; secondaries tipped with white: tail cuneiform, of twelve feathers; the two middle ones black; the two next tipped with white; the others with the white portion gradually increasing to the outermost feather, which is almost wholly white: bill and feet black: irides blackish brown. (*Female.*) Upper plumage resembling that of the male, but the colours more obscure: under parts dirty white, with numerous crescent-shaped dusky lines. (*Egg.*) Light bluish white; the larger end nearly covered with spots of two shades of light brown and ash : long. diam. one inch one line; trans. diam. nine lines and a half.

A migratory, and not abundant species: visits this country sparingly towards the end of Autumn, departing in the Spring. Feeds on small birds, mice, insects, &c. Transfixes its prey when killed upon a thorn before devouring it. Builds in trees, and lays from five to seven eggs.

** *Tail short; even, or slightly rounded.*

30. L. *rufus*, Briss. (*Wood-Chat*.)—Upper plumage variegated with black, white, and ferruginous: under parts white: second qüill equal to the fifth: tail slightly rounded.

L. rufus, *Temm. Man. d'Orn.* tom. I. p. 146. Wood-Chat, *Bew. Brit. Birds*, vol. I. p. 377. *Selb. Illust.* vol. I. p. 153. pl. 100. f. 1. Wood-Shrike, *Mont. Orn. Dict. Supp.*

DIMENS. Entire length seven inches five lines.

DESCRIPT. Forehead, region of the eyes and ears, and upper part of the back, black: occiput and nape ferruginous: scapulars white: wings black; quills white at the base; some of them also edged and tipped with white: rump gray: under parts pure white: tail with the two outermost feathers wholly white, excepting a black spot at the base of the inner web; the two next with the black spot of larger size, and on both webs; the rest black, tipped with white; the two middle ones wholly black. (*Female.*) Colours more obscure: occiput and nape streaked with brown: the black on the back approaching to that colour: under parts dirty white; the breast with transverse brown lines. (*Egg.*) Pale bluish white; the larger end spotted with hair-brown and ash-gray: long. diam. eleven lines; trans. diam. eight lines.

Extremely rare in this country. Has been killed in the neighbourhood of Canterbury, and near Swaffham in Norfolk. In the latter locality, it has been observed to breed more than once. Feeds on insects. Places its nest in the forked branches of bushes.

31. L. *Collurio*, Linn. (*Red-backed Shrike*.)—Back, scapulars, and wing-coverts, ferruginous: breast and flanks rose-red: second quill longer than the fifth: tail even at the extremity.

L. Collurio, *Temm. Man. d'Orn.* tom. I. p. 147. Red-backed Shrike, *Mont. Orn. Dict. Selb. Illust.* vol. I. p. 150. pl. 43. f. 2. and pl. 43. f. 1, 2. *Bew. Brit. Birds*, vol. I. p. 73.

DIMENS. Entire length seven inches: length of the bill (from the forehead) six lines, (from the gape) nine lines and a half; of the tarsus eleven lines; of the tail three inches one line and a half; from the carpus to the end of the wing three inches seven lines: breadth, wings extended, eleven inches four lines.

DESCRIPT. (*Male.*) Crown of the head and nape ash-gray: a black band through the eyes, commencing at the base of the upper mandible, and reaching half-way down the neck: back, scapulars, and wing-coverts, ferruginous brown: rump ash-gray: quills dusky, edged with reddish brown: throat and vent white; breast, belly, and flanks, rose-red: tail nearly even at the extremity; the two middle feathers wholly black; the rest white from the base through two-thirds of their length, the remaining portion black, except the extreme tip which is also white: bill and feet black: irides chestnut-brown. (*Female.*) All the upper parts dull ferruginous brown: nape and rump tinged with ash-gray: under parts grayish white; sides of the neck, breast, and flanks, barred transversely with narrow semicircular dusky lines: tail with the outer webs of the two external feathers edged with white: the four middle ones of a uniform reddish brown. (*Young.*) General plumage resembling that of the adult female, but with less of the grayish tinge on the nape and rump; upper parts here and there barred with blackish brown. (*Egg.*) Pointed: ground colour reddish white, spotted with darker red and ash-gray, the spots forming a band near the large end: long. diam. eleven lines; trans. diam. eight lines. *Obs.* The eggs of this species vary occasionally, and are sometimes greenish white, with spots of light brown and ash-gray intermixed.

A regular summer visitant, arriving in the Spring, and departing in the early part of the Autumn. Not uncommon in some parts of the country. Haunts copses, tall hedges, and furzy commons, where it builds. Nest formed of moss and wool, intermixed with grass, and lined with hair: eggs five or six. Food principally coleopterous insects. *Obs.* The *Wood-Chat* figured at p. 75 (vol. i.) of Bewick's *British Birds*, is only the female of this species.

GEN. 13. MUSCICAPA, *Linn.*

32. M. *grisola*, Linn. (*Spotted Flycatcher.*)—Upper plumage brownish ash : under parts white : head and breast spotted with dusky.

M. grisola, *Temm. Man. d'Orn.* tom. i. p. 152. Spotted Flycatcher, *Mont. Orn. Dict. Selb. Illust.* vol. i. p. 141. pl. 43*. f. 1. *Bew. Brit. Birds*, vol. i. p. 210.

Dimens. Entire length six inches one line : length of the bill (from the forehead) five lines and a half, (from the gape) nine lines ; of the tarsus seven lines ; of the tail two inches eight lines and a half; from the carpus to the end of the wing three inches four lines : breadth, wings extended, ten inches three lines.

Descript. All the upper parts of the plumage brownish ash ; crown of the head obscurely spotted with dusky : throat and middle of the belly white ; sides of the neck and breast with longitudinal streaks of brown; flanks tinged with pale orange-red : legs dusky brown. (*Egg.*) White tinged with blue, and mottled all over with pale reddish yellow brown : long. diam. nine lines; trans. diam. seven lines.

A migratory, and common species. Arrives in this country about the middle, or towards the latter end of May : departs in the Autumn. Frequents gardens. Builds in holes of trees and old walls; often in the branches of trees nailed against walls. Nest formed of bents, moss, &c. interwoven with spider's webs and a little wool; lined with dry grass and feathers. Eggs five in number; hatched about the second week in June. Food insects, taken on the wing. Has no song.

33. M. *luctuosa*, Temm. (*Pied Flycatcher.*)—Upper plumage black; forehead, and under parts of the body, white.

M. luctuosa, *Temm. Man. d'Orn.* tom. i. p. 155. Pied Flycatcher, *Mont. Orn. Dict. and Supp. Selb. Illust.* vol. i. p. 143. pl. 43*. f. 2, 3. *Bew. Brit. Birds*, vol. i. p. 206.

Dimens. Entire length five inches two lines.

Descript. (*Adult male in summer plumage.*) All the upper parts, including the tail, deep black; forehead and under parts pure white :

wings black; the middle and greater coverts white. (*Female.*) Upper parts grayish brown: the white on the forehead and wings less conspicuous: the three outermost tail-feathers edged with white. (*Young male.*) Upper parts grayish brown: wings black; coverts and scapulars broadly edged with white: tail black; the three outermost feathers edged with white on the exterior web, the white portion gradually decreasing: under parts yellowish white, tinged with brown. (*Egg.*) Uniform pale blue: long. diam. eight lines and a half; trans. diam. six lines and a half.

Found in Yorkshire, and in some other parts of the North during the summer months; but is very local. Is probably a bird of passage. Habits similar to those of the last species.

> (2.) *M. albicollis*, Temm. Man. d'Orn. tom. I. p. 153. *White-collared Flycatcher*, Gould, Europ. Birds, part 7.
>
> Mr Gould, in his valuable work above referred to, states that he "has seen this species in a collection of British birds, and was informed that it was supposed to have been killed in England." Its claims, however, to a place in our Fauna require to be confirmed by some further authority. The general character of its plumage is very similar to that of the *M. luctuosa*.

GEN. 14.   CINCLUS, *Bechst.*

34. C. *aquaticus*, Bechst. (*Water-Ouzel.*)—Upper parts brown: throat and breast white: abdomen brownish red.

> C. aquaticus, *Temm. Man. d'Orn.* tom. I. p. 177.   Water-Ouzel, *Mont. Orn. Dict. and Supp. Bew. Brit. Birds*, vol. II. p. 118. European Dipper, *Selb. Illust.* vol. I. p. 172. pl. 45*.

DIMENS. Entire length seven inches six lines.

DESCRIPT. (*Male.*) Upper parts deep brown, passing into black; the feathers on the back and wings edged with ash-gray: throat, eyelids, sides and fore part of the neck, and upper part of the breast, white: lower part of the breast, and belly, brownish red: vent and under tail-coverts dusky brown: bill blackish: irides yellowish brown: legs yellowish. (*Female.*) Upper part of the head and neck cinereous brown: the red and white on the under parts less pure. (*Young of the year.*) Head and upper part of the neck gray: some whitish spots on the wings: the white on the under parts extending almost to the vent, and marked with transverse streaks of grayish 'brown. (*Egg.*) Pointed: white: long. diam. one inch; trans. diam. nine lines.

Inhabits rocky streams, and the banks of rapid rivers, chiefly in the mountainous parts of the country, where it resides all the year. Common in the North of England, and in Scotland: found also in Wales, Devonshire, and occasionally in other parts. Feeds on aquatic insects; and is capable of diving. Nest on the ground, composed of moss and leaves, and arched over. Eggs four to six; hatched in May.

GEN. 15.   TURDUS, *Linn.*

* Plumage brown, and spotted.

35.   T. *viscivorus*, Linn. (*Missel-Thrush.*) — Lore grayish white: upper plumage cinereous brown; wing-coverts edged with white: all the under parts spotted.

T. viscivorus, *Temm. Man. d'Orn.* tom. i. p. 161.   Missel-Thrush,
Mont. Orn. Dict. Selb. Illust. vol. i. p. 158. pl. 44. f. 1.   *Bew.
Brit. Birds*, vol. i. p. 112.

Dimens. Entire length eleven inches: length of the bill (from the
forehead) nine lines, (from the gape) one inch two lines; of the tarsus
one inch three lines; of the tail four inches; from the carpus to the end
of the wing six inches: breadth, wings extended, eighteen inches.

Descript. Upper parts brown, with a tinge of ash-colour: space
between the bill and the eye grayish white: all the under parts white,
lightly shaded here and there with reddish yellow, and marked with deep
brown spots, which are triangular on the throat and fore part of the neck,
oval on the abdomen: upper wing-coverts edged and tipped with whitish:
the three outermost tail feathers tipped with white: bill dusky: feet pale
brown. (*Young of the year.*) Upper parts cinereous brown; head, back,
and scapulars, distinctly spotted with yellow and dusky, the former colour
occupying the central portion, the latter the extreme tip of each feather:
wing-coverts broadly tipped and edged with yellow; quills brown, with a
narrow edging of yellow: tail brown, all the feathers, except the two
middle ones, tipped with white: under parts spotted much as in the
adult bird: legs very pale yellowish brown. *Obs.* This species is subject
to considerable variation of plumage. (*Egg.*) Greenish white, spotted
with red brown; sometimes reddish white, spotted with dark red brown:
long. diam. one inch three lines; trans. diam. eleven lines.

Not uncommon in some parts of the country; in others less frequent.
Haunts woods and orchards. Commences its song very early in the
year; continues it till the end of May. Nest placed in the forked
branches of trees; formed of moss, wool, and coarse grass, and lined with
the finer grasses. Eggs from three to five in number, incubation commencing early in April. Food insects, and berries, particularly those of
the misseltoe.

36.   T. *pilaris*, Linn.   (*Fieldfare.*) — Lore black:
back and wing-coverts brown; head, neck, and rump, ash-
gray: abdomen without spots.

T. pilaris, *Temm. Man. d'Orn.* tom. i. p. 163.   Fieldfare, *Mont. Orn.
Dict. Selb. Illust.* vol. i. p. 160. pl. 45. f. 1.   *Bew. Brit. Birds*,
vol. i. p. 116.

Dimens. Entire length nine inches nine lines: breadth, wings extended, seventeen inches two lines.

Descript. Head, nape, and lower part of the back (including the
rump and upper tail-coverts), deep ash-colour: upper part of the back
and wing-coverts chestnut-brown: space between the eye and bill black;
above the eyes a white streak: throat, sides of the neck, and breast,
yellowish, with oblong dusky spots; feathers on the flanks spotted with
black and edged with white; abdomen without spots: quills dusky with
pale edges: tail black, the two outermost feathers tipped with whitish
on their inner webs: bill yellow; the tips of the mandibles black:
irides brown: feet dusky. (*Female.*) Head ash-colour, obscurely spotted
with brown: throat whitish: feet paler than in the male bird. (*Egg.*)
Light blue, mottled all over with spots of dark red brown.

A winter visitant; first appearing in November: remains with us till
very late in the Spring, but does not breed in this country. During its
stay keeps in flocks. Feeds on haws and other berries; also on insects
and worms.

37. T. *musicus,* Linn. (*Song-Thrush.*)—Lore yellowish : all the upper parts brown tinged with olive : wing-coverts edged with reddish yellow : throat without spots.

T. musicus, *Temm. Man. d'Orn.* tom. I. p. 164.   Song-Thrush, *Selb. Illust.* vol. I. p. 162. pl. 45. f. 2.   *Bew. Brit. Birds,* vol. I. p. 114.   Throstle, *Mont. Orn. Dict.*

DIMENS. Entire length eight inches eleven lines : length of the bill (from the forehead) nine lines, (from the gape) one inch one line ; of the tarsus one inch five lines ; of the tail three inches four lines ; from the carpus to the end of the wing four inches four lines : breadth, wings extended, thirteen inches eight lines.

DESCRIPT. Head, and all the upper parts, brown with a slight tinge of olive : greater wing-coverts edged with orange-yellow : space between the eye and the bill yellowish : throat pure white, without spots ; sides of the neck, and upper part of the breast, reddish yellow, with triangular dusky spots ; belly and flanks white, with oval spots : under wing-coverts pale orange-yellow : bill dusky ; the under mandible yellowish at the base : feet light brown. (*Egg.*) Light blue, with a few black spots at the larger end : long. diam. one inch one line ; trans. diam. ten lines.

Common throughout the country ; remaining with us the whole year. Feeds on berries, insects, and snails. In song from the beginning of February till the end of July. Has frequently two broods in a season, the first of which is hatched about the beginning of April, sometimes earlier. Nest placed in shrubs or low trees ; composed of moss and grass externally, plastered over within with clay and cow-dung. Eggs three to five in number. This species does not collect in flocks during the winter.

38. T. *iliacus,* Linn. (*Redwing.*)—A broad white streak above the eyes : upper parts olive-brown : under wing-coverts and flanks deep orange-red : middle of the abdomen without spots.

T. iliacus, *Temm. Man. d'Orn.* tom. I. p. 165.   Redwing, *Mont. Orn. Dict.   Selb. Illust.* vol. I. p. 165. pl. 45. f. 3.   *Bew. Brit. Birds,* vol. I. p. 118.

DIMENS. Entire length eight inches four lines : length of the bill (from the forehead) eight lines and a half, (from the gape) one inch and half a line ; of the tarsus one inch and half a line ; of the tail three inches three lines ; from the carpus to the end of the wing four inches seven lines : breadth, wings extended, thirteen inches.

DESCRIPT. All the upper parts olive-brown : a broad whitish streak above the eye, extending from the base of the bill backwards for an inch or more : space between the eye and the bill black mixed with yellow : sides of the neck, breast, and sides of the abdomen, white, with numerous large oblong dusky spots ; middle of the abdomen pure white, without spots : under wing-coverts and flanks deep orange-red : feet pale brown : bill dusky ; base of the under mandible yellowish. (*Egg.*) Resembling that of the *Fieldfare,* but smaller, with the spots less numerous.

A migratory species, arriving in large flocks about the beginning of October, or rather later, but generally before the Fieldfare. Haunts pastures and hedges. Feeds principally on insects, and the smaller *helices;* or, when these cannot be obtained, on berries. Suffers more from very severe weather than the Fieldfare. Retires northward in the Spring to breed.

(3.) *T. varius,* Horsf. Zoolog. Research. in Java. Linn. Trans. vol. XIII. p. 149.

> A bird, supposed to be of this species, was shot by Lord Malmesbury in a small gorze-covert, in the parish of Heron Court in Hampshire, Jan. 24, 1828. It being a native of Java, it is not easy to conjecture by what accident it could have reached this country. The plumage, however, was in very fine condition, and did not exhibit any marks of confinement.

** *Plumage uniform ; the ground colour black.*

39. T. *Merula,* Linn. (*Blackbird.*)—Plumage wholly black.

> T. Merula, *Temm. Man. d'Orn.* tom. I. p. 168. Blackbird, *Mont. Orn. Dict. Selb. Illust.* vol. I. p. 167. pl. 45. f. 4. and pl. 43. f. 4. *Bew. Brit. Birds,* vol. I. p. 120.

Dimens. Entire length ten inches eleven lines; length of the bill (from the forehead) eleven lines, (from the gape) one inch three lines; of the tarsus one inch six lines; of the tail four inches eleven lines; from the carpus to the end of the wing five inches: breadth, wings extended, sixteen inches.

Descript. (*Male.*) The whole of the plumage deep black: orbits, bill, and inside of the mouth, bright yellow : irides and feet dusky brown. (*Female.*) Upper parts dusky brown: throat pale brown, with spots of a darker tint; breast reddish brown; abdomen and under tail-coverts dark cinereous brown: bill and legs dusky brown. (*Young newly fledged.*) Upper parts dusky brown, as in the female, but with reddish streaks on the head, nape, upper part of the back, and lesser wing-coverts, occupying the shafts of the feathers : throat and breast reddish brown, obscurely spotted with dusky. (*Egg.*) Light blue, speckled and spotted with light red brown; occasionally uniform blue without spots: long. diam. one inch two lines; trans. diam. ten lines.

Equally common with the Thrush, and like that bird a constant resident in this country. Food the same. Commences its song in the Spring, a little later than that species, and ceases a little earlier. Habits solitary. Nest formed of moss, twigs, and roots, plastered internally, and afterwards lined with the finer grasses. Eggs four or five in number; hatched the beginning of April.

40. T. *torquatus,* Linn. (*Ring-Ouzel.*) — Plumage black, edged with gray : a crescent-shaped white spot on the breast.

> T. torquatus, *Temm. Man. d'Orn.* tom. I. p. 166. Ring-Ouzel, *Mont. Orn. Dict. and Supp. Selb. Illust.* vol. I. p. 169. pl. 44. f. 2. *Bew. Brit. Birds,* vol. I. p. 122.

Dimens. Entire length eleven inches.

Descript. (*Male.*) The whole plumage black: all the feathers edged with ash-gray : on the upper part of the breast a large crescent-shaped gorget of pure white: bill and legs dusky. (*Female.*) Plumage more clouded with gray: the gorget on the breast smaller, not so well defined, and tinged with reddish brown and gray. In the *young* female the gorget is scarcely visible: in the *young male* it is reddish white. (*Egg.*) Light blue, mottled with spots of ash-gray and light brown: long. diam. one inch two lines; trans. diam. ten lines.

Not a common species. Occasionally observed in small flocks about Spring and Autumn, on its passage to and from the northern and mountainous parts of the kingdom where it breeds. Does not winter in England. Nest placed on the ground; in form and texture resembling that of the Blackbird. Food insects and berries.

GEN. 16. ORIOLUS, *Linn.*

41. O. *Galbula*, Linn. (*Golden Oriole.*) — Golden yellow; lore, wings, and tail, black.

O. Galbula, *Temm. Man. d'Orn.* tom. i. p. 129. Golden Oriole, *Mont. Orn. Dict. Selb. Illust.* vol. i. p. 176. pl. 35. f. 1 and 2. *Bew. Brit. Birds,* vol. i. p. 103.

DIMENS. Entire length nine inches six lines: length of the bill (from the forehead) eleven lines and a half, (from the gape) one inch one line and a half; of the tarsus ten lines; of the tail three inches one line; from the carpus to the end of the wing six inches.

DESCRIPT. (*Male.*) Bright golden yellow: space between the eye and the bill black: wings black; the outer webs of the quills edged with white; primary coverts tipped with yellow: two middle tail-feathers wholly black; the others black, tipped with yellow: bill reddish: irides red: feet black. (*Female and young.*) Upper parts olive-green: throat, breast, and abdomen, grayish white, tinged with yellowish, with dusky streaks on the shafts of the feathers: wings brownish black, the quills edged with pale olive-gray: tail olive with a tinge of dusky, the tips of all the feathers, except the two middle ones, yellowish white. (*Egg.*) White, tinged with pink, and sparingly spotted with ash-colour and liver-brown: long. diam. one inch two lines; trans. diam. ten lines.

An occasional visitant in this country, but not often met with. Has been killed in Suffolk, Norfolk, Essex, Hampshire, Devonshire, Cheshire, Lancashire, Northumberland, as well as in Scotland and Ireland. Many of the individuals occurred in the Spring. Frequents wooded districts: feeding on berries and coleopterous insects. Nest purse-shaped, suspended from the forked branches of tall trees. Eggs four or five in number.

GEN. 17. ACCENTOR, *Bechst.*

42. A. *alpinus*, Bechst. (*Alpine Accentor.*)—Cinereous gray, with large brown spots on the back: throat white with brown scales: abdomen grayish white mixed with reddish.

A. alpinus, *Temm. Man. d'Orn.* tom. i. p. 248. Alpine Accentor, *Selb. Illust.* vol. i. p. 247. pl. D. f. 3.

DIMENS. Entire length six inches eight lines. TEMM.

DESCRIPT. Head, breast, neck and back, cinereous gray, with large brown spots between the shoulders: wings and tail dusky brown, variegated with ash-colour; middle and lesser wing-coverts tipped with white: throat white with brown scales; abdomen and sides grayish white, mixed with reddish: bill yellow at the base, black at the tip: feet yellowish. (*Egg.*) Of a uniform bluish green, like that of the next species, but rather larger.

A native of the European Alps, but has twice occurred in this country. In the first instance, a pair were observed at Cambridge by Dr Thackeray, Nov. 23, 1822, and one, a female, killed, which is at present in his collection. In the second, a single individual was shot (a few years since) on the borders of Epping Forest. 'Said to build in the clefts of rocks, and under the eaves of houses. Food insects and seeds.

43.   A. *modularis*, Cuv.· (*Hedge Accentor.*)—Crown of the head cinereous, with brown streaks: back and wing-coverts with large brown spots: throat and breast bluish gray: middle of the abdomen whitish.

> A. modularis, *Temm. Man. d'Orn.* tom. I. p. 249.   Hedge Accentor, *Selb. Illust.* vol. I. p. 248. pl. 43*. f. 4.   Hedge Warbler, *Mont. Orn. Dict. Bew. Brit. Birds*, vol. I. p. 244.

Dimens. Entire length five inches eight lines: length of the bill (from the forehead) five lines, (from the gape) seven lines; of the tarsus ten lines and a half; of the tail two inches six lines; from the carpus to the end of the wing two inches seven lines: breadth, wings extended, eight inches five lines.

Descript. Top of the head cinereous, with brown streaks: back and wing-coverts yellowish brown, with a large darker-coloured spot in the centre of each feather; middle wing-coverts tipped with whitish: sides of the neck, throat, and breast, bluish gray: rump and flanks yellowish brown; middle of the abdomen whitish: under tail-coverts brown, edged with white: bill dusky: irides hazel: feet yellowish brown. (*Egg.*) Of a uniform bluish green: long. diam. nine lines and a half; trans. diam. six lines and a half.

Common in hedges throughout the country; and in song the greater part of the year. Feeds principally on insects. Nest formed of moss and wool, and lined with hair: placed in shrubs and low bushes. Eggs four to six in number; hatched early in April.

GEN. 18.   SYLVIA, *Lath.*

(1. Erithaca, *Swains.*)

44.   S. *Rubecula*, Lath. (*Redbreast.*) — Upper parts brownish gray, tinged with olive: throat and breast bright ferruginous.

> S. Rubecula, *Temm. Man. d'Orn.* tom. I. p. 215.   Redbreast, *Mont. Orn. Dict.   Selb. Illust.* vol. I. p. 188. pl. 46. f. 2.   *Bew. Brit. Birds*, vol. I. p. 235.

Dimens. Entire length five inches nine lines: length of the bill (from the forehead) six lines, (from the gape) eight lines and a half; of the tarsus one inch; from the carpus to the end of the wing two inches eleven lines: breadth, wings extended, nine inches two lines.

DESCRIPT. (*Male.*) Head, and all the upper parts of the body, grayish brown with a tinge of olive: forehead, cheeks, throat, and breast, bright ferruginous red, edged round with ash-gray: belly white: greater and middle wing-coverts sometimes tipped with pale reddish orange: irides black: feet yellowish brown. (*Female.*) Upper parts cinereous brown; the red on the breast not so bright, with the surrounding margin of ash-gray less conspicuous. (*Young of the year.*) Upper parts olive-gray, speckled with pale reddish spots: the breast faintly tinged with reddish yellow, and marked with fine transverse streaks of olive-brown. (*Egg.*) Ground-colour white, spotted with pale yellow brown: long. diam. nine lines and a half; trans. diam. seven lines and a half.

A common and generally diffused species. Frequents gardens and woods. Feeds on insects and worms. In song nearly all the year. Nest formed of moss, dead leaves, and stalks of plants, and lined with hair; placed on the ground, occasionally in the holes of trees and old buildings. Has often two broods in the season, the first of which is sometimes hatched as early as the end of March. Eggs five to seven in number.

(2. PHŒNICURA, *Swains.*)

45. S. *Suecica*, Lath. (*Blue-throated Warbler.*)— Upper parts ash-brown tinged with olive: throat, and fore part of the neck blue, with a central white spot.

S. Suecica, *Temm. Man. d'Orn.* tom. I. p. 216. Blue-throated Warbler, *Lath. Syn.* vol. II. p. 444. Blue-throated or Swedish Redbreast, *Shaw, Nat. Misc.* vol. XVI. pl. 661. Blue-throated Redstart, *Selb. Illust.* vol. I. p. 195. pl. 100. f. 2, 3.

DIMENS. Entire length five inches ten lines: length of the bill (from the forehead) five lines and a half, (from the gape) nine lines; of the tarsus one inch and half a line; of the tail two inches two lines; from the carpus to the end of the wing two inches eleven lines.

DESCRIPT. (*Adult male.*) Upper parts brownish ash, tinged with olive; a whitish streak above the eye; throat, and fore part of the neck, bright azure, in the centre of which is a pure white spot, disappearing in advanced age; beneath the blue a border of black, and beyond this a broader one of red: belly, thighs, and vent, white: basal half of the tail rust-red, the remainder black; the two middle feathers similar to the back, and of one colour throughout. (*Female.*) Upper parts resembling those of the male: throat, and fore part of the neck, white; upper part of the breast dusky, tinged with cinereous and azure, beneath which is a faint indication of the transverse red band; the rest of the under parts dirty white. (*Young.*) Plumage brown, spotted with whitish; a large white space on the throat. (*Egg.*) Uniform greenish blue: long. diam. eight lines; trans. diam. five lines and a half.

Extremely rare in this country. A single individual killed on New-castle Town Moor, in May 1826, is now in the Museum at that place. Said to reside in forests, and to breed in the holes of decayed trees.

46. S. *Phœnicurus*, Lath. (*Redstart.*)—Bluish gray above: throat black; breast, rump, and under tail-coverts, red: second quill equal to the sixth.

S. Phœnicurus, *Temm. Man. d'Orn.* tom. i. p. 220.   Redstart, *Mont. Orn. Dict. Selb. Illust.* vol. i. p. 191. pl. 46. f. 3.   *Bew. Brit. Birds,* vol. i. p. 239.

Dimens.   Entire length five inches seven lines: length of the bill (from the forehead) five lines and a half, (from the gape) eight lines and a half; of the tarsus eleven lines; of the tail two inches five lines; from the carpus to the end of the wing three inches: breadth, wings extended, nine inches four lines.

Descript.   (*Adult male.*)   Forehead white: base of the bill, lore, cheeks, throat, and a portion of the under surface of the neck, black: head, upper part of the neck, and back, deep bluish gray: breast, rump, flanks, and under tail-coverts, bright rust-red; tail the same, excepting the two middle feathers, which are brown: abdomen whitish: feet black. (*Adult female.*)   Upper parts reddish gray: quills and coverts edged with reddish yellow: throat white; in very old individuals dusky, mixed with reddish: breast, flanks, and under tail-coverts pale reddish orange: abdomen whitish: rump rust-red: tail the same, the two middle feathers excepted, which are brown edged with red.   (*Young males of the year.*) Upper parts reddish ash; no white on the forehead; the black on the throat, and the red on the breast, variegated with white.   (*Egg.*) Uniform grayish blue: long. diam. eight lines and a half; trans. diam. six lines and a half.

A summer visitant, first appearing about the second week in April. Common in most parts of the kingdom, but said to be rare in some of the western counties.   Frequents gardens and woods, and builds in the holes of trees and old walls.   Nest composed of moss, and lined with hair and feathers.   Eggs five to seven in number.   Young broods fledged about the second week in June.

47.   S. *Tithys*, Scop.   (*Black Red-tail.*) — Upper parts bluish ash: throat and breast black: rump and under tail-coverts red: second quill equal to the seventh.

S. Tithys, *Temm. Man. d'Orn.* tom. i. p. 218.   Phœnicura Tithys, *Jard. and Selb. Orn.* pl. 86. figs. 1 and 2.   Black Red-tail, *Lath. Syn.* vol. ii. p. 426.   Tithys Redstart, *Selb. Illust.* vol. i. p. 193. pl. D. f. 1, 2.

Dimens.   Entire length five inches three lines.   Temm.

Descript.   (*Male.*)   Upper parts bluish ash: space between the eye and the bill, cheeks, throat and breast, deep black, passing into bluish ash on the belly and sides: rump and under tail-coverts rust-red; the two middle tail feathers brown; the others red: lower part of the abdomen whitish: greater wing-coverts edged with pure white.   (*Female.*) Upper and under parts cinereous, dull above, brighter beneath, passing into whitish on the lower part of the abdomen: quills and wing-coverts dusky, edged with cinereous gray: rump, tail, and under tail-coverts reddish, not so bright as in the male.   (*Egg.*) White: long. diam. ten lines; trans. diam. seven lines.

First added to the British Fauna by Mr Gould, who obtained a specimen which was shot at Kilburn, near London, Oct. 25, 1829.   (See *Zool. Journ.* vol. v. p. 102.)   Two other individuals were procured during the Summer of 1830, one near Bristol, the other at Brighton.   Common on some parts of the Continent.   Habits said to resemble those of the last species.

(3. Salicaria, *Selby.*)

48. S. *Locustella*, Lath. (*Grasshopper Warbler.*)—
Upper plumage olivaceous brown, spotted with dusky:
tail of one colour, long, and very much cuneated.

S. Locustella, *Temm. Man. d'Orn.* tom. i. p. 184. Grasshopper
Warbler, *Mont. Orn. Dict. and Supp. Selb. Illust.* vol. i. p. 199.
pl. 45**. f. 1. *Bew. Brit. Birds*, vol. i. p. 247.

Dimens. Entire length five inches six lines : length of the bill (from
the forehead) five lines, (from the gape) seven lines and a half; of the
tarsus nine lines and a half; of the tail two inches three lines; from the
carpus to the end of the wing two inches four lines and a half.

Descript. All the upper parts olivaceous brown, with a dusky spot
in the centre of each feather, most conspicuous on the head and back :
throat white, bounded by a circle of small oval brown spots : breast and
belly yellowish white; under tail-coverts pale yellowish brown, with dusky
streaks occupying the shafts of the feathers : quills and tail dusky brown,
with pale olive edges : feet yellowish brown : hind claw shorter than the
toe. (*Egg.*) Pale reddish white, speckled all over with darker red brown:
long. diam. eight lines; trans. diam. six lines.

Visits this country in April, but is not generally distributed, and no
where very plentiful. Haunts thickets and furzy commons, principally
in damp situations, and is of shy habits, seldom exposing itself to view.
Note resembling that of the *gryllidæ*. Nest artfully concealed, placed
on the ground, or in thick bushes of furze and bramble; composed of dried
stalks and goose-grass, and lined with fibrous roots. Eggs six in number.
Food principally small coleopterous insects.

49. S. *Phragmitis*, Bechst. (*Sedge Warbler.*)—Upper
plumage olivaceous brown, tinged with yellow, and spotted
with dusky : above the eye a broad white streak : tail
moderate.

S. Phragmitis, *Temm. Man. d'Orn.* tom. i. p. 189. Sedge Warbler,
Mont. Orn. Dict. Selb. Illust. vol. i. p. 201. pl. 45**. f. 2. Reed
Warbler, *Bew. Brit. Birds*, vol. i. p. 246.

Dimens. Entire length five inches five lines : length of the bill (from
the forehead) five lines, (from the gape) seven lines and a half; of the
tarsus nine lines and a half; of the tail two inches one line; from the
carpus to the end of the wing two inches four lines : breadth, wings
extended, seven inches five lines.

Descript. Top of the head, back, scapulars and wing-coverts, olivaceous brown, with a dark spot in the centre of each feather : neck, lower
part of the back, rump and upper tail-coverts, plain yellowish brown : a
broad and very distinct white streak above the eye, arising from the base
of the bill, and extending the length of the head : all the under parts
yellowish white, with a reddish tinge on the breast, sides, and under tail-
coverts : quills brown with pale edges : tail deep yellowish brown : bill
dusky above, whitish beneath : irides dark hazel : feet yellowish brown ;
soles yellow. (*Egg.*) Greenish white; mottled all over with yellow
brown: long. diam. eight lines; trans. diam. six lines.

Very abundant in marshy districts, and by the sides of rivers, visiting
this country about the third week in April. Song much varied, imitative

of that of other birds; kept up unceasingly during the breeding season, and often heard in the night. Nest generally suspended between the stems of reeds at a little distance from the ground; composed of dried stalks, with the addition of a little moss, and lined with hair and the finer grasses. Eggs five or six in number; hatched towards the end of May, or beginning of June.

50.   S. *arundinacea*, Lath.   (*Reed-Wren.*) — Upper parts plain olivaceous brown, without spots : between the bill and the eye a white streak.

S. arundinacea, *Temm. Man. d'Orn.* tom. i. p. 191.   Reed-Wren, *Mont. Orn. Dict. Selb. Illust.* vol. i. p. 203. pl. 45**. f. 3.

Dimens. Entire length five inches four lines : length of the bill (from the forehead) five lines and a half, (from the gape) eight lines; of the tarsus ten lines; of the tail two inches three lines; from the carpus to the end of the wing two inches four lines : breadth, wings extended, seven inches five lines.

Descript. Upper parts of a plain uniform olive brown without spots ; rump and tail faintly tinged with reddish : quills brown with cinereous edges : between the bill and the eye a narrow white streak, but not continued *above* the eye as in the last species : under parts yellowish white, lightest on the throat and down the middle of the belly ; sides inclining to rufous : bill broader throughout, but especially at the base, and also somewhat longer, than in the S. *Phragmitis;* in colour dusky above, yellowish beneath, and along the margin of the upper mandible : irides light hazel : feet pale brown; soles yellow. (*Egg.*) Greenish white, spotted and speckled with ash-green and light brown ; the markings darkest, and most numerous at the larger end: long. diam. nine lines; trans. diam. six lines and a half.

Found in the same situations with the last species, but is much less plentiful, and not so generally diffused. Habits similar. Food the smaller species of *libellulæ*, and other insects. Nest of an oblong panier-shaped form, suspended like that of the S. *Phragmitis;* composed of grasses and the seed-branches of the reed. Eggs four or five in number.

(4.)   *S. Cetti*, Temm. Man. d'Orn. tom. i. p. 194.

According to Temminck, this species has been killed in England. It does not, however, appear to be known to our own naturalists, nor am I aware that it is to be found in any of our British collections.

(4. Philomela, *Swains.*)

51.   S. *Luscinia*, Lath. (*Nightingale.*)—Upper plumage reddish brown ; beneath, cinereous white.

S. Luscinia, *Temm. Man. d'Orn.* tom. i. p. 195.   Nightingale, *Mont. Orn. Dict. and Supp.   Selb. Illust.* vol. i. p. 206. pl. 46. f. 1. *Bew. Brit. Birds,* vol. i. p. 231.

Dimens. Entire length six inches two lines : length of the bill (from the forehead) six lines, (from the gape) nine lines and a half; of the tarsus one inch one line; of the tail two inches eight lines; from the carpus to the end of the wing three inches two lines : breadth, wings extended, nine inches seven lines.

DESCRIPT. All the upper parts reddish brown : tail rust-red : breast
and sides pale ash, passing into yellowish white : throat and belly whitish :
quills pale dusky, edged with reddish brown : bill brown : irides hazel :
feet yellowish brown.   (*Egg.*) Uniform olive-brown : long. diam. ten
lines ; trans. diam. eight lines and a half.
   Common in the southern, midland, and eastern counties ; rare in the
western and northern.   Has not been observed in Cornwall or in Devonshire, excepting on the eastern borders.   First heard about the middle
or end of April : song continued till the first or second week in June.
Frequents woods, copses, and tall hedges.   Nest placed on or near the
ground, sometimes suspended between the stems of herbaceous plants ; of
a very deep, oval form ; composed principally of withered leaves, and lined
with fine grasses.   Eggs four to six in number ; hatched towards the end
of May, or beginning of June.

(5. CURRUCA, *Bechst.*)

52.   S. *Atricapilla*, Lath. (*Black-cap Warbler.*)—Top of the head and occiput deep black (*male*), or reddish brown (*female*) ; rest of the upper plumage greenish ash.

   S. Atricapilla, *Temm. Man. d'Orn.* tom. I. p. 201.   Black-cap, *Mont.
      Orn. Dict.   Selb. Illust.* vol. I. p. 209. pl. 46. f. 5.   *Bew. Brit. Birds,*
      vol. I. p. 249.

   DIMENS. Entire length five inches eight lines : length of the bill
(from the forehead) four lines and a half, (from the gape) seven lines
and a half ; of the tarsus nine lines and a half ; of the tail two inches
four lines ; from the carpus to the end of the wing three inches : breadth,
wings extended, eight inches six lines.
   DESCRIPT. (*Male.*) Forehead, crown, and occiput, deep black : rest
of the upper parts cinereous, with a slight tinge of olive-green : neck and
breast grayish ash ; belly and vent grayish white : bill and feet dusky.
(*Female.*) Crown of the head reddish brown ; rest of the plumage resembling that of the male but rather darker.   (*Egg.*) Pale greenish
white, mottled with light brown and ash-colour, with a few spots and
streaks of dark brown : long. diam. nine lines ; trans. diam. seven lines.
   A common and widely dispersed species, visiting this country about
the same time as the Nightingale.   Song melodious, and rather powerful,
heard till the middle, or occasionally the end of July.   Is partial to
orchards and gardens.   Nest placed in a low bush ; loosely put together ;
constructed of bents and dried stalks, patched on the outside with threads
and cobwebs, and lined with fibrous roots and a few long hairs.   Eggs
five in number, on which the male sits occasionally, as well as the female.

53.   S. *hortensis*, Lath. (*Greater Pettychaps*)—Upper parts grayish brown, tinged with olive : orbits white : tail of one colour.

   S. hortensis, *Temm. Man. d'Orn.* tom. I. p. 206.   Greater Pettychaps,
      Mont. Orn. Dict.   Selb. Illust. vol. I. p. 211. pl. 46. f. 4.   Passe-
      rine Warbler, *Bew. Brit. Birds,* vol. I. p. 243.

   DIMENS. Entire length five inches ten lines : length of the bill (from
the forehead) five lines, (from the gape) eight lines ; of the tarsus ten
lines ; of the tail two inches two lines : from the carpus to the end of the

wing two inches eleven lines: breadth, wings extended, eight inches six lines.

Descript. All the upper parts cinereous brown, tinged with olive: orbits white: below the ear on each side of the neck a patch of ash-gray: throat whitish; breast and sides yellowish gray inclining to brown: belly, vent and under tail-coverts, nearly pure white: quills and tail dusky, with pale edges: feet bluish gray. (*Egg*.) Greenish white ground, speckled with two shades of ash-green: long. diam. nine lines; trans. diam. six lines and a half.

Not uncommon in many parts of the country, making its first appearance about the beginning of May. Frequents gardens, copses, and thick hedges. Song resembling that of the Blackcap, but less powerful; continued till the middle of July. Nest of flimsy structure; formed of the decayed stems of goose-grass, and other fibrous plants. Eggs four or five in number, very like those of the last species.—*Obs.* The *Garden Warbler* and the *Passerine Warbler* of Bewick are probably both referable to the present species. Montagu, however, thought that the *Passerine Warbler* of the author just mentioned was the same as the *Reed Wren*.

54. S. *cinerea*, Lath. (*White-Throat*.)—Top of the head cinereous; back and wing-coverts grayish brown inclining to rufous: under parts white; the breast tinged with rose-red : tail particoloured.

S. cinerea, *Temm. Man. d'Orn.* tom. i. p. 207. White-Throat, *Mont. Orn. Dict. Selb. Illust.* vol. i. p. 213. pl. 46. f. 7. *Bew. Brit. Birds,* vol. i. p. 251.

Dimens. Entire length five inches ten lines: length of the bill (from the forehead) four lines and a half, (from the gape) seven lines; of the tarsus nine lines and a half; of the tail two inches eight lines; from the carpus to the end of the wing two inches eight lines: breadth, wings extended, eight inches seven lines.

Descript. (*Male.*) Crown of the head, space between the eye and the bill, cheeks, and ear-coverts, cinereous: rest of the upper parts cinereous brown, inclining to rufous on the back: quills dusky; secondaries and greater coverts broadly edged with brownish red: throat and middle of the abdomen white; breast faintly tinged with rose-red; flanks and thighs reddish gray: tail deep brown; outer feather white, except at the base of the inner web; the next tipped with white: bill dusky above, paler beneath: irides yellowish: legs yellowish brown. (*Female and young.*) Upper parts more inclining to reddish brown; breast white, without the rosy tinge; the outer tail-feather reddish, instead of white. (*Egg.*) Greenish white ground, spotted and speckled with ash-colour and two shades of ash-green: long. diam. nine lines; trans. diam. six lines and a half.

Common in hedges and thickets throughout the kingdom. First seen about the third week in April. Song often exerted on wing, accompanied by peculiar jerks and gesticulations of the body. Nest similar to that of the last species; placed in low bushes, where there is thick covert. Eggs four or five in number. Young broods fledged about the end of May.

55. S. *Curruca*, Lath. (*Lesser White-Throat.*)— Crown of the head cinereous; back and wing-coverts

grayish ash, tinged with brown: under parts silvery white: tail particoloured.

S. Curruca, *Temm. Man. d'Orn.* tom. I. p. 209. Lesser White-Throat, *Lath. Syn. Supp.* p. 185. pl. 113. *Mont. Orn. Dict. Bew. Brit. Birds*, vol. I. p. 253. *Selb. Illust.* vol. I. p. 215. pl. 100. f. 4.

DIMENS. Entire length five inches three lines.

DESCRIPT. Crown of the head cinereous; space between the eye and the bill, and ear-coverts, the same, but darker-coloured: rest of the upper parts grayish ash, inclining to brown: quills dusky brown, edged with cinereous: all the under parts, from the throat to the vent, silvery white: tail dusky; outer feather white, except at the base of the inner web; the two next faintly tipped with white: bill dusky; the lower mandible pale at the base: irides yellowish: feet dusky lead-colour. (*Egg.*) White, sparingly spotted and speckled, principally at the larger end, with ash-colour and light brown: long. diam. eight lines; trans. diam. six lines.

Common in the eastern, and in some of the midland counties, making its first appearance about the same time as the last species, but apparently rare, or altogether unknown, in many parts of the kingdom. Song soft and inward; generally ending in a shrill shivering cry, which last note is continued at intervals till near the middle of July. Chiefly frequents gardens and copses, concealing itself in the thickest covert. Nest placed in a low bush; of very flimsy structure, composed of dry bents mixed with wool, and lined with finer grasses and a few hairs. Eggs five in number; incubation commencing about the 20th of May.

(6. SYLVIA, *Selb.*)

56. S. *sibilatrix*, Bechst. (*Wood-Wren.*) — Upper parts bright yellowish green: belly and under tail-coverts pure white: wings with the second quill equal to the fourth; third and fourth with their outer webs sloped off at the extremity.

S. sibilatrix, *Temm. Man. d'Orn.* tom. I. p. 223. Wood-Wren, *Lamb in Linn. Trans.* vol. II. p. 245. pl. 24. *Mont. in Linn. Trans.* vol. IV. p. 35. *Mont. Orn. Dict. Selb. Illust.* vol. I. p. 224. pl. 47. f. 2. Yellow Wren, *Bew. Brit. Birds*, vol. I. p. 255.

DIMENS. Entire length five inches three lines: length of the bill (from the forehead) four lines and a half, (from the gape) six lines and a half; of the tarsus nine lines; of the tail two inches one line; from the carpus to the end of the wing three inches: breadth, wings extended, eight inches eight lines.

DESCRIPT. All the upper parts bright yellow green: above the eye a broad streak of bright sulphur-yellow, arising on the forehead at the base of the upper mandible and terminating on the temples: sides of the head, throat, fore part of the neck, axillæ, and thighs, pale primrose-yellow; rest of the under parts pure white: quills dusky, edged with green: tail slightly forked, extending eight lines and a half beyond the tips of the folded wings; dusky, with each feather, except the outermost, likewise edged with bright yellow-green: bill and legs pale brown. (*Egg.*) White, spotted and speckled all over with dark purple, red, and ash-colour: long. diam. eight lines; trans. diam. six lines.

Common in the southern, western, midland, and some of the northern counties, but not generally diffused. In Cambridgeshire very rare. Haunts woods and tall trees; being first heard towards the end of April or beginning of May. Note very peculiar, resembling the word *twee* repeated several times in succession, at first slowly, but afterwards in a hurried manner, and accompanied by a singular shake of the wings. Nest placed on the ground, of an oval form, constructed of dry grass, dead leaves, and a little moss, lined with finer grass and a few hairs. Eggs six in number.

57. S. *Trochilus*, Lath. (*Willow-Wren.*) — Upper parts pale olive-green, tinged with yellow: under tail-coverts yellowish white: wings with the second quill equal to the sixth; the third, fourth, and fifth, with their outer webs sloped off at the extremity.

S. Trochilus, *Temm. Man. d'Orn.* tom. i. p. 224. Yellow Wren, *Mont. Orn. Dict. Selb. Illust.* vol. i. p. 226. pl. 47. f. 3. Willow-Wren, *Bew. Brit. Birds*, vol. i. p. 257.

Dimens. Entire length four inches eleven lines: length of the bill (from the forehead) four lines, (from the gape) six lines; of the tarsus nine lines; of the tail two inches three lines; from the carpus to the end of the wing two inches six lines and a half: breadth, wings extended, eight inches.

Descript. Distinguished from the last principally by its paler colour, and shorter wings; the difference in size not very obvious. Upper parts pale olive-green, with a yellowish tinge, inclining in some specimens to cinereous: from the base of the bill a bright streak of primrose-yellow extending over each eye to the temples, but narrower than in the *S. sibilatrix*: all the under parts yellowish white; the yellow tinge darkest on the breast, sides, axillæ, and ridges of the wings; palest on the middle of the belly: quills and tail brownish ash, faintly edged with olive-green; the latter extending eleven lines beyond the tips of the folded wings, with the two middle feathers shorter than the others: base of the under mandible yellowish: feet light brown. *Obs.* The colour of this species, especially the intensity of the yellow tinge, varies much in different individuals. The variation, however, results more from a difference of age than of sex. The male and female are equally pale in the adult state. The yellowest birds, which are also generally the smallest, occur in the early part of the Spring, and appear to be the young males of the preceding year. (*Egg.*) White, with numerous small specks of pale red: long. diam. seven lines and a half; trans. diam. six lines.

Of frequent occurrence in gardens, plantations, hedges, willow-grounds, and a variety of other situations. On the whole a much more generally diffused species than the last, but, according to Montagu, does not extend so far westward. Song pleasing, but simple, consisting of several plaintive notes in a regularly descending scale; heard from the middle of April to the beginning of July. Nest placed on the ground, nearly spherical, with a small opening at the side near the top; composed of dry grass, stalks, and other herbage, and profusely lined with feathers. Eggs five to seven; hatched the end of May or beginning of June.

58. S. *Hippolais*, Lath. (*Chiff-Chaff.*)—Upper parts olive-green, tinged with yellow; beneath yellowish white:

wings with the second quill equal to the seventh; third, fourth, fifth, and sixth, with their outer webs sloped off at the extremity.

> S. Hippolais, *Lath. Ind. Orn.* vol. ii. p. 507.   S. rufa, *Temm. Man. d'Orn.* tom. i. p. 225.   Lesser Pettychaps, *Mont. Orn. Dict. and Supp. Selb. Illust.* vol. i. p. 222. pl. 47. f. 1.   Chiff-Chaff, *Bew. Brit. Birds*, vol. i. p. 258.

Dimens. Entire length four inches six lines: length of the bill (from the forehead) four lines, (from the gape) six lines and a half; of the tarsus nine lines; of the tail one inch seven lines; from the carpus to the end of the wing two inches three lines: breadth, wings extended, seven inches three lines.

Descript. Usually somewhat smaller than the *S. Trochilus*, with the wings and tail, *relatively* considered, still shorter than in that species; also to be distinguished from it by the characters of the quills above pointed out, but in colour and general appearance almost absolutely the same.   Upper parts olive-green, tinged with yellow and ash-gray: between the bill and the eye, and over each eye, a narrow, faint yellowish white streak: quills cinereous brown, the outer webs edged with yellowish green: all the under parts, including the under tail-coverts, whitish tinged with yellow; the yellow having a tendency on the breast to appear in streaks: axillæ and under wing-coverts bright primrose-yellow: tail extending an inch beyond the tips of the folded wings: feet rather darker than in the last species.   (*Egg.*) White, with a few specks of dark purplish red: long. diam. seven lines; trans. diam. five lines and a half.

Very abundant in some parts of England, but in others much less plentiful than the last species.   Is one of the earliest of our summer visitants, making its first appearance about the middle, or towards the end of March.   Chiefly frequents woods and tall trees, and is of restless habits, being always in motion in search of insects.   Song consisting of only two, rather loud, hollow notes, resembling the words *chip-chop*, or *chiff-chaff*, which are occasionally heard till near the end of September.   Nest similar to that of the last species, placed on the ground, or in very low bushes.   *Obs.* Having compared our English *S. Hippolais* with the *S. rufa* of Temminck*, and found them agreeing closely in all their characters, I have no hesitation in considering them as the same species.

GEN. 19.   MELIZOPHILUS, *Leach.*

59.   M. *provincialis*, Leach. (*Dartford Warbler.*)— Upper plumage deep grayish brown: throat, neck, and breast, ferruginous red.

> Sylvia provincialis, *Temm. Man. d'Orn.* tom. i. p. 211.   S. Dartford-iensis, *Mont. in Linn. Trans.* vol. ix. p. 191.   Dartford Warbler, *Mont. Orn. Dict. and Supp. Selb. Illust.* vol. i. p. 219. pl. 46. f. 6. *Bew. Brit. Birds*, vol. i. p. 234.

Dimens. Entire length five inches six lines.   Mont.

Descript. (*Male.*) All the upper parts dark grayish brown: cheeks cinereous: throat, neck, and breast, deep ferruginous, inclining to purple;

---

* The comparison was made with a specimen of *S. rufa* named by Temminck himself, and sent by him to Mr Gould, to whose kindness I am indebted for it.

sides the same, but not so bright; middle of the belly white: quills dusky; the outer webs edged with dark cinereous, the inner with dark reddish brown: tail blackish brown; the outer feather tipped, and edged externally, with white: bill yellowish white at the base, black at the tip: feet yellowish. (*Female and young.*) Colours paler: throat more or less speckled and streaked with white. (*Egg.*) Greenish white ground, speckled all over with olivaceous brown and cinereous; the markings most numerous at the larger end: long. diam. eight lines; trans. diam. five lines and a half.

A local species; found in Kent, Middlesex, Hampshire, Sussex, Surrey, Devonshire, Cornwall, and a few other parts of England. Remains in this country throughout the year. Haunts chiefly downs and furzy commons. Feeds on insects. Nest somewhat similar to that of the Whitethroat; artfully concealed in the thickest bushes. Young broods fledged in May.

GEN. 20.　REGULUS, *Cuv.*

60. R. *aurocapillus*, Selby. (*Gold-crested Regulus.*)— Upper parts greenish yellow: cheeks and region of the eyes cinereous, without any white bands: crown-feathers elongated, forming a bright yellow orange crest.

> Sylvia Regulus, *Temm. Man. d'Orn.* tom. i. p. 229.　Gold-crested Regulus, *Selb. Illust.* vol. i. p. 229. pl. 47. f. 4 & 5. Golden-Crested Wren, *Mont. Orn. Dict. Bew. Brit. Birds*, vol. i. p. 260.

DIMENS. Entire length three inches six lines: length of the bill (from the forehead) three lines and a half, (from the gape) five lines; of the tarsus seven lines and a half; of the tail one inch five lines; from the carpus to the end of the wing one inch eleven lines: breadth, wings extended, five inches six lines.

DESCRIPT. Upper parts of the plumage olive-green, tinged with yellow: feathers on the top of the head long and narrow, forming a crest, of a rich orange-yellow, the sides bordered with black: base of the bill, cheeks, throat, and region of the eyes, cinereous; the rest of the under parts the same, but with a tinge of yellow: quills dusky, edged with yellowish green; secondaries crossed by a black bar; wing-coverts dusky brown, edged with yellow and tipped with white: tail blackish brown; the feathers edged outwardly with yellowish green, inwardly with whitish: bill black: irides hazel: feet brown. (*Female.*) Colours generally paler: crest on the head smaller, and of a pale lemon-yellow; the lateral edging of black narrower, and less conspicuous. (*Egg.*) Dull pale reddish white: long. diam. six lines; trans. diam. five lines.

Not uncommon, and generally dispersed throughout the kingdom. Does not migrate. Frequents woods and plantations, especially those of fir. Song weak and inward; heard as early as February. Nest generally suspended from the under surface of the branches of spruce firs; formed principally of moss, and lined with feathers. Eggs seven to ten in number. Breeds early in the year.

61. R. *ignicapillus*, Nob. (*Fire-crested Regulus.*)— Upper parts yellowish green, passing into pure yellow on the sides of the neck; cheeks with three longitudinal streaks, one black and two white: crown-feathers elongated, of a brilliant flame-red.

H

Sylvia ignicapilla, *Temm. Man. d'Orn.* tom. i. p. 231.   Fire-crested
Wren, *Gould, Europ. Birds*, part 3.

DIMENS.   Entire length four inches: length of the bill (from the forehead) four lines and a half, (from the gape) five lines and a half; of the
tarsus seven lines and a half; of the tail one inch nine lines; from the
carpus to the end of the wing two inches and half a line.

DESCRIPT.   Closely resembling the last species, but somewhat larger,
with the bill longer and broader at the base.   (*Adult male.*) Upper parts
yellowish tinged with green, passing into nearly pure yellow on the sides
of the neck; crest of a brilliant flame-red, bordered in front and at the
sides of the crown with deep black; cheeks particularly distinguished by
three longitudinal streaks, of which there is no indication in the *R. aurocapillus*; one of these, which is black, passes directly through the eyes
in a line from the base of the upper mandible to the ear-coverts; the
other two are white, and are situate one above, the other beneath the
eyes; forehead cinereous, with a pale reddish tinge : wings and tail the
same as in the last species: under parts more approaching to white: bill
black: irides deep brown: tarsi brown; toes and claws yellowish. (*Female.*)
The longitudinal streaks on the sides of the face the same as in the male,
but the colours more obscure : crest of a dull orange : sides of the neck
olivaceous green, instead of yellow: in other respects the plumage of
the two sexes is similar. (*The young of the year before the first moult.*)
Only to be distinguished from those of the last species, by the longer and
broader bill: cheeks cinereous, without any appearance of the longitudinal streaks: crest of a pale lemon-yellow, scarcely developed: forehead,
and sides of the neck, cinereous: upper parts not so bright as in the adult;
under parts cinereous, tinged with yellow.

A single individual of this species (a young bird of the year) was killed
in a garden at Swaffham Bulbeck in Cambridgeshire, in August 1832.
Since then others have been observed at Brighton by Mr J. E. Gray.   Is
probably to be met with in other parts of the country, though from its
general resemblance to the *R. aurocapillus*, it is easily overlooked.   According to Temminck, this last species resides chiefly at the tops of trees;
the *R. ignicapillus*, more on the lower branches and in small bushes.
The habits of the two in other respects, and nidification, are similar.

GEN. 21.   MOTACILLA, *Linn.*

* *Hind claw moderate; much curved.*

62. M. *alba*, Linn. (*Pied Wagtail.*)—Plumage variegated with black and white.

M. alba, *Temm. Man. d'Orn.* tom. i. p. 255.   Pied Wagtail, *Selb.
Illust.* vol. i. p. 251. pl. 49. f. 1.   *Bew. Brit. Birds*, vol. i. p. 226.
White Wagtail, *Mont. Orn. Dict.*

DIMENS.   Entire length seven inches five lines: length of the bill
(from the forehead) six lines, (from the gape) eight lines: of the tarsus
eleven lines; of the hind toe, claw included, seven lines; of the tail three
inches four lines and a half; from the carpus to the end of the wing three
inches four lines: breadth, wings extended, ten inches nine lines.

DESCRIPT. (*Adult in summer plumage.*) Forehead, cheeks, and sides
of the neck, white; crown of the head, nape, throat, breast, and upper
part of the body, black; belly and vent white: quills and greater wing-
coverts edged with white: the two outer tail-feathers nearly all white:

the others black : bill black: irides dusky. (*Winter plumage.*) Throat
and fore part of the neck white: a crescent-shaped black patch on the
upper part of the breast: all the upper parts inclining to blackish gray.
(*Young of the year.*) Crown of the head, and all the upper parts, ash-
gray, inclining to bluish gray; wing-coverts broadly edged and tipped
with white: on the upper part of the breast a crescent-shaped patch of
dusky ash, the horns of the crescent extending upwards to the ears in a
narrow line on each side of the neck; throat, cheeks, under part of the
neck, and lower part of the breast, dirty white: abdomen nearly pure
white: quills blackish, with an oblong white spot on their inner webs.
(*Egg.*) White, speckled with ash-colour: long. diam. nine lines; trans.
diam. seven lines.

A common and well-known species, frequenting pastures and newly
ploughed lands, more especially those in the vicinity of water. Runs
with great celerity, and occasionally wades in shallow streams in pursuit
of aquatic insects. Migrates in the North of England, but is stationary
in the southern counties. Nest on the ground, or in the holes of trees
and old buildings; composed of moss, fibres, and wool, and lined with
hair. Eggs four or five.

63. M. *Boarula*, Linn. (*Gray Wagtail.*)—Head and
back bluish gray ; rump and under parts bright yellow :
tail very long.

M. Boarula, *Temm. Man. d'Orn.* tom. i. p. 257. Gray Wagtail,
Mont. Orn. Dict. and Supp. Selb. Illust. vol. i. p. 253. pl. 49. f. 2.
Bew. Brit. Birds, vol. i. p. 228.

Dimens. Entire length seven inches nine lines: length of the bill
(from the forehead) five lines, (from the gape) seven lines and a half;
of the tarsus ten lines; of the hind toe, claw included, six lines; of the
tail three inches ten lines; from the carpus to the end of the wing three
inches one line and a half.

Descript. (*Summer plumage.*) Head, neck, back, and scapulars,
bluish gray: above the eyes a pale streak: throat black: rump, and
under parts, bright yellow, the former tinged with green: quills and
six middle tail-feathers grayish black, more or less edged with yellowish
white; outer tail-feather entirely white; the two next white, with a part
of the outer web black: feet dusky brown. (*Winter plumage.*) Throat
yellowish white: under parts of a paler yellow. *Obs.* Female at all times
paler than the male. (*Egg.*) White, nearly covered with reddish yellow
brown, varying in intensity: long. diam. eight lines and a half; trans.
diam. seven lines.

A regular *winter* visitant in the southern parts of the kingdom, but in
the northern a *summer* visitant. Habits and nidification very similar to
those of the last species. Is said to have two broods in the year, the first
of which is generally fledged about the end of May.

** *Hind claw elongated ; but slightly curved.*

64. M. *flava*, Ray. (*Yellow Wagtail.*)—Head, and all
the upper parts of the body, olive-green : streak above the
eye, and under parts, bright yellow : tail moderate.

M. flava, *Ray, Syn. Av.* p. 75. Yellow Wagtail, *Mont. Orn. Dict. and Supp. Selb. Illust.* vol. I. p. 255. pl. 49. f. 3. *Bew. Brit. Birds*, vol. I. p. 229.

DIMENS. Entire length six inches eight lines: length of the bill (from the forehead) five lines and a half, (from the gape) eight lines; of the tarsus eleven lines; of the hind toe, claw included, eight lines and a half; of the tail two inches eleven lines; from the carpus to the end of the wing three inches: breadth, wings extended, nine inches nine lines.

DESCRIPT. (*Male.*) Crown of the head, nape, and ear-coverts, pale olive-green; back, rump, and scapulars, of the same colour but darker: over the eyes a bright yellow streak: all the under parts bright yellow: quills and coverts blackish brown, edged and tipped with yellowish white: tail dusky; the two middle feathers edged with pale olive; the two outer ones with the whole of their external, and a large part of their inner webs white: bill and feet black. (*Female.*) Upper plumage inclining more to cinereous: throat yellowish white; the rest of the under parts yellow, but not so bright as in the male. (*Young of the year.*) Resembling the female, but still paler: upper parts dull cinereous, very faintly tinged with olivaceous: all the under parts yellowish white: greater and lesser coverts broadly tipped with white, forming two bars across the wings: tail as in the adult. (*Egg.*) White, mottled nearly all over with yellow and ash-brown: long. diam. eight lines and a half; trans. diam. six lines and a half.

A migratory species, visiting this country about the end of March, or beginning of April. Frequents downs, and arable lands, and is rather less attached to water than either of the foregoing species. Nest placed on the ground; composed of dried stalks and fibres, and lined with hair. Eggs four or five in number.

65.   M. *neglecta*, Gould. (*Blue-headed Wagtail.*) —
Head bluish lead-colour; streak above the eye white: body above olivaceous green, beneath yellow.

M. neglecta, *Gould in Proceed. of Zool. Soc.* 1832, p. 129. *Id. Europ. Birds*, part 3. M. flava, *Temm. Man. d'Orn.* tom. I. p. 260.

DIMENS. Entire length six inches. GOULD.

DESCRIPT. Closely resembling the last species, from which it scarcely differs excepting in the colour of the head and streak above the eye: the former, instead of being olivaceous like the rest of the upper parts, is of a fine lead gray, approaching to blue; the latter, as well as a second streak beneath the eyes, is white. The *female* has the colours paler, and the eye-streaks indistinctly marked. (*Egg.*) "Greenish olive, with light flesh-coloured blotches." GOULD.

Mr Gould has the merit of having first distinguished this from the *M. flava*, with which it had been previously confounded. The differences between them, he observes, do not depend upon the season, he having obtained specimens of each killed in the same month. The present species, which appears to be the only one known to continental authors, in our own country has hitherto occurred but once. This individual, which was an old bird, was killed near Colchester by Mr H. Doubleday, in the Autumn of 1834. A pair were seen. Food and nidification similar to those of the *M. flava*.

GEN. 22. ANTHUS, *Bechst.*

* Hind claw longer than the toe; very little curved.

66. A. Richardi, Vieill. (*Richard's Pipit.*) — Upper plumage deep brown, with lighter edges : breast spotted : tarsi very long.

A. Richardi, *Temm. Man. d'Orn.* tom. i. p. 263. *Vigors in Zool. Journ.* vol. i. p. 411. pl. 14.   *Bew. Brit. Birds,* vol. i. p. 379. Richard's Pipit, *Selb. Illust.* vol. i. p. 264. pl. 100. f. 5.

DIMENS. Entire length six inches seven lines : length of the hind toe, claw included, one inch. TEMM.

DESCRIPT. Plumage on the crown of the head, back, and scapulars, deep olive-brown ; the edges of the feathers somewhat paler : a broad streak above the eye, temples, throat, belly and vent, pure white ; breast and sides reddish yellow, the former marked with lanceolate dusky spots : wings blackish brown ; quills and coverts broadly edged with yellowish white : tail dusky, edged in the same manner ; outer feather entirely white ; the next with the greater portion white : bill somewhat stronger, and rather more elongated, than in the other species of this genus : upper mandible brown ; lower mandible yellow : irides brown : feet yellowish : hind claw very long, almost straight. (*Egg.*) Reddish white ground, speckled with darker red and light brown : long. diam. ten lines and a half ; trans. diam. seven lines and a half.

Very rare in this country. Has hitherto only occurred in two instances, both times in the neighbourhood of London. Habits unknown.

67. A. pratensis, Bechst. (*Meadow Pipit.*) — Upper plumage olivaceous green, with dusky spots : neck, breast, and sides, spotted : tarsi moderate.

A. pratensis, *Temm. Man. d'Orn.* tom. i. p. 269. Meadow Pipit, *Selb. Illust.* vol. i. p. 260. pl. 49. f. 4. Titlark, *Mont. Orn. Dict. Bew. Brit. Birds,* vol. i. p. 220.

DIMENS. Entire length five inches nine lines : length of the bill (from the forehead) five lines and a half, (from the gape) eight lines ; of the tarsus ten lines and a half ; of the hind toe, claw included, ten lines and a half ; of the tail two inches four lines and a half ; from the carpus to the end of the wing three inches : breadth, wings extended, nine inches ten lines.

DESCRIPT. (*Summer plumage.*) Upper parts olivaceous green tinged with ash-gray ; feathers on the head, back, and scapulars, spotted with dusky : under parts yellowish white ; throat, in the *male*, during the breeding season only, with a deep reddish tinge ; sides of the neck, breast, and upper part of the abdomen, marked with blackish brown spots ; flanks with long dusky streaks : tail dusky ; outer feather edged externally, and broadly tipped, with white ; the next with a small white spot at the extremity : bill slender, dusky ; base of the under mandible yellowish : irides hazel : legs yellowish brown : hind claw not so long as in the last species, slightly curved. (*Winter plumage.*) Upper parts of a brighter and more decided olive-green, with the dusky spots not occupying so large a portion of the feather : under parts ferruginous yellow, and not so much spotted. (*Egg.*) Reddish white ground, mottled all over with darker reddish brown : long. diam. nine lines ; trans. diam. seven lines.

Very abundant on moors, barren heaths, and extensive fenny districts. Stays the whole year. Sings in its descent, after rising to a considerable height in the air, always returning to the ground, or to some low bush, with motionless wings and expanded tail. Nest on the ground; composed of dry grass, bents, and stalks of plants, patched on the outside with a small quantity of green moss, and lined with fine grasses and long horse-hair. Eggs five or six, hatched towards the end of May. *Obs.* The *Pipit Lark* of Montagu is this species in its winter plumage, after the autumnal moult.

**   *Hind claw about equal to the toe in length, or shorter; much curved.*

68.   A. *arboreus,* Bechst. (*Tree Pipit.*)—Upper parts olivaceous green, with dusky spots: breast and sides spotted: wings with two transverse bars of yellowish white.

A. arboreus, *Temm. Man. d'Orn.* tom. I. p. 271.   Tree Pipit, *Selb. Illust.* vol. I. p. 262. pl. 49. f. 5.   Field Lark, *Mont. Orn. Dict.* Tree Lark, *Bew. Brit. Birds,* vol. I. p. 218.

DIMENS. Entire length six inches: length of the bill (from the forehead) five lines and a half, (from the gape) eight lines; of the tarsus nine lines; of the tail two inches six lines; from the carpus to the end of the wing three inches six lines.

DESCRIPT. Strongly resembling the last species, but always to be distinguished by the hind claw, which is shorter than the toe, and so much curved as to form the quadrant of a circle; the bill is also somewhat stronger, and more dilated at the base. Upper parts olivaceous green, tinged with cinereous; the centre of each feather, more especially those on the head and back, dusky brown: lesser and middle wing-coverts tipped with yellowish white, so as to shew a double transverse bar on the wings when closed: throat almost white; fore part of the neck, breast, and flanks, ochre-yellow, marked with dusky spots, large and oval on the breast, but assuming the form of long narrow streaks on the sides; middle of the abdomen pure white; under tail-coverts tinged with yellow, but free from spots: tail dusky brown; the outer feather with the greater portion white; the next only tipped with white: upper mandible wholly dusky, under mandible dusky at the tip, flesh-coloured at the base: irides hazel: legs flesh-coloured, inclining to yellowish brown. (*Egg.*) Grayish white ground, spotted and streaked with ash-colour and dark brown; sometimes pale purple, spotted with darker purple and dark red brown: long. diam. nine lines; trans. diam. seven lines.

A migratory species, visiting this country about the third week in April. Not uncommon in wooded districts, but rarely or never to be found in open country. Sings in its descent like the last species, but always rises from the top of some tall tree, to which it returns gradually, with expanded wings and tail. Song heard till the middle of July. Nest placed on the ground; formed of dry grass, patched externally with moss, and lined with the finer grasses. Eggs four or five.

69.   A. *petrosus,* Flem. (*Rock Pipit.*)—Upper parts greenish brown with darker spots: breast and sides streaked with brown: above the eye a yellowish white streak.

A. aquaticus, *Temm. Man. d'Orn.* tom. I. p. 265. Alauda petrosa, *Mont. in Linn. Trans.* vol. IV. p. 41. Rock Pipit, *Selb. Illust.* vol. I. p. 258. pl. 49. f. 6. Rock Lark, *Mont. Orn. Dict.* Field Lark, *Bew. Brit. Birds,* vol. I. p. 216.

DIMENS. Entire length six inches eight lines: length of the bill (from the forehead) six lines and a half, (from the gape) nine lines; of the tarsus ten lines and a half; of the hind toe, claw included, ten lines; of the tail two inches nine lines; from the carpus to the end of the wing three inches five lines: breadth, wings extended, eleven inches three lines.

DESCRIPT. (*Adult.*) Upper parts dark brown, with a tinge of olive-green; the feathers on the back, and scapulars, obscurely marked with dusky streaks: quills and coverts dusky, edged with greenish white: above the eye a pale streak: all the under parts whitish, with longitudinal brown streaks on the sides of the neck, breast, and flanks: the two middle tail-feathers cinereous brown; the others dusky; the outer one being white in its greater portion, the next tipped with white: bill dusky: irides hazel: feet brown: hind claw curved, about equal to the toe in length. (*Young.*) The olivaceous tinge of the upper parts more conspicuous: under parts yellowish, spotted as in the adult bird: wing-coverts more broadly edged with pale cinereous brown: the white on the outer tail-feathers less pure, inclining to pale dusky. (*Egg.*) Greenish white ground, speckled with ash-brown, darkest at the larger end: long. diam. nine lines and a half; trans. diam. seven lines and a half.

Common on rocky shores, in many parts of the kingdom, where it resides the whole year, but is never found inland. Feeds on marine insects. Song and habits somewhat similar to those of the two last species. Breeds in the clefts of rocks, sometimes at a considerable height from the ground. Nest composed of marine plants, with the addition of a little moss externally, lined with fine grass and a few long hairs. Eggs four or five; hatched early in the Spring.

GEN. 23.   SAXICOLA, *Bechst.*

70. S. *Œnanthe,* Bechst. (*Wheat-Ear.*)—Upper parts ash-gray: forehead, throat, streak above the eyes, and basal portion of the tail, white.

S. Œnanthe, *Temm. Man. d'Orn.* tom. I. p. 237. Wheat-Ear, *Mont. Orn. Dict. Selb. Illust.* vol. I. p. 181. pl. 48. f. 1. *Bew. Brit. Birds,* vol. I. p. 264.

DIMENS. Entire length six inches six lines: length of the bill (from the forehead) six lines, (from the gape) ten lines; of the tarsus eleven lines and a half; of the tail two inches six lines; from the carpus to the end of the wing four inches: breadth, wings extended, twelve inches.

DESCRIPT. (*Male in summer.*) Upper parts of the body bluish ash: from the corners of the bill a black band, passing across the eyes, and extending to the ear-coverts; above the black, a white streak; forehead and throat white: wings brownish black: fore part of the neck, and upper part of the breast, reddish yellow; the rest of the under parts white: tail, with the exception of the two middle feathers which are wholly black, white for two thirds of its length from the base; the remaining portion black: bill and feet black. (*Female.*) Upper parts cinereous brown, tinged with yellow: the band across the eyes brown: quills with

pale edges: less white on the tail. (*Male, after the autumnal moult.*)
All those parts above, which in summer are bluish gray, tinged with
rufous: rump, and streak above the eye, pure white: lore pure black;
but the black on the ear-coverts tinged with rufous: wing-coverts, and
secondary quills, broadly edged with rufous; primaries tipped with reddish white: chin white; throat, breast, and sides of the neck, decided
rufous; rest of the under parts paler, but still tinged with rufous. The
female, at this season, resembles the male, but is more rufous, with the
colours not so well-defined: the black streak through the eye indistinct;
the white one above it dull and inclining to rufous. (*Young of the year.*)
General plumage closely resembling that of the female in autumn: the
band across the eyes very indistinct. (*Egg.*) Uniform delicate pale grey
blue: long. diam. ten lines and a half; trans. diam. seven lines and a half.

A migratory species, making its first appearance about the end of
March, and remaining till September. Common on open downs, and in
uninclosed districts. Builds on the ground, under stones, or in old
rabbit-burrows. Nest composed of moss and grass, and lined with wool
or hair. *Obs.* Previously to quitting this country, it assumes the
autumnal plumage, in which state it does not appear to have been
noticed by Ornithologists.

71. S. *Rubetra*, Bechst. (*Whin-Chat.*)—Upper plumage
dusky brown, edged with reddish yellow: streak above the
eyes, a spot on the wings, and base of the tail, white.

S. Rubetra, *Temm. Man. d'Orn.* tom. I. p. 244. Whin-Chat, *Mont.
    Orn. Dict. Selb. Illust.* vol. I. p. 183. pl. 48. f. 2. *Bew. Brit.
    Birds,* vol. I. p. 266.

DIMENS. Entire length five inches: length of the bill (from the forehead) five lines, (from the gape) eight lines; of the tarsus ten lines; of
the tail two inches; from the carpus to the end of the wing two
inches ten lines and a half; breadth, wings extended, eight inches
eleven lines.

DESCRIPT. (*Male.*) Crown of the head, nape, back, and scapulars, dusky
brown, the feathers broadly edged with reddish yellow: from the base of
the upper mandible a white streak, passing above the eye, and extending
considerably beyond it: cheeks and ear-coverts blackish brown: throat
and breast pale yellowish red, edged laterally with white; rest of the
under parts white, with a faint tinge of yellow: wings brown, with a
large oblong white patch near the shoulder, and another smaller one of
the same colour on the greater coverts of the primaries: rump and upper
tail-coverts yellowish brown streaked with dusky: tail short; the basal
half white, the remainder dusky brown; the two middle feathers entirely
dusky: bill and feet black. (*Female.*) The rufous edging of the upper
plumage broader and more conspicuous: less white on the wing: all the
under parts yellowish white, inclining to rust-red on the breast: upper
tail-coverts rust-red. (*Egg.*) Uniform bluish green, with specks (some
very minute) of dull reddish brown: long. diam. nine lines; trans. diam.
six lines and a half.

Migratory like the last species: seldom appears before the middle
of April. Haunts moors and commons. Nest placed on the ground,
artfully concealed, constructed almost wholly of dried grasses and
stalks, the coarser parts being on the outside, and the finer blades
forming a lining within. Eggs five or six in number; hatched towards
the end of May.

72. S. *Rubicola*, Bechst. (*Stone-Chat.*) — Head and throat black; back black, the feathers with reddish edges: sides of the neck, a spot on the wings, and rump, white.

> S. Rubicola, *Temm. Man. d'Orn.* tom. i. p. 246.   Stone-Chat, *Mont. Orn. Dict. Selb. Illust.* vol. i. p. 185. pl. 48. f. 3 & 4. *Bew. Brit. Birds*, vol. i. p. 268.

DIMENS. Entire length five inches one line: length of the bill (from the forehead) five lines, (from the gape) eight lines; of the tarsus ten lines; of the tail two inches; from the carpus to the end of the wing two inches six lines: breadth, wings extended, eight inches five lines.

DESCRIPT. (*Male in Summer.*) Head and throat black; on each side of the neck a large white spot; back and nape black, some of the feathers on these parts faintly edged with reddish; quills dusky with pale edges; a large white spot on the greater coverts next the body: breast rust-red; the rest of the under parts white, tinged with yellow: rump white: tail black: bill and feet black. In the *winter plumage*, the black on the head and throat is mixed with cinereous brown. (*Female.*) Head and upper parts dusky brown; the feathers edged with yellowish red: quills and tail deep brown, with paler edges: throat dusky, with whitish and reddish spots: the white on the sides of the neck and on the wings not so conspicuous as in the male; the red on the breast paler. (*Young of the year before the first moult.*) Crown of the head dusky brown, with fine yellowish white streaks on the shafts of the feathers; rest of the upper parts as in the adult female, but with a few whitish spots on the nape and top of the back: wing-coverts and scapulars broadly edged with reddish yellow; upper tail-coverts pale rust-colour: throat whitish, speckled with dusky brown. (*Egg.*) Pale gray blue; the larger end speckled with dull reddish brown: long. diam. eight lines and a half; trans. diam. seven lines.

Found in the same places with the preceding species, but remains throughout the year. Habits similar. Nest on the ground; composed of moss and dry bents, and lined with hair or feathers. Eggs five or six; hatched early in the season.

GEN. 24.   PARUS, *Linn.*

(1. PARUS, *Leach.*)

73. P. *major*, Linn. (*Great Titmouse.*)—Head, throat, lower part of the neck, and mesial line of the abdomen, black; cheeks, and a spot on the nape, white.

> P. major, *Temm. Man. d'Orn.* tom. i. p. 287.   Great Titmouse, *Mont. Orn. Dict. Selb. Illust.* vol. i. p. 233. pl. 51. f. 1. *Bew. Brit. Birds*, vol. i. p. 272.

DIMENS. Entire length five inches eleven lines: length of the bill (from the forehead) five lines and a half, (from the gape) seven lines; of the tarsus nine lines and a half; of the tail two inches nine lines; from the carpus to the end of the wing three inches one line; breadth, wings extended, nine inches three lines.

DESCRIPT. Head, throat, fore part of the neck, and a longitudinal streak down the middle of the abdomen, black: cheeks and ear-coverts, and a spot on the nape, white: back olive-green: rump bluish ash:

breast and belly, on each side of the black line, yellow: quills dusky;
wing-coverts bluish ash, the larger tipped with white: tail dusky ash;
the outer feather with the exterior web white; the next tipped with
white: under tail-coverts white: bill black: legs lead-colour. In the
female, the colours are not so vivid, and the abdominal black streak does
not extend so low down. (*Egg.*) White, spotted and speckled with pale
red: long. diam. nine lines and a half; trans. diam. seven lines.

Common in woods and gardens throughout the country. Has a great
variety of notes, which are heard very early in the year. Feeds principally on insects, but will also devour grain, as well as flesh. Breeds in
the hollows of decayed trees, or in old walls. Nest constructed chiefly of
moss, and lined with feathers. Eggs very numerous.

74. P. *cæruleus*, Linn. (*Blue Titmouse.*)—Crown blue, encircled with white: cheeks white, bordered with black, and dark blue.

P. cæruleus, *Temm. Man. d'Orn.* tom. i. p. 289. Blue Titmouse,
Mont. Orn. Dict. Selb. Illust. vol. i. p. 235. pl. 51. f. 2. *Bew.
Brit. Birds*, vol. i. p. 276.

Dimens. Entire length four inches four lines: length of the bill four
lines; of the tarsus eight lines; of the tail one inch eleven lines; from
the carpus to the end of the wing two inches four lines; breadth, wings
extended, seven inches eight lines.

Descript. Forehead white, the white extending backwards in the
form of a narrow band above the eyes, and reaching to the occiput; crown
of the head bright blue: cheeks white, bounded above by a narrow black
line commencing at the bill and passing through the eyes, beneath by
another line of dark blue extending from the chin to the hinder part of
the neck where it becomes broader: back yellowish green: throat, and
longitudinal streak on the middle of the abdomen, deep blue; breast and
sides bright yellow: wings and tail blue; secondaries and greater coverts
tipped with white: bill and legs bluish gray. In the *female*, the blue
colours are more obscure, and the abdominal streak not so well defined.
(*Egg.*) White, spotted with pale red: long. diam. seven lines and a half;
trans. diam. six lines.

Equally common with the last species. Habits and food similar. Nest
placed in the holes of trees; formed of moss, and lined with hair and
feathers.

75. P. *cristatus*, Linn. (*Crested Titmouse.*)—Crown-feathers elongated, forming a crest, black, edged with white: cheeks and sides of the neck dull white: throat, a streak on the temples, and collar, black.

P. cristatus, *Temm. Man. d'Orn.* tom. i. p. 290. Crested Titmouse,
Mont. Orn. Dict. Selb. Illust. vol. i. p. 243. pl. 43. f. 5. *Bew.
Brit. Birds*, vol. i. p. 274.

Dimens. Entire length four inches eight lines: length of the bill
(from the forehead) three lines and a half, (from the gape) five lines;
of the tarsus nine lines; of the tail one inch eight lines; from the carpus
to the end of the wing two inches six lines.

Descript. Feathers on the crown and forehead long and pointed, and capable of erection; of a black colour, edged with white: cheeks and sides of the neck dull white, speckled with black; beneath the eye, and across the ear-coverts, a black streak; throat, upper part of the neck, and a narrow line bounding the white on the sides, deep black: back, and other upper parts, greenish brown: breast and belly white, tinged with buff: bill black: feet bluish gray. The *female* has less black on the throat, and the crown-feathers not so much elongated. (*Egg.*) White, with a few spots and specks of pale red: long. diam. seven lines; trans. diam. six lines.

Confined to Scotland; but stated to be not uncommon amongst the large tracts of pines in the northern parts of that country, particularly in the forest of Glenmoor. Is said to be of solitary habits, and to build in the holes of trees.

76. P. *palustris*, Linn. (*Marsh Titmouse.*)—Crown, throat, and nape, deep black : cheeks yellowish white.

P. palustris, *Temm. Man. d'Orn.* tom. I. p. 291. Marsh Titmouse, *Mont. Orn. Dict. Selb. Illust.* vol. I. p. 237. pl. 51. f. 4. *Bew. Brit. Birds*, vol. I. p. 282.

Dimens. Entire length four inches six lines; length of the bill (from the forehead) four lines, (from the gape) four lines and a half; of the tarsus eight lines; of the tail two inches two lines; from the carpus to the end of the wing two inches four lines and a half: breadth, wings extended, seven inches ten lines.

Descript. Crown of the head, nape, and throat, deep black; cheeks yellowish white: upper parts of the body gray tinged with brown: wings and tail bluish gray, the edges of the feathers somewhat paler: breast and belly white, faintly tinged with grayish brown: bill black, stronger and more conical than in the next species: feet bluish gray. (*Egg.*) White, with a few dark red spots: long. diam. seven lines and a half; trans. diam. six lines.

Not an uncommon species in woods and thickets. Has a loud shrill note, repeated quickly several times in succession, heard very early in the year. Food insects and seeds, occasionally flesh. Builds in the holes of decayed trees: nest composed of moss and the seed-down of the willow, and lined with the last mentioned material. Eggs from six to eight in number.

77. P. *ater*, Linn. (*Cole Titmouse.*) — Head, throat, and fore part of the neck, black ; cheeks, and a spot on the nape, white: two transverse white bars on the wings.

P. ater, *Temm. Man. d'Orn.* tom. I. p. 288. Cole Titmouse, *Mont. Orn. Dict. Selb. Illust.* vol. I. p. 239. pl. 51. f. 3. *Bew. Brit. Birds*, vol. I. p. 278.

Dimens. Entire length four inches three lines: length of the bill (from the forehead) three lines and a half, (from the gape) four lines and a half; of the tarsus seven lines and three quarters; of the tail one inch seven lines and a half; from the carpus to the end of the wing two inches three lines.

Descript. Crown of the head, and nape of the neck, black; the latter with a central white spot: cheeks and sides of the neck white:

throat and fore part of the neck black: upper parts of the body cine-reous, inclining to yellowish on the rump: greater and lesser coverts tipped with white, forming two transverse bars across the wings when closed: belly and abdomen grayish white: tail relatively shorter than in the last species, and slightly forked. The *female* has less white on the sides of the neck, and less black on the throat. (*Egg.*) White, with numerous pale red spots: long. diam. seven lines and a half; trans. diam. six lines.

Not so abundant as the last species. Confined to woods and exten-sive plantations. Feeds principally on insects. Builds in holes. Nest constructed of moss and wool, and lined with hair.

(2. MECISTURA, *Leach.*)

78. P. *caudatus*, Linn. (*Long-tailed Titmouse.*) — Forehead, crown, cheeks, and throat, white : a broad streak across the eye, nape, and back, black ; scapulars reddish.

> P. caudatus, *Temm. Man. d'Orn.* tom. I. p. 296. Long-tailed Tit-mouse, *Mont. Orn. Dict. Selb. Illust.* vol. I. p. 241. pl. 51. f. 5. *Bew. Brit. Birds*, vol. I. p. 279.

DIMENS. Entire length five inches nine lines: length of the bill (from the forehead) two lines and a half, (from the gape) three lines and a half; of the tarsus eight lines; of the tail three inches six lines; from the carpus to the end of the wing two inches four lines: breadth, wings extended, six inches eight lines.

DESCRIPT. Forehead, crown, cheeks and throat, pure white; over each eye a broad black streak, which extends backward to the nape, where it unites with its fellow to form a broader band of the same colour down the middle of the back: sides of the back, and scapulars, rose-red: quills black; greater coverts dusky ash; these last, as well as the secondary quills, edged with white: belly, sides, and abdomen, whitish, tinged with red: tail cuneated, and very long; the six middle feathers black; the others tipped, and obliquely edged on their outer webs, with white: bill black, very short: irides brown: feet dusky. *Obs.* According to Temminck, it is only the *female* which possesses the black streak over the eyes, but as far as my observation goes it is common to both sexes, nor did I ever see a specimen in which it was wanting. *Young* birds have the white parts more or less mottled with brown and dusky; and the black on the back not so deep and well-defined. (*Egg.*) White, with a few pale red specks; frequently quite plain : long. diam. seven lines; trans. diam. five lines.

Far from uncommon in woods and shrubberies, in most parts of the kingdom. Feeds entirely on insects. Nest placed generally in thick bushes; of a very elegant form; long and oval, covered at top, with a small aperture at the side; constructed principally of moss and wool, studded externally with lichens, and lined with a profusion of soft fea-thers. Eggs ten to twelve in number. The young, when fledged, follow the parent birds till the ensuing spring.

GEN. 25.   CALAMOPHILUS, *Leach.*

79. C. *biarmicus*, Leach. (*Bearded Titmouse.*)—Crown and occiput bluish ash ; throat white ; beneath the eye a black pointed moustache.

Parus biarmicus, *Temm. Man. d'Orn.* tom. I. p. 298.   Bearded Titmouse, *Mont. Orn. Dict.   Selb. Illust.* vol. I. p. 244. pl. 51. f. 6. *Bew. Brit. Birds,* vol. I. p. 283.

DIMENS.   Entire length six inches four lines : length of the bill (from the forehead) three lines and a half, (from the gape) five lines ; of the tarsus nine lines and a half ; of the tail three inches one line ; from the carpus to the end of the wing two inches three lines.

DESCRIPT. (*Male.*) Head and occiput bluish gray : between the bill and the eye a tuft of loose pendent black feathers, forming a pointed moustache on each side of the face ; throat and fore part of the neck white ; breast and belly tinged with rose-red : nape, back, and rump, light rufous orange ; scapulars paler : quills dusky, the primaries edged with white ; greater coverts deep black, edged with red on the outer webs, and with reddish white on the inner : under tail-coverts deep black : tail long and cuneated, orange-brown ; the outer feathers tipped, and edged externally, with grayish white : bill orange : irides yellow : feet black. (*Female.*) No black moustache on the sides of the neck ; crown, and upper parts of the body, yellowish brown, spotted with dusky : under tail-coverts of the same colour with the abdomen. (*Egg.*) White, with a few minute specks and streaks of dark red : long. diam. eight lines and a half ; trans. diam. six lines and a half.

A local species confined to fenny districts, and extensive marshes, where there is abundance of reeds.   Found near Winchelsea in Sussex, between Erith and London, and also in some parts of Gloucestershire, Lancashire, Cambridgeshire, and Norfolk.   Nest placed near the ground, fixed to the stems of coarse grasses ; composed entirely of dried bents, the finer ones forming the lining.   Eggs four to six in number.   Said to feed on land mollusca, especially the *Succinea amphibia,* as well as insects.

GEN. 26.   BOMBYCILLA, *Briss.*

80. B. *garrula*, Bonap. (*Bohemian Wax-Wing.*)—Head and upper part of the body purplish red ; the feathers on the crown elongated ; throat black : secondaries and greater coverts tipped with white.

Bombycivora garrula, *Temm. Man. d'Orn.* tom. I. p. 124.   Bohemian Wax-Wing, *Selb. Illust.* vol. I. p. 268. pl. 34*.   Bohemian Chatterer, *Mont. Orn. Dict.*   Chatterer, *Bew. Brit. Birds,* vol. I. p. 98.

DIMENS.   Entire length seven inches eleven lines.

DESCRIPT.   Crown-feathers elongated, forming a pendent crest, of a pale brown purplish red colour : nostrils covered with small black feathers : throat and region of the eyes deep black : all the upper and under parts of the body purplish red, with a tinge of ash-gray, being darkest above : greater wing-coverts black, tipped with white : quills black ; some of the primaries with a *yellow*, others, and all the secondaries, with a *white* spot at the extremity of the outer web: six or eight of the secondaries with the shafts terminating in a small, oblong, flat, cartilaginous process, of a bright red colour, and having the appearance of wax: vent and under tail-coverts reddish orange: tail black, tipped with yellow: bill and feet black: irides purplish red.　In the *female*, there is less black on the throat, and the wax-like appendages are not so numerous: the yellow on the wings, and extremity of the tail, is also not so bright.　In *young* birds the waxen tags do not appear at all till after the first moult; from that time they gradually increase in number as the age of the individual advances.　(*Egg.*) Pale blue, with a few specks of ash-gray and black: long. diam. eleven lines; trans. diam. seven lines.

A very uncommon visitant in this country; making its appearance at irregular intervals.　Occasionally observed in small flocks during the winter season.　Feeds on heps, and the berries of the mountain-ash.

GEN. 27. ALAUDA, *Linn.*

81. A. *alpestris*, Linn. (*Shore Lark.*)—Upper parts reddish brown ; streak above the eye, moustache, and a broad transverse bar on the breast, black.

A. alpestris, *Temm. Man. d'Orn.* tom. I. p. 279.　Shore Lark, *Wils. Amer. Orn.* vol. I. p. 85. pl. 5. f. 4.　*Penn. Arct. Zool.* vol. II. p. 84.

DIMENS.　Entire length six inches ten lines.　TEMM.

DESCRIPT.　(*Male.*) Forehead, throat, and sides of the head behind the eyes, pale yellow; a streak above the eye, continued over the forehead and bounding the yellow on that part, a moustache from the corners of the bill across the cheeks, and a broad transverse bar on the upper part of the breast, deep black: upper parts of the body, wing-coverts, and sides of the breast, reddish brown: quills dusky, edged internally with white: tail with the two middle feathers brown; the rest black; the outer one edged externally with white: lower part of the breast and flanks pale red; belly and abdomen white: bill and feet black. (*Female.*) Forehead yellowish; crown of the head varied with black and brown; the black parts with fine yellowish streaks: the transverse bar on the breast smaller than in the other sex; the black feathers in the tail tipped with whitish. (*Egg.*) White, spotted with black.

A single specimen of this Lark, the only one that has hitherto occurred in this country, was killed on the beach near Sherringham in Norfolk, in March 1830.　The species inhabits the northern parts of Europe, Asia, and America.　Said to frequent extensive plains, and to feed on insects and seeds.　Nest placed on the ground: eggs four or five in number.

(5.) *A. rubra*, Gmel. Syst. vol. II. p. 794.　*Red Lark*, Lewin, Brit. Birds, vol. III. pl. 93.　Mont. Orn. Dict. and Supp. App.

An obscure species, if indeed it be a species, of which very little is known.　Montagu says, "Taken in the Winter of 1812, near Woolwich, in a net with other larks."　The specimen, however, is not in his collection in the British Museum, nor does one exist in any other with which I am acquainted.

82.   A. *arvensis*, Linn.   (*Sky-Lark.*) — Upper parts
reddish brown ; neck and breast reddish white with dusky
spots.

A. arvensis, *Temm. Man. d'Orn.* tom. I. p. 281.   Sky-Lark, *Mont.
Orn. Dict.   Selb. Illust.* vol. I. p. 273. pl. 50. f. 1.   *Bew. Brit.
Birds,* vol. I. p. 213.

DIMENS.   Entire length seven inches three lines : length of the bill
(from the forehead) five lines and a half, (from the gape) eight lines; of
the tarsus eleven lines and a half; of the hind toe, claw included, one
inch one line and a half; of the tail two inches eight lines; from the
carpus to the end of the wing four inches four lines : breadth, wings
extended, thirteen inches six lines.

DESCRIPT.   All the upper parts reddish brown, with a dusky spot in
the middle of each feather; these colours palest on the nape, darkest on
the head and upper part of the back : a whitish streak above the eyes;
cheeks pale yellowish brown : throat yellowish white; fore part of the
neck, and breast, tinged with reddish yellow, and spotted with blackish
brown; sides with dark streaks of this last colour; middle of the abdomen
white : quills edged externally with reddish; the secondaries notched at
their extremities : outer tail-feathers with the tip and exterior web white;
the two next simply edged with white : bill dusky ; the base of the lower
mandible yellowish : feet yellowish : hind claw nearly straight, longer
than the toe.   (*Egg.*) Greenish white ground, spotted all over with
darker green and ash-brown : long. diam. eleven lines; trans. diam. eight
lines and a half.

An abundant, and widely dispersed species.   Frequents open country,
more especially arable and cultivated lands.   Sings in its ascent, and
also whilst suspended at a height in the air.   Commences its song, in the
middle of Summer, as early as two A.M.   Breeds in May.   Nest placed
on the ground; composed of stalks and dry grasses, and lined with the
finer blades.   Eggs four or five in number.   Occasionally two broods in
the season.   Congregates in large flocks during the Winter.

83.   A. *arborea*, Linn.   (*Wood-Lark.*)—Upper parts
reddish brown : neck and breast yellowish white, with
dusky spots : a distinct yellowish white band above the
eyes, reaching from the bill to the occiput.

A. arborea, *Temm. Man. d'Orn.* tom. I. p. 282.   Wood-Lark, *Mont.
Orn. Dict.   Selb. Illust.* vol. I. p. 276. pl. 50. f. 2.   *Bew. Brit.
Birds,* vol. I. p. 222.

DIMENS.   Entire length six inches five lines : breadth, wings extended,
twelve inches six lines.

DESCRIPT.   Distinguished from the last species, by its somewhat
smaller size, shorter tail, and very distinct yellowish white streak passing from the bill above the eyes and reaching nearly to the occiput.
Upper plumage light reddish brown, with dusky spots : crown-feathers
long, and capable of erection : ear-coverts pale brown ; cheeks yellowish
white : neck and breast yellowish white, with longitudinal dusky spots ;
belly dirty white : quills dusky, slightly edged with brown ; wing-coverts
tipped with white : the two middle tail-feathers brown ; the next two

dusky; the others black, with white tips; the outermost of all white on the exterior web: bill brownish black; the base of the lower mandible yellowish white: feet flesh-colour, tinged with yellowish brown; hind claw very long, and nearly straight. (*Egg.*) Light greenish brown ground, mottled with darker brown, with a few dusky streaks at the larger end: long. diam. ten lines; trans. diam. seven lines.

Much less generally distributed than the last species: not uncommon in the southern and western parts of England, but rare in the eastern. Habits similar to those of the Sky-Lark. Sings at a great height in the air, flying round and round in large irregular circles. In song nearly the whole year. Nest on the ground, constructed principally of dry grasses, and lined with a few hairs. Breeds very early in the season. Does not congregate in Winter.

GEN. 28. EMBERIZA, *Linn.*

(1. PLECTROPHANES, *Meyer.*)

84. E. *Lapponica*, Nilss. (*Lapland Bunting.*)—Crown of the head, and upper part of the breast, black; above the eye a white band, prolonged on the sides of the neck: nape, back, and scapulars, varied with brown and red.

E. calcarata, *Temm. Man. d'Orn.* tom. I. p. 322. Plectrophanes Lapponica, *Selb. in Linn. Trans.* vol. xv. p. 156. pl. 1. (Young.) *Faun. Bor. Amer.* part 2. p. 248. pl. 48. (Adult.) Lapland Finch, *Lath. Syn.* vol. II. p. 263. Lapland Lark-Bunting, *Selb. Illust.* vol. I. p. 283. pl. 100. f. 6.

DIMENS. Entire length six inches eight lines: length of the bill (from the forehead) six lines, (from the gape) six lines and a quarter; of the tarsus ten lines; of the hind toe, claw included, nine lines and a quarter; of the tail two inches nine lines; of the folded wing three inches nine lines. *Faun. Bor. Amer.*

DESCRIPT. (*Young.*) "Bill yellowish brown, palest towards the base of the under mandible: head, and all the upper parts of the body, pale wood-brown, tinged with yellowish gray; the shafts of the feathers blackish brown: greater wing-coverts, and secondary quills, blackish brown, deeply margined with chestnut or orange-brown; the tips white: quills dusky brown, paler at the edge: above the eyes a broad streak of pale wood-brown: cheeks and ear-coverts wood-brown, the latter mixed with black: from the corners of the under mandible, on each side of the throat, a streak of blackish brown: throat yellowish white: lower part of the neck and breast dirty white, with numerous dusky spots: belly and vent white: flanks with oblong dusky streaks: tail dusky, the outer feather with the exterior web, and half of the interior, dirty white: the second with a small wedge-shaped white spot near the tip: legs and toes brown: claws not much hooked, the posterior nearly straight, and longer than the toe." SELBY. In the *adult male*, which has not yet occurred in this country, the head, chin, throat, and upper part of the breast, are velvet black, with a broad whitish band down the sides of the neck, commencing above the eyes and dilating into an open space in about the middle of its course: nape bright chestnut; rest of the upper plumage pale reddish brown, with a blackish streak in the middle of each feather: wing-coverts with two obsolete white bands; primaries hair-brown, their exterior edges whitish : belly and under tail-coverts dusky white ; sides of the breast, and flanks, spotted with black : bill bright lemon-yellow, tipped with black : legs pitch-black. The *female* differs in having the chin grayish, the black plumage of the head and breast edged with pale brown and gray, and the chestnut feathers of the nape fringed with white : the white band duller. (*Egg.*) Pale ochre-yellow, spotted with brown. *Faun. Bor. Amer.*

Of this species, which is a native of high northern latitudes, only four individuals have hitherto occurred in this country : the first of these was taken in Cambridgeshire; the second in the neighbourhood of Brighton : the third near London, in Sept. 1828; the fourth near Preston in Lanca-shire, in Oct. 1833. All the specimens were in immature plumage. Habits said to resemble those of the Larks. Food principally seeds. Nest composed externally of dry stems of grass, and lined with hair. Eggs usually seven in number.

85. E. *nivalis*, Linn. (*Snow-Bunting.*)—Head, neck, a large space on the wings, and all the under parts, white.

E. nivalis, *Temm. Man. d'Orn.* tom. i. p. 319. Snow-Bunting, *Mont. Orn. Dict. and Supp. Selb. Illust.* vol. i. p. 279. pl. 52. f. 5. *Bew. Brit. Birds,* vol. i. p. 178.

DIMENS. Entire length six inches eight lines: length of the bill (from the forehead) four lines and a half, (from the gape) five lines and a half : of the tarsus ten lines; of the hind toe, claw included, eight lines; of the tail two inches eight lines ; from the carpus to the end of the wing four inches four lines and a half: breadth, wings extended, twelve inches four lines.

DESCRIPT. (*Male in Summer.*) Head, neck, rump, wing-coverts, upper half of the quills, and all the under parts, pure white : back, scapulars, bastard wing, the lower portion of the quills, and the three secondaries nearest the body, black : the three outer tail-feathers white, with a black spot near the tip; the next with the upper portion of the outer web white : the rest black : bill yellow, dusky at the tip : irides deep brown : feet black. (*Female, and male in Winter.*) Head and neck white, but deeply tinged on the occiput, nape, and ear-coverts, with brownish red ; feathers on the back and scapulars, and three secondaries nearest the body, black, all broadly edged and tipped with reddish ash : greater wing-coverts white, stained with dusky towards the tips of the feathers : breast, rump, and upper tail-coverts, tinged more or less deeply with brownish red : the middle tail-feathers edged, chiefly towards their extremities, with red-dish white : belly and abdomen white. (*Young of the year.*) Crown of the head yellowish brown : forehead, ear-coverts, and a transverse band on the breast, deep reddish brown ; flanks the same, but paler : nape, throat, and fore part of the neck, cinereous, with a tinge of red-dish; the feathers on the back, rump, and scapulars, dusky, with a broad edging of yellowish red ; middle of the wing, and under parts, whitish ; quills and middle tail-feathers dusky, edged with reddish ; the three outer tail-feathers whitish, with a large black spot towards the tip : bill yellowish : feet black. (*Egg.*) Pale reddish white, slightly speckled and spotted with pale red and purple brown : long. diam. eleven lines ; trans. diam. seven lines and a half.

A regular winter visitant in the northern parts of the island, first appearing about the middle of October; but very rare, and only of accidental occurrence, in the southern and midland counties. Breeds within the arctic circle: nest said to be placed within the fissures of rocks. Habits terrestrial: does not perch on trees. Food seeds and insects. *Obs.* The *Emberiza mustelina* and *E. montana* of authors are this species, either in its winter plumage, or in an immature state.

(2. EMBERIZA, *Meyer.*)

86. E. *Miliaria*, Linn. (*Common Bunting.*)—Upper parts yellowish brown, with darker spots : beneath yellowish white, spotted and streaked with black.

E. Miliaria, *Temm. Man. d'Orn.* tom. I. p. 306. Common Bunting, *Mont. Orn. Dict. Selb. Illust.* vol. I. p. 286. pl. 52. f. 1. *Bew. Brit. Birds,* vol. I. p. 168.

DIMENS. Entire length seven inches six lines: length of the bill (from the forehead) six lines and one-third, (from the gape) seven lines and one-third ; of the tarsus one inch one line ; of the tail two inches eleven lines ; from the carpus to the end of the wing three inches nine lines : breadth, wings extended, twelve inches three lines.

DESCRIPT. Upper parts yellowish brown, inclining to olive, the shafts of the feathers with longitudinal dusky streaks : under parts yellowish white, spotted on the breast, and streaked on the sides, with dark brown : quills and tail-feathers dusky, with pale edges ; wing-coverts tipped with yellowish white : upper mandible dusky, under one pale : irides dark hazel : feet pale brown. *Obs.* This species is occasionally found white, or pied. (*Egg.*) Reddish white, or purple red ground, streaked and spotted with dark purple brown : long. diam. one inch ; trans. diam. eight lines and a half.

Common throughout the kingdom, particularly in open cultivated lands. Note harsh and inharmonious, heard at intervals during a great part of the year. Food principally seeds. Nest placed on the ground, formed of straw and fibrous roots, and lined with fine grass and hairs. Congregates in the winter season.

87. E. *Schœniclus*, Linn. (*Reed-Bunting.*) — Head, throat, and fore part of the neck, black : back and scapulars reddish brown, with dusky streaks.

E. Schœniculus, *Temm. Man. d'Orn.* tom. I. p. 307. Reed-Bunting, *Mont. Orn. Dict. Selb. Illust.* vol. I. p. 290. pl. 52. f. 3. *Bew. Brit. Birds,* vol. I. p. 176.

DIMENS. Entire length six inches: length of the bill (from the forehead) four lines and a half, (from the gape) five lines and a half ; of the tarsus ten lines ; of the tail two inches seven lines ; from the carpus to the end of the wing three inches one line : breadth, wings extended, nine inches eleven lines.

DESCRIPT. (*Male in Summer.*) Crown of the head, occiput, cheeks, throat, and upper part of the breast, deep black : a white streak from the corner of the bill downwards joining a collar of the same colour encircling the nape and sides of the neck : back, scapulars, and wing-coverts, reddish brown, with a dusky spot in the middle of each feather : rump and upper tail-coverts bluish ash, tinged with brown : sides of the breast, belly, and under tail-coverts, white ; flanks with a few longitudinal dusky streaks : quills dusky, edged with reddish : tail black ; the two middle feathers broadly edged with rufous; the two outer ones on each side obliquely marked with white towards their extremities, the shafts and tips black: bill black; feet dusky. In *Winter* the black feathers on the head, throat, and breast, are edged with reddish brown. (*Female.*) Crown of the head reddish, with dusky spots: above the eye a pale streak of yellowish brown: throat whitish, bordered on each side by a black line: breast and flanks tinged with reddish, and spotted with dusky: nape and sides of the neck brownish ash: the rest much as in the male bird. The *young* of both sexes are very similar to the adult female. (*Egg.*) Purple white ground, sparingly streaked with dark purple brown: long. diam. nine lines and a half; trans. diam. seven lines.

Common in marshy districts, and by the sides of rivers, in most parts of the kingdom. Feeds on the seeds of aquatic plants, and occasionally on aquatic insects. Song simple and inharmonious. Nest generally placed on the ground, and concealed amongst rushes; constructed of stalks, and other dry vegetable substances, and lined with fine grass, and a scanty supply of long hairs. Eggs four or five in number; incubation commencing about the first week in May. Occasionally a second brood in July.

88.   E. *Citrinella*, Linn. (*Yellow Bunting.*) — Head, neck, and upper part of the breast, yellow : back and scapulars reddish brown, with darker spots.

E. Citrinella, *Temm. Man. d'Orn.* tom. i. p. 304.   Yellow Bunting, *Mont. Orn. Dict.   Selb. Illust.* vol. i. p. 288. pl. 52. f. 2.   *Bew. Brit. Birds,* vol. i. p. 172.

Dimens. Entire length six inches four lines: length of the bill (from the forehead) five lines, (from the gape) seven lines; of the tarsus nine lines and a half; of the tail two inches eleven lines; from the carpus to the end of the wing three inches three lines: breadth, wings extended, ten inches three lines.

Descript. (*Male.*) Head, cheeks, fore part of the neck, abdomen, and under tail-coverts, gamboge-yellow: breast and sides streaked with brownish orange: back and scapulars reddish brown, tinged with olive, with a dusky spot in the centre of each feather; rump bright chestnut, the feathers edged with grayish white: tail dusky; the two outer feathers with a large white conical spot on their inner webs: feet yellowish brown. (*Female.*) The yellow colours much less vivid; that on the head, neck, and throat, marked with olivaceous brown spots; the under parts more clouded and streaked than in the male bird. (*Egg.*) Pale purplish white, streaked and speckled with dark red brown: long. diam. ten lines and a half; trans. diam. eight lines.

Abundant, and generally distributed. Feeds principally on grain. Song scarcely more varied than that of the last species; heard from the first week in February to the beginning or middle of August. Nest placed on or near the ground; composed of straw and dried herbage, and lined with fine grass and long hair. Eggs three to five in number. Breeds late; incubation seldom commencing before the beginning or middle of May. Young fledged about the second week in June. Congregates in Winter.

89.   E. *Cirlus*, Linn. (*Cirl Bunting.*)—Head and nape olivaceous green, with dusky spots : throat, and a narrow band across the eye, black : a yellow gorget on the neck.

E. Cirlus, *Temm. Man. d'Orn.* tom. I. p. 313.   Cirl Bunting, *Mont. Orn. Dict. and Supp. with fig. Id. in Linn. Trans.* vol. VII. p. 276.   *Selb. Illust.* vol. I. p. 292. pl. 52. f. 4.   *Bew. Brit. Birds,* vol. I. p. 174.

DIMENS. Entire length six inches six lines. MONT.
DESCRIPT. (*Male in Summer.*) Crown of the head, and nape of the neck, olivaceous green, with a dusky streak in the centre of each feather; throat, and a narrow band through the eye reaching from the bill to the ear-coverts, black; streak above the eye, and another below it, bright primrose-yellow; a gorget of the same colour encircling the lower part of the neck: breast, immediately below the gorget, yellowish gray, inclining to olive green: back and scapulars fine chestnut brown, the former streaked with dusky; rump olivaceous brown: belly and vent bright yellow; sides of the breast and abdomen brownish red: tail with the two middle feathers chestnut brown; the rest black, the two outer ones obliquely marked with white: bill bluish gray: feet brown, with a tinge of flesh red. In *Winter*, the black on the throat is more dull, the feathers edged with pale yellow. (*Female.*) Head and nape olivaceous brown, with numerous spots of a darker colour; above the eye a dull yellow streak, passing down the sides of the head: chin whitish; the rest of the under parts pale yellow; the breast spotted with reddish; the sides, and under tail-coverts, with large dusky streaks: upper parts as in the male bird, but the colours not so bright, and the spots on the back larger. (*Egg.*) Dirty white, streaked and speckled with dark liver-brown: long. diam. ten lines; trans. diam. eight lines.
Not uncommon in Devonshire, and in one or two of the adjoining counties, where it was first observed by Montagu. Is also found in the Isle of Wight, but is scarcely known in other parts of England. Song and habits somewhat similar to those of the last species. Nest placed in a low bush; composed of dry stalks, roots, and a little moss, and lined with long hair and fibrous roots. Eggs four or five; laid early in May. Congregates in the winter season with Chaffinches, and Yellow Buntings. Said to feed on the berries of the *Solanum.*

90. E. *Hortulana,* Linn. (*Ortolan Bunting.*)—Head and neck olive-gray, obscurely spotted with dusky brown; throat, orbits, and a narrow streak from the corner of the bill downwards, greenish yellow.

E. Hortulana, *Temm. Man. d'Orn.* tom. I. p. 311.   Green-headed Bunting, *Brown, Illust. of Zool.* p. 74. pl. 30.   *Lewin, Brit. Birds,* vol. II. pl. 76.   *Bew. Brit. Birds,* vol. I. p. 170.   Ortolan Bunting, *Selb. Illust.* vol. I. p. 294. pl. 100. f. 7.

DIMENS. Entire length six inches six lines: length of the bill (from the forehead) four lines and three-quarters, (from the gape) five lines and a half; of the tarsus eight lines and a half; of the tail two inches eight lines; from the carpus to the end of the wing three inches five lines.
DESCRIPT. (*Male.*) Head and neck olivaceous green, with a tinge of ash-gray, obscurely spotted with dusky brown; throat, circle round the eyes, and a broadish streak passing downwards from the corner of the bill, greenish yellow: feathers on the back, rump, scapulars, and wing-coverts, deep brown in the middle, reddish at the edges; those on the breast, belly, and abdomen, chestnut red, edged with ash-gray: quills and tail dusky brown; the former with pale edges; the latter with the

two outer feathers white on their inner webs: irides brown: bill and feet flesh red. (*Female.*) Smaller; the yellow on the throat, and round the eyes, paler; breast spotted with brown; the rest of the under parts reddish white: above, as in the male bird, but the colours not so deep, and the spots on the head and nape larger, and more numerous. (*Egg.*) Pale reddish white, streaked and speckled with purple brown.

Only two examples of this species have hitherto occurred in this country: the first taken near London many years ago, and described by Brown, Lewin, and others, under the name of the *Green-headed Bunting* *; the second killed near Manchester, in November 1827. Said to build in bushes, hedges, or among corn, and to lay from four to five eggs. Food grain and insects.

> (6.) *E. Ciris*, Linn. Syst. Nat. tom. i. p. 313. *Painted Finch*, Edwards, Nat. Hist. pl. 130. Gleanings, pl. 273. *Painted Bunting*, Lath. Syn. vol. ii. p. 206.
>
> A single individual of this species is recorded by Montagu (*Orn. Dict. Supp.* Art. *Grosbeak-White-Winged.*) as having been taken alive on Portland Island, in the year 1802; but considerable doubts are entertained, whether it had not escaped from some vessel going up channel. It is a native of some parts both of N. and S. America.

GEN. 29.   FRINGILLA, *Linn.*

(1. Fringilla, *Cuv.*)

91. F. *Cœlebs*, Linn. (*Chaffinch.*) — Head and nape grayish blue: back and scapulars chestnut brown, tinged with olive; rump greenish.

> F. Cœlebs, *Temm. Man. d'Orn.* tom. i. p. 357. Chaffinch, *Mont. Orn. Dict. Selb. Illust.* vol. i. p. 303. pl. 54. f. 4. & 4 *. *Bew. Brit. Birds*, vol. i. p. 188.

Dimens. Entire length five inches eleven lines: length of the bill (from the forehead) five lines, (from the gape) six lines and a half; of the tarsus nine lines; of the tail two inches seven lines; from the carpus to the end of the wing three inches three lines: breadth, wings extended, ten inches.

Descript. (*Male in Spring.*) Forehead black; crown of the head, and nape, deep grayish blue: back and scapulars chestnut brown, with a tinge of olive-green; rump greenish yellow: cheeks, throat, and neck, reddish brown; breast and flanks the same, but somewhat paler; lower part of the abdomen white: lesser wing-coverts white; those of the primary quills, and the bastard wing, entirely black; secondary coverts black, tipped with primrose-yellow: the three first quills black, edged externally with yellowish white; the rest of the primaries, and all the secondaries, with a white spot at the base, with part of the inner web white, and with a portion of the outer web edged with pale yellow: the two middle tail-feathers ash-gray; the others black; the two outer ones on each side having a large white spot on the inner web: bill bluish; the tip black: feet brown. In *Winter*, the colours are paler; and the feathers on all the upper and under parts tipped with ash-gray. (*Female.*) Head, nape, back, and scapulars, pale olive-green, tinged with cinereous brown: cheeks, and under parts, cinereous white: the transverse bars on the wings not so distinct as in the male bird: bill yellowish

---

* See some remarks on this subject by Mr Yarrell, in *Zool. Journ.* vol. iii. p. 498.

gray. *Obs.* This species is subject to much variation of plumage, especially about the wings, which are occasionally almost entirely white. (*Egg.*) Pale purplish white, sparingly streaked and spotted with red brown: long. diam. nine lines and a half; trans. diam. seven lines.

Common throughout the country. In song from the first week in February, to the end of June, or beginning of July. Nest placed against the side of a tree, or in the forked branch of a bush; constructed principally of moss, elegantly studded on the outside with lichens and wool, and lined with feathers and hair. Eggs four or five in number; laid about the third week in April. Collects in flocks, at the approach of Winter.

92. F. *Montifringilla*, Linn. (*Mountain Finch.*)—
Head, cheeks, nape, and upper part of the back, black: rump white: axillæ bright gamboge-yellow.

F. Montifringilla, *Temm. Man. d'Orn.* tom. i. p. 360.   Mountain Finch, *Selb. Illust.* vol. i. p. 306. pl. 54. f. 5. & 5*. *Bew. Brit. Birds,* vol. i. p. 190. Brambling, *Mont. Orn. Dict.*

DIMENS. Entire length six inches five lines: length of the bill (from the forehead) five lines, (from the gape) six lines and a half; of the tarsus ten lines; of the tail two inches seven lines; from the carpus to the end of the wing three inches eight lines: breadth, wings extended, ten inches seven lines.

DESCRIPT. (*Male in Winter.*) Head, cheeks, nape, sides of the neck, and upper part of the back, black; the feathers edged and tipped with reddish brown and ash-gray: throat, fore part of the neck, breast, scapulars, and lesser wing-coverts, pale orange-brown; greater coverts black, tipped with orange-red; quills black, edged with yellow on their outer webs; some of the primaries with a white spot at the base, forming an oblique bar of that colour when the wing is closed: axillæ bright gamboge-yellow: rump, belly, and under tail-coverts, white: sides inclining to reddish, with a few dusky spots: tail black; the two middle feathers edged with reddish ash; the outer one white on the exterior web: bill lemon-yellow at the base, black at the tip: irides dusky: feet grayish brown. In *Summer,* the plumage on the head, neck, and back, is deep black, without the edging of reddish brown: the bill is also bluish. (*Female.*) Colours not so bright: crown of the head reddish brown, the feathers edged with ash-gray; a black streak above the eye; cheeks, nape, and sides of the neck, cinereous; back dusky brown, the feathers broadly margined with reddish ash: the rufous tinge on the breast and wings much fainter than in the male bird. In *young* birds the throat is white. (*Egg.*) "Spotted with yellowish." TEMM.

A native of the northern parts of Europe. In this country only a winter visitant, making its appearance towards the end of Autumn, and departing early in the Spring. Is more plentiful, and more generally distributed, some seasons than others, according to the state of the weather. Habits very similar to those of the last species. Said to build in fir and pine-forests, and to construct a nest of moss and wool, lined with hair and feathers: eggs five in number.

(2. PYRGITA, *Cuv.*)

93. F. *domestica*, Linn. (*House Sparrow.*) — Crown
and occiput bluish ash ; cheeks, and sides of the neck, grayish white : a transverse whitish bar on the wing.

F. domestica, *Temm. Man. d'Orn.* tom. I. p. 350. House Sparrow, *Mont. Orn. Dict. Selb. Illust.* vol. I. p. 298. pl. 54. f. 2. & 2*. *Bew. Brit. Birds,* vol. I. p. 184.

Dimens. Entire length six inches two lines: length of the bill (from the forehead) five lines and a half, (from the gape) seven lines; of the tarsus nine lines; of the tail two inches four lines; from the carpus to the end of the wing two inches eleven lines: breadth, wings extended, nine inches four lines.

Descript. (*Male.*) Crown of the head, and occiput, bluish ash; space between the bill and the eye, throat, and fore part of the neck, deep black; above the eyes, and behind the ears, a band of reddish brown: cheeks, sides of the neck, breast, and abdomen, grayish white: plumage on the back and wings dusky brown, edged with reddish; the latter with one transverse whitish bar: tail brown, edged with yellowish gray: bill bluish black: irides hazel: feet brown. (*Female.*) Head and nape cinereous brown: above and behind the eye a yellowish streak: upper parts plain brown, darkest in the middle of the feathers: under parts grayish white, without any black on the throat and neck: bill much paler than in the male. *Obs.* White, and other varieties of this species are not unfrequent. (*Egg.*) White; spotted and streaked with ash-colour and dusky brown: varies considerably in the number and intensity of the markings: long. diam. ten lines; trans. diam. seven lines.

Plentiful in all parts of the kingdom. Attached to the habitations of man. Food, grain and insects. Nest generally placed under the eaves of buildings, or in the holes of old walls; composed of hay and straw in large quantities and loosely put together, lined principally with feathers. Eggs five or six.

94. F. *montana*, Linn. (*Tree Sparrow.*)—Crown and occiput chestnut-brown; sides of the neck, and a collar on the nape, white: wings with two transverse white bars.

F. montana, *Temm. Man. d'Orn.* tom. I. p. 354. Tree Sparrow, *Mont. Orn. Dict. Selb. Illust.* vol. I. p. 300. pl. 55. f. 2. Mountain Sparrow, *Bew. Brit. Birds,* vol. I. p. 187.

Dimens. Entire length five inches six lines. Mont.

Descript. Crown of the head, and nape of the neck, deep chestnut-brown, tinged with cinereous: space between the eye and the bill, ear-coverts, and throat, black: cheeks, sides of the neck, and an interrupted collar on the nape, white: feathers on the back, and scapulars, dusky in the middle, rufous brown at the edges; wing-coverts chestnut-brown, tipped with white, forming two distinct bars of that colour across the wings: quills black, with reddish edges: tail rufous brown: bill black: feet pale brown. In the *female*, the colours are more obscure; the black on the throat and ear-coverts hardly visible. (*Egg.*) Dirty white; speckled all over with light ash-brown: long. diam. eight lines and a half; trans. diam. six lines.

Much less abundant, and more partially distributed, than the last species. Met with sparingly in Yorkshire, Lancashire, Lincolnshire, and some other of the northern and eastern counties. Does not frequent buildings, but is partial to old trees, in the holes of which it builds. Nest formed of the same materials as that of the House-Sparrow. Number of eggs the same.

(3. Coccothraustes, *Briss.*)

95. F. *Coccothraustes*, Temm. (*Common Grosbeak.*)
—Base of the bill, lore, and throat, black : crown, cheeks, and rump, reddish brown.

> F. Coccothraustes, *Temm. Man. d'Orn.* tom. i. p. 344.   Haw Grosbeak, *Mont. Orn. Dict.*   Haw-Finch, *Selb. Illust.* vol. i. p. 324. pl. 55. f. 1.   Grosbeak, *Bew. Brit. Birds*, vol. i. p. 159.

DIMENS.   Entire length seven inches : length of the bill (from the forehead) nine lines, (from the gape) ten lines ; of the tarsus ten lines and a half ; of the tail one inch ten lines ; from the carpus to the end of the wing three inches ten lines.

DESCRIPT.   Base of the bill, space between the bill and the eye, chin and throat, deep black : a broad collar of ash-gray on the nape : crown of the head, cheeks, rump, and upper tail-coverts, reddish, or chestnut-brown ; back, scapulars, and lesser wing-coverts, the same, but of a deeper tint ; greater coverts grayish white, forming a broad bar of that colour across the wings ; secondary quills, and some of the primaries, glossy black, with an oblong white spot on the middle of their inner webs, the former truncated at their extremities ; the rest of the quills entirely black : breast and belly vinous red ; vent and under tail-coverts white : tail with the two middle feathers like the back ; the others dusky brown with a large oblong white spot at the extremities of their inner webs : bill blue in Summer, whitish in Winter.   In the *female*, the colours are paler, but similarly disposed to those of the other sex.   (*Young of the year before the first moult.*)   Throat yellow ; crown of the head, cheeks, and all the upper parts of the body, yellowish brown : under parts yellowish white ; the breast, belly and flanks, spotted with brown, this last colour occupying the tip of each feather.   (*Egg.*)   White tinged with blue, spotted and streaked with ash-gray and dark brown : long. diam. eleven lines ; trans. diam. eight lines and a half.

Only an occasional visitant in this country during the winter months. Principally observed in the southern counties.   In a few instances, has been known to remain and breed.   Feeds on haws, and other stone fruits.   Builds in hedges and tall trees : nest composed of twigs, lichens, and vegetable fibres ; lined with feathers or horse-hair, and other soft materials.   Eggs three to five in number.

96. F. *Chloris*, Temm. (*Green Grosbeak.*)—All the upper and under parts of the body yellowish green.

> F. Chloris, *Temm. Man. d'Orn.* tom. i. p. 346.   Green Grosbeak, *Mont. Orn. Dict.   Bew. Brit. Birds*, vol. i. p. 163.   *Selb. Illust.* vol. i. p. 326. pl. 54. f. 3.

DIMENS.   Entire length six inches : length of the bill (from the forehead) six lines, (from the gape) seven lines ; of the tarsus nine lines ; of the tail two inches five lines ; from the carpus to the end of the wing three inches four lines : breadth, wings extended, ten inches four lines.

DESCRIPT.   (*Male.*)   All the upper and under parts of the body, scapulars, and lesser wing-coverts, yellowish green ; greater coverts and secondary quills ash-gray ; primaries dusky, edged on the outer web with gamboge-yellow : tail a little forked ; the two middle feathers blackish gray, edged with yellowish gray ; the four outer feathers on each side with the greater part of the exterior web yellow : bill and feet pale flesh-red : irides dark hazel. (*Female.*) Upper parts cinereous, tinged with green : under parts much paler than in the male bird : the yellow edging on the outer webs of the primaries and tail-feathers not so bright. (*Egg.*) White, tinged with light blue ; the larger end speckled and spotted with purplish gray and dark brown : long. diam. nine lines and a half; trans. diam. six lines and a half.

Common in all parts of the country, and stationary throughout the year. Feeds on seeds and grain. Has a harsh monotonous note, heard from the end of February to the middle of August or September. Nest generally placed in a thick bush ; composed of slender twigs, bents, and moss, interwoven with wool, and lined with hair and feathers. Eggs four or five. Is a late breeder, and seldom hatches before the middle of May. Collects into flocks in Winter.

(4. CARDUELIS, *Briss.*)

97. F. *Carduelis*, Linn. (*Goldfinch.*)—Base of the bill, occiput, and nape, black : forehead and throat, blood-red.

F. Carduelis, *Temm. Man. d'Orn.* tom. I. p. 376. Goldfinch, *Mont. Orn. Dict. Selb. Illust.* vol. I. p. 312. pl. 55. f. 8, 9. *Bew. Brit. Birds,* vol. I. p. 192.

DIMENS. Entire length five inches : length of the bill (from the forehead) six lines, (from the gape) six lines and a half; of the tarsus seven lines ; of the tail two inches one line ; from the carpus to the end of the wing three inches one line : breadth, wings extended, nine inches three lines.

DESCRIPT. Base of the bill, lore, occiput and nape, deep black : forehead and throat arterial blood-red : cheeks, ear-coverts, fore part of the neck, belly and abdomen, white : back, scapulars, and sides of the breast, deep yellowish brown : lesser wing-coverts black ; greater coverts, and basal half of the quills, bright yellow ; the remaining portion of the quills black, with a white spot at the tips : tail black ; the two outer feathers on each side with a large oblong white spot on the middle of the inner web ; the others tipped with white : bill yellowish white ; the tip dusky : feet pale flesh-red. In the *female*, there is less red on the forehead and throat ; the lesser wing-coverts are brown, and the colours in general not so bright as in the male bird. (*Egg.*) Pale bluish white, with a few spots and lines of pale purple and brown : long. diam. eight lines and a half; trans. diam. six lines.

An abundant and generally diffused species. Feeds on seeds, especially those of the thistle. Song pleasing and varied; heard from April till towards the end of the Summer. Builds in orchards, gardens, and plantations : nest placed in trees or shrubs, composed of moss and wool with the addition of a few bents or lichens, and lined with wool, hair, and thistle-down. Eggs four or five in number. Is rather a late breeder.

98. F. *Spinus*, Linn. (*Siskin.*)—Crown of the head, and throat, black : nape dusky green : above and behind the eye a broad yellow streak.

F. Spinus, *Temm. Man. d'Orn.* tom. I. p. 371. Siskin, *Mont. Orn. Dict. & Supp. Selb. Illust.* vol. I. p. 309. pl. 55. f. 6, 7. *Bew. Brit. Birds,* vol. I. p. 194.

DIMENS. Entire length four inches seven lines: length of the bill (from the forehead) four lines and a half, (from the gape) five lines and a half; of the tarsus six lines and a half; of the tail one inch nine lines; from the carpus to the end of the wing two inches nine lines.

DESCRIPT. (*Male.*) Throat, crown, and occiput, deep black: nape variegated with dusky and greenish yellow: above and rather behind the eye a broad streak of sulphur-yellow: neck, breast, and belly, yellow: back, scapulars, and lesser wing-coverts, yellowish green, inclining to ash-gray, with a longitudinal dusky spot in the centre of each feather: greater coverts black, tipped with yellow: quills dusky with yellow edges: tail dusky, all the feathers except the two middle ones with a large yellow spot near the base extending over both webs: sides and lower part of the abdomen whitish, with longitudinal dusky streaks: bill whitish, the tip black: feet pale brown. (*Female.*) All the upper parts, cheeks, and sides of the neck, cinereous, with longitudinal dusky spots: under parts yellowish white; breast, flanks, and under tail-coverts, with dusky green streaks: the transverse bars on the wings, and the yellow edging of the quills, paler than in the other sex. (*Egg.*) Pale bluish white, spotted with purple brown.

A winter visitant, but not making its appearance regularly. Occasionally observed in large flocks in Cambridgeshire about the commencement of the new year, during its stay in which parts, it appears to feed almost entirely on the seeds of the alder. Is known also in Norfolk, Suffolk, Sussex, Middlesex, and Northumberland. Breeds in the North of Europe, and is said to build in the highest branches of the pine. Lays five eggs.

(5. LINARIA, *Steph.*)

99. F. *Linaria*, Linn. (*Lesser Redpole.*) — Forehead, lore, and throat, black: bill much compressed towards the tip, yellowish: feet brown.

F. Linaria, *Temm. Man. d'Orn.* tom. I. p. 373. Lesser Redpole, *Mont. Orn. Dict. Bew. Brit. Birds,* vol. I. p. 200. *Selb. Illust.* vol. I. p. 320. pl. 54. f. 6.

DIMENS. Entire length five inches three lines: length of the bill (from the forehead) three lines, (from the gape) four lines and a quarter; of the tarsus six lines and a half; of the tail two inches one line; from the carpus to the end of the wing two inches eight lines.

DESCRIPT. (*Male in Spring.*) Forehead, space between the bill and the eye, chin and throat, black: crown of the head deep crimson: occiput, nape, and upper part of the back, blackish brown, the feathers edged with pale reddish brown: sides of the throat, fore part of the neck, breast, sides of the abdomen, and rump, carmine-red, the latter tinged with grayish; middle of the belly, vent, and under tail-coverts, white tinged with rose-colour: wings and tail dusky, the feathers edged with pale reddish brown; wing-coverts tipped with pale yellowish brown, so as to shew two transverse bars: bill more compressed than in the next species; yellow, the culmen and tip dusky: feet brown. (*Female.*) Less red on the crown of the head; forehead brown, mixed with yellowish white: rump and under parts only slightly tinged with rose-red: throat dusky; breast and middle of the belly whitish; flanks, vent, and under tail-coverts, marked with large longitudinal spots of blackish brown. (*Egg.*) Pale blue, speckled and streaked with pale purplish red, and reddish brown: long. diam. seven lines and a half; trans. diam. five lines and a half.

Var. β. Linaria borealis, *Selb. in Newcast. Nat. Hist. Trans.* vol. I. p. 263. Lesser Redpole, (a large variety,) *Selb. Illust.* vol. I. pl. 53**. f. 2. Mealy Redpole, (Lin. canescens,) *Gould, Europ. Birds,* part XI. Larger than the more common variety: general plumage paler: rump grayish white: tail rather longer: the two cross-bars on the wings broader and more strongly marked.

Resident in Scotland and the North of England throughout the year. In the southern counties only a winter visitant, appearing at that season in large flocks, and frequently associating with other species. Nest, according to Selby, placed in a bush or low tree; constructed of moss and the stalks of dry grass, intermixed with down from the catkin of the willow, which also forms the lining. Eggs four or five in number. Young seldom fledged before the end of June or beginning of July. Feeds on the catkins of the birch and alder, as well as on the seeds and buds of other trees. The *Var. β.* is considered by some as a distinct species, but this point requires further investigation. It is known to the bird-catchers in the neighbourhood of London, by the name of *Mealy-backed Redpole,* the more common variety being called by them *Stone Redpole.*

100. F. *cannabina,* Linn. (*Common Linnet.*)—Throat yellowish white, with dusky streaks: scapulars and wing-coverts, chestnut-brown: bill thick at the base; bluish gray: feet brown.

> F. cannabina, *Temm. Man. d'Orn.* tom. I. p. 364. Common or Brown Linnet, *Selb. Illust.* vol. I. p. 315. pl. 55. f. 3, 4. Greater Redpole, and Linnet, *Mont. Orn. Dict. Bew. Brit. Birds,* vol. I. pp. 198 & 202.

DIMENS. Entire length five inches five lines: length of the bill (from the forehead) five lines, (from the gape) five lines and a half; of the tarsus seven lines and a half; of the tail two inches three lines; from the carpus to the end of the wing three inches one line: breadth, wings extended, nine inches six lines.

DESCRIPT. (*Male in Spring.*) Forehead and breast bright carmine-red: throat and fore part of the neck yellowish white, with longitudinal dusky streaks: crown, nape, and sides of the neck, cinereous: back, scapulars, and wing-coverts, chestnut-brown: middle of the belly and abdomen grayish white; flanks pale reddish brown: quills black, edged with white, so as to exhibit a longitudinal bar of the latter colour when the wing is closed: tail forked; the two middle feathers wholly black, and pointed; the rest black, narrowly edged with white on the outer web, and more broadly so on the inner: bill bluish gray; not so much compressed as in the last species: feet brown. In *young* birds, the red on the breast and forehead is not so intense, or so widely extended. (*Male in Winter.*) Crown of the head with large dusky spots: back and scapulars chestnut-brown, broadly edged with pale yellowish brown: breast cinereous brown, faintly tinged with red, the tips of the feathers yellowish white: flanks with large oblong brown streaks. (*Female.*) Smaller in size. All the upper parts pale cinereous brown, with dusky spots; wing-coverts reddish brown: under parts pale reddish, inclining to white on the belly: breast and flanks with streaks of dusky brown. (*Egg.*) Pale bluish white, speckled with pale purple and red-brown: long. diam. nine lines; trans. diam. six lines and a half.

Common throughout the country. Song commenced in March or April, and continued during the greater part of the Summer. Nest

placed in hedges and low bushes; formed of moss and bents interwoven
with wool, and lined with wool and hair. Eggs four or five in number;
hatched the beginning of May. Occasionally a second brood. In Winter,
collects into large flocks. *Obs.* The *Fringilla Linota* of authors is this
species in its winter plumage.

101. F. *Montium*, Gmel. (*Mountain Linnet.*)—
Throat reddish brown, without spots; greater wing-coverts
edged with white; rump purplish red : bill thick and short;
yellow: feet black.

F. Montium, *Temm. Man. d'Orn.* tom. i. p. 368.   Mountain Linnet
    or Twite, *Selb. Illust.* vol. i. p. 318. pl. 55. f. 5.   *Bew. Brit. Birds,*
    vol. i. p. 204.   Twite, *Mont. Orn. Dict.*

DIMENS. Entire length five inches three lines: length of the bill
(from the forehead) three lines, (from the gape) four lines; of the tarsus
seven lines and a half; of the tail two inches three lines; from the
carpus to the end of the wing two inches eleven lines: breadth, wings
extended, eight inches nine lines.
DESCRIPT. (*Male in Spring.*) Throat, fore part of the neck, and sides
of the head, pale reddish brown: the feathers on the crown, nape, and
back, deep black in the middle, reddish at the edges: rump fine purplish
red: greater wing-coverts edged and tipped with whitish: quills dusky;
the primaries edged with pale brown, the secondaries with white, on their
outer webs: breast and sides, yellowish brown, with large dusky spots;
middle of the belly and vent whitish: tail forked; deep brown, edged
with dirty white: bill smaller than in the last species, wax-yellow:
irides brown: feet black. In *Winter*, the black on the upper parts
assumes more of a brownish tinge. (*Female.*) All the upper parts brown,
the feathers edged with pale yellowish red; rump the same as the back,
without the tinge of purplish red: bill yellowish, tipped with brown.
(*Egg.*) Pale bluish white, speckled with pale purple-red: long. diam.
eight lines; trans. diam. six lines and a half.
Not uncommon in the neighbourhood of London, and in many of the
eastern counties, during the winter season, flocking with the last species.
At the approach of Spring, retires to the northern and mountainous parts
of England and Scotland, where it breeds. Nest placed amongst heath,
and formed of that plant with the addition of dry grass and wool, lined
with this latter material, fibres of roots, and the finer parts of the heath.
Eggs four or five in number.

GEN. 30. PYRRHULA, *Briss.*

102. P. *vulgaris*, Temm. (*Common Bullfinch.*)—Base
of the bill, crown, throat, wings and tail, black : cheeks,
and under parts, red.

P. vulgaris, *Temm. Man. d'Orn.* tom. i. p. 338.   Bullfinch, *Mont.*
    Orn. Dict.   Bew. Brit. Birds, vol. i. p. 165.   *Selb. Illust.* vol. i.
    p. 336. pl. 54. f. 1. & 1*.

DIMENS. Entire length six inches four lines: length of the bill (from
the forehead) four lines and a half, (from the gape) five lines; of the
tarsus nine lines; of the tail two inches six lines; from the carpus to the

end of the wing three inches two lines: breadth, wings extended, nine inches nine lines.

DESCRIPT. (*Male*.) Crown and occiput, base of the bill, throat, wings, and tail, velvet black tinged with purple: nape and back bluish ash: cheeks, sides and under part of the neck, breast, belly, and flanks, bright tile-red: rump, vent, and under tail-coverts, white: greater wing-coverts tipped with grayish white, shewing a transverse bar across the wing: bill and feet dusky brown. (*Female*.) Upper parts bluish ash, with a tinge of yellowish brown: under parts reddish brown: less white on the rump. (*Egg*.) Pale blue, speckled and streaked with purplish gray and dark purple: long. diam. nine lines and a half; trans. diam. seven lines.

Common in most parts of the country. Frequents woods, gardens, and orchards. Food seeds and the buds of trees. Nest generally placed in thick bushes; composed of dry twigs, and lined with fibrous roots. Eggs four or five; hatched towards the end of May.

103. P. *Enucleator*, Temm. (*Pine Bullfinch.*) — General plumage reddish orange, spotted on the back and scapulars with dusky brown.

> P. Enucleator, *Temm. Man. d'Orn.* tom. i. p. 333. Pine Grosbeak, *Mont. Orn. Dict. Bew. Brit. Birds*, vol. i. p. 161. Pine Bullfinch, *Selb. Illust.* vol. i. p. 334. pl. 53*. f. 1, 2.

DIMENS. Entire length seven inches four or five lines. TEMM.

DESCRIPT. (*Adult male*.) Head, throat, neck, and under parts, fine reddish orange, inclining to yellowish orange on the breast and belly: the feathers on the back, scapulars, and rump, deep brown, broadly bordered with yellowish orange: wings and tail black; the former with two transverse white bars: secondaries edged with white; primaries and tail-feathers, with orange. (*Young male*.) Head, neck, throat, breast, and rump, bright crimson; feathers on the back and scapulars broadly bordered with the same colour: the transverse bars on the wings, and the edging of the secondary quills, likewise crimson: flanks, belly, and vent, cinereous, tinged with red. (*Female*.) Head, nape, and rump, yellowish brown, more or less inclining to orange: back and scapulars cinereous brown: under parts ash-gray very faintly tinged with yellowish orange: the transverse bars on the wings grayish white, and not so broad as in the other sex; quills edged with greenish orange. (*Egg*.) White: long. diam. one inch one line; trans. diam. ten lines.

An occasional visitor in the northern districts of Scotland, but very rare. According to Messrs. C. & J. Paget (*Nat. Hist. of Yarm.* p. 6.) a flight was once observed near Yarmouth. Partial to pine-forests. Food seeds and berries. Breeds within the Arctic Circle. Nest placed in trees, not far from the ground, formed of twigs and sticks, and lined with feathers. Eggs four in number.

GEN. 31. LOXIA, *Briss.*

104. L. *curvirostra*, Linn. (*Common Cross-Bill.*)— Bill as long as the middle toe, moderately hooked; the crossing point of the lower mandible passing beyond the ridge of the upper.

L. curvirostra, *Temm. Man. d'Orn.* tom. i. p. 328.   Common Cross-Bill, *Selb. Illust.* vol. i. p. 329. pl. 53.   Cross-Bill, *Mont. Orn. Dict. Bew. Brit. Birds,* vol. i. p. 153.

DIMENS. Entire length six inches four lines.

DESCRIPT. (*Adult male.*) General colour of the plumage cinereous, deeply tinged with greenish yellow: rump yellow: lower part of the abdomen ash-gray with dusky spots: quills and tail-feathers dusky, with greenish edges; great and middle wing-coverts edged with yellowish white: bill yellowish brown: irides and feet brown. (*Male after the first moult.*) All the upper and under parts brick-red, tinged with yellowish gray: quills and tail-feathers dusky, edged with yellowish green: under tail-coverts white, spotted with dusky. (*Female, and young of the year.*) Brownish gray, more or less tinged with greenish: rump yellow: under parts whitish, with longitudinal dusky streaks. (*Egg.*) Pale bluish white, speckled with red-brown: long. diam. ten lines and a half; trans. diam. eight lines.

An occasional visitant in this country, at irregular intervals. Generally observed during the summer and autumnal months, in larger or smaller flocks. Breeds in the northern parts of Europe during the Winter, or very early in the Spring. According to Sheppard has been known to breed in Suffolk in one or two instances. Nest placed in the forked branches of pines, composed of moss and lichens, and lined with feathers. Eggs four or five in number. Food principally the seeds of the pine and other firs.

105.   L. *Pytiopsittacus,* Bechst. (*Parrot Cross-Bill.*) —Bill shorter than the middle toe, very strong and broad at the base, much hooked; the crossing point of the lower mandible not reaching so high as the ridge of the upper.

L. Pytiopsittacus, *Temm. Man. d'Orn.* tom. i. p. 325.   Parrot Cross-Bill, *Selb. Illust.* vol. i. p. 332. pl. 53**. f. 1.   *Bew. Brit. Birds,* vol. i. p. 157.

DIMENS. Entire length seven inches. TEMM.

DESCRIPT. (*Adult male.*) All the upper and under parts of the body olive-ash: cheeks, throat, and sides of the neck, cinereous: head spotted with brown: rump greenish yellow; breast and belly the same, tinged with gray; sides streaked with dusky: quills and tail-feathers dusky brown, edged with olive-ash: bill dusky horn-colour: irides, and feet, brown. (*Male after the first moult.*) Upper and under parts bright red: wings and tail blackish brown; the quills edged with reddish. (*Female.*) "Upper parts greenish ash, with large spots of cinereous brown: throat and neck grayish, tinged with brown; rest of the under parts cinereous, slightly tinged with greenish yellow: rump yellowish: abdomen whitish; under tail-coverts spotted with brown." TEMM.   The *young of the year* very much resemble the female. (*Egg.*) Similar to that of the last species.

This species, which inhabits the northern parts of Europe and America, has been occasionally met with in this country, but is much more rare than the last. Habits similar. Breeds, according to Temminck, in some climates, during the Winter; in others, in May. Eggs four or five in number.

106. L. *leucoptera*, Gmel. (*White-winged Cross-Bill.*) —Bill longer than the middle toe : wings with two broad transverse white bars.

L. leucoptera, *Bonap. Amer. Orn.* vol. ii. pl. 15. f. 3. (Female.) *Faun. Bor. Am.* vol. ii. p. 263. L. falcirostra, *Lath. Ind. Orn.* p. 371. White-winged Cross-bill, *Lath. Syn.* vol. ii. p. 108. *Wilson, Amer. Orn.* vol. iv. pl. 31. f. 3.

Dimens. Entire length six inches three lines : length of the bill (from the forehead) seven lines, (from the gape) seven lines and a half; of the tarsus seven lines and a half; of the tail two inches seven lines; of the wing three inches six lines. *Faun. Bor. Am.*

Descript. (*Immature male.*) Head, nape, back, rump, and all the under parts, crimson-red, partially spotted with dusky : wings and tail black, the former with two broad, white, transverse bars : region of the bill brown. "As the bird acquires its *mature* plumage, the red parts change to greenish yellow, the rump assuming a purer yellow. The *female and young, before their first moult*, are greenish, with yellowish rumps; their bellies whitish, streaked with blackish brown." *Faun. Bor. Am.* (*Egg.*) White, marked with yellowish spots.

A native of North America, but has been killed in Ireland, in one instance, within two miles of Belfast. Partial to thick forests of white spruce, feeding on the seeds of the cones. Nest generally fixed about half way up a pine-tree, composed of grass, mud, and feathers. Eggs five in number.

GEN. 32.   STURNUS, *Linn.*

107. S. *vulgaris*, Linn. (*Common Starling.*)—Plumage black, with purple and green reflections; the feathers tipped with yellowish white.

S. vulgaris, *Temm. Man. d'Orn.* tom. i. p. 132. Common Starling, *Selb. Illust.* vol. i. p. 340. pl. 36. f. 1. Starling, *Mont. Orn. Dict.* Bew. *Brit. Birds*, vol. i. p. 105.

Dimens. Entire length eight inches four lines : length of the bill (from the forehead) one inch, (from the gape) one inch four lines; of the tarsus one inch one line; of the tail two inches seven lines; from the carpus to the end of the wing four inches ten lines : breadth, wings extended, fifteen inches.

Descript. (*Male.*) General colour of the plumage black, with brilliant purple and golden-green reflections; the feathers on the upper parts tipped with small triangular yellowish white spots : quills and tail-feathers dusky brown, with pale reddish edges : under tail-coverts edged with white : bill yellow : feet flesh-red, inclining to brown. In the *Female*, the spots are more numerous, and diffused over the under as well as the upper parts. In the *young of the year previously to the autumnal moult*, the plumage is of a uniform cinereous brown without any spots; the throat and lower part of the abdomen whitish. In this state it is the *Solitary Thrush* of Montagu. The perfect plumage is probably not attained till the third year. (*Egg.*) Of a uniform delicate pale blue : long. diam. one inch two lines; trans. diam. ten lines.

A plentiful and widely dispersed species. Partial to old trees, church steeples, and ruinous buildings, in the holes of which it breeds. Nest formed of dry grass. Eggs four or five in number. In the Autumn, congregates in immense flocks. Food insects and worms, occasionally grain.

GEN. 33. PASTOR, *Temm.*

108. P. *roseus*, Temm. (*Rose-coloured Pastor.*) — Head and neck violet-black, the feathers on the crown elongated, forming a crest: back and belly rose-red.

> P. roseus, *Temm. Man. d'Orn.* tom. i. p. 136. Rose-coloured Pastor, *Selb. Illust.* vol. i. p. 343. pl. 36. f. 2, 3. Rose-coloured Ouzel, *Mont. Orn. Dict.* Rose-coloured Starling, *Bew. Brit. Birds*, vol. i. p. 110.

DIMENS. Entire length eight inches six lines: breadth, wings extended, fourteen inches six lines.

DESCRIPT. (*Male.*) Head and neck velvet-black, with violet and green reflections; the feathers on the crown very much elongated, forming a pendent crest: back, rump, breast, and belly, rose-red: wings and tail blackish brown, with violet reflections: under tail-coverts, and thighs, black: upper mandible, and tip of the lower one, yellowish red, the remaining portion black: irides deep brown: feet yellowish red. (*Female.*) General colours of the plumage much paler; the red tinged with brown: the feathers on the crown shorter. (*Young of the year.*) "No crest: all the upper parts of the body isabella-brown: wings and tail brown, all the feathers fringed with white and ash-colour: throat and middle of the abdomen pure white; the rest of the under parts cinereous brown: bill yellow at the base, brown at the tip: feet brown." TEMM. (*Egg.*) Colour unknown.

An occasional, but very rare, visitant in this country. Has been observed in Sussex, Suffolk, Norfolk, Cambridgeshire, Oxfordshire, Dorsetshire, Northumberland, Lancashire, and Ireland. Habits similar to those of the Starling. Food insects and their larvæ; also seeds, and occasionally cherries. Said to build in the holes of trees and old walls, and to lay six eggs.

GEN. 34. FREGILUS, *Cuv.*

109. *Graculus*, Selb. (*Cornish Chough.*)—Black: bill and feet coral-red.

> Pyrrhocorax Graculus, *Temm. Man. d'Orn.* tom. i. p. 122. Cornish Chough, *Selb. Illust.* vol. i. p. 365. pl. 33. Chough, *Bew. Brit. Birds*, vol. i. p. 90. Red-legged Crow, *Mont. Orn. Dict.*

DIMENS. Entire length sixteen inches: length of the bill (from the forehead) two inches two lines and a half, (from the gape) two inches two lines and a half; of the tarsus two inches; of the tail five inches seven lines; from the carpus to the end of the wing ten inches eight lines: breadth, wings extended, two feet eight inches.

DESCRIPT. The whole plumage black, with purple and green reflections: bill and feet bright coral-red: claws black. (*Egg.*) Yellowish white, spotted with ash-gray and light brown: long. diam. one inch eight lines; trans. diam. one inch one line.

A local species; principally met with on the rocky coasts of Cornwall, Devonshire, Dorsetshire, and Wales. Found also in some of the Scotch Islands. Never observed inland, but breeds on the cliffs, or in old buildings near the sea. Nest formed of sticks, and lined with wool and hair. Eggs four or five in number. Food insects, grain, and berries.

GEN. 35.　CORVUS, *Linn.*

(1. Corvus, *Cuv.*)

110.　C. *Corax*, Linn.　(*Raven.*) — Plumage black, glossed with blue: nostrils covered by bristly feathers half the length of the bill: tail considerably rounded, extending two inches beyond the folded wings.

> C. Corax, *Temm. Man. d'Orn.* tom. i. p. 107.　*Id.* tom. iii. p. 56.
> Raven, *Mont. Orn. Dict. Selb. Illust.* vol. i. p. 346. pl. 27*.
> *Bew. Brit. Birds,* vol. i. p. 79.

Dimens. Entire length two feet one inch: length of the bill two inches nine lines: breadth, wings extended, four feet.

Descript. The whole plumage black; the upper parts with purple and blue reflections, beneath less glossy: throat feathers long, loose, and acuminated: bill very strong, black: irides with two circles, the outer one brown, the inner gray: tail very much rounded, much longer than in the next species, reaching about two inches beyond the tips of the folded wings: feet black. According to Low, *white varieties* have been met with in the Orkneys. (*Egg.*) Pale green ground, spotted and speckled with darker greenish brown: long. diam. two inches; trans. diam. one inch four lines.

The largest species in the genus. Widely dispersed over the country, but not very plentiful. Breeds very early in the year, on steep cliffs, or in large and lofty trees. Nest formed of sticks, and lined with wool, hair, and other substances. Eggs five or six in number. Food small quadrupeds, poultry, game, as well as carrion, and other animal substances. The male and female pair for life, and generally haunt the same spot every year for the purpose of nidification.

111.　C. *Corone*, Linn.　(*Carrion Crow.*)—Plumage bluish black: nostrils covered by bristly feathers one-third the length of the bill: tail moderately rounded, extending an inch and a quarter beyond the wings.

> C. Corone, *Temm. Man. d'Orn.* tom. i. p. 108.　*Id.* tom. iii. p. 58.
> Carrion Crow, *Mont. Orn. Dict. Selb. Illust.* vol. i. p. 349. pl. 28.
> The Crow, *Bew. Brit. Birds,* vol. i. p. 81.

Dimens. Entire length nineteen inches: length of the bill (from the forehead) one inch ten lines, (from the gape) two inches one line; of the tarsus two inches one line; of the tail seven inches six lines; from the carpus to the end of the wing twelve inches: breadth, wings extended, three feet.

Descript. Much smaller than the last species, but very similar in general appearance: plumage entirely black; the upper parts reflecting green and violet: tail relatively shorter than in the *C. Corax,* and not so

K

much rounded at the extremity: bill and feet black: irides dark hazel. (*Egg.*) Ground colour pale bluish green, spotted and speckled with two shades of ash-colour and clove-brown: long. diam. one inch eight lines; trans. diam. one inch two lines.

Common throughout the kingdom, but most abundant in wooded districts. Food and habits similar to those of the Raven. Like that species found in pairs all the year. Breeds as early as February. Nest generally placed in the forked branch of a lofty tree, composed of sticks, and lined with wool, hair, and other soft materials. Eggs four or five in number.

112. C. *Cornix*, Linn. (*Hooded Crow*.) — Head, throat, wings and tail, black, glossed with blue; the rest of the plumage ash-gray.

C. Cornix, *Temm. Man. d'Orn.* tom. I. p. 109. *Id.* tom. III. p. 59.
Hooded Crow, *Mont. Orn. Dict. Selb. Illust.* vol. I. p. 351. pl. 29.
Bew. Brit. Birds, vol. I. p. 83.

DIMENS. Entire length nineteen inches six lines: length of the bill (from the forehead) two inches two lines, (from the gape) two inches two lines and a half; of the tarsus two inches three lines; of the tail seven inches six lines; from the carpus to the end of the wing twelve inches ten lines: breadth, wings extended, three feet two inches.

DESCRIPT. Head, throat, wings, and tail, black, with blue and green reflections: all the rest of the plumage cinereous gray: bill strong, shaped like that of the *Raven*, black: nostrils covered by reflected bristly feathers, as in the last species: irides brown: feet black. (*Egg.*) Mottled all over with dark greenish brown on a lighter green ground: long. diam. one inch ten lines; trans. diam. one inch three lines.

Resident all the year in the western and northern parts of Scotland, and in some parts of Ireland, but only a winter visitant in England, where it arrives about the end of October, and remains till the approach of Spring. During its stay chiefly frequents the sea-coast, or open and extensive downs. Builds in trees and ·rocky cliffs, constructing a nest similar to that of the Carrion Crow. Breeds early in the year, and lays four or five eggs. Feeds indiscriminately on all kinds of animal substances.

113. C. *frugilegus*, Linn. (*Rook*.)—Plumage black, glossed with blue: base of the bill (in the *adult* bird) denuded of feathers, and covered with a white scurf.

C. frugilegus, *Temm. Man. d'Orn.* tom. I. p. 110. *Id.* tom. III. p. 59.
Rook, *Mont. Orn. Dict. Selb. Illust.* vol. I. p. 353. pl. 30. *Bew. Brit. Birds*, vol. I. p. 85.

DIMENS. Entire length eighteen inches: length of the bill (from the forehead) two inches two lines, (from the gape) two inches four lines; of the tarsus two inches; of the tail seven inches five lines; from the carpus to the end of the wing twelve inches six lines: breadth, wings extended, three feet.

DESCRIPT. Strongly resembling the *C. Corone*, but may always be distinguished in the adult state by the naked, scabrous, whitish skin surrounding the base of the bill, and by the entire absence of nasal feathers: plumage wholly black as in that species, but rather more glossy, reflecting rich tints of blue and violet-purple: bill not quite so strong, and of not so deep a black: colour of the feet similar. *Young* birds, on leaving the nest, have the base of the bill feathered as in the *C. Corone*, and the nostrils likewise covered by reflected bristles: in the course of the autumn these feathers fall off, and are not replaced by others. *Varieties* are occasionally met with entirely white, or pied, or with the tips of all the feathers whitish. (*Egg.*) Ground colour pale green, nearly covered with blotches of dark greenish brown: long. diam. one inch eight lines; trans. diam. one inch two lines.

Common throughout the country, especially in cultivated districts. Of gregarious habits, breeding together, and likewise seeking food in large companies. Subsists principally on the grub of the cockchaffer, wire-worm, and other insects; but will occasionally devour corn, and, during the winter season, is very destructive to turnips. Builds early in March, and hatches in April. Nests crowded together at the tops of the tallest trees, composed principally of fresh twigs forcibly detached from the branches of the neighbouring trees, and lined with grass and fibrous roots. Eggs four or five in number.

114. C. *Monedula*, Linn. (*Jackdaw.*)—Crown of the head, and upper parts of the body, bluish black; occiput and nape ash-gray.

C. Monedula, *Temm. Man. d'Orn.* tom. I. p. 111. *Id.* tom. III. p. 60.
Jackdaw, *Mont. Orn. Dict. Selb. Illust.* vol. I. p. 356. pl. 31. f. 1.
Bew. Brit. Birds, vol. I. p. 88.

DIMENS. Entire length twelve inches nine lines: length of the bill (from the forehead) one inch two lines, (from the gape) one inch five lines; of the tarsus one inch eight lines: of the tail four inches eleven lines; from the carpus to the end of the wing nine inches: breadth, wings extended, two feet three inches three lines.

DESCRIPT. Much smaller than any of the foregoing species: crown of the head, and all the upper parts of the body, black, glossed with violet-blue; occiput and nape ash-gray: wings and tail the same as the back: under parts deep black, not so much glossed as the upper: bill and feet black: irides grayish white. *Varies* occasionally like the last species. (*Egg.*) Pale bluish white, spotted with ash-colour and clove-brown: long. diam. one inch seven lines; trans. diam. one inch and half a line.

Common in all parts of the country. Much attached to churches and other buildings, especially such as are in a ruinous and deserted state. Builds in such situations, as well as in the holes of decayed trees; occasionally in chimnies, and even rabbit-burrows. Nest composed of sticks mixed up occasionally with horse-dung, and lined with wool and other soft substances. Eggs four to six in number. Gregarious like the last species, with which it often associates. Feeds on a great variety of animal and vegetable substances.

(2. PICA, *Cuv.*)

115. C. *Pica*, Linn. (*Magpie.*)—Head, throat, back, and breast, black; scapulars and belly white.

C. Pica, *Temm. Man. d'Orn.* tom. I. p. 113. Garrulus Picus,
　　Id. tom. III. p. 63. Magpie, *Mont. Orn. Dict. Selb. Illust.*
　　vol. I. p. 358. pl. 31. f. 2. *Bew. Brit. Birds*, vol. I. p. 92.

DIMENS. Entire length eighteen inches: length of the bill (from the
forehead) one inch four lines, (from the gape) one inch eight lines; of
the tarsus two inches one line; of the tail nine inches ten lines; from
the carpus to the end of the wing seven inches eight lines: breadth,
wings extended, twenty-three inches.

DESCRIPT. Head, throat, neck, upper part of the breast, back, vent,
under tail-coverts, and thighs, deep velvet-black; the feathers on the
throat rather loose and open, each tipped with a short bristle: scapulars,
lower part of the breast, and belly, pure white: wings and tail greenish
black, reflecting rich tints of purple, bronze, and blue, according to the
position of the eye; the inner webs of the primaries with a large oblong
white spot: tail very much cuneated, the length of the feathers rapidly
decreasing: irides dark brown: bill and feet black. (*Egg.*) Pale bluish
white, spotted all over with ash-colour and two shades of greenish brown:
long. diam. one inch four lines and a half; trans. diam. one inch.

A common and generally dispersed species. Frequents woods and
thickets, but prefers those near adjoining to villages. Generally observed
in pairs. Nest placed in a thick bush, or on the top of a lofty tree; of
an oval shape, constructed of sticks closely interwoven with each other,
the bottom plastered with clay, and lined with fibrous roots. Eggs six or
seven in number, laid early in the Spring. Omnivorous.

GEN. 36.　GARRULUS, *Briss.*

116.　G. *glandarius*, Flem. (*Jay.*)—A black moustache
descending from the corners of the bill: greater wing-
coverts barred with black and blue.

Corvus glandarius, *Temm. Man. d'Orn.* tom. I. p. 114. Garrulus
　　gland. *Id.* tom. III. p. 65. Jay, *Mont. Orn. Dict. Selb. Illust.*
　　vol. I. p. 362. pl. 32. *Bew. Brit. Birds*, vol. I. p. 94.

DIMENS. Entire length thirteen inches eight lines: length of the
bill (from the forehead) one inch two lines, (from the gape) one inch five
lines and a half; of the tarsus one inch eight lines and a half; of the tail
five inches nine lines; from the carpus to the end of the wing seven
inches three lines: breadth, wings extended, twenty-two inches.

DESCRIPT. The feathers on the forehead and crown narrow and
elongated, forming a crest capable of erection, of a whitish colour streaked
with black; chin whitish; a broad black moustache, about an inch in
length, descending from the corners of the lower mandible: general
colour of the plumage above and below reddish or purple ash; darkest on
the nape and upper part of the back; rump, and lower part of the abdomen, whitish: quills and tail dusky, the former edged with white;
bastard wing and greater coverts bright blue, with several narrow transverse bars of black; secondaries with an oblong white spot on their outer
webs: bill black: irides bluish white: feet livid brown. White and other
varieties sometimes occur. (*Egg.*) Yellowish white ground, minutely
and thickly speckled all over with light brown, presenting the appearance
of a uniform yellow gray-brown: long. diam. one inch four lines; trans.
diam. one inch.

Found in extensive woods in most parts of the kingdom. Like the last species omnivorous. Has a harsh disagreeable cry. Nest placed in close thickets, formed of sticks and lined with fibrous roots. Lays five or six eggs.

GEN. 37. NUCIFRAGA, *Briss.*

117. N. *Caryocatactes*, Temm. (*Nutcracker.*) — General plumage brown, spotted on the back and under parts with white.

> N. Caryocatactes, *Temm. Man. d'Orn.* tom. I. p. 117. *Id.* tom. III.
> p. 67. Nutcracker, *Mont. Orn. Dict. Selb. Illust.* vol. I. p. 368.
> pl. 33* *Bew. Brit. Birds*, vol. I. p. 97.

DIMENS. Entire length thirteen inches: length of the bill two inches. MONT.

DESCRIPT. Crown of the head and nape of the neck blackish brown: the rest of the plumage above and below the body rusty brown, spotted with white; the spots on the back guttiform; those beneath disposed longitudinally on each feather, and occupying a larger portion of the ground-colour: vent white: quills and tail black; the latter with a broad white bar at the extremity: irides brown: bill and feet dusky. (*Egg.*) Pale buff-coloured white: long. diam. one inch three lines; trans. diam. ten lines and a half.

Very rare in this country. Has been only observed in three or four instances. One killed near Moyston in Flintshire in October 1753; a second in Kent; a third in Northumberland in the autumn of 1819. Common in some of the mountainous parts of the North of Europe, residing principally in woods. Feeds on seeds and nuts; also on insects. Said to build in the holes of trees, and to lay five or six eggs.

GEN. 38. PICUS, *Linn.*

118. P. *viridis*, Linn. (*Green Woodpecker.*)—Plumage above, green; beneath, greenish ash; crown and occiput bright red.

> P. viridis, *Temm. Man. d'Orn.* tom. I. p. 391. *Id.* tom. III. p. 280.
> Green Woodpecker, *Mont. Orn. Dict. Selb. Illust.* vol. I. p. 372.
> pl. 38. f. 2. *Bew. Brit. Birds*, vol. I. p. 136.

DIMENS. Entire length thirteen inches three lines: length of the bill (from the forehead) one inch seven lines, (from the gape) one inch nine lines and a half; of the tarsus one inch two lines; of the tail three inches ten lines and a half; from the carpus to the end of the wing six inches four lines; breadth, wings extended, twenty-one inches.

DESCRIPT. Crown and occiput bright crimson-red; base of the bill, and region of the eyes, black; a broad crimson moustache edged with black descending from the corner of the bill: hind part of the neck, back, and wing-coverts, bright green: rump gamboge-yellow: under parts

grayish white, with a tinge of yellowish green : primaries dusky, with a
regular series of pale yellow spots on the outer webs : tail-feathers presenting alternate bars of green and dusky-brown; the extreme tips
black : bill dusky; the base of the lower mandible yellowish : irides
white : feet pale brown with a tinge of green. In the *female*, there is
less red on the head, and less black about the eyes; the moustache is
entirely black. In *very young* birds, the red on the head is mixed with
yellowish gray; the green tint on the upper parts paler, and irregularly
spotted with ash-gray; the moustache incomplete; and the under parts
marked with transverse brown bars. (*Egg.*) Smooth and shining : pure
white : long. diam. one inch two lines and a half; trans. diam. ten lines
and a half.

Common in wooded districts throughout the country. Feeds on ants,
and other insects; more especially the larvæ of the timber-eating species,
which it extracts by means of its long tongue, after having perforated the
wood with its bill. Breeds in the holes of trees. Eggs four or five in
number, deposited on the rotten wood.

119. P. *major*, Linn. (*Great Spotted Woodpecker.*)

—Crown, and upper part of the body, black ; scapulars
white ; a crimson patch on the occiput.

> P. major, *Temm. Man. d'Orn.* tom. i. p. 395. *Id.* tom. iii. p. 281.
> Greater Spotted Woodpecker, *Mont. Orn. Dict.   Selb. Illust.*
> vol. i. p. 376. pl. 38. f. 3.   Pied Woodpecker, *Bew. Brit. Birds*,
> vol. i. p. 138.

DIMENS. Entire length nine inches six lines : length of the bill (from
the forehead) one inch, (from the gape) one inch three lines; of the tarsus
eleven lines; of the tail three inches seven lines; from the carpus to the
end of the wing five inches : breadth, wings extended, fourteen inches.

DESCRIPT. A transverse band of dirty white on the forehead; crown
of the head black; occiput crimson-red : cheeks and ear-coverts white,
bounded beneath by a black line, proceeding from the corner of the
mouth towards the nape, from whence arises another line of the same
colour that passes down the side of the neck and terminates on the
breast : a white spot on each side of the back part of the neck : back and
lesser wing-coverts velvet-black; scapulars, and some of the greater
coverts nearest the body, pure white : quills black, with a series of white
spots on each web : throat, fore part of the neck, breast and belly, dirty
white; abdomen and under tail-coverts rich crimson : tail black; the
three lateral feathers on each side spotted and tipped with dirty white :
irides red : bill and feet dusky lead-colour. In the *female*, there is no
red on the occiput. In *young* birds, until the first moult, the crown
of the head is red, and the patch on the occiput black; the black on the
upper parts is also tinged with brown. (*Egg.*) Smooth; shining white :
long. diam. one inch; trans. diam. nine lines.

Not so abundant as the last species, but hardly to be esteemed rare.
Haunts, food, and habits, very similar. Makes a loud jarring noise in the
Spring, (probably a call-note to the other sex) by striking its bill very
rapidly and many times in succession against the branch of a tree.
Makes no nest, but deposits its eggs, which are four or five in number,
in the holes of decayed trees.   *Obs.* The *Middle Spotted Woodpecker* of
Montagu and other English authors is only the young of this species :
the *Picus medius* of Temminck is distinct; but not hitherto found in this
country.

120. P. *minor*, Linn. (*Lesser Spotted Woodpecker.*) —Occiput and nape black : middle of the back, and scapulars, with white and black bars : crown of the head red.

P. minor, *Temm. Man. d'Orn.* tom. i. p. 399. *Id.* tom. iii. p. 283. Lesser Spotted Woodpecker, *Mont. Orn. Dict. Selb. Illust.* vol. i. p. 379. pl. 38. f. 4. Barred Woodpecker, *Bew. Brit. Birds*, vol. i. p. 140.

DIMENS. Entire length five inches six lines: length of the bill seven lines and a half: breadth, wings extended, twelve inches.

DESCRIPT. Forehead dirty white; crown of the head bright red; streak over the eyes, occiput and nape, black; cheeks and sides of the neck white; from the corners of the lower mandible a black streak directed towards the shoulders : upper part of the back, rump, and lesser wing-coverts, glossy black; the rest of the upper parts, including the middle region of the back, scapulars, and quills, transversely barred with black and white: all the under parts of a dirty brownish white, with a few fine longitudinal dark streaks on the breast and sides : the four middle tail-feathers glossy black; the three outer ones on each side tipped with white and barred with black : bill and feet dusky lead-colour : irides reddish brown. In the *female*, the crown is dirty white instead of red. (*Egg*.) Smooth, delicate white : long. diam. nine lines and a half; trans. diam. seven lines.

Much less frequent than either of the preceding species, and only partially distributed. Met with in the counties of Gloucester, Wilts, Hereford, Cambridge, Norfolk, and Suffolk. Said to be very rare in the North of England. Habits and nidification similar to those of the last.

(7.) *P. martius*, Linn. (*Great Black Woodpecker.*) Temm. Man. d'Orn. tom. i. p. 390. Gould, Europ. Birds, part i. Selb. Illust. vol. i. p. 375. pl. D. f. 4.

This species, which is not uncommon in the northern parts of Europe, has been included in the British lists principally on the authority of Drs. Latham and Pulteney. The former author states (*Syn. Supp.* p. 104.) that it has been sometimes seen in Devonshire; the latter (*Cat. Dors.* p. 6.) that it has been more than once shot in Dorsetshire. No specimen, however, known to have been certainly killed in this country, exists in any of our museums, and there is strong reason to doubt the reality of its claims to a place in the British Fauna.

(8.) *P. villosus*, Linn. (*Hairy Woodpecker.*) Lewin, Brit. Birds, vol. ii. pl. 50. Mont. Orn. Dict.

A pair of this species, in the collection of the late Dowager Duchess of Portland, were said to have been shot near Halifax in Yorkshire. It is supposed, however, that this was an error, and that the above locality had been confounded with Halifax in North America, where the species is not uncommon.

(9.) *P. tridactylus*, Linn. (*Three-toed Woodpecker.*) Temm. Man. d'Orn. tom. i. p. 401. Don. Brit. Birds, vol. vi. pl. 143.

A native of the northern parts of Europe and America. According to Donovan, has been shot in the North of Scotland. This, however, is probably a mistake.

GEN. 39. YUNX, *Linn.*

121. Y. *Torquilla*, Linn. (*Wryneck.*)—Upper parts yellowish gray, irregularly spotted and lined with brown and black ; a broad black mesial stripe on the back of the neck.

Y. Torquilla, *Temm. Man. d'Orn.* tom. i. p. 403.   *Id.* tom. iii.
  p. 284. Wryneck, *Mont. Orn. Dict.   Selb. Illust.* vol. i. p. 381.
  pl. 38. f. 1.   *Bew. Brit. Birds,* vol. i. p. 129.

Dimens. Entire length seven inches four lines: length of the bill (from the forehead) six lines, (from the gape) nine lines and a half; of the tarsus eight lines and a half; of the tail two inches eight lines : from the carpus to the end of the wing three inches six lines : breadth, wings extended, eleven inches.

Descript.   General colour of the upper plumage reddish or yellowish gray, irregularly spotted and marked with various shades of brown and black ; more particularly a broad black mesial streak extending from the occiput to the upper part of the back : throat and fore part of the neck reddish yellow, with transverse undulating dusky lines; rest of the under parts whitish, with arrow-shaped black spots : quills marked on the outer webs with oblong red spots : tail-feathers rounded at the end, mottled like the back, with four transverse black bars : irides yellowish brown : bill and feet brown. (*Egg.*) Smooth, delicate white : long. diam. nine lines and a half; trans. diam. seven lines.

A summer visitant, first appearing about the second week in April. Not very uncommon in wooded districts throughout the southern, mid-land, and eastern counties: said to be more rare in the West of England. Utters a loud and oft-repeated cry during the breeding season, somewhat resembling that of the Kestril Hawk. Food principally ants. Eggs six to ten in number, deposited in the holes of trees on the rotten wood, without a nest.

GEN. 40. CERTHIA, *Linn.*

122. C. *familiaris*, Linn. (*Common Creeper.*)—Upper parts yellowish brown ; variegated with dusky, and white spots : rump reddish yellow.

C. familiaris, *Temm. Man. d'Orn.* tom. i. p. 410.   *Id.* tom. iii. p. 288.
  Common Creeper, *Mont. Orn. Dict.   Selb. Illust.* vol. i. p. 388.
  pl. 39. f. 2.   *Bew. Brit. Birds,* vol. i. p. 148.

Dimens. Entire length five inches: length of the bill (from the fore-head) five lines and three quarters, (from the gape) seven lines; of the tarsus seven lines ; of the tail two inches one line and a half ; from the carpus to the end of the wing two inches four lines : breadth, wings extended, seven inches three lines.

Descript.   Head, and all the upper parts, yellowish brown, mixed with dusky, with an oblong whitish spot in the centre of each feather; rump rust-red; above the eyes a whitish streak : all the under parts white, tinged with reddish on the lower part of the abdomen: quills tipped with white ; the four first dusky, the rest with a broad transverse reddish white bar about the middle: tail-feathers long, stiff, and acuminated; brownish gray, tinged with red: upper mandible brown; lower mandible yellowish: feet yellowish brown. (*Egg.*) White, with pale red spots often confined to the larger end: long. diam. eight lines; trans. diam. five lines and a half.

Common and generally dispersed throughout the country. Climbs trees with great facility. Feeds entirely on insects. Has a peculiar, but rather monotonous song, heard early in the Spring, and continued during the breeding season. Nest placed in the holes, or under the loose bark, of decayed trees; formed of small sticks, wool, and mosses packed rudely together, and lined with feathers and fine shreds of wood. Eggs six or eight in number, laid towards the end of April.

GEN. 41.   TROGLODYTES, *Cuv.*

123.   T. *Europæus*, Selb. (*Common Wren.*)—Upper plumage reddish brown, with transverse dusky lines: over the eye a narrow white streak.

Sylvia Troglodytes, *Temm. Man. d'Orn.* tom. i. p. 233.   Troglodytes vulgaris, *Id.* tom. iii. p. 160.   Common Wren, *Mont. Orn. Dict. Selb. Illust.* vol. i. p. 390. pl. 47. f. 6.   *Bew. Brit. Birds*, vol. i. p. 262.

DIMENS. Entire length three inches nine lines: length of the bill (from the forehead) three lines and a half, (from the gape) five lines and a half; of the tarsus nine lines; of the tail one inch four lines; from the carpus to the end of the wing one inch nine lines and a half: breadth, wings extended, six inches three lines.

DESCRIPT. Upper parts of the body deep reddish brown, faintly marked with transverse dusky lines: over each eye a pale narrow streak: quills barred alternately on their outer webs with blackish brown and reddish: tail dusky, with transverse black bars: under parts light rufous-brown; the sides and thighs streaked with darker lines; under tail-coverts obscurely spotted with white: bill brown: irides hazel: feet yellowish brown. (*Egg.*) White, with a few specks of pale red: long. diam. seven lines and a half; trans. diam. six lines.

Abundant in all parts of the country, remaining with us the whole year. Frequents gardens and out-houses. Song shrill and loud. Nest often fixed against the thatch of buildings, or placed in the holes of trees, more rarely on the ground; of an oval form, covered over at top, with the entrance on one side; formed of moss, hay, leaves, and other materials, and lined with feathers occasionally mixed with hair. Eggs five to eight in number.

GEN. 42.   UPUPA, *Linn.*

124.   U. *Epops*, Linn. (*Hoopoe.*)—Head, neck, and breast, purplish red; wings black, barred with white.

U. Epops, *Temm. Man. d'Orn.* tom. i. p. 415.   *Id.* tom. iii. p. 291. Hoopoe, *Mont. Orn. Dict. Selb. Illust.* vol. i. p. 393. pl. 40. f. 2. *Bew. Brit. Birds*, vol. i. p. 144.

DIMENS. Entire length twelve inches: length of the bill two inches six lines: breadth, wings extended, nineteen inches.

DESCRIPT. Crest composed of a double row of elongated feathers, orange-red tipped with black; a small white patch intervening between these two colours: head, neck, and breast, purplish red: upper part of

the back pale grayish brown; middle and lower regions barred with
black and white: wings black, with five transverse bars of yellowish
white: tail black, with one broad, white, crescent-shaped bar: abdomen
white; flanks with a few longitudinal brown streaks: bill flesh-red at
the base, black at the tip: irides and feet brown. In the *female*, the
colours are paler, and the crest not so long. (*Egg.*) Uniform pale
lavender-gray: long. diam. one inch and half a line; trans. diam. eight
lines.

Visits this country most years, but in very small numbers. Occasionally met with in different parts of England: rarely breeds with us.
Habits somewhat terrestrial. Food coleopterous, and other insects.
Builds in the holes of trees. Nest formed of bents, and lined with
feathers and other soft materials. Eggs four or five in number.

GEN. 43.   SITTA, *Linn.*

125.   S. *Europœa*, Linn. (*Nuthatch.*)—Upper parts
bluish ash; breast and belly reddish yellow; a black
streak across the eye and ear-coverts.

S. Europæa, *Temm. Man. d'Orn.* tom. I. p. 407.   *Id.* tom. III. p. 285.
    Nuthatch, *Mont. Orn. Dict.   Selb. Illust.* vol. I. p. 385. pl. 39.
    f. 1.   *Bew. Brit. Birds*, vol. I. p. 142.

Dimens. Entire length six inches ten lines.

Descript. Crown of the head, and all the upper parts of the plumage
bluish ash: a black streak from the corner of the bill across each eye,
and extending down the sides of the neck as far as the shoulder; cheeks
and throat white; breast and belly dull orange-red; sides and thighs
ferruginous chestnut: under tail-coverts white, edged with ferruginous:
tail consisting of twelve short flexible feathers; the two middle ones
gray; the four outer ones black, with a white spot near the extremity,
the tips themselves gray: bill blackish above, white at the base of the
lower mandible: irides hazel: feet yellowish gray.   (*Egg.*) White,
spotted and speckled with pale red: long. diam. nine lines; trans.
diam. seven lines.

Common in wooded districts in many parts of the country, but not in
all. Rare in some of the northern and western counties. Remains with
us the whole year. Climbs trees with great facility, and equally well in
all directions. Feeds principally on nuts, which it breaks with its bill
after having firmly fixed them in the crevices of old trees: occasionally
on insects. Nest placed in the holes of trees, and formed of dead leaves.
Eggs six or seven in number.

GEN. 44.   CUCULUS, *Linn.*

126.   C. *canorus*, Linn. (*Common Cuckow.*)—Head,
neck, breast, and upper parts, bluish ash: abdomen whitish,
with transverse dusky bars.

C. canorus, *Temm. Man. d'Orn.* tom. I. p. 381.   *Id.* tom. III. p. 272.
    Common Cuckow, *Mont. Orn. Dict. and Supp.   Selb. Illust.* vol. I.
    p. 397. pl. 37, and pl. 43. f. 3.   *Bew. Brit. Birds*, vol. I. p. 124.

Dimens. Entire length thirteen inches six lines: length of the bill
(from the forehead) ten lines, (from the gape) one inch two lines: of the

tarsus ten lines and a half; of the tail six inches ten lines; from the carpus to the end of the wing eight inches nine lines: breadth, wings extended, twenty-three inches four lines.

DESCRIPT. (*Adult.*) Head, and all the upper parts, bluish ash; throat, fore part of the neck, and breast, the same, but rather paler; belly, thighs, and under tail-coverts, whitish, with transverse streaks of dusky brown: quills dusky, barred on the inner webs with oval white spots: tail black, with a series of small white spots on the shafts of the feathers; the tips also white: bill dusky, yellowish at the base and edges; inside of the mouth, and orbits, orange-yellow: irides and feet yellow. (*Young of the year.*) All the upper parts of a deep clove-brown, with transverse bars of pale ferruginous brown, the feathers tipped with whitish; a patch of white on the occiput: throat, and under parts, yellowish white, with transverse black bars: quills spotted with reddish brown on their inner webs: tail with alternate oblique bars of red and brown, the brown bar nearest the extremity broader than the others; the shafts of the feathers with a series of white spots; the tips white: irides liver-brown. (*Egg.*) White, speckled all over with ash-brown; or reddish white, speckled with nutmeg-brown: long. diam. eleven lines; trans. diam. eight lines and a half.

Visits this country early in April, and leaves it again about the beginning of July; the young of the year remaining till September. Feeds principally on caterpillars, and other insects. Makes no nest, but commits its eggs (five or six in number) to the nests of other birds, generally selecting those of the Hedge Sparrow, Wagtail, or Tit-Pipit.

GEN. 45. COCCYZUS, *Vieill.*

127. C. *Americanus*, Bonap. (*Carolina Cuckow.*)

Cuculus Americanus, *Linn. Syst. Nat.* tom. I. p. 170. *Lath. Ind. Orn.* vol. I. p. 219. Cuc. Carolinensis, *Wils. Amer. Orn.* vol. IV. p. 13. pl. 28. f. 1. Cuc. cinerosus, *Temm. Man. d'Orn.* tom. III. p. 277. Carolina Cuckow, *Lath. Syn.* vol. II. p. 537.

DIMENS. Entire length eleven inches four lines: length of the bill (from the forehead) one inch, (from the gape) one inch four lines; of the tarsus eleven lines and a half; of the tail five inches seven lines; from the carpus to the end of the wing five inches nine lines.

DESCRIPT. All the upper parts of the head and body, wings, and two middle tail-feathers, cinereous brown, with a slight tinge of olivaceous; the other tail-feathers black, with a broad white space at the extremity of each of the three outermost; the fourth just tipped with white; primaries and wing-coverts bright rufous: throat, sides of the neck, and all the under parts white: upper mandible black, edged with yellow at the base; lower mandible yellow, tipped with black: legs black: tarsi long, naked. (*Egg.*) "Of a uniform greenish blue colour". WILS.

The above description of this species is taken from a specimen in the collection of the Zoological Society, which was killed in the preserves of Lord Cawdor in Wales, in the autumn of 1832. Three other individuals have occurred in this country, two in Ireland, and one in Cornwall. Inhabits the northern parts of America. Habits said to be essentially different from those of the *Common Cuckow.* Constructs its own nest, and rears its own young. Eggs three or four in number. Food, according to Wilson, principally caterpillars.

GEN. 46. CORACIAS, *Linn.*

128. C. *garrula*, Linn. (*Garrulous Roller.*)—Head, neck, and under parts, light bluish green : back and scapulars reddish brown.

C. garrula, *Temm. Man. d'Orn.* tom. I. p. 127. *Id.* tom. III. p. 72.
Garrulous Roller, *Mont. Orn. Dict. Selb. Illust.* vol. I. p. 117.
pl. 34. Roller, *Bew. Brit. Birds*, vol. I. p. 100.

DIMENS. Entire length twelve inches six lines. MONT.

DESCRIPT. Head, neck, breast, and belly, verditer blue, inclining to sea-green: back and scapulars reddish brown: lesser wing-coverts rich violet-blue; greater coverts pale green: rump, and basal portion of the quills, purplish blue; the tips of the quills dusky: tail somewhat forked; the outer feather pale ultra-marine blue, tipped with black; the others dusky green, tipped with pale bluish green, the two middle ones excepted, which are dusky throughout: bill black at the tip, yellowish brown at the base: feet yellowish. (*Egg.*) Smooth, shining white: long. diam. one inch five lines; trans. diam. one inch one line.

Very rare in this country, and only an accidental visitant. First noticed as a British bird by Pennant, who describes one killed in Cornwall. Solitary individuals have been since met with in Sussex, Suffolk, Norfolk, Scotland, Northumberland, and Yorkshire. Said to be common in Germany and Sweden, inhabiting large forests. Feeds principally on insects. Builds in the holes of decayed trees, and lays from four to seven eggs.

GEN. 47. MEROPS, *Linn.*

129. M. *Apiaster*, Linn. (*Common Bee-eater.*)—Forehead greenish white; nape, and upper part of the back, deep chestnut : throat yellow, bounded by a black line.

M. Apiaster, *Temm. Man. d'Orn.* tom. I. p. 420. *Id.* tom. III. p. 293.
Sow. Brit. Misc. pl. 69. Common Bee-eater, *Mont. Orn. Dict.
Selb. Illust.* vol. I. p. 114. pl. 41. *Bew. Brit. Birds*, vol. I. p. 146.

DIMENS. Entire length eleven inches. TEMM.

DESCRIPT. Forehead white, passing into bluish green; crown of the head, nape, and upper part of the back, deep chestnut; middle and lower regions of the back of a paler chestnut, passing into reddish yellow: from the corner of the bill a black streak passing over the eyes and across the ear-coverts, where it meets another line of black encircling the lower portion of the neck; throat golden yellow; the rest of the under parts bluish green: lesser wing-coverts grass-green; middle portion of the wing brownish red: quills dusky, passing into bluish green on their outer webs: tail greenish blue; the two middle feathers somewhat darker, pointed, and longer than the others by more than an inch: bill black: irides red: feet brown. In the *female*, the colours are for the most part similar, only paler: the two middle tail-feathers are relatively shorter. (*Egg.*) Smooth, and white; nearly round: long. diam. one inch and half a line; trans. diam. eleven lines.

Very rarely seen in this country. A flight of about twenty was observed near Mattishall in Norfolk, and one killed, in June 1794: since which period, other specimens have been met with on different occasions in Cornwall, Devonshire, and Ireland. Common in some parts of the Continent. Feeds principally on Hymenopterous insects. Nest placed in deep holes excavated in the banks of rivers for that purpose. Eggs five to seven in number.

GEN. 48.   ALCEDO, *Linn.*

130.   A. *Ispida*, Linn.   (*Common King-Fisher.*) — Crown of the head, and wing-coverts, deep green, spotted with azure-blue: behind the eye a patch of orange-brown passing into white.

A. Ispida, *Temm. Man. d'Orn.* tom. I. p. 423.   *Id.* tom. III. p. 296.
Common King-Fisher, *Mont. Orn. Dict.   Selb. Illust.* vol. I.
p. 136. pl. 40. f. 1.   *Bew. Brit. Birds*, vol. II. p. 121.

DIMENS.   Entire length seven inches three lines: length of the bill (from the forehead) one inch six lines, (from the gape) two inches; of the tarsus four lines and a half; from the carpus to the end of the wing three inches: breadth, wings extended, ten inches.
DESCRIPT.   Crown of the head deep olive-green, the feathers tipped with bright azure-blue: from the upper mandible to the eye a dusky streak; behind the eye a band of orange-brown, passing on the sides of the neck into a white patch; below this, extending from the base of the lower mandible to near the insertion of the wing, a broad streak of azure-green: middle of the back, rump, and upper tail-coverts, fine bright azure: throat, and fore part of the neck, yellowish white; rest of the under parts ferruginous orange: wing-coverts and quills deep greenish blue, spotted like the crown of the head, but more sparingly: tail greenish blue; the shafts of the feathers black: bill dusky brown, reddish at the base: irides hazel: feet orange-red.   (*Egg.*) Smooth, and white; nearly round: long. diam. ten lines and a half; trans. diam. nine lines.
Generally diffused over the country, and resident all the year.   Haunts the banks of rivers and clear streams, feeding on fish and aquatic insects. Is rapid on wing, and frequently utters a shrill cry in its flight.   Nest placed in holes in the ground near the water's edge, consisting simply of fish bones and other indigestable parts of the food rejected from the stomach after eating.   Eggs six or seven in number.

GEN. 49.   HIRUNDO, *Linn.*

131.   H. *rustica*, Linn.   (*Chimney Swallow.*)—Upper parts, and a transverse bar on the breast, bluish black; forehead and throat chestnut-red.

H. rustica, *Temm. Man. d'Orn.* tom. I. p. 427.   *Id.* tom. III. p. 297.
Chimney Swallow, *Mont. Orn. Dict.   Selb. Illust.* vol. I. p. 120.
pl. 42. f. 1.   Swallow, *Bew. Brit. Birds*, vol. I. p. 287.

DIMENS.   Entire length seven inches: length of the bill (from the forehead) three lines and a half, (from the gape) eight lines; of the tarsus five lines: of the tail four inches: the same, excluding the long lateral feathers, two inches six lines; from the carpus to the end of the wing four inches eleven lines: breadth, wings extended, thirteen inches four lines.

DESCRIPT. Forehead and throat chestnut-red: all the upper parts, sides of the neck, and a broad transverse bar on the breast, black, with purple and blue reflections: belly and vent reddish white: tail long, and very much forked; the two middle feathers plain; the rest with a large white spot on the inner web: bill and feet black. The *female* has less red on the forehead, and less black on the breast: the under parts are also whiter, and the outer tail-feathers not so long as in the male bird. In *young* birds, the long tail-feathers do not appear till after the first moult. White *varieties* are sometimes met with. (*Egg.*) White, spotted and speckled with ash-colour and dark red-brown: long. diam. nine lines and a half; trans. diam. six lines and a half.

A summer visitant, making its first appearance about the second or third week in April, and staying till towards the middle or end of October. Feeds entirely on insects, taken on the wing. Builds generally in chimnies: nest formed of mud plastered together, and lined with feathers. Eggs four or five in number: two broods in the season. Previous to migration, congregates in large flocks.

132. H. *urbica*, Linn. (*House Martin.*)—Head, nape, and back, bluish black; rump white: tarsi and toes feathered.

H. urbica, *Temm. Man. d'Orn.* tom. I. p. 428. *Id.* tom. III. p. 300. Martin, *Mont. Orn. Dict. Selb. Illust.* vol. I. p. 123. pl. 42. f. 2. *Bew. Brit. Birds*, vol. I. p. 293.

DIMENS. Entire length five inches six lines: length of the bill (from the forehead) three lines, (from the gape) six lines; of the tarsus six lines; of the tail two inches six lines and a half; from the carpus to the end of the wing four inches four lines.

DESCRIPT. Head, nape, and all the upper region of the back, glossy bluish black: wings and tail dusky brown; the latter forked, but not so long as in the last species: rump, and all the under parts from the chin to the vent, pure white: tarsi and toes covered with a white down. (*Egg.*) Smooth, delicate white: long. diam. nine lines and a half; trans. diam. six lines.

Rather later in its arrival than the last species, being seldom observed in abundance before the end of April or beginning of May. Departs in October; but a few stragglers may occasionally be seen on to November. Food and habits similar to those of the Swallow: flight not so rapid. Nest usually fixed under the eaves of houses or in the angles of windows; formed entirely of mud externally, and lined with feathers. Lays four or five eggs, and has frequently two broods in the Summer. Congregates at the approach of Autumn.

133. H. *riparia*, Linn. (*Bank Martin.*)—All the upper parts, and a transverse band on the breast, cinereous brown: tarsi and toes naked.

H. riparia, *Temm. Man. d'Orn.* tom. I. p. 429. *Id.* tom. III. p. 300. Sand Martin, *Mont. Orn. Dict. Selb. Illust.* vol. I. p. 125. pl. 42. f. 3. *Bew. Brit. Birds*, vol. I. p. 295.

Dimens. Entire length five inches two lines: length of the bill (from the forehead) three lines, (from the gape) six lines; of the tarsus five lines and a half; of the tail two inches one line; from the carpus to the end of the wing four inches.

Descript. All the upper parts, cheeks, and a transverse band on the breast, cinereous or mouse-coloured brown: wings and tail inclining to dusky brown: throat, fore part of the neck, belly, and under tail-coverts, white: tarsi and toes naked, with the exception of a few small feathers near the origin of the hind toe. In *young* birds, all the upper parts of the plumage are edged with pale reddish brown; the tail-feathers with yellowish white. (*Egg.*) White: long. diam. eight lines; trans. diam. six lines.

First seen about the beginning of April. More locally distributed than either of the preceding species. Found only in the neighbourhood of sand-pits and the high banks of rivers, in which situations it builds, excavating a horizontal hole in the loose soil to the depth of two or three feet. Nest placed at the extremity of the hole, formed of dry grass and straw, and lined with feathers. Eggs four or five in number.

GEN. 50.　CYPSELUS, *Illig.*

134.　C. *Apus*, Flem. (*Common Swift.*)—Chin white; all the rest of the plumage black.

> C. murarius, *Temm. Man. d'Orn.* tom. I. p. 434. *Id.* tom. III. p. 303. Common Swift, *Selb. Illust.* vol. I. p. 127. pl. 42. f. 4. Swift, *Mont. Orn. Dict. Bew. Brit. Birds*, vol. I. p. 296.

Dimens. Entire length eight inches: length of the bill (from the forehead) three lines and a half, (from the gape) eight lines and a half; of the tarsus six lines; of the tail three inches four lines; from the carpus to the end of the wing five inches ten lines and a half: breadth, wings extended, seventeen inches.

Descript. Throat grayish white: all the rest of the plumage above and below sooty black, with greenish reflections: wings and tail extremely long; the latter much forked: tarsi covered with small feathers: irides deep brown: bill and feet black. (*Egg.*) White: long. diam. one inch; trans. diam. eight lines.

Late in its arrival, seldom shewing itself before the beginning, and in some parts of the country not till near the end, of May. Departs also much sooner than the other species of this family; generally about the middle of August. Haunts churches, towers, and other lofty buildings, in the holes of which it breeds. Nest formed of dried grass, and lined with feathers. Eggs two in number. Only one brood in the season. Feeds entirely on insects, taken on the wing. Flight high, and extremely rapid.

135.　C. *alpinus*, Temm. (*Alpine Swift.*)—General colour of the plumage grayish brown: throat, and middle of the abdomen, white.

> C. alpinus, *Temm. Man. d'Orn.* tom. I. p. 433. *Id.* tom. III. p. 303. Greatest Martin, *Edw. Nat. Hist.* vol. I. pl. 27. White-bellied Swift, *Lath. Syn.* vol. II. p. 586.

DIMENS. Entire length nearly nine inches. TEMM.

DESCRIPT. All the upper parts of a uniform grayish brown; breast, sides of the abdomen, and under tail-coverts the same: throat, and middle of the belly, pure white: tarsi covered with brown feathers. (*Egg.*) White: of the same size as that of the last species.

Shot within eight or ten miles of the south coast of Ireland in 1829, about Midsummer. A second individual has been since killed at Kingsgate, in the Isle of Thanet, and a third in Norfolk. Common in alpine and rocky districts in the southern parts of Europe. Habits somewhat similar to those of the Common Swift. Said to build in the clefts of rocks and old buildings, and to lay three or four eggs.

GEN. 51. CAPRIMULGUS, *Linn.*

136. C. *Europæus*, Linn. (*European Goatsucker.*)— General colour of the plumage ash-gray, variegated with black, brown, and ferruginous.

C. Europæus, *Temm. Man. d'Orn.* tom. I. p. 436. *Id.* tom. III. p. 304. European Goat-sucker, *Mont. Orn. Dict. Selb. Illust.* vol. I. p. 131. pl. 42*. Night-Jar, *Bew. Brit. Birds*, vol. I. p. 302.

DIMENS. Entire length ten inches four lines: length of the bill (from the forehead) four lines and a half, (from the gape) one inch two lines; of the tarsus eight lines; of the tail four inches eight lines; from the carpus to the end of the wing seven inches nine lines: breadth, wings extended, twenty-two inches.

DESCRIPT. General colour of the plumage ash-gray, beautifully variegated with black, brown, ferruginous, and white, disposed in bars, spots and specks, of different shades and sizes: on the head, and down the middle of the back, some longitudinal black streaks: a little white on the throat, as well as beneath each eye: under parts in general yellowish brown, with transverse undulating lines of black: quills dusky, with ferruginous spots on their outer webs; the three first with also a large white patch on the inner web about midway: tail nearly square, yellowish gray, with transverse zigzag bars of black; the two outer feathers on each side tipped with white: bill and irides dusky: feet yellowish brown. The *female* wants the white spots on the tips of the quills and two outer tail-feathers. (*Egg.*) Of an oval form: ground colour white, mottled with cinereous; this last colour including spots of two shades of brown: long. diam. one inch two lines; trans. diam. ten lines and a half.

One of our latest summer visitants, not appearing before the middle or end of May. Stays till September or October. Generally distributed over the kingdom, but not equally plentiful in all parts. Found principally in wild uncultivated districts, where there is wood. Habits crepuscular. Food insects; particularly the *Melolontha* tribe, and the larger *Lepidoptera.* Makes no nest, but lays two eggs on the bare ground amongst fern, heath, or long grass. During the season of incubation the male utters a peculiar noise somewhat resembling that of a spinning-wheel.

ORDER III.   RASORES.

GEN. 52.   COLUMBA, *Linn.*

(1. Columba, *Swains.*)

137.   C. *Palumbus*, Linn. (*Ring-Dove.*)—Bluish ash;
a white space on the sides of the neck, and on the edge
of the wing; extremity of the tail black.

C. Palumbus, *Temm. Man. d'Orn.* tom. II. p. 444. *Id. Pig. et Gall.*
tom. I. p. 78.   Ring-Dove, *Mont. Orn. Dict. Selb. Illust.* vol. I.
p. 406. pl. 56. f. 1.   *Bew. Brit. Birds*, vol. I. p. 307.

DIMENS.   Entire length sixteen inches six lines: length of the bill
(from the forehead) nine lines and a half, (from the gape) one inch two
lines; of the tarsus one inch four lines; of the tail six inches six lines;
from the carpus to the end of the wing ten inches: breadth, wings
extended, twenty-nine inches four lines.

DESCRIPT.   Head, cheeks, throat, rump, and basal portion of the tail,
bluish ash; on each side of the neck a patch of white; a narrow line of
the same colour running longitudinally down the wing near the edge;
breast and upper part of the belly vinaceous red, with glossy green reflections; back and wing-coverts deep bluish gray; quills dusky, edged with
white; a broad black bar at the extremity of the tail; vent, thighs, and
under tail-coverts, grayish white: bill orange; the soft portion at the base
covered with a white mealy substance: irides light yellow: feet red.
(*Egg.*) White: long. diam. one inch eight lines; trans. diam. one inch
two lines.

Common in wooded districts throughout the country, remaining the
whole year.   Utters a cooing note, heard from February to the beginning
of October.   Builds in April: nest placed in trees, formed of twigs loosely
put together, and very shallow.   Eggs always two in number.   Frequently
a second, or even a third brood.   Food, grain and seeds of various kinds;
during severe weather, the leaves of turnips, and other vegetables.   Collects in large flocks at the approach of Winter.

138.   C. *Œnas*, Linn. (*Stock-Dove.*) —Bluish ash;
sides of the neck with green and purple reflections; some
of the wing-coverts spotted with black, forming an irregular transverse bar; rump bluish gray; tip of the tail
black.

C. Œnas, *Temm. Man. d'Orn.* tom. II. p. 445. *Id. Pig. et Gall.*
tom. I. p. 118.   Stock-Dove, *Selb. Illust.* vol. I. p. 408. pl. 56*.
f. 1.

DIMENS.   Entire length twelve inches six lines: length of the bill
(from the forehead) eight lines and a half, (from the gape) eleven lines

L

and a half; of the tarsus one inch; of the tail four inches; from the carpus to the end of the wing eight inches six lines.

Descript. Head and throat deep bluish gray; sides of the neck glossed with metallic hues of green and purple; back brownish gray; rump and upper tail-coverts pale bluish ash; breast pale vinaceous red; belly, thighs, and under tail-coverts, bluish gray; quills dusky, passing into bluish gray at the base of the feathers; wing-coverts of the same colour as the back, but rather paler; some of the greater ones, as well as the three last secondary quills, spotted with black, forming an irregular bar across the wings: tail bluish ash passing into black at the extremity; the outer web of the external feather with an oblong white spot towards the base: irides deep reddish brown: feet red. (*Egg.*) White: oval: long. diam. one inch six lines and a half; trans. diam. one inch two lines.

Inhabits woods with the preceding species, but is less plentiful and more local. Not uncommon in some of the midland and eastern counties, where it remains the whole year. Builds in the hollows of old pollard trees, and lays two eggs. Does not cooe like the Ring-Dove, but utters a hollow rumbling note, heard at intervals throughout the spring and summer months. Flocks with that species in the Winter, and supports itself in the same manner.

139. C. *Livia*, Briss. (*Rock-Dove.*) — Bluish ash; sides of the neck glossed with green reflections; wings with two distinct transverse bars; rump white; tip of the tail black.

C. Livia, *Temm. Man. d'Orn.* tom. ii. p. 446.   *Id. Pig. et Gall.* tom. i. p. 125.   Rock-Dove, *Selb. Illust.* vol. i. p. 410. pl. 56*. f. 2. Wild Pigeon, *Bew. Brit. Birds*, vol. i. p. 309.

Dimens. Entire length twelve inches eight lines.
Descript. Head, throat, upper part of the back, wing-coverts, and under parts, bluish ash; sides of the neck, and upper part of the breast, glossed with shades of green and purple-red; rump, and lower part of the back, white: wings with two distinct black transverse bars; quills tipped with black: tail deep bluish gray at the base, black at the extremity; outer feather with the external web white: irides pale reddish orange: feet red. (*Egg.*) White: of a sub-oval form, and rather pointed: long. diam. one inch five lines; trans. diam. one inch two lines and a half.

Found on rocky cliffs, principally those in the neighbourhood of the sea. Met with in various parts of England, but is most abundant on the eastern coast. Breeds in caverns, and the recesses of rocks, and lays two eggs. Has two or three broods in the year. Feeds on grain, seeds, and some of the smaller *Helices*. *Obs.* The *Domestic Pigeon*, with its numerous varieties, is descended from this species.

140. C. *Turtur*, Linn. (*Turtle-Dove.*) — Back and rump cinereous brown; a black space on the sides of the neck; tip of the tail white.

C. Turtur, *Temm. Man. d'Orn.* tom. ii. p. 448.   *Id. Pig. et Gall.* tom. i. p. 305.   Turtle-Dove, *Mont. Orn. Dict.   Selb. Illust.* vol. i. p. 413. pl. 56. f. 2.   *Bew. Brit. Birds*, vol. i. p. 312.

Dimens. Entire length eleven inches three lines: length of the bill (from the forehead) eight lines, (from the gape) ten lines; of the tarsus ten lines; of the tail four inches four lines; from the carpus to the end of the wing six inches nine lines: breadth, wings extended, twenty inches.

Descript. Head and nape cinereous, with a tinge of vinaceous red; on each side of the neck a patch of black feathers tipped with grayish white; back and rump cinereous brown; inside and edge of the wings bluish ash; greater quills brownish black; secondaries bluish gray; scapulars and wing-coverts ferruginous brown inclining to rust-red, with a black spot in the middle of each feather: fore part of the neck and breast pale vinaceous; belly and under tail-coverts pure white: the two middle tail-feathers wholly brown; the rest bluish black, tipped with white; the outer one also white on the external web: bill brown: irides reddish orange: orbits and feet red. (*Egg.*) White: rather more pointed than in any of the former species: long. diam. one inch two lines and a half; trans. diam. ten lines.

Var. β. Spotted-necked Turtle. *Lath. Syn.* vol. II. p. 645. *Don. Brit. Birds*, vol. VII. pl. 149. Characterized principally by having the whole side of the neck black, each feather having a round white spot near the extremity, instead of being tipped with white.

A migratory species, visiting this country in May, and departing at the approach of Autumn. Not generally diffused. Said to be most abundant in Kent and Buckinghamshire. Occurs sparingly in some of the eastern and western counties, but is rare northward. Frequents thick woods, and builds in trees, constructing a nest like that of the Ring-Dove, but smaller. Eggs two in number, laid about the middle of June. Utters a peculiar plaintive note during the breeding season, which is sometimes continued at intervals till near the middle of August. Food all kinds of grain and seeds. The *Spotted-necked* variety, first noticed by Dr Latham, appears to have been only met with in Buckinghamshire.

(2. Ectopistes, *Swains.*)

141. C. *migratoria,* Linn. (*Passenger Pigeon.*) — Cinereous; sides of the neck glossed with green and purple; wing-coverts spotted with black.

C. migratoria, *Temm. Pig. et Gall.* tom. I. p. 346. Passenger Pigeon, *Lath. Syn.* vol. II. p. 661. (Male.) Canada Turtle, *Id.* vol. II. p. 658. (Female.) Pass. Pig., *Wilson, Amer. Orn.* vol. V. p. 102. pl. 44. f. 1.

Dimens. Entire length sixteen inches nine lines and a half: length of the bill one inch; of the tarsus one inch one line and a half: breadth, wings extended, twenty-four inches six lines. Flem.

Descript. "Chin, cheeks, head, back, and rump, bluish gray; shoulders with a tinge of yellowish brown: side of the neck, and behind, rich reddish purple, iridescent: fore-neck deep chestnut, becoming paler on the breast, or rather salmon-coloured, and passing to white on the belly and vent: thighs like the breast: quills brownish black, the gray colour of the margin of the outer web increasing at the base of the secondaries, and towards the ends of the inner ones: bastard wing, and greater coverts of the primaries, brownish black; greater coverts of the secondaries gray: lesser coverts and outer scapulars tinged with yellowish brown, with black spots: tail of twelve feathers, the two middle ones produced, the rest decreasing to the exterior: the two middle ones dusky black, the next gray, the inner margin white towards the extremity, with a black and brown spot near the base; the fourth and third gray, with the black spot; the second gray, with the black and brown spot; the outer web and tip of the first white, lower half of the inner web gray, with a black and brown spot: upper tail-coverts long, produced; lower ones white: bill black: bare space round the eyes livid: irides reddish orange: feet reddish, paler behind than before; claws black." FLEM. The *female*, according to Wilson, is somewhat smaller, with the colours in general less vivid and more tinged with brown: the gold spot on the sides of the neck smaller, and less brilliant.

A native of North America. The only individual which has hitherto occurred in this country is recorded by Dr Fleming (*Brit. An.* p. 145.) to have been "shot in the neighbourhood of a pigeon-house at Westhall, in the parish of Monymeal, Fifeshire, Dec. 31, 1825. The feathers were quite fresh and entire, like those of a wild bird."

---

* GEN. 53. MELEAGRIS, *Linn.*

* 142. M. *Gallopavo*, Linn. (*Turkey.*)—Head, and upper part of the neck, almost naked, with a bluish papillose skin: a tuft of black hairs on the breast.

M. Gallopavo, *Temm. Pig. et Gall.* tom. II. p. 374. and tom. III. p. 677.
Turkey, *Bew. Brit. Birds*, vol. I. p. 325.

DIMENS. Entire length three feet six inches: breadth, wings extended, about four feet.

DESCRIPT. Head, and upper portion of the neck, bare of feathers; skin bluish, rough with numerous fleshy papillæ of various hues and sizes; throat furnished with a pendulous carunculated wattle, of a bright scarlet colour, increasing in intensity when the bird is under excitement: also an elongated fleshy appendage arising from the base of the upper mandible, in its contracted state about an inch long, but when relaxed hanging down considerably below the bill: a tuft of long pendent hair from the middle of the breast: colour of the plumage very variable ; generally dark gray, inclining to black, or black with transverse whitish bars ; occasionally pure white, or pied; more rarely of a fine deep copper-colour, with the greater quills pure white, and the tail dirty white. In the *female* the pectoral tuft is wanting ; the frontal caruncle is smaller, and remains always contracted; there is also no power of erecting and expanding the tail, as in the male bird. (*Egg.*) Yellowish white, spotted and speckled all over with reddish yellow: long. diam. three inches; trans. diam. one inch eleven lines.

Found in a wild state in North America. Generally supposed to have been introduced into England about the year 1524. Lays early in the Spring, and produces from fifteen to seventeen in a brood. Seldom hatches more than once in the season in this climate.

---

* GEN. 54. PAVO, *Linn.*

* 143. P. *cristatus*, Linn. (*Crested Peacock.*)—Crest on the head compressed.

P. cristatus, domesticus, *Temm. Pig. et Gall.* tom. II. p. 35. and tom. III. p. 651.
Peacock, *Bew. Brit. Birds*, vol. I. p. 328.

DIMENS. Entire length three feet eight inches: length of the train four feet five inches ; of the tail one foot seven inches.
DESCRIPT. (*Male.*) Crest of twenty-four feathers, the shafts slender, and scarcely webbed except at the tips, which are of a golden green colour : head, throat, neck, and breast, of a rich blue, glossed with green and gold ; above and

beneath the eye a white streak ; back and rump golden green, glossed with copper, the feathers edged and tipped with velvet black ; train, or upper tail-coverts, very much elongated, reaching considerably beyond the tail, capable of being erected and expanded at will ; each of the feathers composing it having a conspicuous ocellated spot at the extremity ; the true tail concealed beneath the train, consisting of eighteen feathers, of a grayish brown colour: scapulars and lesser wing-coverts variegated with black, and reddish cream-colour ; middle coverts deep blue, glossed with golden green ; greater coverts, bastard wing, and primary quills, rufous ; the other quills dusky, some of them being variegated with red, and tinged with golden green : belly and sides greenish black ; thighs tawny yellow : irides yellow : bill whitish : feet grayish brown. (*Female.*) Smaller ; the train shorter than the tail, and without the ocellated spots ; sides of the head with more white ; throat of that colour ; neck green ; general colour of the body, and wings, cinereous brown, the feathers on the breast being tipped with white : irides lead gray. White and pied *varieties* sometimes occur in both sexes. (*Egg.*) Yellowish white, sparingly speckled with reddish yellow : long. diam. two inches eight lines ; trans. diam. two inches one line.

A native of India, where it is still met with in an unreclaimed state. The period of its first introduction into this country, not very well ascertained. Lays from five to eight eggs. Time of incubation from twenty-seven to thirty days. The approach of the breeding season announced by the loud discordant screams of the male bird, first heard towards the end of March, and continued at intervals through the Summer. The two sexes are similar in plumage during the first year, and the train of the male does not appear till the third.

* GEN. 55. GALLUS, *Briss.*

* 144. G. *domesticus,* Briss. (*Domestic Cock.*) — Caruncle on the head compressed, denticulated ; throat with two pendulous wattles : neck-feathers linear and elongated.

G. domesticus, *Temm. Pig. et Gall.* tom. II. p. 92. and tom. III. p. 654. The Cock, *Bew. Brit. Birds,* vol. I. p. 316.

DIMENS. Extremely variable.

DESCRIPT. Crest or comb with eight or nine serratures, of a bright coral-red ; two long wattles beneath the lower mandible of the same colour ; throat, and space round the eyes, naked, the skin on these parts red ; beneath the ears a naked white spot : neck-feathers very much elongated, and linear throughout their whole length : tail ascending, compressed, forming two planes inclined to one another at an acute angle ; of fourteen feathers, the two middle ones considerably the longest, and bending gracefully over the others : thighs strong and muscular : tarsi with long bent spurs : colours of the plumage very variable ; more commonly, the head, neck, back, and wing-coverts, orange-red ; under parts whitish, or velvet black ; the sickle-shaped feathers of the tail blackish blue : sometimes the whole plumage pure white. The *Hen* is always smaller, with the comb and wattles less developed ; the colours are also less brilliant, and the tail wants the long pendent feathers. (*Egg.*) White ; varying in size and shape according to the breed.

Var. β. cristatus. (Crested Cock.) *Temm. Pig. et Gall.* tom. II. p. 239. Distinguished by having a tuft of feathers on the head instead of a comb.

Var. γ. pusillus. (Bantam Cock.) *Temm. Pig. et Gall.* tom. II. p. 242. A small variety, with the feet and toes feathered.

Var. δ. Pumilio. (Dwarf Cock.) *Temm. Pig. et Gall.* tom. II. p. 244. Very small, scarcely larger than a pigeon : feet short, generally feathered.

Var. ε. pentadactylus. (Dorking Cock.) *Temm. Pig. et Gall.* tom. III. p. 658. Differs from the others in having five toes on each foot, three in front, and two behind. Generally considered as most abundant in the neighbourhood of Dorking, in Surrey.

Known in a state of domestication from the earliest times. The original stock very uncertain ; referred by Temminck to the *G. Bankiva,* Temm. (*Pig. et Gall.* tom. II. p. 87.) a species met with at the present day in a wild state in the Island of Java. In the domestic varieties, the cock is polygamous. The hen is very prolific, and continues to lay during a great part of the year. Period of incubation about three weeks.

Besides the above, the following species are domesticated in some parts of England.

* (10.) *G. Morio*, Temm. (*Negro Cock.*) Pig. et Gall. tom. ii. p. 253.

    Originally from India. Remarkable for having the comb, wattles, skin, and periosteum black.

* (11.) *G. lanatus*, Temm. (*Silk Cock.*) Pig. et Gall. tom. ii. p. 256.

    A native of Japan. Has the webs of the feathers disunited, and of a very silky texture. Plumage white.

* (12.) *G. crispus*, Briss. (*Frizzled Cock.*) Temm. Pig. et Gall. tom. ii. p. 259.

    From Asia. All the feathers reflexed, and as it were curled. Smaller than the common poultry.

* (13.) *G. ecaudatus*, Temm. (*Rumpless Cock.*) Pig. et Gall. tom. ii, p. 267.

    Inhabits Ceylon. No tail or tail-coverts: wants the last dorsal vertebra.

* GEN. 56. PHASIANUS, *Linn.*

* 145. P. *Colchicus*, Linn. (*Common Pheasant.*)—Head and neck metallic green, glossed with blue; breast and flanks reddish orange with purple reflections, the feathers edged and tipped with violet black.

P. Colchicus, *Temm. Man. d'Orn.* tom. ii. p. 453. *Id. Pig. et Gall.* tom. ii. p. 289. Common Pheasant, *Mont. Orn. Dict. Selb. Illust.* vol. i. p. 417. pl. 57.

DIMENS. Entire length three feet: length of the bill (from the forehead) one inch two lines, (from the gape) one inch four lines; of the tarsus three inches; of the tail one foot nine inches; from the carpus to the end of the wing ten inches: breadth, wings extended, two feet seven inches six lines. (*Male.*)
DESCRIPT. (*Male.*) Head and neck of a rich metallic green, passing beneath into blue and violet; on each side of the occiput a tuft of dark golden green feathers capable of being erected at will, and most conspicuous in the breeding season: a large naked space on the cheeks, thickly studded with scarlet papillæ intermixed with minute black specks: lower part of the neck, breast, belly, and sides, of a brilliant orange red, with a faint tinge of purple, each feather being edged and tipped with violet black; lower part of the abdomen, and thighs, blackish brown: scapulars, and feathers on the back, dusky brown in the middle, broadly edged with purplish orange, within which is a yellowish white band; lower part of the back, tail-coverts, and saddle-hackle feathers, exhibiting different shades of green, intermixed with orange and purple: primary and secondary quills dusky brown, with yellowish white bars: tertials and wing-coverts reddish yellow, stained and spotted with dark purple-red: tail long, and very much cuneated, of an olive-gray, or grayish brown colour, with transverse black bars, each feather being fringed with purplish red: bill pale horn-colour: irides yellowish orange: feet grayish black: spurs sharp and pointed, half an inch or more in length. (*Female.*) Smaller: general colour of the plumage yellowish brown, variegated with gray and rufous; head, neck, and upper part of the body, with the central portion of each feather black: region of the eyes feathered: tail much shorter than in the male bird, but barred in a similar manner. (*Young of the year, till after the first moult.*) Of a uniform gray colour, somewhat resembling the adult female: spur of the male short and blunt. *Varieties*, white and pied, are not unfrequent. (*Egg.*) Uniform olive-brown: long. diam. one inch ten lines; trans. diam. one inch five lines.
Var. β. Ring Pheasant. *Bew. Brit. Birds*, vol. i. p. 321. With a collar of white round the neck. This variety, which is not uncommon, has originally proceeded from a cross between this and the following species, with which last it must not be confounded.

Var. γ. Hybrid Pheasant. (*Male.*) Head, neck, and breast, deep brownish black, with a slight gloss of bottle-green ; the rest of the plumage above and below the body, scapulars and wing-coverts, deep red brown, the feathers edged with glossy black, and many of them streaked with yellow down the shafts : tail shorter than in the common Pheasant, compressed, and slightly arched, of a deep brown, approaching to black, variegated with specks and transverse undulating lines of ochre-yellow. (*Female.*) Head, neck, and breast, dull brown without the gloss of green ; a greater portion of the upper plumage reddish or yellowish brown ; many of the feathers without the black border : tail brown, barred with reddish yellow ; more cuneated than in the male bird. This variety, which is occasionally met with, is an hybrid production between the *Common Pheasant* and the *Domestic Fowl.* The plumage is of course subject to considerable variation. The above descriptions were taken from a pair in the museum of the Cambridge Philosophical Society. Is sometimes called a *Pero.* The *female* has been known to breed again with the *male Game-Fowl.*

Supposed to have been brought originally from the banks of the Phasis, a river of ancient Colchis. Common at present throughout the greater part of Asia and Europe. Period of its first introduction into Britain uncertain. As a naturalized species, is generally distributed throughout England, but considered rare in Scotland. Frequents woods. Is polygamous, like the Domestic Cock. Commences laying in April, and hatches towards the end of May. Number of eggs from ten to fourteen, deposited on the ground, amongst long grass, or in cornfields ; sometimes in thick copses. Barren hens, which have partially assumed the spurs and plumage of the cock, are not unfrequently met with, and are termed by sportsmen *Mule-Birds.*

* 146. P. *torquatus,* Temm. (*Ring-necked Pheasant.*)—Head and neck bluish green, the latter encircled by a white collar; a line of white over each eye; lower part of the neck and breast copper-red, the feathers deeply divided by a black line.

P. torquatus, *Temm. Pig. et Gall.* tom. ii. p. 326. and tom. iii. p. 670. Ring Pheasant, *Leach, Zool. Misc.* vol. ii. p. 14. pl. 66. *Shaw, Gen. Zool.* vol. xi. p. 228.

Dimens. Rather less than those of the last species.

Descript. (*Male.*) Head and upper part of the neck deep green, with a tinge of violet blue ; below this colour, a collar of pure white, broadest on the sides of the neck, narrow behind, entirely interrupted in front : over each eye a narrow white line, not always present : plumage on the upper part of the back, between the shoulders, of a paler colour than in the *P. Colchicus,* approaching to orange-yellow, the feathers of a more pointed form, and deeply divided by a blackish blue line ; all the lower portion of the back varied with different shades of green, passing off at the sides into bluish ash ; saddle-hackle feathers pale rust-colour ; upper tail-coverts the same, passing into greenish yellow ; feathers on the fore part of the neck below the white collar, and those on the breast, of a fine bright copper-red, with a much narrower edging of black than in the last species, but deeply indented at the tip by a lanceolate black line : wings and tail paler than in the Common Pheasant. (*Female.*) General colour similar to that of the female of *C. Colchicus;* breast not so much spotted ; transverse bars on the tail more distinctly marked. (*Egg.*) Bluish green, with small spots of a darker tint. Temm.

A native of China, from whence it has been introduced into England, and naturalized in many parts of the country. In consequence, however, of its breeding freely with the Common Pheasant*, it has become so intermixed with that species, as very rarely to occur at present exhibiting the pure plumage which characterizes it in its wild state. The above description of the cock bird was taken from a fine specimen in the possession of Mr Leadbeater.

* (14.) *Bohemian Pheasant.*

This name is employed in many parts of England to distinguish a species or variety of the Pheasant, which is met with in several preserves, but which does not appear to have received the notice of ornithologists. The

---

* This does not happen in its native country, where both species are equally plentiful, but, according to Temminck, keep perfectly separate.

head and neck are coloured much the same as in the *P. colchicus,* but all
the rest of the plumage is of a uniform pale brownish yellow, the feathers
being edged with black, and indented at the tips, as in the species last
described : tail rather darker than the body, but paler than in the common
Pheasant. The history of this peculiar breed, together with the origin of
its name, does not appear to be well ascertained.

* GEN. 57. NUMIDA, *Linn.*

* 147. N. *Meleagris,* Linn. (*Guinea Pintado.*)—Head and upper
part of the neck naked, the skin bluish : plumage on the body bluish
gray, with white spots.

N. Meleagris, *Temm. Pig. et Gall.* tom. ii. p. 431. and tom. iii. p. 680. Guinea
Pintado, *Lath. Syn.* vol. ii. p. 685. *Bew. Brit. Birds,* vol. i. p. 332.

Dimens. Entire length twenty-two inches : length of the bill one inch three
lines.
Descript. Head bare of feathers, and covered with a naked bluish skin ; on the
crown a callous conical protuberance, compressed at the sides and directed backwards, of a bluish red colour ; at the base of the upper mandible, on each side,
a loose pendulous wattle, bluish in the male, red in the female : upper part of the
neck nearly naked, being thinly furnished with hair-like feathers, which on the nape
are directed upwards ; the skin bluish ash ; lower part of the neck feathered, inclining to purple : general colour of the plumage on the other parts, dusky, or dark
bluish gray, sprinkled all over, the breast alone excepted, with round white spots of
various sizes : back very much rounded : tail short, and bent down : feet brownish
red. (*Egg.*) Yellowish white ground, mottled all over with reddish yellow : long.
diam. two inches ; trans. diam. one inch seven lines.
Brought originally from Africa, but has long been domesticated in Britain. Is
very prolific ; but the young are difficult to rear. Of a restless disposition, and very
clamorous.

GEN. 58. TETRAO, *Linn.*

(1. Tetrao, *Steph.*)

† 148. T. *Urogallus,* Linn. (*Wood Grous.*)—Chin and throat-feathers
elongated : breast glossed with dark green : bill white : tail rounded at
the extremity.

T. Urogallus, *Temm. Man. d'Orn.* tom. ii. p. 457. *Id. Pig. et Gall.* tom. iii.
pp. 114, & 696. Wood Grous, *Lewin, Brit. Birds,* vol. iv. pl. 132.
Mont. Orn. Dict. Bew. Brit. Birds, vol. i. p. 335. (Trachea) *Linn.
Trans.* vol. xvi. pl. 21. f. 1.

Dimens. Entire length (*male*) two feet ten inches, (*female*) two feet one inch :
breadth, wings extended, three feet six inches.
Descript. (*Male.*) Head and neck dusky ash, passing into black on the chin
and throat where the feathers are elongated ; a bare red skin above the eye, and
beneath it a small spot of white feathers ; breast of a fine dark glossy green ; rest of
the under parts black, with white spots : wings and scapulars chestnut brown, finely
speckled with dusky ; lower part of the back, rump, sides, and upper tail-coverts,
marked with numerous small undulating lines and specks of black, upon an ash-
coloured ground : tail very much rounded, black, some of the feathers having a
white spot on each side near the extremity : tarsi thickly clothed with brown hair-
like feathers : bill yellowish white : irides hazel. (*Female.*) Very much smaller :
head, neck, and back, barred transversely with black and tawny red ; throat tawny
red, without spots ; breast deep red, with a few white spots ; belly barred like the
back, some of the feathers tipped with white : quills dusky, mottled on their outer
webs with light rufous brown : tail dark red with black bars ; the tip white : bill
dusky. (*Young male after the first moult.*) Upper plumage not so deep coloured
as in the adult, inclining to cinereous gray ; green on the breast more dull ; a few
red feathers shewing themselves in different parts of the body ; tail often tipped with

white. *Before the first moult,* both sexes resemble the adult female. (*Egg.*) Light reddish yellow brown, spotted all over with two shades of darker brown: long. diam. two inches three lines ; trans. diam. one inch eight lines.

Formerly abundant in the mountainous forests of Scotland and Ireland, but now extirpated. The last specimen killed in the former country is said to have been shot near Inverness, about fifty years ago. Ceased to exist in Ireland at a considerably earlier period. Occurs plentifully at the present day in many parts of the Continent, and is much attached to pines, birch, and juniper. Said to feed on the berries of the latter, and on the buds and tender sprays of the two former. The males are polygamous. The females build on the ground, and lay from six to sixteen eggs.

† (15.)   *T. medius,* Meyer. (*Hybrid Grous.*) Temm. Man. d'Orn. tom. II. p. 459. Id. Pig. et Gall. tom. III. pp. 129, & 698. *Spurious Grous,* Lath. Syn. Supp. p. 214.

> Said to have been formerly found in Scotland, on the authority of Mr Tunstall; but there is no existing evidence by which the truth of this assertion can be proved. Is principally distinguished from the last species by having the neck and breast of a rich bronzed purple hue, the bill black, and the tail slightly forked, with the outermost feathers the longest. The *egg,* which is figured by Klein, is very similar in colour to that of the *Wood Grous,* but in size a little smaller.

149. T. *Tetrix,* Linn. (*Black Grous.*) — Throat-feathers not elongated : the general plumage violet-black, with a white bar on the wings : tail very much forked ; the lateral feathers bending outwards.

T. Tetrix, *Temm. Man. d'Orn.* tom. II. p. 460. *Id. Pig. et Gall.* tom. III. pp. 140, & 699.   Black Grous, *Mont. Orn. Dict. Selb. Illust.* vol. I. p. 423. pls. 58, & 58*. *Bew. Brit. Birds,* vol. I. p. 338.

Dimens. Entire length twenty-three inches: breadth, wings extended, three feet.

Descript. (*Male.*) Head, neck, breast, back, and rump, black, with purple and blue reflections: eyebrows naked, vermilion-red ; beneath the eye a white spot: abdomen, wing-coverts, and tail, deep black . secondary quills tipped with white, forming, with the adjacent coverts, a broad bar of that colour across each wing: under tail-coverts pure white : the lateral feathers of the tail much longer than the middle ones, curling outwards: tarsi clothed with blackish gray hair-like feathers : bill black: irides bluish. (*Female.*) Smaller ; general colour of the upper plumage ferruginous yellow, barred and mottled with black : greater wing-coverts tipped with white: breast orange-brown, with black bars : belly dusky brown, with whitish and red bars : tail very slightly forked, variegated with ferruginous and black; the tip grayish white. *The young of the year, till after the first moult,* resemble the adult female. (*Egg.*) Yellowish white, spotted and speckled with orange-brown : long. diam. two inches ; trans. diam. one inch five lines.

Most abundant in Scotland, and the northern parts of England. Occurs more or less sparingly in some parts of North Wales, as well as in the counties of Stafford, Somerset, Devon, Sussex, and Hants. Partial to woody, heathy, and mountainous situations. Feeds on berries, and on the tops of heath and birch. Is polygamous. Nest placed on the ground, generally under the shelter of a low bush, composed of a few dried stems of grass. Eggs from six to ten in number, laid in May. *Obs.* The *Hybrid Bird* figured in White's *Nat. Hist. of Selborne,* is probably a young male of this species, having nearly completed the first moult.

(2. LAGOPUS, *Vieill.*)

**150. T. *Scoticus*, Temm. (*Red Ptarmigan.*) — The
entire plumage chestnut-brown, variegated with black:
tail of sixteen feathers; the six outer ones on each side
dusky.**

> T. Scoticus, *Temm. Man. d'Orn.* tom. II. p. 465. Red Grous, *Mont.*
> *Orn. Dict. Selb. Illust.* vol. I. p. 427. pl. 59. f. 1. *Bew. Brit.*
> *Birds*, vol. I. p. 341.

DIMENS. Entire length sixteen inches: length of the bill (from the
forehead) nine lines and a half, (from the gape) eleven lines; of the
tarsus one inch five lines; of the tail four inches; from the carpus to
the end of the wing eight inches three lines.

DESCRIPT. (*Male.*) The whole plumage of a deep chestnut-brown,
nearly plain on the head and neck, but marked on the back and wing-
coverts with black spots of different sizes, and beneath the body with
numerous undulating black lines: orbits, and a small patch at the base
of the lower mandible, white: naked space above the eyes fringed, of a
bright scarlet colour: frequently some of the feathers on the abdomen
tipped with white: tail of sixteen feathers; the four middle ones reddish
brown, with transverse black lines; the rest of a uniform dusky brown:
bill black, half concealed by the nasal feathers: irides chestnut-brown:
tarsi and toes thickly clothed with grayish white feathers; claws light
horn-colour. (*Female.*) Colours not so dark as in the male; the brown
varied with reddish yellow, and marked with a greater number of black
spots and lines: naked space above the eyes less conspicuous. (*Egg.*)
Reddish white, nearly covered with blotches and spots of umber-brown:
long. diam. one inch nine lines; trans. diam. one inch three lines.

Peculiar to the British Islands. Found plentifully in Scotland, as well
as in some of the mountainous parts of England and Wales. Frequents
moors, heaths, and extensive uncultivated wastes: never resorts to woods.
Feeds on berries, and the tender tops of heaths. Is monogamous, and
pairs in January. Commences laying in March or April. Eggs eight
to twelve in number, deposited on the ground.

**151. T. *Lagopus*, Sab. (*Common Ptarmigan.*)—Cine-
reous, with transverse undulating lines and spots of black,
(*summer*); or wholly white, (*winter*); shafts of the quills,
and lateral tail-feathers, always black.**

> T. Lagopus, *Sab. Supp. Parry's First Voy.* p. cxcvii. *Richards.*
> *App. Parry's Second Voy.* p. 350. Ptarmigan, *Mont. Orn. Dict.*
> *Selb. Illust.* vol. I. p. 430. pl. 59. f. 2. & pl. 60. *Shaw, Gen. Zool.*
> vol. XI. p. 287. pl. 21. *Bew. Brit. Birds*, vol. I. p. 343.

DIMENS. Entire length fifteen inches: length of the bill (from the
forehead) eight lines, (from the gape) ten lines; of the tarsus one inch
four lines; of the tail four inches three lines; of the wing seven inches
six lines: breadth, wings extended, twenty-four inches six lines.

DESCRIPT. (*Male in winter plumage.*) A streak from the corner of
the bill across the eyes, lateral tail-feathers, and shafts of the quills,
black; the rest of the plumage pure white: above the eyes a scarlet
fringed membrane: tarsi and toes thickly clothed with woolly feathers:

bill and claws black. In the *female*, the naked membrane above the eyes is less conspicuous, and the black on the lore altogether wanting. (*Male in summer plumage.*) All the upper parts of the body, scapulars, tertiaries, neck, breast, and sides under the wings, cinereous brown, with transverse undulating black lines, and minute dusky spots: a few reddish orange bars on the head and neck only : primary and secondary quills, (with the exception of the black shafts,) greater part of the wing-coverts, belly, under tail-coverts, and legs, white: two middle tail-feathers nearly all white; the others black, some of them slightly tipped with white. The *female in summer* does not show so much of the rufous tint on the head and neck. (*Egg.*) Yellowish white, sparingly blotched and spotted with black brown: long. diam. one inch eight lines; trans. diam. one inch two lines.

Found only in the Highlands of Scotland, and the adjacent isles. Was formerly met with in some of the mountainous parts of Cumberland and Westmoreland, but is supposed now to be extinct in England. Frequents the summits of the loftiest hills, from whence it rarely descends into the plains. Feeds on the berries and tender shoots of alpine plants. Pairs early in the Spring, and lays its eggs on the bare ground, from eight to fourteen in number. *Obs.* This species is also met with in North America; but it is not found on the Continent, where it has been always confounded with the following.

152. T. *rupestris*, Sab. (*Rock Ptarmigan.*)—Brownish yellow, with transverse black bars, (*summer*) ; or pure white, (*winter*) ; shafts of the quills, and lateral tail-feathers, always black.

> T. rupestris, *Sab. Supp. Parry's First Voy.* p. cxcv. *Richards. App. Parry's Second Voy.* p. 348. *Faun. Bor. Amer.* pt. 2. p. 354. pl. 64. T. Lagopus, *Temm. Man. d'Orn.* tom. II. p. 468. Rock Ptarmigan, *Shaw, Gen. Zool.* vol. XI. p. 290.

DIMENS. Entire length fourteen inches: length of the bill (from the forehead) seven lines, (from the gape) one inch; of the tarsus one inch four lines and a half; of the tail four inches; of the wing seven inches. *Faun. Bor. Am.*

DESCRIPT. Smaller than the last species, with the bill longer and narrower. *Winter plumage* exactly similar. *Summer plumage* characterized by its brownish yellow colour, with rather broad blackish brown bars, exhibiting none of the cinereous tint which predominates in the *Common Ptarmigan :* on the *upper* parts the black markings prevail over the yellow; on the *under* the yellow ground is most conspicuous: primary and secondary quills, with some of the coverts, white; the shafts of the quills black: tail with the two middle feathers barred like the back; the rest black, faintly tipped with white. The *male* is at all seasons distinguished from the *female* by a black band across the eye, as in the last species: in its *summer plumage*, it is furthermore characterized by having the middle of the belly white. (*Egg.*) Buffy white, nearly covered with spots of two shades of dark red brown: long. diam. one inch nine lines; trans. diam. one inch two lines.

This species, originally described by Pennant, (*Arct. Zool.* vol. II. p. 364. no. 184.) occurs in North America, and very plentifully in some parts of the Continent, but has been confounded by Temminck and others with the Common Ptarmigan. In Great Britain, it has hitherto only occurred once. This specimen was killed in Perthshire in Scotland, and is now in the collection of Lord Derby Is said to frequent dry rocky grounds, and to feed on the tops of small birch. Hatches in June.

GEN. 59. PERDIX, *Briss.*

(1. PERDIX, *Steph.*)

153. P. *cinerea*, Briss. (*Common Partridge.*)—Upper parts cinereous, variegated with brown and black; a deep chestnut crescent-shaped spot on the breast.

> P. cinerea, *Temm. Man. d'Orn.* tom. II. p. 488. *Id. Pig. et Gall.* tom. III. pp. 373, & 728. Common Partridge, *Mont. Orn. Dict. Selb. Illust.* vol. I. p. 433. pl. 61. *Bew. Brit. Birds,* vol. I. p. 348.

DIMENS. Entire length thirteen inches: length of the bill (from the forehead) seven lines and a half, (from the gape) eleven lines; of the tarsus one inch nine lines; of the tail three inches four lines; from the carpus to the end of the wing six inches two lines: breadth, wings extended, twenty inches six lines.

DESCRIPT. (*Male.*) Sides of the face, throat, and eyebrows, bright rust-colour: behind the eye a naked red skin: neck and breast bluish gray with fine zigzag black lines; on the lower part of the breast a large patch of deep chestnut-brown in the shape of a horse-shoe; flanks cinereous, with undulating black lines, and a large rust-coloured bar towards the tip of each feather: back, rump, and upper tail-coverts, cinereous brown, with transverse zigzag black lines and a few narrow bars of reddish brown; scapulars and wing-coverts of a deeper hue than the back, with the shafts of the feathers yellowish white; quills blackish gray, spotted and barred with pale yellowish red: tail of eighteen feathers; the four middle ones marked like the back; the others bright rust-colour: bill and feet bluish gray: irides hazel. (*Female.*) Less of the rust-coloured tinge on the head and throat; feathers on the crown of the head spotted with white: upper plumage generally darker, with a greater number of black bars and spots: the horse-shoe mark on the breast very indistinct, or altogether wanting. White and pied *varieties* are occasionally met with. (*Egg.*) Of a uniform olive-brown: long. diam. one inch five lines; trans. diam. one inch and half a line.

Abundant in all the cultivated parts of Great Britain, but seldom found at a distance from arable land, which is its favourite haunt. Feeds on seeds and insects, and especially on the pupæ of ants. Pairs in February, and commences laying about the middle or end of May. Eggs twelve to twenty in number, deposited on the ground, amongst brush-wood and long grass, or in fields of clover and standing corn. Period of incubation three weeks. After the young are hatched, they flock together in *coveys* till the following Spring.

* 154. **P. *rubra*, Briss. (*Red-legged Partridge.*)** — Upper parts reddish ash: throat and cheeks white; bounded by a collar of black, expanding on the breast, and spotted with white.

> P. rubra, *Temm. Man. d'Orn.* tom. II. p. 485. *Id. Pig. et Gall.* tom. III. pp. 361, and 726. Guernsey Partridge, *Mont. Orn. Dict. Lewin, Brit. Birds,* vol. IV. pl. 137. Red-legged Partridge, *Bew. Brit. Birds,* vol. I. p. 345.

Dimens. Entire length thirteen inches six lines: length of the bill (from the forehead) eight lines, (from the gape) eleven lines; of the tarsus one inch seven lines; of the tail four inches one line; from the carpus to the end of the wing six inches six lines: breadth, wings extended, twenty inches ten lines.

Descript. Forehead cinereous; crown and occiput reddish brown; throat, cheeks, and a broad band above the eyes reaching nearly to the nape, white; the white space on the throat bounded by a black collar, which spreads itself out on the upper part of the breast and sides of the neck, in the form of black spots on a light reddish ground; intermixed with the black spots are a few white ones; lower part of the breast bluish ash; belly and abdomen rust-red; feathers on the sides cinereous, each with two transverse bars across the middle, the first white and the second black, beyond which last is a large semilunar rust-coloured spot, occupying the tip: all the upper parts of the body reddish ash: quills grayish brown, with the outer webs ochre-yellow: the four middle tail-feathers like the back; the others rust-colour: orbits, bill, and feet, bright red. In the *female* the colours are paler, but in other respects similar to those of the male. (*Egg.*) Reddish yellow white, spotted and speckled with reddish brown: long. diam. one inch seven lines and a half; trans. diam. one inch three lines.

A native of France, Italy, and the Islands of Guernsey and Jersey, from whence it has been imported into England, and naturalized in some parts of the country*. Is very common in Suffolk, frequenting waste heaths and extensive barren lands, more than cornfields. Occasionally met with in other counties. On the Continent, is said to lay from fifteen to eighteen eggs.

* (2. Ortyx, *Steph.*)

* 155. P. *Virginiana*, Lath. (*Virginian Partridge.*) — Body above brownish chestnut, variegated with black and rufous; beneath whitish, with transverse undulating black lines: throat, and a broad band above the eye, white.

> Tetrao Virginianus, *Wils. Amer. Orn.* vol. vi. p. 21. pl. 47. f. 2.   Perdix borealis, *Temm. Pig. et Gall.* tom. iii. pp. 436, and 735. Virginia *and* Maryland Partridge, *Lath. Syn.* vol. ii. pp. 777, and 778.   Northern Colin, *Shaw, Gen. Zool.* vol. xi. p. 377.

Dimens. Entire length nine inches: length of the bill six lines and a half; of the tarsus one inch three lines.

Descript. (*Male.*) Forehead black; a broad white band above the eye, reaching to the nape, above which is another narrow one of black; throat white, bounded by a black collar, which expands over the front and sides of the neck in the form of black spots, mixed with others white and red: crown of the head chestnut-red, spotted with black; upper part of the body brownish chestnut; the edges of the feathers cinereous, with fine black streaks; in the middle of the back some large black spots edged with rufous: scapulars and greater wing-coverts spotted with black and red on their inner webs, with cinereous and red on their outer; lesser coverts red with fine undulating black lines: breast reddish white, with transverse black lines; belly pure white, with black lines of a semicircular form; feathers on the sides rufous, with an edging of white spots surrounded with black; under tail-coverts red, spotted with black; tail bluish ash; the middle feathers rufous towards their extremities, and marked with brown undulating lines: bill black, reddish at the base: feet brownish red. (*Female.*) Throat, and band above the eyes, pale red, the latter without the black streak above; the red on the throat surrounded by spots of black, brown, and white: nape, and upper part of the head, spotted with red: feathers on the back more deeply edged with cinereous; rest of the upper parts paler than in the male bird: breast bright red, with two little white spots towards the extremity of each feather: tail bluish ash, with fine undulating lines of brown and whitish towards the tips of all the feathers: bill with more red at the base than in the male. The *young of the year* very much resemble the adult female, but the transverse undulating lines on the back and tail-feathers are more numerous. (*Egg.*) White: of a pointed form: long. diam. one inch two lines and a half; trans. diam. one inch.

---

* Pulteney observes that this species has been killed near Weymouth in Dorsetshire, and suggests the probability of its sometimes reaching this country from the Islands of Jersey and Guernsey. If this be ever the case, which is rather doubtful, it has a claim to be considered as one of our *native* species.

A native of North America, from whence it has been recently introduced into this country. Is now naturalized in Suffolk. Said to frequent woody situations, and to perch in trees. Lays from twenty to twenty-five eggs, and has two broods in the year.

(3. COTURNIX, *Briss.*)

156. P. *Coturnix*, Lath. (*Common Quail.*)—Upper parts cinereous brown, variegated with black: over each eye, and on the crown of the head, a longitudinal whitish streak.

P. Coturnix, *Temm. Man. d'Orn.* tom. II. p. 491. Coturnix dacty-lisonans, *Temm. Pig. et Gall.* tom. III. pp. 478, & 740. Common Quail, *Selb. Illust.* vol. I. p. 437. pl. 62. Quail, *Mont. Orn. Dict. and Supp. Bew. Brit. Birds,* vol. I. p. 351.

DIMENS. Entire length eight inches: length of the bill (from the forehead) five lines, (from the gape) seven lines and a half; of the tarsus one inch two lines; of the tail one inch six lines; from the carpus to the end of the wing four inches six lines: breadth, wings extended, fourteen inches four lines.

DESCRIPT. (*Male.*) Crown and occiput black, the feathers edged with rufous brown; over each eye a yellowish white streak extending to a considerable distance down the neck; a similar streak along the top of the head running parallel with the above; throat rufous, bounded by a double crescent of dusky brown: upper part of the neck, back, scapulars, and wing-coverts, cinereous brown, variegated with black; the shafts and central portions of the feathers yellowish white: lower part of the neck, breast, and sides, pale rufous brown with longitudinal white streaks; belly yellowish white: tail blackish brown; the shafts and tips of the feathers whitish: bill and feet yellowish brown. (*Female.*) Throat white, without the double crescent-shaped markings: breast yellowish white, spotted with blackish brown: general colours of the plumage paler. *Obs.* This species is subject to some variation of colours: in *very old males*, the cheeks and throat are dusky brown, and the whole plumage of a more vivid hue. Occasionally found pure white; or of a uniform deep brown. (*Egg.*) Yellowish white, blotched and spotted with umber-brown: long. diam. one inch one line; trans. diam. eleven lines.

A migratory species visiting this country in May, and usually departing towards the end of the Autumn. Occasionally remains through the Winter. Chiefly frequents corn-fields and open lands. Feeds on seeds, grain, and insects. The males are polygamous, and utter during the breeding season a peculiar whistling note. Eggs eight or ten in number, deposited on the bare ground.

GEN. 60.   OTIS, *Linn.*

157.   O. *Tarda*, Linn. (*Great Bustard.*) — Upper part of the body reddish orange, with transverse black bars; beneath whitish: a tuft of long filiform feathers (*male*) from the corners of the lower mandible.

O. Tarda, *Temm. Man. d'Orn.* tom. II. p. 506.   Great Bustard,
   Mont. Orn. Dict. and Supp. Selb. Illust. vol. I. p. 442. pls. 64,
   & 64*.   *Bew. Brit. Birds*, vol. I. p. 355.

DIMENS.   Entire length nearly four feet: breadth nine feet.   PENN.
DESCRIPT.   (*Male.*)   Head and neck cinereous; a longitudinal black
streak on the crown reaching to the occiput; feathers on the chin loose,
and somewhat elongated, with the barbs disunited; a long moustache
of similar feathers on each side of the lower mandible:* back, scapulars,
lesser wing-coverts, rump, and tail-coverts, yellowish or reddish orange,
barred and variegated with black; greater wing-coverts, and some of the
secondaries, pale cinereous; primaries black, slightly tipped with white:
upper part of the breast reddish orange; rest of the under parts white:
tail brownish orange, some of the lateral feathers with the basal half and
tip white; the whole crossed by two or three black bars near the extremity: bill dusky, the under mandible somewhat paler than the upper:
irides light hazel: feet dusky brown.   (*Female.*)   Much smaller, and
without the tuft of long feathers on each side of the lower mandible;
the longitudinal streak on the crown not so obvious; fore part of the
neck of a deeper gray: in other respects similar to the male.   (*Egg.*)
Olive-brown, sparingly and indistinctly blotched with greenish brocoli-
brown: long. diam. two inches eleven lines; trans. diam. two inches two
lines.

Formerly met with in great plenty on the plains of Wiltshire and
Dorsetshire, the Wolds of Yorkshire, as well as in many other parts of
England, and also in Scotland.   Is now become extremely rare, and
almost confined to the open parts of Norfolk and Suffolk, where the
species still continues to breed in small quantities.   Single individuals
are occasionally observed in Cambridgeshire, and a fine male specimen
was killed near Ickleton in that county in January 1831.   Where plentiful, usually found in flocks.   Food green corn, seeds, insects, as well as
various other vegetable and animal substances.   Lays two eggs early in
the Spring, which are deposited on the bare ground: rarely, however,
hatches more than one in this country.   Period of incubation about four
weeks.   The male of this species possesses a capacious pouch, situate
along the fore part of the neck, said to be capable of holding several
quarts of water.

158.   O. *Tetrax*, Linn.   (*Little Bustard*.) — Upper
part of the body yellowish orange, with fine zigzag dusky
lines; neck black, encircled by a double collar of white.

O. Tetrax, *Temm. Man. d'Orn.* tom. II. p. 507. Little Bustard,
   Mont. Orn. Dict. and Supp. Selb. Illust. vol. I. p. 447. pl. 65.
   Bew. Brit. Birds, vol. I. p. 359.

DIMENS.   Entire length seventeen inches and a half: breadth, wings
extended, two feet ten inches and a half.   SELBY.
DESCRIPT.   (*Adult male.*)   Crown and occiput yellowish orange, with
brown spots: sides of the head and throat deep cinereous, bounded by a
white collar encircling the upper part of the neck; lower part of the neck
black; on the breast another, and somewhat broader, white collar, below
which is a narrow one of black: upper parts yellowish orange, crossed by
numerous fine zigzag dusky lines, and mottled with large spots more

---

* Bewick's figure represents this species with a longitudinal black streak or band on the sides
of the neck.   It is possible that such a character may exist during the short period of the breeding
season only, but it is not usually present.

sparingly distributed: the edge of the wing, belly, and upper tail-coverts, pure white: irides orange: bill and feet gray. (*Female and young male.*) Throat white; sides of the head, and upper part of the neck, yellowish orange, with longitudinal dusky streaks; lower part of the neck and upper part of the breast the same, with transverse crescent-shaped dusky bars; lower breast and flanks whitish, spotted with dusky, the latter with a few fine longitudinal streaks besides on the shafts of the feathers: belly and abdomen white and unspotted: upper parts of the body much as in the male. *Obs.* It is not certain whether the *adult male* does not lose the black neck and double white collar in Winter, and resemble the female during that season. (*Egg.*) Of a uniform olive-brown: long. diam. two inches; trans. diam. one inch six lines.

A very rare, and only occasional, visitant in this country. Has been taken alive on the edge of Newmarket Heath, and more recently near Caxton in Cambridgeshire. Other specimens have been killed in Sussex, Kent, Devonshire, Northumberland, and Suffolk. Frequents plains, and large tracts of open country. Said to be graminivorous. Eggs three to five in number, deposited on the ground.

ORDER IV. GRALLATORES.

GEN. 61.   CURSORIUS, *Lath.*

159. C. *isabellinus*, Meyer. (*Cream-coloured Courser.*)— General plumage reddish cream-colour: behind the eyes a double black streak.

> C. isabellinus, *Temm. Man. d'Orn.* tom. II. p. 513.   Cream-coloured Plover, *Lath. Syn.* vol. III. p. 217.   *Id. Supp.* p. 254. pl. 116. *Mont. Orn. Dict.*   Cream-coloured Swiftfoot, *Selb. Illust.* vol. II. p. 217. pl. 33**.

DIMENS. Entire length ten inches: length of the bill nine lines. MONT.

DESCRIPT. General colour of the plumage reddish cream or buff orange, the feathers in some places with pale edges; throat whitish; behind the eye a double black streak directed towards the occiput: quills black: lateral tail-feathers black towards their extremities, the black including a small white spot: abdomen whitish: bill black: legs yellowish brown. In *immature birds*, the scapulars and wing-coverts are crossed by numerous, fine, zigzag bars of a darker tint, more particularly towards the tips of the feathers: the black streak behind the eyes very faint. (*Egg.*) Unknown.

A native of Africa, very rarely occurring in Europe. In England only three specimens have been hitherto observed. The first of these was shot near St Albans, in Kent, many years back; the second was killed in October 1827, at Charnwood Forest, Leicestershire; the third in the year following, at Freston, near Aldborough, in Suffolk. Very little is known of its habits. Said to inhabit dry plains, and to run very swiftly.

GEN. 62. OEDICNEMUS, *Temm.*

160.   OE. *crepitans*, Temm. (*Common Thick-knee.*)—
Reddish ash, with longitudinal dusky spots: quills black;
the first two primaries with a broad white bar across each
web.

> OE. crepitans, *Temm. Man. d'Orn.* tom. II. p. 521. Common Thick-
> knee, *Selb. Illust.* vol. II. p. 250. pl. 40. Great Plover, *Bew. Brit.*
> *Birds*, vol. I. p. 362. Thick-kneed Bustard, *Mont. Orn. Dict.*

DIMENS. Entire length eighteen inches.

DESCRIPT. Upper parts reddish ash, with a longitudinal dusky spot
down the middle of each feather: space between the eye and the bill,
cheeks, throat, belly, and thighs, white: neck and breast tinged with
reddish, and marked with fine longitudinal dusky streaks: a pale bar
across the wing-coverts: quills black; the first with a large and conspicuous white spot near the middle; the second with one somewhat
smaller: under tail-coverts red: all the tail-feathers, the two middle
ones excepted, tipped with black: bill yellowish at the base, black at
the tip: irides, orbits, and feet, yellow. In *young birds*, the markings
are less distinct. (*Egg.*) Stone-colour, blotched, spotted, and streaked,
with ash-blue and dark brown: long. diam. two inches two lines; trans.
diam. one inch seven lines.

A migratory species visiting this country about the latter end of April
or beginning of May, and departing in the Autumn. Is most abundant
in the southern, midland, and eastern counties. Frequents heaths, extensive corn-lands, and other open districts. Feeds on insects, worms, and
reptiles. Makes no nest, but lays its eggs, which are two in number, on
the bare ground. During the breeding season the male utters a loud
shrill cry, heard more particularly in the dusk of the evening. The name
of this species is derived from a peculiar enlargement of the upper part
of the tarsus, and of the joint immediately above it, most conspicuous in
the young birds of the year.

GEN. 63. CHARADRIUS, *Linn.*

161.   C. *pluvialis*, Linn. (*Golden Plover.*)—Upper parts
dusky brown, spotted with yellow; cheeks, neck, and breast,
variegated with brown and ash-colour; throat and abdomen
white.

> C. pluvialis, *Temm. Man. d'Orn.* tom. II. p. 535. Golden Plover,
> *Mont. Orn. Dict. Selb. Illust.* vol. II. p. 231. pl. 37. *Bew. Brit.*
> *Birds*, vol. I. p. 367.

DIMENS. Entire length eleven inches: length of the bill (from the
forehead) ten lines and a half, (from the gape) one inch one line; of the
tarsus one inch seven lines; of the tail three inches; from the carpus to
the end of the wing seven inches six lines: breadth, wings extended,
twenty-three inches.

DESCRIPT. (*Winter plumage.*) Crown of the head, and all the upper
parts, dusky brown, thickly spotted with king's yellow, the spots being
disposed along the edges and at the tips of the feathers; cheeks, neck,

and breast, somewhat variegated with streaks and spots of ash-gray and yellowish brown ; throat, belly, and abdomen, white : quills, and greater coverts, dusky ; the former white along the shafts ; the latter tipped with white : bill dusky : irides dark brown : legs deep ash-colour. (*Summer plumage.*) Upper parts deep black ; the edges of the feathers spotted with bright yellow : forehead, and space above the eyes, white : sides of the neck white, with large black and yellow spots : throat, fore part of the neck, and all the under parts, black. (*Young of the year.*) Upper parts cinereous brown, with spots of yellowish ash. (*Egg.*) Yellowish stone-colour, blotched and spotted with brownish black : long. diam. two inches ; trans. diam. one inch four lines.

Not uncommon ; migrating from the southern to the northern parts of the kingdom at the approach of the breeding season. Haunts moors, heaths, and other open districts ; occasionally, during severe weather, the sea-coast. Food worms and insects. Eggs four in number ; laid in May.

162.   C. *Morinellus*, Linn. (*Dotterel Plover*.)—Upper plumage brownish ash, edged with reddish : breast reddish ash : a broad streak above the eyes, and a gorget on the breast, white.

C. Morinellus, *Temm. Man. d'Orn.* tom. II. p. 537. Dotterel, *Mont. Orn. Dict. Selb. Illust.* vol. II. p. 236. pl. 39. *Bew. Brit. Birds,* vol. I. p. 369.

Dimens.   Entire length nine inches five lines : length of the bill (from the forehead) seven lines and a half, (from the gape) one inch ; of the tarsus one inch six lines ; from the carpus to the end of the wing six inches.

Descript.   (*Winter plumage.*) Forehead, throat, and sides of the face, whitish, dotted and streaked with black ; above each eye a broad band of reddish white, reaching to the nape : crown of the head, and occiput, dusky ash ; rest of the upper parts cinereous, tinged with olivaceous brown ; the feathers on the back, wing-coverts, and scapulars, edged with rust-red : fore part of the neck, breast, and flanks, reddish ash ; a gorget on the upper part of the breast extending on each side to the bend of the wing, and middle of the belly, pure white : shaft of the first primary white ; the rest dusky : tail cinereous, passing into dusky towards the end ; the extreme tip white : bill black : irides brown : feet greenish ash. (*Male in summer plumage.*) Cheeks, throat, and band above the eyes, pure white ; crown of the head, and occiput, deep brown ; back, wing-coverts, and scapulars, olivaceous ash, the feathers edged with deep red ; nape, also the sides and lower part of the neck in front, cinereous ; on the upper part of the breast a white gorget bounded above by a blackish line ; lower part of the breast, and flanks, bright rust-colour ; middle of the belly deep black ; vent and under tail-coverts reddish white.   In the *female,* the white band above the eyes is less conspicuous ; the gorget on the breast has a reddish tinge ; the black patch on the abdomen is smaller, and much intermixed with white.   In *young birds,* the crown of the head is reddish, with longitudinal dusky spots ; the upper parts are more cinereous ; and the tail is tipped with pale red. (*Egg.*) Yellowish olive, blotched and spotted with dark brownish black : long. diam. one inch seven lines and a half ; trans. diam. one inch two lines and a half.

Common in some parts of England, appearing twice in the year, Spring and Autumn, in its passage to and from more northern latitudes, in which it breeds. Winters in the South of Europe. With us generally observed in small flocks on heaths, moors, and other open districts. Feeds on insects and worms. This and the preceding species are in high estimation for the table.

163. C. *Hiaticula*, Linn. (*Ringed Plover*.) — Upper parts cinereous brown: a broad gorget of black on the breast: bill orange and black: feet orange.

> C. Hiaticula, *Temm. Man. d'Orn.* tom. ii. p. 539. Ringed Plover, *Mont. Orn. Dict. Selb. Illust.* vol. ii. p. 240. pl. 38. f. 1, 2. *Bew. Brit. Birds*, vol. i. p. 371.

DIMENS. Entire length seven inches six lines: length of the bill (from the forehead) seven lines, (from the gape) seven lines and a half; of the tarsus eleven lines and a half; of the tail two inches five lines; from the carpus to the end of the wing five inches: breadth, wings extended, sixteen inches.

DESCRIPT. (*Adult male in Summer and Winter*.) Forehead, space between the eye and the bill, and sides of the face, black; across the forehead, and through the eyes, a white band, passing backwards to the occiput: on the upper part of the breast a broad gorget of black, the ends of which unite on the nape becoming narrower: throat, a collar round the neck immediately above the black gorget, and the rest of the under parts, pure white: crown of the head, and all the upper parts, cinereous brown: quills dusky, with an oval white spot about the middle of each feather; shafts partly white: outermost tail-feather wholly white; the next white, with a small brown spot on the inner web; the rest dusky, tipped with white, the two middle ones excepted, which are dusky throughout: bill orange at the base, black at the tip: irides hazel: feet yellowish orange. (*Adult female.*) Lore and cheeks cinereous brown; forehead white, surmounted by a narrow transverse band of dusky ash; streak through the eye faint; gorget dusky: in other respects similar to the male. (*Young of the year.*) Between the bill and the eye a dusky streak; forehead dirty white, without the coronal black band: upper parts cinereous brown, the feathers edged with yellowish white: gorget on the breast cinereous brown: bill dusky: feet yellowish. (*Egg.*) Stone-colour, spotted and streaked with ash-blue and black: long diam one inch five lines; trans. diam. one inch and half a line.

Common on most parts of the sea-coast, resorting occasionally to inland marshes and the banks of rivers. Remains with us the whole year. Food insects and worms. Pairs in May, and towards the end of that month deposits its eggs, which are four in number, in a small cavity in the sand, just above high-water mark. In the Autumn becomes gregarious, and keeps in small flocks throughout the Winter.

164. C. *minor*, Meyer. (*Little Ringed Plover*.)—Upper parts cinereous brown : a gorget of black on the breast: bill entirely black: feet flesh-colour.

> C. minor, *Temm. Man. d'Orn.* tom. ii. p. 542. Little Ring-Dotterel, *Gould, Europ. Birds*, part xi. pl. 2.

Dimens. Entire length five inches: length of the bill (from the fore-head) five lines and a half; of the tarsus eleven lines and a half; of the tail two inches four lines; from the carpus to the end of the wing four inches four lines.

Descript. (*Adult male and female.*) " Bill black; a band of the same colour passing from the bill to the eye, and extending over the ear-coverts; the forehead pure white, above which on the crown a black band passes from eye to eye; the occiput gray, beneath which a white circle spreads from the throat round the neck; this is succeeded by a black band, broad on the chest, but narrowing until it meets at the back of the neck; the whole of the upper plumage, with the exception of the rump, which is white, of a fine brownish gray; under surface white; feet and legs flesh-colour; irides hazel. (*Young.*) Wants the black collar and facial markings, the crown of the head and face being brownish gray; in every other respect resembles the adult, except that a brownish tint pervades the whole of the upper plumage, and that every feather is edged with a lighter margin. (*Egg.*) Yellowish white, marked with blotches of black and brown." Gould.

A single individual of this species, a young bird of the year, has been recently killed at Shoreham in Sussex. It is now in the possession of Mr Henry Doubleday of Epping. No other has hitherto occurred in Britain, although from its close resemblance to the *C. Hiaticula*, it is possible that the species may have been overlooked. Common on some parts of the Continent. According to Temminck, resorts by preference more to the banks of rivers than to the shores of the sea. Food and nidification similar to those of the Ringed Plover.

165.   C. *Cantianus*, Lath. (*Kentish Plover.*)—Upper parts cinereous brown: a large patch of black on each side of the breast: bill and feet black.

C. Cantianus, *Temm. Man. d'Orn.* tom. ii. p. 544.   Kentish Plover, *Lewin, Brit. Birds,* vol. v. pl. 85.   *Lath. Syn. Supp.* ii. p. 316.   *Mont. Orn. Dict. Supp.   Selb. Illust.* vol. ii. p. 243. pl. 38. f. 3.

Dimens. Entire length six inches ten lines: length of the bill (from the forehead) seven lines and a half, (from the gape) nine lines; of the tarsus one inch and half a line; of the tail two inches one line; from the carpus to the end of the wing four inches three lines: breadth, wings extended, thirteen inches four lines.

Descript. (*Male.*) Forehead, a broad streak over each eye, a ring round the neck, and all the under parts, pure white: space between the eye and the bill, and ear-coverts, black; an angular black patch imme-diately behind the forehead, advancing towards the crown; a larger patch of the same colour on each side of the breast: crown of the head, and nape of the neck, light brownish red; rest of the upper parts cinereous brown: shafts of the primaries white: the two outermost tail-feathers white; the next whitish; the rest cinereous brown: bill and feet black. (*Female.*) The white on the forehead of less extent; behind it a small transverse black streak instead of the large angular spot; space between the eye and the bill, ear-coverts, and the patch on each side of the breast, cinereous brown: the red on the head and nape tinged with ash-gray. (*Young before the first moult.*) The white on the forehead, above the eyes, and across the nape of the neck, very indistinct: no black any

where; the large spot on the sides of the breast light brown, and very
faintly indicated: all the upper parts light cinereous brown. (*Egg.*)
Stone-colour, spotted with black: long. diam. one inch three lines;
trans. diam. eleven lines.

A rare, and only occasional visitant in this country. Observed many
years ago near Sandwich in Kent, by the late Mr Boys. Has more
recently occurred in two or three instances on the coasts of Norfolk
and Sussex. Habits similar to those of the last two species. Feeds
on marine insects and worms. Lays from three to five eggs on the
bare sand.

GEN. 64.   VANELLUS, *Briss.*

(1. SQUATAROLA, *Cuv.*)

166. V. *griseus*, Briss. (*Gray Plover.*)—Upper plumage
dusky brown, variegated with white and ash-colour; breast
and belly white with dark spots (*winter*), or of a uniform
deep black (*summer*).

V. melanogaster, *Temm. Man. d'Orn.* tom. II. p. 547.   Gray Plover,
  Selb. Illust. vol. II. p. 227. pl. 35.   *Bew. Brit. Birds*, vol. II. p. 83.
  Gray Sandpiper, *Mont. Orn. Dict.*

DIMENS. Entire length eleven inches six lines: length of the bill
(from the forehead) one inch and half a line, (from the gape) one inch
three lines and a half; of the tarsus one inch eleven lines; of the naked
part of the tibia ten lines; of the tail three inches; from the carpus to
the end of the wing seven inches eleven lines: breadth, wings extended,
twenty-five inches six lines.

DESCRIPT. (*Winter plumage.*) Forehead and chin white; streak over
the eyes, fore part of the neck, sides of the breast, and flanks, white,
variegated with spots of brown and ash-colour: head, and all the upper
parts of the body, dusky brown, the feathers edged and tipped with
grayish white: belly, abdomen, thighs, and upper tail-coverts, pure
white: beneath the wing some long black feathers arising from the
axilla: tail white, towards the tip reddish, with transverse brown bars
which become paler and less numerous on the lateral feathers: bill
black: irides dusky: feet blackish gray. (*Summer plumage.*) Space
between the eye and the bill, throat, sides and fore part of the neck,
middle of the breast, belly and flanks, deep black; forehead, a broad
streak above the eyes continued down the sides of the neck and breast
and bounding the black on those parts, thighs and abdomen, pure white:
crown and occiput cinereous brown, with the shafts of the feathers black;
nape variegated with brown, black, and white; back, scapulars, and wing-
coverts, black, all the feathers broadly tipped and barred with white.
(*Young of the year.*) Somewhat similar to the adult bird in winter
plumage: the spots on the sides of the breast and flanks larger, but
of a paler tint: upper parts of a uniform light gray, variegated with
whitish spots; the transverse bars on the tail gray. (*Egg.*) Light
clay-brown, spotted and streaked all over with dark brown: long. diam.
one inch ten lines; trans. diam. one inch four lines.

Found on many parts of the British coast during the autumn and
winter months, but not in very great abundance. Retires northward to
breed. Feeds on worms and marine insects. Lays four eggs.

(2. Vanellus, *Cuv.*)

167. V. *cristatus*, Meyer. (*Crested Lapwing.*)—Crown, fore part of the neck and breast, greenish black; occipital feathers very much elongated, slightly recurved : back and scapulars olive-green.

V. cristatus, *Temm. Man. d'Orn.* tom. ii. p. 550. Crested or Green Lapwing, *Selb. Illust.* vol. ii. p. 221. pl. 34. Lapwing, *Mont. Orn. Dict. Bew. Brit. Birds,* vol. ii. p. 79.

DIMENS. Entire length twelve inches six lines: length of the bill (from the forehead) ten lines and a half, (from the gape) one inch and half a line; of the tarsus one inch ten lines; of the tail four inches six lines; from the carpus to the end of the wing eight inches seven lines: breadth, wings extended, twenty-eight inches five lines.
DESCRIPT. (*Winter plumage.*) Occipital feathers very much elongated, filiform, and turned upwards at the extremities: forehead, crown, crest, fore part of the neck, and upper part of the breast, of a shining greenish black; throat, region of the eyes, and sides of the neck, white: back, scapulars, and wing-coverts, olive-green, glossed with blue and purplish red: quills black; the first four primaries tipped with white: lower part of the breast, and all the belly, white; upper and under tail-coverts pale ferruginous: basal half of the tail white; the remaining portion black, tipped with white; outer feather almost entirely white: bill dusky: feet brownish red. (*Summer plumage.*) The black on the neck of a deeper tint, and extending over the throat; the occipital crest somewhat longer; the green and blue reflections on the upper parts more brilliant. (*Young of the year.*) The crest shorter; throat variegated with white and cinereous brown; beneath the eyes a dusky streak; all the feathers, as well on the upper as under parts, tipped with reddish yellow: feet olivaceous ash. (*Egg.*) Olive ground, blotched and spotted nearly all over with blackish brown: long. diam. one inch eleven lines; trans. diam. one inch four lines.
Common in most parts of the kingdom, frequenting fens and moist fields, as well as heaths, warrens, and upland situations. Has a peculiar note resembling the word *pee-wit.* Feeds on insects, worms, and snails. Breeds early in the Spring, depositing its eggs, which are four in number, on the bare ground. Collects in vast flocks at the approach of Autumn.

GEN. 65.   STREPSILAS, *Illig.*

168. S. *Interpres*, Leach. (*Common Turnstone.*) — Upper parts variegated with white, black, and ferruginous; breast and abdomen white.

S. collaris, *Temm. Man. d'Orn.* tom. ii. p. 553. Common Turnstone, *Selb. Illust.* vol. ii. p. 204. pl. 33*. Turnstone, *Mont. Orn. Dict. Bew. Brit. Birds,* vol. ii. pp. 108, and 110.

DIMENS. Entire length nine inches three lines: length of the bill (from the forehead) ten lines, (from the gape) eleven lines and a half; of the tarsus eleven lines and a half; of the tail two inches eight lines:

from the carpus to the end of the wing five inches ten lines: breadth, wings extended, eighteen inches eight lines.

DESCRIPT. (*Adult male.*) Forehead, space between the eye and the bill, throat, and a broad collar on the nape of the neck, white: across the forehead a narrow black streak, which, passing over the eyes, becomes somewhat enlarged on the sides of the face; from thence it branches off on one side to the base of the lower mandible, on the other to the sides of the neck, where it again enlarges before uniting with a large gorget of the same colour on the lower part of the neck and breast: rest of the under parts pure white: crown of the head reddish white, with longitudinal black streaks; upper part of the back, scapulars, and wing-coverts, ferruginous brown, with large irregularly distributed black spots; lower back and upper tail-coverts white; a broad black bar on the rump: quills dusky; the shafts of the primaries, and tips of the secondaries, white: tail black, tipped with white; the outer feather entirely white: bill black: feet orange-yellow. The *female* differs only in having the colours not so distinct; the white on the head and neck less pure; the black not so deep. (*Young of the year.*) Head and nape cinereous brown, with darker variegations; some white spots on the cheeks and sides of the neck; throat and fore part of the neck whitish; sides of the breast deep brown, the feathers tipped with whitish; rest of the under parts, and a considerable portion of the back, pure white; upper part of the back, scapulars, and wing-coverts, deep brown, all the feathers broadly edged with yellowish; the transverse bar on the rump deep brown, bordered with red: feet yellowish red. (*Egg.*) Reddish white, blotched and spotted with dark chestnut-brown: long. diam. one inch six lines; trans. diam. one inch.

A regular winter visitant on many parts of the coast, appearing in August and departing in the Spring. Goes northward to breed. Derives its name from its habit of turning over the small stones on the shore in search of marine insects and worms on which it feeds. Is said to lay three or four eggs, in a small hollow in the sand. *Obs.* The *Tringa Morinella* of Linnæus is this bird in immature plumage.

GEN. 66. CALIDRIS, *Illig.*

169. C. *arenaria*, Illig. (*Sanderling.*) — Cinereous above, with darker streaks; all the under parts white (*winter*): or ferruginous and white spotted with black; belly and abdomen white (*summer*).

C. arenaria, *Temm. Man. d'Orn.* tom. II. p. 524. Sanderling, *Mont. Orn. Dict. Bew. Brit. Birds*, vol. I. p. 375. Common Sanderling, *Selb. Illust.* vol. II. p. 208. pl. 36.

DIMENS. Entire length eight inches: length of the bill (from the forehead) one inch and half a line, (from the gape) one inch one line; of the tarsus eleven lines and a half; of the middle toe, claw included, eight lines and a half; from the carpus to the end of the wing four inches nine lines.

DESCRIPT. (*Winter plumage.*) All the upper parts cinereous, with the shafts of the feathers blackish brown: forehead, cheeks, throat, sides of the neck, and all the under parts, pure white: the bend and edge of the wing blackish gray; primaries dusky, the edges and tips inclining to brown; wing-coverts broadly edged with white: tail deep gray, the feathers edged with white; the two middle ones darkest: bill, irides, and feet, black. (*Summer plumage.*) Cheeks and upper part of the head black, the feathers edged and variegated with ferruginous and white: throat, neck, breast, and upper part of the flanks, reddish ash; the central portion of each feather black, the tip whitish: the rest of the under parts pure white: back and scapulars deep ferruginous, with large irregular patches and spots of black; all the feathers edged and tipped with whitish: wing-coverts dusky brown, margined and tipped with white, forming a transverse bar: the two middle tail-feathers blackish brown with ferruginous edges. (*Young of the year before the first moult.*) " Feathers on the crown of the head, back, scapulars, and wing-coverts, black, edged and spotted with yellowish; between the bill and the eye a cinereous brown streak; nape, sides of the neck, and sides of the breast, pale gray, with fine undulating streaks; forehead, throat, fore part of the neck, and all the under parts, pure white: wings and tail as in the adult." TEMM. (*Egg.*) Olive ground, spotted and speckled with black: long. diam. one inch four lines; trans. diam. one inch.

Not uncommon in small flocks on many parts of the coast, but rarely observed inland. Is partial to sandy shores, feeding upon marine insects. Breeds in high northern latitudes. Said to construct a rude nest of grass in the marshes, and to lay four eggs; incubation commencing in the middle of June.

GEN. 67.   HÆMATOPUS, *Linn.*

170.   H. *ostralegus*, Linn. (*Pied Oyster-Catcher.*) — Head, neck, and upper parts black : breast and abdomen white: bill and feet red.

H. ostralegus, *Temm. Man. d'Orn.* tom. II. p. 531. Pied Oyster-Catcher, *Mont. Orn. Dict. and Supp.* Oyster-Catcher, *Bew. Brit. Birds,* vol. II. p. 113. Common Oyster-Catcher, *Selb. Illust.* vol. II. p. 200. pl. 33.

DIMENS. Entire length sixteen inches ten lines : length of the bill (from the forehead) two inches nine lines, (from the gape) two inches ten lines and a half; of the tarsus one inch eleven lines; of the tail four inches two lines; from the carpus to the end of the wing ten inches nine lines: breadth, wings extended, thirty-two inches.

DESCRIPT. (*Winter plumage.*) Head, neck, upper part of the breast, back, wings, and extremity of the tail, deep black; a collar on the throat, and a small spot beneath the eye, white: under parts, rump, basal portion of the tail and quill feathers, and a transverse bar on the wings, pure white: bill, and orbits, bright orange-red: irides crimson: feet purplish red. The *summer plumage* is distinguished by the absence of the white collar; in other respects similar. (*Young of the year.*) "The black parts of the plumage shaded with brown, the feathers being edged with this last colour; the white dull and soiled: bill and orbits dusky brown: irides brown : feet livid gray." TEMM. (*Egg.*) Yellowish stone-colour, spotted with ash-gray and dark brown: long. diam. two inches two lines; trans. diam. one inch six lines.

Common on the coast, assembling in small flocks during the winter season. Is never found inland. Feeds principally on marine insects and the bivalve mollusca. Makes no nest, but deposits its eggs, two to four in number, on the bare ground above high-water mark. The male bird has a loud screaming note during the season of incubation.

(16.) *Psophia crepitans*, Linn. (*Gold-breasted Trumpeter.*) Lath.
Syn. vol. II. p. 793. pl. 68.

An individual of this species is recorded by Montagu, in the Supplement
to his Ornithological Dictionary, as having been taken in Surrey; it had,
however, probably escaped from confinement, being a native of South
America.

GEN. 68.　GRUS, *Pall.*

171.　G. *cinerea*, Bechst. (*Common Crane.*) — Cinereous; occiput, throat, and fore part of the neck, dusky
gray.

G. cinerea, *Temm. Man. d'Orn.* tom. II. p. 557. Common Crane,
Selb. Illust. vol. II. p. 4. pl. 1. *Bew. Brit. Birds*, vol. II. p. 3.
Crane, *Mont. Orn. Dict.* (Trachea,) *Linn. Trans.* vol. IV. pl. 12.
f. 4.

DIMENS. Entire length five feet: length of the bill above four inches.
MONT.

DESCRIPT. Forehead, and space between the bill and the eyes, covered
with black bristly hairs; crown of the head naked, the skin of an orange-
red colour: occiput, throat, and fore part of the neck, of a deep blackish
gray; all the upper and under parts of the body dark ash-colour: quills,
and greater coverts, black; secondaries and tertials elongated, and of an
arched form, with the barbs of the feathers disunited, forming an elegant
tuft of floating plumes capable of being erected or depressed at will: bill
greenish black, the tip horn-colour: feet black. "*Young birds, previously to the second autumnal moult,* are without the naked patch on
the crown of the head, or have it very small; the deep blackish tint on
the occiput and fore part of the neck is simply indicated by a few
longitudinal spots of that colour. In *old individuals,* a large whitish
space is found behind the eyes, and on each side of the upper part
of the neck." TEMM. (*Egg.*) Pale greenish olive ground, blotched and
spotted with darker green and olive-brown: long. diam. four inches;
trans. diam. two inches six lines.

According to Ray, this species was formerly met with in Cambridgeshire, in large flocks, during the winter months. In still earlier times
it is stated by Turner to have bred in some of our fens. It must now be
considered as an extremely rare and accidental visitant. Pennant makes
mention of a single specimen which was killed in his time near Cambridge. One or two others are said to have occurred in Kent; and so
recently as in the year 1828, an individual is recorded to have been killed
in Cornwall. This last is now in the collection of Mr Drew, Devonport.
(*Loud. Mag.* vol. III. p. 177.) Is a common inhabitant of marshy plains
in the eastern parts of Europe. Feeds on grain, aquatic plants, worms,
and small reptiles. Said to build amongst rushes and other thick herbage, sometimes on the roofs of solitary houses. Lays two eggs.

GEN. 69.   ARDEA, *Linn.*

(1. ARDEA, *Steph.*)

* *A large portion of the tibia naked; tarsi long.*

172.   A. *cinerea*, Lath.   (*Common Heron.*)—General
plumage bluish ash: middle toe, claw included, much
shorter than the tarsus.

> A. cinerea, *Temm. Man. d'Orn.* tom. II. p. 567. Common Heron,
> *Mont. Orn. Dict. and Supp.   Selb. Illust.* vol. II. p. 11. pl. 2.
> Heron, *Bew. Brit. Birds*, vol. II. p. 9.

DIMENS.   Entire length three feet two inches nine lines; length of
the bill (from the forehead) four inches seven lines, (from the gape) five
inches eleven lines; of the tarsus five inches eight lines; of the naked
part of the tibia two inches eight lines; of the middle toe three inches
ten lines; of the tail eight inches; from the carpus to the end of the
wing one foot five inches seven lines: breadth, wings extended, five feet
six inches.

DESCRIPT.   (*Adult, after the age of three years.*) A pendent crest
of narrow, elongated, black feathers at the back of the head: forehead,
crown, neck, middle of the breast, belly, edge of the wing, thighs, and
under tail-coverts, pure white; fore part of the neck with large longitudinal black spots; the feathers towards the bottom loose and very
much elongated, falling over the breast in an elegant manner: occiput, sides of the breast, and flanks, deep black: back and wings of a
fine bluish gray: quills black: scapulars long and pointed, forming
graceful plumes of a silvery gray colour: tail bluish ash: bill deep
yellow: irides bright gamboge-yellow: lore, and naked skin round the
eyes, greenish: feet brown; the naked part of the tibia red. (*Young,
and until the age of three years.*) Occipital crest very short, almost
wanting: without the long plumes on the scapulars and lower part of
the neck: forehead and crown cinereous; throat white; neck pale ash-
colour, inclining to whitish underneath, with large longitudinal dusky
spots: rest of the upper parts bluish ash; ridge of the wing spotted with
white and reddish brown; quills of a deep bluish black: breast, belly,
and vent, white, spotted with black: upper mandible dusky brown, variegated with yellowish spots; lower mandible wholly yellow: irides yellow:
lore and orbits greenish yellow: feet dark ash-colour; the lower part of
the tibia yellowish. (*Egg.*) Of a uniform sea-green: long. diam. two
inches three lines; trans. diam. one inch nine lines.

A common and generally distributed species. Congregates in large
societies during the breeding season, building on the loftiest trees. At
other times of the year, of rather solitary habits, residing principally in
marshy districts, or in the neighbourhood of streams and rivers. Feeds
principally on fish and small reptiles. Lays three or four eggs.

173.   A. *purpurea*, Linn. (*Purple Heron.*)—General
plumage reddish ash, inclining to purple: middle toe, claw
included, as long as, or longer than the tarsus.

> A. purpurea, *Temm. Man. d'Orn.* tom. II. p. 570. Crested Purple
> Heron, *Lath. Syn.* vol. III. p. 95.   *Shaw, Zool.* vol. XI. p. 556.
> *Selb. Illust.* vol. II. p. 15. pl. 3.

DIMENS. Entire length two feet eleven inches : length of the bill (from the forehead) four inches eight lines, (from the gape) five inches seven lines; of the tarsus four inches seven lines; of the naked part of the tibia two inches seven lines; of the middle toe, claw included, four inches nine lines; from the carpus to the end of the wing thirteen inches eight lines.

DESCRIPT. (*Adult in perfect plumage.*) Crown, and occipital crest, black, with green reflections; throat white; cheeks, and sides of the neck, reddish brown, with three longitudinal narrow black bands, two lateral reaching from the eyes to the breast, the third commencing at the nape and running down the back of the neck for two-thirds of its length: front of the neck variegated with red, black, and purple; the feathers on the lower part long and acuminated, of a purplish white colour: back, wings, and tail, reddish ash; scapulars long and subulate as in the last species, of a rich brilliant purple red: breast and flanks of a deep brownish red, tinged with purple; belly and thighs red: bill and orbits bright yellow: irides yellowish orange: soles of the feet, posterior part of the tarsus, and naked space on the tibia, yellow; fore part of the tarsus, and upper surface of the toes, greenish brown. (*Immature plumage.*) No occipital crest, or simply a few red feathers somewhat longer than the others: without the long subulate plumes on the lower part of the neck and scapulars: forehead black; nape and cheeks pale red; throat white; fore part of the neck, and sides of the breast, yellowish white, with numerous longitudinal black spots: back, wings, and tail, dusky ash, all the feathers edged with reddish ash: belly and thighs whitish: a large portion of the upper mandible dusky; lower mandible, orbits, and irides, pale yellow. (*Egg.*) Pale asparagus-green : long. diam. two inches four lines; trans. diam. one inch seven lines.

A very rare, and only occasional visitant in this country. Has been killed in Berkshire, Cambridgeshire, Norfolk, and in one or two other parts of the kingdom. A specimen is also said to have flown on board a fishing boat off the coast of Cornwall in May 1822, and to have been taken. Frequents the same kind of situations as the last species. Food similar. Is said to breed amongst reeds and thick underwood, rarely in trees. Lays three eggs. *Obs.* The *African Heron* of Montagu's *Ornithological Dictionary* (A. Caspica, *Gmel.*) is this species in immature plumage.

(17.) *A. alba*, Linn. Syst. Nat. tom. I. p. 239. *Great White Heron*, Mont. Orn. Dict. & Supp. Selb. Illust. vol. II. p. 18. *A. Egretta*, Temm. Man. d'Orn. tom. II. p. 572. *White Heron*, Penn. Brit. Zool. vol. II. p. 427. pl. 62.

> Recorded as a British species by Ray. Asserted by Latham, on the authority of Dr Heysham, to have been shot in Cumberland : supposed also to have been seen in Devonshire and Suffolk. There is, however, no well authenticated instance of its having been met with in this country of late years, nor any British-killed specimen in existence. It is distinguished by its pure white plumage, greenish yellow bill, and long slender legs : the *adult, in the breeding season,* possesses an occipital crest and long dorsal plumes like the next species. Is common in some of the eastern parts of Europe. Food and habits resembling those of its congeners.

174. A. *Garzetta*, Linn. (*Little Egret Heron.*) — Whole plumage white : bill black.

A. Garzetta, *Temm. Man. d'Orn.* tom. II. p. 574. Egret, *Penn. Brit. Zool.* vol. II. *Append.* p. 631. pl. 7. *Mont. Orn. Dict.* Little Egret, *Lath. Syn.* vol. III. p. 90. *Bew. Brit. Birds*, vol. II. p. 17. Little Egret Heron, *Selb. Illust.* vol. II. p. 21. pl. 5.

DIMENS. Entire length one foot ten or eleven inches. TEMM.

DESCRIPT. (*Mature plumage in Summer.*) The whole plumage pure white: a pendent crest of long narrow feathers on the occiput: a tuft of similar feathers, the webs silky and disunited, on the lower part of the neck: dorsal plumes likewise of the same character, being of a soft silky texture, with the shafts very much elongated, and floating loosely over the back and tail from between the shoulders: bill black: lore and orbits greenish: irides bright yellow: feet dusky green; toes greenish yellow. (*Immature plumage, and adult in Winter.*) The long silky plumes on the occiput, back, and lower part of the neck, wanting. In *very young* birds, the white is of a dull tint; the bill, lore, orbits, irides, and feet, black. (*Egg.*) White. TEMM.

An extremely rare and accidental visitant in this country, though supposed to have been more plentiful formerly. Said to have been shot in Anglesey, and also in Ireland, many years ago. More recently, in April 1824, two specimens are recorded to have been killed near Penzance in Cornwall, and one of them to have been preserved. Common on some parts of the Continent. Said to breed in marshes, and to lay four or five eggs.

175. A. *russata*, Wagler. (*Buff-backed Heron.*) — Occiput, nape, and hind-neck, saffron-yellow: wings, tail, and under parts, pure white.

A. russata, *Wagler, Syst. Av.* part i. sp. 12. A. æquinoctialis, *Mont. in Linn. Trans.* vol. IX. p. 197. Little White Heron, *Mont. Orn. Dict. Supp. with fig.* Buff-backed Heron, *Selb. Illust.* vol. II. p. 24.

DIMENS.* Entire length twenty inches six lines: length of the bill (from the forehead) two inches two lines, (from the gape) two inches ten lines; of the tarsus three inches; of the naked part of the tibia one inch one line; of the middle toe, claw included, two inches nine lines; of the tail four inches; from the carpus to the end of the wing ten inches.

DESCRIPT. (*Mature plumage.*) " Occiput, nape, and hinder part of the neck, clothed with rigid open feathers of a saffron yellow: throat, front part of the neck and breast, white, tinged with sienna-yellow: long flowing plumes of the back ochre-yellow: wings, tail, and under parts of the body, pure white." SELBY. (*Immature plumage.*) "The whole plumage snow white, excepting the crown of the head, and the upper part of the neck before, which are buff: the skin of a very dark colour, almost black, so that on the cheeks and sides of the neck, where the feathers are thin, it is partly seen, giving a dingy shade to the white plumage of those parts: feathers on the back of the head somewhat elongated, but scarcely forming a crest: those on the lower part of the neck before, rather more elongated, hanging loosely over the upper part of the breast: tail slightly forked, and so short as to be entirely covered by the wings when folded: bill, lore, and orbits, orange-yellow: irides pale yellow: legs nearly black, with a tinge of green." MONT.

A single individual of this species, in immature plumage, and probably a young bird of the year, is recorded by Montagu to have been shot near Kingsbridge in Devonshire, the latter end of October 1805. This specimen, which is now in the British Museum, is the only one that has

---

* These measurements were taken from Montagu's original specimen.

hitherto occurred in this country.   It is a native of the warmer parts of
Europe, as well as of Asia and Africa, but is not found in America, "the
Ardea æquinoctialis, with which it has been confounded, being a distinct
species, and confined to that continent." Habits unknown.

** *A small portion of the tibia naked; tarsi short.*

176. A. *Ralloides*, Scop. (*Squacco Heron.*) — Occipital feathers very long, white edged with black: neck,
back, and scapulars, pale buff-orange.

> A. Ralloides, *Temm. Man. d'Orn.* tom. ii. p. 581. Squacco Heron,
> *Selb. Illust.* vol. ii. p. 25. pl. 6. *Mont. Orn. Dict. Supp. Shaw,
> Zool.* vol. xi. p. 574.

DIMENS. Entire length seventeen inches: length of the bill (from the
forehead) two inches six lines, (from the gape) three inches one line; of
the tarsus two inches; of the naked part of the tibia eleven lines and
a half; of the middle toe, claw included, two inches seven lines; of the
tail three inches one line; from the carpus to the end of the wing eight
inches eleven lines.

DESCRIPT. (*Mature plumage.*) Forehead and crown yellowish, streaked
with black; a pendent crest from the occiput, consisting of eight or ten
very long narrow white feathers edged at the sides with black: throat
white; neck, breast, upper part of the back, and scapulars, pale buff
orange; middle and lower regions of the back inclining to ferruginous
chestnut, the feathers on those parts very long, with disunited webs, and
floating loosely over the tail: all the rest of the plumage pure white: bill
bluish green at the base, the tip black: lore and orbits gray tinged with
green: irides yellow: feet yellow tinged with green. (*Plumage during
the first and second year.*) "Without the long occipital feathers; head,
neck, and wing-coverts, ferruginous brown, with large longitudinal
spots of a darker tint; throat, rump, and tail, pure white; upper part of
the back and scapulars of a more or less deep brown: upper mandible
greenish brown; lower mandible greenish yellow: orbits green: irides
very pale yellow: feet greenish ash." TEMM. (*Egg.*) Unknown.

A very rare, and only occasional visitant. Shot at Boyton in Wiltshire,
in 1775. A second specimen taken at Ormsby in Norfolk, in 1820; a
third killed in Cambridgeshire; and a fourth, a female, at Bridgewater,
in the Summer of 1825. Common in some parts of the Continent. Frequents marshes and the borders of rivers, &c. Food small fish, mollusca,
and insects. Said to build in trees. *Obs.* This species is the *A. comata*
of Pallas and some other authors.

*** *Tibia entirely feathered; tarsi short.*

177. A. *minuta*, Linn. (*Little Heron.*) — Crown,
back, scapulars, and secondaries, black, glossed with green;
neck, wing-coverts, and under parts, reddish yellow.

> A. minuta, *Temm. Man. d'Orn.* tom. ii. p. 584. Little Bittern,
> *Penn. Brit. Zool.* vol. ii. *App.* p. 633. pl. 8. *Mont. Orn. Dict.
> and Supp. Selb. Illust.* vol. ii. p. 36. pl. 6*. *Bew. Brit. Birds,*
> vol. ii. pp. 25, & 27.

DIMENS. Entire length fourteen inches three lines: length of the
bill two inches; of the tarsus one inch nine lines; of the middle toe,
claw included, two inches; from the carpus to the end of the wing five
inches six lines.

DESCRIPT. (*Adult.*) Crown of the head, occiput, back, scapulars,
secondary quills, and tail, black, with green reflections; cheeks, neck,
wing-coverts, and all the under parts, reddish yellow; flanks, and sides of
the breast, with a few brownish streaks; primary quills dusky: bill
yellow; the tip brown: orbits and irides yellow: feet greenish yellow.
(*Young of the year.*) Crown of the head brown; fore part of the neck
whitish with longitudinal dark streaks; cheeks, nape, breast, back, and
wing-coverts, chestnut-brown more or less deep, some of the feathers
with pale rufous edges: greater quills and tail-feathers deep brown.
(*Egg.*) White: long. diam. one inch six lines; trans. diam. eleven lines.

Occasionally met with, but not common in this country. Montagu
records that three specimens were shot in Devonshire during the Spring
and Summer of 1808. Others have been killed at different times in the
Orkneys, in Somersetshire, Shropshire, Northumberland, Hampshire,
Suffolk, and, more recently, on Uxbridge Moor, and on the banks of the
Thames near Windsor. Frequents marshes, and rushy places on the
borders of lakes and rivers. Breeds in such situations, constructing a
large nest of leaves and rushes. Lays five or six eggs. Food small
reptiles, fish, worms, &c.

(2. BOTAURUS, *Steph.*)

178. A. *stellaris*, Linn. (*Common Bittern.*)—Upper
parts ochre-yellow, variegated with black; beneath paler,
with oblong dusky streaks.

A. stellaris, *Temm. Man. d'Orn.* tom. II. p. 580. Bittern, *Mont.
Orn. Dict. Selb. Illust.* vol. II. p. 30. pl. 8. *Bew. Brit. Birds,*
vol. II. p. 22.

DIMENS. Entire length two feet six inches six lines: length of the
bill (from the forehead) three inches, (from the gape) four inches; of the
tarsus three inches eleven lines; of the middle toe four inches six lines;
of the tail four inches six lines; from the carpus to the end of the wing
one foot: breadth, wings extended, three feet nine inches.

DESCRIPT. Crown of the head, and a broad streak or moustache from
the corners of the mouth, black; neck-feathers loose and elongated,
capable of being raised and depressed at will, ochre-yellow, with brown
zigzag transverse lines on the sides, and long streaks and spots of reddish
brown in front: all the upper parts of the body ochre-yellow tinged with
orange-red, with a large dusky spot in the middle of each feather;
primary and secondary quills, primary coverts, and spurious winglet,
ferruginous, with transverse bars of blackish brown; rest of the coverts
and scapulars, mottled like the back with ochre-yellow and zigzag dusky
lines: under parts of the same colour as above but paler, with large
oblong longitudinal dusky streaks: upper mandible brown, passing into
yellow at the edges; lower mandible, lore, orbits, and feet, greenish
yellow; irides bright gamboge-yellow. (*Egg.*) Of a uniform light olive-
brown: long. diam. two inches two lines; trans. diam. one inch six
lines.

Occasionally met with in extensive marshes, as well as on the borders
of rivers and lakes, but not so abundant as formerly. Is partial to sedge

and reedy situations. Breeds in such spots, constructing a nest of rushes and other coarse plants. Lays four or five eggs. During the Spring the male utters a singular bellowing note which may be heard to a considerable distance. Feeds on small fish and reptiles, as well as on aquatic worms and insects.

179. A. *lentiginosa*, Mont. (*American Bittern.*) — Back and scapulars chocolate-brown glossed with purple, the edges of the feathers paler : under parts ochre-yellow, marked on the neck, breast, and belly, with broad chestnut streaks.

> A. lentiginosa, *Faun. Amer. Bor*. vol. II. p. 374. Freckled Heron, *Mont. Orn. Dict. Supp. with fig.* American Bittern, *Selb. Illust.* vol. II. p. 34.

Dimens. Entire length about twenty-three inches : length of the bill (from the forehead) two inches nine lines : of the tarsus three inches nine lines. Mont.

Descript. Crown of the head chocolate brown, passing into dull yellow at the nape, where the feathers are much elongated; chin and throat white, with a row of brown feathers down the middle; from the base of the lower mandible a black streak increasing on the upper part of the neck on each side; cheeks yellowish, with an obscure dusky line at the corner of the eye; the feathers on the neck long and broad, with their webs partly unconnected; those in front pale dull yellow with broad chestnut streaks; hind neck bare; the feathers on the breast long, of a fine chocolate brown, glossed with purple, and edged with dull yellow; belly and sides the same, but not so bright, the brown marks becoming speckled; vent and under tail-coverts yellowish white: back and scapulars chocolate-brown, minutely speckled and glossed with purple, the edges of the feathers paler: wing-coverts dull yellow, darkest in the middle of each feather, the margins prettily speckled; first and second order of quills, their greater coverts, and the spurious wing, dusky lead-colour, with a tinge of cinereous; primaries very slightly tipped with brown; secondaries and the greater coverts tipped more deeply with the same, and prettily speckled on the light part: the closed wings not reaching to the end of the tail: upper mandible dusky above, greenish yellow at the sides; lower mandible greenish yellow: legs greenish. Mont. (*Egg.*) Cinereous green. *Faun. Am. Bor.*

A single individual of this species, which is common in North America, was shot by Mr Cunningham, in the parish of Piddletown, in Dorsetshire, in the Autumn of 1804. This specimen was procured by Montagu, and, with the rest of his collection, is now in the British Museum. No other has occurred since in this country. Has a loud booming note similar to that of the last species. Said to frequent marshes and willow thickets, and to lay four eggs.

(3. Nycticorax, *Steph.*)

180. A. *Nycticorax*, Linn. (*Common Night-Heron.*) —Crown, back, and scapulars, black, glossed with green : under parts white.

A. Nycticorax, *Temm. Man. d'Orn.* tom. ii. p. 577. Night-Heron, *Mont. Orn. Dict. Selb. Illust.* vol. ii. p. 39. pls. 7, & 7*. *Bew. Brit. Birds*, vol. ii. p. 14.

Dimens. Entire length twenty-one inches: length of the bill (from the forehead) two inches eight lines, (from the gape) three inches eight lines; of the tarsus two inches eleven lines; of the naked part of the tibia ten lines; of the middle toe, claw included, three inches one line and a half; from the carpus to the end of the wing twelve inches.

Descript. (*Adult.*) Crown of the head, nape, upper part of the back, and scapulars, black glossed with green: occiput ornamented with three very narrow white feathers, measuring six or seven inches in length: lower part of the back, wings, and tail, of a fine pearl-gray: forehead, a narrow streak above the eye, throat, and all the under parts, pure white; bill black, yellowish at the base: lore and orbits green: irides deep orange: feet yellowish green. (*Young of the year.*) Without the long subulate feathers on the occiput: upper part of the head, nape, back, and scapulars, dull brown, with longitudinal streaks of yellowish white, one in the middle of each feather: throat white, spotted with brown; feathers on the sides and fore part of the neck, yellowish, broadly edged with brown; rest of the under parts variegated with brown, white, and ash-colour; middle of the belly whitish: wing-coverts, and primary quills, cinereous brown, with large pisciform yellowish white spots at the tips of the feathers: culmen and tip of the bill dusky brown, the remaining portion greenish yellow: irides brown: feet olivaceous brown. (*Egg.*) Pale blue: long. diam. two inches three lines; trans. diam. one inch nine lines.

Very rare in this country. Has been killed in Norfolk, Suffolk, Bedfordshire, Oxfordshire, Buckinghamshire, and, in more than one instance, near London. Common in the South of Europe; inhabiting the borders of rivers and lakes where there is covert. Feeds on fish, reptiles, worms, &c. Said to build on the ground, and to lay three or four eggs. *Obs.* The *Ardea Gardeni* of Gmelin and Latham is the young of the year of this species. The supposed *A. Cayennensis*, recorded by Mr Youell to have been taken near Yarmouth, May 23, 1824, (*Linn. Trans.* vol. xiv. p. 588.) is probably only a variety of the common Night-Heron, having double the usual number of long occipital feathers.

GEN. 70.   CICONIA, *Briss.*

181.   C. *alba*, Ray.   (*White Stork.*)—White; scapulars and wings black.

C. alba, *Temm. Man. d'Orn.* tom. ii. p. 560.   Stork, *Mont. Orn. Dict. & Supp. Bew. Brit. Birds*, vol. ii. p. 6.   White Stork, *Selb. Illust.* vol. ii., p. 45. pl. 11.

Dimens. Entire length three feet eight inches: length of the bill (from the forehead) seven inches six lines; of the tarsus twelve inches; of the naked part of the tibia six inches.

Descript. Head, neck, and all the body, pure white; scapulars and wings black: bill and legs red: naked skin round the eyes black: irides brown. In *young birds*, the black on the wings inclines to brown, and the bill is dusky tinged with reddish. (*Egg.*) White: long. diam. two inches ten lines; trans. diam. one inch eleven lines.

Very common in Holland, France, and other parts of the Continent, but in this country a rare, and only accidental visitant. Montagu mentions one which was killed near Salisbury, in February 1790; a second shot at Sandwich in Kent, in 1805; and a third in Hampshire, in the Autumn of 1808. Has occurred in Suffolk more recently, and in three or four instances. Food, reptiles and insects, as well as small quadrupeds and birds. Builds on the tops of houses, or in old trees. Lays generally three eggs.

182. C. *nigra*, Ray. (*Black Stork.*)—Black; lower part of the breast, and belly, white.

C. nigra, *Temm. Man. d'Orn.* tom. ii. p. 561.   Black Stork, *Mont. in Linn. Trans.* vol. xii. p. 19.   *Selb. Illust.* vol. ii. p. 48. pl. 11*.

Dimens. Entire length nearly three feet. Temm.
Descript. (*Adult.*) Head, neck, all the upper parts of the body, wings, and tail, black, with purple and green reflections; lower part of the breast and belly dingy white: bill and orbits bright orange: irides hazel: legs and toes deep red. (*Young.*) Head, and upper part of the neck, pale reddish brown, with the central portion of the feathers dusky; back, scapulars, wings, and tail, dusky brown, very slightly glossed with greenish hues: bill, orbits, and legs, olivaceous green. (*Egg.*) Of a uniform greenish buff: long. diam. two inches seven lines; trans. diam. one inch eleven lines.

A single individual of this species is recorded by Montagu as having been shot in West Sedge-Moor, in Somersetshire, May 13, 1814. The specimen is now in the British Museum. A second is stated to have been shot in October 1832, in the parish of Otley, about eight miles from Ipswich. (*Loud. Mag. of Nat. Hist.* vol. vii. p. 53.) Inhabits various parts of Europe, but is less plentiful than the last species. Said to frequent wooded swamps and extensive forests. Food, small fish, reptiles, and insects. Builds in lofty trees, and lays two or three eggs.

GEN. 71. PLATALEA, *Linn.*

183. P. *Leucorodia*, Linn. (*White Spoonbill.*)—Whole plumage white: a pendent crest of long subulate feathers on the occiput.

P. Leucorodia, *Temm. Man. d'Orn.* tom. ii. p. 595.   Spoonbill, *Penn. Brit. Zool.* vol. ii. *App.* p. 634. pl. 9. *Bew. Brit. Birds,* vol. ii. p. 29.   White Spoonbill, *Mont. Orn. Dict. & Supp. Selb. Illust.* vol. ii. p. 51. pl. 10.   (Trachea,) *Linn. Trans.* vol. xvi. pl. 19.

Dimens. Entire length thirty-one inches: length of the bill (from the forehead) seven inches three lines; breadth of the spoon one inch ten lines; length of the tarsus four inches eleven lines; of the naked part of the tibia two inches eight lines; from the carpus to the end of the wing fourteen inches six lines.
Descript. (*Old male.*) The whole plumage pure white, with the exception of a large patch of buff yellow on the upper part of the breast, from whence a narrow band of the same colour ascends on each side towards the top of the back: lore, orbits, and naked space on the throat, orange-yellow: bill black; the tip ochre-yellow: irides red: legs black.

N

(Old female.) Somewhat smaller; the occipital crest shorter; the buff-coloured patch on the breast not so distinct. *(Young of the year.)* No occipital crest; the feathers on the head short and rounded: shafts of all the quills, and tips of the primaries, deep black; the rest of the plumage white: bill deep ash-colour, soft, and very flexible: irides cinereous: lore and orbits dingy white. " The patch of buff yellow on the breast does not shew itself till the second or third year." Temm. *(Egg.)* White, spotted with pale reddish brown: long. diam. two inches nine lines; trans. diam. one inch nine lines.

A rare visitant in England. Pennant mentions a flock which migrated into the marshes near Yarmouth in Norfolk, in April 1774. Since then, specimens have been killed in Somersetshire, Devonshire, Dorsetshire, Lincolnshire, and Suffolk. In the latter county, three were shot out of a flight of seven, which appeared at Thorpe, in the Autumn of 1828. Common in Holland, frequenting the banks of rivers near their junction with the sea. Food, small fish, mollusca, and aquatic insects. Builds in trees or amongst rushes. Eggs two or three in number.

GEN. 72.   IBIS, *Lacép.*

184. I. *Falcinellus*, Temm. (*Glossy Ibis.*)—Neck, breast, and under parts, bright chestnut-red; lower back, rump, wings and tail, glossy green with purple reflections.

I. Falcinellus, *Temm. Man. d'Orn.* tom. ii. p. 598. Glossy Ibis, *Mont. Orn. Dict. & Supp.   Selb. Illust.* vol. ii. p. 56. pl. 12. *Bew. Brit. Birds,* vol. ii. p. 33. Bay Ibis, *Sow. Brit. Misc.* pl. 17. Brazilian Curlew, *Shaw, Nat. Misc.* vol. xvii. pl. 705.

Dimens. Entire length twenty-two inches: length of the bill (from the forehead) four inches, (from the gape) four inches three lines; of the tarsus three inches six lines; of the naked part of the tibia one inch ten lines; from the carpus to the end of the wing ten inches eight lines.

Descript. *(Old bird in perfect plumage.)* Head chestnut-red passing into dusky brown; neck, breast, upper part of the back, and all the under parts, bright chestnut-red; lower part of the back, rump, wing-coverts, primary quills, and tail-feathers, deep dusky green, with purple and bronze reflections: bill and legs greenish black: lore and orbits green: irides brown. *(Young till the age of three years.)* Head, throat, and upper part of the neck, dusky brown, the feathers with whitish edges; occasionally a few large irregular spots and transverse bars of this last colour: lower part of the neck, breast, belly, and thighs, cinereous black, with a few greenish reflections on the breast: back and scapulars greenish brown; wings and tail as in the perfect plumage, but with the gloss of green and purple much less brilliant. The *young of the year* have the whole plumage inclining more to cinereous brown; and the feathers on the head and neck more broadly edged with white. *(Egg.)* Unknown.

An occasional but rare visitant in this country. Has been killed in the counties of Cornwall, Devon., Kent, Berks., Norfolk, and Northumberland; also in Anglesea and Ireland. Most of the specimens have occurred during the autumnal and winter months. Common in the eastern parts of Europe, inhabiting the borders of lakes and rivers. Feeds on insects, worms, shell-fish, and aquatic plants. Nidification unknown. *Obs.* The *Green Ibis* of Latham and other authors is this species in immature plumage.

GEN. 73. NUMENIUS, *Briss.*

185. N. *arquata,* Lath. (*Common Curlew.*)—Head, neck, breast, and all the upper parts, rufous ash ; belly and abdomen white ; the whole variegated with dusky spots.

N. arquata, *Temm. Man. d'Orn.* tom. ii. p. 603. Common Curlew, *Mont. Orn. Dict. & Supp. Selb. Illust.* vol. ii. p. 62. pl. 13. *Bew. Brit. Birds,* vol. ii. p. 36.

Dimens. Entire length twenty-two to twenty-five inches: length of the bill from four to six inches and a half; of the tarsus three inches three lines; of the naked part of the tibia one inch two lines; of the tail four inches six lines; from the carpus to the end of the wing from eleven to eleven inches and a half: breadth, wings extended, from thirty-eight to forty-two inches.

Descript. Head, neck, and breast, pale ash tinged with rufous, the shafts and central portion of the feathers dusky; upper part of the back and scapulars blackish brown, the feathers broadly edged with pale rufous brown; lower part of the back inclining to white, spotted with black; tail yellowish white, with transverse brown bars; belly and abdomen white, with longitudinal dusky spots: upper mandible blackish brown; lower one flesh-colour: irides brown: legs and toes pale bluish gray. The *female* has less of the rufous, and more of the cinereous, tinge. The *young of the year* have the bill considerably shorter, and nearly straight. (*Egg.*) Olive green; blotched and spotted with darker green and dark brown: long. diam. two inches seven lines; trans. diam. one inch eleven lines.

Common on most parts of the coast during the winter season. Retires inland, and to more elevated spots, to breed. Deposits its eggs amongst heath, rushes, or long grass, to the number of four or five. Feeds on worms, insects, and molluscous animals.

186. N. *Phæopus,* Lath. (*Whimbrel.*)—Head, neck, breast, and upper parts, pale ash-colour, spotted with dusky: lower back, belly and abdomen, white.

N. Phæopus, *Temm. Man. d'Orn.* tom. ii. p. 604. Whimbrel, *Mont. Orn. Dict. Bew. Brit. Birds,* vol. ii. p. 38. Whimbrel Curlew, *Selb. Illust.* vol. ii. p. 65. pl. 14.

Dimens. Entire length seventeen inches: length of the bill three inches three lines: breadth, wings extended, twenty-nine inches.

Descript. Forehead and crown cinereous brown, the latter divided in the middle by a longitudinal pale streak; over each eye, another but broader streak of white variegated with brown; sides of the head, neck, and breast, pale ash, with longitudinal brown streaks : upper part of the back, scapulars, and wing-coverts, deep brown, the feathers with pale edges; lower part of the back white; rump barred with cinereous brown; tail cinereous brown, with oblique bars of a darker shade: belly and abdomen white; sides barred with brown: bill dusky, reddish towards the base: irides brown: legs deep bluish gray. In *young birds,* the bill is shorter, and not so much curved. (*Egg.*) Dark olive-brown, blotched with darker brown: long. diam. two inches five lines; trans. diam. one inch eight lines.

Found on many parts of the coast during the autumnal and winter months, but is less common than the last species. Generally observed in small flocks. Retires to the northern parts of the country to breed. Nest said to be placed on open and exposed heaths. Food, worms and insects.

GEN. 74.   TOTANUS, *Bechst.*

* *Bill slender, straight.*

187. T. *fuscus*, Leisl. (*Dusky Sandpiper.*)—Upper plumage ash-gray; rump, and all the under parts, pure white: legs, and base of the lower mandible, red.

T. fuscus, *Temm. Man. d'Orn.* tom. ii. p. 639. Cambridge Godwit, *Penn. Brit. Zool.* vol. ii. p. 447. Dusky Sandpiper, *Shaw, Zool.* vol. xii. p. 132. *Selb. Illust.* vol. ii. p. 69. pl. 15. f. 1, & 2.

DIMENS. Entire length twelve inches: length of the bill two inches five lines; of the tarsus two inches three lines; of the naked part of the tibia one inch three lines and a half; of the tail two inches seven lines; from the carpus to the end of the wing six inches ten lines.

DESCRIPT. (*Adult in winter plumage.*) Crown of the head, nape, and back, cinereous gray, with fine dusky streaks on the shafts of the feathers; wing-coverts and scapulars cinereous gray edged with white*; between the bill and the eye a blackish patch, above which is a white streak; cheeks, sides and fore part of the neck, variegated with white and ash-colour; throat, breast, rump, belly and abdomen, pure white; flanks whitish, passing into ash-gray: upper tail-coverts, and three or four outermost tail-feathers, with alternate transverse bars of white and dusky brown; central tail-feathers of a uniform ash-gray narrowly edged with white: bill black; base of the lower mandible red: legs bright orange-red. (*Adult in summer plumage.*) All the upper parts deep blackish brown, the feathers on the back, scapulars, and wing-coverts, marked on the edges with white spots mostly of a triangular form; head, neck, and under parts, dusky gray; the neck without spots, but the breast and belly sometimes with a narrow crescent-shaped white edging at the tip of each feather: vent, and under tail-coverts, barred transversely with white and dusky; tail dusky ash, with little transverse white streaks at the edges of the feathers, not reaching to the shafts: legs reddish brown. (*Young of the year.*) Upper parts olivaceous brown, the feathers on the back with a narrow edging of white; wing-coverts and scapulars with triangular white spots: all the under parts whitish, with indistinct spots and transverse undulating bars of cinereous brown: legs and toes orange-red.

A rare species in Great Britain, but has been killed at different times in various parts of the country. Frequents marshes, and the borders of lakes and rivers. Food, principally molluscous animals. Retires to high northern latitudes to breed. *Obs.* The *Cambridge Godwit* of Pennant is this species in the adult winter plumage: the *Spotted Redshank* of the same author (the *Spotted Snipe* of Latham and Montagu) is the young bird of the year.

188. T. *Calidris*, Bechst. (*Redshank Sandpiper.*)— Upper parts cinereous brown; rump, belly, and greater portion of the secondary quills, white: legs, and basal half of both mandibles, red.

---

* Probably, *in very old birds*, this white edging disappears.

T. **Calidris**, *Temm. Man. d'Orn.* tom. ii. p. 643.   Redshank Sandpiper, *Selb. Illust.* vol. ii. p. 72. pl. 16. f. 1.   Redshank, *Mont.*
Orn. Dict. & Supp.   Bew. Brit. Birds, vol. ii. p. 71.

Dimens.   Entire length twelve inches two lines: length of the bill
(from the forehead) one inch seven lines and a half, (from the gape) one
inch ten lines; of the tarsus one inch ten lines; of the naked part of the
tibia nine lines; of the middle toe one inch four lines and a half; of the
tail two inches ten lines; from the carpus to the end of the wing six
inches eleven lines: breadth, wings extended, twenty-one inches nine
lines.

Descript.   (*Winter plumage.*) Crown of the head, space between the
eye and the bill, back of the neck, upper part of the back, scapulars, and
wing-coverts, cinereous brown, darkest on the shafts of the feathers;
throat, and streak above the eye, white; sides of the head, fore part of
the neck, and breast, grayish white, the shafts of the feathers brown;
lower part of the back, belly and abdomen, white: greater quills with
their coverts dusky brown; secondaries white for a considerable portion
of their length: tail and upper tail-coverts with transverse zigzag bars of
white and deep brown: bill red; the tip black: irides brown: feet pale
red.   (*Summer plumage.*) A white streak from the base of the upper
mandible to the eye: all the upper parts cinereous inclining to olivaceous
brown, with a broad longitudinal dusky streak in the middle of each
feather; lower part of the back white: sides of the head, throat, front of
the neck, breast, and belly, white, spotted and streaked with blackish
brown: tail barred with black and white, and tipped with pure white, the
white passing into ash-colour on the four middle feathers: legs, and
basal half of the bill, bright vermilion-red.   (*Young, till after the first
moult.*) Upper parts dark cinereous brown, all the feathers edged with
yellowish white; on the back and scapulars this edging assumes the
form of angular spots; wing-coverts dusky brown, edged and tipped with
yellowish white: throat whitish, dotted with brown; neck and breast
cinereous, with longitudinal narrow streaks; belly white; flanks, abdomen, and under tail-coverts, spotted with brown: tip of the tail reddish:
bill livid flesh-colour; the tip brown: legs orange-yellow.   (*Egg.*) Pale
reddish white, tinged with green; blotched, spotted, and speckled, with
dark red brown: long. diam. one inch six lines and a half; trans. diam.
one inch two lines.

Not uncommon on the coast during the autumnal and winter months.
Retires inland to breed.   Nest placed in marshes and wet pastures.
Eggs four in number.   Food, insects and worms, and bivalve mollusca.
Obs. The *Gambet* of Pennant is either this species, or a *Ruff* in winter
plumage.

189.   T.   *Ochropus*,   Temm.   (*Green Sandpiper.*) —
Upper parts greenish brown: tail barred with brown and
white; the two outer feathers on each side almost entirely
white: bill and legs greenish.

T. **Ochropus**, *Temm. Man. d'Orn.* tom. ii. p. 651.   T. Glareola,
Markwick in Linn. Trans. vol. i. p. 128. tab. 11.   Green Sandpiper, *Mont. Orn. Dict. & Supp.   Bew. Brit. Birds,* vol. ii.
p. 85.   *Selb. Illust.* vol. ii. p. 75. pl. 16. f. 2.

Dimens.   Entire length nine inches eight lines: length of the bill
(from the forehead) one inch three lines and a half, (from the gape) one

inch six lines; of the naked part of the tibia nine lines; of the tarsus one
inch four lines; of the middle toe one inch two lines; of the tail two
inches five lines; from the carpus to the end of the wing five inches six
lines.

DESCRIPT. (*Winter plumage.*) Crown of the head, nape, sides of the
neck, and all the upper parts, cinereous brown, tinged with olivaceous
green; the feathers on the back, scapulars, and wing-coverts, with an
edging of small whitish spots : from the bill to the eye a brown streak,
above which is another one of white : upper tail-coverts, and all the under
parts, pure white; the breast, and front of the neck, variegated with
numerous longitudinal dusky streaks : quills deep brown : under wing-
coverts marked with transverse undulating white bars : tail white for one-
third of its length from the base; the remaining portion barred with
brown and white; the two middle feathers with five broad bars of the
former colour; the two next with four; the two next with three; the two
next with two; the two next with only one bar, but with a small spot
higher up on the exterior web; the two outer feathers almost entirely
white, having only a single small spot on the exterior web : bill dusky
green passing into black at the tip : irides deep brown : legs and toes
greenish gray. The *summer plumage* is distinguished by the deeper
tint of the upper parts, which are also more glossed with green, and more
spotted on the edges of the feathers; the streaks on the fore part of the
neck are likewise more distinct. (*Young of the year.*) "All the upper
parts paler; the spots fewer, and of a yellowish colour; nape tinged with
cinereous : sides of the breast of the same colour as the back, and marked
with white spots; all the fore part of the neck, and middle of the breast,
with lanceolate brown streaks : less white at the base of the tail; the
bars of brown on the middle feathers broader." TEMM. (*Egg.*) Yellowish
clay-colour, blotched and spotted with ash-gray and dark brown : long.
diam. one inch four lines and a half; trans. diam. one inch.

An occasional visitant in this country, principally during the autumnal
and winter months, but scarcely to be called common. Frequents the
borders of streams, rivers, and other pieces of fresh water, but rarely
resorts to the coast. Said to breed in the central parts of Europe, constructing its nest in the sand or grass by the water-side. Lays from three
to five eggs. Food, insects and worms.

190. T. *Glareola*, Temm. (*Wood Sandpiper.*)—Upper
parts deep brown, spotted with white : all the tail-feathers
barred with brown and white; the two outer feathers on
each side with the inner web entirely white : bill and legs
greenish.

T. Glareola, *Temm. Man. d'Orn.* tom. II. p. 654. Wood Sandpiper,
Mont. Orn. Dict. & Supp. with fig. Selb. Illust. vol. II. p. 77.
pl. 16. f. 3. Long-legged Sandpiper, *Mont. Orn. Dict. Supp.
App.*

DIMENS. Entire length nine inches : length of the bill not quite one
inch three lines. MONT.

DESCRIPT. (*Winter plumage.*) Between the bill and the eye a dusky
streak, above which is a broader one of white continued over the eye :
forehead, crown, back, and wings, deep brown ; the edges of the feathers
on the back, scapulars, and wing-coverts, marked with white and grayish
white spots : ear-coverts dusky ; cheeks, nape, fore-neck, and breast, dirty

white, irregularly variegated with spots and streaks of cinereous brown;
flanks with transverse undulating bars of the same colour: throat, middle
of the abdomen, upper and under tail-coverts, pure white; some of the
tail-coverts with fine dusky streaks on the shafts of the feathers: tail with
alternate black and white bars, narrower and more numerous than in the
last species; the two outer feathers with their inner webs pure white:
legs greenish; much longer than in the *T. Ochropus*, with a greater portion of the tibia naked: bill black, towards the base greenish. (*Summer
plumage.*) "Crown of the head, and nape, streaked longitudinally with
brown and whitish; cheeks, fore part of the neck, breast, and flanks,
nearly pure white, streaked longitudinally with deep brown: all the
feathers of the back with a very large black spot in the centre, and
with two whitish spots on each side of the webs; scapulars, when raised,
appearing barred with broad dusky bands: the rest as in winter." Temm.
(*Young of the year.*) "All the deep brown part of the plumage covered
with small, closely approximating, red spots: the whole of the breast
waved with cinereous, with irregular brown spots: legs, and base of
the bill, dirty yellowish green: tail-feathers irregularly barred." Temm.
(*Egg.*) Reddish white; the larger end nearly covered with dark red
brown: long. diam. one inch five lines; trans. diam. one inch one line.

First noticed as British by Montagu, who has accurately pointed out
the distinctions between this and the last species. The individual described by him was shot on the coast of South Devon early in August.
He mentions another which was shot at Woolwich, at the same period of
the year. Since his time individuals have been killed in the counties of
Durham, Northumberland, Norfolk, and Cambridge. The species, however, must be considered as a rare and accidental visitant in this country.
Common on many parts of the Continent. Breeds, according to Temminck, within the arctic circle. Constructs its nest in marshes, and lays
four eggs. Food, insects and worms.

(18.) *T. Macularia*, Temm. Man. d'Orn. tom. ii. p. 656. Wils.
Amer. Orn. vol. vii. p. 60. pl. 59. f. 1. *Spotted Sandpiper*, Mont.
Orn. Dict. & Supp. Selb. Illust. vol. ii. p. 84. pl. 17. *Spotted
Tringa*, Edw. Glean. pl. 277.

There is no well-authenticated instance in which this species has occurred in Great Britain, although by all our authors it has been included
in the British lists. The bird shot in Essex described by Edwards, and
supposed by him to be the same as his *Spotted Tringa* from America, is
evidently nothing more (as Mr Selby has suggested) than the common
Totanus Hypoleucos. The same may be said of the *Spotted Sandpiper* of
Bewick. (*Brit. Birds*, vol. ii. p. 97.) The true *T. Macularia* is abundant in the United States, where it represents the Common Sandpiper of this
country. Nevertheless, as (according to Temminck) it has certainly been
killed in two or three instances on the Continent, it is still possible that it
may occasionally occur in this country.

191. T. *Hypoleucos*, Temm. (*Common Sandpiper.*)
—All the upper parts cinereous brown glossed with olive ;
under parts pure white, without spots.

T. Hypoleucos, *Temm. Man. d'Orn.* tom. ii. p. 657. Common Sandpiper, *Mont. Orn. Dict. Bew. Brit. Birds*, vol. ii. p. 93. *Selb.
Illust.* vol. ii. p. 81. pl. 15. f. 3, & 4.

Dimens. Entire length seven inches nine lines: length of the bill
(from the forehead) one inch, (from the gape) one inch one line and

a half; of the tarsus eleven lines; of the naked part of the tibia four
lines; of the tail two inches four lines; from the carpus to the end of the
wing four inches three lines: breadth, wings extended, thirteen inches
three lines.

Descript. (*Summer and winter plumage.*) All the upper parts cinereous brown, glossed with olivaceous green; the shafts of the feathers
being of a darker tint; back and wing-coverts marked with fine transverse undulating lines of dusky brown: over the eye a whitish streak:
under parts pure white, streaked on the breast and sides of the neck with
cinereous brown: quills brown, with a large white spot on their inner
webs, the two first excepted : the four middle tail-feathers like the back;
the two next on each side tipped with white; the outer one tipped with
white, and barred on the exterior web with white and brown: bill and
legs grayish brown, the latter tinged with green. (*Young of the year.*)
Throat, and fore part of the neck, pure white, spotted only on the sides;
the white streak above the eyes broader and more distinct; the feathers
on the back edged with reddish and dusky; wing-coverts tipped with
red and black bars. (*Egg.*) Reddish white, spotted and speckled with
umber brown: long. diam. one inch four lines; trans. diam. one inch.

A regular summer visitant, making its appearance in the Spring and
departing in the Autumn. Chiefly frequents the banks of rivers and
lakes. Nest constructed of moss and dry leaves, and placed in the grass
by the water side. Eggs four or five in number. Utters a clear piping
note in its flight. Food, worms and insects.

** Bill rather strong, slightly recurved.

192. T. *Glottis*, Bechst. (*Greenshank.*)—Back, scapulars, and wing-coverts, dusky brown, the feathers edged
with yellowish white; lower back, and under parts, white:
legs greenish.

T. Glottis, *Temm. Man. d'Orn.* tom. ii. p. 659. Greenshank, *Mont.
Orn. Dict. & App. to Supp. Bew. Brit. Birds,* vol. ii. p. 67.
Selb. Illust. vol. ii. p. 86. pl. 19.

Dimens. Entire length fourteen inches. Mont.

Descript. (*Winter plumage.*) From the upper mandible to the eye
a white streak; head, cheeks, sides and back part of the neck, and sides
of the breast, cinereous white, with longitudinal brown streaks; upper
part of the back, scapulars, and wing-coverts, dusky brown, all the feathers broadly edged with yellowish white: lower part of the back, upper
tail-coverts, throat, fore part of the neck, middle of the breast, belly, and
other under parts, pure white: quills dusky; some of them spotted with
white on their inner webs: tail white, with transverse bars of brown; the
outer feathers entirely white, with the exception of a narrow longitudinal
streak on the exterior web : bill cinereous brown : legs yellowish green;
long and slender. (*Summer plumage.*) Crown, nape, cheeks, and sides
of the neck, with dusky streaks; throat white; fore-neck, breast, upper
part of the abdomen, and flanks, white, with oval black spots; rest of the
under parts pure white: upper part of the back and scapulars deep black,
the feathers on the back edged with white, those on the scapulars with
reddish white spots; greater coverts reddish ash with black shafts, edged
with white: the two middle tail-feathers cinereous, with transverse brown
zigzag bars. (*Egg.*) Olive-brown, covered all over with dusky spots.

Occasionally observed in small flocks on the coast during the winter months, as well as in marshes and on the banks of rivers, but not a very common species. Doubtful whether it ever remains through the Summer. According to Temminck, breeds within the arctic circle. Food, small fish, worms, and molluscous animals. *Obs.* The *Cinereous Godwit* of Pennant is this species in its winter plumage.

GEN. 75.   RECURVIROSTRA, *Linn.*

193.   R. *Avocetta,* Linn.   (*Scooping Avocet.*)—Crown, nape, scapulars, and lesser wing-coverts, black : cheeks, and under parts, pure white.

R. Avocetta, *Temm. Man. d'Orn.* tom. II. p. 590.   Scooping Avocet, *Mont. Orn. Dict.   Bew. Brit. Birds,* vol. II. p. 116.   *Selb. Illust.* vol. II. p. 90. pl. 20.

Dimens. Entire length eighteen inches: length of the bill (from the forehead) three inches, (from the gape) three inches two lines; of the tarsus two inches nine lines; of the tail two inches ten lines; from the carpus to the end of the wing eight inches six lines.
Descript. (*Adult male and female.*) Crown of the head, nape, two-thirds of the hind part of the neck, scapulars, lesser wing-coverts, and primary quills, deep black; all the rest of the plumage pure white: bill black: irides reddish brown: legs bluish gray. (*Young of the year, before the first moult.*) " All the black parts of the plumage tinged with brown; the brownish black on the head not extending below the occiput; scapulars edged with red, and tipped with reddish ash: legs cinereous." Temm. (*Egg.*) Olive-brown, blotched and spotted with brownish black: long. diam. two inches three lines; trans. diam. one inch five lines.
Common in small flocks on some parts of the coast during Winter. Breeds in the fens of Lincolnshire, and in other extensive marshy districts. Lays two eggs. Food, worms and marine insects.

GEN. 76.   HIMANTOPUS, *Briss.*

194.   H. *melanopterus,* Temm.   (*Black-winged Long-shanks.*)—Occiput, nape, back, and wings, black: rest of the plumage white.

H. melanopterus, *Temm. Man. d'Orn.* tom. II. p. 528.   Long-legged Plover, *Penn. Brit. Zool.* vol. II. p. 476.   *White, Nat. Hist. Selb.* vol. II. p. 84. *with fig. Mont. Orn. Dict.   Shaw, Nat. Misc.* vol. VI. pl. 195.   *Bew. Brit. Birds,* vol. I. p. 365. Black-winged Stilt, *Selb. Illust.* vol. II. p. 247. pl. 39*.

Dimens. Entire length thirteen inches: length of the bill two inches six lines; of the naked part of the tibia three inches six lines; of the tarsus four inches six lines: breadth, wings extended, two feet five inches.
Descript. Occiput and nape black, the latter sometimes whitish with dusky streaks; back and wings glossy black, with greenish reflections: forehead, sides of the face, neck, and all the under parts, pure white, slightly tinged on the breast and belly with rose-red: rump white: tail cinereous gray; the outer feathers white: bill black: irides crimson:

legs vermilion red. In *very old males*, according to Temminck, the nape, and sometimes the occiput also, is perfectly white. (*Female.*) Smaller; the black on the back and wings inclining more to brown, and without the greenish gloss. (*Young.*) Crown, occiput, and nape, dusky ash; back and wings brown; all the feathers with whitish edges: legs orange-red. (*Egg.*) Pale blue, blotched and streaked with ash green and red brown: long. diam. one inch nine lines; trans. diam. one inch three lines.

A very rare and accidental visitant in this country. Sibbald mentions a pair which were killed in Scotland. Pennant has recorded another which was shot near Oxford; and White, in his *Nat. Hist. of Selborne*, speaks of five which were shot out of a flock of six on the verge of Frinsham pond, near Farnham in Surrey, in April 1779. Has also been killed in Anglesea, and more recently in Lincolnshire and Norfolk. Said to be not uncommon in the eastern parts of Europe, frequenting the borders of rivers and lakes. Breeds in the salt marshes of Hungary and Russia. Food, tadpoles, aquatic and other insects, &c.

GEN. 77.   LIMOSA, *Briss.*

195. L. *rufa*, Briss. (*Bar-tailed Godwit.*)—Tail with eight or nine transverse dusky bars: bill curving considerably upwards: middle claw short, and without serratures: tarsi moderate.

> L. rufa, *Temm. Man. d'Orn.* tom. II. p. 668. Godwit, and Red Godwit, *Bew. Brit. Birds*, vol. II. pp. 59, & 61. *Shaw, Gen. Zool.* vol. XII. p. 77. Red Godwit, *Selb. Illust.* vol. II. p. 98. pl. 22.

DIMENS. Entire length sixteen inches: length of the bill (from the forehead) three inches, (from the gape) three inches one line; of the tarsus one inch eleven lines and a half: of the tail two inches ten lines; from the carpus to the end of the wing eight inches.

DESCRIPT. (*Male and female in winter plumage.*) Crown of the head, space between the bill and the eyes, cheeks, and neck, pale ash-colour, with longitudinal streaks of brown; a broad line above the eyes, throat, breast, belly, and abdomen, white; flanks, and some of the under tail-coverts, streaked with dusky brown: upper part of the back, and scapulars, cinereous gray, the shafts of the feathers, together with a small portion of the webs on each side, brownish black; wing-coverts white, the middle of each feather deep brown: quills dusky: lower part of the back, and rump, white with dusky spots: tail with alternate transverse bars of white and blackish brown: basal half of the bill purplish red; the tip black: irides brown: legs and toes black. (*Male in summer plumage.*) Crown and nape reddish orange, with brown streaks; line above the eyes, throat, sides of the neck, and all the under parts, bright ferruginous; the feathers on the sides of the breast, flanks, and under tail-coverts, longitudinally streaked with dusky brown: back and scapulars deep black, with oval spots of pale orange-red at the edges of the feathers; wing-coverts dusky ash, edged with white: quills black with white shafts: rump white spotted with black: tail with alternate bars of brown and reddish white. In the *female*, the ferruginous tint is not so bright: the upper parts are deep brown, with cinereous undulating lines, and yellowish spots on the edges of the feathers: under parts pale reddish yellow, with the exception of the middle of the belly, which is pure white; sides of the breast, flanks, and under tail-coverts, streaked with black. Dimensions always larger than those of the male.

(*Young of the year.*) " Head, nape, back, and scapulars, deep brown,
the feathers with an interrupted border of yellowish white ; wing-coverts
broadly edged with white: neck, breast, and flanks, reddish ash, with fine
longitudinal streaks of brown ; band above the eyes, throat, and belly,
pure white ; rump and under tail-coverts white, with large lanceolate
dusky spots : tail with broad zigzag bars of brown and reddish white ;
the tip white: legs dusky ash." TEMM. (*Egg.*) Unknown.

Not uncommon on our coasts during the winter months, and occasionally observed in marshes inland. Said to resort to higher latitudes to
breed. Food, worms, insects, and the smaller bivalve mollusca. *Obs.* The
Common Godwit of Montagu and Bewick is this species in its winter
plumage. The *Red Godwit* of this last author (but not of Montagu) is
the same in its summer plumage.

196. L. *melanura*, Leisl. (*Black-tailed Godwit.*) —
Tail of a uniform black, the base white: wings with a
transverse white bar : bill very slightly curved : middle
claw long and serrated : tarsi very long.

> L. melanura, *Temm. Man. d'Orn.* tom. II. p. 664. Black-tailed
> Godwit, *Selb. Illust.* vol. II. p. 94. pl. 21. Jadreka Snipe, *Mont.
> Orn. Dict. Supp. with fig.*

DIMENS. Entire length seventeen inches six lines : length of the bill
(from the forehead) four inches seven lines, (from the gape) four inches
ten lines ; of the tarsus three inches four lines ; of the naked part of the
tibia one inch eleven lines ; of the tail three inches eight lines ; from the
carpus to the end of the wing nine inches six lines.

DESCRIPT. (*Winter plumage.*) All the upper parts of a uniform
cinereous brown, with the shafts of the feathers of a somewhat deeper
tint : streak over the eyes, throat, fore part of the neck, breast, and
flanks, grayish white ; belly, abdomen, rump, and under tail-coverts,
white: quills dusky ; the basal portion of some of the primaries white:
greater wing-coverts ash-gray, broadly edged with white : tail white for
one-third of its length from the base ; the remaining portion black ; the
two middle feathers tipped with white : bill orange-red at the base, the
tip black: legs dusky gray ; the tarsi much longer than in the last
species. (*Summer plumage.*) Crown of the head black, the feathers
edged with bright red : eye-streak reddish white ; space between the
eye and the bill brown ; throat and neck bright ferruginous, with brownish specks ; breast and flanks the same, with transverse crescent-shaped
bars of brownish black: upper part of the back, and scapulars, deep black,
all the feathers tipped with reddish brown, and edged at the sides with
small spots of the same colour ; wing-coverts much as in winter : middle
of the belly, abdomen, and base of the tail, white ; lower part of the back,
and remainder of the tail, deep black: base of the bill bright orange : legs
black. (*Young of the year, before the first moult.*) Eye-streak, throat,
base of the tail, belly, and abdomen, white: upper part of the head brown,
the feathers with reddish edges ; neck and breast pale red, tinged with
cinereous ; feathers on the back and scapulars dusky brown, surrounded
by a red border ; wing-coverts cinereous, broadly edged and tipped with
reddish white ; tips of the tail-feathers edged with white: extremity of
the bill brown. (*Egg.*) Light olive-brown, blotched and spotted with
darker brown: long. diam. two inches two lines ; trans. diam. one inch
six lines.

Much less frequent than the last species. Haunts marshes and low
pastures, and is not often observed on the sea-coast. Is occasionally known
to breed in the fenny districts of Cambridgeshire and Norfolk. Lays four
eggs. Food, insects and their larvæ, worms, &c.

GEN. 78.   SCOLOPAX, *Linn.*

(1. RUSTICOLA, *Vieill.*)

197. S. *Rusticola*, Linn. (*Woodcock.*)—Plumage, above,
variegated with black, gray, and ferruginous; beneath, yellowish white, with transverse undulating bars: occiput
barred transversely.

S. Rusticola, *Temm. Man. d'Orn.* tom. II. p. 673. Woodcock, *Mont.
Orn. Dict. & Supp. Bew. Brit. Birds,* vol. II. p. 43. *Selb. Illust.*
vol. II. p. 107. pl. 23. f. 1.

DIMENS. Entire length thirteen inches nine lines: length of the bill
(from the forehead) two inches nine lines, (from the gape) two inches
seven lines and a half; of the tarsus one inch four lines and a half; of
the middle toe one inch eight lines and a half; of the tail three inches
three lines; from the carpus to the end of the wing seven inches ten
lines: breadth, wings extended, twenty-four inches six lines.

DESCRIPT. Forehead and crown ash-gray, tinged with rufous; from
the corner of the bill to the eyes a dusky streak; occiput with four broad
transverse bars of blackish brown; rest of the upper parts variegated with
chestnut-brown, ochre-yellow, and ash-gray, with zigzag lines and large
irregular spots of black: throat plain white; rest of the under parts yellowish white, passing into rufous on the breast and fore part of the neck,
with transverse undulating bars of dusky brown: quills barred with ferruginous and black on their outer webs: tail of twelve feathers, black,
the outer webs edged with rufous; the tips ash-gray above, silvery white
beneath: bill flesh-colour, tinged with gray: legs livid. The *female* is
somewhat larger, with the colours more dull. (*Egg.*) Pale yellowish
white; the larger end blotched and spotted with ash-gray and two shades
of reddish yellow brown: long. diam. one inch nine lines; trans. diam. one
inch four lines.

A winter visitant, appearing regularly about the beginning of October,
and departing in February or March. Frequents moist woods, and is of
common occurrence in most parts of the kingdom. Occasionally known
to breed with us. Nest placed on the ground, in a small hollow. Eggs
four in number. Food, insects and worms.

(2. SCOLOPAX, *Vieill.*)

198.   S. *Sabini*, Vig. (*Sabine's Snipe.*) — Whole
plumage variegated with black and chestnut; paler beneath: tail of twelve feathers.

S. Sabini, *Vigors in Linn. Trans.* vol. XIV. p. 556. pl. 21. *Jard.
and Selb. Orn.* vol. I. pl. 27. Sabine's Snipe, *Selb. Illust.* vol. II.
p. 118. pl. 24. f. 1. *Bew. Brit. Birds,* vol. II. p. 416.

DIMENS. Entire length nine inches and three-tenths: length of the
bill two inches and seven-tenths; of the tarsus one inch three lines; from
the carpus to the end of the wing five inches and one-tenth. VIG.

Descript. Cheeks, throat, neck, and breast, brownish black, speckled with chestnut-brown; belly and abdomen grayish black, with transverse bars of chestnut-brown: back and scapulars deep black, with chestnut-brown bars and spots; under wing-coverts blackish gray: tail of twelve feathers, black at the base, ferruginous at the tip, with transverse black bars: bill brownish black, with a tinge of chestnut at the base of the upper mandible: legs brownish black. (*Egg.*) Unknown.

A rare and little known species first described by Mr Vigors from a specimen shot in the Queen's County in Ireland, in August 1822. Since then, two other individuals have occurred in Kent, a third in another part of Ireland, and a fourth near Morpeth in Northumberland. This last is in the collection of Mr Selby. Has not hitherto been met with in any other part of the world.

199. S. *major*, Gmel. (*Great Snipe.*)—All the under parts spotted and barred with black : tail of sixteen feathers.

S. major, *Temm. Man. d'Orn.* tom. ii. p. 675. Great Snipe, *Mont. Orn. Dict. & Supp. Selb. Illust.* vol. ii. p. 115. pl. 23. f. 2. *Bew. Brit. Birds*, vol. ii. p. 415.

Dimens. Entire length twelve inches six lines: length of the bill two inches nine lines: breadth, wings extended, nineteen inches.

Descript. Crown of the head black, divided in the middle by a longitudinal streak of yellowish white; between the bill and the eye a dusky brown streak: cheeks, eyelids, and throat, yellowish white, speckled with brown: hind part of the neck pale yellowish brown, spotted with black: back and scapulars variegated with black and chestnut-brown, many of the feathers edged on their outer webs with yellowish white; greater coverts tipped with white: under parts white, tinged with yellowish brown, spotted and barred all over with blackish brown: quills dusky: tail of sixteen feathers, black for two-thirds of its length from the base, the remaining portion bright chestnut with black bars, the tip yellowish white: bill pale brown, inclining to flesh-colour at the base: legs greenish gray. (*Egg.*) Yellow olive-brown, spotted with two shades of reddish brown: long. diam. one inch nine lines; trans. diam. one inch two lines.

A rare and only occasional visitant in this country. Inhabits extensive marshes in the North of Europe. Nest said to be placed in such situations, amongst rushes and other herbage. Eggs three or four in number. Food, worms, insects, and small shells.

200. S. *Gallinago*, Linn. (*Common Snipe.*) — Neck and breast mottled with black and pale ferruginous ; belly and abdomen pure unspotted white: tail of fourteen feathers.

S. Gallinago, *Temm. Man. d'Orn.* tom. ii. p. 676. Common Snipe, *Mont. Orn. Dict. Selb. Illust.* vol. ii. p. 121. pl. 23. f. 3. Snipe, *Bew. Brit. Birds*, vol. ii. p. 50.

Dimens. Entire length eleven inches five lines: length of the bill (from the forehead) two inches ten lines and a half, (from the gape) two inches ten lines; of the tarsus one inch two lines; of the tail two inches

three lines; from the carpus to the end of the wing five inches four lines:
breadth, wings extended, seventeen inches five lines.

DESCRIPT. Upper parts very similar to those of the last species:
crown black, divided by a yellowish white line; a similar streak from the
base of the upper mandible over each eye; between the bill and the eye
a dusky line: back and scapulars velvet-black, with transverse bars of
chestnut-brown, and longitudinal streaks of ochre-yellow; wing-coverts
dusky brown, edged with reddish white; quills black: chin and throat
white; cheeks, neck, and upper part of the breast, mottled with black
and pale ferruginous brown; flanks with white and dusky transverse
bars; lower part of the breast, belly and abdomen, pure white without
spots: tail of fourteen feathers, black for two-thirds of its length from the
base, the remaining portion reddish brown, with black bars, the tip
reddish white: bill brown, paler at the base: legs dusky gray, tinged
with green. (*Egg.*) Pale yellowish white; the larger end spotted with
three shades of brown: long. diam. one inch six lines; trans. diam. one
inch one line.

A common inhabitant of marshes and low meadows throughout the
kingdom. Is generally considered as migratory, appearing early in the
Autumn and departing in the Spring. In certain districts, however,
many remain annually through the breeding season. Nest placed on the
ground, concealed amongst rushes and coarse grass. Eggs four or five
in number. Food, worms, insects, &c.

201. S. *Gallinula*, Linn. (*Jack Snipe.*) — Crown
divided longitudinally by a black band reaching to the
nape: neck and breast spotted ; belly and abdomen pure
white: tail of twelve feathers.

S. Gallinula, *Temm. Man. d'Orn.* tom. II. p. 678.   Jack Snipe, *Mont.
Orn. Dict.   Selb. Illust.* vol. II. p. 125. pl. 23. f. 5.   Judcock, *Bew.
Brit. Birds*, vol. II. p. 54.

DIMENS. Entire length eight inches six lines: length of the bill
(from the forehead) one inch eight lines and a half, (from the gape) one
inch seven lines; of the tarsus eleven lines; of the tail one inch eleven
lines; from the carpus to the end of the wing four inches three lines.

DESCRIPT. Crown divided by a black band slightly edged with reddish brown, extending from the forehead to the nape; beneath this, and
parallel with it, are two streaks of yellowish white, separated by another
of black; between the bill and the eye a dusky line; throat white; front
of the neck, and upper part of the breast, pale yellowish brown tinged
with ash, with longitudinal spots of a deeper tint: back and scapulars
black, glossed with green and purple reflections; the latter with the
outer webs cream-yellow, forming two conspicuous longitudinal bands
extending from the shoulders to the tail; quills dusky; wing-coverts
black, edged with pale brown and white: lower part of the breast, belly
and abdomen, pure white: tail of twelve feathers, dusky, edged with pale
ferruginous brown: bill bluish at the base, black towards the tip: legs
greenish gray. (*Egg.*) Yellowish olive; the larger end spotted with two
shades of brown: long. diam. one inch three lines; trans. diam. ten
lines.

Found with the last species, but not quite so plentiful. Arrives early
in October and departs in March. Very rarely remains with us to breed.
Nest said to resemble that of the Common Snipe. Eggs four or five.
Food similar.

(3. Macroramphus, *Leach.*)

202. S. *grisea*, Gmel. (*Brown Snipe.*) — Cinereous
brown (*winter*), or reddish brown variegated with black
(*summer*); rump and tail white, with numerous dusky
transverse bars.

S. grisea, *Temm. Man. d'Orn.* tom. ii. p. 679. S. Noveboracensis,
Wils. Amer. Orn. vol. vii. p. 45. pl. 58. f. 1. (Summer.) Brown
Snipe, *Mont. Orn. Dict.* vol. ii. *with fig. in Supp.* Brown Long-
beak, *Selb. Illust.* vol. ii. p. 103. pl. 24. f. 2. *Shaw, Zool.* vol. xii.
p. 61. pl. 9.·

Dimens. Entire length eleven inches: length of the bill two inches
six lines. Mont.

Descript. (*Winter plumage.*) Between the bill and the eye a dusky
streak; above that, passing over the eye, a white one; cheeks and throat
white, the former with a few brown streaks; crown of the head, neck,
and upper part of the breast, cinereous brown, with the shafts of the
feathers somewhat darker; back and scapulars dark brown, the feathers
edged with cinereous and rufous brown; primary quills dusky, the inner
ones and largest coverts immediately over them, slightly tipped with
white; smaller coverts above and below the bastard winglet dusky and
white; the rest of the coverts cinereous brown, darkest in the middle of
the feathers; secondary quills dusky brown, tipped and edged with white:
lower part of the breast, belly and thighs, pure white; flanks whitish
with dusky bars; lower part of the back white; rump and under tail-
coverts white, with transverse crescent-shaped dusky spots; upper tail-
coverts and tail thickly barred with black and white: tail of twelve
feathers: bill dusky, lightest at the base: legs yellow olivaceous green.
Mont. (*Summer plumage.*) Crown of the head, nape, back and scapulars, reddish brown, variegated with black and yellowish; cheeks, and
streak over the eye, pale reddish; fore part of the neck and breast reddish brown; wing-coverts cinereous, with whitish edges; belly, rump
and tail, as in winter. Temm. (*Young of the year.*) All the upper parts
blackish brown, the feathers broadly edged with bright ferruginous;
eyestreak and under parts dirty white tinged with reddish, the ferruginous
tint most distinct on the breast; breast and flanks spotted with dusky
brown; the middle tail-feathers tipped with red. Temm. (*Egg.*) Unknown.

A native of North America. In this country an extremely rare and
accidental visitant. The specimen described by Montagu was shot in the
beginning of October, on the coast of Devonshire. Since then, a second,
in summer plumage, has been killed at Yarmouth. Wilson states that it
differs in its habits from the Common Snipe, keeping on the sea-coast,
and being seldom or never seen inland. Food, worms and small bivalve
mollusca. Nidification unknown.

GEN. 79. TRINGA, *Briss.*

(I. Machetes, *Cuv.*)

203. T. *pugnax*, Linn. (*Ruff.*)—Tail rounded; the
two middle feathers barred; the three outer ones on each
side always of one colour.

T. pugnax, *Temm. Man. d'Orn.* tom. ii. p. 631.   Ruff, *Mont. Orn.
Dict. & Supp. Bew. Brit. Birds,* vol. ii. p. 75.   *Selb. Illust.*
vol. ii. p. 130. pl. 25.

Dimens. Entire length twelve inches five lines: length of the bill
(from the forehead) one inch six lines, (from the gape) one inch six
lines; of the tarsus one inch ten lines; of the tail two inches eight
lines; from the carpus to the end of the wing seven inches five lines.
Descript. (*Male, during the breeding season.*) Face naked, covered
with warty pimples of a reddish yellow colour: sides of the occiput
adorned with two tufts of elongated feathers: beneath the throat a large
frill or ruff of similar feathers standing out in a very conspicuous manner;
colours of the ruff extremely variable, changing occasionally even in the
same individual in different seasons; yellowish white, barred with black;
or entirely black, glossed with purple; in some, of a uniform yellowish
brown; in others, pure white; or varied with black, white, and yellow:
upper parts of the body likewise variable, in general presenting a mixture of cinereous brown, yellowish white, reddish brown, and black: sides
of the breast and flanks pale reddish brown, with transverse black bars,
sometimes entirely black; middle of the belly, abdomen, and under tail-
coverts, white: quills dusky: the four middle tail-feathers barred with
black; the rest of a uniform colour: bill yellowish orange: legs yellow.
(*Male in autumn and winter.*) Face feathered: no elongated tufts on
the occiput, or frill on the neck: throat, fore part of the neck, belly, and
other under parts, pure white; or stained and mottled with black; sometimes black, with transverse undulating white lines: breast reddish
brown, with spots of a deeper tint: upper plumage variable, generally
brown with black spots, the feathers edged with reddish; greater coverts,
and middle tail-feathers, barred with black and reddish brown: bill
brownish: legs yellowish brown.   (*Female* or *Reeve.*) Smaller: at all
times without the ruff and occipital tufts: upper parts of the body
cinereous brown, mixed with black, the black glossed with steel blue:
neck and breast the same, but paler: belly and abdomen white: bill
black: legs yellowish.   (*Young of the year.*) Very much resembling
the female in winter plumage, but with the fore part of the neck and
breast of a dull reddish ash; the feathers on the head, back, scapulars,
and greater wing-coverts, dusky brown, broadly edged with reddish
yellow; lesser coverts edged with reddish white: throat, belly, and abdomen, pure white: bill black: feet greenish.   (*Egg.*) Olive, blotched
and spotted with clove and liver-brown: long. diam. one inch seven
lines; trans. diam. one inch one line and a half.
A migratory species, arriving early in the Spring, and departing in
September.   Rather locally distributed.   Principally confined to the
marshes of Lincolnshire, Norfolk, and the Isle of Ely.   Males very
pugnacious during the breeding season.   Nest usually placed on a has-
sock of grass, in the most swampy situations.   Eggs four in number, laid
in the first or second week of May.   Towards the end of June, or beginning of July, the ruff on the neck of the male bird begins to fall.   Food,
insects and worms.   *Obs.* The *Shore Sandpiper, Greenwich Sandpiper,*
and *Equestrian Sandpiper* of Latham, are all referable to this species in
different states of plumage.

(2. Tringa, *Selb.*)

204.   T. *subarquata,* Temm.   (*Pigmy Curlew.*)—Bill
slightly bent down, much longer than the head: the two

middle tail-feathers longer than the others : a considerable part of the tibia naked : tarsus fourteen lines in length.

T. subarquata, *Temm. Man. d'Orn.* tom. II. p. 609. Pigmy Curlew, *Mont. Orn. Dict. & Supp. with fig. Bew. Brit. Birds,* vol. II. p. 40. Curlew Tringa, *Selb. Illust.* vol. II. p. 157. pl. 26. f. 4, & 5.

DIMENS. Entire length eight inches: length of the bill one inch seven lines ; of the tarsus one inch two lines ; of the naked part of the tibia six lines ; of the middle toe, claw included, eleven lines ; from the carpus to the end of the wing five inches two lines.

DESCRIPT. (*Winter plumage.*) Face, streak above the eye, throat, belly, and other under parts, pure white; from the bill to the eye a brown streak; upper part of the head, back, scapulars, and wing-coverts, cine-reous brown, with the shafts of the feathers somewhat darker; feathers on the nape streaked longitudinally with brown, and edged with whitish; fore part of the neck and breast the same, but paler: upper tail-coverts white: tail cinereous gray, the feathers edged with white: bill black: irides brown: legs dusky gray. (*Summer plumage.*) Face, streak over the eye, and throat, white, speckled with brown: crown of the head black, the feathers edged with pale reddish brown; hind part of the neck red-dish brown, with longitudinal black streaks; fore part of the neck, breast, belly, and abdomen, chestnut red, with a few white feathers and brown spots sometimes intermixed, according to the period of the season: back, scapulars, and greater coverts, deep black, the edges of the feathers marked with a series of reddish brown spots, the tips yellowish gray; rest of the coverts deep brown, margined with grayish white, or (during the season of incubation) with reddish yellow: upper and under tail-coverts white, with transverse black bars. (*Young of the year, before the first moult.*) Back, scapulars, and wing-coverts, dusky ash, all the feathers broadly edged and tipped with yellowish white: breast tinged with white and pale yellowish brown, faintly streaked, but without any distinct spots: feet brown. (*Egg.*) Yellowish, with brown spots. TEMM.

A rare visitant in this country. Occasionally met with on the eastern and southern coasts, but not often observed inland. Common in Holland in Spring and Autumn at the periods of its migration. Sometimes breeds in that country. Nest placed on the sea-shore, or on the borders of exten-sive waters. Eggs four or five in number. Food, insects and worms. *Obs.* The *Red Sandpiper* of Latham and Montagu is this species in its summer plumage.

205. T. *variabilis,* Meyer. (*Dunlin.*) — Bill nearly straight, a little longer than the head: the two middle tail-feathers longer than the others, and pointed : a small part of the tibia naked : tarsus eleven lines and a half.

T. variabilis, *Temm. Man. d'Orn.* tom. II. p. 612. Dunlin and Purre, *Mont. Orn. Dict. Bew. Brit. Birds,* vol. II. pp. 102, & 104. Dunlin, *Selb. Illust.* vol. II. p. 153. pl. 26.

DIMENS. Entire length seven inches nine lines: length of the bill (from the forehead) one inch three lines, (from the gape) one inch two lines and a half; of the tarsus eleven lines and a half; of the naked part of the tibia three lines; of the tail two inches one line; from the carpus

O

to the end of the wing four inches eight lines: breadth, wings extended,
fourteen inches seven lines.

DESCRIPT. (*Winter plumage.*) Throat, and a streak from the base of
the upper mandible to the eye, pure white: head, back of the neck, and
all the upper parts, cinereous brown, the shafts of the feathers being of a
darker tint; quills dusky; greater wing-coverts tipped with white: lower
part of the neck and breast grayish white, mottled with pale cinereous
brown; belly, vent, and under tail-coverts, pure white: rump and two
middle tail-feathers dusky brown; the other tail-feathers brownish ash,
edged with white: bill black: irides and legs dusky brown. (*Summer
plumage.*) Throat white; cheeks, sides and fore part of the neck, and
breast, reddish white, with fine longitudinal streaks of black; belly and
abdomen, during the period of incubation, deep black, but afterwards
mixed with white: crown of the head black, the feathers edged with reddish brown; back, scapulars, and lesser wing-coverts, deep black, all the
feathers broadly edged with reddish brown, and tipped with grayish
white; quills and greater coverts dusky brown, the latter tipped with
white: two middle tail-feathers blackish brown; the rest cinereous edged
with white. (*Spring and autumn, at the time of change.*) Throat and
eye-streak white; upper parts dusky brown, the feathers edged with reddish yellow; amongst them several cinereous brown, as in winter: wing-
coverts wholly cinereous brown, without any red: neck and breast reddish
yellow, spotted with brown; belly white, more or less spotted and blotched
with dusky brown; vent and under tail-coverts pure white. *Obs.* The
female is somewhat larger than the *male*, and has a longer bill. In
young birds, the bill is nearly straight. (*Egg.*) Greenish white, blotched
and spotted with two shades of dark red brown: long. diam. one inch four
lines and a half; trans. diam. eleven lines and a half.

A common species on all parts of the coast during the greater part
of the year. At the approach of the breeding season, retires to inland
marshes and the banks of rivers. Lays four eggs. Food, worms and
insects. *Obs.* The *Purre* of authors is this species in its winter plumage.

206. T. *pectoralis*, Bon. (*Pectoral Sandpiper.*) —
" Bill shorter than the head, compressed and reddish
yellow at the base: rump black: middle tail-feathers
longest: feet greenish yellow: tarsus one inch." BON.

T. pectoralis, *Bon. Syn.* p. 318. *Id. Amer. Orn.* vol. IV. p. 43.
pl. 23. f. 2.* Pectoral Sandpiper, *Nuttall, Orn. of Unit. States,*
vol. II. p. 111.

DIMENS. Entire length about ten inches: length of the bill (from the
rictus) rather more than one inch; of the tarsus one inch. NUTT.
DESCRIPT. " *Summer plumage,* varied with black and rufous, beneath
white; breast cinereous, strongly lineated with blackish. *Winter plumage,*
cinereous brown, beneath white." BON.

Mr Yarrell informs me that a single individual of this species, which is
not uncommon in the United States, has been killed at Yarmouth. It is
at present in the possession of Mr Hoy of that place. Its identity with
the *T. pectoralis* of America was confirmed by Mr Audubon, to whom
the specimen was submitted for examination.

* The above reference to Bonaparte's *American Ornithology* is on the authority of Nuttall.
I have not been able to get sight of the fourth volume of that work myself.

207. T. *maritima*, Brunn. (*Purple Sandpiper.*) —
Bill slightly curved at the tip, longer than the head :
tibia almost entirely feathered : feet, and base of the bill,
coloured.

T. maritima, *Temm. Man. d'Orn.* tom. II. p. 619. T. nigricans,
Mont. in Linn. Trans. vol. IV. p. 40. pl. 2. Purple Sandpiper,
Mont. Orn. Dict. & Supp. Purple or Rock Tringa, *Selb. Illust.*
vol. II. p. 150. pl. 26. f. 6.

DIMENS. Entire length eight inches three lines : length of the bill
one inch three lines ; of the tarsus ten lines and a half ; of the tail two
inches three lines ; from the carpus to the end of the wing five inches two
lines.

DESCRIPT. (*Winter plumage.*) Crown of the head, cheeks, sides and
nape of the neck, dusky brown, with a tinge of ash-gray ; throat white ;
orbits, and a spot near the anterior angle of the eye, grayish white : back
and scapulars black, with purple and violet reflections, the feathers edged
with deep ash-colour ; wing-coverts black, edged with white : breast gray,
the feathers fringed with white ; rest of the under parts white, streaked
on the flanks with dark ash-colour ; under tail-coverts with lanceolate
dusky spots : rump, and two middle tail-feathers, deep black ; the other
tail-feathers dusky ash, fringed with white : bill reddish orange at the
base, the tip dusky : irides dusky : legs ochre-yellow. (*Summer plumage.*)
" Crown of the head, back, and scapulars, violet black, the feathers with
a broad edging of white mixed with a little ferruginous : fore part of the
neck, breast, and belly, cinereous white, with lanceolate dusky spots ;
flanks, and sides of the neck, with oval spots ; under tail-coverts with
longitudinal streaks ; middle of the belly pure white." TEMM. (*Young
of the year.*) " Crown, back, scapulars, secondary quills, and middle tail-
feathers, dull black, all the feathers edged and tipped with pale red ;
wing-coverts broadly tipped with white : sides and fore part of the neck
with longitudinal streaks, the feathers edged with cinereous ; some large
spots disposed longitudinally on the flanks and under tail-coverts : legs
and base of the bill pale yellowish." TEMM. (*Egg.*) Yellowish olive, spotted and speckled with reddish brown : long. diam. one inch three lines ;
trans. diam. eleven lines.

Inhabits rocky shores, and is found on many parts of the British coast
during the winter months. Retires to higher latitudes to breed. A few
have been observed by Mr Selby to remain for that purpose on the Fern
Islands. Food, marine insects and small bivalve mollusca. *Obs.* The
Knot of Bewick is the young of the year of this species.

208. T. *Temminckii*, Leisl. (*Temminck's Stint.*) —
Bill very slightly curved at the tip, shorter than the head :
tail cuneiform, the outer feathers pure white : tarsus eight
lines and a half.

T. Temminckii, *Temm. Man. d'Orn.* tom. II. p. 622. Little Sandpiper, *Mont. Orn. Dict. App.* Temminck's Tringa, *Selb. Illust.*
vol. II. p. 144. pl. 27*. f. 1, & 2.

DIMENS. Entire length six inches : length of the bill (from the forehead) eight lines and a half, (from the gape) eight lines ; of the tarsus

eight lines and a half or nine lines; of the naked part of the tibia four
lines and a half; of the tail one inch nine lines; from the carpus to the
end of the wing three inches nine lines: breadth, wings extended, eleven
inches eight lines.

DESCRIPT. (*Winter plumage*.) All the upper parts deep brown, with
dusky streaks on the shafts of the feathers: throat white; breast and
front of the neck cinereous brown; belly, vent, and under tail-coverts,
pure white: tail cuneated; the four middle feathers cinereous brown; the
others whitish; the two outer ones on each side nearly pure white: bill
dusky: legs olivaceous brown. (*Summer plumage*.) "Upper parts black,
all the feathers broadly edged with deep ferruginous: throat white; forehead, breast, and fore part of the neck, reddish ash, with very small longitudinal black spots; the rest of the under parts, and outer tail-feathers,
pure white; the two middle tail-feathers dusky brown, edged with deep
ferruginous." TEMM. (*Young of the year, before the first moult.*) Upper
parts cinereous brown, palest on the nape; all the feathers, those on the
nape excepted, with a double edging of dusky brown and yellowish white:
throat white; over the eyes a whitish streak; breast and sides of the
neck cinereous gray, with a faint tinge of reddish; rest of the under parts
white; tail-feathers, the outer one excepted, tipped with reddish: bill and
legs somewhat paler than in the adult. (*Egg.*) Unknown.

A rare, and only an occasional, visitant in this country. Montagu's
specimen was killed on the south coast of Devonshire. Others have been
since met with in Sussex and Norfolk. Frequents the borders of lakes
and rivers. Probably breeds in high northern latitudes. Food, insects
and worms.

209. T. *minuta*, Leisl. (*Little Stint*.)——Bill straight, shorter than the head: tail doubly forked, the outer feathers cinereous brown edged with whitish: tarsus ten lines and a half.

T. minuta, *Temm. Man. d'Orn.* tom. II. p. 624. Little Stint, *Bew.
Brit. Birds,* vol. II. p. 107. Minute Tringa, *Selb. Illust.* vol. II.
p. 147. pl. 27*. f. 3, & 4.

DIMENS. Entire length six inches: length of the bill eight lines and
a half; of the tarsus ten lines and a half.

DESCRIPT. (*Winter plumage*.) All the upper parts cinereous brown,
with the shafts of the feathers dusky: a brown streak between the bill
and the eye; above the eyes another indistinct one of white; sides of the
breast cinereous, tinged with brown; throat, fore part of the neck, middle
of the breast, and all the other under parts, pure white: the outer tail-
feathers cinereous brown, edged with whitish; the two middle ones brown:
bill and legs black. (*Summer plumage*.) Crown black, the feathers edged
with reddish brown; forehead and eye-streak somewhat paler than the
rest of the head; throat white; cheeks, sides of the neck, and sides of the
breast, pale reddish spotted with brown; middle of the breast, and all the
under parts, pure white: feathers on the back, scapulars, wing-coverts,
rump, and two middle tail-feathers, deep black, all broadly edged and
tipped with bright red brown; the outer tail-feathers cinereous brown,
edged lighter. (*Young of the year, before the first moult.*) "Crown of
the head dusky, the feathers edged with yellowish red; forehead, eye-
streak, throat, front of the neck, middle of the breast, and other under

parts, pure white; between the bill and the eye a brown streak; sides of the breast reddish, mixed with cinereous brown; nape and sides of the neck cinereous and brown: back, scapulars, and wing-coverts, dusky brown; the feathers on the upper part of the back broadly edged with red, the scapulars with yellowish white; wing-coverts with a narrow edging of yellowish red: two middle tail-feathers dusky, edged with reddish ash; the others edged with white." TEMM. (*Egg.*) Reddish white ground, spotted and speckled with dark red brown: long. diam. one inch one line and a half; trans. diam. nine lines.

Like the last, a rare and occasional visitant. Habits similar. Supposed to retire northward to breed.

210. T. *Canutus*, Linn. (*Knot.*) — Bill straight, a little longer than the head, very much dilated at the tip: tail even at the extremity: a small part of the tibia naked.

T. cinerea, *Temm. Man. d'Orn.* tom. II. p. 627.　Knot and Ash-coloured Sandpiper, *Mont. Orn. Dict. & Supp.*　Knot, *Selb. Illust.* vol. II. p. 138. pl. 27. f. 2, 3, & 4.

DIMENS. Entire length ten inches: length of the bill (from the forehead) one inch four lines; of the tarsus one inch two lines; of the tail two inches six lines; from the carpus to the end of the wing six inches three lines.

DESCRIPT. (*Winter plumage.*) Crown of the head, hind part of the neck, back, and scapulars, light ash-gray, with the shafts of the feathers darker; wing-coverts the same, but tipped with white, forming a transverse bar: from the bill to the eye a dusky streak; over the eye a white one: forehead, throat, and all the under parts, white; sides and front of the neck longitudinally streaked with brown; breast and flanks with transverse bars of cinereous brown: rump and upper tail-coverts white, with transverse crescent-shaped black bars: tail-feathers cinereous, edged with white: bill and legs blackish gray: irides hazel. (*Summer plumage.*) Throat, eye-streak, sides and fore part of the neck, breast, belly, and flanks, ferruginous brown: crown of the head, and nape, ferruginous, with small longitudinal black streaks; back and scapulars deep black, all the feathers edged with bright ferruginous; on the scapulars some large spots of the same colour: abdomen white, mixed with ferruginous: upper tail-coverts white, barred with black and spotted with ferruginous: tail dusky ash, the feathers edged with whitish. (*Young of the year, before the first moult.*) Crown of the head, and nape, ash-gray, with large brown spots longitudinally disposed; back and scapulars deep cinereous gray, all the feathers with a double edging of black and yellowish white, the latter colour being outermost: a brown streak between the bill and the eye; breast with a faint tinge of reddish gray: bill greenish gray: legs greenish yellow: in other respects like the adult in winter. (*Egg.*) Pale yellowish olive, blotched and spotted with two shades of dark red brown: long. diam. one inch eight lines; trans. diam. one inch one line and a half.

A winter visitant, arriving in September and departing in April. Frequents the coast in large flocks; and, during certain parts of the year, inland fens and marshy districts. Retires northward to breed. Said to lay four eggs on a tuft of withered grass. Food, worms, marine insects, and small bivalve mollusca.

211. T. *rufescens*, Vieill. (*Buff-breasted Sandpiper.*)
—Bill slender, very slightly curved, not longer than the
head: tail cuneated: the outer feathers light brown, edged
with white: a large portion of the tibia naked.

> T. rufescens, *Vieill. Gal. Ois.* tom. ii. p. 105. pl. 238.   *Yarr. in
> Linn. Trans.* vol. xvi. p. 109. pl. 11.   Buff-breasted Tringa, *Selb.
> Illust.* vol. ii. p. 142. pl. 27. f. 1.

Dimens. Entire length eight inches: length of the bill (from the
forehead) nine lines, (from the gape) one inch; of the naked part of the
tibia six lines; of the tarsus one inch three lines. Yarr.

Descript. Feathers on the top of the head dark brown, approaching
to black, edged with very light brown, giving a mottled appearance; hind
part of the neck light brown, with a minute dark spot in the centre of
each feather; back, scapulars, and tertials, blackish brown, the edges of
the feathers paler; primaries nearly black, tipped with white; the shafts
white; tail-coverts brown, with lighter-coloured edges; tail cuneiform,
the middle feathers black, the shafts and edges lighter; the feathers on
each side light brown, inclosed by a zone of black, and edged with white:
chin, sides of the neck, throat, and breast, light brown tinged with buff;
abdomen, flanks, and under tail-coverts, white, but pervaded also by the
buff colour of the higher parts; sides of the neck with darker coloured
spots: anterior portion of the under surface of the wing rufous brown;
the outer portion spotted; under coverts pure white; shafts of the primaries on their under surface pearl-white, the outer web dusky, the inner
web also dusky, and plain on the part nearest the shaft, the other inner
half of the web beautifully mottled with dark specks; secondary quills
also mottled at their bases, and ending in sabre-shaped points, presenting
a regular series of lines formed by alternating shades of white, black, and
dusky bands, well defined in the adult bird, and presenting a beautifully
variegated appearance, peculiar to the species: bill black: legs brown.
Yarr. (*Egg.*) Unknown.

A single individual of this species, which is a native of Louisiana in
America, was shot early in September 1826, in the parish of Melbourne
in Cambridgeshire, in company with some dotterel. It is now in the
possession of Mr Yarrell. A second was killed at Sheringham in
Norfolk, July 28, 1832. This last, which proved to be a female, is in
the Norwich Museum. It appears to be a very rare species even in its
native country. Habits and nidification unknown.

GEN. 80. LOBIPES, *Cuv.*

212. L. *hyperboreus*, Steph. (*Red Lobefoot.*)—Crown,
nape, and sides of the breast, deep ash; sides of the neck
ferruginous: legs greenish gray.

> Phalaropus hyperboreus, *Temm. Man. d'Orn.* tom. ii. p. 709. P.
> Williamsii, *Simm. in Linn. Trans.* vol. viii. p. 264. Red Phala-
> rope, *Mont. Orn. Dict. & Supp.* Red-necked Phalarope, *Sow.
> Brit. Misc.* pl. 10. *Bew. Brit. Birds,* vol. ii. p. 149. Red Lobe-
> foot, *Selb. Illust.* vol. ii. p. 166. pl. 28.

Dimens. Entire length seven inches three lines: length of the bill
(from the forehead) ten lines and a half, (from the gape) eleven lines; of

the tarsus nine lines and a quarter; of the naked part of the tibia three lines; of the tail one inch eleven lines; from the carpus to the end of the wing four inches two lines.

DESCRIPT. (*Adult male in spring plumage.*) Crown of the head, nape, sides of the breast, space between the eye and the bill, as well as a little streak behind the eye, deep ash-gray; sides and front of the neck bright ferruginous; throat, middle of the breast, and all the other under parts, pure white; flanks marked with large ash-coloured spots: back, and scapulars, deep black, the feathers with broad ferruginous edges; wing-coverts black, tipped with white, forming a transverse bar: two middle tail-feathers deep black; the lateral feathers ash-gray bordered by a narrow white band: bill black: irides brown: legs greenish gray. (*Adult female.*) The ash-gray about the eyes tinged with reddish; the red on the front of the neck of less extent, and mixed with cinereous; the spots on the flanks larger and more numerous. (*Young of the year.*) "Crown of the head, occiput, nape, and a spot behind the eyes, dusky brown; back, scapulars, and two middle tail-feathers, the same, but all the feathers broadly edged with pale red: primaries and wing-coverts dusky, edged and tipped with whitish; the transverse bar on the wing not so broad as in the adult: forehead, throat, front of the neck, breast, and other under parts, white, passing into pale ash on the sides of the breast and on the flanks; a faint yellowish tinge on the sides of the neck: inside of the tarsi yellow; the outer portion, and toes, yellowish green." TEMM. The *adult in winter* is nearly similar to the young. (*Egg.*) Olive ground, blotched and streaked all over with dark red-brown: long. diam. one inch two lines; trans. diam. ten lines.

Abundant in the Orkneys, Hebrides, and northern parts of Scotland, but very rarely met with in England. Has been shot in Yorkshire, and in Norfolk. Frequents fresh-water lakes, on the banks of which it breeds. Lays three eggs. Food, worms and insects.

GEN. 81.　PHALAROPUS, *Briss.*

213.　P. *lobatus*, Flem. (*Gray Phalarope.*) — Gray above; beneath white (*winter*): or brown above, the feathers edged with rufous; beneath red (*summer*): a transverse white bar on the wings.

P. platyrhinchus, *Temm. Man. d'Orn.* tom. II. p. 712.　Gray Phalarope, *Mont. Orn. Dict. Selb. Illust.* vol. II. p. 162. pl. 28.　Red and Gray Phalaropes, *Bew. Brit. Birds,* vol. II. pp. 146, & 147.

DIMENS.　Entire length eight inches six or nine lines.　TEMM.

DESCRIPT. (*Winter plumage.*) Forehead and crown white; occiput, ear-coverts, and a streak down the nape of the neck, dusky gray; sides of the breast, back, scapulars, and rump, bluish ash, with the shafts of the feathers dusky; some of the scapulars tipped with white; wing-coverts the same, forming a transverse bar on the wing: throat, sides of the neck, middle of the breast, and all the other under parts, pure white: tail dusky gray, the feathers edged with cinereous: bill yellowish red at the base, dusky brown at the tip: irides reddish yellow: legs greenish gray. (*Summer plumage.*) Forehead, throat, crown, and occiput, deep grayish ash; region of the eyes, and sides of the occiput, white: sides and fore part of the neck, breast, belly, and abdomen, brick-red: nape, back, scapulars, and upper tail-coverts, dusky brown, all the feathers

broadly edged with reddish yellow; wing-coverts black, tipped with
white; rump white, spotted with black. (*Young of the year, before
the first moult.*) "Occiput, and a streak over the eyes, dusky; nape,
back, scapulars, and upper tail-coverts, cinereous brown; the feathers
on the back and scapulars broadly edged with yellowish; rump white,
variegated with brown; wing-coverts edged and tipped with yellowish
white; tail cinereous brown, the middle feathers with a broad yellowish
edging: forehead, throat, sides and front of the neck, breast, and other
under parts, pure white: bill cinereous brown: feet greenish yellow."
TEMM. (*Egg.*) Stone-colour, tinged with green; blotched and spotted
all over with blackish brown: long. diam. one inch two lines; trans. diam.
ten lines and a half.

A rare and occasional visitant in this country, but has been killed
in several instances and in widely separated localities. Said to be not
uncommon in some seasons on the coast of Cornwall. Very abundant in
the arctic regions, and in the eastern parts of Europe and Asia, where
it is supposed to breed. Inhabits the borders of rivers and large lakes.
Swims with great facility, and preys upon marine and other aquatic
insects. *Obs.* The *Red Phalarope* of Bewick is this species in summer
plumage.

GEN. 82.   GLAREOLA, *Briss.*

214.   G. *Pratincola*, Leach. (*Collared Pratincole.*)—
Above brown; throat and front of the neck reddish white,
bounded by a black line: tail very much forked.

G. Pratincola, *Leach in Linn. Trans.* vol. XIII. p. 131. pl. 12. G.
torquata, *Temm. Man. d'Orn.* tom. II. p. 500. Collared Pratincole, *Selb. Illust.* vol. II. p. 213. part i. pl. 63. Austrian Pratincole,
Mont. Orn. Dict. Supp. with fig.

DIMENS. Entire length ten inches. MONT.
DESCRIPT. (*Adult.*) All the upper parts brown, more or less inclining
to gray, occasionally with a reddish tinge on the crown and nape; back
and scapulars with faint reflections of greenish bronze: throat and front
of the neck reddish or yellowish white, bounded by a black line passing upwards to the corners of the bill; space between the bill and the
eye black; breast clouded with brown; under wing-coverts bright ferruginous; belly, abdomen, upper and under tail-coverts, white, the two
former sometimes tinged with reddish: quills dusky brown: tail very
much forked, dusky, with more or less of white at the base: bill black,
the base reddish: irides reddish brown: eyes surrounded by a naked
red circle: legs rufous brown. (*Young.*) Upper parts cinereous brown,
shaded with brown of a darker tint; the feathers edged with reddish
white: throat whitish, the surrounding black line simply indicated by
a few spots: breast and belly deep gray, occasionally spotted with brown:
tail but little forked; the outer feathers much shorter than in the adult
bird. *Obs.* This species is subject to considerable variation of plumage:
the colours are more or less intense even in individuals of the same age.
(*Egg.*) White: long. diam. one inch two lines; trans. diam. ten lines.

This species, which is a native of some parts of Europe and Asia, must
in this country be regarded as an extremely rare and accidental visitant.
The first recorded British-killed specimen was shot near Liverpool on the
18th of May 1804, and is now in the collection of Lord Derby. A second

individual was killed by Mr Bullock in the Isle of Unst, about three miles from the northern extremity of Britain, on the 16th of August 1812. More recently (May 1827) a pair are recorded to have been shot on the Breydon-wall, Yarmouth*. Said to frequent the banks of rivers, lakes, inland seas, and extensive marshes. Builds in the reeds and other thick herbage growing in such situations. Lays three or four eggs. Flight extremely rapid. Food, principally insects, which are taken on the wing.

GEN. 83. RALLUS, *Linn.*

215. R. *aquaticus*, Linn. (*Water-Rail.*) — Above, olive-brown spotted with dusky ; beneath, bluish gray ; flanks with black and white bars.

> R. aquaticus, *Temm. Man. d'Orn.* tom. ii. p. 683. Rail, *Mont. Orn. Dict. & Supp. Selb. Illust.* vol. ii. p. 172. pl. 29. Water-Rail, *Bew. Brit. Birds,* vol. ii. p. 126.

DIMENS. Entire length eleven inches five lines : length of the bill (from the forehead) one inch seven lines, (from the gape) one inch eight lines and a half: of the tarsus one inch eight lines ; of the naked part of the tibia six lines ; of the middle toe, claw included, two inches ; of the tail two inches three lines ; from the carpus to the end of the wing four inches eight lines : breadth, wings extended, sixteen inches.

DESCRIPT. Crown of the head, and all the upper parts, olive-brown, with a dusky spot in the middle of each feather : chin whitish ; sides of the head, sides and fore part of the neck, breast and belly, deep ash or lead-gray ; flanks with alternate transverse bars of black and white ; under tail-coverts white : quills dusky brown : tail blackish ; the feathers edged with olive-brown: bill red ; the culmen, and tips of both mandibles, brown : irides red : legs and toes brownish flesh-colour. A *white variety* is said to have occurred not long since in Berkshire†. (*Egg.*) Buffy white: the larger end speckled with ash-gray and orange-brown: long. diam. one inch four lines ; trans. diam. one inch.

Not uncommon in most parts of the country, frequenting marshes, streams, and the banks of rivers. Remains with us the whole year. Nest placed amongst rushes and other aquatic herbage. Eggs six to ten in number. Food, insects, and molluscous animals ; also vegetables.

GEN. 84. CREX, *Bechst.*

216. C. *pratensis*, Bechst. (*Corn-Crake.*) — Above, rufous brown, the centres of the feathers dusky ; beneath, yellowish white : wing-coverts ferruginous.

> Gallinula Crex, *Temm. Man. d'Orn.* tom. ii. p. 686. Crake Galli-nule, *Mont. Orn. Dict.* Land-Rail or Corn-Crake, *Bew. Brit. Birds,* vol. ii. p. 130. Meadow or Corn-Crake, *Selb. Illust.* vol. ii. p. 176. pl. 30.

DIMENS. Entire length ten inches : length of the bill (from the fore-head) nine lines and a half, (from the gape) one inch one line ; of the tarsus one inch five lines and a half ; of the tail two inches and half a line ; from the carpus to the end of the wing five inches four lines.

---

* Paget's *Nat. Hist. of Yarm.* p. 10.   † Loudon's *Mag. of Nat. Hist.* vol. v. p. 384.

DESCRIPT. All the upper parts dusky brown, the feathers broadly margined with reddish ash; over each eye a broad cinereous streak, prolonged down the sides of the head and neck; wing-coverts ferruginous brown; quills reddish brown, tinged with gray: under parts yellowish white, tinged with cinereous, passing into pale yellowish brown on the breast; flanks and under tail-coverts with transverse bars of pale orange-brown: upper mandible of the bill brown; lower mandible whitish: irides pale brown: legs yellowish brown, tinged with gray. (*Egg.*) Pale reddish white, spotted and speckled with ash-gray and pale red brown: long. diam. one inch six lines; trans. diam. one inch one line.

A regular summer visitant, appearing about the latter end of April, and departing in October. Is pretty generally distributed throughout the kingdom, though said to be most plentiful in the northern parts of it, and in Ireland. Frequents cornfields and meadows, particularly such as are in the neighbourhood of water. Nest on the ground, generally in high grass, rudely constructed of moss and a few dry plants. Eggs from seven to twelve in number, laid the middle of June. During the breeding season, the male is heard to utter a singular noise resembling the word *Crex*, frequently repeated. Food, insects, worms, and snails; also seeds and aquatic vegetables.

217. C. *Porzana*, Selby. (*Spotted Crake*.) — Above olive-brown, with dusky streaks and white spots; beneath cinereous olive, spotted with white.

Gallinula Porzana, *Temm. Man. d'Orn.* tom. II. p. 688.   Spotted Crake, *Selb. Illust.* vol. II. p. 179. pl. 30*. f. 1, & 2.   Spotted Gallinule, *Mont. Orn. Dict. Bew. Brit. Birds*, vol. II. p. 132.

DIMENS. Entire length nine inches: length of the bill (from the forehead) nine lines and a quarter, (from the gape) ten lines and a half; of the tarsus one inch four lines and a half; of the naked part of the tibia six lines; of the middle toe, claw included, one inch eight lines; of the tail two inches; from the carpus to the end of the wing four inches eight lines: breadth, wings extended, fifteen inches two lines.

DESCRIPT. (*Adult male.*) Crown of the head, back, scapulars and rump, olive-brown, with a dusky spot in the middle of each feather, all except the first elegantly spotted and streaked with pure white: forehead, throat, and a broad streak above the eyes, ash-gray, the latter speckled with white; nape thickly spotted with black and white; cheeks cinereous, speckled with black; fore part of the neck and breast pale olivaceous, tinged with ash-gray, and spotted with white: belly and vent cinereous white; flanks marked with transverse bars of white, black, and olivaceous brown: quills brown: wing-coverts olivaceous brown, sparingly spotted with white: bill greenish yellow, passing into orange-yellow at the base: irides reddish hazel: legs and toes yellowish green. In the *female*, the cheeks have a reddish tinge, and are speckled with brown. (*Egg.*) Pale reddish white, spotted and speckled with dark red brown: long. diam. one inch three lines; trans. diam. eleven lines.

Sparingly distributed over many parts of the kingdom, frequenting marshes, the banks of rivers, and other watery places. Is usually considered as a migratory species, retiring at the approach of Winter, but it is highly probable that many individuals remain with us throughout the year. It is certainly to be met with from early in March to the middle of November. Is fond of concealing itself in the thickest covert, amongst reeds and rushes, in which situations it is not easily roused. Nest rudely constructed of aquatic plants, and said to float upon the water. Eggs eight to twelve in number. Food, similar to that of the last species.

218. C. *Baillonii*, Selby. (*Baillon's Crake.*)—Upper parts olivaceous red, with numerous white spots on the back and wings: these last reaching to half the length of the tail: bill green: legs flesh-colour.

> Gallinula Baillonii, *Temm. Man. d'Orn.* tom. II. p. 692. *Jard. and Selb. Orn.* pl. 15. Baillon's Crake, *Selb. Illust.* vol. I. p. 182. pl. 30*. f. 3. *Shaw, Gen. Zool.* vol. XII. p. 228. pl. 27.

DIMENS. Entire length seven inches: length of the bill (from the forehead) eight lines, (from the gape) eight lines and a half; of the naked part of the tibia five lines; of the tarsus one inch and half a line; of the middle toe, claw included, one inch six lines; of the tail one inch eight lines; from the carpus to the end of the wing three inches five lines.

DESCRIPT. (*Adult mule.*) Crown of the head, and back of the neck, wood-brown; throat, cheeks, front and sides of the neck, breast and belly, bluish gray: back, scapulars, and wing-coverts, yellowish brown, tinged with olivaceous, and marked with numerous irregular white spots, each surrounded by a black border: flanks, abdomen, and under tail-coverts, black, with narrow transverse white bars: bill dark green, thicker and shorter than in the next species: irides reddish: legs and toes flesh-colour. In the *female* the colours are the same, but of a paler tint. (*Young.*) Throat, and middle of the belly, white, with transverse undulations of cinereous and olivaceous brown; flanks olivaceous, spotted with white: upper parts as in the adult, but the white spots fewer in number: bill greenish brown. (*Egg.*) Light olive-brown, spotted with darker brown: long. diam. one inch one line; trans. diam. nine lines and a half.

A rare and accidental visitant in this country. A specimen, caught alive at Melbourne in Cambridgeshire, in January 1823, is now in the collection of Dr Thackeray. According to Sheppard, it has also occurred in Suffolk. Common in the southern and eastern parts of Europe. Frequents the same situations as the last species. Food similar. Nest always placed in the vicinity of water. Eggs seven or eight in number.

219. C. *pusilla*, Selby. (*Little Crake.*)—Upper parts olivaceous with a few white streaks; wings without spots, reaching to the extremity of the tail: bill and legs sap-green.

> Gallinula pusilla, *Temm. Man. d'Orn.* tom. II. p. 690. Little Crake, *Selb. Illust.* vol. II. p. 185. pl. 30*. f. 4. Little and Olivaceous Gallinules, *Mont. Orn. Dict. Supp. with figs.* Little Gallinule, *Bew. Brit. Birds*, vol. II. p. 134.

DIMENS. Entire length seven inches nine lines: length of the bill (from the forehead) eight lines and a half, (from the gape) nine lines; of the naked part of the tibia five lines; of the tarsus one inch one line and a half; of the middle toe, claw included, one inch seven lines; of the tail one inch seven lines and a half; from the carpus to the end of the wing three inches ten lines and a half.

DESCRIPT. (*Adult male.*) Throat, sides of the head and neck, breast and belly, deep bluish gray, without spots: crown of the head, and all the upper parts, olivaceous green, with the centres of the feathers dusky; on the back a large blackish patch, marked with a few longitudinal white streaks: abdomen and flanks with indistinct transverse bars of white and brown; under tail-coverts black, barred with white: bill longer and more slender than in the last species, of a fine sap-green, somewhat reddish at the base: irides red: legs and toes sap-green. (*Adult female.*) Sides of the head pale cinereous; throat whitish; front of the neck, breast and belly, reddish ash; thighs and abdomen cinereous; under tail-coverts tipped with white; all the upper parts reddish brown: on the back a large dusky patch, sparingly spotted with white; wing-coverts olivaceous ash. (*Young.*) Colours paler; the whole of the throat whitish; fore-neck and upper breast light buff; the white marks on the upper part of the back very few in number; flanks brown, with transverse white bars. (*Egg.*) Light olive-brown, spotted with darker brown: long. diam. one inch two lines; trans. diam. nine lines and a half.

Like the last, a very rare and only occasional visitant in this country. Montagu's specimen was shot near Ashburton, in Devonshire, in 1809. Others have been since killed in Yorkshire, Derbyshire, Cambridgeshire, and Norfolk. Common in the eastern parts of Europe. Haunts and food similar to those of the last species. Nest said to be placed amongst reeds and other aquatic herbage. Eggs seven or eight in number. *Obs.* The *Olivaceous Gallinule* of Montagu is supposed to be the adult state, and the *Little Gallinule* of the same author the young, of this species.

GEN. 85. GALLINULA, *Lath.*

220. G. *chloropus*, Lath. (*Common Gallinule.*) — Plumage above deep olive-brown; beneath blackish gray; ridge of the wing, and under tail-coverts, white.

G. chloropus, *Temm. Man. d'Orn.* tom. II. p. 693. Common Gallinule, *Mont. Orn. Dict. Bew. Brit. Birds*, vol. II. p. 137. *Selb. Illust.* vol. II. p. 188. pl. 31.

DIMENS. Entire length thirteen inches: length of the bill (from the base of the frontal disk) one inch five lines, (from the gape) one inch one line and a half; of the tarsus one inch eleven lines; of the middle toe two inches eight lines and a half; of the tail two inches eleven lines; from the carpus to the end of the wing six inches nine lines: breadth, wings extended, twenty inches ten lines.

DESCRIPT. Head, throat, neck, and upper part of the breast, dusky gray; rest of the under parts deep bluish gray, the feathers on the belly and abdomen edged with grayish white: upper parts dark olive: flanks with large longitudinal streaks of white: ridge of the wing, and under tail-coverts, pure white, the latter with a few black feathers intermixed: frontal disk, and base of the bill, red; tip of the bill yellow: irides red: legs and toes olive-green: on the naked part of the tibia a red circle. (*Young, till after the second autumnal moult.*) Crown of the head and nape, as well as the rest of the upper parts, olivaceous brown: throat, front of the neck, and a spot beneath the eye, whitish; breast, belly, and abdomen, pale gray ; flanks tinged with olivaceous : tip of the bill greenish, passing into olivaceous brown at the base : the frontal disk of small size, of a deep olivaceous brown : irides brown : legs olivaceous : the naked part of the tibia yellowish. (*Egg.*) Reddish white, sparingly spotted and speckled with orange-brown : long. diam. one inch eight lines and a half ; trans. diam. one inch three lines and a half.

A common inhabitant of marshy places and the banks of rivers throughout the country. Runs swiftly, and is also a good swimmer. Nest constructed of rushes and other dry herbage ; generally placed on the ground near the water's edge, but occasionally in trees. Eggs from five to eight in number, laid early in April. Food, insects, seeds, and aquatic vegetables.

GEN. 86.   FULICA, *Linn.*

221.   F. *atra*, Linn.   (*Common   Coot.*) — Head   and neck black ;   back black,   tinged   with   cinereous :   beneath paler.

F. atra, *Temm. Man. d'Orn.* tom. ii. p. 706.   Common Coot, *Mont. Orn. Dict. & Supp.   Selb. Illust.* vol. ii. p. 193. pl. 32.   Coot, *Bew. Brit. Birds,* vol. ii. p. 141.

Dimens.   Entire length sixteen inches : length of the bill (from the base of the frontal disk) one inch ten lines, (from the gape) one inch five lines ; of the tarsus two inches two lines ; of the middle toe three inches three lines ; of the tail two inches ; from the carpus to the end of the wing seven inches nine lines.

Descript.   Head and neck deep black ; back, wings, and all the upper parts, black, with a tinge of ash-gray : breast, belly, flanks, thighs, and under tail-coverts, deep bluish gray : frontal disk large, of a pure white : bill white, with a tinge of rose-red : irides crimson red : legs and toes ash-gray, tinged with greenish ; naked part of the tibia orange-yellow.   In *young birds*, the frontal disk is very small ; the under parts of the plumage pale gray.   (*Egg.*) Stone-colour, speckled with nutmeg-brown : long. diam. two inches one line ; trans. diam. one inch six lines.

Common in most parts of the country, frequenting lakes, rivers, and extensive ponds.   Makes a large floating nest of flags, and other aquatic plants, and lays from eight to twelve eggs.   Feeds on aquatic insects and vegetables. *Obs.* The *Greater Coot* (F. aterrima, *Linn.*) is not distinct from the present species.

ORDER V. NATATORES.

GEN. 87.   ANSER, *Briss.*

(1. Anser, *Steph.*)

222. A. *ferus*, Steph. (*Wild Goose.*)—Bill strong and elevated ; orange, the nail whitish : legs flesh-colour : wings not reaching to the extremity of the tail.

> Anas Anser ferus, *Temm. Man. d'Orn.* tom. ii. p. 818. Gray-Lag Goose, *Mont. Orn. Dict. & Supp. Bew. Brit. Birds,* vol. ii. p. 282. Gray-Lag Wild Goose, *Selb. Illust.* vol. ii. p. 261. pl. 41.

Dimens. Entire length two feet ten inches.

Descript. Head and neck clove-brown, tinged with gray, the feathers on this last part loose and furrowed : back, scapulars, greater and middle wing-coverts, clove-brown, the feathers deeply edged with grayish white ; lesser wing-coverts, and base of the primary quills, bluish gray : breast and belly grayish-white, undulated with bars of a deeper tint : rump, abdomen, and under tail-coverts, pure white : tail deep clovebrown ; the middle feathers edged with white ; the outer one on each side almost wholly white : bill large and elevated ; orange-red, the nail whitish : irides deep brown : legs flesh-colour. (*Egg.*) Ivory white : smooth and shining : long. diam. three inches one line ; trans. diam. two inches one line.

Said to have been formerly very plentiful in this country ; and resident in the fens of Lincolnshire all the year. Now only met with in small flocks during the winter months, and that not very frequently. Breeds in marshes, and lays from five to ten eggs. Food, aquatic vegetables, and grain of all kinds. *Obs.* The *Domestic Goose* is usually considered as having been derived from this species, but such a circumstance is rendered highly improbable from the well known fact that the Common Gander after attaining a certain age is invariably white. Montagu also observes* that a specimen of the *Anser ferus,* which was shot in the wing by a farmer in Wiltshire, and kept alive many years, would *never associate* with the tame geese. In fact the origin of these last is unknown.

223. A. *Segetum*, Steph. (*Bean Goose.*)—Bill long, and somewhat depressed ; orange, the base and nail black : legs orange : wings extending beyond the extremity of the tail.

> Anas Segetum, *Temm. Man. d'Orn.* tom. ii. p. 820. Bean Goose, *Mont. Orn. Dict. & Supp. Selb. Illust.* vol. ii. p. 263. pl. 42.

Dimens. Entire length two feet six inches : length of the tarsus three inches three lines ; from the carpus to the end of the wing eighteen inches six lines.

---

* *Orn. Dict.* vol. i. Art. *Goose-Gray-Lag.*

Descript. Head, and upper part of the neck, dark cinereous brown, the feathers of the latter having a furrowed appearance; lower part of the neck, breast and belly, ash-gray, clouded with shades of a deeper tint: back, scapulars and wing-coverts, deep clove-brown, the feathers edged with whitish: quills dark gray, passing into black at the tips; the shafts white: rump blackish brown: tail dusky, tipped with white: vent, and under tail-coverts, pure white: bill black at the base and on the nail, the middle part orange-red: irides deep brown: legs orange. *Obs.* This species is subject to some little variation of plumage: the brown tint on the upper parts prevails over the ash-colour, or the ash-colour over the brown: the secondary quills are also sometimes edged and tipped with white: occasionally, though very rarely, the nail of the bill is white, in which case it becomes extremely difficult to distinguish it from the last species, which at all times it closely resembles; it is however smaller than the *A. ferus*, and the wings when closed extend nearly half an inch beyond the tip of the tail. (*Egg.*) Dull yellowish white: long. diam. three inches five lines; trans. diam. two inches five lines.

Much more plentiful than the last species. Visits this country in large flocks at the approach of Winter, resorting to fens, and also to upland corn-fields for the sake of the green wheat. Clamorous in its flight, which, when the flock is numerous, generally assumes a figured form. Retires in May to more northern latitudes to breed. Said to make a nest in marshes, amongst the coarse herbage, and to lay from ten to twelve eggs.

224. A. *albifrons*, Steph. (*White-fronted Goose.*) — Bill and legs orange; the former with the nail white: wings reaching a little beyond the tail: a large white space on the forehead.

Anas albifrons, *Temm. Man. d'Orn.* tom. ii. p. 821. White-fronted Goose, *Mont. Orn. Dict. Bew. Brit. Birds*, vol. ii. p. 294. *Selb. Illust.* vol. ii. p. 266. pl. 43.

Dimens. Entire length two feet three inches: length of the bill (from the gape) two inches; of the tarsus three inches; from the carpus to the end of the wing sixteen inches.

Descript. Region of the bill, and a large space on the forehead, pure white, the latter bordered with deep brown; a small white spot also under each eye: rest of the head, and neck, brownish ash: back, scapulars, and flanks, clove-brown, the feathers edged with reddish ash: quills black; greater wing-coverts tipped with white: breast, belly and abdomen, dirty white, irregularly marked with patches and transverse bars of black: tail blackish gray; the middle feathers edged with white, the outer one on each side almost wholly white: upper and under tail-coverts white: bill flesh-red, tinged with orange; the nail white: irides brown: legs yellowish orange. In *young birds*, there is no white patch on the forehead, but occasionally two or three small white spots at the base of the bill: the head and neck are tinged with reddish; the ash-coloured tints on the rest of the plumage paler than in the adult. (*Egg.*) White, tinged with buff: long. diam. two inches ten lines; trans. diam. one inch eleven lines.

Appears in small flocks during the winter months. Not uncommon in the fenny districts in the eastern parts of England. Said to breed in the Arctic Regions. Food, grain, seeds and vegetables.

(2. Bernicla, *Steph.*)

225. A. *Leucopsis*, Bechst. (*Common Bernicle.*) —
Forehead, cheeks, and throat, white : crown, occiput, neck
and breast, black.

> Anas Leucopsis, *Temm. Man. d'Orn.* tom. ii. p. 823. Bernicle
> Goose, *Mont. Orn. Dict. & Supp. Selb. Illust.* vol. ii. p. 268.
> pl. 44. Bernacle, *Bew. Brit. Birds*, vol. ii. p. 302.

Dimens. Entire length two feet one inch.

Descript. Forehead, cheeks and throat, white; between the bill and
the eyes a dark streak; crown, occiput, neck, and upper part of the breast,
black; back, scapulars and wing-coverts, undulated with transverse
bars of ash-gray, black, and grayish white, this last colour occupying the
tips of the feathers: quills blackish gray: under parts of the body pure
white, passing into ash-gray on the flanks: tail black: irides dusky
brown : bill and legs black. The *young* have the white on the forehead
mottled with dusky; the feathers on the back and wings tipped with
reddish; the flanks of a darker colour than in the adult: legs dusky
brown. (*Egg.*) Greenish white: long. diam. three inches; trans. diam.
one inch eleven lines.

Common as a winter visitant on the western coast of England, and in
some parts of Ireland. Rarely seen in the southern and eastern districts. Retires early in the year to high latitudes, where it breeds.
Food, seeds, grain, and aquatic and other vegetables.

226. A. *torquatus*, Frisch. (*Brent Bernicle.*)—Head,
neck, and breast, black ; a white patch on each side of
the neck.

> Anas Bernicla, *Temm. Man. d'Orn.* tom. ii. p. 824. Brent Goose,
> *Mont. Orn. Dict. & Supp. Bew. Brit. Birds*, vol. ii. p. 300.
> *Selb. Illust.* vol. ii. p. 271. pl. 45.

Dimens. Entire length twenty-nine inches.

Descript. (*Adult.*) Head, neck, and upper part of the breast, black;
on each side of the neck about half-way down, a patch of white, mottled
with black: back, scapulars, and wing-coverts, clove-brown, the feathers
edged with ash-gray : middle of the belly dark gray, tinged with brown :
flanks barred transversely with gray and white; vent, upper and under
tail-coverts, pure white: rump, quills, and tail, black : irides dusky brown :
bill and legs black. (*Young of the year.*) Head and neck dusky ash,
the latter without the white patch on the sides ; the feathers on the back
and breast tipped with reddish brown; flanks gray, with transverse bars
of reddish white. (*Egg.*) Dull white, slightly tinged with green: long.
diam. two inches eleven lines ; trans. diam. one inch nine lines.

Like the other species of this family, a regular winter visitant. Most
abundant in the eastern and southern parts of the kingdom. Frequents
the sea coast, and also inland marshes. Retires northward in February
and March to breed. Said to lay from eight to ten eggs. Food, similar
to that of the last.

227. A. *ruficollis*, Pall. (*Red-breasted Bernicle.*)—
Crown and throat black ; fore-neck and breast ferruginous ;
lore, and a double longitudinal line on the sides of the
neck, white.

Anas ruficollis, *Temm. Man. d'Orn.* tom. ii. p. 826. Red-breasted
Goose, *Mont. Orn. Dict. Bew. Brit. Birds,* vol. ii. p. 280.
Selb. Illust. vol. ii. p. 275. pl. 46.

DIMENS. Entire length twenty-one inches. MONT.

DESCRIPT. Forehead, crown, list down the back of the neck, throat,
and a band extending upwards to the eyes, black: between the bill and
the eye, a large oval white spot ; behind the eye, and surrounding a large
patch of orange-brown on the side of the neck, a list of white, which is
further extended, and forms a line of division between the orange-brown
and black of the lower neck : fore part of the neck, and breast, fine
orange-brown; the latter margined by a list of black, and another of
white; a second bar of white immediately before the shoulders: back,
belly, quills and tail, black : abdomen, vent, thighs, upper and under
tail-coverts, pure white : greater wing-coverts black, edged with white:
bill reddish brown; the nail black: legs blackish brown, tinged with
red. SELB. (*Egg.*) Unknown.

A very rare and accidental visitant in this country. According to
Montagu, one was shot near London in 1766, and another taken alive in
Yorkshire about the same time. Other specimens are stated to have
been killed in Cambridgeshire during the severe winter of 1813. It has
also occurred in Norfolk. Inhabits the North of Asia, about the shores of
the Frozen Sea. Appears in Russia as a regular bird of passage. Said
to breed in the northern parts of that country.

* 228. A. *Ægyptiacus*, Briss. (*Egyptian Goose.*)—Forehead, crown,
and throat, white; a large patch encircling the eyes, and another on the
breast, chestnut-red.

Chenalopex Ægyptiaca, *Steph. in Shaw's Gen. Zool.* vol. xii. part 2. p. 43.
pl. 42. Egyptian Goose, *Lath. Syn.* vol. iii. p. 453. *Shaw, Nat. Misc.*
vol. xv. pl. 605. *Bew. Brit. Birds,* vol. ii. p. 298.

DIMENS. Entire length two feet three inches: length of the bill two inches.
LATH.

DESCRIPT. (*Male.*) Forehead, crown, and throat, white ; this last somewhat
spotted with chestnut: on the sides of the head, surrounding the eyes, a large patch
of chestnut-red : upper part of the neck pale chestnut, becoming deeper at bottom
where it unites with the back : this last, and scapulars, brownish red, with numerous
transverse undulating dusky lines : wing-coverts white ; the greater ones barred
with black near their tips : primary quills black, the first five entirely so, the rest
edged with glossy green : secondaries tinged with reddish bay, and edged with chestnut: lower back, rump and tail, black : middle of the belly white ; all the rest of
the under parts pale rufous ash, with narrow undulating dusky lines ; on the breast
a large deep chestnut-coloured spot : bill red, the tip black : eyelids red : irides
yellowish : legs red. (*Female.*) "Chestnut patch round the eye smaller : chin
white : the chestnut patch on the breast smaller, if not wholly wanting : lesser
wing-coverts white ; the others pale ash-colour, with darker edges ; the lower
order fringed with white, forming a bar on the wing : scapulars and secondary quills
much inclined to chestnut: the rest as in the male." LATH. (*Egg.*) Dull
white, tinged with buff: long. diam. two inches nine lines ; trans. diam. two
inches.

A native of Egypt, and other parts of Africa. Has been introduced into this
country, and become partly naturalized in some places. A small flock is recorded

to have visited the banks of the Tweed, at Carham, in February 1832, and two to have been killed; but according to the conjectures of Mr Selby, they had probably made their escape from Gosforth, the seat of the Earl of Wemyss, upon the Firth of Forth, where great numbers of these birds are kept in the artificial pieces of water.

(3. PLECTROPTERUS, *Leach.*)

229. A. *Gambensis,* Briss. *(Spur-winged Goose.)* — Neck, and all the upper plumage, black, glossed with purple on the back; cheeks, throat, and under parts, white: bill and legs red.

Plectropterus Gambensis, *Steph. in Shaw's Gen. Zool.* vol. XII. part ii. p. 6. pl. 36. Spur-winged Goose, *Lath. Syn.* vol. III. p. 452. pl. 102. *Bew. Brit. Birds,* vol. II. p. 296.

DIMENS. Size of the *Common Goose,* but standing higher on its legs: length of the bill more than two inches. LATH.
DESCRIPT. Upper part of the head and neck dingy brown; ear-coverts, and sides of the throat, white, spotted with brown; the lower part of the neck, sides of the breast, and all the upper plumage, appear black, but this colour is lost, particularly in the scapulars and tertials, which are most resplendently bronzed and glossed with brilliant green, and most of the outer webs of the other feathers partake of the same hue: on the bend of the wing or wrist a strong white horny spur, turning upwards, about five-eighths of an inch (Latham says an inch and a half) in length, and pointing rather inwards: the whole of the edges of the wing, from the alula spuria to the elbow and shoulder, white: all the under parts the same: bill reddish yellow: legs and toes somewhat longer than those of the *Wild Goose;* of a red or orange-yellow. BEW. (*Egg.*) Unknown.
A single individual of this rare species, shot near St Germain's, Cornwall, June 20th, 1821, is now in the Newcastle Museum. No other has hitherto occurred in this country. Inhabits Gambia, and other parts of Africa. Habits and nidification unknown.

GEN. 88.   CYGNUS, *Meyer.*

* 230. C. *Guineensis,* Nob. (*Guinea Swan.*) — An elevated knob at the base of the upper mandible: upper parts of the body brownish gray, the edges of the feathers paler: a black list down the nape of the neck.

Anser Guineensis, *Briss. Orn.* vol. II. p. 435. Chinese Goose, *Lath. Syn.* vol. III. p. 447. Swan Goose, *Bew. Brit. Birds,* vol. II. p. 274.

DIMENS. Entire length three feet four inches.
DESCRIPT. (*Male.*) A white streak from the corners of the mouth surrounding the base of the upper mandible: down the nape a longitudinal black stripe reaching from the occiput to the back; this last, and all the rest of the upper parts, brownish gray, with the edges of the feathers somewhat paler: fore part of the neck and breast yellowish brown; belly, abdomen, and under tail-coverts, white; flanks brownish gray, the feathers edged with white: bill with a large protuberance at the base of the upper mandible; black, the base and protuberance sometimes orange: under the throat a loose skin, forming a kind of pouch and almost bare of feathers: legs black. (*Female.*) Smaller: the frontal protuberance not so much developed. (*Egg.*) White: long. diam. three inches three lines; trans. diam. two inches two lines.
A domesticated species, common in many parts of England on artificial pieces of water. Native country somewhat doubtful. Said by some to have been brought

originally from Guinea. Of restless habits, and very clamorous. Spends much of its time on land, seldom taking to the water except for safety in times of danger. Feeds on grain and vegetables. Lays in March.

* 231. C. *Canadensis*, Steph. (*Canada Swan.*)—No protuberance at the base of the bill: head and neck black; under the throat a crescent-shaped white patch.

> C. Canadensis, *Steph. in Shaw's Gen. Zool.* vol. xii. part 2. p. 19. Anser Canadensis, *Faun. Bor. Am.* part ii. p. 468. Canada Goose, *Lath. Syn.* vol. iii. p. 450. *Wils. Amer. Orn.* vol. viii. p. 53. pl. 67. f. 4. Cravat Goose, *Bew. Brit. Birds*, vol. ii. p. 276.

Dimens. Entire length forty-one inches: length of the bill (above) two inches two lines, (from the gape) two inches six lines; of the tarsus three inches seven lines; of the tail nine inches; of the wing nineteen inches six lines. *Faun. Bor. Am.*

Descript. Head, and greater part of the neck, black; under the throat a crescent-shaped white patch, passing upwards behind the eyes, and reaching nearly to the occiput: back, scapulars, wings and flanks, grayish brown, the tips of the feathers paler: lower part of the neck in front, breast, belly, and under parts, pure white: primary quills, rump and tail, black: bill and legs black. (*Egg.*) Dull dirty white: long. diam. three inches four lines; trans. diam. two inches four lines.

A native of North America, from which country it has been introduced into Europe. Is not uncommon in some parts, and may be considered as in a great measure naturalized. Small flocks are occasionally observed in England in a state of liberty and independance.

232. C. *ferus*, Ray. (*Whistling Swan.*)—Bill black; the base and sides, these last to beyond the nostrils, bright yellow: plumage white: tail of twenty feathers.

> Anas Cygnus, *Temm. Man. d'Orn.* tom. ii. p. 828. Whistling Swan, *Mont. Orn. Dict. & Supp. Selb. Illust.* vol. ii. p. 278. pl. 47. Wild Swan, *Bew. Brit. Birds*, vol. ii. p. 265. (Trachea,) *Linn. Trans.* vol. iv. pl. 12. f. 1.

Dimens. Entire length five feet: length of the bill (to the forehead) four inches four lines and a half, (to the eye) five inches three lines; of the tarsus four inches; of the middle toe six inches six lines; from the carpus to the end of the wing twenty-five inches six lines: breadth, wings extended, seven feet ten inches. Yarr.

Descript. The whole plumage pure white, with the exception of the head and nape, which have a faint tinge of orange-yellow: bill black; cere at the base, and a portion of the sides extending beyond the line of the nostrils, lemon-yellow; the same cere passes backward, and forms a yellow space round the eyes: irides brown: legs black. In the *female*, the bill is of a paler yellow than in the male. In *young birds*, the whole plumage is pale gray; the cere, and naked skin surrounding the eyes, pale flesh-colour; the legs reddish gray. The plumage is probably not perfected till the fourth or fifth year. (*Egg.*) Dull white, faintly tinged with greenish: long. diam. four inches one line; trans. diam. two inches eight lines.

A periodical winter visitant in the northern parts of Britain. Seldom observed southward except in very severe seasons. According to Low, some few pairs remain and breed in the Orkneys; but the greater number retire to the Arctic Regions at the approach of Spring. Builds on the ground near water, and lays from five to seven eggs. Has a loud harsh note. Food, aquatic plants and insects.

233. C. *Bewickii*, Yarr. (*Bewick's Swan*.)—Bill black; the base orange-yellow: plumage white: tail of eighteen feathers.

> C. Bewickii, *Yarrell in Linn. Trans.* vol. xvi. p. 445. *Selby in Newcastle Nat. Hist. Trans.* vol. i. p. 17. *Faun. Bor. Amer.* part ii. p. 465. Bewick's Swan, *Selb. Illust.* vol. ii. p. 284. pl. 47*. (Trachea,) *Linn. Trans.* vol. xvi. pls. 24, & 25.

Dimens. Entire length three feet nine inches: length of the bill (to the forehead) three inches six lines, (to the eye) four inches four lines and a half; of the tarsus three inches nine lines; of the middle toe five inches three lines; from the carpus to the end of the wing twenty inches six lines: breadth, wings extended, six feet one inch. Yarr.

Descript. Closely resembling the last species, but one third smaller. The whole plumage pure white, tinged, in immature specimens, on the crown, nape, and belly, with reddish orange: bill black; the cere at the base and sides orange-yellow; but no part of this last colour advancing beyond the nostrils, (as in the *C. ferus*): irides orange-yellow: legs black: tail cuneiform, of only eighteen feathers*. The *young* are gray, and the adult plumage is not perfected till the third or fourth year. (*Egg.*) Brownish white, slightly clouded with a darker tint: long. diam. three inches seven lines; trans. diam. two inches six lines.

Like the last,· a winter visitant. First distinguished as a peculiar species by Mr Yarrell, and, about the same time, by Mr Richard Wingate of Newcastle. Breeds in the Arctic Regions. Said to possess a weaker voice than that of the *C. ferus*. Habits similar.

* **234. C. *Olor*, Steph. (*Mute Swan*.)**—Bill red; the edges of the mandibles, nail, and a protuberance at the base, black: plumage white: tail of twenty-four feathers.

> C. Olor, *Steph. in Shaw's Gen. Zool.* vol. xii. part ii. p. 15. pl. 38. Anas Olor, *Temm. Man. d'Orn.* tom. ii. p. 830. Mute Swan, *Lewin, Brit. Birds,* vol. vii. pl. 237. *Bew. Brit. Birds,* vol. ii. p. 270.

Dimens. Entire length five feet or upwards.
Descript. The whole plumage, without exception, pure white: bill red; the edges of the mandibles, nail at the tip, nostrils, and a large protuberance on the forehead, as well as the space round the eyes, deep black: irides brown: legs black; sometimes tinged with reddish. The *female* is smaller, with the frontal protuberance not so much developed. The *young of the year* are ash-brown, with the bill and legs dusky gray. *In the second year* the bill assumes a yellowish tint, and white feathers appear intermixed with the gray ones. *In the third year*, the plumage is perfected. (*Egg.*) Dull greenish white: long. diam. four inches; trans. diam. two inches nine lines.

Well known in a domesticated or half-reclaimed state on many of our rivers and artificial pieces of water. Found wild in the eastern parts of Europe. Makes a nest amongst aquatic herbage in February or March, and lays from six to eight eggs. Food, aquatic vegetables and insects; occasionally small fish. Is destitute of the shrill voice which distinguishes the two preceding species.

---

* According to Mr Thompson there are twenty tail-feathers. See *Lond. & Edinb. Phil. Mag.* Oct. 1834. p. 299.

GEN. 89.   TADORNA, Leach.

235. T. rutila, Steph. (*Ruddy Shieldrake.*) — Back, scapulars, and under parts, ferruginous yellow; a black collar on the neck; a large white space on the wings: bill and legs black.

Anas rutila, *Temm. Man. d'Orn.* tom. II. p. 832. Ruddy Goose, *Lath. Syn.* vol. III. p. 456. Ruddy or Casarka Shieldrake, *Selb. Illust.* vol. II. p. 293. pl. 48**. Ferruginous Duck, *Bew. Brit. Birds*, vol. II. p. 313.

DIMENS. Entire length about twenty-three inches. SELB.

DESCRIPT. Forehead, cheeks and chin, pale ochreous yellow; region of the eyes, crown and nape, grayish white; neck, as far as the collar, ochreous yellow, tinged with orange; collar about half an inch in width, black, glossed with green: breast, back, scapulars, and under parts of the body, deep gallstone-yellow, tinged with orange, deepest upon the breast; the feathers on the upper parts with paler edges, and the ends of the long tertials passing into sienna-yellow: lesser and middle wing-coverts white: secondary quills, and greater coverts, green, glossed with purple, forming a very large speculum: greater quills black: tail-coverts black, glossed with duck-green: bill, legs and feet, black. SELB. The *female*, according to Temminck, wants the black collar, and has more white about the head and neck. (*Egg.*) White. TEMM.

An extremely rare and accidental visitant in this country. One killed at Bryanston in Dorsetshire, in the Winter of 1776, is now in the Newcastle Museum. Mr Selby has figured a second individual which was killed in the South of England, and which is in his own collection. Inhabits the eastern parts of Europe. Said to make a nest in the banks of rivers, or in burrows excavated for the purpose in deserted hillocks; occasionally in hollow trees. Lays eight or nine eggs. Food, aquatic plants and insects.

236. T. Bellonii, Steph. (*Common Shieldrake.*) — Back, wing-coverts, and flanks, white; a broad ferruginous band on the breast: bill, frontal protuberance, and legs, red.

Anas Tadorna, *Temm. Man. d'Orn.* tom. II. p. 833. Shieldrake, *Mont. Orn. Dict. & Supp.* *Bew. Brit. Birds*, vol. II. p. 341. Common Shieldrake, *Selb. Illust.* vol. II. p. 289. pl. 48. (Trachea,) *Linn. Trans.* vol. IV. pl. 15. f. 8, & 9.

DIMENS. Entire length twenty-three inches six lines: length of the bill (from the forehead) two inches, (from the gape) two inches three lines; of the tarsus one inch eleven lines; of the tail three inches eight lines; from the carpus to the end of the wing twelve inches three lines.

DESCRIPT. (*Male.*) Head, and upper part of the neck, deep blackish green with glossy reflections; lower part of the neck, back, wing-coverts, flanks, rump, and base of the tail, pure white: across the breast a broad band of ferruginous brown, the ends passing upwards and uniting between the shoulders: scapulars, a mesial abdominal list dilating at the vent, and tip of the tail, black; under tail-coverts pale reddish brown: primaries black: speculum rich bronzed-green: three or four of the

secondaries next the body with their outer webs rich orange-brown:
bill, and a fleshy protuberance on the forehead, blood-red: irides brown:
legs flesh-red, inclining to crimson. (*Female.*) Smaller: no frontal protuberance, but a whitish spot instead: the colours in general not so
bright as in the male: the pectoral band, and abdominal list, narrower;
the latter often interrupted by large white spots. (*Young of the year.*)
Forehead, cheeks, front of the neck, back, and all the under parts, white:
crown and back of the neck dusky brown, with whitish specks: the pectoral band faintly indicated by a very pale tinge of reddish on that part:
scapulars dusky gray, edged with pale ash: lesser wing-coverts white,
with dusky tips, having a mottled appearance: tail tipped with cinereous brown: bill reddish brown: legs livid gray. (*Egg.*) Smooth;
shining white: long. diam. two inches nine lines; trans. diam. one inch
eleven lines.

A common species on many parts of the British coast, where it resides
all the year. Prefers flat and sandy shores. Is rarely observed inland.
Breeds in rabbit-burrows, or in holes in sand-banks excavated for the
purpose. Eggs eight to ten in number. Food, sea-weed, marine insects,
and small bivalve mollusca.

* GEN. 90. CAIRINA, *Flem.*

* 237. C. *moschata*, Flem. (*Musk Duck.*)—Bill, legs, and naked skin
surrounding the eyes, red: plumage various.

Anas moschata, *Linn. Syst. Nat.* tom. I. p. 199. Muscovy Duck, *Lath. Syn.*
vol. III. p. 476. Musk Duck, *Bew. Brit. Birds*, vol. II. p. 317. (Trachea,)
Linn. Trans. vol. xv. pl. 15. f. *a.*

DIMENS. Entire length twenty-four inches six lines.
DESCRIPT. Crown-feathers slightly elongated, forming a kind of tuft capable of
being erected: colours of the plumage extremely variable in the domesticated state;
often wholly white, the tuft on the crown excepted, which is black; or slate-gray,
with more or less white on the under parts: tail of twenty feathers; the outer one
on each side always white: bill, a fleshy protuberance at its base, and a naked
papillose skin surrounding the eyes, red: legs short and thick; red, or reddish
yellow. In the *female*, the naked papillose skin is of a paler hue, and does not
cover so large a part of the face: she is also smaller than the male. (*Egg.*)
Yellowish white: long. diam. two inches six lines; trans. diam. one inch nine
lines.
Common in a domesticated state, throughout Europe, but not indigenous.
Native country said to be America, where they are still found wild. A prolific
species, laying often, and producing from eight to twelve in a brood. Nest occasionally placed in the holes of trees. Has a musky smell arising from the gland on
the rump. *Obs.* The *Anas bicolor* of Donovan* is a hybrid between this species and
the common *Domestic Duck*, which not unfrequently breed together.

GEN. 91. ANAS, *Linn.*

(1. ANAS, *Swains.*)

238. A. *clypeata*, Linn. (*Common Shoveller.*)—Head
and neck glossy green; back brown; belly and abdomen
brownish red; lesser wing-coverts pale blue.

A. clypeata, *Temm. Man. d'Orn.* tom. II. p. 842. Shoveller, *Mont.*
Orn. Dict. & Supp. Bew. Brit. Birds, vol. II. p. 345. Common
Shoveller, *Selb. Illust.* vol. II. p. 297. pl. 48*. (Trachea,) *Linn.*
Trans. vol. IV. pl. 13. f. 4, & 5.

* *Brit. Birds*, vol. IX. pl. 212.

DIMENS. Entire length twenty inches: length of the bill (from the forehead) two inches eight lines, (from the gape) two inches eleven lines; of the tarsus one inch five lines; of the tail four inches; from the carpus to the end of the wing nine inches: breadth, wings extended, two feet six inches.

DESCRIPT. (*Male*.) Head, and upper half of the neck, deep green, with glossy reflections; lower part of the neck, breast, and scapulars, white; belly, abdomen, and flanks, brownish red: back, and primary quills, umber-brown: lesser wing-coverts pale blue; greater ones tipped with white, forming an oblique bar across the wings and an upper border to the speculum, which last is of a brilliant grass-green: rump, upper and under tail-coverts, brown, glossed with blackish green; sides of the rump white: tail brown, the feathers edged with white; the outer one wholly white: bill black: irides yellow: legs orange. (*Female*.) Head pale reddish brown, with fine dusky streaks; rest of the upper parts dusky brown, the feathers edged with reddish white: under parts reddish,. with large brown spots: lesser wing-coverts slightly glossed with pale blue: speculum not so bright as in the male. *Obs*. During the breeding season the *male* has a red breast, in which state it is the *Red-breasted Shoveller* of Pennant: after the expiration of that season, he partially assumes the *female* plumage, which is retained till the period of the autumnal moult. (*Egg*.) White, tinged with green: long. diam. two inches two lines; trans. diam. one inch six lines.

Not a very abundant species, but met with occasionally during the winter months, principally in the eastern parts of the country. A few pairs are said to remain and breed in the marshes in Norfolk. Nest placed near the water's edge amongst aquatic herbage. Eggs twelve to fourteen in number. Food, worms and aquatic insects.

(2. CHAULIODUS, *Swains*.)

239. A. *Strepera*, Linn. (*Gadwall*.)—Back, breast, scapulars and flanks, dusky brown, marked with undulating white lines: speculum white.

A. Strepera, *Temm. Man. d'Orn.* tom. II. p. 837. Gadwall, *Mont. Orn. Dict. Bew. Brit. Birds*, vol. II. p. 348. Common Gadwall, *Selb. Illust.* vol. II. p. 301. pl. 51, & pl. 49*. f. 1. *Gould, Europ. Birds*, part viii. (Trachea,) *Linn. Trans.* vol. IV. pl. 13. f. 7, & 8.

DIMENS. Entire length twenty inches: length of the bill (from the forehead) one inch nine lines, (from the gape) two inches; of the tarsus one inch six lines; of the tail three inches five lines; from the carpus to the end of the wing ten inches seven lines.

DESCRIPT. (*Male*.) Head and neck grayish white, speckled with brown; lower part of the neck, breast, and back, clove-brown, marked with crescent-shaped white lines; scapulars and flanks undulated with white and blackish brown; middle wing-coverts chestnut-brown; greater coverts, rump, and upper and under tail-coverts, black, glossed with purplish blue: speculum white: belly and abdomen white, minutely speckled with grayish brown: tail cinereous, edged with white: bill brownish black, pale beneath: legs orange. (*Female*.) Not very dissimilar to the male, but with the undulating lines and crescent-shaped bars less distinctly marked. (*Young of the year*.) " Of a uniform rusty brown above, each feather having a central mark of dusky black; the under surface white." GOULD. (*Egg*.) " Greenish ash-colour." TEMM.

A winter visitant, but not of very common occurrence. According to Temminck, is very abundant in Holland, frequenting the same situations as the common Wild Duck. Breeds in marshes, and lays eight or nine eggs. Food, fish, and aquatic insects and vegetables.

(3. DAFILA, Leach.)

240. A. acuta, Linn. (Pintail.) — A longitudinal white line on each side of the occiput; back and flanks undulated with black and grayish white; two central elongated tail-feathers black.

> A. acuta, *Temm. Man. d'Orn.* tom. II. p. 838. Pintail Duck, *Mont. Orn. Dict. & Supp. Bew. Brit. Birds,* vol. II. p. 356. Common Pintail, *Selb. Illust.* vol. II. p. 311. pl. 49, & pl. 49*. f. 2. (Trachea,) *Linn. Trans.* vol. IV. pl. 13. f. 6.

DIMENS. Entire length twenty-six inches; the same, central tail-feathers excluded, twenty-four inches: length of the bill (from the forehead) two inches, (from the gape) two inches three lines; of the tarsus one inch seven lines; of the middle toe, nail included, two inches four lines; of the tail six inches; from the carpus to the end of the wing ten inches six lines.

DESCRIPT. *(Male.)* Forehead and crown umber-brown, the feathers with paler edges; rest of the head, chin, and throat, dark hair-brown, slightly glossed behind the ears with purplish green: fore part of the neck, and two lateral streaks passing upwards to the occiput, white; neck above deep blackish brown: the whole of the back, flanks, and sides of the breast, beautifully marked with transverse undulating lines of black and grayish white: scapulars black; tertials long and acuminated, velvet black, with a broadish edging of grayish white: wing-coverts and primaries hair-brown; speculum blackish green, glossed with purple, bordered above by a pale ferruginous bar, below by a white one: breast, belly, and abdomen, white; the latter minutely speckled with gray towards the vent: tail, and upper coverts, dark cinereous brown, the edges of the feathers paler; two central elongated feathers, and under coverts, black: bill black, the sides of the upper mandible bluish gray: legs blackish gray. *(Female.)* Smaller than the male: head and neck reddish brown, speckled and streaked with dusky: all the upper plumage blackish brown, the feathers edged with reddish white: under parts reddish yellow, obscurely spotted with brown: speculum dull, without the green gloss; bordered above with yellowish, beneath with whitish: tail conical, but the two middle feathers scarcely longer than the others. *Obs.* The *male* of this species (as in the case of the *Shoveller*) partially assumes the *female* plumage after the expiration of the breeding season: it is, however, not retained beyond the autumnal moult. *(Egg.)* Greenish white, tinged with buff: long. diam. two inches three lines; trans. diam. one inch seven lines and a half.

Not of unfrequent occurrence during the winter months. Breeds in higher latitudes. Said to lay eight or nine eggs. Food, aquatic insects and vegetables, fish, and molluscous animals.

(4. BOSCHAS, Swains.)

241. A. glocitans, Pall. (Bimaculated Duck.)—Before and behind the eyes an irregular patch of chestnut-brown;

back and flanks undulated with black: two middle tail-
feathers somewhat elongated, straight.

Querquedula glocitans, *Vigors in Linn. Trans.* vol. XIV. p. 559.
Bimaculated Duck, *Penn. Brit. Zool.* vol. II. p. 602. pl. 100.
no. 287. *Bew. Brit. Birds*, vol. II. p. 362. Bimaculated Teal,
Selb. Illust. vol. II. p. 321. pls. 55, & 55*.

DIMENS. Length of the body, bill included, fifteen inches nine lines;
of the bill (to the forehead) one inch and nine-tenths, (to the gape) two
inches and one-tenth; of the tarsus one inch and six lines; from the
carpus to the end of the wing eight inches and two-fifths. VIG.

DESCRIPT. (*Male.*) Forehead, crown, and occiput, very deep reddish
brown, glossed with purplish black, passing, on the hinder part of the
neck, into deep violet purple: between the bill and the eye, and behind
the ear-coverts, two large irregular patches of chestnut-brown, margined
and varied with white: cheeks, and sides of the neck, glossy duck-green,
the rest of the upper neck, and throat, greenish black: front of the lower
neck, and sides of the breast, reddish brown, with oval black spots;
middle part of the breast pale reddish brown, spotted with black: ground-
colour of the back pale sienna-yellow, undulated with black lines; scapulars the same, tipped with glossy Scotch blue; wing-coverts hair-brown,
the lower row tipped with pale wood-brown; speculum dark green, glossed
with purple: upper and under tail-coverts greenish black, glossed with
purple: tail cuneiform, the two middle feathers black, narrow and acuminated, considerably longer than the others, which are hair-brown, edged
with white: belly and abdomen yellowish white, with undulating black
lines, darkest and most distinct upon the flanks: bill blackish gray, passing towards the base and margins, into dirty orange-yellow: legs pale
orange. SELB. (*Female.*) Chin and throat pale buff; head and neck
the same, but spotted and streaked with black, the spots largest and most
distinct upon the crown; lower neck, and sides of the breast, pale yellowish brown, with blackish brown spots; middle of the breast, and belly,
white; abdomen white, with faint hair-brown spots; flanks variegated
with yellowish brown and blackish brown: upper parts blackish brown,
the feathers deeply edged with reddish white and yellowish brown; lesser
wing-coverts hair-brown, the lower tier deeply tipped with pale reddish
brown: upper half of the speculum green, with purple reflections; the
lower half velvet-black, with white tips to the feathers: quills and tail
hair-brown, the latter edged with white and reddish white: legs orange-
yellow. SELB.

Very little is known of this extremely rare species, which was first
described by Pennant from a male specimen taken in a decoy in 1771.
Two other individuals, a male and female, occurred near Maldon in
Essex, in the Winter of 1812-13. These last, which are described in
the Linnæan Transactions by Mr Vigors, are now in the Museum of
the Zoological Society. Habits and nidification unknown. Stated by
Pallas to be a native of Siberia, frequenting Lake Baikal and the
River Lena.

242. A. *Boschas*, Linn. (*Mallard.*)—Head, and upper
part of the neck, deep green; below the green a white
collar: four middle tail-feathers (in the *male*) recurved.

A. Boschas, *Temm. Man. d'Orn.* tom. II. p. 835. Common Duck, *Mont. Orn. Dict. & Supp.* Common Wild Duck, *Selb. Illust.* vol. II. p. 305. pls. 50 & 50*. Mallard, *Bew. Brit. Birds*, vol. II. p. 325. (Trachea,) *Linn. Trans.* vol. IV. pl. 13. f. 10.

DIMENS. Entire length twenty-four inches: length of the bill (from the forehead) two inches three lines, (from the gape) two inches seven lines; of the tarsus one inch eleven lines; of the middle toe, nail included, two inches five lines; of the tail three inches five lines; from the carpus to the end of the wing ten inches nine lines: breadth, wings extended, three feet.

DESCRIPT. (*Male.*) Head, and upper half of the neck, deep emerald-green, approaching to black on the cheeks and forehead; beneath the green a white collar; rest of the neck, and breast, dark chestnut; upper part of the back, wing-coverts, and primary quills, cinereous brown of different tints, the first with fine transverse lines of gray; rump, upper and under tail-coverts, blackish green; scapulars, flanks, abdomen, and sides of the rump, grayish white, with fine transverse undulating lines of clove-brown; some of the outer scapulars chestnut, with the transverse lines darker: speculum deep Prussian-blue, with purple and green reflections; bounded above and below by a double border, the inner one velvet-black, the outer one white: tail grayish brown, all the feathers bordered with white: bill greenish yellow: irides reddish brown: legs orange. (*Female.*) Smaller: all the upper parts umber-brown, of different shades, the feathers edged with pale reddish brown: head and neck with dusky streaks; throat whitish; breast and under parts yellowish gray, obscurely spotted and streaked with brown: speculum as in the male: four central tail-feathers straight: bill greenish gray. The *young males, till after the first moult,* resemble the females. (*Egg.*) Smooth; greenish white: long. diam. two inches three lines and a half; trans. diam. one inch seven lines.

A common species in most parts of the country during the winter months, some few remaining to breed. Frequents lakes, marshes, and rivers. Nest generally placed on the ground amongst aquatic herbage; sometimes in trees. Eggs ten or twelve in number. The young are called *Flappers.* Food, insects, slugs, grain, and aquatic vegetables. *Obs.* The *Domestic Duck* derives its origin from this species.

* (19.) *A. adunca,* Linn. (*Hook-billed Duck.*) Lath. Syn. vol. III. p. 495. Don. Brit. Birds, vol. IX. pl. 218.

Differs from the *Common Mallard* in the bill being broader, longer, and inclined more downwards at the tip. In other respects similar. It does not seem to be satisfactorily determined, whether it be a peculiar species, or only a variety of that last described; probably, however, the latter. Not uncommon in the domestic state.

243. A. *Querquedula,* Linn. (*Garganey.*) — Crown, occiput, and list on the nape, dark brown; a white stripe from the eye down each side of the neck: speculum grayish green.

A. Querquedula, *Temm. Man. d'Orn.* tom. II. p. 844. Garganey, *Mont. Orn. Dict. & Supp. Bew. Brit. Birds,* vol. II. p. 372. Garganey Teal, *Selb. Illust.* vol. II. p. 318. pl. 53. (Trachea,) *Linn. Trans.* vol. IV. pl. 13. f. 2, & 3.

DIMENS. Entire length sixteen inches six lines.

DESCRIPT. (*Male.*) Crown, occiput, and a list down the back part of the neck, dark umber-brown; throat black; over each eye a band of pure white, prolonged down the sides of the neck; cheeks, and upper part of the neck, chestnut-brown, with fine longitudinal streaks of white; lower part of the neck, and breast, pale yellowish brown, beautifully marked with crescent-shaped black bars: back grayish black, the feathers edged with ash-colour and yellowish brown; scapulars long and narrow, black, with a broad central white streak; wing-coverts bluish ash: speculum grayish green, bordered above and below by a white bar: belly white; flanks marked with transverse undulating black lines: vent, upper and under tail-coverts, yellowish white, spotted with black: tail dusky gray, the edges of the feathers lighter: irides light hazel: bill and legs blackish gray. (*Female.*) Smaller: throat white: head, neck, and upper parts of the body, brown, approaching to dusky, the feathers with pale whitish edges: the streak behind the eye very faint and ill-defined: lower part of the breast, and belly, white; flanks and abdomen spotted with brown: wing-coverts dark ash-gray; speculum dull, the green tinge almost wanting. The *young males* resemble the females: as they advance to maturity, the plumage assumes a mixed character. (*Egg.*) Buff-colour: long. diam. one inch nine lines; trans. diam. one inch three lines.

Met with in small numbers principally during the winter and spring months, and probably remains to breed in some parts of the country, but this last point does not appear to have been fully ascertained. Frequents marshes and the reedy banks of rivers. Nest placed on the ground in situations of the above nature. Eggs ten or twelve. Food, slugs, insects, seeds, and aquatic plants.

244. A. *Crecca*, Linn. (*Teal.*) — Crown, cheeks, and neck, ferruginous brown; from the eyes to the nape a broad longitudinal band of green: speculum, half green and half black.

A. Crecca, *Temm. Man. d'Orn.* tom. II. p. 846. Teal, *Mont. Orn. Dict. Bew. Brit. Birds*, vol. II. p. 374. Common Teal, *Selb. Illust.* vol. II. p. 315. pl. 54. (Trachea,) *Linn. Trans.* vol. IV. pl. 13. f. 1.

DIMENS. Entire length fourteen inches six lines: length of the bill (from the forehead) one inch five lines, (from the gape) one inch eight lines; of the tarsus one inch two lines; of the tail three inches two lines; from the carpus to the end of the wing seven inches five lines.

DESCRIPT. (*Male.*) Crown of the head, cheeks, front and sides of the neck, ferruginous brown; on the sides of the head, inclosing the eye, a large patch of deep green, passing off backwards to the nape in the form of a broad band; sides of the lower part of the neck, back, scapulars, and flanks, beautifully marked with transverse undulating lines of black and white; some of the longer scapulars cream-yellow, with a portion of their outer webs velvet-black: lower part of the neck in front, and breast, reddish white, with round black spots: wing-coverts brown, tinged with gray; speculum deep green in the middle, velvet-black at the sides, bordered above by a broad white bar: belly and abdomen white: under tail-coverts blackish brown, bordered at the sides with yellowish white: tail cuneiform, brown, the feathers edged with white: bill dusky gray: irides light hazel: legs brown, with a tinge of ash-gray. (*Female.*) Extremely similar to that of the last species, but may always be distinguished by the speculum, which is of the same dark black and green colours as in the male: head, neck, and all the upper parts, dusky brown, the feathers more or less broadly edged with pale reddish brown; throat, cheeks, and a band behind the eyes, yellowish white, spotted with black: under parts yellowish white. The *young males* resemble the female. (*Egg.*) White, tinged with buff: long. diam. one inch nine lines; trans. diam. one inch four lines.

A common species during the winter months, appearing in small flocks, and frequenting fresh waters. Remains to breed in some parts of the country. Nest (according to Selby) composed of rushes and other aquatic grasses, and lined with down. Eggs ten to twelve in number. Food, similar to that of the Garganey. *Obs.* The name of *Summer Teal* appears to be applied in some places indiscriminately to this and the last species, when met with during the Spring of the year.

GEN. 92. MARECA, *Steph.*

245. M. *Penelope*, Selb. (*Wigeon.*) — Forehead yellowish white; rest of the head, and neck, chestnut-red: back and flanks undulated with black and white.

Anas Penelope, *Temm. Man. d'Orn.* tom. II. p. 840. Wigeon, *Mont. Orn. Dict. & Supp. Bew. Brit. Birds*, vol. II. p. 350. Common Wigeon, *Selb. Illust.* vol. II. p. 324. pl. 52. (Trachea,) *Linn. Trans.* vol. IV. pl. 13. f. 9.

DIMENS. Entire length twenty inches: length of the bill (from the forehead) one inch six lines, (from the gape) one inch nine lines; of the tarsus one inch six lines; of the tail three inches ten lines; from the carpus to the end of the wing ten inches three lines.

DESCRIPT. (*Male.*) Forehead and crown cream-yellow; rest of the head, and upper part of the neck, chestnut-red, the cheeks speckled with black; throat black; lower part of the neck, and breast, vinaceous red: back and flanks marked with transverse undulating lines of black and white; scapulars black, edged with white; wing-coverts white, some of the lesser ones nearest the body pale grayish brown: quills cinereous brown: speculum composed of three bars, the middle one glossy green, the upper and under ones black: belly and abdomen white: tail cunei-form; the two middle feathers pointed, and considerably longer than the others; of a blackish gray colour: under tail-coverts black: bill bluish gray, the tip black: irides hazel: legs dusky lead-colour. (*Female.*) Head and neck rufous brown, speckled with dusky; back and scapulars dusky brown, the feathers with reddish edges; wing-coverts brown, edged with whitish; speculum without the gloss of dark green: breast, belly, and abdomen, much as in the male: flanks rufous brown, the tips of the feathers inclining to ash-gray: bill and legs dusky gray. *Obs.* This species appears to be subject to considerable variation of plumage. In *very old males*, according to Temminck, the forehead alone is yellowish white, this colour not extending over the crown. *Young males of the year* resemble the female. (*Egg.*) Brownish white: long. diam. two inches; trans. diam. one inch seven lines.

A winter visitant of common occurrence in most parts of the country. Frequents rivers and marshes, as well as inlets of the sea. Keeps in flocks, and has a peculiar whistling note. Breeds in the north of Europe, and is said to lay eight or nine eggs. Food, aquatic insects and vegetables, mollusca, and small fish. In confinement, has been known to breed with the Pintail and the Common Duck.

* GEN. 93. DENDRONESSA, *Swains.*

* 246. D. *Sponsa*, Swains. (*Summer Duck.*)—A pendent occipital crest, variegated with green, purple, and white.

D. Sponsa, *Faun. Bor. Amer.* part ii. p. 446. Summer Duck, *Lath. Syn.* vol. iii. p. 546. *Wils. Amer. Orn.* vol. viii. p. 97. pl. 70. f. 3. (Trachea,) *Mont. Orn. Dict. Supp. App.* last plate, f. 5.

Dimens. Entire length twenty-one inches : length of the bill (above) one inch four lines and a half, (to gape) one inch ten lines ; of tarsus one inch six lines ; of middle toe and claw two inches ; of tail four inches six lines ; of folded wing nine inches. *Faun. Bor. Am.*

Descript. (*Male.*) Head above, and space between the eye and the bill, glossy dark green ; cheeks, and a large patch on the sides of the throat, purple, with blue reflections ; pendent occipital crest of green and auricula purple, marked with two narrow white lines, one of them terminating behind the eye, the other extending over the eye to the bill ; sides of the neck purplish red, changing on the front of the neck and sides of the breast to brown, and there spotted with white : scapulars, wings and tail, exhibiting a play of duck-green, purple, blue, and velvet-black, colours ; intèrscapulars, posterior part of the back, rump, and upper tail-coverts, blackish green and purple ; several of the lateral coverts reddish orange ; a hair-like, splendent, reddish purple tuft on each side of the rump ; the under coverts brown : chin, throat, a collar round the neck, a crescentic bar on the ears, the middle of the breast, and whole of the abdomen, white : flanks yellowish gray, finely undulated with black ; the tips of the long feathers, and also of those on the shoulder, broadly barred with white and black : inner wing-coverts white, barred with brown : almost all the coloured plumage shows a play of colours with metallic lustre : bill red ; a space between the nostrils, its tip, margins, and lower mandible, black : legs orange. *Faun. Bor. Am.*   (*Female.*) Without the fine lines on the flanks, and the hair-like tufts on the sides of the rump : crest shorter : the plumage less vivid, particularly about the head, where it is mostly brown. *Faun. Bor. Am.* (*Egg.*) Smooth ; yellowish white : long. diam. one inch ten lines ; trans. diam. one inch six lines.

A pair of these birds, male and female, were shot about two or three years since near Dorking in Surrey, but as the species is strictly a native of America, and is often kept in aviaries, its claims to a place in the British Fauna must be considered as extremely doubtful. It is probable that the above individuals had escaped from confinement. Common in Mexico, and other parts of North America ; breeding in hollow trees. Said to lay thirteen eggs. Food, according to Wilson, principally acorns, seeds of the wild oats, and insects.

GEN. 94. SOMATERIA, *Leach.*

247. S. *mollissima*, Leach. (*Eider Duck.*) — Lateral prolongations at the base of the bill in the form of two narrow flat lamellæ : bill and legs greenish gray.

Anas mollissima, *Temm. Man. d'Orn.* tom. ii. p. 848. Eider Duck, *Mont. Orn. Dict. & Supp. Bew. Brit. Birds*, vol. ii. p. 305. Common Eider, *Selb. Illust.* vol. ii. p. 338. pls. 70, & 70 *. (Trachea,) *Linn. Trans.* vol. xii. pl. 30. f. 1, & 2.

Dimens. Entire length twenty-four inches : length of the bill (from the forehead) two inches three lines, (from the gape) two inches ten lines ; of the tarsus two inches ; of the tail four inches ; from the carpus to the end of the wing eleven inches six lines.

Descript. (*Male.*) Contour of the frontal lamellæ, forehead and crown, taking in the eyes, black, glossed with violet ; a longitudinal band on the top of the head, cheeks, throat, and front of the neck, whitish ; occiput, nape, and a large patch beneath the temples, green-ish white ; lower part of the neck, back, scapulars, lesser wing-coverts,

curved tertiaries, and sides of the rump, pure white; breast white, with
a tinge of vinaceous red: belly, abdomen, and rump, deep black: quills
and tail-feathers dark ash-gray: bill dull green: irides brown: legs dull
greenish gray. (*Female.*) The whole plumage reddish brown, with transverse black bars: head and back part of the neck marked with dusky
streaks: wing-coverts black, edged with ferruginous: across the wing
two rather indistinct white bars: belly and abdomen dark reddish brown,
with obscure transverse black bars. *Young males* are said not to attain
to maturity till the *fourth year.* According to Montagu, during the *first
year*, the back is white, and the usual parts, except the crown, black:
but the rest of the body variegated with black and white. In the *second
year*, the crown of the head is black, and the neck and breast spotted with
black and white. (*Egg.*) Smooth, shining, olive-green: long. diam. three
inches; trans. diam. two inches one line.

Common in the northern parts of Britain, particularly in the Scotch
Islands, where it breeds. Rarely observed southwards; but has been
killed in a single instance on the coast of Devon. Nest placed on the
ground, near the edge of the sea, or on projecting rocks, formed of sea-
weed, and copiously lined with down, which the female plucks from her
body. Eggs five or six in number. Food, marine plants and insects,
small fish, and bivalve mollusca.

248. S. *spectabilis*, Leach. (*King Duck.*) — Lateral
prolongations at the base of the bill in the form of two
elevated, compressed tubercles: bill red: legs ochre-
yellow.

> Anas spectabilis, *Temm. Man. d'Orn.* tom. ii. p. 851. King Duck,
>   *Mont. Orn. Dict. Bew. Brit. Birds*, vol. ii. p. 310. King Eider,
>   *Selb. Illust.* vol. ii. p. 343. pl. 71. (Trachea,) *Linn. Trans.* vol. xv.
>   pl. 15. f. c & d.

DIMENS. Entire length twenty-four inches six lines: length of the
bill (from the forehead) one inch two lines, (from the gape) two inches
seven lines; of the tarsus one inch ten lines and a half; of the tail four
inches three lines; of the wing eleven inches six lines: height of the frontal
tubercles one inch six lines; breadth of ditto one inch. *Faun. Bor. Am.*

DESCRIPT. (*Male.*) Contour of the frontal tubercles, under eyelid,
edge of the upper one, and two converging bands on the throat, meeting on the chin, rich velvet-black: crown, occiput, and nape, fine bluish
gray: cheeks and ear-coverts glossy pea-green, passing into dull white
towards the chin and neck; a whitish line from the corners of the eye
down the sides of the occiput reaching to the nape; lower part of the
neck, and breast, white tinged with buff; upper part of the back, wing-
coverts, and a large space on each side of the rump, pure white: scapulars, secondaries, curved tertiaries, lower part of the back, rump, tail-
coverts, belly, and abdomen, deep black: greater quills and tail blackish
brown: bill vermilion-red; frontal tubercles, and base of the lower mandible, orange: legs ochre-yellow. (*Female.*) "Exactly resembling the
female *Eider*, except that the frontal plates of the upper mandible, instead of being almost horizontal, are more nearly vertical. The bill is
also shorter than that of the Eider." *Faun. Bor. Am.* (*Young male.*)
"Head and neck dusky yellowish gray, crowded with blackish spots:
upper plumage mostly pitch-black, with yellowish brown edgings: breast
and flanks yellowish brown, spotted and barred with black; belly the

same colours intimately mixed: bill as in the female." *Faun. Bor. Am.* (*Egg.*) Greenish yellow white: long. diam. two inches nine lines; trans. diam. one inch nine lines.

A native of high northern latitudes, but met with occasionally in small numbers in some of the Scotch Islands. Mr Bullock is stated by Montagu to have found it breeding in Papa Westra, one of the Orkneys, in the latter end of June. The nest was on a rock impending the sea. Eggs six in number, covered, as in the case of the last species, with the down of the female. Very rare on the English coast, but has been killed at Aldborough in Suffolk.

GEN. 95. OIDEMIA, *Flem.*

249. O. *fusca*, Flem. (*Velvet Scoter.*)——Black; a white speculum on the wings : legs red.

> Anas fusca, *Temm. Man. d'Orn.* tom. ii. p. 854. Velvet Duck, *Mont. Orn. Dict. Bew. Brit. Birds*, vol. ii. p. 320. Velvet Scoter, *Selb. Illust.* vol. ii. p. 333. pl. 67. (Trachea,) *Linn. Trans.* vol. xvi. pl. 21. f. 2, & 3.

DIMENS. Entire length twenty-three inches: length of the bill (from the forehead) one inch eight lines, (from the gape) two inches seven lines; of the tarsus one inch nine lines; of the tail three inches ten lines; from the carpus to the end of the wing ten inches ten lines.

DESCRIPT. (*Male.*) A small crescent-shaped spot beneath the eyes, and speculum on the wings, white; all the rest of the plumage deep velvet-black: bill orange; a protuberance at the base, nostrils, and margins of both mandibles, black; legs and toes scarlet-red; membranes black. (*Female.*) Similar to the male, but smaller, with the upper plumage more inclining to brown; under parts gray, spotted and streaked with dusky brown; between the bill and the eye, and on the orifice of the ears, a grayish white spot: bill blackish brown; slightly inflated at the base: legs and toes dull red. (*Egg.*) Buff-colour: long. diam. two inches ten lines; trans. diam. one inch ten lines.

A periodical winter visitant, but not very numerous in the southern parts of the country. Frequents the sea-coast, and is very rarely observed inland. Breeds, according to Temminck, in the Arctic Regions, and lays from eight to ten eggs. Dives well in search of its food, which consists principally of bivalve mollusca.

250. O. *nigra*, Flem. (*Black Scoter.*) —— The whole plumage black; no speculum on the wings : legs dusky gray.

> Anas nigra, *Temm. Man. d'Orn.* tom. ii. p. 856. Scoter, *Mont. Orn. Dict. Bew. Brit. Birds*, vol. ii. p. 322. Black Scoter, *Selb. Illust.* vol. ii. p. 329. pl. 68.

DIMENS. Entire length eighteen inches: length of the bill (from the forehead) one inch eight lines, (from the gape) two inches; of the tarsus one inch nine lines; of the tail three inches four lines; from the carpus to the end of the wing eight inches nine lines: breadth, wings extended, two feet nine inches. (*Female.*)

DESCRIPT. (*Male.*) The whole plumage, without exception, deep velvet-black, with glossy reflections: bill, and a globular protuberance at the base, black; the latter divided longitudinally by a mesial band of

orange yellow, which, passing onwards, spreads over half the bill, but
does not reach the tip by half an inch: irides brown: orbits yellow:
tarsi and toes dusky gray; membranes black.   (*Female.*) Crown, occiput
and nape, dusky brown; cheeks, throat and breast, dull brown, with a
mixture of ash-gray: back, wings, and under parts of the body, deep
brown, the feathers edged at the tips with whitish brown: bill somewhat
elevated at the base, but without the globular protuberance; of a blackish
colour; the nostrils, and a spot towards the tip, yellowish: orbits brown.
(*Young male, during the first year.*) " Resembles the adult female, but the
colours are somewhat paler: space between the eye and the bill, crown,
occiput, nape and breast, deep brown; beneath the eyes, sides and fore
part of the neck, pure white; the rest of the plumage sooty-brown: bill
elevated at the base; livid brown; the nostrils flesh-colour: legs dirty
yellowish green; membranes dusky." Temm.   (*Egg.*) Pale buff-colour:
long. diam. two inches six lines; trans. diam. one inch nine lines.

Like the last a winter visitant. Not uncommon on most parts of the
British coast, but seldom observed inland. Breeds in high northern
latitudes. Dives well, and feeds principally on shell-fish.

(20.) *O. perspicillata,* Steph. (*Surf Scoter.*) Faun. Bor. Amer.
part ii. p. 449. Selb. Illust. vol. ii. p. 335. pl. 69. *Anas persp.,*
Temm. Man. d'Orn. tom. ii. p. 853. Wils. Amer. Orn. vol. viii.
p. 49. pl. 67. f. 1.

According to Temminck, this species, which is plentiful in North America, has occurred occasionally in the Orkneys. It does not, however,
appear to have been met with by any of our own naturalists, nor am I
aware that any native specimens exist in our collections. It is distinguished
from the *O nigra,* which it most nearly resembles, by a frontal band, and
a large patch on the nape, of pure white: rest of the plumage black: no
speculum.

(21.) *O. leucocephala,* Steph. (*White-headed Scoter.*) Flem. Brit.
An. p. 119. *Anas leuc.,* Temm. Man. d'Orn. tom. ii. p. 859.

Dr Fleming is of opinion that the *female Scoter* described by Montagu in
the *Supp.* to the *Orn. Dict.* and the *White-throated Duck* of Pennant (*Brit.
Zool.* vol. ii. pl. 98.) are referable to this species, to which he has accordingly given a place in the British Fauna. It is, however, very probable,
that the individuals above alluded to may have been only immature specimens of the *O nigra,* which has generally more or less white on the head
and neck during the first year. There is no good authority for considering
the *O leucocephala* as British. It inhabits the eastern parts of Europe, and,
according to Temminck, is never found in Holland.

GEN. 96. FULIGULA, *Ray.*

251. F. *rufina,* Steph. (*Red-crested Pochard.*)—Head,
throat, and crest, reddish brown; lower part of the neck,
breast, belly and abdomen, deep black; flanks, a large
spot on the shoulders, and speculum on the wing, white:
bill and legs red.

F. rufina, *Steph. in Shaw's Gen. Zool.* vol. xii. part ii. p. 188. pl. 54.
Anas rufina, *Temm. Man. d'Orn.* tom. ii. p. 864. *Yarrell in Zool.
Journ.* vol. ii. p. 492. Red-crested Duck, *Lath. Syn.* vol. iii.
p. 544. Red-crested Pochard, *Selb. Illust.* vol. ii. p. 350. (Trachea,)
Linn. Trans. vol. xv. pl. 15. f. e.

DIMENS. Entire length twenty-one inches: length of the bill (from the forehead) two inches two lines, (from the gape) two inches three lines; of the tarsus one inch four lines and a half; from the carpus to the end of the wing ten inches.

DESCRIPT. (*Male.*) Head, cheeks, throat, and upper part of the neck, reddish brown or bay; the feathers on the crown elongated, and of a silky texture, forming a crest, somewhat paler than the rest of the head; lower part of the neck, breast, belly, and abdomen, deep black: back, wings, and tail, pale brown: flanks, bend of the wing, a large spot on the sides of the back, speculum, and basal part of the primary quills, white: bill, tarsi, and toes, bright red: nail of the bill white: membranes of the feet black: irides bright red. (*Female.*) " Crown, occiput and nape, deep brown; the crest less tufted; cheeks, throat, and sides of the neck, ash-brown; breast and flanks yellowish brown; belly and abdomen gray: back, wings, and tail, brown, with a slight tinge of ochre; no white spot on the sides of the back; speculum half grayish white, the other half pale brown; base of the quills white tinged with brown: bill, tarsi, and toes, reddish brown." TEMM. (*Immature male.*) Nape, fore part of the neck, and breast, dark brown; abdomen of a lighter brown; flanks white, tinged with pink: tail ash-brown; upper and under tail-coverts dark brown: legs and toes orange: the rest as in the adult. (*Egg.*) Uniform olive-brown: long. diam. two inches two lines; trans. diam. one inch six lines.

An immature male of this species, supposed to be in the second year, was shot near Boston, in January 1826, while feeding on fresh water in company with some Wigeons. It is now in the collection of Mr Yarrell. Since then a second individual has occurred in the London markets; a third has been shot at Yarmouth; and a fourth killed near Colchester. This last is in the Museum of the Cambridge Philosophical Society. Said to inhabit the north-eastern parts of Europe. Food, according to Temminck, shell-fish and aquatic vegetables.

252. F. *ferina*, Steph. (*Common Pochard.*)—Head and neck bright chestnut; breast black; flanks and scapulars with undulating lines of black and grayish white: no speculum: bill black, with a broad transverse blue band.

Anas ferina, *Temm. Man. d'Orn.* tom. II. p. 868.   Pochard, *Mont. Orn. Dict. & Supp. Bew. Brit. Birds*, vol. II. p. 353.   Red-headed Pochard, *Selb. Illust.* vol. II. p. 347. pl. 63. f. 1.   (Trachea,) *Linn. Trans.* vol. IV. pl. 14. f. 5.

DIMENS. Entire length nineteen inches: length of the bill (from the forehead) two inches, (from the gape) two inches three lines; of the tarsus one inch nine lines; of the tail two inches six lines; from the carpus to the end of the wing eight inches three lines: breadth, wings extended, twenty-nine inches seven lines.

DESCRIPT. (*Male.*) Head and neck bright chestnut-red: upper part of the back, breast, and rump, black; middle and lower regions of the back, scapulars, wing-coverts, flanks, and thighs, grayish white, with numerous fine undulating black lines; belly and abdomen whitish, very faintly undulated in the same manner, the lines becoming darker towards the vent: secondary quills bluish gray: primaries, and tail, dark cine-reous brown, approaching to dusky : upper and under tail-coverts black: bill black, with a broad transverse band in the middle of deep bluish gray: irides orange-yellow : legs bluish gray; membranes black. (*Female.*) Crown, nape, sides of the neck, and upper part of the back, reddish brown; throat, and fore part of the neck, white, mixed with reddish; breast reddish brown, the feathers edged and mottled with reddish white; flanks with large brown spots ; middle of the belly grayish white : back and wings much as in the male, but the undulating lines less distinct. *Young males of the year* resemble the female. In those of *one and two years*, the breast is dusky brown, and the head and neck not so bright as in the adult. (*Egg.*) Greenish white: long. diam. two inches three lines and a half; trans. diam. one inch seven lines and a half.

Not uncommon on the British coasts during the winter season. Has been known to breed in Norfolk, but the greater part retire northwards in the Spring for that purpose. Eggs twelve to fourteen in number. Food, small fish, marine plants, and insects.

253. F. *Nyroca*, Steph. (*Nyroca Pochard.*) — Head, neck, breast, and flanks, ferruginous: irides, and a spot on the chin, white: speculum white, edged with black beneath: bill and legs bluish gray.

F. Nyroca, *Steph. in Shaw's Gen. Zool.* vol. XII. part ii. p. 201. pl. 55. Anas leucophthalmos, *Temm. Man. d'Orn.* tom. II. p. 876. Olive-tufted Duck, *Sow. Brit. Misc.* pl. 21. Ferruginous Duck, *Penn. Brit. Zool.* vol. II. p. 601. pl. 99.? *Mont. Orn. Dict. Supp. with fig.* Castaneous Duck, *Mont. Orn. Dict. Supp. App. Bew. Brit. Birds,* vol. II. p. 315. Nyroca Pochard, *Selb. Illust.* vol. II. p. 352. pl. 63. f. 2. (Trachea,) *Mont. Orn. Dict. Supp. App.* last pl. f. 1.

DIMENS. Entire length (*male*) sixteen inches six lines, (*female*) eighteen inches. MONT.

DESCRIPT. (*Male.*) Head and neck dark ferruginous; on the lower part of the neck a narrow collar of blackish brown; back, scapulars, and wing-coverts, dusky brown, somewhat glossed with green and purple re-flections, the whole finely powdered with pale reddish brown : a whitish spot on the chin; lower part of the neck, breast, and flanks, bright fer-ruginous; belly and under tail-coverts pure white; abdomen and vent blackish gray, finely speckled with yellowish white: speculum white, edged with black at the lower part: primaries dusky: tail, and upper tail-coverts, dusky brown, with a slight dash of ferruginous: bill and legs bluish gray; the nail of the bill, and the membranes of the toes, black: irides white. (*Female.*) Head, neck, breast, and flanks, brown, all the feathers tipped with reddish; the collar on the lower part of the neck indistinct: upper parts dusky, the feathers tipped with pale brown: the rest as in the male. (*Young of the year.*) "Crown of the head dusky brown: all the feathers on the upper parts edged and tipped with reddish brown : the white on the belly clouded with pale brown." TEMM. (*Egg.*) "White, tinged with greenish." TEMM.

A rare species in this country, but met with occasionally during the winter months. Frequents rivers, extensive lakes, and other inland waters. Said to breed in marshes and on the rushy banks of rivers, laying from eight to ten eggs. Food, insects, aquatic vegetables, &c.

Obs. The *Collared Duck* of Donovan*, is either a female or a young male of this species, with a slight variation of plumage. The *Ferruginous Duck* of Pennant is also probably referable to this species in some one of its different states.

254. F. *dispar*†, Steph. (*Western Pochard.*)—White; beneath ferruginous: forehead, and occipital band, green; orbits, throat, collar and back, black.

> F. dispar, *Steph. in Shaw's Gen. Zool.* vol. XII. part ii. p. 206. Anas dispar, *Gmel. Syst.* tom. I. part ii. p. 535. Western Duck, *Penn. Arct. Zool.* vol. II. p. 289. pl. 23. *Lath. Syn.* vol. III. p. 532. *Id. Supp.* p. 275. *Shaw, Nat. Misc.* vol. I. pl. 32. Western Pochard, *Shaw, Gen. Zool.* l. c. *Selb. Illust.* vol. II. p. 360. pl. 66**.

DIMENS. Entire length seventeen inches. LATH.

DESCRIPT. (*Male.*) Crown, sides of the head and neck, and hind part of this last for half-way, white; across the forehead a band of pea-green; a transverse fascia on the nape of the same colour, but deeper, at the lower corner of which, on each side, is a round black spot of the size of a pea; behind the eye, another spot of the same colour, but irregular in shape: chin, throat, and fore part of the neck, black, communicating with a collar of the same which surrounds the neck about the middle: from the hind part of this the black passes down over the back, quite to the tail: breast and sides pale ferruginous, deepening into chestnut at the middle, growing still deeper as it passes on towards the vent, where the colour is black: wing-coverts white; primary quills dusky black: secondaries six inches long, and curving downwards, partly white, partly black, the colours divided obliquely on each feather; scapulars also long, and curving elegantly downwards over the greater coverts, as in the *Garganey;* each of these has the web next the body scarcely broader than the shaft itself, and both of them white; the other web very broad, and black: tail pointed, brown: bill and legs black. LATH. (*Female.*) The whole plumage mixed brown and ferruginous, not unlike that of the *Woodcock:* the quills all straight, and of a dusky colour; some of the secondaries with white tips, forming a spot on the wing; some of the wing-coverts also tipped with white: legs black. LATH. (*Egg.*) Unknown.

Only one individual of this species has hitherto occurred in Britain. This, which was a male specimen, is recorded to have been shot at Caistor near Yarmouth, in February 1830. It is now in the Norwich Museum. Said to inhabit Kamtschatka, and the western coast of America, breeding amongst rocks. Pennant states that it flies in flocks, and never enters the mouths of rivers.

255. F. *Marila*, Steph. (*Scaup Pochard.*) — Head and neck black, glossed with green; back and scapulars whitish, with undulating black lines; belly, flanks, and speculum on the wing, white; bill blue; legs ash-colour.

---

* *Brit. Birds*, vol. VI. pl. 147.

† I have followed Stephens and Selby in referring this species to the present genus, until its true situation be determined. It may be questioned, however, whether it be not more properly an *Oidemia*.

Anas Marila, *Temm. Man. d'Orn.* tom. ii. p. 865. Scaup Duck,
Mont. Orn. Dict. & Supp. Bew. Brit. Birds, vol. ii. p. 339.
Scaup Pochard, *Selb. Illust.* vol. ii. p. 354. pls. 66, & 66 *. (Trachea,)
Linn. Trans. vol. iv. pl. 14. f. 3, & 4.

Dimens. Entire length twenty inches: length of the bill (from the
forehead) one inch ten lines, (from the gape) two inches three lines; of
the tarsus one inch five lines and a half; of the tail two inches seven
lines; from the carpus to the end of the wing eight inches six lines.

Descript. (*Male.*) Head and upper part of the neck greenish black,
with glossy reflections; lower part of the neck, and breast, pitch-black:
upper part of the back, and scapulars, grayish white, with fine, distant,
transverse, undulating, black lines: first three primaries, and greater
wing-coverts, dusky; lesser coverts the same, but finely powdered with
white; the rest of the primaries, and all the secondaries, white, with
dusky tips, forming a bar across the wings when closed: rump, tail, and
upper tail-coverts, deep brown, approaching to black: belly and flanks
pure white, the latter slightly undulated with black: vent, and under
tail-coverts, spotted with brown: bill pale grayish blue; the nail black:
irides bright yellow: legs ash-gray; the membranes dusky. (*Female.*)
A broad, transverse band of yellowish white surrounding the base of the
bill; rest of the head, and neck, chocolate-brown; lower part of the neck,
breast, and rump, deep brown; back and scapulars dusky brown, with
minute specks and close undulating lines of grayish white; flanks spotted
and undulated with brown: irides dull yellow. (*Young male.*) More or
less similar to the adult female: some white feathers at the base of the
bill; the black on the head and neck without gloss, and mixed with feathers of a dusky brown colour: the white on the back variegated with
brown spots; and the undulating lines closer together than in the *adult
male:* belly dirty white, spotted with gray; flanks spotted with dusky
brown. (*Egg.*) Greenish yellow white: long. diam. two inches five lines;
trans. diam. one inch eight lines and a half.

A regular winter visitant, and not uncommon at that season on most
parts of the British coast. Occasionally met with inland on fresh waters.
Retires in the Spring to high northern latitudes to breed. Food, marine
plants and insects, fish, and molluscous animals. *Obs.* The *Anas fræ-
nata* of Sparmann* is the female of this species.

256. F. *cristata*, Steph. (*Tufted Pochard.*) — Head,
neck, crest, and upper parts, black, with glossy reflections;
belly, flanks, and speculum on the wing, white: bill and
legs bluish gray.

Anas Fuligula, *Temm. Man. d'Orn.* tom. ii. p. 873. Tufted Duck,
Mont. Orn. Dict. Bew. Brit. Birds, vol. ii. p. 370. Tufted
Pochard, *Selb. Illust.* vol. ii. p. 357. pl. 65.

Dimens. Entire length seventeen inches: length of the bill (from the
forehead) one inch seven lines, (from the gape) one inch nine lines and a
half; of the tarsus one inch four lines and a half; of the tail two inches
four lines; from the carpus to the end of the wing eight inches two lines:
breadth, wings extended, twenty-eight inches two lines.

Descript. (*Male.*) Head, neck, breast, and a pendent crest of elongated feathers on the occiput, black, with green and purple reflections:
back, wings, and rump, brownish black, slightly glossed with green, and

* *Sow. Brit. Misc.* pl. 62.

finely powdered with grayish white: belly, flanks, and a transverse bar
on the wings, pure white; vent black: bill pale bluish gray; the nail
black: irides gamboge-yellow: legs bluish gray; the membranes dusky.
(*Female.*) Smaller than the male; the crest shorter, and of not so bright
a colour: head, neck, breast, and upper part of the back, dull black,
clouded with deep brown; back and wings dull blackish brown, speckled
with grayish brown; breast and flanks with large spots of reddish brown;
belly whitish, tinged with reddish brown: speculum smaller than in the
male: bill and legs of a deeper colour: irides pale yellow. (*Young
of the year of both sexes.*) "No indication of a crest: a large whitish
spot on the sides of the bill; some white also on the forehead and occasionally behind the eyes: head, neck, and breast, dull brown, this last
variegated with reddish brown: back and wings dusky brown, the feathers edged with paler brown: flanks reddish brown: the bar on the
wings small and whitish: abdomen variegated with brown and ash-
colour: irides dirty yellow. *Young males* have the belly of a purer
white than *young females. After the first moult,* the white at the base
of the bill disappears; the crest begins to shew itself; and the plumage
becomes deeper coloured." TEMM. (*Egg.*) Greenish white, spotted with
light brown: long. diam. two inches one line; trans. diam. one inch five
lines.

Like the last species, a winter visitant. Met with on the coast, and
also not unfrequently on fresh waters. Said to breed within the Arctic
Circle. Food, similar to that of the preceding.

GEN. 97. CLANGULA, *Flem.*

257. C. *chrysophthalmos*, Steph. (*Golden-Eye Garrot.*)
—A large patch before the eye, lower neck, all the under
parts, and speculum on the wing, white: legs orange-
yellow.

Anas Clangula, *Temm. Man. d'Orn.* tom. II. p. 870. Golden-Eye,
Mont. Orn. Dict. & Supp. Bew. Brit. Birds, vol. II. p. 365.
Common Golden-Eye Garrot, *Selb. Illust.* vol. II. p. 367. pl. 62.
(Trachea,) *Linn. Trans.* vol. IV. pl. 15. f. 1, & 2.

DIMENS. Entire length eighteen inches six lines: length of the bill
(from the forehead) one inch three lines, (from the gape) one inch seven
lines; of the tarsus one inch six lines; of the tail three inches seven lines;
from the carpus to the end of the wing eight inches three lines.

DESCRIPT. (*Male.*) A large white space at the corner of the bill,
immediately beneath the lore; rest of the head, and upper half of the
neck, glossy black, with green and violet reflections; lower part of the
neck, breast, belly, abdomen, and flanks, pure white: back, rump, and
upper tail-coverts, deep glossy black: scapulars partly white and partly
black; primary quills, four outer secondaries, and coverts bordering the
wing, brownish black; rest of the secondaries, middle and greater coverts,
white: tail blackish gray: bill black: irides golden yellow: legs orange;
the membranes, and joints of the toes, dusky. (*Female.*) No white space
at the corner of the bill; the whole head, and upper part of the neck,
deep umber-brown; lower part of the neck, belly, and abdomen, pure
white; breast and flanks deep ash-colour, margined with grayish white:
all the upper parts grayish black, the feathers edged and tipped with ash:
greater coverts white, tipped with black; lesser coverts blackish gray,
tipped with white: bill black; the tip yellowish: irides and legs pale

yellow. *Young males of the year* resemble the adult female. *At the end of the first year*, the white patch at the base of the bill begins to shew itself, and the head becomes black, but without any glossy reflections. (*Egg.*) Buff-coloured white: long. diam. two inches four lines; trans. diam. one inch seven lines.

Met with in small flocks during the winter season, but not very plentiful. Frequents principally rivers, lakes, and other inland waters. Retires northward in the Spring to breed. Said to lay as many as fourteen eggs. Food, aquatic insects and vegetables, small fish, and mollusca. *Obs.* The *Anas Glaucion* of Linnæus may be referred either to the female, or to the young, of this species.

258.   C. *histrionica*, Steph. (*Harlequin Garrot.*)—A large patch on the lore, spot on the ears, longitudinal band on the sides of the neck, collar, and pectoral fascia on each side, white: speculum blue: legs dusky.

> C. histrionica, *Steph. in Shaw's Gen. Zool.* vol. XII. part ii. p. 180. pl. 57.   Anas histrionica, *Temm. Man. d'Orn.* tom. II. p. 878. *Sow. Brit. Misc.* pl. 6.   Harlequin Duck, *Mont. Orn. Dict.* Harlequin Garrot, *Selb. Illust.* vol. II. p. 371. pl. 60.

DIMENS. Entire length seventeen inches: length of the bill one inch six lines. MONT.

DESCRIPT. (*Male.*) Head and neck black, glossed with violet: a large patch between the bill and the eye, a spot behind the eyes, a longitudinal band on the sides of the neck, collar lower down, a broad crescent-shaped fascia on the sides of the breast, and part of the scapulars, all pure white: lower part of the neck, and breast, bluish ash; flanks ferruginous; belly brown: back, wings, and rump, glossy black, with blue and violet reflections: speculum on the wing very deep violet: bill black: irides brown: legs, and membranes of the toes, dusky blue. TEMM. (*Female.*) Smaller: all the upper plumage deep brown, clouded with ash-colour: towards the forehead, and a little before the eyes, a small white spot; a large patch of the same colour on the orifice of the ears: throat whitish; breast and belly whitish, clouded and spotted with brown; flanks reddish brown. TEMM. (*Egg.*) White, tinged with buff-colour: long. diam. two inches two lines; trans. diam. one inch seven lines.

A very rare occasional winter visitant. A pair, male and female, killed some years back in the north of Scotland, were in the collection of the late Mr James Sowerby. A young female has also been shot in one of the Orkney Islands. According to Messrs. C. and J. Paget*, it has also been obtained in one instance in the market at Yarmouth. Common in the northern parts of Europe and America. Said to breed on the borders of waters, amongst shrubs and herbage, and to lay from ten to twelve eggs. Food, shell-fish and insects. *Obs.* The *Anas minuta* of Linnæus is the female of this species.

> (22.) C. *albeola*, Steph. (*Buffel-headed Garrot.*) Faun. Bor. Amer. part ii. p. 458. *Anas albeola*, Wils. Amer. Orn. vol. VIII. p. 51. pl. 67. f. 2 & 3. Don. Brit. Birds, vol. X. pl. 226.
>
> Inserted as British by Donovan in his "British Birds," but without any authority being given. A common species in North America.

---

* *Nat. Hist. of Yarm.* p. 12.

GEN. 98. HARELDA, *Leach.*

259. H. *glacialis*, Steph. (*Long-tailed Hareld.*) —
Crown and nape white; a large chestnut patch on the
sides of the neck: bill black, with a transverse orange
band: legs yellow.

Anas glacialis, *Temm. Man. d'Orn.* tom. ii. p. 860. *Sabine in
Linn. Trans.* vol. xii. p. 555. Long-tailed Duck, *Mont. Orn.
Dict. & Supp. Bew. Brit. Birds*, vol. ii. p. 359. Long-tailed
Hareld, *Selb. Illust.* vol. ii. p. 363. pl. 61. (Trachea,) *Linn.
Trans.* vol. xii. pl. 30. f. 3, & 4.

DIMENS. Entire length, central tail-feathers included, twenty-two
inches. MONT.

DESCRIPT. (*Adult male in winter.*) Crown of the head, nape, front
and lower part of the neck, pure white; cheeks and chin ash-gray; on
each side of the upper part of the neck a large patch of chestnut-brown:
breast, back, wings, rump, and the two elongated central tail-feathers,
brownish black: scapulars, and lateral tail-feathers, white; the former
long, narrow, and acuminated: flanks cinereous; belly and abdomen
white: bill black, with a broad transverse band of deep orange before
the nostrils: irides reddish yellow: legs yellowish; the membranes of
the toes dusky. (*Adult female.*) " Forehead, chin, and eyebrows, whitish
ash ; crown of the head, and the large patch on the sides of the neck, dusky
ash: nape, front and lower part of the neck, as well as the belly and
abdomen, pure white: breast variegated with brown and dark ash-colour:
back, scapulars, and wing-coverts, blackish brown, the feathers edged and
tipped with reddish ash : the rest of the upper parts sooty brown: tail short,
the central feathers not elongated; blackish brown, edged with white:
bill bluish; the transverse band yellowish: irides light brown : legs bluish
gray." TEMM. (*Male in summer.*) "The whole under part of the neck,
and breast, black; sides of the head, and a little beyond the eye, brownish
white; round the eye some white feathers; from the bill a black line runs
on the top of the head to the crown, which is black; back of the neck chiefly
black, but at a small distance below the crown with a few white feathers
intermixed; the black of the back of the neck extends down the back, but
in the centre of the upper part of the back near the neck is a patch of
black feathers edged with ferruginous; scapulars long and narrow, black
in the centre, and edged with ferruginous white, the longer ones having
more white; wings brownish black, the quills being palest; lower belly
and sides to the rump, and under tail-coverts, white, a line of black descending between the white from the back to the tail." SAB. (*Young of
the year.*) Not very different from the adult female: face, crown, occiput,
and nape, grayish white, mottled with brown and ash-colour; on the
sides of the neck a large dusky patch; throat and upper part of the neck
in front, belly and abdomen, white; lower part of the neck in front, and
upper part of the breast, spotted and variegated with brownish ash; sides
of the breast with transverse crescent-shaped bars of the same colour;
back, and rest of the upper parts, deep brown; the scapulars edged with
ash-colour, and some of the shorter ones with white feathers intermixed;
tail short, the two middle feathers not elongated. (*Egg.*) Greenish
yellow white: long. diam. two inches two lines; trans. diam. one inch six
lines and a half.

A regular winter visitant in the Orkneys, and on the northern coasts
of England and Scotland. Seldom met with in the southern parts of

Britain; such individuals generally young birds. Breeds in the Arctic
Regions. Nest said to be placed amongst grass, near the edge of the
sea. Eggs, according to Temminck, five in number; according to others,
from ten to fourteen. Food, marine insects, and bivalve mollusca. *Obs.*
The *Anas hyemalis* of Linnæus is the male of this species in summer
plumage.

GEN. 99.   MERGUS, *Linn.*

260.   M. *Merganser,* Linn. (*Goosander.*) — Breast,
and speculum on the wing, white; the latter without
transverse bars: bill and legs red: crest (in the *adult
male*) short and bushy.

> M. Merganser, *Temm. Man. d'Orn.* tom. ii. p. 881.   Goosander,
> *Mont. Orn. Dict. & Supp. Selb. Illust.* vol. ii. p. 375. pl. 57.
> Goosander and Dun-Diver, *Bew. Brit. Birds,* vol. ii. pp. 250, &
> 253. (Trachea,) *Linn. Trans.* vol. xv. pl. 15. f. *h.*

DIMENS. Entire length twenty-four inches: length of the bill (from
the forehead) two inches four lines, (from the gape) three inches one
line; of the tarsus two inches one line; from the carpus to the end of the
wing ten inches six lines. (*Young male.*) The *adult* sometimes attains a
length of twenty-nine inches.

DESCRIPT. (*Adult male.*) Head and upper part of the neck glossy
greenish black; the feathers on the crown and occiput elongated, forming
a short crest: lower part of the neck, breast, belly, and abdomen, cream-
yellow, fading after death to pure white: upper part of the back, inner
scapulars, humeral wing-coverts, bastard winglet, basal halves of the
greater coverts, and fourteen outer quills, black; outer scapulars, all
the lesser coverts except the humeral ones, six of the secondary quills,
and tips of the greater coverts immediately above them, white: lower
back, rump, and tail, deep ash-gray: bill vermilion red; the ridge and
nail black: irides red: legs vermilion. (*Adult female.*) Crown, and occipital crest (the latter longer and more slender than in the male), ferruginous brown; rest of the head, and upper part of the neck, bright ferruginous; chin and throat pure white; lower part of the neck before, sides
of the breast, flanks, and thighs, white and ash-colour mixed; belly and
abdomen yellowish white: back, scapulars, wing-coverts, and tail, deep
ash-colour: primaries dusky; six of the secondaries, and tips of the impending coverts, white, forming a large speculum: bill dull red; the
ridge and nail blackish: irides brown: legs yellowish red. *The young
males of the year* resemble the adult female. *At the end of the first
year,* the white on the throat begins to be spotted with black; the crown
becomes dusky, and the ferruginous brown on the neck is edged at
bottom with black; the white also appears on the wing-coverts. (*Egg.*)
Pale olive white, tinged with buff: long. diam. two inches six lines; trans.
diam. one inch eight lines.

Found in the Orkneys, and some other of the Scotch Islands, throughout the year. In England, only a winter visitant, and seldom seen at all
in the more southern districts except in severe seasons. Frequents lakes,
large rivers, and estuaries. Nest said to be placed amongst loose stones
near the edge of the water. occasionally in bushes, or in hollow trees.

Eggs twelve to fourteen in number. Food, fish and amphibious reptiles.
Obs. The *Dun-Diver* of English authors (*M. Castor*, Linn.) is the female
of this species.

261. M. *Serrator*, Linn. (*Red-breasted Merganser.*)
—Breast (in the *male*) reddish brown: speculum white,
divided by one or two transverse bars: bill and legs red:
crest (in the *adult male*) long, slender, pendent.

> M. Serrator, *Temm. Man. d'Orn.* tom. ii. p. 884. Red-breasted
> Merganser, *Mont. Orn. Dict. Bew. Brit. Birds*, vol. ii. p. 257.
> *Selb. Illust.* vol. ii. p. 379. pls. 58, & 58*. (Trachea,) *Linn. Trans.*
> vol. iv. pl. 16. f. 1, & 2.

DIMENS. Entire length twenty-two inches.

DESCRIPT. (*Adult male.*) Head, a pendent crest on the occiput, and
upper part of the neck, greenish black, with glossy reflections; lower
part of the neck white, the mesial line behind black; breast reddish
brown, spotted and variegated with black; near the insertion of the
wing, several large white spots bordered with black; upper part of the
back, scapulars, edge of the wing and adjoining coverts, and primary
quills, black; the rest of the coverts, and secondary quills, white, but
these last, as well as the greater coverts, black at the base, the whole
together forming a large speculum divided by two transverse bars; some
of the secondaries nearest the body white edged with black: lower part
of the back, rump, and flanks, grayish white, with fine transverse undulating black lines; belly and abdomen white; tail brown: bill orange-
red; the ridge and nail dusky: irides red: legs orange. (*Adult female.*)
Head, crest, and neck, dull ferruginous; throat white; fore part of the
neck and breast variegated with white and ash-gray; lower part of the
neck behind, back, rump, wings, and scapulars, dark ash-colour; speculum white, with one transverse bar, by which it may always be distinguished from the female of the last species, which in other respects
it greatly resembles; belly and abdomen white; flanks dark ash-colour:
bill and legs dull orange: irides brown. (*Young male of the year.*)
Head deep brown; throat grayish white; bill pale red; irides yellowish: in other respects like the adult female. (*Egg.*) Smooth and
shining; buff-colour: long. diam. two inches six lines; trans. diam.
one inch nine lines.

A winter visitant in England, and not often met with in the southern
counties. More abundant northward. Breeds in some of the Scotch lakes.
Nest, according to Selby, composed of small sticks and grass, lined with
the down of the female; built in the thick herbage on the edge of the
main land, or on small islets in the lake. Eggs nine in number. Food
and habits similar to those of the last species.

262. M. *cucullatus*, Linn. (*Hooded Merganser.*) —
Head, neck, and upper parts, dusky brown ; beneath white :
bill and legs reddish brown: crest (in the *adult male*)
semicircular, compressed.

> M. cucullatus, *Faun. Bor. Amer.* part ii. p. 463. Hooded Mer-
> ganser, *Lath. Syn.* vol. iii. p. 426. pl. 101. *Wils. Amer. Orn.*
> vol. viii. p. 79. pl. 69. f. 1. *Selb. in Trans. of Nat. Hist. Soc.*
> *of Newcast.* vol. i. p. 292. *Id. Illust.* vol. ii. p. 383. pl. 58**.

Dimens. Entire length eighteen inches: length of the bill (from the forehead) one inch eight lines, (from the gape) two inches; of the tarsus one inch three lines; of the tail three inches ten lines; from the carpus to the end of the wing seven inches six lines.

Descript. (*Adult male.*) Head and neck brownish black; the former furnished with a large, erect, compressed, semicircular crest, extending from the forehead to the occiput; frontal portion of the crest, space above the eyes, and a narrow edging all round, black; the remaining part of the crest, and a space on the sides of the occiput commencing five lines behind the eyes, white: back, rump, tail, primaries, and lesser wing-coverts, dark cinereous brown, inclining in some places to dusky: greater coverts, and secondary quills, black at the base, the tips of the former and outer borders of the latter white; the whole together forming a speculum consisting of four bars, black and white alternately: scapulars long and pointed, greenish black, streaked longitudinally down the middle with white: sides of the breast with two or three crescent-shaped transverse black bars; flanks, and sides of the body near the bend of the wing, finely undulated with black and yellowish brown; rest of the under parts pure white: bill and legs reddish brown. (*Adult female.*) Forehead, sides of the head, and all the neck, dark ash-colour; crest shorter than in the male, reddish ash, without any white: all the upper plumage dark brown, approaching to dusky, with a tinge of cinereous: upper part of the breast, and flanks, lead-gray, the feathers edged with white: the rest much as in the male. (*Young female?*) "Chin grayish white, speckled with grayish brown; the whole of the face, cheeks, and neck, of an uniform grayish brown, or mouse-colour; crown of the head darker; occipital crest pale reddish brown, tinged with gray; breast gray, the margins of the feathers paler; upper back and wing-coverts grayish black, the feathers margined with obscure grayish brown; scapulars and lower back black; margins of four or five of the secondary quills white, forming a small spot in the middle of the wings; quills and tail grayish black; lower part of the breast, belly, and abdomen, pure white, with a silken lustre; sides and flanks deep grayish brown; bill reddish brown at the base, the tip black; legs reddish brown." Selb. (*Egg.*) White. Lath.

A single individual of this species, supposed to be a young female, of which the above is a description borrowed from Selby, was killed near Yarmouth in Norfolk, in the Winter of 1829. It is a native of North America, and had never before been seen in Europe. According to Latham, appears at Hudson's Bay as a summer visitant at the end of May, and builds close to the lakes. Nest composed of grass, lined with feathers from the breast. Eggs four to six in number. Frequents fresh waters.

263. M. *albellus*, Linn. (*Smew.*)—Head, neck, and all the under parts, white; cheeks and occiput greenish black: bill and legs bluish gray: crest (in the *adult male*) bushy, moderately elongated.

M. albellus, *Temm. Man. d'Orn.* tom. ii. p. 887. Smew, *Mont. Orn. Dict. & Supp. Bew. Brit. Birds*, vol. ii. p. 260. *Selb. Illust.* vol. ii. p. 385. pl. 59. (Trachea,) *Linn. Trans.* vol. iv. pl. 16. f. 3, & 4.

Dimens. Entire length seventeen inches: length of the bill (from the forehead) one inch three lines, (from the gape) one inch eight lines; of the tarsus one inch six lines.

DESCRIPT. (*Adult male.*) A large patch on each side of the base of the bill enclosing the eyes, and another longitudinal one on the occiput, black glossed with green; rest of the head, occipital crest, neck, some of the lesser wing-coverts, and all the under parts, pure white: two crescent-shaped fasciæ advancing forwards from the shoulders on each side, and partly encircling, one the lower part of the neck, the other the upper part of the breast, black; back, lesser coverts bordering the wing, and primary quills, black; scapulars white, edged on the outer webs with black: secondary quills and greater coverts black, tipped with white: tail and upper tail-coverts bluish gray: flanks and thighs with transverse undulating lines of black: bill and legs deep bluish gray; the membranes of the toes dusky. (*Female.*) Crown, cheeks, and occiput, reddish brown; crest shorter than in the male; throat, sides and front of the upper part of the neck, belly, and abdomen, white; lower part of the neck, breast, and flanks, clouded with ash-colour: back, tail, and upper tail-coverts, deep ash-gray: wings much as in the male, only the dark parts gray instead of black. *The young of the year* resemble the adult female. *After the second moult*, the male begins to shew a few black feathers on the sides of the face, the first indication of the large patch which characterizes that part in the adult; some white appears on the crown and occiput; the back becomes partially black; and the two crescent-shaped fasciæ are faintly traced out on the sides of the breast. (*Egg.*) " Whitish." TEMM.

A winter visitant, but not of very frequent occurrence. Principally met with in the neighbourhood of fresh waters. Breeds in high latitudes on the borders of lakes and rivers. Eggs, according to Temminck, from eight to twelve in number. Food, fish, and aquatic vegetables. *Obs.* The *Minute Merganser* of Montagu and other English authors (*M. minutus,* Linn.) is referable to the female and young of this species.

GEN. 100.   PODICEPS, *Lath.*

264.   P. *cristatus*, Lath. (*Great Crested Grebe.*) — Bill longer than the head, reddish, the tip white; distance from the nostrils to the tip about eighteen lines.

P. cristatus, *Temm. Man. d'Orn.* tom. II. p. 717.   Crested Grebe, *Mont. Orn. Dict. Selb. Illust.* vol. II. p. 394. pl. 73.   Great Crested Grebe, *Bew. Brit. Birds,* vol. II. p. 153.

DIMENS.   Entire length twenty-one inches: length of the bill, from the forehead, two inches.

DESCRIPT. (*Adult in perfect plumage.*) Streak above the eye, cheeks, and throat, white: crown of the head, occipital crest, and a large ruff of feathers standing out round the neck, glossy black, passing into reddish yellow on the sides of the head: fore part of the neck, and all the under parts of the body, silvery white: hind part of the neck, back, scapulars, and wing-coverts, dusky brown: secondaries pure white, forming an oblique bar across the wings: naked space between the bill and the eye red: bill reddish, dusky brown on the ridge, the tip grayish white: irides crimson red: legs dusky; internally yellowish white. (*From two to three years of age.*) " Occipital crest very short, edged with white; face white, without the reddish tinge; the feathers of the ruff very short: a dusky streak passing from the bill beneath the eyes, and reaching to the occiput. (*Young before the age of two years.*) No occipital crest or ruff round the neck: forehead and face white; these parts as well as the upper part of the neck with zigzag streaks of dusky brown disposed in every direction: irides pale yellow. *The young of the year before the first moult* have the head and upper part of the neck deep brown." Temm. (*Egg.*) White: long. diam. two inches two lines; trans. diam. one inch six lines.

Not an uncommon species, remaining with us the whole year. Frequents lakes and extensive tracts of fen; during severe weather in Winter, the sea-coast and the estuaries of large rivers. Is said to breed in the meres of Shropshire and Cheshire, and also in Lincolnshire. Makes a large floating nest of aquatic plants. Eggs four in number. Food, principally fish. Dives well, and is rarely seen on land. *Obs.* The *Tippet Grebe* of Pennant (*C. Urinator*, Linn.) is this species in its immature state.

265. P. *rubricollis*, Lath. (*Red-necked Grebe.*)—Bill of the length of the head, black, towards the base yellow; distance from the nostrils to the tip thirteen or fourteen lines: no ruff; occipital crest very short.

P. rubricollis, *Temm. Man. d'Orn.* tom. ii. p. 720. Red-necked Grebe, *Mont. Orn. Dict. & Supp. Bew. Brit. Birds*, vol. ii. p. 161. *Selb. Illust.* vol. ii. p. 392. pl. 72.

Dimens. Entire length seventeen inches six lines: length of the bill (from the forehead) one inch nine lines, (from the gape) two inches two lines; of the tarsus two inches two lines; from the carpus to the end of the wing seven inches two lines: breadth, wings extended, twenty-nine inches three lines.

Descript. (*Adult.*) Forehead, crown, and occipital tuft, blackish brown: cheeks and throat ash-gray: back part of the neck, and all the upper parts of the body, deep blackish gray: front of the neck, sides and upper part of the breast, bright orange-brown; rest of the under parts white; flanks and thighs with a few dusky streaks: lower half of the secondary quills white: bill black, with the base of both mandibles bright gamboge-yellow: irides reddish brown: legs externally black; internally yellowish green. (*Young, at the age of two years.*) Throat and cheeks white: forehead, crown, and rest of the upper parts, dusky brown, without any elongated feathers on the occiput: sides of the neck, and upper part of the breast, pale reddish, mixed with dusky and brown; rest of the under parts white, faintly spotted with dusky on the flanks and towards the vent: base of the bill livid yellow. (*Egg.*) Dirty white: long. diam. one inch ten lines; trans. diam. one inch four lines.

Of rare occurrence in this country, particularly in the adult state. Abundant in the eastern parts of Europe. Frequents rivers and lakes, and also the sea-coast. Food, small fish, aquatic insects and vegetables. Nest said to resemble that of the last species. Eggs three or four in number.

266. P. *cornutus*, Lath. (*Sclavonian Grebe.*) — Bill strong, shorter than the head, compressed throughout its length; black, the tip red: eyes with a double iris: distance from the nostrils to the tip of the bill seven lines.

P. cornutus, *Temm. Man. d'Orn.* tom. ii. p. 721. Sclavonian Grebe,
Mont. Orn. Dict. & Supp. with fig. Horned Grebe, *Selb. Illust.*
vol. ii. p. 397. pl. 74.

Dimens. Entire length thirteen inches six lines: length of the bill
(from the forehead) ten lines and a half, (from the gape) one inch four
lines; of the tarsus one inch eight lines: breadth, wings extended, twenty-
two inches.

Descript. (*Adult.*) Forehead, crown, and an ample ruff surrounding
the upper part of the neck, deep glossy black; above and behind the eyes
two large spreading tufts of feathers of an orange-red colour, resembling
horns: nape, and all the upper parts of the body, dusky brown: secondaries
white: front of the neck, and breast, bright ferruginous chestnut; rest
of the under parts pure white, with the exception of the flanks, which are
tinged with ferruginous: base of the bill, and naked space between it and
the eyes, rose-red; rest of the bill black, the extreme tip red: irides with
a double circle; the inner one yellow; the outer and broader one bright
red: legs externally black, internally gray. (*Young, during the first
year.*) Head, nape, and all the upper parts, dusky ash, deepest on the
head, where it approaches to black, with a slight gloss of green; no indications of a ruff or tufts above the eyes: throat and cheeks pure white;
a narrow line of the same colour extends from behind the ears on each
side to the back of the head: middle of the fore part of the neck ash-
colour; sides of the breast, and flanks, tinged with dusky; rest of the
under parts, as well as the secondary quills, pure white: bill dusky, with
the sides of the upper, and base of the under, mandible, of a livid flesh-
colour, the tip pale brown: the inner circle of the iris white; the outer
one red: legs externally dusky; internally bluish ash. (*Egg.*) Dirty
greenish white: long. diam. one inch nine lines; trans. diam. one inch
three lines.

More common than the last species, but not very plentiful, at least in
the adult state. Found on fresh waters, as well as on the sea-coast. Food,
similar to that of the two preceding species. Constructs a floating nest
amongst reeds and rushes, and lays three or four eggs. *Obs.* The second
of the two states of plumage described above is that of the *Dusky Grebe*
of Pennant, (*P. obscurus,* Lath.) which Temminck considers as the young
of this species. Not being prepared to prove to the contrary, I have followed him in this last instance. Yet the correctness of such an opinion
may reasonably be called in question on the ground of anatomical differences observed by Mr Yarrell. This last gentleman remarked that in
an adult male specimen of the *P. cornutus,* the stomach was membrano-
muscular, and the cæcal appendages only an inch and a half in length;
whereas in a specimen of the *P. obscurus* dissected by him, the stomach
was more decidedly muscular, and the cæcal appendages nearly five inches
in length. These facts appear to indicate a difference of species.

267. P. *auritus,* Lath. (*Eared Grebe.*)—Bill shorter
than the head, depressed at the base, slightly recurved at
the tip, black: iris single: distance from the nostrils to the
tip of the bill six lines and a half.

P. auritus, *Temm. Man. d'Orn.* tom. ii. p. 725. Eared Grebe,
Mont. Orn. Dict. & Supp. with fig. Bew. Brit. Birds, vol. ii.
p. 157. *Selb. Illust.* vol. ii. p. 399. pl. 74.

DIMENS. Entire length twelve inches: length of the bill (from the forehead) eleven lines, (from the gape) one inch one line and a half; of the tarsus one inch six lines and a half; from the carpus to the end of the wing five inches.

DESCRIPT. (*Adult.*) Cheeks, forehead, crown, a short occipital crest, and upper part of the neck, deep black: sides of the head ornamented with two tufts of long slender feathers, of a pale yellow passing into deep orange; these tufts arise from behind the upper part of the eye, and passing backwards cover the orifice of the ears: throat, lower part of the neck, sides of the breast, and all the upper parts, dull black: secondary quills white: flanks and thighs deep chestnut-red, mixed with dusky; the rest of the under parts pure white: bill black, reddish at the base: irides vermilion: legs externally dusky ash, internally greenish ash. (*Young of the year.*) Closely resembling the young of the last species, but may always be distinguished by the peculiar shape of the bill, which is depressed at the base, and slightly recurved at the tip: the iris is also of one colour, and the white on the cheeks more extended, reaching down the sides of the neck. (*Egg.*) Dirty yellowish white: long. diam. one inch nine lines; trans. diam. one inch three lines.

A rare species in this country, particularly in the adult state. Met with principally in the winter months, but, according to Montagu, breeds in the fens of Lincolnshire. Chiefly frequents lakes and inland waters, occasionally the estuaries of rivers. Nest placed amongst reeds and other aquatic herbage. Eggs three or four in number.

268.   P. *minor*, Lath. (*Little Grebe.*) — Bill very short, compressed; distance from the nostrils to the tip six lines: no ruff or occipital crest: tarsi furnished posteriorly with a double row of serratures.

P. minor, *Temm. Man. d'Orn.* tom. II. p. 727. Colymbus Hebridicus, *Sow. Brit. Misc.* pl. 70. Little and Black-chin Grebes, *Mont. Orn. Dict. & Supp. Bew. Brit. Birds,* vol. II. p. 163. Little Grebe, *Selb. Illust.* vol. II. p. 401. pl. 75.

DIMENS. Entire length nine inches ten lines: length of the bill one inch: of the tarsus one inch three lines; from the carpus to the end of the wing four inches three lines.

DESCRIPT. (*Adult in summer.*) Throat, crown of the head, and nape, deep black; cheeks, sides and fore part of the neck, bright chestnut: upper parts of the body dark brown, tinged with olivaceous green; rump inclining to reddish brown: primary quills brownish ash; secondaries white at the base and on their inner webs: under parts dusky ash, darkest on the breast, vent, and flanks; thighs tinged with reddish: bill black; the extreme tip, and base of the lower mandible, together with the lore, yellowish white: irides reddish brown: legs externally dusky olive, internally livid flesh-colour. (*Young of the year.*) Crown of the head, nape, and all the upper parts of the body, a mixture of brown and ash-colour, with a slight tinge of rufous: throat pure white: sides of the neck reddish ash; fore part of the neck, upper part of the breast, and flanks, the same but paler; belly and abdomen white: bill shorter than in the adult; the lower mandible, and edges of the upper mandible, yellowish white; rest of the bill deep brown: irides brown. (*Egg.*) Smooth; white: long. diam. one inch seven lines; trans. diam. one inch one line.

The most abundant species in the genus. A frequent inhabitant of ponds, marshes, lakes, and the banks of rivers. Rarely observed on the sea-coast. Feeds on *Dyticidæ* and other aquatic insects. Is an excellent diver. Constructs a large floating nest, which it attaches to the stems of rushes. Eggs four to six in number.

GEN. 101.　COLYMBUS, *Lath.*

269. C. *glacialis*, Linn. (*Northern Diver.*) — Head and neck violet-black, the latter with a double interrupted white collar : bill upwards of four inches in length ; the upper mandible nearly straight.

C. glacialis, *Temm. Man. d'Orn.* tom. II. p. 910.　Northern Diver, *Mont. Orn. Dict. Selb. Illust.* vol. II. p. 406. pl. 76.　Great Northern Diver, *Bew. Brit. Birds*, vol. II. p. 174.

DIMENS.　Entire length thirty-three inches : length of the bill (from the forehead) two inches eleven lines, (from the gape) four inches one line ; of the tarsus three inches ; of the middle toe, nail included, four inches two lines ; of the tail three inches six lines ; from the carpus to the end of the wing thirteen inches six lines.

DESCRIPT.　(*Adult.*) Head and neck velvet black, with green and purple reflections ; beneath the throat a transverse semilunar white band with black streaks ; on the lower part of the neck an interrupted broad collar, streaked longitudinally with black and white : all the upper parts black, elegantly spotted with white ; the spots on the back and rump small and round ; those on the scapulars larger and of an oblong-square form, disposed in rows, two at the extremity of each feather : quills and tail without spots : breast and other under parts white ; flanks and sides of the breast with a few black streaks : bill black ; the ridge of the upper mandible very slightly arched above ; lower mandible channelled beneath, appearing deepest in the middle ; the gonys sloping upwards to the tip ; tomia of both mandibles, but particularly of the lower one, inflected : irides brown : feet externally dusky brown ; the inner portions, and membranes between the toes, whitish. (*Young of the year.*) Head, occiput, and all the back part of the neck, cinereous brown ; cheeks speckled with white and ash-colour ; throat, front of the neck, and all the other under parts, pure white ; the feathers on the back, wings, rump, and flanks, deep brown in the middle, the edges and tips ash-colour : upper mandible cinereous gray ; lower mandible whitish. *At the age of one year*, a transverse bar of dusky brown appears on the middle of the neck, forming a sort of collar ; the back assumes a dusky tint, and the small white spots begin for the first time to shew themselves. *At the end of the second year*, the dusky bar on the neck extends further ; the brown on the head and nape becomes mixed with greenish black ; the spots on the back and wings increase in number, and the band on the throat, and collar on the lower part of the neck, are more distinctly indicated by longitudinal streaks of brown and white. *At the age of three years*, the plumage is perfect. (*Egg.*) Dark olive-brown, with a few spots of umber-brown : long. diam. three inches six lines ; trans. diam. two inches three lines.

A native of high northern latitudes.　Abundant in the Orkneys, Hebrides, and other Scotch islands, but very rare in the southern parts of Britain. Has been taken in Cornwall, Bedfordshire, and occasionally, in the immature state, on different parts of the English coast. Seldom flies, but dives well. Feeds on fish, particularly herrings. Nest said to be placed on the borders of fresh-water lakes, or in small islands. Eggs two in number. *Obs.* In its immature state, this species is the *C. Immer* of authors, but it is probable that in some cases this name has been also applied to the young of the next species.

270. C. *arcticus*, Linn. (*Black-throated Diver.*) — Occiput and nape cinereous brown; front of the neck violet-black: bill upwards of three inches in length; the upper mandible slightly curved.

C. arcticus, *Temm. Man. d'Orn.* tom. II. p. 913. Black-throated Diver, *Mont. Orn. Dict. Bew. Brit. Birds,* vol. II. p. 181. *Selb. Illust.* vol. II. p. 411. pl. 77.

DIMENS. Entire length from twenty-four to twenty-six inches. TEMM.

DESCRIPT. (*Adult plumage.*) Head and nape cinereous brown, passing into black on the forehead; throat and fore part of the neck black, with purple and green reflections; immediately beneath the throat a narrow transverse white band longitudinally streaked with black; on the sides of the neck a broader band extending downwards from the ears, streaked longitudinally with black and white; lower part of the neck, and sides of the breast, streaked in a similar manner: back, rump, and flanks, deep black, without spots; scapulars with twelve or thirteen transverse bars of pure white; wing-coverts black, speckled with white: breast and under parts of the body pure white: quills dusky: tail grayish black; under tail-coverts barred with black: bill dusky; the upper mandible more curved than in the last species; the lower mandible not thickened in the middle, and without the groove: irides brown: feet externally brown; internally whitish; membranes whitish. (*Young of the year.*) Only to be distinguished from those of the last species by their inferior size, and characters of the bill as above pointed out: plumage almost entirely similar: there is generally, however, a dusky band running longitudinally down the sides of the neck, which is not present in the young of *C. glacialis. At the age of one year,* a few black feathers begin to appear on the throat, as well as on the back, rump, and flanks; there is also a slight indication of the broad band of black and white streaks on the sides of the neck. *At the end of the second year,* the colour of the forehead begins to deepen; the violet-black extends itself over the throat and fore part of the neck, still however mixed with a few white feathers; the band of longitudinal streaks appears more distinct; and the black and white on the upper parts becomes tolerably well-defined. *At the age of three years,* the plumage is generally matured. (*Egg.*) Dark olive-brown, thinly spotted with umber-brown: long. diam. two inches nine lines; trans. diam. one inch ten lines.

Inhabits the Arctic Regions, but appears occasionally as a winter visitant on different parts of the British coast, as well as on lakes and rivers inland. Feeds on fish, frogs, insects, and aquatic vegetables. Nest placed amongst reeds and rushes in marshy situations, or on the banks of rivers. Eggs two in number. *Obs.* The *Lesser Imber* of Bewick is probably the young of this species.

271. C. *septentrionalis*, Linn. (*Red-throated Diver.*)
—Occiput and nape with black and white streaks; front
of the neck deep orange-brown : bill scarcely, or not exceeding, three inches; slightly recurved; tomia very much
inflexed.

C. septentrionalis, *Temm. Man. d'Orn.* tom. II. p. 916.   Red-throated
    Diver, *Mont. Orn. Dict. & Supp.   Bew. Brit. Birds*, vol. II. p. 177.
    Selb. Illust. vol. II. p. 414. pls. 78, & 78*.

Dimens. Entire length twenty-six inches three lines : length of the
bill (from the forehead) one inch eleven lines, (from the gape) two inches
eleven lines and a half; of the tarsus two inches ten lines; of the tail
three inches three lines; from the carpus to the end of the wing eleven
inches three lines : breadth, wings extended, three feet four inches seven
lines.

Descript. (*Adult.*) Sides of the head, throat, and sides of the neck,
mouse-gray; crown spotted with black; occiput, nape and lower part of
the neck, marked with longitudinal streaks of black and white; on the
front of the neck a large patch of deep orange-brown; breast and belly
silvery white; flanks and thighs streaked with dusky : back, and all the
other upper parts dusky brown, without spots in very old individuals, but
in birds of three and four years of age, with small indistinct whitish spots :
bill black; the commissure quite straight, but the upper mandible being
considerably depressed above the nostrils, the ascending gonys of the
lower one causes the bill, when closed, to appear slightly recurved; tomia
of both mandibles greatly inflexed : irides orange-brown; feet externally
greenish gray; internally livid white; membranes the same. (*Young of
the year*.) Forehead, crown and nape, dusky ash, finely streaked with
grayish white; back, scapulars and rump, dusky brown, copiously
sprinkled with small oval white spots arranged along the edges of the
feathers; wing-coverts bordered with white towards their extremities :
throat, sides and fore part of the neck, breast and belly, white; flanks
streaked with dusky : quills and tail blackish gray : bill grayish white;
the ridge of the upper mandible dusky : irides brown : legs externally
dark greenish gray; internally, as well as the webs, greenish white. *At
the end of the first year*, a few red feathers begin to show themselves on
the fore part of the neck; and the white spots on the upper parts become
smaller and less distinct. *After the second moult*, the patch of orange-
brown is nearly complete, but still mixed with a little white : the spots on
the upper parts gradually disappear as the bird advances to maturity.
(*Egg.*) Chestnut-brown, rather thickly spotted with dark umber-brown :
long. diam. two inches eight lines; trans. diam. one inch ten lines.

A winter visitant on the English coast of not unfrequent occurrence,
particularly in the immature state. Remains all the year in the Orkney
and Shetland Islands, where it breeds. Nest placed on the extreme
borders of lakes, composed of rushes and dry grass. Eggs two in number.
Food similar to that of the last species. Flies well, and is also an
excellent diver. *Obs.* The *Speckled Diver* of Pennant and Montagu
(*C. stellatus*, Gmel.) is the young of the year of this species. The
Second Speckled Diver of Bewick is perhaps the same bird a little
nearer advanced to maturity.

R

GEN. 102. URIA, *Briss.*

272. U. *Troile*, Lath. (*Foolish Guillemot.*)—Above brownish black; secondaries tipped with white: bill as long as, or somewhat longer than, the head; very much compressed throughout: feet dusky.

U. Troile, *Temm. Man. d'Orn.* tom. II. p. 921. Foolish Guillemot, *Mont. Orn. Dict. & Supp. Bew. Brit. Birds,* vol. II. p. 166. *Selb. Illust.* vol. II. p. 420. pl. 79.

DIMENS. Entire length seventeen inches nine lines: length of the bill (from the forehead) one inch eleven lines, (from the gape) two inches eleven lines; of the tarsus one inch seven lines; of the tail two inches; from the carpus to the end of the wing six inches two lines: breadth, wings extended, two feet.

DESCRIPT. (*Adult in winter.*) Forehead, crown, space between the eye and the bill, a longitudinal streak running backwards from the eyes to the distance of an inch and a half, and all the upper parts, including the wings, brownish black with a slight tinge of cinereous: occiput grayish white, opening on each side into a white patch above the longitudinal streak behind the eyes: secondaries tipped with white: all the under plumage pure white; the black on the sides of the neck advancing in front towards the lower part, and forming a faint collar of grayish ash: bill black; very much compressed throughout its whole length: inside of the mouth dull yellow: irides brown: feet and toes yellowish brown; posterior part of the tarsus and membranes black. (*Summer plumage.*) The whole of the head, including the region of the eyes, throat, and all the upper part of the neck, pitchy brown: inside of the mouth bright yellow: the rest as in winter. (*Young of the year.*) Resemble the adult in winter, but distinguished by the shorter bill: the longitudinal streak behind the eyes is also indistinct; and the white on the under parts less pure, with more of cinereous brown on the lower part of the neck: tarsi and toes dull yellowish; membranes brown. (*Egg.*) Very variable in colour, scarcely two being found precisely alike; generally bluish green, more or less blotched and streaked with black: long. diam. three inches two lines; trans. diam. one inch eleven lines.

Met with in great abundance on various parts of the British coast throughout the year. Breeds on cliffs and rocky islands, generally in large companies. Lays a single egg. Food, fish, marine insects, and small bivalve mollusca. *Obs.* The *Lesser Guillemot* of English authors is the young of the year of this species.

(23.) *U. Brunnichii*, Sab. in Linn. Trans. vol. XII. p. 538. Temm. Man. d'Orn. tom. II. p. 924. *Franks' Guillemot,* Shaw, Gen. Zool. vol. XII. p. 243. pl. 62.

According to the statement of a writer in the *Quarterly Review,* (vol. XLVII. p. 354.) this species has been killed off one of the Shetland Isles. It is distinguished from the last, principally by the form of the bill, which is shorter and more dilated at the base.

273. U. *Grylle*, Lath. (*Black Guillemot.*) — Above black; a large white patch on the middle of the wings: legs red.

U. Grylle, *Temm. Man. d'Orn.* tom. ii. p. 925. Black Guillemot, *Mont. Orn. Dict. & Supp. Bew. Brit. Birds,* vol. ii. p. 170. *Selb. Illust.* vol. ii. p. 426. pl. 80.

Dɪᴍᴇɴs. Entire length thirteen inches six lines: length of the bill (from the forehead) one inch two lines, (from the gape) one inch five lines; of the tarsus one inch two lines; of the tail one inch nine lines; from the carpus to the end of the wing six inches seven lines.

Dᴇsᴄʀɪᴘᴛ. (*Adult in winter.*) Crown, nape, and all the rest of the upper plumage, sooty black, with the exception of the middle and greater coverts, which are white, forming a large patch of this last colour on the middle of the wing: sides of the face, and all the under parts, white: bill black: irides brown: inside of the mouth, and legs, reddish. (*Summer plumage.*) The whole plumage above and below, with the exception of the large white space in the middle of the wing, deep sooty black: bill black: inside of the mouth, and legs, bright red. (*Young of the year.*) Crown of the head and nape blackish, the feathers edged with white; lower part of the neck, and sides of the breast, gray and white mixed: throat, breast, and rest of the under parts, white: upper back and scapulars black, many of the feathers edged with white; lower back black; rump white mixed with black: wings as in the adult, but the white speculum marked with cinereous and dusky spots: irides dusky brown: inside of the mouth, and legs, reddish. (*Egg.*) White, tinged with green: blotched, spotted and speckled, with ash-gray, red brown, and very dark brown: long. diam. two inches three lines: trans. diam. one inch six lines.

Met with principally in the northern parts of the kingdom. Common in the Hebrides and Orkney Islands, where they remain all the year, breeding in the crevices of rocks. Like the last species, lays but one egg*. Food, small fish, and marine crustacea. *Obs.* The *Spotted Guillemot* of authors is this species in its immature state.

GEN. 103.　MERGULUS, *Ray.*

274. M. *Alle*, Selby. (*Common Rotche.*) — Black; breast, belly, a spot above the eyes, and tips of the secondaries, white.

Uria Alle, *Temm. Man. d'Orn.* tom. ii. p. 928. Little Auk, *Mont. Orn. Dict. & Supp. Bew. Brit. Birds,* vol. ii. p. 408. Common Rotche, *Selb. Illust.* vol. ii. p. 430. pl. 81.

Dɪᴍᴇɴs. Entire length eight inches ten lines: length of the bill (from the forehead) seven lines, (from the gape) eleven lines; of the tarsus nine lines and a half; of the tail one inch seven lines; from the carpus to the end of the wing four inches nine lines.

Dᴇsᴄʀɪᴘᴛ. (*Winter plumage.*) Crown, region of the eyes, nape, sides of the breast, and all the upper parts of the body, deep black: wings pitchy brown; secondaries tipped with white; tertials broadly edged with the same colour: throat, front and sides of the neck, and all the under parts, pure white: sides of the head also white, variegated with dusky streaks, the white passing off towards the occiput in the form of an indistinct narrow band: bill black: irides dusky brown: tarsi and toes

---

* A writer in Loudon's *Magazine of Natural History,* (vol. v. p. 418.) asserts that in a large number of instances he has invariably found the eggs *two* in number.

yellowish brown; membranes greenish brown. (*Summer plumage.*)
Head, cheeks, throat, and neck, deep sooty black; a small white spot
above the eyes: the rest as in winter. (*Young of the year.*) Distinguished by their shorter and smaller bill: plumage similar to that of
the adult in winter. (*Egg.*) Of a uniform pale blue: long. diam. one
inch seven lines; trans. diam. one inch one line.

Very abundant in the Arctic Regions, where it breeds. Met with occasionally on the British coasts during the winter season. Lays two eggs,
which are deposited in the holes and crevices of the steepest rocks. Food,
marine insects and crustacea.

GEN. 104. FRATERCULA, *Briss.*

275. F. *arctica*, Steph. (*Puffin.*)—Crown, collar and
back, black; cheeks, throat, breast and belly, white.

Mormon Fratercula, *Temm. Man. d'Orn.* tom. II. p. 933. Puffin,
Mont. Orn. Dict. & Supp. Bew. Brit. Birds, vol. II. p. 404.
Selb. Illust. vol. II. p. 439. pl. 83*

DIMENS. Entire length thirteen inches six lines: length of the bill
(from the forehead) one inch eleven lines, (from the gape) one inch six
lines; height of the bill at the base one inch seven lines; length of the
tarsus one inch one line; of the tail two inches two lines; from the carpus
to the end of the wing six inches six lines.

DESCRIPT. (*Adult at all seasons.*) Crown of the head, a broad collar
surrounding the neck, and all the upper parts of the body, deep glossy
black: quills dusky brown: cheeks, throat, and a broad band above the
eyes, grayish white; breast, belly, and rest of the under parts, pure
white: bill bluish gray at the base, orange-yellow in the middle, bright
red at the tip; upper mandible with three obliquely transverse furrows;
lower mandible with two: irides, and horny appendages to the eyelids,
grayish white: orbits red: legs orange-red; claws black. (*Young of the
year.*) Bill much smaller and narrower; the sides smooth, with the transverse furrows very indistinctly marked; colour yellowish brown: space
between the eye and the bill dusky ash; cheeks and throat of a deeper
gray than in the adult bird; collar inclining to dusky gray: legs dull
red. (*Egg.*) White; indistinctly spotted with ash-gray: long. diam. two
inches three lines; trans. diam. one inch seven lines.

Found in great abundance on many parts of the British coast during
the spring and summer months. First seen about the middle of April.
Breeds in the crevices of rocks, and, in some places, in holes in the ground
burrowed out for the purpose. Deposits a single egg about the middle of
May. Towards the end of August or beginning of September, leaves the
British kingdom, and retires southward for the Winter. Food, small fish,
particularly sprats, marine crustacea and insects, occasionally sea-weed.

GEN. 105. ALCA, *Linn.*

276. A. *Torda*, Linn. (*Razor-Bill.*) — Wings reaching to the rump: bill black, with a transverse white band:
a narrow white line in advance of the eye.

A. Torda, *Temm. Man. d'Orn.* tom. II. p. 936. Razor-Bill, *Mont.*
  Orn. Dict. Bew. Brit. Birds, vol. II. p. 399. Razor-Bill Auk,
  Selb. Illust. vol. II. p. 435. pl. 83.

DIMENS. Entire length seventeen inches: length of the bill (from the forehead) one inch six lines, (from the gape) one inch eleven lines; of the tarsus one inch three lines; of the tail three inches four lines; from the carpus to the end of the wing seven inches six lines.

DESCRIPT. (*Adult in winter.*) Crown of the head, nape, sides of the neck, and all the upper parts, deep black; from the eye to the middle of the bill a narrow longitudinal interrupted white line: secondaries tipped with white, forming a transverse bar on the wings: throat, front of the neck, breast, and all the under parts, pure white; sides of the occiput white mixed with cinereous, bounded beneath by a narrow black band reaching backwards from the eyes: bill black, with three or four transverse furrows, the middle one pure white: irides hazel: legs dusky. (*Adult in summer.*) The narrow streak from the eye to the middle of the bill pure uninterrupted white: head, throat, and all the upper and fore part of the neck, deep brownish black: the rest as in winter. (*Young of the year.*) Similar to the adult in winter plumage, but easily distinguished by the smaller and narrower bill, without the transverse white furrow: the white line from the bill to the eye obscure and ill-defined: crown and nape cinereous black; sides of the neck, and towards the occiput, white, tinged with ash-colour: under parts pure white. (*Egg.*) White; blotched and spotted with dark red brown and blackish brown: long. diam. two inches nine lines; trans. diam. one inch ten lines.

Like the Puffin, a regular summer visitant. Common on all the rocky parts of the coast during the breeding season, and generally found in large companies. Deposits a single egg, about the beginning of May, on the tops and projecting shelves of the highest cliffs. Food, sprats and other small fish, and marine crustacea. *Obs.* The *Black-billed Auk* of English authors (*A. Pica,* Gmel.) is this species in its immature state.

277. A. *impennis,* Linn. (*Great Auk.*)—Wings very short, not reaching to the rump : bill black : an oval white patch in advance of the eye.

A. impennis, *Temm. Man. d'Orn.* tom. II. p. 939. Great Auk,
  Mont. Orn. Dict. Bew. Brit. Birds, vol. II. p. 397. *Selb. Illust.*
  vol. II. p. 433. pl. 82.

DIMENS. Entire length thirty-two inches: length of the bill (from the forehead) three inches six lines, (from the gape) four inches two lines; of the tarsus two inches five lines; of the tail three inches three lines; from the carpus to the end of the wing seven inches.

DESCRIPT. (*Summer plumage.*) Head, nape, back, wings, and tail, deep glossy black; front and sides of the neck black tinged with brown : immediately before the eye a large oval white patch: flanks dark ash-colour; rest of the under parts pure white: wings very short, incapacitated for flight; secondary quills tipped with white: bill black; crossed obliquely by several transverse furrows and ridges: irides and legs black. (*Winter plumage.*) "Unknown." TEMM. (*Egg.*) Dirty white, tinged with yellow: blotched and streaked, principally at the larger end, with black: long. diam. four inches ten lines; trans. diam. two inches nine lines.

A native of high northern latitudes, very rarely visiting the British coast. Occasionally seen in the Orkneys, and off the island of St. Kilda. Resides entirely in the open sea, never coming on land except for the purpose of breeding. Swims and dives with the greatest rapidity. Lays one egg in the clefts of rocks close to the water's edge. Food, fish and marine vegetables.

GEN. 106.   PHALACROCORAX, *Briss.*

278. P. *Carbo,* Steph. (*Common Cormorant.*) — A white gorget under the throat: tail of fourteen feathers.

> Carbo Cormoranus, *Temm. Man. d'Orn.* tom. II. p. 894.   Corvorant, *Mont. Orn. Dict.* Cormorant, *Bew. Brit. Birds,* vol. II. p. 379. *Leach, Zool. Misc.* vol. III. pl. 123.   Common Cormorant, *Selb. Illust.* vol. II. p. 446. pl. 84.

DIMENS. Entire length three feet: length of the bill (from the fore-head) two inches ten lines, (from the gape) four inches.

DESCRIPT. (*Adult in winter.*) Crown of the head, neck, breast, and all the under parts, black, with blue and green reflections; throat dusky, mixed with whitish; on the neck a few indistinct whitish streaks: upper part of the back, scapulars, and wing-coverts, cinereous brown, glossed with purple and bronze, all the feathers broadly edged with deep black: lower back and rump bluish black: quills and tail dusky: bill dark horn-colour: naked skin encircling the eyes greenish yellow: irides green: guttural pouch yellowish: legs black. (*Summer plumage.*) The feathers on the occiput and nape slender and elongated, forming a crest of a bluish black colour, glossed with green: a broad gorget of pure white on the throat: crown and upper part of the neck variegated with numerous long slender white feathers of a silky texture: upon the thighs a large patch of pure white: the rest as in winter. (*Young of the year.*) "Crown, nape, and back, deep brown, with a slight gloss of green: gorget whitish gray; front of the neck, and all the under parts, grayish brown, with whitish spots, more particularly on the breast and middle of the belly, intermixed: upper part of the back, scapulars, and wing-coverts, ash-gray, the feathers broadly edged with deep brown: bill pale brown: irides brown." TEMM. (*Egg.*) White, tinged with pale blue; the outer surface of a rough calcareous texture: long. diam. two inches nine lines; trans. diam. one inch seven lines.

Common on most parts of the British coast throughout the year. Occasionally observed inland on the banks of rivers. Swims and dives with great ease. Breeds on the summits of lofty rocks, sometimes in trees. Constructs a nest of sea-weed, and lays three or four eggs. Feeds entirely on fish. *Obs.* The *Crested Cormorant* of Bewick is this species in its summer plumage.

279. P. *cristatus,* Steph. (*Crested Shag.*) — The whole of the plumage, throat included, deep green: tail short, of twelve feathers.

> Carbo cristatus, *Temm. Man. d'Orn.* tom. II. p. 900.   Shag, and Crested Shag, *Mont. Orn. Dict. & Supp.*   Shag, *Bew. Brit.*

Birds, vol. ii. p. 387. Crested Shag, or Green Cormorant, *Selb.*
Illust. vol. ii. p. 450. pl. 86. *Gould, Europ. Birds*, part x.

DIMENS. Entire length twenty-eight inches : length of the bill (from
the forehead) two inches six lines, (from the gape) three inches six lines ;
of the tarsus two inches two lines ; of the tail five inches two lines ; from
the carpus to the end of the wing ten inches nine lines.

DESCRIPT. (*Adult in winter.*) The whole plumage of a rich deep
glossy green : upper part of the back, scapulars, and wing-coverts, with
bronze and purple reflections, each feather surrounded with a narrow
edging of velvet-black : tail short and rounded ; black : base of the bill
and guttural pouch gamboge-yellow : bill brown : irides green : legs
black. (*Summer plumage.*) Remarkable for an elegant crest of long
spreading feathers rising from the forehead between the eyes, capable
of erection : no white feathers on the neck and thighs as in the last
species. (*Young of the year.*) " Distinguished by the whole of the upper
part of the plumage being brown, slightly tinted with green ; the under
surface brownish ash, more or less inclining to white." GOULD. (*Egg.*)
White, tinged with pale greenish blue in patches ; the outer surface
rough and calcareous : long. diam. two inches five lines ; trans. diam.
one inch five lines.

Found principally in the rocky islands of the North of England and
Scotland. Common in the Orkneys. Breeds on the shelves of steep
rocks. Nest formed of sea-weed. Eggs from three to five in number.
Feeds on fish. Is never seen inland.

(24.) *P. Graculus*, Steph. in Shaw's Gen. Zool. vol. XIII. part i.
p. 82. *Carbo Graculus*, Temm. Man. d'Orn. tom. ii. p. 897. *Black
Cormorant*, Gould, Europ. Birds, part x.

This species has been considered as British by nearly all our own Ornithologists, though upon very insufficient authority. It appears to have
been confounded with the *P. cristatus*, from which it differs in the darker
colour of its plumage, as well as in other characters which will be found
pointed out in Mr Gould's work above referred to.

GEN. 107. SULA, *Briss.*

280. S. *Bassana*, Briss. (*Common Gannet.*)—Crown
and occiput pale buff ; quills black ; the rest of the plumage
white.

S. alba, *Temm. Man. d'Orn.* tom. ii. p. 905. Gannet, *Mont. Orn.*
Dict. & Supp. Bew. Brit. Birds, vol. ii. p. 390. Solan Gannet,
Selb. Illust. vol. ii. p. 455. pls. 86*, & 87.

DIMENS. Entire length three feet.

DESCRIPT. (*Adult in perfect plumage.*) Crown of the head, and
occiput, pale buff-yellow ; all the rest of the plumage milk-white, with
the exception of the quills and bastard winglet, which are black : bill
bluish gray at the base, the tip white : naked skin surrounding the eyes
pale blue : at the corners of the mouth a dusky membrane having the
appearance of being a prolongation of the gape : chin destitute of feathers, and of a dusky colour ; capable of great distention, forming a kind
of pouch : irides yellow : acrotarsia and acrodactyla with longitudinal
streaks of pale green ; membranes dusky ; claws white : tail-feathers
strong and pointed : twelve in number : the middle ones longest. (*Young*

during the first year.) All the upper plumage dusky brown; under parts brown, varied with ash-gray: bill, naked membranes, and irides, brown: tail simply rounded. *After the second moult*, the head, neck and breast, are grayish black, with a small triangular white spot at the tip of each feather: back, rump, and wings, the same, but the spots on these parts larger and more scattered: abdomen whitish, the feathers margined with cinereous brown: quills and tail grayish black; the latter more conical, with the shafts of the feathers white: bill grayish brown, whitish towards the tip: acrotarsia and acrodactyla greenish brown; the longitudinal streaks grayish white: membranes cinereous brown: claws whitish. *At each succeeding moult* the plumage becomes whiter, till the commencement of the fourth year, when it is fully matured. (*Egg*.) Dull white, tinged with green: long. diam. three inches three lines; trans. diam. one inch ten lines.

Plentiful during the breeding season on some parts of the northern coasts, particularly in the Isle of Bass in the Frith of Forth, St. Kilda, and some other of the Scotch Islands. Migrates southward in the Autumn, and, during Winter, may be observed in most parts of the British channel. At such times generally keeps far out at sea, but has been killed inland in a few rare instances. Feeds on herrings, pilchards, and other fish, on which it darts with great force and velocity. Nest placed on steep inaccessible rocks. Lays a single egg.

(25.) *Pelecanus Onocrotalus*, Temm. Man. d'Orn. tom. ii. p. 891.
Pelican, Mont. Orn. Dict. Supp. Gould, Europ. Birds, part xii.

An individual of this species, which is a native of Eastern Europe, is recorded (according to Montagu) to have been shot in England, in May 1663, at Horsey Fen. It is conjectured, however, that it might have escaped from confinement.

GEN. 108. STERNA, *Linn*.

(1. Sterna, *Steph*.)

281. S. *Caspia*, Pall. (*Caspian Tern*.) — Bill thick and strong, bright red: tarsus one inch six lines in length: tail short, forked.

S. Caspia, *Temm. Man. d'Orn.* tom. ii. p. 733. Caspian Tern, *Lath. Syn.* vol. iii. p. 350. *Selb. Illust.* vol. ii. p. 463.

Dimens. Entire length twenty-one inches: length of the bill (from the forehead) two inches nine lines, (from the gape) three inches two lines; of the tarsus one inch six lines; of the tail six inches; from the carpus to the end of the wing sixteen inches six lines.

Descript. (*Summer plumage*.) Forehead, crown, and long occipital feathers, deep black: nape, back, scapulars and wing-coverts, bluish ash: quills brown, tinged with ash-gray: sides of the head, fore part of the neck, and all the under parts of the plumage, pure white: tail pale ash-gray: bill coral-red: legs black. Wings reaching about four inches and a half beyond the tail. (*Winter plumage*.) Forehead, and a part of the crown, pure white; occiput variegated with black and white: the rest as in summer. (*Young of the year*.) "Forehead and crown as in the adult in winter plumage; rest of the upper parts ash-coloured brown, with large dusky spots and transverse bars: tail with a large dusky space at the extremity: quills almost entirely of the same colour: under parts white: bill dull red; the tip dusky." Temm. (*Egg*.) Yellowish stone-colour, spotted with ash-gray and dark red-brown; long. diam. two inches six lines; trans. diam. one inch eight lines.

Inhabits the borders of the Baltic, the Caspian Sea, and Archipelago. Is also occasionally seen on the coasts of France and Holland. In England, a few specimens have occurred at Yarmouth and Aldborough. One killed at this last place is in the Museum of the Cambridge Philosophical Society. Feeds on fish. Eggs two or three in number; deposited in a hole in the sand, or on the rocks by the edge of the sea.

282. S. *Cantiaca*, Gmel. (*Sandwich Tern.*) — Bill long ; black, the tip yellowish : legs short ; black : tarsus one inch : tail long, very much forked, shorter than the wings.

S. Cantiaca, *Temm. Man. d'Orn.* tom. ii. p. 735. Sandwich Tern, *Mont. Orn. Dict. & Supp. with fig. Bew. Brit. Birds*, vol. ii. p. 189. *Selb. Illust.* vol. ii. p. 464. pl. 88. f. 3.

Dimens. Entire length eighteen inches: length of the bill two inches : breadth, wings extended, two feet nine inches. Mont.

Descript. (*Summer plumage*.) Forehead, crown, and long occipital feathers, deep black; nape and upper part of the back pure white; middle and lower regions of the back, scapulars and wing-coverts, pale bluish ash: all the under parts pure white, tinged on the breast and fore part of the neck, with rose-red: quills deep ash-gray, having a velvety appearance, bordered on their inner webs with a broad band of pure white; some of the primaries tipped with black: tail white; considerably forked: bill black for two-thirds of its length from the base, the tip straw-yellow: legs black; the under surface of the toes and membranes yellow: claws black. (*Winter plumage*.) Forehead and crown white, spotted with black towards the occiput; the long occipital feathers deep black, but fringed with white: under parts pure white, without the rosy tinge on the neck and breast: the rest as in winter. (*Young of the year*.) Crown and occiput a mixture of black, white, and pale reddish brown: all the under parts pure white: upper part of the back and scapulars reddish, with transverse bars of dusky brown; some of the larger scapulars broadly bordered with brown; tips of the wing-coverts with a crescent-shaped edging of this colour: primary and secondary quills dusky ash, edged and tipped with white: tail ash-coloured at the base; the remaining portion dusky, with the tips of the feathers white: bill livid black; the extreme tip yellowish. (*Egg*.) Yellowish stone-colour; thickly spotted with ash-gray, orange-brown, and deep red-brown: long. diam. two inches; trans. diam. one inch five lines.

A summer visitant. Frequents the coasts of Kent, Sussex, and Suffolk. Occasionally met with in other parts of the country, but not a generally diffused species. Breeds in large societies. Eggs two or three, laid on sandy shores or on rocks. Feeds on fish.

283. S. *Dougallii*, Mont. (*Roseate Tern*.)—Bill slender ; black, the base reddish : legs orange : tarsus nine lines : tail greatly forked, extending far beyond the wings.

S. Dougalli, *Temm. Man. d'Orn.* tom. II. p. 738.　Roseate Tern,
　　Mont. Orn. Dict. Supp. with fig.　Bew. Brit. Birds, vol. II. p. 192.
　　Selb. Illust. vol. II. p. 470. pl. 89. f. 1, & 2.

DIMENS.　Entire length fifteen inches six lines: length of the bill
(from the forehead) one inch seven lines and a half; of the tarsus nine
lines; of the tail seven inches; from the carpus to the end of the wing
nine inches three lines.
　DESCRIPT.　(*Summer plumage.*) Forehead, crown, and long occipital
feathers, deep black; cheeks, throat, neck, and all the under parts, pure
white, tinged on the fore part of the neck, breast, and belly, with rose-
red: back, scapulars, and wing-coverts, pale ash-gray: first quill with
the outer web hoary black; the rest gray; all of them with a deep
border of white on their inner webs: tail white, greatly forked; the outer
feathers very long and subulate, extending upwards of two inches
beyond the wings: bill long and slender, black, passing into orange-red
at the base: legs bright orange: claws small and black. (*Young.*) "Bill
brownish black; the base orange-yellow: forehead and crown cream-
yellow, tinged with gray: region of the eyes, ear-coverts, and nape,
grayish black, mixed with yellowish white: throat, sides of the neck,
and under parts, white: ridge of the wings blackish gray, margined
paler: back and wing-coverts bluish gray, marbled with grayish black
and yellowish white: tail-feathers with the exterior webs gray, the
interior and tips white: quills gray, margined with white: legs pale
gallstone-yellow." SELB. (*Egg.*) Yellowish stone-colour; spotted and
speckled with ash-gray and dark brown: long. diam. one inch nine lines
and a half; trans. diam. one inch two lines and a half.
　First noticed on the Cumbrey Islands in the Frith of Clyde, by
Dr Macdougall, who communicated the species to Montagu.　Since observed on other parts of the Scotch coast, and also in the Fern Islands, to
which last locality, according to Selby, they resort annually to breed.
Eggs two or three in number.　Food, small fish.

284.　S. *Hirundo*, Linn.　(*Common Tern.*) — Bill moderate; red, the tip black: legs red: tarsus nine lines and a half: tail very much forked; as long as, or a little shorter than, the wings.

S. Hirundo, *Temm. Man. d'Orn.* tom. II. p. 740.　Common Tern,
　　Mont. Orn. Dict. & Supp.　Bew. Brit. Birds, vol. II. p. 185.
　　Selb. Illust. vol. II. p. 468. pl. 90. f. 1.

DIMENS.　Entire length thirteen inches six lines: length of the bill
(from the forehead) one inch seven lines and a half, (from the gape) two
inches two lines: of the tarsus nine lines and a half; of the tail four
inches six lines; from the carpus to the end of the wing ten inches four
lines: breadth, wings extended, two feet six inches eight lines.
　DESCRIPT.　(*Summer plumage.*) Forehead, crown, and long occipital
feathers, deep black; hind part of the neck, back, scapulars, and wing-
coverts, bluish ash: all the under parts white, the breast and belly faintly
tinged with ash-colour: quills ash-gray passing into dusky brown at the
tips, with a large oblong white space on their inner webs: tail whitish;
the outer webs of the two exterior feathers dusky ash: bill red, passing
into dusky at the tip: feet bright red.　The *winter plumage* differs
simply in having the forehead and crown of a more dull black. (*Young
of the year.*) Forehead and anterior part of the crown dirty white: rest

of the crown towards the occiput marked with large dusky spots; the long occipital feathers brownish black with whitish tips: rest of the upper parts dull bluish ash, irregularly spotted with pale reddish brown, all the feathers whitish at the tips: under parts whitish: many of the quills tipped with white: lesser wing-coverts near the top of the wing dusky gray with whitish edges: tail ash-gray, passing into white on the inner webs, and towards the tips of the feathers: bill much shorter than in the adult bird, dusky at the tip, dull orange at the base: legs orange. (*Egg.*) Yellowish stone-colour, blotched and spotted with ash-gray and dark. red-brown: long. diam. one inch eight lines; trans. diam. one inch two lines.

Common on most parts of the British coast during the spring and summer months. Retires in the Autumn. Occasionally observed inland on fresh waters. Breeds on flat shingly shores, and lays from two to four eggs. Of noisy and restless habits; constantly on wing. Food, insects and small fish. *Obs.* The *Brown Tern* of English authors is probably this species in immature plumage.

285. S. *arctica*, Temm. (*Arctic Tern.*)—Bill slender; red throughout its whole length: legs red: tarsus seven lines: tail very much forked; longer than the wings.

S. arctica, *Temm. Man. d'Orn.* tom. II. p. 742. Arctic Tern, *Shaw, Gen. Zool.* vol. XIII. p. 152. *Selb. Illust.* vol. II. p. 473. pl. 90. f. 2.

DIMENS. Entire length fifteen inches: length of the bill (from the forehead) one inch four lines, (from the gape) one inch ten lines; of the tarsus seven lines; of the tail seven inches nine lines; from the carpus to the end of the wing eleven inches.

DESCRIPT. Very similar to the last species, but may always be distinguished by the much shorter tarsus, and the bill which is also shorter and generally of one colour throughout. (*Summer plumage.*) Forehead, crown, and long occipital feathers, deep black; rest of the upper parts bluish ash: throat, front of the neck, breast and belly, of the same colour as the back; vent, under tail-coverts, and a band beneath the eyes, pure white: tail very much forked; always longer than in the last species, extending (in the *adult male*) one inch two lines beyond the wings: tarsi and toes very short; of a deep crimson-red: bill crimson-red. (*Egg.*) Colour the same as in the last species: long. diam. one inch seven lines; trans. diam. one inch one line.

Inhabits the Arctic Regions; resorting to the northern coasts of Britain during the summer months. According to Selby, very abundant in the Fern Islands, where they arrive about the middle of May and stay till August. Eggs laid on the gravelly beach or bare ground; three or four in number. Food, small fish, particularly the fry of the sand-eel (*Ammodytes Tobianus.*)

286. S. *minuta*, Linn. (*Lesser Tern.*)—Bill orange, the tip black: legs orange: tarsus seven lines and a half: tail very much forked, shorter than the wings: forehead white.

S. minuta, *Temm. Man. d'Orn.* tom. ii. p. 752. Lesser Tern, *Mont. Orn. Dict. & Supp. Bew. Brit. Birds*, vol. ii. p. 187. *Selb. Illust.* vol. ii. p. 475. pl. 89. f. 3, & 4.

DIMENS. Entire length nine inches six lines: length of the bill (from the forehead) one inch two lines, (from the gape) one inch eight lines; of the tarsus seven lines and a half; of the middle toe, claw included, nine lines; of the tail three inches three lines; from the carpus to the end of the wing seven inches one line: breadth, wings extended, twenty inches ten lines.

DESCRIPT. (*Summer and winter plumage.*) Forehead, and a streak above the eyes, pure white: between the bill and the eye a broad black streak; crown, occiput, and nape, deep black; back, scapulars, and wing-coverts, pale bluish ash: first three quills blackish gray, their inner webs broadly edged with white: cheeks, all the under parts, and tail, pure white: bill yellowish orange; the tip black: irides black: legs and toes orange-yellow. (*Young of the year.*) Forehead yellowish white; crown, occiput and nape, brownish, with dusky streaks; before and behind the eyes a black spot; back, scapulars and wing-coverts, yellowish white, with a crescent-shaped dusky bar near the extremity of each feather: quills cinereous gray, edged and tipped with yellowish white: tail nearly even at the end, white tinged with cinereous, the tips of the feathers yellowish: throat, and all the under parts, white: bill shorter than in the adult bird, pale yellow, the tip blackish brown: legs dull yellow. (*Egg.*) Stone-colour, spotted and speckled with ash-gray and dark chestnut-brown: long. diam. one inch four lines; trans. diam. eleven lines.

Common on some parts of the British coast during the summer months, but not so generally diffused as the *S. Hirundo.* Habits similar to those of that species. Eggs two or three in number, deposited on the shingly beach a little above high water-mark. Food, marine worms and winged insects.

287. S. *nigra*, Linn. (*Black Tern.*) — Bill black: legs reddish brown: membranes of the toes deeply emarginated: tarsus seven lines and a half: tail not much forked, shorter than the wings.

S. nigra, *Temm. Man. d'Orn.* tom. ii. p. 749. Black Tern, *Mont. Orn. Dict. & Supp. Bew. Brit. Birds*, vol. ii. p. 195. *Selb. Illust.* vol. ii. p. 477. pl. 91.

DIMENS. Entire length ten inches three lines: length of the bill (from the forehead) one inch two lines, (from the gape) one inch seven lines and a half; of the tarsus seven lines and a half; of the tail three inches three lines; from the carpus to the end of the wing eight inches eight lines: breadth, wings extended, twenty-three inches six lines.

DESCRIPT. (*Summer plumage.*) Head and neck deep black: breast and abdomen dark ash-colour: all the upper parts, including the rump and tail, deep bluish gray: vent and under tail-coverts pure white: wings, when closed, extending one inch seven lines beyond the tail: bill black: irides brown: legs purplish brown. (*Winter plumage.*) Forehead, space between the bill and the eyes, throat and all the fore part of the neck, white: the rest as in summer. *In the intermediate seasons,*

these parts are often found mottled with black upon a white ground.
(*Young of the year.*) " Forehead, space between the eye and the bill,
sides and front of the neck, as well as all the under parts, pure white:
on the sides of the breast a large patch of dusky ash: before the eyes
a crescent-shaped spot of the same colour: crown, occiput and nape,
black: back and scapulars brown, the feathers edged and tipped with
reddish white: wings, rump and tail, ash-colour; the wing-coverts
tipped with reddish white: bill brown at the base: irides brown : legs
livid brown." TEMM. (*Egg.*) Dark olive-brown, blotched and spotted
with black, principally at the larger end : long. diam. one inch five lines;
trans. diam. one inch.

  A more inland species than any of the foregoing. Found principally
in marshes, and on the banks of rivers and lakes. First seen towards the
latter end of April: departs in October. Common in Romney Marsh in
Kent, and also in the fens of Cambridgeshire and Lincolnshire, where
they breed. Eggs two to four in number, deposited on the bare grass
in swampy and sedgy spots. Young hatched in the beginning of July.
Food, worms and aquatic insects. *Obs.* The *S. nævia* of Gmelin is the
young of this species.

  288. S. *Anglica*, Mont. (*Gull-billed Tern.*) — Bill
short and thick; entirely black : legs long; black : claws
nearly straight : tarsus one inch three or four lines : tail
not much forked, considerably shorter than the wings.

  S. Anglica, *Temm. Man. d'Orn.* tom. II. p. 744. Gull-billed Tern,
    Mont. Orn. Dict. Supp. with fig. Selb. Illust. vol. II. p. 480.
    pl. 88. f. 1.

  DIMENS. Entire length thirteen inches six lines: length of the bill
(from the forehead) one inch six lines, (from the gape) one inch ten lines
and a half; of the tarsus one inch two lines and a half to four lines and
a half*.
  DESCRIPT. (*Summer plumage.*) Forehead, crown, occiput and nape,
deep black; the feathers on this last part long and silky: all the rest of
the upper plumage pale bluish ash: quills dusky gray, having a hoary
appearance; the tips of the first five black : under parts white : wings
extending three inches beyond the extremity of the tail: bill deep black;
remarkably thick and strong; gonys of the lower mandible ascending ;
mental angle very prominent, as in the genus *Larus :* irides brown : legs
long; black, with a reddish tinge: toes long; the claws unusually
straight; membranes deeply emarginated. (*Winter plumage.*) Forehead
and crown white; anterior angle of the eye, and spot behind the ears,
grayish black : the rest as in summer, with the exception of the primary
quills, which are not quite so deeply coloured at the tips. (*Young of the
year.*) "Crown white, with small longitudinal brownish streaks: back
and wings bluish ash, mixed with gray and yellowish brown: all the
under parts pure white: quills brownish ash: tail very little forked, ash-
colour, the tips of the feathers white: base of the bill yellowish ; the rest,
towards the tip, dusky brown." TEMM. (*Egg.*) Dark olive-
brown, spotted with ash-colour and two shades of dark red-brown : long.
diam. one inch eleven lines; trans. diam. one inch four lines.

  * A difference of two lines was found in the length of the tarsus in two British-killed specimens
in the British Museum.

A rare species in this country, first described by Montagu, who obtained specimens from the coast of Sussex. According to Temminck, very abundant in Hungary. Said to breed on the marshy borders of salt-water lakes, and to lay three or four eggs. Food, winged insects.

(2. Anous, *Steph.*)

289. S. *stolida*, Linn. (*Black Noddy.*)—Forehead and crown whitish; the rest of the plumage, as well as the bill and legs, black.

S. stolida, *Lath. Ind. Orn.* vol. ii. p. 805. Noddy, *Lath. Syn.* vol. iii. p. 354. Black Noddy, *Steph. in Shaw's Gen. Zool.* vol. xiii. p. 140. pl. 17.

Dimens. Entire length fifteen inches: length of the bill (from the forehead) one inch nine lines, (from the gape) two inches; of the tarsus eleven lines; of the tail five inches; from the carpus to the end of the wing ten inches.

Descript. Forehead and crown whitish, passing into ash-gray on the occiput and nape: all the rest of the plumage brownish black, deepest on the quills and tail-feathers: wings just reaching to the extremity of the tail, which is even, or somewhat rounded, at its tip: bill and legs black: the membranes of the toes entire, without any indication of a notch.

Two adult individuals of this species were taken a few years since during Summer on the Wexford coast, between the Tasker Lighthouse and Dublin Bay. From one of these, which was obligingly submitted to my examination by Mr W. Thompson of Belfast, the above measurements and description were taken. Not previously known as British. Frequently met with at sea, but generally within the Tropics. Stated by Latham to breed in the Bahama Islands, laying its eggs on the bare rocks.

GEN. 109.   LARUS, *Linn.*

(1. Xema, *Leach.*)

290. L. *Sabini*, Sabine. (*Sabine's Gull.*)

Xema Sabini, *Leach in Ross's Voy. App.* p. lvii. *with fig.* Larus Sabini, *Sab. in Linn. Trans.* vol. xii. p. 520. pl. 29. Sabine's Xeme, *Steph. in Shaw's Gen. Zool.* vol. xiii. part i. p. 177. pl. 20.

Dimens. Entire length from twelve inches six lines to fourteen inches: length of the bill one inch; of the tarsus one inch six lines; of the tail five inches: breadth, wings extended, about thirty-three inches. Sab.

Descript. (*Adult male and female in summer.*) "The whole of the head, and upper part of the neck, a very dark ash or lead-colour; the remainder of the neck behind and before, as well as the breast and belly, pure white; a narrow black collar surrounding the neck at the meeting of the ash-colour and white: back, scapulars, and wing-coverts, ash-coloured, very much lighter than the head, but darker than the corresponding parts of the *Larus ridibundus;* the lower ends of the scapulars tipped with white: the first five primary quills with black shafts, the whole outer webs of these black, the edge of their upper webs white to within an inch and a half of the tips, the white sometimes continued to the tip; the tips of the first and second of these quills in some white, in others black; the tips of the third, fourth, and fifth, white, giving the wing when closed a spotted appearance; the sixth primary quill with a white shaft, having the web more or less black, but principally white, with sometimes a black spot near the end; the other primaries, the secondaries, and tertials, white; the whole under parts of the wings white: wings extending an inch or more beyond the longest feather of the tail: upper and under tail-coverts white; the tail feathers twelve, the outer narrower than the central ones; the depth of the fork nearly an inch: bill with the base of both mandibles black as far as the angular projection of the lower mandible, the remainder yellow; inside of the mouth bright vermilion: irides dark, surrounded by a naked circle of the same colour as the inside of the mouth; a small white speck beneath the eye, scarcely perceptible: legs black." SAB. In the *immature* state and in the *adult winter* plumage, the black head is wanting. (*Egg*.) Of a regular shape, and not much pointed: colour olive, much blotched with brown: long. diam. one inch six lines. SAB.

Two individuals of this species, which was first observed in Greenland by Captain Sabine, are recorded by Mr Thompson* to have been killed in Ireland. The specimens, which are both in the plumage of the first year, are now in the Museums of the Natural History Society of Belfast and the Royal Society of Dublin. Not much is known respecting its habits. Said to lay two eggs on the bare ground, which are hatched in the last week in July. Food, marine insects.

(2. LARUS, Steph.)

* Small; less than twenty inches in length.

291. L. *minutus*, Pall. (*Little Gull.*)—Bill brownish red: legs bright vermilion: tarsus eleven or twelve lines: hind toe very small; the claw straight: all the quills tipped with white; the shafts of the primaries brown.

L. minutus, *Temm. Man. d'Orn.* tom. II. p. 787. Little Gull, *Mont. Orn. Dict. App. to Supp. with fig. Lath. Syn.* vol. III. p. 391. *Selb. Illust.* vol. II. p. 484. pl. 92. *Gould, Europ. Birds,* part xi.

DIMENS. Entire length rather exceeding ten inches: length of the bill (from the forehead) rather more than nine lines; of the tarsus rather more than one inch; from the carpus to the end of the wing eight inches six lines. MONT.

DESCRIPT. (*Adult in winter.*) Forehead, cheeks, and a space behind the eyes, throat, and all the under parts, pure white: occiput, nape, a spot at the anterior angle of the eye, and ear-coverts, deep blackish gray: upper parts of the body and wings pale bluish ash: primary and secondary quills broadly tipped with white: inside of the wings deep blackish gray: tail white: bill and irides dusky brown: legs bright vermilion-red. (*Adult in summer.*) "All the head, and upper part of the neck, black; a white crescent-shaped spot behind the eyes; lower part of the neck, and all the under parts, white, with a faint blush of rose-red: back, scapulars, and wings, very pale bluish ash; rump and tail pure white: primary quills cinereous; all of them, as well as the secondaries, tipped with white: bill deep lake-red: irides dark brown: legs crimson." TEMM. (*Young of the year.*) "Forehead and crown white, tinged with gray: nape, and back part of the neck, and upper parts of the body, blackish gray, tinged with clove-brown, the margins of the feathers fringed with grayish white : scapulars edged with white : greater coverts pale bluish gray, deeply edged with white : first four quills with the outer webs and tips black, the inner webs grayish white : under plumage white : tail white for two-thirds of its length, terminated by a broad black bar : legs livid or flesh-red." SELB. (*Egg.*) Unknown.

A very rare occasional visitant in this country. First noticed by Montagu, who describes a specimen shot on the Thames near Chelsea. Has since occurred at Yarmouth, and on one or two other parts of the eastern coast, as well as in Cornwall and Scotland. Said to be common in the East of Europe, frequenting rivers and lakes. Food, insects and worms. Nidification unknown.

292. L. *capistratus*, Temm. (*Brown-headed Gull.*)— Bill and legs reddish brown : tarsus one inch seven lines : shafts of the outer primaries white : *in summer*, a brown hood reaching to the occiput.

L. capistratus, *Temm. Man. d'Orn.* tom. II. p. 785. *Flem. Brit. An.* p. 142. *Yarr. in Proceed. of Zool. Soc.* 1831. p. 151. Masked Gull, *Shaw, Gen. Zool.* vol. x$\overset{\text{ge}}{\text{i}}$.i. p. 204.

DIMENS. Entire length fifteen inches : length of the bill (above, from the first feathers) one inch and half a line, (from the gape) one inch ten lines ; of the tarsus one inch seven lines ; of the middle toe, nail included, one inch six lines ; from the carpus to the end of the wing eleven inches eight lines, YARR.

DESCRIPT. (*Winter plumage.*) According to Temminck, exactly similar to that of the next species, from which however the present may always be distinguished by its smaller size, its more slender as well as shorter bill, shorter tarsi, and smaller feet : the colour of the feet is also reddish brown. (*Summer plumage.*) " Head, and upper part of the neck, brocoli-brown, bounded by blackish brown, descending lowest at the fore part, some of the dark feathers at the margin in front tipped with white ; the remaining portion of the neck, breast, abdomen, vent and tail, pure white : upper surface of the wings pale ash-gray ; under surface grayish white : primaries white, edged and tipped with black, broadest on the inner web ; shafts white : bill, legs, and toes, brownish red ; membranes chocolate-brown." YARR. (*Egg.*) Unknown.

But little is known of the habits of this species, which was first characterized as distinct from the two following, by Temminck. Inhabits the Shetland and Orkney Islands. Has been also shot in Ireland, in the neighbourhood of Belfast, by Mr Thompson. Food and nidification unknown.

293. L. *ridibundus*, Linn. (*Black-headed Gull.*)— Bill and legs bright red : tarsus one inch ten lines : a large white space in the middle of the primaries : *in summer*, a deep brown hood reaching below the occiput.

L. ridibundus, *Temm. Man. d'Orn.* tom. II. p. 780. Black-headed Gull, *Mont. Orn. Dict. & Supp. Selb. Illust.* vol. II. p. 486. pl. 92. Black-headed and Red-legged Gulls, *Bew. Brit. Birds,* vol. II. pp. 222, & 225.

Dimens. Entire length seventeen inches: length of the bill (from the forehead) one inch four lines, (from the gape) two inches one line; of the tarsus one inch ten lines; of the middle toe, nail included, one inch eight lines; of the tail four inches eight lines; from the carpus to the end of the wing twelve inches nine lines: breadth, wings extended, three feet three inches five lines.

Descript. (*Winter plumage.*) Head, neck, and tail, pure white, with the exception of a small black spot at the anterior angle of the eye, and a dusky patch on the ear-coverts: back, scapulars, secondary quills, and greater portion of the wing-coverts, pale bluish ash; some of the long coverts annexed to the primaries, as well as the spurious winglet, white: primaries themselves white, passing into dusky ash towards the margins of the inner webs, and broadly tipped with black; (according to Temminck, *in very old birds*, the extreme tips of the primaries are *white*;) first primary with the outer web black from the base nearly to the tip, the inner edging of dusky ash very narrow; under surface of the wings dark ash-gray: breast, belly, and abdomen, white, in some individuals faintly tinged with flesh-red: irides deep brown: bill and legs bright vermilion-red. (*Summer plumage.*) Head, and upper part of the neck, deep brown; at the posterior angle of the eye a white spot; lower part of the neck, and all the under plumage, fine white, tinged with rose-red: bill and legs deep purplish red: the rest as in winter. (*Young of the year.*) Head and occiput very pale brown; a large white spot behind the eyes: under parts, and a collar on the nape, white; the white on the front of the neck with a faint reddish tinge; flanks with brown crescent-shaped marks: back, scapulars, and middle coverts, deep brown, the feathers with yellowish edges; upper edge of the wing, rump, and greater part of the tail-feathers, white, these last tipped with dusky brown: primaries white at the base and on the inner webs, black on the outer, and at the tips; greater coverts bluish ash: bill livid at the base; the tip black · legs yellowish. *After the first autumnal moult*, the back and wings are bluish ash, with a few brown feathers intermixed, some of those on the wings still with yellowish edges: forehead, and all the under parts, pure white; head white, spotted with pale ash-colour; ear-coverts, and a spot at the anterior angle of the eye, brown: tail still with a dark bar at its tip: bill reddish at the base; the tip brown. *In this state they remain till the following spring, when they assume the perfect summer plumage.* (*Egg.*) Yellowish olive-brown, spotted with two shades of darker brown: long. diam. two inches one line and a half; trans. diam. one inch six lines.

A common species; frequenting inland marshes and the banks of rivers. Resorts to the sea-coast in Winter. Breeds in fens and low swampy meadows, laying three eggs on tufts of grass previously trodden down. Food, insects, worms, and small fish. *Obs.* The *Red-legged Gull* (L. cinerarius, *Gmel.*) is this species in its winter plumage. The *Brown-headed Gull* (L. erythropus, *Gmel.*) is the young of the year after the first autumnal moult.

294. L. *Atracilla*, Linn. (*Laughing Gull.*)—Bill and legs deep brownish red: tarsus one inch ten lines: primaries entirely black; extending considerably beyond the tail: *in summer* a lead-coloured hood reaching below the occiput.

L. Atracilla, *Temm. Man. d'Orn.* tom. II. p. 779. L. ridibundus, *Wils. Amer. Orn.* vol. IX. pl. 74. f. 4. Laughing Gull, *Mont. Orn. Dict.*

S

Dimens. Entire length eighteen inches. Mont.

Descript. (*Winter plumage.*) According to Temminck, unknown.
(*Summer plumage.*) " Head, and upper part of the neck, deep lead-gray,
the hood extending lower in front than behind; a white spot above and
beneath the eyes; lower part of the neck, breast, belly, and tail, pure
white: back, wings, and secondary quills, lead-gray; tips of the secondaries white; the whole of the primaries, which extend two inches beyond
the tail, deep black: bill and legs very deep lake-red." Temm. (*Egg.*)
Dun clay-colour, thinly marked with small irregular touches of pale
purple and pale brown: long. diam. two inches and a quarter; trans.
diam. one inch and a half. Wils.

But little known as a British species. Montagu records having seen
five individuals together feeding in a pool upon the Shingly Flats near
Winchelsea in Sussex, and two others near Hastings. One was shot.
According to Temminck, not uncommon in the South of Europe. Food,
the remains of fish and crustacea, and also insects. Said by Wilson to
breed in marshes, and to lay three eggs.

295. L. *tridactylus*, Lath. (*Kittiwake Gull.*) — Bill greenish yellow: legs dusky: tarsus one inch four lines: a small knob instead of a back toe: primaries bluish ash, the outer one edged externally with black.

L. tridactylus, *Temm. Man. d'Orn.* tom. ii. p. 774. Kittiwake,
Mont. Orn. Dict. & Supp. Bew. Brit. Birds, vol. ii. pp. 218,
& 220. *Selb. Illust.* vol. ii. p. 493. pl. 94.

Dimens. Entire length fifteen inches six lines: length of the bill
(from the forehead) one inch four lines, (from the gape) two inches one
line; of the tarsus one inch four lines; of the tail five inches five lines;
from the carpus to the end of the wing twelve inches: breadth, wings
extended, three feet one inch four lines.

Descript. (*Adult in winter.*) Crown, occiput, nape, and sides of the
neck, pale bluish ash, with a few fine dusky streaks before the eyes;
forehead, region of the eyes, and all the under parts, pure white: back,
scapulars, and wing-coverts, bluish ash: quills the same; the first with
the whole of the outer web black; first four primaries tipped with black,
two or three of them with a small white spot at the extreme point; the
fifth white at the extremity, with a black bar near the tip: rump and
tail pure white: bill greenish yellow: inside of the mouth, and eye-lids,
orange-red: irides brown: legs dusky brown, tinged with olivaceous.
(*Summer plumage.*) The whole head and neck pure white, without any
bluish ash on the nape, or dusky streaks before the eyes: the rest as in
winter. (*Young of the year.*) Head, neck, and all the under parts, white,
with the exception of a small spot at the anterior angle of the eye,
a patch on the ear-coverts, and a larger crescent-shaped patch on the
nape, which are dusky ash: back and wings deep bluish ash, the feathers
tipped with dusky brown; the bend, and upper edge, of the wing, black;
scapulars and secondary quills with large dusky spots; primaries black:
tail black at the end, the extreme tips of the feathers whitish; the outer
feather entirely white: bill, irides, and eye-lids, black. *After the first
moult*, the back assumes the bluish ash, with still a few feathers intermixed spotted with brown; the spots on the head, and crescent-shaped
patch on the nape, appear of a dark bluish ash; the wing-coverts also
still spotted with dark brown; tail black at the extremity: bill greenish

yellow spotted with dusky: forehead, and all the under parts, white.
The plumage is not matured till after the second autumnal moult.
(*Egg.*) Stone-colour, thickly spotted with ash-gray and two shades of
light brown: long. diam. two inches two lines and a half; trans. diam.
one inch seven lines.

Not uncommon during the summer months on rocky coasts in the
northern parts of the island. More rarely observed southwards. Breeds
on the narrow ledges of steep cliffs, and lays three eggs. Food, small
fish and marine insects. *Obs.* The *Tarrock Gull* of English authors is
this species in its immature plumage.

296. L. *canus*, Linn. (*Common Gull.*)—Bill greenish,
yellow at the tip: legs greenish gray: tarsus two inches
one line: wings reaching beyond the tail; first two primaries black, with a large white spot near the tips.

> L. canus, *Temm. Man. d'Orn.* tom. II. p. 771.   Common Gull,
> *Mont. Orn. Dict. & Supp.   Bew. Brit. Birds,* vol. II. p. 216.
> *Selb. Illust.* vol. II. p. 490. pl. 93.

DIMENS. Entire length seventeen inches three lines: length of the
bill (from the first feathers) one inch three lines, (from the gape) two
inches two lines; of the tarsus two inches one line; of the naked part
of the tibia seven lines; of the middle toe, nail included, one inch nine
lines; of the tail five inches four lines; from the carpus to the end of
the wing thirteen inches ten lines.

DESCRIPT. (*Adult in winter.*) Head, occiput, nape, and sides of the
neck, white, spotted with blackish gray: all the under parts, rump and
tail, pure white: back, scapulars and wing-coverts, fine bluish ash: first
two primaries black, with a large white space near the tips; the rest
black towards the extremities, the tips themselves, as well as those of
the scapulars and secondary quills, white: bill greenish at the base, the
tip ochre-yellow: inside of the mouth orange: irides brown: naked
circle round the eyes reddish brown: legs greenish gray; the webs
blotched with yellowish. (*Summer plumage.*) Head, occiput, nape and
sides of the neck, pure white, without the dusky spots: bill ochre-yellow:
naked circle round the eyes bright vermilion: legs pale ochre-yellow,
spotted with bluish ash: the rest as in winter. (*Young of the year.*)
A black spot before the eyes; all the upper plumage grayish brown, the
feathers on the back and wings edged and tipped with yellowish white;
those on the upper part of the back with fine streaks of this colour:
forehead, and all the under parts, whitish, the breast and flanks spotted
with ash-gray; throat, and middle of the belly, pure white: quills of
a uniform dusky brown, neither tipped nor spotted with white: tail white,
with a broad black bar near the extremity: base of the bill livid dirty
white; the tip black: naked circle round the eyes brown: legs livid
white. *In the second year,* the head, neck, rump, and under parts, are
white, the two former streaked with dusky brown: back and scapulars
bluish ash; wing-coverts still mottled with brown and white feathers:
quills dusky brown, sometimes tipped with white: tail white, with a black
bar towards the tip. *The plumage is not matured till after the second
spring moult.* (*Egg.*) Dark olive-brown, spotted with two shades of
darker brown: long. diam. two inches three lines; trans. diam. one
inch six lines.

Common on all parts of the coast, and, during the winter months,
occasionally observed inland at a considerable distance from the sea.

Breeds on the ledges of rocks.   Nest constructed of sea-weed.   Eggs two or three in number.   Food, worms, marine insects, and small fish. *Obs.* The *Winter Gull* of English authors (*L. hybernus*, Gmel.) is this species in its immature plumage.

297. L. *eburneus*, Gmel. (*Ivory Gull.*)—Bill strong; bluish gray, the tip yellow : legs black : naked part of the tibia very small : tarsus one inch six lines : membranes of the toes somewhat abbreviated : *adult* plumage pure white.

> L. eburneus, *Temm. Man. d'Orn.* tom. II. p. 769.   *Sabine in Linn.* *Trans.* vol. XII. p. 548.   *Edmondst. in Wern. Mem.* vol. IV. p. 501. Ivory Gull, *Bew. Brit. Birds*, vol. II. p. 214.   *Selb. Illust.* vol. II. p. 497. pl. 94*.

DIMENS.   Entire length sixteen inches, (according to *Temminck*, nineteen inches): breadth, wings extended, three feet three inches and a half. EDMONDST.

DESCRIPT.   (*Adult plumage.*) The whole plumage pure white, without spots of any kind: bill thick and strong; deep bluish gray at the base, the rest ochre-yellow : irides deep brown : wings extending an inch and a half beyond the tail: legs black; covered with a rough skin : tibia feathered nearly to the tarsal joint: membranes of the toes short, and slightly hollowed out : claws much hooked. In *young birds*, the plumage is pale grayish white, more or less spotted and barred with blackish brown ; the tips of the quills, and a transverse bar at the extremity of the tail, of this last colour : bill blackish, the tip yellow : legs blackish gray. (*Egg.*) Unknown.

Inhabits the Arctic Regions.   A rare and accidental visitant in this country.   A solitary individual in its second year's plumage is recorded by Mr. Edmondston to have been killed in Balta Sound, Zetland, Dec. 13, 1822.   It is now in the Edinburgh Museum.   According to Selby, it has been also killed, in an immature state, in the Frith of Clyde.   Food, according to Captain Sabine, blubber and the flesh of whales.   Nidification unknown.

** *Large ; exceeding twenty inches in length.*

298. L. *argentatus*, Brunn. (*Herring Gull.*) — Bill yellow : legs livid flesh-colour : tarsus two inches six lines : wings reaching very little beyond the tail : shafts of the primaries black : mantle (in the *adult*) bluish ash.

> L. argentatus, *Temm. Man. d'Orn.* tom. II. p. 764.   Herring Gull, *Mont. Orn. Dict. & Supp.   Bew. Brit. Birds*, vol. II. p. 207. *Selb. Illust.* vol. II. p. 504. pls. 96, & 96*.

DIMENS.   Entire length twenty-three inches: length of the bill (from the forehead) two inches three lines, (from the gape) two inches nine lines ; of the tarsus two inches six lines ; from the carpus to the end of the wing seventeen inches three lines.

DESCRIPT.   (*Adult in winter.*) Crown, region of the eyes, occiput, nape and sides of the neck, white, with longitudinal streaks of pale brown on the shafts of the feathers: back, scapulars and wing-coverts, fine bluish ash : primaries dusky, passing into deep black towards their extremities, all terminated by a large white spot; scapulars and secondaries edged and tipped with white: forehead, throat, and all the other under parts, as well as the tail, pure white: bill ochre-yellow; the angle of the lower mandible orange-red: orbits orange: irides pale yellow: legs livid flesh-colour. (*Summer plumage.*) The whole head and neck pure white, without any brown streaks: the rest as in winter. (*Young of the year.*) Head, neck, and all the under parts, ash-gray, spotted and variegated with light brown: upper parts brown and ash-colour, all the feathers edged with reddish: quills dusky brown, without any white at the tips: tail whitish at the base, becoming browner towards the extremity, the tips of the feathers reddish yellow: bill grayish black: irides and orbits brown: legs livid brown. *In the second year*, the colours are similar, but somewhat paler. *After the second autumnal moult*, the bluish ash begins to appear upon the back mottled with brown, and the irides get lighter, inclining to yellow. *The following spring*, the colour of the bill changes; the bluish ash extends further and assumes a purer tint; and the bar on the tail gradually disappears. *At the age of three years, or after the third autumnal moult*, the plumage is matured. (*Egg.*) Light olive-brown, spotted with two shades of dark brown: long. diam. two inches six lines; trans. diam. one inch nine lines.

Common on all parts of the coast throughout the year. In the young or immature state, occasionally observed inland in the vicinity of rivers and fresh-water lakes. Breeds on rocky cliffs, constructing a nest of long dry grass. Eggs two or three in number. Feeds on fish, worms, and marine *rejectamenta*. *Obs.* The name of *Wagel Gull* has been applied indiscriminately to the young of this, and the two following species.

299.　L. *fuscus*, Linn. (*Lesser Black-backed Gull*.)— Bill and legs yellow: tarsus two inches three lines: wings reaching two inches beyond the tail: shafts of the primaries black : mantle (in the *adult*) grayish black.

L. fuscus, *Temm. Man. d'Orn.* tom. ii. p. 767.　Lesser Black-backed Gull, *Mont. Orn. Dict. & Supp. Bew. Brit. Birds*, vol. ii. p. 205. *Selb. Illust.* vol. ii. p. 509. pl. 95.

DIMENS.　Entire length twenty-four inches. MONT.

DESCRIPT. (*Adult in winter.*) Crown, region of the eyes, occiput, nape and sides of the neck, white, with longitudinal streaks of pale brown on the shafts of the feathers: upper part of the back, scapulars and wing-coverts, deep blackish gray: primaries black, deepening in tint towards their extremities; the first two with an oval white spot near their tips; the rest finely tipped with white; secondaries, and some of the longer scapulars, likewise tipped with white: forehead, throat, and all the under parts, lower part of the back, rump and tail, pure white: bill ochre-yellow; the angle of the lower mandible orange-red: irides gamboge-yellow: orbits bright red: legs yellow. (*Summer plumage*.) The whole head and neck pure white, without any brown streaks: the rest as in winter. (*Young of the year.*) Scarcely to be distinguished from that of the last species, excepting by the shorter tarsus and some- what longer wings: plumage extremely similar: throat, and front of the neck, whitish, with longitudinal streaks of pale brown: rest of the under parts grayish white, almost entirely covered with large deep brown spots: upper plumage dusky brown, all the feathers edged with a narrow yel-

lowish band: primaries black, without any white at the tips: tail grayish
white at the base, variegated with black; the remaining portion deep
blackish brown, the feathers tipped with white: bill blackish, brown at
the base: legs dull yellow. (*Egg.*) Yellow stone-colour, thickly spotted
with ash-gray and two shades of brown: long. diam. two inches ten
lines; trans. diam. one inch eleven lines.

Not uncommon on the northern coasts of Britain: less abundant southward, though generally diffused. Like the last species, occasionally
observed inland on the banks of rivers. Breeds on rocks and steep
shores, and lays from two to four eggs. Food, fish, worms, and marine
rejectamenta.

300. L. *marinus*, Linn. (*Great Black-backed Gull.*)— Bill yellow: legs pale flesh-colour: tarsus three inches: wings reaching very little beyond the tail: shafts of the primaries black: mantle (in the *adult*) grayish black.

> L. marinus, *Temm. Man. d'Orn.* tom. II. p. 760. Great Black-backed
> Gull, *Mont. Orn. Dict. Selb. Illust.* vol. II. p. 507. pl. 97. Black-
> backed Gull, *Bew. Brit. Birds*, vol. II. p. 201.

DIMENS. Entire length near thirty inches: length of the bill three
inches six lines: breadth, wings extended, five feet nine or ten inches.
MONT.

DESCRIPT. Distinguished from the last species, which it closely
resembles, by its superior size, longer tarsus, and by the colour of the
legs. (*Adult in winter.*) Crown, region of the eyes, occiput and nape,
white, with longitudinal streaks of pale brown on the shafts of the
feathers: back, scapulars and wing-coverts, deep grayish black: forehead, throat, neck, all the under parts, rump and tail, pure white:
primaries black towards their extremities, all terminated by a large
white space; secondaries and scapulars tipped with white: bill thick
and strong; bright yellow; the angle of the lower mandible blood-red,
enclosing a dusky spot not found in the last species: irides gamboge-
yellow: orbits red: legs livid white, or pale flesh-colour. (*Summer
plumage.*) The whole head and neck pure white, without any streaks of
brown: orbits orange: the rest as in winter. (*Young of the year.*)
Similar to that of the two last species, but always larger, with the bill
stronger. Head, and fore part of the neck, grayish white, with numerous
brown spots largest on the neck: upper plumage dusky brown, all the
feathers edged and tipped with reddish white: wing-coverts marked with
transverse bars of this last colour: under parts dirty gray, with broad
zig-zag streaks and spots of brown: primaries dusky, sometimes finely
tipped with white: middle tail-feathers almost wholly dusky; lateral
feathers black towards their extremities; all edged and tipped with
whitish: bill black: irides and orbits brown: legs livid gray. *During
the second year*, the colours of the plumage undergo no material change:
only the gray on the head and under parts gradually gives place to a
purer white, while the brown spots decrease in size and number; the
base and tip of the bill also assume a livid tinge. *After the second
autumnal moult*, the mantle of grayish black begins to show itself,
though still variegated with large irregular brown spots; the white on
the under parts is now nearly perfected; the tail white, variegated in parts
with black; the bill becomes livid yellow spotted with black; and the
red spot, enclosing a black one, appears at the corner of the lower

mandible. *At the age of three years, or after the third autumnal moult,*
the plumage is matured. (*Egg.*) Yellow-brown tinged with green,
sparingly spotted with slate-colour and dark brown : long. diam. three
inches two lines; trans. diam. two inches four lines.

Not so plentiful as the two last species, but occasionally met with on
most parts of the British coast. Observed by Montagu in considerable
abundance on the extensive sandy flats of the coast of Caermarthenshire.
In the immature state, it is sometimes seen far inland. Breeds in the
Orkneys, and on the northern shores of Scotland. Nest placed on the
shelves of insulated rocks. Eggs three or four in number. Food, fish,
carrion, and any animal matter.

301.   L. *Islandicus*, Edmondst. (*Iceland Gull.*)—Bill
yellow : legs livid : tarsus two inches three lines : wings
reaching a little beyond the tail : shafts of the primaries
white : mantle (in the *adult*) pale bluish ash.

> L. Islandicus, *Edmondst. in Wern. Mem.* vol. iv. p. 506. *Flem.
> Brit. An.* p. 139. L. argentatus, *Sab. in Linn. Trans.* vol. xii.
> p. 546.? Iceland Gull, *Selb. Illust.* vol. ii. p. 501. pl. 98.

Dimens. Entire length twenty-two inches, (Edmondst.): length
of the bill (from the forehead) two inches; of the tarsus two inches
three lines, (Selb.): breadth, wings extended, four feet four inches,
(Edmondst.)

Descript. Distinguished from the next species by its smaller size,
and from the *L. argentatus*, which it also closely resembles, by the
entire absence of black on the primary quills. (*Adult in winter.*) " Back,
and upper wing-coverts, very pale blue; all the rest of the plumage
white, except the head and upper part of the neck, which are streaked
with gray, as occurs in the winter dress of the other large gulls; wing-
feathers and scapulars tipped with a more brilliant and pure white tinge
than that which occurs on the rest of the plumage : irides pale yellow :
bill smaller and more slender than in the *Herring Gull*: feet deep flesh-
colour." Edmondst. (*Immature Bird.*) " Ground-colour of the entire
plumage pale yellowish gray ; the feathers being barred and mottled with
pale broccoli-brown: quills grayish white, with a slight tinge of broccoli-
brown: tail pale broccoli-brown, marbled with white: bill pale flesh-
red, or livid, at the base, with the tip blackish, or dark horn-colour:
irides pale yellowish gray: legs and toes pale livid flesh-red." Selb.

A winter visitant in the Shetland Isles, and the northern parts of
Scotland, where it was first observed by Mr. Edmondston, who has
pointed out the distinctions between this and the next species. Mr.
Selby has also obtained a few specimens on the coast of Northumberland,
though all in immature plumage. By this last gentleman it is considered to be the same as the *L. argentatus* of Sabine, *l. c.* (the *L.
arcticus* of Macgillivray, *Wern. Mem.* vol. v. p. 268), which is described
as being plentiful in Baffin's Bay, Davis's Straits, and Melville Island.
Mr. Selby observes that it is also common upon the Iceland coast, and
thinks it probable that many of those which winter with us retire there
to breed. Feeds on fish, the flesh of whales, and other carrion.

302.   L. *glaucus*, Brunn. (*Glaucous Gull.*) — Bill
yellow : legs livid : tarsus three inches : wings barely

reaching to the end of the tail: primaries terminated by
a large white space; the shafts entirely white: mantle (in
the *adult*) pale bluish ash.

> L. glaucus, *Temm. Man. d'Orn.* tom. II. p. 757. *Sab. in Linn.*
> *Trans.* vol. XII. p. 543. Glaucous Gull, *Bew. Brit. Birds*, vol. II.
> pp. 209, & 212. *Selb. Illust.* vol. II. p. 498. pl. 99. Iceland Gull,
> *Edmondst. in Wern. Mem.* vol. IV. pp. 176, & 182.

DIMENS. Entire length twenty-nine inches: length of the bill
(above) three inches, (from the gape) four inches; of the tarsus three
inches; of the tail ten inches; of the wing nineteen inches: breadth,
wings extended, five feet two inches. *Faun. Bor. Am.*

DESCRIPT. Similar to the last, but larger, with the bill more robust,
and the tarsus longer. (*Adult in winter.*) Head and neck white,
streaked and mottled with gray: back, scapulars, and wing-coverts, pale
bluish ash, not so dark as in the *L. argentatus:* primaries terminated by
a large white space, extending upwards more than two inches; the shafts
entirely white: tips of the secondaries, tail, and all the under parts, also
pure white: bill very strong; wax-yellow; the angular projection of the
lower mandible bright arterial blood-red: orbits red: irides yellow: legs
livid flesh-colour. (*Summer plumage.*) Head and neck pure white, without the gray streaks: the rest as in winter. (*Young of the year.*) Whole
plumage mottled throughout with white, gray, and light brown, as in
the young of the *L. marinus* and *L. argentatus*, but the colours are
always paler than in those species, especially those of the quills, which
are brown instead of dusky, with the shafts entirely white: bill horn-
coloured at the base, the tip brownish black; the angular projection of
the lower mandible not so strongly defined as in the adult: legs flesh-
colour. From the young of the last species, they are only to be distinguished by their greater size. (*Egg.*) Stone-colour, spotted with ash-
gray, and two shades of red-brown: long. diam. two inches nine lines;
trans. diam. one inch eleven lines.

A regular winter visitant in the Shetland Islands, where it arrives
about the middle of Autumn. In the southern parts of Britain rare, and
only occasionally met with: such individuals generally young birds in
immature plumage. A bold and voracious species, preying on fish and
other birds, as well as on carrion. Retires to the Arctic Regions to
breed.

GEN. 110.   LESTRIS, *Illig.*

303. **L.** *Cataractes*, Temm. (*Common Skua.*)—Central
tail-feathers but slightly elongated; of equal breadth
throughout; square at the extremities: tarsus two inches
eight lines.

> L. Cataractes, *Temm. Man. d'Orn.* tom. II. p. 792. Cataractes
> vulgaris, *Flem. in Edinb. Phil. Journ.* vol. I. p. 99. Skua Gull,
> *Mont. Orn. Dict. & Supp. Bew. Brit. Birds*, vol. II. p. 229.
> Common Skua, *Selb. Illust.* vol. II. p. 514. pl. 100.

DIMENS. Entire length twenty-five inches; the same, central tail-
feathers excluded, twenty-two inches: length of the bill one inch nine
lines; of the tarsus two inches eight lines.

Descript. Head, and region of the eyes, deep brown: neck, and all
the under parts, reddish ash, tinged with brown: back and scapulars
dark ferruginous brown, the feathers edged at the sides with dusky
brown; wing-coverts, secondary quills, and tail-feathers, brown: basal
half of the primaries white; the remaining portion deep brown; the
first with the whole of the outer web brown; shafts of the quills, as
well as those of the tail-feathers, white: bill black, the base brownish:
irides brown: orbits black: legs covered with large black scales; back
part of the tarsus with very little indication of the projecting asperities
which characterize the next species: claws strong and black. *Obs.* The
plumage of this species is not subject to any important variation, either
from age or season. One, which lived ten years in confinement, ex-
hibited no tendency towards the light colour which characterizes the
adult under plumage of the other species of the genus. (*Egg.*) Olive-
brown, blotched with darker brown: long. diam. two inches nine lines;
trans. diam. two inches.

Met with in small numbers in the Shetland Islands, particularly in
those of Foulah and Unst, where they remain the whole year. Of
very rare and accidental occurrence in the southern parts of Britain.
Montagu mentions one that was shot at Sandwich in Kent, in the Winter
of 1800. Another has been since killed in Somersetshire. A bold and
rapacious species, obtaining its food principally by pursuing the larger
kinds of Gulls, and compelling them to disgorge the fish which they
have obtained. Flight very impetuous. Breeds in large companies
on high hills and unfrequented moors. Nest constructed of a few dried
weeds. Eggs two in number.

304. L. *pomarinus*, Temm. (*Pomarine Skua.*) —
Central tail-feathers projecting three inches; of equal
breadth throughout; rounded at the tips: tarsus two
inches and half a line; covered posteriorly with rough
angular scales.

L. pomarinus, *Temm. Man. d'Orn.* tom. II. p. 793. *Faun. Bor.
Amer.* part ii. p. 429. Pomarine Skua, *Shaw, Gen. Zool.* vol. XIII.
p. 216. pl. 24. *Selb. Illust.* vol. II. p. 517. pl. 101**.

Dimens. Entire length twenty-one inches; the same, central tail-
feathers excluded, eighteen inches: length of the bill (above) one inch
seven lines, (from the gape) one inch ten lines; of the tarsus two inches
and half a line; of the central tail-feathers nine inches six lines; of the
wing fifteen inches. *Faun. Bor. Am.*

Descript. (*Male and female in perfect plumage.*) Face, crown of the
head, occiput, back, wings, and tail, of a uniform deep brown: neck
straw-yellow; the feathers on the nape long and subulate: ear-coverts,
throat, breast and belly, white; vent, and under tail-coverts, blackish
brown; flanks, and sides of the breast, blotched with the same: shafts
of the quills and tail-feathers white, except at their tips: bill dark
brown, the tip black: legs black: tail slightly rounded, excluding the
central pair of feathers; these last of equal breadth throughout, and
rounded at the tips: tarsus posteriorly very rugose, clothed with angular
projecting scales. (*Middle age.*) "The whole plumage above and below
very deep brown; the feathers on the neck and nape slightly elongated,
tinged with yellow: central tail-feathers shorter than in the adult in

perfect plumage, but always of the same breadth throughout, and
rounded at the tips: bill and legs the same as described above." TEMM.
(*Young of the year*.) Head, neck, and all the upper parts, very dark
cinereous brown, approaching to dusky; the feathers on the back, scapulars, and wing-coverts, edged with reddish yellow: quills and tail dusky:
under parts yellowish ash, with dusky spots and indistinct transverse
undulating bars; upper and under tail-coverts with more distinct bands
of brown and yellowish white: base of the bill greenish blue; the tip
black: legs bluish ash; base of the toes, and membranes, whitish:
central tail-feathers not more than half an inch longer than the others:
the tips broad and rounded. (*Egg*.) Pale green; the larger end spotted
with two shades of red brown: long. diam. two inches three lines; trans.
diam. one inch six lines and a half.

Inhabits the Arctic Regions. In this country, at least in the adult
state, a rare and accidental visitant. Immature birds have been killed
at Dover, Brighton, near London, and in several instances on the eastern
coast of England. Food and habits similar to those of the last species.
Said to breed on low hills in marshy districts, or on rocks. Constructs
a nest of grass and moss coarsely put together, and lays two or three
eggs.

305. L. *Richardsonii*, Swains. (*Richardson's Skua*.)
—Central tail-feathers projecting three inches; gradually
tapering from the base; the tips acute: tarsus one inch
nine lines; slightly rugose posteriorly.

> L. Richardsonii, *Faun. Bor. Amer*. part ii. p. 433. pl. 73. Cataractes
> parasiticus, *Flem. in Edinb. Phil. Journ*. vol. I. p. 101. *Id. Brit.
> An*. p. 138. Larus parasiticus, *Edmondst. in Edinb. Phil. Journ*.
> vol. v. p. 169, & vol. VII. p. 91. Arctic and Black-toed Gulls, *Mont.
> Orn. Dict. & Supp. Bew. Brit. Birds*, vol. II. pp. 232, & 235.
> Arctic Skua, *Selb. Illust*. vol. II. p. 520. pls. 101, & 101*. Rich-
> ardson's Lestris, *Gould, Europ. Birds*, part iv.

DIMENS. Entire length twenty-one inches; the same, central tail-
feathers excluded, eighteen inches: length of the bill (from the forehead)
one inch two lines and a half, (from the gape) one inch nine lines and
a half; of the tarsus one inch nine lines; of the middle toe, nail included, one inch eight lines and a half; of the central tail-feathers nine
inches; from the carpus to the end of the wing thirteen inches.

DESCRIPT. (*Adult male and female*.) Forehead, crown, occiput, wings
and tail, deep black; the shafts of the quills and tail-feathers white to
near their tips: rest of the upper plumage deep dusky brown tinged with
cinereous: nape, and ear-coverts, straw-yellow: throat, breast, and belly,
yellowish white: flanks, vent, and under tail-coverts, deep dusky gray:
bill and legs black. (*Young of the year*.) Extremely similar to those
of the last species, from which they are scarcely to be distinguished
except by their inferior size, less robust bill, shorter and smoother tarsi,
and the sharp-pointed central tail-feathers, which in length, however, do
not at this age exceed the others by more than half an inch: legs, and
basal portion of the webs as well as toes, flesh-colour; the remaining
portion of the membranes black, in which state it is the *Black-toed Gull*
of English authors. *Obs*. In the *adult state* this species is subject to
considerable variation of plumage; in some instances all the under parts
being of a uniform dark brownish gray without any of the yellow and

yellowish white tints, which are probably not assumed till after a certain number of years. This variety is common to both sexes, and individuals may be found shewing every intermediate shade of colouring. (*Egg.*) Olive-brown, spotted with two shades of darker brown: long. diam. two inches four lines; trans. diam. one inch eight lines.

Of much more frequent occurrence than the next species, with which it appears to have been confounded by many authors. It is, however, principally confined to the Orkney and Shetland Islands, where it breeds in considerable abundance. Immature birds are occasionally met with on different parts of the English coast. Habits predatory, like those of the Common Skua. Nest constructed of dry grass, on unfrequented heaths. Eggs two in number. Is gregarious during the breeding season.

306. L. *parasiticus*, Temm. (*Arctic Skua.*)—Central tail-feathers projecting six inches or more; gradually tapering from the base; the tips acute: tarsus one inch six lines; slightly rugose.

> L. parasiticus, *Temm. Man. d'Orn.* tom. ii. p. 796. *Faun. Bor.
> Amer.* vol. ii. p. 430. Parasitic Gull, *Gould, Europ. Birds,*
> part iv.

Dimens. Entire length twenty-one inches: length of the bill one inch and six-eighths; of the tarsus one inch six lines; of the central tail-feathers twelve inches; of the wing eleven inches nine lines. Gould*.

Descript. Extremely similar to the last species, from which it is principally distinguished by its smaller size, longer central tail-feathers, and shorter tarsi. (*Adult male and female.*) "Top of the head, and the space between the bill and the eyes, of a deep blackish brown, terminating at the occiput; the whole of the upper surface of a clear brownish gray; quills and tail-feathers much darker; the throat, neck, and under surface, of a pure white, with the exception of the cheeks and sides of the neck, which are tinged with a delicate straw-yellow; legs and feet black. The *young* resemble in colouring the other species of the genus of the same age." Gould.

Apparently of rare occurrence in this country; nor is it known whether it ever breeds with us like the last species. A British-killed specimen is in the British Museum, but I am ignorant from what locality it was obtained. According to Mr. Thompson, it has repeatedly occurred in the Bays of Dublin and Belfast. It is possible that the *Arctic Gull* noticed by Montagu in the *Supplement* to his "Ornithological Dictionary," may have been of this species, but in all other cases the bird described under that name by English authors is referable to the *L. Richardsonii*. According to Mr. Gould, the present species inhabits the shores of the Baltic, the rugged coasts of Norway, and the Polar Regions. Habits similar to those of the rest of the genus.

---

* The following are the dimensions of a *British specimen*, in the British Museum, as given in the *Faun. Bor. Amer.* by Dr. Richardson. "Length, total, twenty inches; the same, excluding central tail-feathers, fourteen inches; of central pair eleven inches; of wing eleven inches six lines; of bill above one inch; of bill to rictus one inch seven lines and a half; of tarsus one inch seven lines; of middle toe one inch three lines; of hind toe two lines; of hind nail two lines; transverse diameter of the bill at the front four lines."

GEN. 111. PROCELLARIA, *Linn.*

(1. PROCELLARIA, *Vig.*)

307. P. *glacialis*, Linn. (*Northern Fulmar.*)—White; back and wings hoary ash: bill and legs yellow.

> P. glacialis, *Temm. Man. d'Orn.* tom. II. p. 802. Fulmar, *Mont. Orn. Dict. & Supp. Bew. Brit. Birds*, vol. II. p. 239. Fulmar Petrel, *Selb. Illust.* vol. II. p. 525. pl. 102.

DIMENS. Entire length seventeen inches: length of the bill (from the forehead) one inch seven lines, (from the gape) one inch nine lines and a half; of the tarsus one inch ten lines; of the tail five inches; from the carpus to the end of the wing twelve inches.

DESCRIPT. (*Adult.*) Head, neck, rump, tail, and all the under parts, pure white: back, scapulars, wing-coverts, and secondary quills, bluish ash; primaries dusky gray: tail rounded at the tip: bill yellow; the nasal tube tinged with orange: irides and legs yellowish. (*Young of the year.*) "All the parts of the body pale gray tinged with brown; the feathers on the back and wings tipped with deeper brown: primary quills and tail-feathers of one uniform tint of brownish gray: before the eyes an angular dusky spot: bill and legs yellowish ash." TEMM. (*Egg.*) Dirty yellowish white: long. diam. two inches ten lines; trans. diam. two inches.

Not uncommon in some of the islands off the North of Scotland, but very rarely seen in the southern parts of Britain. Has been killed on the coast of Durham, as well as, in a single instance, in South Wales. Breeds in St. Kilda, and is said to lay one egg in the holes of rocks. Habits very oceanic: seldom resorts to land, except for the purpose of nidification. Food, fish, the flesh of whales, and other animal matter.

(2. PUFFINUS, *Ray.*)

308. P. *Puffinus*, Linn. (*Cinereous Shearwater.*)— Bill two inches long: tarsus one inch ten lines: tail conical. TEMM.

> P. Puffinus, *Temm. Man. d'Orn.* tom. II. p. 805. Cinereous Shearwater, *Shaw, Gen. Zool.* vol. XIII. part i. p. 227. *Selb. Illust.* vol. II. p. 528. pl. 102*.

DIMENS. Length eighteen inches. TEMM.

DESCRIPT. "Head, back part of the neck, and the upper plumage, blackish brown, with the margins and tips of the feathers of the scapulars lighter: throat, lower part of the neck, and the whole of the under plumage, deep ash-gray, with a tinge of broccoli-brown: quills and tail brownish black: legs having the outer part of the tarsus deep gray; the inner part, and webs, yellowish." SELB.

Mr. Selby informs us that he has lately obtained an individual of this species on the coast of Northumberland. It has not been noticed by any other British author. Possibly, however, it may have been confounded with the *P. Anglorum*, which it is said closely to resemble. According to Temminck, it is common in the Mediterranean and on the coast of Spain. Habits unknown.

309. P. *Anglorum*, Temm. (*Manks Shearwater.*) —
Bill very slender ; one inch seven or eight lines in length :
tail rounded ; wings extending a little beyond its extremity :
tarsus one inch nine lines.

P. Anglorum, *Temm. Man. d'Orn.* tom. ii. p. 806.   Shearwater,
Mont. Orn. Dict. Bew. Brit. Birds, vol. ii. p. 241.   Manks
Shearwater, *Selb. Illust.* vol. ii. p. 529. pl. 102.

Dimens. Entire length fourteen inches : length of the bill (from the
forehead) one inch six lines, (from the gape) one inch nine lines ; of the
tarsus one inch nine lines ; of the tail three inches seven lines ; from
the carpus to the end of the wing nine inches six lines.

Descript. Forehead, crown, nape, and all the upper parts, including
the wings and tail, deep grayish black, with a slight gloss : sides of the
neck mottled with indistinct transverse bars of gray and white : chin,
throat, and all the other under parts, pure white : some of the under
tail-coverts black on their outer webs : bill yellowish brown at the base,
dusky towards the tip : legs gray ; the front of the tarsus, and part of
the webs, yellowish white, tinged with flesh-red. (*Egg.*) White : long.
diam. two inches five lines ; trans. diam. one inch nine lines.

Formerly very abundant during the breeding season in the Calf of
Man, but less so of late years.   Common in the Orkneys, and some other
of the Scotch Islands, to which they resort in February or March.   Builds
in the interstices of rocks, sometimes in rabbit-burrows, laying but one
egg.   Food, fish, and marine insects and worms.   Habits crepuscular ;
seldom appearing abroad during the bright sunshine.

310.   P. *fuliginosa*, Strickland.

Puffinus fuliginosus, *Proceed. of Zool. Soc.* 1832. p. 129.

Dimens.   Entire length eighteen inches : length of the bill (from the
forehead) one inch and seven-eighths, (from the gape) two inches and
a half ; of the tarsus two inches and a quarter ; of the middle toe two
inches and a half ; of the wing twelve inches. Strickl.

Descript.   Brown above and below ; the wings of a deeper tint :
throat faintly tinged with gray : bill of the same colour : tarsi externally, and the outer toes, brown ; tarsi internally, and membranes,
ochraceous brown. Strickl.

A single individual of this species, supposed to be the *Procellaria
fuliginosa* of Kuhl, was shot in the middle of August 1828, at the mouth
of the Tees, and is now in the collection of Mr. Arthur Strickland, of
Boynton in Yorkshire.   It is the only specimen hitherto obtained.
Native country and habits unknown.

(3. Thalassidroma, *Vig.*)

311.   P. *pelagica*, Linn. (*Stormy Petrel.*) — Black :
tail even ; the wings extending a little beyond its tip :
tarsus ten lines and a half.

P. pelagica, *Temm. Man. d'Orn.* tom. ii. p. 810.   Stormy Petrel,
Mont. Orn. Dict. & Supp.   Bew. Brit. Birds, vol. ii. p. 246.
Selb. Illust. vol. ii. p. 533. pl. 103. f. 2.

DIMENS. Entire length five inches ten lines.

DESCRIPT. Head, back, quills, and tail, ink-black: wing-coverts brownish black; a broad transverse bar of white on the rump; some of the scapulars and secondary quills tipped with white: all the under parts brownish black: bill and legs black: irides brown. (*Egg.*) White; of a roundish-oval form: long. diam. one inch one line; trans. diam. ten lines.

A common species on some parts of the British coast, especially northwards. Very abundant in the Orkneys and Hebrides, where it breeds in holes in the ground or in the clefts of rocks, laying one, sometimes two eggs. Generally keeps far out at sea, though occasionally driven inland by storms. Individuals under such circumstances, have been met with in various parts of England. Skims the surface of the ocean, feeding on the marine insects which float on the waves. When approached, ejects from its mouth an oily substance of a very rancid smell.

312. P. *Leachii*, Temm. (*Leach's Petrel.*) — Black : tail forked ; the wings not reaching beyond its tip : tarsus eleven lines and a half.

P. Leachii, *Temm. Man. d'Orn.* tom. II. p. 812. Leach's Petrel, *Steph. in Shaw's Gen. Zool.* vol. XIII. part i. p. 219. pl. 25. Forktailed Petrel, *Bew. Brit. Birds,* vol. II. p. 244. *Selb. Illust.* vol. II. p. 537. pl. 103. f. 1.

DIMENS. Entire length seven inches six lines.

DESCRIPT. Head, neck, and under parts, grayish black: back and scapulars pitch-black: quills black; the coverts lighter, forming a bar of dusky brown: upper and lateral under tail-coverts white, the shafts and tips black: tail black, of twelve feathers, forked; the depth of the fork half an inch: bill and legs black. (*Egg.*) White; roundish-oval: long. diam. one inch four lines; trans. diam. eleven lines.

First discovered by Mr. Bullock in St. Kilda. Since ascertained to be not uncommon in that island, though rare elsewhere. Solitary individuals have been occasionally met with in different parts of England. Breeds in the clefts of rocks, and lays one egg. Food and habits similar to those of the last species.

(26.)   P. *Wilsoni*, Bonap. — Vigors in Zool. Journ. vol. I. p. 425. P. *pelagica*, Wils. Amer. Orn. vol. VII. p. 90. pl. 60. f. 6.

I am informed by Mr. Yarrell that this species has been killed in the British Channel, though at some distance from land. It inhabits the western shores of the Atlantic, and is principally distinguished by a large oblong yellow spot on the membranes of the toes. Length of the tarsus nearly an inch and a half.

CLASS III. REPTILIA.

ORDER I. TESTUDINATA.

Body enclosed in a double shield; the head, neck, limbs, and tail, alone free; the upper shield formed by the union of the ribs and dorsal vertebræ, the lower one by the pieces of the sternum: jaws horny, without teeth: four feet.

I. CHELONIADÆ.—*Feet, especially the anterior pair, elongated; compressed, fin-shaped.*

1. SPHARGIS. — Shell covered with a continuous coriaceous skin: claws obsolete.

2. CHELONIA.—Shell covered with horny plates: feet with claws.

ORDER II. SAURIA.

Skin covered with scales: body and tail elongated: jaws furnished with teeth: generally four feet.

I. LACERTIDÆ. — *Tongue slender, extensile, bifid: all the feet with five toes; these last*

*separate, unequal, armed with claws: scales dis-
posed, under the belly and round the tail, in
transverse parallel bands.*

3. LACERTA.——Palate armed with two rows of teeth:
upper part of the head protected by large squamous plates,
terminating posteriorly in a line with the orifices of
the ears: a collar on the under side of the neck formed
by a transverse row of flat broad scales, separated from
those of the breast by a space covered with small granulated scales*: scales on the abdomen much broader than
those on the back, and not keeled: one row of femoral
pores on each thigh.

ORDER III. OPHIDIA.

Skin covered with scales: body cylindrical, very much
elongated: jaws furnished with teeth: no feet.

I. ANGUIDÆ.——*A third eyelid: body entirely
covered with imbricated scales: jaws not dilatable: rudimentary scapular and clavicular bones
beneath the skin.*

4. ANGUIS.——No appearance of extremities visible
externally: tympanum concealed beneath the skin:
maxillary teeth compressed and hooked; no teeth on
the palate.

---

* This last character, though applicable to the few species met with in this country, must be
received with some limitation in the case of two or three others found on the Continent, in which
the collar, though still free at the sides, is interrupted in the middle.

II. SERPENTIDÆ.—*No third eyelid: abdomen covered with broad transverse plates: jaws dilatable: no vestiges of bones of the sternum and shoulder.*

5. NATRIX. — Subcaudal plates arranged in pairs: four nearly equal rows of imperforate teeth above, and two below: no poison-fangs.

6. VIPERA. — Subcaudal plates arranged in pairs: maxillaries armed with poison-fangs, but without ordinary teeth.

T

ORDER I.   TESTUDINATA.

GEN. 1.   SPHARGIS, *Merr.*

1. S. *coriacea*, Gray. (*Coriaceous Turtle.*) — Shell oval, pointed behind, with three longitudinal ridges.

> S. coriacea, *Gray, Syn. Rept.* part i. p. 51. Testudo coriacea, *Linn. Syst. Nat.* tom. i. p. 350. *Turt. Brit. Faun.* p. 78. Coriudo coriacea, *Flem. Brit. An.* p. 149. Turtle, *Borl. Nat. Hist. of Cornw.* p. 285. pl. 27. f. 4. Coriaceous Tortoise, *Penn. Brit. Zool.* vol. iii. p. 7. pl. 1. Coriaceous Turtle, *Shaw, Gen. Zool.* vol. iii. part i. p. 77. pl. 21.

Dimens. Said to attain the length of eight feet.

This species, which is a native of the Mediterranean, has been occasionally taken in our seas. Borlase mentions two which were caught in the mackerel-nets off the coast of Cornwall, in July 1756. The largest measured six feet nine inches from the tip of the nose to the end of the shell, and was adjudged to weigh eight hundred pounds. Pennant speaks of a third individual, of equal weight with that just alluded to, which was taken on the coast of Dorsetshire. *Obs.* The anterior extremities in this species are said to be proportionably longer in the young animal than in the adult.

GEN. 2.   CHELONIA, *Brongn.*

2. C. *imbricata*, Gray. (*Imbricated Turtle.*) — Shell elliptic, carinated, with the plates of the disk imbricated.

> C. imbricata, *Gray, Syn. Rept.* part i. p. 52. Testudo imbricata, *Linn. Syst. Nat.* tom. i. p. 350. *Turt. Brit. Faun.* p. 78. Chelona imbricata, *Flem. Brit. An.* p. 149. Imbricated Turtle, *Shaw, Gen. Zool.* vol. iii. part i. p. 89. pl. 26.

Dimens. General length, from the tip of the bill to the end of the shell, about three feet: has been known to measure five feet. Shaw.

Descript. Body roundish-ovate, slightly heart-shaped, slightly carinated down the back: head small, prominent; with the upper mandible curved over the lower: two claws on each foot: plates of the disk imbricated, thirteen in number, rather square, semi-transparent, variegated; of the circumference twenty-five, pointed and incumbent on each other in a serrated manner: tail a mere notch. Turt.

A native of the American seas: has occurred, however, in a few instances, as a straggler, on the British coasts. The first individual (according to Dr. Fleming) is recorded by Sibbald, as having appeared in Orkney. Dr. Fleming himself states that he has "credible testimony of its having been taken at Papa Stour, one of the West Zetland Islands." Dr. Turton has also mentioned one which was taken in the Severn in the Spring of 1774. This last was placed in a fish pond, where it lived till the Winter.

ORDER II. SAURIA.

GEN. 3. LACERTA, *Cuv.*

3. L. *Stirpium*, Daud. (*Sand Lizard.*) — Occipital plate rudimentary; frontal large, nearly as broad behind as before: temples covered with small plates: abdominal lamellæ in six longitudinal rows: fore feet with the third toe longest: femoral pores from twelve to fifteen.

Lézard des Souches, *Edwards in Ann. des Sci. Nat.* (1829) tom. xvi. p. 65. *Dugès, Id.* p. 377.

DIMENS. The following are those of an English specimen. Entire length seven inches: length of the head (measured above to the posterior margin of the occipital plate) nine lines, (underneath from the extremity of the lower jaw to the posterior margin of the collar) one inch and half a line; of the body (from the collar to the anus) two inches one line; of the tail three inches ten lines; of the hind leg one inch three lines and a half; of the fore leg ten lines and a half. These measurements are, however, probably often exceeded.

DESCRIPT. (*Form.*) Larger than the common species; the body and limbs thicker and stronger in proportion to the entire length. Occipital plate rudimentary, very much smaller than the parietal plates; frontal large, and nearly as broad at its posterior, as at its anterior margin; space between the eye and the meatus auditorius covered with small plates of various sizes: collar composed of eleven lamellæ; the margin irregularly toothed or notched: pectoral triangle well-defined: abdominal lamellæ in six longitudinal rows; the two middle rows much narrower than the adjoining ones, with the lamellæ of a parabolic form: ante-anal lamella single*, large, somewhat pentagonal: dorsal scales small, of an irregular form, approaching to square or hexagonal, with a distinct longitudinal keel directed somewhat obliquely; those on the sides of the body

* In one specimen it was observed to be double. but this is probably accidental.

larger, with the keel obsolete: caudal scales oblong, but becoming longer
and narrower as they approach the tip of the tail, each terminating below
in an obtuse point, and furnished with a longitudinal keel, which also
becomes more strongly marked towards the extremity: tail itself moderately stout at its origin, but gradually tapering to a fine sharp point;
with fifty-three (Dugès says from fifty to eighty) whorls of the scales last
described: fore legs not reaching beyond the eyes, when placed against
the sides of the head; strong, with the third toe a little longer than the
fourth; all the claws strong and sharp, and more developed than those
on the hind feet: hind feet reaching to the carpus of the fore: thighs
very much compressed; the number of femoral pores varying (according
to Dugès) from twelve to fifteen,—in this specimen, on the right thigh
thirteen, on the left fifteen. (*Colours.*) Said to be very variable. In my
specimen, the upper parts dark green, thickly spotted with black; a
broad interrupted fascia of dark greenish brown down the middle, containing interrupted lines of yellow spots: under parts light bluish green,
with small black spots much less numerous than above.

Of this species I have seen but two indigenous specimens, which
were obtained by W. Yarrell, Esq. from the neighbourhood of Poole
in Dorsetshire. For one of these I am indebted to the kindness of
that gentleman. It is common in France, and will probably be met
with in other parts of our own country, as soon as our native Reptiles
shall have received more attention from naturalists. It is very distinct
from the *L. agilis*, though at first sight, and without close examination,
it might pass for a large variety of that species. With its habits I am
unacquainted. *Obs.* The *L. arenicola* of Daudin is a variety of this
species.

(1.) L. *viridis*, Daud. Hist. Nat. des Rept. tom. III. pl. 34.
Lacertus viridis, Ray, Syn. Quad. p. 264. *Lézard piqueté*.
Edwards in Ann. des Sci. Nat. (1829) tom. XVI. p. 64. *Lézard
vert*, Dugès, Id. p. 373.

> Larger, with the tail much longer in proportion to the body, than the
> *L. Stirpium*. Occipital plate small, and triangular; interparietal lozenge-
> shaped; frontal very large, quite as broad at its posterior as at its anterior
> margin: collar consisting of eight lamellæ; the alternate ones smaller, and
> of a triangular form: abdominal lamellæ in six rows, the two middle ones
> much narrower than the others: tail with upwards of a hundred whorls of
> scales: toes long and slender; the third and fourth on the fore feet of
> equal length: femoral pores from fifteen to eighteen. *Colour* generally a
> brilliant green variegated with black specks on the back, flanks, and limbs:
> abdominal lamellæ plain yellowish green. Attains a length of eighteen
> inches.
>
> This species, which is well known on the Continent and in the Island of
> Guernsey, is said by Ray to be found in Ireland, but its existence in this
> last country does not appear to have been confirmed by any subsequent
> observer. It is also doubtful whether it be indigenous in any part of
> England, though possibly the "beautiful green *Lacerti*" observed by
> Mr. White "on the sunny sand-banks near Farnham, in Surrey,"[*] may
> have belonged to this species.

4. L. *agilis*, Berkenh. (*Common Lizard.*)—Occipital
plate rudimentary; frontal large, as broad behind as

---

[*] *Nat. Hist. of Selborne:* seventeenth letter to Mr. Pennant.

before: temples covered with small plates: abdominal lamellæ in six rows: fore feet with the fourth toe longest: femoral pores from nine to eleven.

> L. agilis, *Berkenh. Syn.* vol. i. p. 56. *Sheppard in Linn. Trans.* vol. vii. p. 49. *Turt. Brit. Faun.* p. 79. *Flem. Brit. An.* p. 150.
> Scaly Lizard, *Penn. Brit. Zool.* vol. iii. p. 21. pl. 2. no. 7.

Dimens. Entire length from six inches to six inches nine lines. Relative proportions very variable.

Descript. (*Form.*) In every respect smaller, and more slender, than the *L. Stirpium:* snout rather sharper: head more depressed, with the superciliary plates raised above the level of the crown: occipital plate very small; frontal large, as broad at its posterior as at its anterior margin: temples covered with small plates, more numerous than in *L. Stirpium:* collar composed of nine nearly equal lamellæ, with the posterior margin entire: pectoral triangle ill-defined, the lamellæ crowded together in an irregular manner: abdominal lamellæ in six longitudinal rows; the two middle rows a little narrower than the adjoining ones, with the lamellæ in these rows approaching to square or rectangular: dorsal scales rather narrower than in the *L. Stirpium;* not carinated, or with the keel very obsolete: caudal scales similar, but the keel of these also less strongly marked; the terminal point of each scale is likewise more obtuse, causing the whorls to appear less crenated: feet much slenderer than in the above species; the fore feet with the fourth toe a little longer than the third; claws small, and not more developed before than behind: thighs scarcely compressed; the number of femoral pores tolerably constant, generally nine, sometimes ten, rarely eleven. The following are *sexual distinctions.* In the *male*, the tail and legs are longer in proportion to the body; the former is nearly (in some specimens quite) two-thirds of the entire length; the hind leg, applied to the side of the abdomen, reaches to, or passes beyond, the carpus of the fore foot: the ante-anal lamella is shorter and broader, or more transverse: the under side of the base of the tail is flattened, with a slight longitudinal depression in the middle just behind the vent; during the season of sexual excitement the base of the tail is much dilated at the sides, appearing swollen. In the *female*, the abdomen is longer, and the tail shorter, the latter being often not more than half the entire length: the hind leg barely reaches to the tips of the claws of the fore foot: the ante-anal lamella is longer in proportion to its breadth, and of a more decided hexagonal or pentagonal form: the base of the tail is rounded, and convex underneath, and never dilated at the sides* (*Colours.*) Extremely variable. Upper parts generally cinereous brown, more or less dark, often tinged with bluish green; a dark list down the middle of the back, with parallel fasciæ at the sides; these last broader than the former, commencing behind the eyes, and sometimes extending to near the extremity of the tail; between the mesial list and lateral fasciæ, are one or more rows of black spots, and sometimes the same number of yellow ones: under surface of the body and base of the tail, and sides of the abdomen, in the *male*, bright orange, more or less spotted with black; in the *female*, generally pale yellowish green without spots. *Obs.* In some individuals, the whole of the upper parts are plain cinereous brown, without any markings whatever.

---

* Some of the above distinctions were first pointed out by Mr. Gray in a communication made to the Zoological Society, in May 1832. (See *Proceed. of Zool. Soc.* 1832. p. 112.) I have myself since examined a large number of individuals, and confirmed the accuracy of them.

An extremely abundant species in all parts of the country, frequenting heaths, moors, woods, sand-banks, &c. Is fond of basking in the sunshine, and in warm weather is extremely active. Forms a retreat under ground, in which it resides wholly during Winter. Is first seen in March, or early in April. Feeds principally on insects. Is ovoviviparous; the young broods appearing in June or July. Tail extremely brittle, but, when broken, gradually reproduced. The renewed part, however, according to Dugès, never acquires vertebræ.

(2.) *L. œdura*, Sheppard in Linn. Trans. vol. VII. p. 50.

This supposed species is principally characterized by the circumstance of the "tail bulging out a little below the base, which gives it the appearance of having been cut off and set on again." I am indebted to Mr. Gray for a suggestion, which he has since published in the *Proceedings of the Zoological Society*, (1834. p. 101.) that it is nothing more than the *male* of the common species in summer, when under the full influence of sexual excitement. I think this extremely probable.

(3.) *L. anguiformis*, Sheppard in Linn. Trans. vol. VII. p. 51.

"Head very light brown above, with four dark spots; yellowish white beneath: back with a black line along the middle, reaching from the head to about half an inch beyond the hind legs; on each side of this a broader one of dark brown (these beyond the black line unite, and reach to the end of the tail); next to these succeeds a fine yellow stripe that extends to the end of the tail; then a black one, which reaches no further than the middle line, and afterwards a dark brown stripe mixed with a few yellow spots extending to the end of the tail: a little above the hind legs, in some specimens, is a slight division of the scales, forming a transverse line: belly yellowish white, with a few black spots: tail, under part dirty white, spotted with black as far as within an inch of the end; the remainder marked lengthways with long bars of black: legs dark brown spotted with black. Length seven inches and upwards." Shepp.
Another species instituted by Mr. Sheppard, but too imperfectly characterized to rank as certainly distinct from those already described. Mr. Sheppard states that he once saw a specimen above a foot long, a length to which, I believe, the common *L. agilis* never attains. Unfortunately, however, this gentleman has in his description almost entirely confined himself to noticing the colours, than which, in these Reptiles, nothing can be more variable *.

Obs. Before concluding this account of our British *Lacertæ*, it may be stated that several other allied species, formerly confounded under the general name of *L. agilis*, are known on the Continent, some of which may possibly occur in this country, although hitherto overlooked by naturalists. Pennant speaks of a Lizard, "which was killed near Woscot, in the parish of Swinford, Worcestershire, in 1741, which was two feet six inches long, and four inches in girth†." He adds, that "another was killed at Penbury, in the same county." It is very possible that these may have been the *L. ocellata* of Daudin, which is found in the South of Europe, and which, according to Dugès, sometimes exceeds two

---

* Mr. Sheppard thinks that this species may be the *Lacerta anguiformis* of Ray. It is clear, however, that Ray, in his enumeration of the British species of "*Eft or Swift*," as he terms them, has only copied from Merrett, (*Pinax*, p. 169.) who, I suspect, by the *Lacertus terrestris anguiformis in Ericetis*, meant nothing more than our *Common Lizard*, which he calls *anguiformis*, in order to distinguish it from the *scale-less* Efts, belonging to the modern genus *Triton*, between which and the true *Lacertæ*, the writers of that day did not sufficiently discriminate. Merrett's other species, viz. 1. *Terrestris vulg. ventre nigro-maculato*, 2. *Parvus terrestris fuscus oppido rarus*, 3. *Aquat. fuscus*, and 4. *Aquat. niger*, are probably all referable to one or other of our two well-known British species of *Triton*, being called *terrestres* or *aquatici*, according as they may happen to have been found on land or in water.

† No further light is thrown upon this species, in the *Illustrations of the Natural History of Worcestershire*, lately published by Dr. Hastings, who simply alludes to the circumstance, as mentioned by Pennant.

feet in length. In the event of their occurring to any future observer, it may be useful to mention that the *L. ocellata*, independently of its great size, may be easily distinguished by the circumstance of its having the occipital plate very much developed, and at least quite as large as either the frontal or parietal plates* : it also possesses eight or ten longitudinal rows of abdominal lamellæ.

Another species, which may be briefly alluded to, is the *L. muralis* of Latreille, very common on the Continent, and apparently closely resembling our own *L. agilis*, from which, however, it would seem to differ in having the temples covered with very small granulated scales, resembling those of the back, in the middle of which is *one* circular plate † : the number of femoral pores is also much greater, varying from eighteen to twenty-five. It may be stated, that Mr. Gray is of opinion that this species is identical with the *Common Lizard* of this country. Judging, however, from the descriptions of French authors, I cannot but consider this as at present doubtful ‡.

ORDER III. OPHIDIA.

GEN. 4. ANGUIS, *Cuv.*

5. A. *fragilis*, Linn. (*Blind-Worm.*)

A. fragilis, *Linn. Syst. Nat.* tom. I. p. 392. *Turt. Brit. Faun.* p. 81. *Flem. Brit. An.* p. 155. *Cuv. Reg. An.* tom. II. p. 70. Cæcilia, *Ray, Syn. Quad.* p. 289. Blind-Worm, *Penn. Brit. Zool.* vol. III. p. 36. pl. 4. no. 15. Common Slow-Worm, *Shaw, Gen. Zool.* vol. III. p. 579.

DIMENS. Length from ten to twelve inches ; rarely more.

DESCRIPT. (*Form.*) Head small; body larger (more bulky in the *female* than in the *male*), cylindrical, and of nearly equal thickness throughout ; tail long, equalling half the entire length, sometimes more, blunt at the extremity : eyes small: gape extending a little beyond the eyes : teeth small, slightly hooked, with the points directed backwards : tongue broad ; the tip deeply notched : upper part of the head covered

---

* See a representation of the plates of the head in this species, as well as in *L. Stirpium* and *L. viridis*, in the *Ann. des Sci. Nat.* tom. XVI. pl. 5. f. 1, 3, & 4.

† This plate Edwards calls *disque massetérin.* See *Ann. des Sci. Nat.* tom. XVI. pl. 7. f. 3., where is a representation of the side of the head in this species.

‡ For a more detailed account of the above species, as well as of some others found on the Continent, which may possibly occur in England, I refer the reader to two valuable memoirs, already alluded to, one by Milne Edwards the other by Dugès, in the 16th volume of the *Annales des Sciences Naturelles.* In the same memoirs will be found an explanation of the nomenclature employed in designating the different external parts of these animals, more particularly of the plates on the upper part of the head, which furnish important characters for distinguishing some nearly allied species.

I would also recommend to our own naturalists, in drawing up descriptions of these Reptiles in future, to pay more attention to *form*, as opposed to *colour*. This last can scarcely ever be depended upon. It not only varies to a very great extent in the same species, but in the same individual, according to age, season, and the period of time which may have elapsed since the last moult of the cuticle. Edwards has observed that in general the spots are more regular and better defined in *young*, than in *adult* specimens.

with squamous plates; frontal large; parietal and interparietal plates
moderately developed, the latter of a triangular form, with the apex
directed backwards : sides of the head, throat, and all the upper as
well as under surface of the body and tail, covered with small imbricated scales of a rounded form and not keeled; those on the sides set
obliquely with respect to the axis of the body. (*Colours.*) Glistening
brownish gray above, inclining to reddish on the sides; bluish black
beneath : along the back several parallel rows of small dark spots :
sometimes all the upper surface light yellowish brown without spots;
the sides only marked with a dusky fascia, commencing behind the eyes,
and reaching to the extremity of the tail.   *Obs.* The markings are most
distinct in young specimens.

Common in most parts of the country. Frequents woods and gardens.
Feeds on worms and insects. Is ovoviviparous. Motion slow.

(4.) *A. Eryx*, Linn. Syst. Nat. tom. I. p. 392.   *Aberdeen Snake*,
Penn. Brit. Zool. vol. III. p. 35.

" Length fifteen inches : tongue broad and forked : nostrils small, round,
and placed near the tip of the nose : eyes lodged in oblong fissures above
the angle of the mouth : belly of a bluish lead-colour, marked with small
white spots irregularly disposed : the rest of the body grayish brown, with
three longitudinal dusky lines, one extending from the head along the
back to the point of the tail ; the others broader, and extending the whole
length of the sides : no *scuta* ; but entirely covered with small scales ;
largest on the upper part of the head." PENN.

The above is a description of a Snake, communicated to Linnæus and
Pennant by the late Dr. David Skene, and said to inhabit Aberdeenshire.
It is probably nothing more than a variety of the common *A. fragilis*.

GEN. 5.   NATRIX, *Flem.*

6.   N. *torquata*, Ray. (*Ringed Snake.*) — Dorsal
scales carinated : a lunulate yellow spot on each side of
the nape, with a black one behind.

N. torquata, *Ray, Syn. Quad.* p. 334.   *Flem. Brit. An.* p. 156.
Coluber Natrix, *Linn. Syst. Nat.* tom. I. p. 380.   *Turt. Brit.
Faun.* p. 81.   Ringed Snake, *Penn. Brit. Zool.* vol. III. p. 33.
pl. 4. no. 13.   *Shaw, Gen. Zool.* vol. III. p. 446.   Couleuvre à
collier, *Cuv. Reg. An.* tom. II. p. 83.

DIMENS. Length from three to four feet; sometimes more.   *Obs.*
The *female* is always much larger than the *male.*
DESCRIPT. (*Form.*) Head depressed, and broader than the neck;
body slender, elongated, thickest in the middle, gradually tapering posteriorly; tail about one-fifth of the entire length, rather sharp-pointed at
the extremity : gape the length of the head, arched, ascending upwards
behind : teeth very small, serrated, arranged in two rows on each side of
the jaws : upper part of the head protected by large squamous plates;
the frontal and fronto-parietal plates of considerable size ; seven plates
on each side of the upper jaw : dorsal scales imbricated, oval, with an
elevated keel down the middle; becoming broader and larger at the sides,
with the keel obsolete : plates of the belly broad, transverse, oblong, in
number about one hundred and seventy; subcaudal plates arranged in
pairs, from sixty to sixty-five on each side. (*Colours.*) Upper parts cine-

reous brown, tinged with green: at the back of the head a double lunulate spot, of a bright yellow colour, behind which is a double one of black, larger and more triangular: two rows of small black spots disposed longitudinally down the middle of the back, besides which are some larger ones on the sides, uniting to form short transverse undulating bars: throat, and beneath the neck, yellowish white; abdomen, and under surface of the tail, dusky blue, mottled in some places with yellowish white; edges of the abdomen with a series of yellowish white spots.

A common species; met with in woods and hedges, as well as in marshes. Is particularly abundant in the fens of Cambridgeshire, where it sometimes attains a large size. Often takes to the water, especially when alarmed; and swims easily: will occasionally remain at the bottom for a considerable time. Feeds on frogs, mice, insects, &c. Is oviparous. Eggs from sixteen to twenty in number, often deposited on dunghills, or under hedges. Hybernates during Winter: reappears in March, or early in April. When irritated, voids a fœtid substance.

> (5.) *N. Dumfrisiensis,* Flem. Brit. An. p. 156. *Coluber Dumfrisiensis,* Sow. Brit. Misc. p. 5. pl. 3. Loud. Mag. of Nat. Hist. vol. ii. p. 458. (Copied.)
>
>> An obscure species, of which little is known. Said to be particularly characterized by having "the scales of the back extremely simple, not carinated: plates on the belly one hundred and sixty-two: scales under the tail about eighty. Of a pale brown colour, with pairs of reddish brown stripes from side to side, over the back, somewhat zigzag; with intervening spots on the sides." Sow. Only one specimen known, which was discovered by T. W. Simmons, near Dumfries. According to Sowerby's figure, which is said to be of the natural size, its length does not exceed three or four inches. Probably an immature variety of the common species.

GEN. 6. VIPERA, *Daud.*

7. V. *communis,* Leach. (*Common Viper.*) — Three plates on the upper part of the head, larger than the surrounding scales; dorsal scales carinated: a series of confluent rhomboidal black spots down the back.

> V. communis, *Leach, Zool. Misc.* vol. iii. p. 7. *Flem. Brit. An.* p. 156. Vipera, *Ray, Syn. Quad.* p. 285. Coluber Berus, *Turt. Brit. Faun.* p. 80. Viper, *Penn. Brit. Zool.* vol. iii. p. 26. pl. 4. no. 12. Common Viper, *Shaw, Gen. Zool.* vol. iii. p. 365. pl. 101.

Dimens. Length from one and a half to two feet; rarely more.

Descript. (*Form.*) Shorter, and, in proportion to its length, thicker, than the *Natrix torquata.* Head depressed, widening behind the eyes; neck somewhat contracted; gape as long as the head, slightly ascending posteriorly; jaws very dilatable; two rows of fine teeth on the palatines, but none on the maxillaries, besides the poison-fangs: body gradually increasing in thickness to about the middle of the entire length, from that point scarcely diminishing to the vent, beyond which it tapers quite suddenly: tail very short, not one-ninth of the entire length, terminating in a sharp point: upper part of the head covered with small squamous plates, different from the imbricated scales of the back; of these plates three are larger than the rest, one situate in the middle between the eyes, the two others immediately behind the first: dorsal scales imbri-

cated, oval approaching triangular, carinated; increasing in size towards
the sides of the body, where the longitudinal keel becomes lost: beneath
the lower jaw some imbricated scales without a keel: plates of the belly
transverse, oblong, about one hundred and forty-three in number; subcaudal plates about thirty-three on each side. (*Colours.*) Extremely
variable: ground of the back and upper parts, in some, dirty yellow; in
others olive, or pale cinereous brown: space between the eyes, and an
oval patch on each side of the occiput, black or dark brown; a zigzag
dorsal fascia of the same colour commencing at the nape and reaching to
the extremity of the tail (in some the fascia assumes rather the appearance of a longitudinal row of confluent diamond-shaped spots); also a
row of small triangular black spots along each side parallel to the dorsal
fascia: belly, and beneath the tail, steel-blue, stained in some places with
yellowish; sometimes almost wholly black. *Obs.* The markings above
vary much in intensity of colouring, but always preserve, those on the
head especially, nearly the same form. The following are some of the
principal varieties noticed by authors.

Var. β. Red Viper. *Rackett in Linn. Trans.* vol. xii. p. 349. *Strickland in Loud. Mag. of Nat. Hist.* vol. vi. p. 399. *Gray in Proceed. of
Zool. Soc.* (1834) p. 101. Coluber Chersea, *Linn. Syst. Nat.* tom. i.
p. 377. Petite Vipère, *Cuv. Reg. An.* tom. ii. p. 92. "Of a bright ferruginous red, with zigzag markings down the back, resembling in form
those of the *Common Viper;* but instead of being black or dark brown,
they are of a deep mahogany colour: also a series of irregular spots of
the same colour along each side: the zigzag line terminates at the back
of the head in a heart-shaped spot, placed between two converging dark-
coloured bands, which meet on the top of the head, and again diverge
towards the eyes: belly ferruginous, like the back. Head much broader
and shorter than in the *Common Viper.*" Strickl.

Var. γ. Blue-bellied Viper. (Coluber cæruleus,) *Shepp. in Linn.
Trans.* vol. vii. p. 56.

Var. δ. Black Viper. *Leach, Zool. Misc.* vol. iii. pl. 124. Coluber
Prester, *Linn. Syst. Nat.* tom. i. p. 377. Wholly black, or very dark
brown; the markings hardly distinguishable.

Common in many parts of the kingdom, frequenting thickets, old chalk-
pits, and other waste places, more especially where the soil is dry. Said
to be most abundant in the Western Islands. In Cambridgeshire very
rare. Feeds on mice, frogs, and insects. Brings forth its young alive.

Var. β. was first obtained by the Rev. T. Rackett from Cranborne
Chase, in Dorsetshire. It has been since met with in Suffolk, Worcestershire, Somersetshire, and Berkshire. By some it is considered as a distinct species; I have, however, no hesitation myself in regarding it as a
mere variety of the common kind. The fact of its being always found of
a small size is probably due to the circumstance of the colours changing
in advanced life.

Var. γ. was described by Mr Sheppard, who considered it as another
distinct species. He does not state whence his specimen was obtained.

Var. δ. has been found in Suffolk, and a few other parts of England,
but is very rare.

Obs. The *Coluber Berus* of Linnæus is thought by Cuvier to be
the same as his *Vipère Commune*, a species perfectly distinct from the
Common Viper of England*.

---

* *Obs.* According to Mr. Lyell, (*Prin. of Geol.* vol. ii. p. 103.) none of the above three species
of *Ophidian Reptiles* have been observed hitherto in Ireland. According, however, to another
author, (*Edinb. New Phil. Journ.* vol. xviii. p. 373.) Snakes have been lately imported into
that country, and, "are at present (1835) multiplying rapidly within a few miles of the tomb of
St. Patrick."

CLASS IV. AMPHIBIA.

ORDER I. CADUCIBRANCHIA.

Gills deciduous.

I. RANIDÆ. — *No tail in the adult state: fore feet with four toes; hind feet with five, or with the rudiment of a sixth.*

1. RANA.——Skin smooth : hind feet very long, adapted for leaping; more or less palmated : upper jaw with a row of small fine teeth ; also a transverse interrupted row on the middle of the palate.

2. BUFO.——Body swollen : skin warty ; a porous protuberance behind the ears: hind feet of moderate length : jaws without teeth.

II. SALAMANDRIDÆ. — *Body elongated; tail always present: fore feet with four toes; hind feet with five.*

3. TRITON.——Tail compressed : jaws furnished with numerous small teeth ; two longitudinal rows of similar teeth on the palate.

ORDER I.   CADUCIBRANCHIA.

GEN. 1.   RANA, *Laurent.*

1. R. *temporaria,* Linn. (*Common Frog.*)—Reddish
or yellowish brown, spotted with black; an elongated
black patch behind the eyes.

> R. temporaria, *Linn. Syst. Nat.* tom. I. p. 357.   *Turt. Brit. Faun.*
> p. 80.   *Flem. Brit. An.* p. 158.   R. aquatica, *Ray, Syn. Quad.*
> p. 247.   R. fusca, *Ræs. Ran.* tabb. 1-3.   Common Frog, *Penn.*
> *Brit. Zool.* vol. III. p. 9.   *Shaw, Gen. Zool.* vol. III. p. 97. pl. 29.
> *Id. Nat. Misc.* vol. XX. pl. 864. (Variety.)   Grenouille rousse,
> *Cuv. Reg. An.* tom. II. p. 105.

DIMENS. (*Average.*) Length (from the end of the snout to the anus)
two inches seven lines; hind leg (from its union with the body to the
extremity of the longest toe) four inches; fore leg (measured in the same
way) one inch five lines and a half.   *Obs.* Often attains a larger size.

DESCRIPT. (*Form.*) Body slender, compared with that of the *Toad:*
head approaching triangular, the snout a little pointed: gape wide,
extending to a vertical line from the posterior part of the orbit: teeth
minute, forming a single row in the upper jaw; none in the lower; also
an interrupted row across the front of the palate: tongue soft, fleshy,
spatula-shaped, emarginated at the tip, folded back upon itself when not
in use: eyes somewhat elevated above the forehead: back generally flat;
sometimes a little gibbous behind: fore feet moderate, with four divided
toes; third toe longest; second shortest; first and fourth nearly equal:
hind feet more than half as long again as the body; the thighs strong
and muscular; toes on these feet palmated, five in number, with scarcely
the rudiment of a sixth; fourth toe considerably longer than any of the
others; third and fifth equal: skin naked, every-where smooth, excepting
between the thighs, where it is a little rugose. (*Colours.*) Variable:
above brown, yellowish brown, or reddish brown, more or less spotted
with black; the spots forming transverse fasciæ on the legs: beneath
whitish, or yellowish white; generally plain, but sometimes spotted like
the back.   The most constant mark is an elongated patch of brown or
brownish black behind the eyes, on each side of the occiput: there is also
generally more or less indication of a whitish line running longitudinally
down each side of the back, and enclosing a space paler than the adjoining regions.

Common and generally distributed in England and Scotland: said,
however, to have been unknown in Ireland previously to 1696, in which

year the species was introduced, for the first time, into that country*.
Frequents the water during its *larva* state; afterwards, only resorts
to it occasionally, or for the purpose of spawning. *Ova* deposited in
clusters, in ditches and shallow ponds, about the middle of March: young,
or *Tadpoles*, hatched a month or five weeks afterwards, according to the
season: by the eighteenth of June, these are nearly full-sized, and
begin to acquire their fore feet: towards the end of that month or the
beginning of the next (varying in different years), the young frogs come
on land, but the tail is still preserved for a short time afterwards.
During the breeding season, the thumb of the *male* is much swollen.
Food, principally insects.

(1.) *R. esculenta*, Linn. Syst. Nat. tom. I. p. 357. Turt. Brit.
Faun. p. 80. Flem. Brit. An. p. 159. *R. viridis*, Rœs. Ran.
tabb. 13, 14. *Edible Frog*, Penn. Brit. Zool. vol. III. p. 13. *Green
Frog*, Shaw, Gen. Zool. vol. III. p. 103. pl. 31. Id. Nat. Misc.
vol. xx. pl. 871.

Larger than the common species. Colour olive-green, spotted with
black : three longitudinal streaks of yellow down the back : belly yellowish.

This species, which is common in France and in other parts of the
Continent, has been included in the British Fauna upon rather doubtful
authority. In the late Mr. Don's account of the plants and animals
found in Forfarshire, it is asserted (p. 37.) that a few are occasionally to
be met with about the lakes in that district, although rather rare. More
recently, Dr. Stark is said † to have found it in the neighbourhood of
Edinburgh. I cannot but think, however, that, in both these instances,
some other species, possibly a new one, has been mistaken for it, since it
seems hardly probable that an animal so common in the South of Europe,
should be found in Scotland, and not in any part of England. Although
represented as indigenous by all our British authors, none, with the exception of those above mentioned, have assigned any locality for it. It is
much to be desired that Dr. Stark would investigate the subject more
thoroughly, and compare the specimens, which he finds in his neighbour-
hood‡, with the true *R. esculenta* of the Continent.

GEN. 2. BUFO, *Laurent.*

2. B. *vulgaris*, Flem. (*Common Toad.*) — Lurid
brownish gray, with reddish brown tubercles: body large
and swollen.

B. vulgaris, *Flem. Brit. An.* p. 159. B. terrestris, *Rœs. Ran.* tab. 20.
Rana Bufo, *Linn. Syst. Nat.* tom. I. p. 354. *Turt. Brit. Faun.*
p. 80. Bufo, *Ray, Syn. Quad.* p. 252. Toad, *Penn. Brit. Zool.*
vol. III. p. 14. Common Toad, *Shaw, Gen. Zool.* vol. III. p. 138.
pl. 40. Crapaud commun, *Cuv. Reg. An.* tom. II. p. 109.

DIMENS. Length three inches three lines; hind leg three inches six
lines; fore leg two inches.
DESCRIPT. (*Form.*) Body broad, thick, and very much swollen: head
large, with the crown much flattened, the snout obtuse and rounded:
gape extremely wide: no teeth either in the jaws or on the palate:

---

* See *Edinb. New Phil. Journ.* vol. xviii. p. 372.

† *Proceed. of Zool. Soc.* (1833) p. 88.

‡ That they are not simple varieties of the *R. temporaria*, is probable from the circumstance
of Dr. Stark's having observed osteological differences between them and the species just alluded
to. But I think it remains to be shewn that they are really the *R. esculenta*.

tongue with the apex entire: eyes moderately projecting; above each
a slight protuberance studded with pores; a larger protuberance of the
same kind on each side of the head behind the ears, with pores more
numerous and secreting a fœtid humour: fore feet with four divided
toes; third toe longest; first and second equal, both a little shorter
than the fourth: hind legs moderate, scarcely longer than the body;
the toes on these feet semi-palmated, five in number with the rudiment
of a sixth; fourth toe much the longest; third a little longer than
the fifth: skin every-where covered with warts and pimples of
various sizes; largest on the back, but most crowded beneath. (*Colours.*)
Upper parts of a lurid brownish gray, sometimes inclining to olive, at
other times to black; the colour of the tubercles rufous brown: beneath
yellowish white; either plain, or irregularly spotted with black.

Common in most parts of Great Britain: rare, however, in Ireland, if
not an introduced species in that country. Frequents the shady parts of
woods and gardens, cellars, and other damp places. Always a few days
later in spawning than the Frog; the difference, in some seasons, amounting to more than a fortnight. *Ova* deposited in long necklace-like chains.
Feeds on worms and insects, but is capable of remaining a long time
without nourishment. Said to be very long-lived. *Obs.* The *Great Frog*
of Pennant* is evidently nothing more than a large variety of this
species†.

3. B. *Calamita*, Laurent. (*Natter-Jack.*)—Olivaceous,
or yellowish brown; a bright yellow line down the middle
of the back : eyes very much elevated.

> Rana Bufo, β, *Gmel. Linn.* tom. I. part iii. p. 1047. R. Rubetra,
> *Turt. Brit. An.* p. 80. Bufo Rubeta, *Flem. Brit. An.* p. 159.
> B. terrest. fœtidus, *Ræs. Ran.* tab. 24. f. 1. Natter-Jack, *Penn.
> Brit. Zool.* vol. III. p. 19. *Jenyns in Camb. Phil. Trans.*
> vol. III. p. 373. Mephitic Toad, *Shaw, Gen. Zool.* vol. III. p. 149.
> pl. 43. *Id. Nat. Misc.* vol. XXIII. pl. 999.

DIMENS. Length two inches seven lines and a half; hind leg two
inches; fore leg one inch three lines.

DESCRIPT. (*Form.*) General appearance similar to that of the last
species; but the eyes more projecting, with the eye-lids very much elevated above the crown: porous protuberance behind the ears not so
large: toes on the fore feet more nearly equal; the third, notwithstanding, a little longer than the others; first and second not shorter than
the fourth: hind legs not so long as the body; the toes on these feet
much less palmated than in the *B. vulgaris;* the sixth toe scarcely at
all developed: skin similarly covered with warts and pimples. (*Colours.*)
Above, yellowish brown, or olivaceous, clouded here and there with
darker shades; a line of bright yellow along the middle of the back;
warts and pimples, especially the porous protuberance behind the eyes,
reddish: beneath, whitish, often spotted with black: legs marked with
transverse black bands.

First observed near Revesby Abbey in Lincolnshire, by the late
Sir J. Banks. Has been since met with in plenty on many of the
heaths about London, as well as on Gamlingay Heath in Cambridge-

---

* *Brit. Zool.* vol. III. p. 20.
† See, on this subject, Leach's *Zoological Miscellany*, vol. III. p. 9. pl. 125.

shire, and in two or three localities in Norfolk. Appears to affect dry sandy districts. Of much more active habits than the Common Toad, its pace being a kind of shuffling run: never leaps. Spawns later in the season.

Obs. Before concluding this family, it may be just stated, that amongst the British species, Merrett has enumerated the *Tree Frog,* (Ranunculus viridis, *Pinax Rer. Nat. Brit.* p. 169.) This, however, is so obviously a mistake, that there is no occasion to dwell longer on the circumstance.

GEN. 3. TRITON, *Laurent.*

4. T. *palustris*, Flem. (*Warty Eft.*)—Body rough ; with scattered pores ; a distinct lateral line of pores : dorsal and caudal crests (in the *male*) separate; the former deeply serrated.

T. palustris, *Flem. Brit. An.* p. 157. Lacerta palustris, *Linn. Syst. Nat.* tom. I. p. 370. *Shepp. in Linn. Trans.* vol. VII. p. 52. *Turt. Brit. Faun.* p. 79. Salamandra aquatica, *Ray, Syn. Quad.* p. 273. S. cristata, *Latr. Hist. Nat. des Sal. de France,* pp. 29, & 43. pl. 3. f. 3. A. Warty Lizard, *Penn. Brit. Zool.* vol. III. p. 23. pl. 3. Warted Newt, *Shaw, Nat. Misc.* vol. VIII. pl. 279. Great Water-Newt, *Id. Gen. Zool.* vol. III. p. 296. pl. 82. Salamandre crêtée, *Cuv. Reg. An.* tom. II. p. 116.

DIMENS. Entire length from five to six inches, rarely more.

DESCRIPT. (*Form.*) Head depressed: snout obtuse and rounded: gape extending a little beyond the eyes: teeth minute, sharp, slightly hooked, forming a single row in each jaw, and two parallel rows on the palate: a collar beneath the neck formed by a loose fold of the skin: fore feet extending a little beyond the snout; each with four flattened toes; third toe longest; second a little longer than the fourth; this last a little longer than the first: hind feet, placed against the sides of the abdomen, reaching to the carpus of the fore; with five toes, more developed than those in front; third and fourth toes equal, and longest; second longer than the fifth; first shortest: tail about two-fifths of the entire length; very much compressed, with its upper and under edges sharply keeled; of a lanceolate form, gradually tapering to an obtuse point: skin warty, uniformly covered with scattered pores; parotids porous; also a row of distinct pores on each side of the body, forming a line between the fore and hind legs. *Obs.* In the *male* the abdomen is rather shorter, compared with the entire length, than in the *female;* the hind feet are somewhat larger and stronger; the back, *during the spring,* is ornamented with an elevated membranous crest, commencing between the eyes, and running longitudinally down the mesial line to near the tail; this last is also furnished with a similar but separate membrane along its upper and under ridges, causing it to appear at the base as broad as the body; both membranes, but the dorsal more especially, are deeply jagged, and serrated. In the *female,* there is only a slight dorsal ridge occupying the place of the membrane in the other sex. (*Colours.*) Upper parts blackish brown, with round spots of a somewhat darker tint: breast and abdomen bright orange, or orange-yellow, with

conspicuous round black spots, sometimes confluent, and forming interrupted transverse fasciæ : sides dotted with white : frequently a silvery
white band along the sides of the tail : membranes dusky, tinged with
violet.

Not uncommon in ditches, ponds, and other stagnant waters, during
the spring months. Late in Summer, is sometimes met with on land, in
damp shady situations : this, however, is probably in consequence of the
drying up of the waters in its accustomed haunts. *Ova* deposited on
aquatic plants.

5. T. *punctatus*, Bonap. (*Common Eft.*) — Body
smooth, without pores; lateral line of pores indistinct ;
top of the head with two porous bands : dorsal and
caudal crests united, and uniformly crenate.

T. punctatus, *Bonap. Faun. Ital.* fasc. i. tab. 4. f. 4.   T. aquaticus,
Flem. Brit. An. p. 158.   Lacerta aquatica, *Linn. Syst. Nat.*
tom. i. p. 370 ?   L. maculata, *Shepp. in Linn. Trans.* vol. vii.
p. 53.   *Turt. Brit. Faun.* p. 79.   Salamandra punctata, *Latr.
Hist. Nat. des Sal. de France*, pp. 31, & 53. pl. 6. f. 6. A. (Male.)
B. (Female.)   Smaller or Common Water-Newt, *Shaw, Nat.
Misc.* vol. xi. pl. 412.   *Id. Gen. Zool.* vol. iii. p. 298. pl. 83.
Salamandre ponctuée, *Cuv. Reg. An.* tom. ii. p. 116.

Dimens.   Entire length from three and a half to four inches.
Descript. (*Form.*) Always much smaller than the last species, from
which it may be further distinguished by its smooth soft skin : tail terminating in rather a sharper point than in the *T. palustris :* fore feet,
relatively, a little longer ; but the disposition of the toes on both fore and
hind feet similar : very little trace of a collar beneath the throat : two
rows of pores on the top of the head, but none on the body ; occasionally
a few distant pores between the legs forming an indistinct lateral line.
In the *male*, the dorsal crest commences at the occiput, and is more elevated than in the *L. palustris ;* it also forms one continuous membrane
with the crest of the tail ; its margin, instead of being serrated, is regularly
crenate, or festooned, throughout its whole length : *during the season of
love*, the hind toes of this sex are also broadly fringed with dilated membranes. (*Colours.*) Above light brownish gray, inclining to olivaceous ;
beneath yellowish, passing into bright orange *in the spring :* every-where
marked with round black spots of unequal sizes : on the head the spots
unite to form longitudinal streaks ; there is generally also a yellowish
white fascia commencing beneath the eyes, and terminating a little beyond them.   *Obs.* The *female* is much less spotted than the *male ;* the
spots are also smaller : sometimes, in this sex, the under parts are quite
plain.

Equally common with the last species, and found in similar situations.

Obs. The above species is subject to considerable variation. It is also
often found on land, a circumstance which tends in some measure to alter
its characters. In such specimens, the skin loses its softness ; becoming
at the same time opaque, and somewhat corrugated : the membranes of
the back and tail entirely disappear, causing this last to appear narrower,
and thicker in proportion to its depth : the toes, from being flattened,
become rounded : the colours also are every-where more obscure. In
this state it is the *Lacerta vulgaris* of Sheppard and Turton (and pro-

bably of Linnæus also), the *Triton vulgaris* of Fleming, the *Brown Lizard* of Pennant, and the *Common Newt* of Shaw. By these authors, the variety in question is considered as a distinct species, an opinion to which I was formerly myself inclined. I am, however, now perfectly satisfied, from the examination of a large number of specimens, that it is identical with the aquatic kind, and that all its peculiarities may be traced to the change of circumstances under which it is placed. Sheppard lays great stress upon the fact of its being observed "of all sizes, from one to four inches in length, but never in any other than a perfect state;" and he considers this "a sufficient proof that, like the rest of the *land* lizards, it undergoes no change." The same circumstance is noticed by Shaw, who regards it as an argument in favour of its being viviparous. I suspect, however, that the period of time during which this species remains in the larva state, although perhaps constant in ordinary cases, is subject to much variation; and that if any thing occur to oblige the young to exchange their native element for another before they would naturally attain their perfect form, the gills are cast prematurely, to enable the animal to accommodate itself to its new circumstances. The fact of such small specimens, as Sheppard has noticed, being found on land is indisputable, but I think I have generally observed some traces of there having *been* gills at no very long period before. I may just add, that Sheppard appears to have confounded, as Pennant had done before him, the *males* of these reptiles, when possessing the dorsal and caudal fins, with the *larvæ*.

6. T. *vittatus*, Gray. (*Striped Eft.*)—Body smooth, without pores; lateral line of pores distinct; top of the head with two porous bands: dorsal and caudal crests irregularly and deeply notched.

T. vittatus, *Gray's Mss.*

DIMENS. Entire length four inches six lines.
DESCRIPT. Skin smooth: top of the head and parotids with scattered pores: a series of distant pores on the lower part of the sides between the fore and hind legs. *Male in summer* with a high, deeply notched, dorsal crest, commencing in front of the eyes, and with a deep notch over the vent; continued into a low entire crest extending the whole length of the tail. *Colour* white (yellow? when alive), with unequal black spots; tail black; belly, under sides of the legs and tail, and a broad streak along each side of the body and tail, white.
Var. β. Throat white, with a few spots; upper part of the tail pale, black-spotted: dorsal crest very low; caudal crest distinct.
Var. γ. Above black, beneath white; throat black-spotted: dorsal crest none.
A new species, discovered in ponds near London, by J. E. Gray, Esq. to whom I am indebted for the above description. It differs remarkably from both the preceding species in the form of the dorsal crest, and in the disposition of the colours. From the *T. palustris*, it may be further distinguished by its smooth skin.

U

CLASS V. PISCES.

(I. OSSEI.)

Skeleton bony; the osseous matter disposed in fibres: sutures of the cranium distinct: maxillary and intermaxillary bones, always one, and generally both, present.

§ *1.* PECTINIBRANCHII. — *Branchiæ in continuous pectinated ridges; furnished with an opercle and branchiostegous membrane: jaws complete, and free.*

ORDER I. ACANTHOPTERYGII.

The first portion of the dorsal fin, or the entire first dorsal when two are present, with simple spinous rays: also the anal and ventrals with one or more of the anterior rays generally spinous.

I. PERCIDÆ. — *Scales generally rough, with ciliated margins: margin of the opercle or preopercle, sometimes both, denticulated, or armed with spines: both jaws, as well as the vomer, and almost always the palatine bones also, armed with teeth.*

1. PERCA. — Body oblong, somewhat compressed: ventrals beneath the pectorals: branchiostegous membrane with seven rays: preopercle with the basal and posterior margins denticulated: jaws, vomer, and palatines, all armed with small teeth: scales rough; not easily detached.

(1. PERCA.) Two dorsals: opercle with the upper half covered with scales, terminating behind in a flattened point: infra-orbitals slightly denticulated: tongue smooth.

(2. LABRAX.) Two dorsals: opercle entirely covered with scales, terminating behind in two spines: infra-orbitals not denticulated: tongue rough with minute teeth.

(3. SERRANUS.) A single dorsal: cheeks and opercle covered with scales; the latter terminating behind in one or more flattened points: jaws with some elongated sharp teeth among the smaller ones.

(4. ACERINA.) A single dorsal: head without scales, pitted with indentations: opercle terminating behind in a single spine: teeth uniform.

2. TRACHINUS.—Head compressed; body elongated: ventrals before the pectorals: two dorsals; the first short; the second, as well as the anal, long: branchiostegous membrane with six rays: opercle with one strong spine directed backwards: two small spines in front of the eye: both jaws, as well as the vomer and palatines, armed with minute teeth.

(1.) *SPHYRÆNA.*—Body elongated: two dorsals remote from each other: head oblong: lower jaw pointed, longer than the upper, with some of the teeth larger than the others: branchiostegous membrane with seven rays: no denticulations on the preopercle, or spines on the opercle.

3. MULLUS.—Body oblong, thick: ventrals a little behind the pectorals: two dorsals widely separated: branchiostegous membrane with four rays: no denticulations on the preopercle, or spines on the opercle: teeth in the lower jaw and on the palatines only: chin with two long barbules: scales large, deciduous*.

---

* *Obs.* The characters of this and the last genus depart rather from those of the rest of the *Percidæ.* With respect to the *Mullets,* Cuvier observes that they might almost be considered as a distinct family.

II. LORICATI. — *Infra-orbitals extending more or less over the cheeks, articulating behind with the preopercle : head mailed, or otherwise armed.*

4. TRIGLA.——Head mailed; in the form of a parallelo-piped : opercle, and bones of the shoulder, armed with spines : body scaly : two dorsals : three detached rays beneath the pectorals : branchiostegous membrane with seven rays : fine velvet-like teeth* in both jaws and on the front of the vomer.

5. COTTUS.——Head broad, depressed ; more or less armed with spines : body naked, without scales : two dorsals, distinct, or very slightly connected : lower rays of the pectorals simple : branchiostegous membrane with six rays : teeth in both jaws, and also on the front of the vomer ; none on the palatines : ventrals small.

6. ASPIDOPHORUS.——Head broad, depressed; armed with spines and tubercles : body attenuated behind, mailed with angular plates : two dorsals, nearly contiguous : branchiostegous membrane with six rays ; furnished, as well as the chin, with small thread-like filaments : teeth in both jaws; none on the vomer or palatines : ventrals small †.

7. SCORPÆNA.——Head compressed, armed more or less with spines and tubercles : body oblong, scaly : a single dorsal : lower rays of the pectorals simple : branch-iostegous membrane with seven rays : velvet-like teeth in both jaws, and on the palatines.

(1. SEBASTES.) Head scaly ; spines on the preopercle and opercle ; no tubercles.

8. GASTEROSTEUS. — Head without spines or tubercles : body generally more or less protected by

* The terms employed to designate the different forms of teeth in fishes have been, in most instances, adopted from Cuvier. See *Hist. Nat. des Poiss.* tom. I. p. 362. See also Yarrell's *British Fishes*, vol. I. p. 99.

† Since the publication of my Catalogue, I have inclined to the opinion that this group should rank higher than as a mere subdivision of the genus *Cottus*.

shield-like plates: several free spines instead of a first dorsal: ventrals reduced nearly to a single spine: branchiostegous membrane with three rays: teeth in both jaws; none on the vomer or palatines.

> (1. GASTEROSTEUS.) Bones of the pelvis united, forming a triangular plate on the abdomen: ventrals with only one soft ray.

> (2. SPINACHIA.) Bones of the pelvis separate: ventrals with two soft rays: lateral line armed with large carinated scales.

III. SCIÆNIDÆ.—*Preopercle denticulated; opercle with spines: mouth but little protractile: no teeth on the vomer or palatines: bones of the face and cranium often cavernous, causing the snout to appear more or less protuberant.*

9. SCIÆNA.—Head protuberant, covered entirely, as well as the body, with scales: two dorsals; the second much longer than the first: anal short: preopercle, except in advanced age, denticulated; opercle terminating behind in two flat spines: branchiostegous membrane with seven rays.

> (1. SCIÆNA.) A row of strong pointed teeth in each jaw, accompanied, in the upper, by smaller ones behind: anal with only one small spine: no cirrus on the chin.

>> (*UMBRINA.*) A broad band of fine small teeth in each jaw without an anterior row of stronger ones: anal with two spines, the second strong: a barbule beneath the symphysis of the lower jaw.

IV. SPARIDÆ.—*No denticulations on the preopercle, or spines on the opercle: palate without teeth: mouth not protractile: body oval, covered with large scales; no scales on the vertical fins.*

10. SPARUS.—Sides of the jaws furnished with rounded molars forming a pavement: cheeks scaly: a single dorsal: branchiostegous membrane with five or six rays.

(1. Chrysophrys.) From four to six conical incisors in each jaw; molars large, in three or more rows.

(2. Pagrus.) Conical incisors in front, with card-like teeth behind; molars of moderate size, in only two rows.

(3. Pagellus.) All the anterior teeth fine and card-like: molars small, in two or more rows.

11. DENTEX.—All the teeth conical, and forming but a single row; some of the anterior ones longer than the others, and hooked: cheeks scaly: a single dorsal: branchiostegous membrane with six rays.

12. CANTHARUS. — All the teeth card-like, and crowded together; the anterior row larger, and more hooked, than the others: cheeks scaly: branchiostegous membrane with six rays.

V. SQUAMIPINNATI. — *Body compressed; scaly: dorsal and anal fins, or at least their soft portions, closely covered with scales.*

13. BRAMA.—Both jaws, as well as the palatine bones, with card-like teeth: dorsal and anal fins long: the spinous rays few in number: branchiostegous membrane with seven rays.

VI. SCOMBRIDÆ. — *Opercular pieces without denticulations: scales very small: body smooth: vertical fins not scaly.*

14. SCOMBER.—Lateral line unarmed: two dorsals; the first continuous; posterior rays of the second, as well as the corresponding rays of the anal, separated into spurious finlets: body fusiform: branchiostegous membrane with seven rays.

(1. Scomber.) First dorsal separated from the second by a wide space: scales small, and every-where uniform: sides of the tail with two small cutaneous ridges.

(2. Thynnus.) First dorsal reaching nearly to the second: some large scales surrounding the thorax, forming a corselet: sides of the tail with a cartilaginous keel between two small cutaneous ridges.

15. XIPHIAS.—Lateral line unarmed: a single dorsal, continuous: body elongated; the snout produced into a sword-like process: jaws without teeth: sides of the tail with a strong projecting keel: ventrals wanting.

16. CENTRONOTUS.—Lateral line unarmed: spines of the first dorsal free, and not connected by a membrane: ventrals always present.

(1. NAUCRATES.) Body fusiform: sides of the tail keeled: two free spines before the anal.

(*LICHIA.*) Body compressed: sides of the tail not keeled: two free spines before the anal: before the spines on the back, a reclined spine directed forwards.

17. CARANX.—Lateral line armed with large, imbricated, spinous plates: two distinct dorsals; before the first a sharp reclined spine directed forwards: some free spines before the anal; sometimes connected, forming a small fin: body fusiform.

18. ZEUS.—One dorsal; the spinous and soft portions separated by a deep notch: body oval, compressed: mouth very protractile: teeth small, few in number: ventrals thoracic.

(1. ZEUS.) Dorsal spines accompanied by filamentous prolongations of the membrane: a series of forked spines along the base of the dorsal and anal.

(2. CAPROS.) No spines at the base of the dorsal and anal fins: body covered with rough scales.

19. LAMPRIS.—Dorsal entire, very much elevated anteriorly: anal also elevated, with one small spine in front of the base: body oval, compressed: no teeth: ventrals, and lobes of the caudal, very much elongated; the former abdominal, with ten rays: sides of the tail keeled.

20. CORYPHÆNA. — Body compressed, elongated: upper part of the head presenting a sharp edge: one dorsal running the whole length of the back; all the

rays nearly equally flexible, but the anterior ones not articulated : branchiostegous membrane with seven rays.

(1. CENTROLOPHUS.) Head oblong: palate destitute of teeth: a space without rays between the occiput and commencement of the dorsal.

VII. TÆNIOIDEI.—*Body very much elongated, as well as compressed: scales small.*

* *Snout elongated : gape wide: teeth strong, sharp, and cutting : lower jaw projecting.*

21. LEPIDOPUS. — Ventrals reduced to two small scales : dorsal extending throughout the whole length : anal narrow : caudal well formed : branchiostegous membrane with eight rays.

22. TRICHIURUS. — Ventrals and caudal wanting : tail produced into a long, slender, compressed filament : anal represented by a series of small, almost invisible, spines : branchiostegous membrane with seven rays.

** *Mouth small, very protractile : teeth small.*

23. GYMNETRUS. — Anal entirely wanting : dorsal long ; the anterior rays prolonged : ventrals very long (but easily broken) : caudal of few rays, attached vertically to the extremity of the tail, which terminates in a small hook or claw : branchiostegous membrane with six rays.

*** *Snout very short ; gape oblique : teeth well developed.*

24. CEPOLA.—Dorsal and anal long, both reaching to the base of the caudal : ventrals moderately developed : branchiostegous membrane with six rays.

VIII. MUGILIDÆ. — *Body oval, approaching to cylindric ; covered with large scales : snout very short : mouth transverse, angular when*

closed, the lower jaw with an eminence in the middle fitting into a corresponding hollow in the upper: teeth extremely minute: two dorsals widely separated: ventrals a little behind the pectorals: branchiostegous membrane with six rays.

25. MUGIL.

26. ATHERINA*. — Body elongated: mouth very protractile: teeth minute: two dorsals widely separate: ventrals behind the pectorals: branchiostegous membrane with six rays: a broad silver band along each side.

IX. GOBIADÆ. — *Body elongated: one or two dorsals: the spinous rays always slender and flexible.*

27. BLENNIUS.—Ventrals before the pectorals, very much reduced: body elongated, compressed, smooth, covered with small scales: gill-opening large; the membrane continued across the breast: a single dorsal, composed almost entirely of simple flexible rays.

(1. BLENNIUS.) Teeth long, even, close-set, forming a single row; the last in the series sometimes longer than the others, and curved: ventrals of two or three rays.

(2. GUNNELLUS.) Teeth short, in more than one row: ventrals extremely small, reduced nearly to a single ray: dorsal extending the whole length; all the rays simple, and without articulations.

28. ZOARCES†.—Ventrals before the pectorals, small, with three rays: body elongated, covered with a mucous

secretion : branchiostegous membrane with six rays; not continued across the breast: teeth conical, in one row at the sides of the jaws, in many in front: fins invested with a thick skin; the dorsal, anal, and caudal, united; all the rays of the dorsal soft and articulated.

29. ANARRHICHAS. — Ventrals wanting: body elongated, smooth : one dorsal, composed entirely of simple rays, not connected with the caudal: branchiostegous membrane with six rays : palatines, vomer, and mandibles, armed with large osseous tubercles; the anterior teeth long, and conical.

30. GOBIUS.—Ventrals thoracic; more or less united at the edges, forming a funnel-shaped cavity : body elongated, scaly : head moderate : gill-opening small; branchiostegous membrane with five rays: two distinct dorsals; the first with the spines extremely flexible.

31. CALLIONYMUS.—Ventrals jugular, widely separate, broader than the pectorals: body naked : head broad and depressed : gill-opening reduced to a small hole on each side of the nape: preopercle terminating behind in several small spines: two dorsals.

X. LOPHIADÆ.—*Bones of the carpus elongated, forming a kind of arm supporting the pectorals: skeleton semicartilaginous.*

32. LOPHIUS.—Ventrals before the pectorals: opercle and branchiostegous rays enveloped in the skin; the branchiostegous membrane forming a large purse-like cavity in the axilla : two distinct dorsals ; in front of which are some free rays, produced into long slender filaments: skin naked: head broad and depressed; extremely large with respect to the rest of the body.

XI. LABRIDÆ. — *Body oblong, scaly : only one dorsal; the spines invested with membranous shreds, extending beyond their tips, and giving them a bifid appearance: lips fleshy: pharyngeans three in number; two above and one below; all armed with strong teeth.*

33. LABRUS. — Lips double: branchiostegous membrane with from four to six rays: maxillary teeth conical; the anterior ones longest: pharyngeans cylindrical, blunt, forming a pavement.

(1. LABRUS.) Preopercle with the margin entire: cheeks and opercle scaly: first dorsal spines not elongated.

(*LACHNOLAIMUS.*) First dorsal spines produced into long flexible threads.

(2. JULIS.) First dorsal spines elongated: head entirely smooth, and without scales.

(3. CRENILABRUS.) Preopercle with the margin denticulated: cheeks and opercle scaly.

XII. CENTRISCIDÆ.—*Mouth at the extremity of a long tube, formed by a prolongation of the rostral and opercular bones.*

34. CENTRISCUS.—Body oblong-oval, compressed at the sides, carinated beneath: mouth extremely small, cleft obliquely: branchiostegous membrane with two or three slender rays: two dorsals; the anterior one placed very backward, with the first spine much longer and stouter than the others: ventrals small, behind the pectorals.

ORDER II.   MALACOPTERYGII.

All the fin-rays, with the exception sometimes of the first in the dorsal and the first in the pectorals, soft and cartilaginous; these rays of an articulated structure, and generally more or less branched at their extremities.

(I. ABDOMINALES.)

Ventrals suspended from the abdomen, and situate far
behind the pectorals.

I. CYPRINIDÆ. — *Mouth small: jaws weak,
generally without teeth, formed by the intermaxillaries: pharyngeans with strong teeth: body
scaly: no adipose fin.*

35. CYPRINUS. — Jaws without teeth: lips simple,
with or without barbules: branchiostegous membrane with
three flat rays: one dorsal: scales generally large.

> (*1. Cyprinus.) Dorsal long; the second ray, as well as that in the anal,
> a serrated spine.

> (2. Barbus.) Dorsal and anal short; the former with the second or
> third ray strongly spinous: upper jaw with four barbules; two
> at the angles, and two at the extremity of the mouth.

> (3. Gobio.) Dorsal and anal short; without spines: upper jaw with
> two barbules.

> (4. Tinca.) Dorsal and anal short; without spines: scales small,
> and slimy: two very short barbules.

> (5. Abramis.) Neither spines nor barbules: dorsal short, placed
> behind the ventrals: anal long.

> (6. Leuciscus.) Dorsal and anal short; without spines: mouth
> without barbules.

36. COBITIS.—Jaws without teeth: lips fleshy, fitted
to act as suckers, furnished with barbules: body elongated:
scales small, and slimy: branchiostegous membrane with
three rays: one dorsal: ventrals very much behind.

II. ESOCIDÆ.—*Mouth with strong teeth: upper
jaw formed by the intermaxillary, or if not
entirely, the maxillary without teeth and concealed in the substance of the lips: no adipose
fin.*

37. ESOX.—Snout broad, oblong, rounded, depressed;
intermaxillaries forming two-thirds of the upper jaw; teeth

in both jaws, as well as on the vomer, palatines, tongue, pharyngeans, and branchial arches: one dorsal; the anal opposite.

38. BELONE.—Snout attenuated, greatly prolonged: intermaxillaries forming the entire margin of the upper jaw: both jaws with small teeth; none on the vomer, palatines, or tongue: body elongated: a row of carinated scales along each side of the abdomen.

(1. BELONE.) Dorsal and anal entire.

(2. SCOMBERESOX.) Last rays of the dorsal and anal detached, forming spurious finlets.

---

39. EXOCŒTUS.—Head depressed, scaly: intermaxillaries without pedicles, and forming the entire margin of the upper jaw: both jaws with small teeth: branchiostegous membrane with ten rays: a row of carinated scales forming a projecting line along the bottom of each flank: pectorals extremely large, almost as long as the body.

SILURIDÆ.—No true scales: skin naked, or covered with osseous plates: adipose fin often present: upper jaw formed by the intermaxillaries: first ray of the dorsal and pectoral fins generally a strong articulated spine.

(2.) *SILURUS.*—Skin naked, covered with a mucous secretion: head depressed: mouth terminal; with several fleshy barbules: card-like teeth in both jaws, as well as on the vomer: dorsal small, without any sensible spine: anal very long.

III. SALMONIDÆ. — *Body scaly: two dorsals; the first with all the rays soft; second small, and adipose.*

40. SALMO. — The greater part of the upper jaw formed by the maxillaries: one row of sharp teeth on the maxillaries, intermaxillaries, palatines, and mandibulars;

two rows on the vomer, tongue, and pharyngeans : ventrals opposite the middle of the first dorsal; the adipose fin opposite the anal : branchiostegous membrane with more than eight rays.

41. OSMERUS.—Two rows of teeth on each palatine, the vomer with only a few teeth in front: ventrals opposite the anterior margin of the first dorsal: branchiostegous membrane with only eight rays: scales minute.

42. THYMALLUS.—Gape small: jaws, tongue, palatines, and front of the vomer, with very fine velvet-like teeth: branchiostegous membrane with ten rays: first dorsal long and high: scales large.

43. COREGONUS.—Mouth as in the last genus, but the teeth still smaller, sometimes entirely wanting : dorsal not so long as it is high in front: scales· very large.

(3.) *SCOPELUS.*—Gape and gill-openings large: both jaws with very small teeth; the margin of the upper formed entirely by the intermaxillaries : tongue and palate smooth : branchiostegous membrane with nine or ten rays: first dorsal answering to the space between the ventrals and the anal; second very small, with vestiges of rays.

IV. CLUPEIDÆ. — *No adipose fin : upper jaw formed in the middle by the intermaxillaries, which are without pedicles ; at the sides by the maxillaries : body scaly.*

44. CLUPEA. — Intermaxillaries short, and narrow ; sides of the jaw formed by the maxillaries in three pieces, and alone protractile: mouth with few teeth, sometimes with none : abdomen compressed, the scales on the margin forming a serrated ridge : gill-opening very large ; branchiostegous membrane with from six to eight rays.

(1. CLUPEA.) Maxillaries arched in front: gape moderate; upper lip entire.

(2. ALOSA.) Upper jaw notched in the middle.

45. ENGRAULIS. — Snout projecting, and sharp-pointed ; intermaxillaries placed beneath, and very small ;

maxillaries straight and long: gape extending to behind
the eyes: both jaws furnished with teeth: gill-opening
extremely large; the membrane with twelve or more rays:
abdomen generally smooth.

(4.) *LEPISOSTEUS.*—Snout prolonged, formed by the union of the
intermaxillaries, maxillaries, palatines, vomer and ethmoid: lower jaw
equalling it: both jaws armed with sharp teeth: branchiostegous membrane with three rays: body covered with hard osseous scales: dorsal and
anal opposite, placed very far back.

(II. SUBBRACHIALES.)

Ventrals immediately beneath the pectorals, the pelvis
being attached to the bones of the shoulder.

V. GADIDÆ*.—*Ventrals jugular, pointed: body
covered with soft scales: all the fins soft: jaws,
and front of the vomer, armed with several rows
of sharp card or rasp-like teeth: gill-opening
large with seven rays.*

46. GADUS.—Body oval, moderately elongated: head
compressed: three dorsals: two anals: one barbule at the
extremity of the lower jaw: ventrals with six rays.

47. MERLANGUS.—Body elongated: three dorsals:
two anals: no barbule on the chin: ventrals with six rays.

48. MERLUCCIUS. — Body elongated: head compressed: two dorsals; the first small: one anal: chin
without barbules.

49. LOTA.—Body slender, elongated, compressed behind: two dorsals; the first short; the second dorsal, as

---

* I have followed Cuvier in the arrangement of the genera belonging to this family. It may be
questioned, however, whether those which he has adopted are all of them groups of equal value,
and whether some might not with more propriety be lowered to a subordinate rank. Without
an extensive acquaintance with foreign species, it would be presumptuous to decide this point.

well as the anal, long: chin with one or more barbules: ventrals with six or seven rays.

50. MOTELLA.——Body elongated, compressed behind: first dorsal but little elevated, and scarcely perceptible; the rays detached and hair-like, all, except the first, very minute: second dorsal, and also the anal, long: ventrals with six or seven rays.

51. BROSMUS.——Body elongated, compressed behind: only one long dorsal extending nearly to the caudal: anal long: chin with a single barbule: ventrals thick and fleshy, consisting of five rays.

52. PHYCIS.——Ventrals consisting of only a single ray; often forked: head large: body elongated: chin with one barbule: two dorsals; the second, as well as the anal, long.

53. RANICEPS.——Head depressed and very broad: body very much compressed behind: two dorsals; the first very small, scarcely perceptible; the second, as well as the anal, long: ventrals with the two first rays elongated.

VI. PLEURONECTIDÆ.—*Body deep, very much compressed; with both the eyes on the same side of the head: sides of the mouth, and generally the pectorals, unequal: dorsal and anal extending the whole length of the back and abdomen respectively: ventrals appearing like a continuation of the anal: branchiostegous membrane with six rays.*

54. PLATESSA.——A single row of obtuse cutting teeth in each jaw; and generally a pavement of teeth on the pharyngeans: dorsal commencing in a line with the upper eye, and leaving, as well as the anal, a space between it and the caudal: form rhomboidal: eyes on the right side.

55. HIPPOGLOSSUS.—Jaws and pharyngeans armed with strong sharp teeth: dorsal commencing in a line with the upper eye, and terminating before the caudal: form oblong: eyes sometimes on the right, sometimes on the left side.

56. PLEURONECTES.—Jaws and pharyngeans with fine card-like teeth: dorsal commencing immediately above the upper lip, and reaching, as well as the anal, to very near the caudal: form rhomboidal: eyes generally on the left side.

57. SOLEA.—Mouth irregular, and as it were twisted on the side opposed to the eyes, and furnished on that side only with fine velvet-like teeth, the upper side being without teeth: form oblong-oval: snout rounded, advancing beyond the mouth: dorsal commencing at the mouth, and reaching, as well as the anal, quite to the caudal.

 (1. SOLEA.) Pectorals of moderate size, and not very unequal.

 (2. MONOCHIRUS.) Pectoral on the side of the eyes extremely small; that on the side opposite rudimentary, or altogether wanting.

VII. DISCOBOLI. — *Ventrals united, forming a concave disk beneath the body: skin without scales.*

58. LEPADOGASTER. — Pectorals large, descending to the inferior surface of the body, then doubling forwards upon themselves, and finally uniting under the throat by a transverse membrane: a second, circular, concave disk behind the disk formed by the united ventrals: head broad and depressed; snout projecting: gill-opening small; branchiostegous membrane with four or five rays: one dorsal.

59. CYCLOPTERUS.—Pectorals large, uniting under the throat, and enclosing the disk of the ventrals: no second

disk : mouth broad ; both jaws, as well as the pharyngeans, armed with small pointed teeth : gill-opening closed at bottom ; branchiostegous membrane with six rays.

> (1. CYCLOPTERUS.) A first dorsal more or less obvious, with simple rays; a second, with branched rays, opposite the anal : body thick.

> (2. LIPARIS.) Only one, moderately long, dorsal: anal long : body smooth; elongated, compressed behind.

VIII. ECHENEIDIDÆ.—*An oval flattened disk on the upper part of the head, composed of several transverse cartilaginous plates directed obliquely backwards, and toothed on their posterior margin : body elongated, covered with small scales.*

60. ECHENEIS.

(III. APODES.)

Ventrals wanting.

IX. ANGUILLIDÆ.—*Body very much elongated : skin thick and soft ; the scales deeply imbedded, and scarcely apparent.*

61. ANGUILLA.—Gills opening by a small aperture on each side beneath the pectoral : dorsal and anal fins prolonged round the end of the tail, forming by their union a pointed caudal.

> (1. ANGUILLA.) Dorsal commencing considerably behind the pectorals: upper jaw shorter than the lower.

> (2. CONGER.) Dorsal commencing a little behind, sometimes immediately above, the pectorals: upper jaw longest.

(5.) *OPHISURUS.*—Gills opening by a small aperture beneath the pectorals : dorsal and anal not reaching to the end of the tail, which terminates in a point, and is itself without a fin.

62. MURÆNA. — Gills opening by a small aperture on each side : pectorals wanting : dorsal and anal uniting at the tail; low, sometimes scarcely distinguishable.

---

63. LEPTOCEPHALUS.—Gill-opening small, before the pectoral: body very much compressed, ribband-shaped : head extremely small; snout short: pectorals scarcely perceptible : dorsal and anal obsolete, uniting at the extremity of the tail.

64. OPHIDIUM.—Gills opening by a moderately large aperture ; furnished with a distinct opercle and branchiostegous membrane : body very much compressed : dorsal and anal uniting to form a pointed caudal ; the dorsal rays articulated, but not branched.

65. AMMODYTES.—Gill-opening very large ; all the pieces of the opercle considerably developed : snout sharp ; upper jaw capable of great extension, but when at rest shorter than the lower: dorsal and anal separated from the caudal by a small space; the dorsal furnished with simple articulated rays: caudal forked.

---

§ *II.* LOPHOBRANCHII.—*Branchiæ in small round tufts disposed in pairs along the branchial arches; opercle large, confined on all sides by a membrane, with only a small hole for the external aperture; branchiostegous rays rudimentary: jaws complete, and free.*

ORDER III.   OSTEODERMI.

Body mailed with transverse angular plates: snout very much produced, formed by a prolongation of

x 2

the bones of the head and gill-covers : generally one dorsal, with simple slender rays: the other fins often wanting.

66. SYNGNATHUS.—Snout prolonged into a tube; mouth placed at the extremity, and cleft nearly vertically : body very much elongated, slender, and of nearly equal thickness throughout: gill-opening towards the nape : ventrals always wanting.

67. HIPPOCAMPUS. — Snout tubular ; the mouth placed at the extremity : trunk of the body laterally compressed, and more elevated than the tail : the joints of the squamous plates raised in ridges ; the projecting angles spinous : ventrals, and also caudal, always wanting.

§ *III.* PLECTOGNATHI. — *Jaws incomplete; maxillary firmly attached to the side of the intermaxillary, which alone forms the jaw; palatine arch united to the cranium by suture, and immoveable : branchiæ with the pectinations continuous ; opercle and rays concealed beneath the skin; external aperture a simple cleft.*

ORDER IV.   GYMNODONTES.

No true teeth ; but the jaws covered with a lamellated substance resembling ivory, either entire, or divided in the middle by a suture: opercle small; branchiostegous rays five on each side; both deeply concealed.

68. TETRODON. —— Jaws divided in the middle, so as to present the appearance of four teeth, two above and two below: skin rough, with small slightly projecting spines: body capable of inflation.

69. ORTHAGORISCUS. —— Jaws undivided: body very much compressed, short, truncated behind; rough, but without spines; not capable of inflation: dorsal and anal fins high and pointed, uniting with the caudal.

ORDER V.   SCLERODERMI.

Snout very much produced, of a conical or pyramidal form; mouth small, with distinct teeth in each jaw: skin generally scabrous, or clothed with large scales.

70. BALISTES.——Eight teeth, forming a single row, in each jaw: body compressed; covered with large, hard, rhomboidal scales, not overlapping one another: first dorsal with three spines; the first much the largest, the third very small, and remote from the others: extremity of the pelvis projecting; armed with prickles.

(II. CARTILAGINEI.)

Skeleton cartilaginous: bones soft, and destitute of fibres: sutures of the cranium indistinct: maxillary and intermaxillary bones either wanting or rudi-mentary; the palatines, or vomer alone, supplying their place.

ORDER VI. ELEUTHEROPOMI.

Branchiæ pectinated, free, with one large external aperture; furnished with an opercle, but without rays in the membrane: upper jaw formed by the palatine bone firmly united to the maxillary: intermaxillary rudimentary.

71. ACIPENSER.—Body elongated; mailed, as well as the head, with osseous tubercles arranged in longitudinal rows: snout conical: mouth placed beneath, very protractile, small, without teeth: nostrils and eyes lateral: four pendent barbules on the under surface of the snout.

ORDER VII. ACANTHORRHINI.

Branchiæ pectinated; adhering by a large portion of their external margin; opening outwards by a single aperture, communicating with five others at the bottom of the general cavity: opercle rudimentary: the vomer alone present to represent the upper jaw.

72. CHIMÆRA.—Body elongated: snout conic, marked with lines of pores: jaws armed with hard plates instead of teeth; four above and two below: between the eyes a fleshy process, bent forwards, and terminating in a cluster of small spines: first dorsal armed with a strong spine; placed above the pectorals: second dorsal commencing immediately behind the first, and reaching to the end of the tail, which terminates in a long filament.

ORDER VIII. PLAGIOSTOMI.

Branchiæ pectinated, fixed, opening outwards by several distinct apertures: no opercle: jaws represented
by the palatine and postmandibular bones, which
alone are armed with teeth: pectorals and ventrals
always present; the latter (in the *male*) furnished
on their internal margin with long appendages.

73. SQUALUS.—Body elongated; tail thick and muscular: pectorals of moderate size: branchial openings at the
sides of the neck: snout more or less produced; with the
mouth and nostrils beneath: eyes lateral.

* *Snout short and obtuse: nostrils near the mouth, with
a groove-like prolongation conducting to the edge of
the lips; more or less closed by two membranous flaps;
caudal truncated at the extremity, not forked.*

(1. SCYLLIUM.) Teeth sharp and pointed, with small denticulations
on each side: temporal orifices, as well as an anal fin, always
present: dorsals very much behind, the first never in advance of
the ventrals: branchial openings partly above the pectorals.

** *Snout produced: nostrils not prolonged in a groove,
nor furnished with valves: caudal with a lobe beneath,
causing it to appear forked.*

(2. CARCHARIAS.) Snout depressed; the nostrils beneath the middle:
teeth cutting, pointed, and generally denticulated at the sides:
first dorsal far in advance of the ventrals; second nearly opposité
to the anal, which is always present: no temporal orifices: the
last of the branchial openings above the pectorals.

(3. LAMNA.) Snout pyramidal; the nostrils beneath the base: teeth
sharp and pointed, denticulated at the sides: first dorsal considerably before the ventrals: anal present: no temporal orifices:
the branchial openings all before the pectorals.

(4. GALEUS.) Snout depressed; the nostrils beneath the middle:
teeth sharp and pointed, with a few denticulations on the outer
side only: temporal orifices, as well as an anal fin, present: the
last branchial opening above the pectoral.

(5. MUSTELUS.) Snout depressed; the nostrils beneath the middle:
teeth small and obtuse, forming a closely-compacted pavement
in each jaw: temporal orifices, and anal, both present.

(6. SELACHE.) Teeth small, conical, not denticulated at the sides : temporal orifices, and anal, both present : branchial openings all before the pectorals ; large, nearly surrounding the neck.

(7. SPINAX.) Snout depressed : teeth small, cutting, in several rows : temporal orifices present : anal wanting : branchial openings all before the pectorals : a sharp strong spine in front of each of the dorsals.

(8. SCYMNUS.) Teeth in two or more rows ; in the upper jaw lanceolate, with the cutting edges rough ; in the lower jaw pyramidal, compressed, with the cutting edges crenated : temporal orifices present : anal wanting : no spines before the dorsal fins.

74. ZYGÆNA.—Head flattened horizontally, truncated in front, with the sides very much produced, appearing hammer-shaped ; eyes placed at the extremities of the lateral prolongations ; the nostrils at their anterior margin ; mouth beneath : pectorals moderate : branchial openings at the sides of the neck.

75. SQUATINA.—Body broad, flattened horizontally : pectorals large, extending forwards, but separated from the neck by a cleft, in which are the branchial openings : head round ; mouth at the extremity of the snout ; eyes above : temporal orifices present : no anal : both dorsals further back than the ventrals.

(6.) *PRISTIS.* — Body elongated, flattened anteriorly ; with the branchial openings placed beneath : snout prolonged into a depressed sword-shaped beak, armed on each side with strong sharp spines resembling teeth : mouth beneath : temporal orifices present : no anal fin.

76. RAIA.—Body very much flattened, resembling a disk : pectorals extremely large, uniting with the snout, or with each other, anteriorly ; extending backwards to near the base of the ventrals : mouth, nostrils, and branchial openings, beneath ; eyes, and temporal orifices, above : dorsals almost always upon the tail.

(1. TORPEDO.) Tail short, and moderately thick : the disk of the body nearly circular ; the part between the pectorals and the head and the branchiæ furnished with an electrical apparatus : teeth small and sharp.

(2. RAIA.) Tail slender ; furnished above, towards its extremity, with two small dorsals, and sometimes the vestige of a caudal : disk rhomboidal : teeth slender, close-set, arranged in a quincuncial order.

(3. TRYGON.) Tail slender; armed with a sharp serrated spine, but without fins: head enveloped in the pectorals, which meet at an obtuse angle: teeth slender, set in a quincuncial order.

(MYLIOBATIS.) Tail long, and extremely slender; armed with one or more serrated spines, and also furnished, near its base, with a small dorsal: head projecting from between the pectorals, which last are much broader transversely than in the preceding sub-genera: jaws with broad flat teeth forming a pavement.

(4. CEPHALOPTERA.) Tail long, and extremely slender; armed with a spine, and also furnished at its base with a small dorsal: head truncated in front, placed between two horn-like prolongations of the pectorals, which are broad as in the last sub-genus: teeth small, and very slender, finely denticulated.

ORDER IX. CYCLOSTOMI.

Branchiæ purse-shaped, fixed, opening outwards by several apertures: jaws represented by an immoveable cartilaginous ring, formed by the union of the palatine and mandibular bones: no pectorals or ventrals: body elongated: the skeleton very imperfectly developed.

77. PETROMYZON. — Seven branchial openings on each side: maxillary ring armed with strong teeth; inside of the mouth furnished with tooth-like tubercles: lip circular: tongue with two longitudinal rows of small teeth.

78. AMMOCŒTES. — Branchial openings seven in number: lip semicircular, and covering only the upper part of the mouth: no teeth; but the opening of the mouth furnished with a row of small branched cirri.

79. MYXINE. — Branchial openings two in number, placed beneath: maxillary ring entirely membranaceous, with a single tooth above: tongue with strong teeth arranged in two longitudinal rows on each side: mouth circular, surrounded by eight barbules; a spiracle above, communicating with the interior.

(I. OSSEI.)

ORDER I.   ACANTHOPTERYGII.

GEN. 1.   PERCA, *Linn.*

(1. Perca, *Cuv.*)

1. P. *fluviatilis*, Linn. (*Common Perch.*) — Back
dusky green ; sides with five transverse dusky bands.

P. fluviatilis, *Linn. Syst. Nat.* tom. I. p. 481. *Bloch, Ichth.* pl. 52.
  Don. Brit. Fish. vol. III. pl. 52. *Flem. Brit. An.* p. 213. Perch,
  Will. Hist. Pisc. p. 291. tab. S. 13. f. 1. *Penn. Brit. Zool.* vol. III.
  p. 254. pl. 48. no. 124. *Id.* (Edit. 1812.) vol. III. p. 345. pl. 59.
  Bowd. Brit. fr. wat. Fish. Draw. 5. *Yarr. Brit. Fish.* vol. I.
  p. 1. Perche commune, *Cuv. et Val. Poiss.* tom. II. p. 14.
  Cuv. Reg. An. tom. II. p. 132.

LENGTH.   From twelve to sixteen inches.
DESCRIPT. (*Form.*) Body compressed ; back much arched, highest
over the ventrals, the depth at that part equalling nearly one-third of
the length, caudal fin excluded ; greatest thickness half the depth : head
about one-fourth of the entire length, including caudal : nape depressed,
the back rising rather suddenly behind ; profile gently falling from the
forehead towards the end of the snout : jaws nearly equal ; teeth in both
jaws, as well as on the vomer, palatines, pharyngeans, and branchial
arches : head and cheeks for the most part smooth, the latter covered
with small scales : preopercle rectangular ; the margin finely serrated :
opercle triangular, terminating behind in a sharp point ; the subopercle
reaching beyond it : gill-opening large : lateral line nearly parallel with
the line of the back, its course at about one-fourth of the depth* : first
dorsal fin commencing in a line with the terminating point of the opercle ; the fourth and fifth rays longest ; those on each side gradually decreasing ; the last, or last two, in the fin small and inconspicuous ; height
of the longest rays nearly equalling half the depth of the body : second

* The course of the lateral line is said to be at half, one-third, one-fourth, &c. of the depth.
when the distance from the line of the back to the lateral line equals half, one-third, one-fourth,
&c. of the depth of the body.

dorsal close behind the first, only a small space intervening; first ray spinous, not half the length of the second; third and fourth longest; the succeeding ones slightly decreasing; all the soft rays, except the first, branched: anal commencing rather more backward than the second dorsal, and not extending so far towards the caudal; two first rays strongly spinous, much shorter than the others, which are all branched and articulated: caudal forked: pectorals about two-thirds of the length of the head; the middle rays longest, those on each side decreasing; the first two and last three simple, the others branched: ventrals placed a little behind the pectorals, equal to them in length; the first ray strongly spinous, shorter than the others, which are all soft and much branched: number of rays in the respective fins,

D. 14 or 15—1/13; A. 2/8; C. 17, and some short ones; P. 14; V. 1/5*:

vent in a line with the commencement of the second dorsal. Number of vertebræ forty-two. (*Colours.*) Back and sides dusky green, with from five to seven dark transverse bands; abdomen white, tinged with red: ventrals bright scarlet; anal and caudal fins somewhat paler; dorsals and pectorals dusky, tinged with red; often a conspicuous black spot on the webs connecting the four last spines in the first dorsal.

Common in lakes, rivers, and streams. Found, according to Cuvier, throughout all the temperate parts of Europe, and a great part of Asia. Often, but not always, gregarious. Feeds on worms, insects, small crustacea, and the fry of other fish. Seldom attains a greater weight than four pounds, but has been known to weigh as many as nine. Spawns in April or May. A monstrous variety, with the back hunched, and the lower part of the back-bone next the tail much distorted, is mentioned by Pennant as found in a lake in Merionethshire.

(2. Labrax, *Cuv.*)

2. P. *Labrax*, Linn. (*Common Basse.*)

P. Labrax, *Linn. Syst. Nat.* tom. i. p. 482. *Don. Brit. Fish.* vol. ii. pl. 43. *Flem. Brit. An.* p. 213. Sciæna Labrax, *Bloch, Ichth.* pl. 301. Labrax Lupus, *Cuv. et Val. Poiss.* tom. ii. p. 41. pl. 11. *Cuv. Reg. An.* tom. ii. p. 133. Basse, *Will. Hist. Pisc.* p. 271. tab. R. 1. *Penn. Brit. Zool.* vol. iii. p. 257. pl. 49. *Id.* (Edit. 1812.) vol. iii. p. 348. pl. 60. *Yarr. Brit. Fish.* vol. i. p. 6.

Length. From one to two feet.

Descript. (*Form.*) Resembling the *Perch*, but more elongated; the back not so high: greatest depth a little behind the ventrals, equalling one-fourth of the length, caudal excluded: line of the back somewhat concave beneath the first dorsal, and convex beneath the second: head nearly one-fourth of the entire length, including caudal: lower jaw longest; strong card-like teeth on the intermaxillary, vomer, and palatines; on the sides, and towards the root, of the tongue, teeth like velvet: head smooth; cheeks covered with small scales: preopercle large; the serratures on the ascending margin more developed than in the *Perch;* the basal margin with three strong spines: opercle triangular, the posterior angle armed with two strong flattened spines: lateral line curved, descending a little from the upper angle of the opercle to

---

* I have adopted throughout the same kind of formula, by which to express the number of fin-rays, as that employed by MM. Cuvier and Valenciennes.

beneath about the middle of the first dorsal, then passing off straight to the caudal: first dorsal commencing a little behind the ventrals: the fourth and fifth rays equal and longest, those on each side gradually decreasing: second dorsal immediately after the first; first ray spinous, not half the length of the succeeding ones; third longest; the others gradually decreasing: space between the second dorsal and the base of the caudal equal to that occupied by the former fin: anal similar to the second dorsal, but placed a little more backward, with the three first rays spinous, gradually increasing in length: caudal a little forked: pectorals somewhat less than half the length of the head, covered at their base with small scales; the first ray simple, and shorter than the others: ventrals a little behind the pectorals, equal to them in length; the first ray spinous: number of rays in the respective fins,

D. 9—1/12; A. 3/11; C. 17; P. 17; V. 1/5.

Number of vertebræ twenty-six. (*Colours.*) Back and sides dusky gray, with a paler spot in the middle of each scale; lower portion of the sides, and abdomen, white, slightly silvery; cheeks and gill-covers with a faint yellowish tinge; posterior portion of the opercle almost black, forming a dark patch on that part: dorsal and caudal fins deep lead-gray; anal the same, tipped with whitish; pectorals pale gray; ventrals nearly white.

Met with occasionally on different parts of the coast, particularly southward, and likewise in the estuaries of rivers, but seldom in any great plenty. Common in the Mediterranean, where it attains a much larger size than in the British seas. Said to be very voracious. *Obs.* Pennant states that in the young of this species the space above the side-line is marked with small black spots: the same remark is made by Cuvier and others; but certainly in the larger number of British specimens there is no indication of these spots whatever.

(3. Serranus, *Cuv.*)

3. P. *Cabrilla*, Linn. (*Smooth Serranus.*) — Jaws without scales: cheeks and gill-covers marked with three or four oblique bands; sides with a few longitudinal bands.

P. Cabrilla, *Linn. Syst. Nat.* tom. I. p. 488.   P. Channus, *Couch in Loud. Mag. of Nat. Hist.* vol. v. p. 19. fig. 6.   Serranus Cabrilla, *Cuv. et Val. Poiss.* tom. II. p. 166. pl. 29.   Smooth Serranus, *Yarr. Brit. Fish.* vol. I. p. 9.   Smooth Perch, *Couch, l.c.* Serran commun, *Cuv. Reg. An.* tom. II. p. 139.

Length.   About ten inches.   Couch.
Descript.   (*Form.*) "Under jaw longest: teeth in both and in the palate, numerous, irregular, sharp, incurved: tongue small, loose: eyes large, near the top of the head: first plate of the gill-covers serrate; the second with two (in the *female* one) obscure spines, scarcely to be distinguished, except in form, from the scales with which the gill-covers and body are thickly and firmly covered, and which are also ciliated: seven rays in the gill-membrane, curved; the superior broad: body compressed, deep: dorsal fin commencing opposite the ventrals: pectorals pointed: caudal slightly curved: number of fin-rays,

D. 10/14; A. 2/7; C. 17; P. 15; V. 6:

lateral line nearer the back." COUCH. According to Cuvier and Valenciennes, this species is distinguished from the *Serranus Scriba*, Cuv. (a
closely allied one found in the Mediterranean) by its shorter snout and
rather more convex forehead*; larger eye; and rather less rounded preopercle, with the denticulations towards the angle a little stronger: the
lower jaw has the under surface of its branches chagrined and vermiculated by little marks in the skin. (*Colours.*) "Colour of the back brown,
having, in some specimens, distinct bars running round to the belly:
sides yellow, reddish, or saffron-coloured, more faint below: two irregular parallel whitish lines pass along the side from head to tail; a third,
more imperfect, on the belly: gill-plates with several faintish blue stripes,
running obliquely downward: fins striped longitudinally with red and
yellow; pectorals wholly yellow." COUCH.

An abundant species in the Mediterranean. In the British seas it has
hitherto been observed only by Mr. Couch, who represents it as common
on the coast of Cornwall, "keeping in the neighbourhood of rocks, not far
from land."

4. P. *Gigas*, Gmel. (*Dusky Serranus.*)—Lower jaw
covered with very small scales.

P. Gigas, *Gmel. Linn.* tom. I. part iii. p. 1315. P. robusta, *Couch
in Loud. Mag. of Nat. Hist.* vol. v. p. 21. fig. 7. Serranus Gigas,
Cuv. et Val. Poiss. tom. II. p. 201. pl. 33. Dusky Serranus,
Yarr. Brit. Fish. vol. I. p. 15. Dusky Perch, *Couch, l. c.*
Merou brun, *Cuv. Reg. An.* tom. II. p. 140.

LENGTH. Three feet. COUCH.

DESCRIPT. (*Form.*) "Depth, exclusive of the fins (in a specimen
measuring three feet in length), seven inches: body thick and solid:
under jaw longest; both jaws, as well as the palate, with numerous
slender incurved teeth: in front of the under one a bed of them: lips
like those of the *Cod-Fish:* two large open nasal orifices, and a large
hole under the projection of the nasal bone: first plate of the gill-covers
serrate, the second with a broad flat spine projecting through the skin,
and pointing backward; the fleshy covering of the gill-cover elongated
posteriorly; seven rays in the gill-membrane: body and head covered
with large scales: lateral line gently curved: dorsal fin single, long,
expanding towards its termination, with eleven spinous rays, the first
short, and seventeen soft rays, the last two from one origin: pectorals
round, with nineteen rays: ventrals fastened down by a membrane
through part of their course, with six rays: vent an inch and a half
from the origin of the anal fin, which last has two spinous and nine
soft rays, the last two from one origin: caudal roundish, with sixteen
rays." COUCH. According to Cuvier and Valenciennes, the number of
fin-rays stands thus:

D. 11/15 or 16; A. 3/8; C. 15; P. 17; V. 1/5.

(*Colours.*) "Back reddish brown, lighter on the belly: two slightly
marked lines on the gill-covers running obliquely downward, one on
each plate." COUCH.

---

* Or rather that part of the face between the forehead and the nostrils termed by the French
chanfrein, for which we have no adequate term in the English language.

A single individual of this species, which is found in the Mediterranean, has been obtained by Mr. Couch from the coast of Cornwall. Cuvier states that nothing is known of its habits, excepting that at Nice it approaches the shores in the months of May and April. Usual weight from ten to twenty pounds.

(1.) *Serranus Couchii*, Yarr. Brit. Fish. vol. I. p. 12.   *Stone-Basse*, Couch in Linn. Trans. vol. XIV. p. 81.

> This fish, which has been observed by Mr. Couch on the Cornish coast, accompanying floating timber covered with barnacles, remains yet to be identified with certainty. It was considered by that gentleman, in his paper in the *Linnæan Transactions*, as the *Pagrus totus argenteus* of Sloane*, a species, however, to which it evidently bears no affinity. Mr. Yarrell, who has received an original drawing of the fish from Mr. Couch, regards it as a new species of *Serranus*, and accordingly names it after its first discoverer. No description of it has been yet published.

(4. ACERINA, *Cuv.*)

5. P. *Cernua*, Linn. (*Common Ruffe*.) — Back and sides olivaceous, spotted with dusky brown.

P. Cernua, *Linn. Syst. Nat.* tom. I. p. 487. *Bloch, Ichth.* pl. 5. f. 2. *Don. Brit. Fish.* vol. II. pl. 39. Cernua fluviatilis, *Flem. Brit. An.* p. 212. Acerina vulgaris, *Cuv. et Val. Poiss.* tom. III. p. 4. pl. 41. *Cuv. Reg. An.* tom. II. p. 144. Ruffe, *Will. Hist. Pisc.* p. 334. tab. X. 14. f. 2. *Penn. Brit. Zool.* vol. III. p. 259. *Id.* (Edit. 1812.) vol. III. p. 350. *Bowd. Brit. fr. wat. Fish.* Draw. 10. *Yarr. Brit. Fish.* vol. I. p. 18.

LENGTH. From six to eight inches.

DESCRIPT. (*Form.*) Not so high in the back as the *Perch*, and less compressed in proportion; head broader, thicker, and more obtuse about the snout: greatest depth at the commencement of the dorsal fin, equalling one-fourth of the entire length, including caudal; thickness two-thirds of the depth: eyes very large and brilliant; their diameter one-fourth of the length of the head: mouth rather small; upper jaw a little the longest; both jaws, as well as the vomer, armed with fine teeth like velvet; pharyngeans card-like: head without scales; the snout, cheeks, and jaws, pitted with numerous excavations: preopercle with the ascending and basal margins strongly spined; posterior angle of the opercle terminating in a small spine: gill-opening very large: lateral line nearly parallel with the back; its course at rather less than one-third of the depth: dorsal commencing in a line with the posterior angle of the opercle; first ray very short; the succeeding ones gradually lengthening to the fourth and fifth, which are longest, then decreasing to the fourteenth, or last of the spinous rays; soft portion of the fin rather more than half the length of the spinous; middle rays longest, the last ray double: anal commencing a little nearer the caudal than the soft dorsal, and not extending quite so far; first two rays strongly spinous and slightly curved: finless portion of the tail about equal in length to the caudal; this last deeply forked: pectorals in a line with the commencement of the dorsal, and two-thirds the length of the head; all the rays soft, and, except the first and last, branched; middle rays longest: vent-

* *Nat. Hist. of Jam.* vol. II. tab. 253. f. 1.

rals about equal to the pectorals, placed immediately beneath them; the first ray spinous, rather more than half as long as the third, which is longest; all the soft rays much branched: number of rays,

D. 14/12; A. 2/5; C. 17; P. 14; V. 1/5.

Number of vertebræ thirty-seven. (*Colours.*) Back and upper half of the sides pale brown, inclining to olivaceous; lower portion of the sides with a slight lustre of golden yellow: belly silvery: cheeks, opercle, and breast, of a pearly hue, with a play of iridescent colours varying according to the light: head, back, and a portion of the sides, sprinkled with brownish spots: dorsal, caudal, and pectorals, yellowish gray, speckled with brown; ventrals and anal pale yellowish white, without spots: irides with the upper portion dusky, the lower half inclining to golden yellow.

Not uncommon in rivers and clear streams. Said to have been first observed by Dr. Caius in the river Yare near Norwich. Habits somewhat resembling those of the Perch. Seldom exceeds a weight of three ounces. Spawns in March and April. Is sometimes called a *Pope*.

GEN. 2.   TRACHINUS, *Linn.*

6. T. *Draco*, Linn. (*Great Weever*.)—Entire length six times the depth of the body: second dorsal with about thirty rays: lower jaw ascending.

>   T. Draco, *Linn. Syst. Nat.* tom. I. p. 435.   *Cuv. et Val. Poiss.*
>   tom. III. p. 178.   *Cuv. Reg. An.* tom. II. p. 152.   T. major, *Don.*
>   *Brit. Fish.* vol. v. pl. 107.   *Flem. Brit. An.* p. 214.   Weever,
>   *Will. Hist. Pisc.* p. 288. tab. S. 10. f. 1.   Great Weever, *Penn.*
>   *Brit. Zool.* vol. III. p. 171. pl. 29.   *Id.* (Edit. 1812.) vol. III.
>   p. 229. pl. 33.   *Yarr. Brit. Fish.* vol. I. p. 20.

Length. From twelve to fifteen inches.

Descript. (*Form.*) Elongated; head and sides much compressed; entire length more than six times the depth of the body, and more than four times the length of the head; greatest thickness half the depth: head and back nearly in a continuous straight line; the profile slightly falling from the forehead; abdomen scarcely more convex than the back: lower jaw longer than the upper, and sloping upwards to meet it; both armed, as well as the vomer, palatines, pterygoidians, and pharyngeans, with fine sharp velvet-like teeth; arch-bones of the gills with a number of serrated tooth-like processes: before and rather above each eye are two short strong spines; there is also a strong sharp spine on the upper part of the opercle, but not projecting beyond the edge of the membrane: supra-scapulars represented by a large denticulated scale: lateral line straight; its course at rather less than one-fourth of the depth: scales small, disposed in oblique transverse rows: first dorsal very short, commencing immediately above the denticulated scale; spines stiff and very sharp; the third longest; those on each side gradually decreasing; the last very small and partly concealed: second dorsal immediately after the first, almost continuous with it, and extending nearly to the caudal: anal thick and fleshy, rather longer than the second dorsal, the ends of the rays reaching beyond the webs, and somewhat hooked: caudal scarcely notched: pectorals two-thirds of the length of the head: the third and six following rays branched, the others simple; ninth ray longest: ventrals before the pectorals, and scarcely more than half their length; first ray short and spinous; the rest articulated, with the last three branched; fourth ray in the fin longest: number of rays,

D. 6—31; A. 1—31; C. 13; P. 16; V. 1/5.

Number of vertebræ forty. (*Colours.*) Back and upper portion of the sides reddish brown, with interrupted lines of black and yellow, running parallel with the oblique rows of scales; lower part of the sides, and abdomen, white, with interrupted yellow lines: first dorsal with the web deep black; second dorsal and caudal pale, more or less spotted with brown.

Met with occasionally at Weymouth, Hastings, Scarborough, and other parts of the coast. Is much apprehended by fishermen on account of its spines, which are sharp, and capable of inflicting a severe wound: they are usually considered as venomous, but, in the opinion of Cuvier, there is no real secretion of any poisonous fluid. Feeds on small fish, crustacea, and marine insects. Spawns in June.

7. T. *Vipera*, Cuv. (*Little Weever.*)—Entire length four times the depth of the body: second dorsal with twenty-four rays: lower jaw nearly vertical.

T. Vipera, *Cuv. et Val. Poiss.* tom. III. p. 189.   *Cuv. Reg. An.* tom. II. p. 152.   T. Draco, *Bloch, Ichth.* pl. 61.   *Don. Brit. Fish.* vol. I. pl. 23.   *Flem. Brit. An.* p. 213.   Common Weever, *Penn. Brit. Zool.* vol. III. p. 169. pl. 28. no. 71.   *Id.* (Edit. 1812.) vol. III. p. 226. pl. 32. Lesser Weever, *Yarr. Brit. Fish.* vol. I. p. 25.

LENGTH. Five or six inches; rarely more.

DESCRIPT. (*Form.*) Much resembling the *T. Draco*, but deeper in proportion to its length, owing to the greater convexity of the abdomen: profile not falling, but in the same horizontal line with the back; lower jaw more nearly vertical; when the mouth is opened wide, the upper jaw becomes exactly vertical, and the lower forms with it a right angle: side very much compressed: greatest depth beneath the first dorsal, equalling one-fourth of the length, caudal excluded: thickness half the depth: teeth (in the lower jaw especially) very sharp, and somewhat longer in proportion than in the last species: the toothed scale, formed by the supra-scapular and part of the omoplat, of a different form, rounded, bilobated, and more deeply denticulated: lateral line commencing at the above scale, and running nearly straight throughout its course: the oblique transverse lines on the sides, formed by the scales, much less strongly marked: first dorsal more distinctly separated from the second; the fifth and sixth spines (the last especially) very small and inconspicuous: pectorals pointed, about equal in length to the depth of the body: caudal rounded: number of fin rays,

D. 6—23 or 24; A. 25; C. 13; P. 14; V. 1/5.

(*Colours.*) Back reddish gray; sides and abdomen much paler than in the last species, approaching to silvery white, with faint indications of transverse yellow lines: the web connecting the four first spines of the first dorsal deep black: a black spot at the extremity of the caudal fin.

Rather more common than the last species, and met with on various parts of the British coast. Said to conceal itself in the loose soil at the bottom of the water, with only its head exposed. It is probably the small species alluded to by Willughby*, under the name of *Otterpike*.

GEN. (1.) *SPHYRÆNA*, Schn.

(2.) *S. vulgaris*, Cuv. et Val. Poiss. tom. III. p. 242. *S. Spet*, Lacép. Hist. Nat. des Poiss. tom. v. p. 326. *Esox Sphyræna*, Linn. Syst. Nat. tom. I. p. 515. Bloch, Ichth. pl. 389. *Sea-Pike*, Couch in Linn. Trans. vol. XIV. p. 84.

> A very doubtful native. Inserted by Mr. Couch in his list of Cornish Fishes, accompanied by a remark that he had been informed that a fish, which he could refer to no other species but this, had been taken some time since near Falmouth. He had never, however, himself met with a specimen. Inhabits the Mediterranean.

GEN. 3.   MULLUS, *Linn.*

8.   M. *Surmuletus*, Linn. (*Striped Surmullet.*)—Red; sides with longitudinal yellow lines: profile descending obliquely from the forehead.

M. Surmuletus, *Linn. Syst. Nat.* tom. I. p. 496. *Bloch, Ichth.* pl. 57. *Don. Brit. Fish.* vol. I. pl. 12. *Flem. Brit. An.* p. 216. Surmullet, *Will. Hist. Pisc.* p. 285. tab. S. 7. f. 1. Striped Surmullet, *Penn. Brit. Zool.* vol. III. p. 274. pl. 53. *Id.* (Edit. 1812.) vol. III. p. 368. pl. 64. Striped Red Mullet, *Yarr. Brit. Fish.* vol. I. p. 27. Surmulet, *Cuv. et Val. Poiss.* tom. III. p. 319. *Cuv. Reg. An.* tom. II. p. 157.

LENGTH. From twelve to fifteen inches.

DESCRIPT. (*Form.*) Thick and blunt-headed, the profile falling abruptly from the forehead: greatest depth contained four times and a half in the entire length; thickness rather more than half the depth: head a little less than one-fourth of the whole length: eyes large; mouth small; jaws equal; the lower one only with fine teeth like velvet; teeth in the middle of the palate forming a pavement: chin with two barbules two-thirds of the length of the head: scales large, particularly those on the head and gill-covers, easily falling: first dorsal commencing at a little less than one-third of the entire length from the end of the snout; its length half the depth of the body; its height somewhat exceeding its length; first three rays nearly equal, the others gradually decreasing: space between the two dorsals equalling the length of the first: height of the second rather less than that of the first; its length somewhat greater; first ray shorter than the second and third, which are equal and longest; the succeeding ones gradually decreasing; all the rays branched: caudal deeply forked: anal similar to the second dorsal, commencing a little more backward, but ending in the same vertical line; all the rays except the first branched: finless portion o: the tail not quite one-fourth of the whole length, caudal excluded: pectorals not quite so long as the head, narrow, and somewhat pointed; the first two rays simple, the others branched; fourth ray longest: ventrals exactly beneath them, and nearly of the same length; first ray spinous; second and third longest; all the soft rays branched: number of rays altogether,

D. 7—8 or 9; A. 8; C. 15, and some short ones; P. 17; V. 1/6.

Number of vertebræ twenty-four. (*Colours.*) General colour of the back and sides vermilion-red, with three longitudinal lines of golden yellow: throat, breast, belly, and beneath the tail, white, tinged with rose-colour: fins pale red, inclining in some places to yellow. When the scales are rubbed off, the skin appears crimson.

Met with principally on the southern coast. Mr. Couch states that it "is a migratory fish, and usually reaches the Cornish shores about Midsummer. Its common habit is to keep close to the ground, but the migration is performed near the surface." Has no air-bladder. Food, according to Bloch, crustacea, small fish, and shelled mollusca. Spawns in Spring.

9. M. *barbatus*, Linn. (*Red Surmullet.*)—Plain red ; the sides without any longitudinal yellow lines : profile nearly vertical.

> M. barbatus, *Linn. Syst. Nat.* tom. I. p. 495. *Bloch, Ichth.* pl. 348. f. 2. Mullus, *Will. Hist. Pisc.* p. 285. tab. S. 7. f. 2. Red Surmullet, *Penn. Brit. Zool.* vol. III. p. 271. *Id.* (Edit. 1812.) vol. III. p. 365. Plain Red Mullet, *Yarr. Brit. Fish.* vol. I. p. 32. Vrai Rouget, ou Rouget-Barbet, *Cuv. et Val. Poiss.* tom. III. p. 325. pl. 70.

LENGTH. Rarely exceeds six inches.

DESCRIPT. (*Form.*) Readily distinguished from the last species by the form of the head, the fall of the profile approaching more nearly to vertical : the pores of the infra-orbitals are also larger and more numerous ; the scales narrower, with fewer indentations on their basal margin, and with the striæ more strongly marked. In other respects the two species are similar. (*Colours.*) Of a deeper red, and of a more uniform colour than the *M. Surmuletus*, without any longitudinal yellow lines* : beneath the body silvery : fins yellow. CUV.

This species, which resides principally in the Mediterranean, appears to be extremely rare in the British seas. Pennant mentions having heard of its being taken on the coast of Scotland. More recently two specimens have been obtained by Mr. Couch on the coast of Cornwall. There is no other recorded instance in which it has been noticed. Habits said to resemble those of the *M. Surmuletus.*

GEN. 4.   TRIGLA, *Cuv.*

* *Body marked with fine transverse thread-like ridges.*

10.   T. *Pini*, Bl. (*Pine-leaved Gurnard.*)—Transverse lines not reaching below the middle of the sides : lateral line smooth : profile oblique.

------

* According to Mr. Couch, there is *one* yellow line, a little below the lateral line.

T. Pini, *Bloch, Ichth.* pl. 355.   T. lineata, *Mont. in Wern. Mem.*
vol. II. p. 460.   *Flem. Brit. An.* p. 215.   Pine-leaved Gurnard,
Shaw, Nat. Misc. vol. XXII. pl. 954.   Red Gurnard, *Yarr. Brit.
Fish.* vol. I. p. 34.   Grondin rouge, ou Rouget commun, *Cuv. et
Val. Poiss.* tom. IV. p. 20.   *Cuv. Reg. An.* tom. II. p. 159.

LENGTH.   From twelve to sixteen inches.

DESCRIPT.   (*Form.*) Head large; body tapering from the nape to
the base of the caudal: greatest depth contained six times and a half
in the entire length; length of the head contained four times; thickness three-fourths of the depth: profile falling obliquely, and making
with the cranium an angle of one hundred and thirty to one hundred
and forty degrees; the descending line slightly concave: sides of the
head flat and nearly vertical; space between the eyes contracted and
concave: cheeks and upper part of the head rough with granulations
disposed in lines radiating from different centres: extremity of the snout
slightly emarginated, with three or four blunt denticulations on each side:
above the anterior angle of the eye two or three short spines: supra-scapulars ending in a sharp point, of a semi-elliptic form, with the inner
margin denticulated: opercle with two short spines, not extending beyond the membrane; the upper one directed obliquely upwards, the lower
one backwards: clavicle likewise terminating in a short, but sharp, point:
lateral line straight, slender, bifurcating at the caudal extremity, and
almost perfectly smooth, as is the rest of the body, with the exception of
the dorsal ridges, which are deeply and strongly serrated: upper part of
the sides marked with a number of transverse parallel lines, intersecting
the lateral line nearly at right angles, and reaching as far below as above
it: both dorsals placed in a groove; the first commencing above the
supra-scapulars, of a triangular form, with the first ray serrated; second
ray longer and stouter than the others, equalling the depth of the body
at this point: second dorsal a very little behind the first, scarcely more
than half as high, but twice as long: anal answering to the second
dorsal, but commencing a little further backward: caudal slightly forked:
pectorals equalling one-fourth of the whole length, reaching a little beyond the vent; first seven rays branched, gradually decreasing; the rest
simple: ventrals somewhat shorter than the pectorals; the spinous ray
half as long as the soft ones, which are all branched:

B. 7; D. 9—18; A. 17; C. 11, and some short ones; P. 11, & 3 free; V. 1/5.

Number of vertebræ thirty-six or thirty-seven. (*Colours.*) The whole
body, fins included, rose-red; the under parts somewhat paler: occasionally the red is more or less clouded with brown and cinereous.

Very abundant on the southern and western coasts, and generally
known by the name of the *Red Gurnard*. Is considered by many
authors as the *T. Cuculus* of Linnæus, but since the characters in the
Systema Naturæ apply equally well to two species, this must remain
doubtful. Feeds principally on crustaceous animals. Spawns in May
or June.

11.   T. *lineata*, Gmel. (*Streaked Gurnard.*)—Transverse lines encircling the whole body : lateral line sharply
serrated : profile nearly vertical.

T. lineata, *Gmel. Linn.* tom. I. pt. 3. p. 1345.   *Bloch, Ichth.* pl. 354.
Don. Brit. Fish. vol. I. pl. 4.   T. Adriatica, *Flem. Brit. An.* p. 215.
Mullus imberbis, *Will. Hist. Pisc.* p. 278. tab. S. 1. f. 1.   Cuculus

lineatus, *Ray, Syn. Pisc.* p. 165. Streaked Gurnard, *Penn. Brit.
Zool.* vol. III. p. 281. pl. 57. *Id.* (Edit. 1812.) vol. III. p. 377. pl. 66.
Yarr. Brit. Fish. vol. I. p. 46. Rouget Camard, *Cuv. et Val. Poiss.*
tom. IV. p. 25. *Cuv. Reg. An.* tom. II. p. 159.

LENGTH. One foot.

DESCRIPT. (*Form.*) Thicker anteriorly than the last species; the
body tapering behind more suddenly: head shorter: profile falling more
abruptly; the descending line inclining to convex: cheeks higher in proportion: snout less emarginated; the denticulations at the sides very
indistinct: opercle broader, the terminating spine shorter and blunter:
clavicular spine not quite so sharp: first dorsal with the second ray
shorter; the first more strongly serrated; all the spinous rays weaker:
pectorals longer: lateral line and dorsal ridges sharply serrated: rest of
the body roughish, marked with elevated transverse lines, which, after
crossing the lateral line, pass onwards to the abdomen, where they ramify,
thus nearly encircling the whole body; these lines correspond in number
with the scales on the lateral line, amounting to about sixty-eight: number of fin-rays,

D. 10—16; A. 16; C. 11 or 13; P. 10, and 3; V. 1/5.

Number of vertebræ thirty-three. (*Colours.*) Dusky green, mottled with
purple, and sprinkled with red and gamboge-yellow spots; lower portion
of the sides silvery white, clouded with flesh-red: pectorals blue on their
under surface, but red at the base; their upper surface yellowish green,
spotted with red: free rays yellowish green tipped with red: ventrals
white: dorsal, anal, and caudal fins, red; the rays of the anal tipped with
white. *Obs.* Young fish are much less variegated, and generally want
the yellow spots.

A rare species; first observed on the Cornish coast by Mr. Jago, and
communicated by him to Petiver. Is occasionally met with at Weymouth,
Hastings, and as far north as the coast of Scotland. At Hastings it
is called the *French Gurnard.* The colours are very variable, but the
transverse lines encircling the whole body will always serve to identify
the species.

** *Body without transverse lines.*

12. T. *Hirundo*, Bl. (*Sapphirine Gurnard.*)—Lateral
line smooth : pectorals very large, reaching beyond the
ventrals : spine on the opercle scarcely projecting beyond
the membrane.

T. Hirundo, *Bloch, Ichth.* pl. 60. *Don. Brit. Fish.* vol. I. pl. 1.
T. lævis, *Mont. in Wern. Mem.* vol. II. p. 455. *Flem. Brit. An.*
p. 214. Tub-Fish, *Will. Hist. Pisc.* p. 280. tab. P. 4. Sapphirine
Gurnard, *Penn. Brit. Zool.* vol. III. p. 280. pl. 56. *Id.* (Edit. 1812.)
vol. III. p. 376. pl. 68. *Yarr. Brit. Fish.* vol. I. p. 41. Le Perlon,
Cuv. et Val. Poiss. tom. IV. p. 29. *Cuv. Reg. An.* tom. II. p. 159.

LENGTH. From eighteen to twenty-five inches.

DESCRIPT. (*Form.*) Somewhat resembling the *T. Pini* in its general
proportions, but thicker in the body, and broader across the head and
snout: inclination of the profile the same: eyes much smaller; the space
between them broader and not so much hollowed out: cheeks smoother :

snout more emarginated, with three or four rather blunt denticulations
on each side: gape more capacious: supra-scapulars triangular, the terminating spine rather more pointed: the opercular and clavicular spines
preserve the same proportions, the former scarcely projecting beyond the
membrane: lateral line straight, slender, and almost perfectly smooth,
like the rest of the body with the exception of the dorsal ridges, which
are strongly serrated: spines in the first dorsal not so strong as in the
T. Pini; the first with very obsolete denticulations; the second scarcely
longer than the adjoining ones, and equalling not above two-thirds of the
depth of the body: pectorals contained three times and a half in the whole
length: number of fin rays,

D. 9—16; A. 16; C. 11 or 13; P. 10, and 3; V. 1/5:

scales very small, oval, smooth, entire; those on the lateral line scarcely
projecting beyond the others. Number of vertebræ thirty-three or thirty-
four. (*Colours.*) General colour of the head and body brownish red, here
and there faintly tinged with yellowish green: pectorals bluish green on
their inner surface, edged and spotted with bright blue; on their outer
surface brownish red with the rays whitish: first dorsal reddish; second,
as well as the anal and ventrals, nearly white.

Common on the southern and western coasts, attaining a larger size
than any other British species, the *T. Lyra* excepted. Is sometimes
called a *Tub-Fish.* According to Mr. Couch, "sheds its spawn about
Christmas." *Obs.* Linnæus and Pennant have erroneously attributed
to this species a rough lateral line, a circumstance which appears to
have misled Montagu, when he established a second under the name
of *T. lævis.* All the individuals which have fallen under my notice
have had this part perfectly smooth, neither does Cuvier seem at all
aware of there being any allied species in which it is otherwise.

13. T. *Lyra*, Linn. (*Piper.*) — Lateral line smooth: pectorals large, reaching beyond the ventrals: humeral and opercular spines extremely long: snout divided into two dentated processes.

T. Lyra, *Linn. Syst. Nat.* tom. i. p. 496. *Bloch, Ichth.* pl. 350.
Don. Brit. Fish. vol. v. pl. 118. *Flem. Brit. An.* p. 215. Piper,
Will. Hist. Pisc. p. 282. tab. S. 1. f. 4. *Penn. Brit. Zool.* vol. iii.
p. 279. pl. 55. *Id.* (Edit. 1812.) vol. iii. p. 374. pl. 67. *Yarr. Brit.
Fish.* vol. i. p. 44. La Lyre, *Cuv. et Val. Poiss.* tom. iv. p. 40.
Cuv. Reg. An. tom. ii. p. 159.

Length. From twenty to twenty-eight inches.

Descript. (*Form.*) Readily distinguished from all the other British
species by the length of the opercular and humeral spines. Head very
large: depth at the nape a little less than one-fifth of the whole length;
length of the head one-fourth: snout deeply emarginated; the lateral
lobes much more produced than in any other species; the margin of
each lobe divided into twelve or fifteen teeth, the middle ones long and
pointed: the whole head finely granulated: only one, rather strong, spine
at the anterior angle of the orbit: the spine on the supra-scapular, and
the large one on the opercle, longer and sharper than in any other species; the humeral spine still more developed; the humeral bone, when
measured to the end of the spine, equalling more than half the length of
the head: first dorsal with the rays very sharp, a little bent, and smooth;
the first and second only with their anterior edges obsoletely denticu-

lated; second and third rays equal; fourth scarcely shorter: pectorals
very large, equalling nearly one-third of the entire length, extending
considerably beyond the ventrals:

D. 9—16 ; A. 16; C. 11, and some short ones ; P. 14, and 3; V. 1/5:

dorsal ridges more strongly serrated than in the other species: lateral
line smooth. Number of vertebræ thirty-three. (*Colours.*) "The general colour pale flesh-colour, rosy or darker on the back, and the belly
white; fins bluish at the base, and tinged with reddish towards the
extremities: irides fine golden yellow." DON.

Far from common; the name of *Piper* being often applied to the
last species, which is of much more frequent occurrence in the London
market. Frequents the western coasts at all seasons of the year, according to information communicated to Pennant. Is also occasionally taken
at Weymouth. Attains a weight of nearly seven pounds. Feeds on
crustacea. This and some other species have the power of uttering a
low grumbling sound when taken out of the water. The English name
of *Piper* is derived from this circumstance.

14. T. *Gurnardus*, Linn. (*Gray Gurnard.*)—Lateral
line sharply serrated : pectorals of equal length with the
ventrals: humeral and opercular spines long : first three
rays of the first dorsal granulated.

> T. Gurnardus, *Linn. Syst. Nat.* tom. I. p. 497. *Bloch, Ichth.* pl. 58.
> *Don. Brit. Fish.* vol. II. pl. 30. *Flem. Brit. An.* p. 215. Gray
> Gurnard, *Will. Hist. Pisc.* p. 279. tab. S. 2. f. 1. *Penn. Brit.
> Zool.* vol. III. p. 276. pl. 54. *Id.* (Edit. 1812.) vol. III. p. 371. pl. 65.
> *Yarr. Brit. Fish.* vol. I. p. 48. Le Grondin gris, *Cuv. et Val. Poiss.*
> tom. IV. p. 45. *Cuv. Reg. An.* tom. II. p. 160.

LENGTH. From fifteen to twenty inches; rarely two feet.
DESCRIPT. (*Form.*) Body more elongated, the snout longer, and the
profile less inclined than in most of the other species: depth at the pectorals less than one-sixth of the whole length: length of the head one-
fourth: cranium very slightly hollowed out between the eyes: descending
line of the profile nearly straight: emargination of the snout moderate;
each lobe with three or four well-marked denticulations: the whole of
the head, as well as the shoulders, granulated: two sharp spines at the
anterior angle of the orbit: supra-scapular and its terminating spine
much as in the *T. Hirundo:* humeral and opercular spines strongly
developed, much more so than in any of the foregoing species, excepting the *T. Lyra;* the latter projecting four or five lines beyond the membrane: spines of the first dorsal very strong, especially the first three,
which are rough with granulations; the second longest, a little exceeding
the depth of the body beneath: pectorals and ventrals of equal length,
both falling short of the vent by two or three lines:

D. 8—20; A. 19; C. 11, besides short ones; P. 10, and 3; V. 1/5:

lateral line broad, and sharply serrated; the scales larger than those on
the rest of the body: scales on the dorsal ridges with little projecting
crests, which are crenated and rough with minute granulations, but not
spinous. Number of vertebræ thirty-eight. (*Colours.*) Above gray,
clouded with brown, and more or less spotted with black and yellowish
white: beneath silvery: lateral line forming a longitudinal whitish band:
sometimes the whole body red, or inclining to that colour.

One of the most common species on the British coasts. Pennant states that it sometimes attains the length of two feet and a half; it is however usually found much less. Said to keep near the bottom, preying on shells and crustacea. According to Bloch, spawns in May and June.

15. T. *Cuculus*, Bl. (*Red Gurnard.*) — Constantly red, with a black spot on the first dorsal : this last with the first three rays smooth, without granulations.

T. Cuculus, *Bloch, Ichth.* pl. 59. *Mont. in Wern. Mem.* vol. ii. p. 457. *Flem. Brit. An.* p. 215.   T. Blochii, *Yarr. Brit. Fish.* vol. i. p. 50.   Red Gurnard, *Penn. Brit. Zool.* vol. iii. p. 278. pl. 57.   *Id.* (Edit. 1812.) vol. iii. p. 373. pl. 66.   Le Grondin Rouge, *Cuv. et Val. Poiss.* tom. iv. p. 48.   *Cuv. Reg. An.* tom. ii. p. 160.

Length. From nine to twelve inches.
Descript. Distinguished from the last species, which it closely re-sembles in all its other characters, by the first three spines in the dorsal fin not being granulated, but simply with a few denticulations on the anterior edges of the first two : the crests likewise of the scales on the dorsal ridges are not crenated as in the *Grey Gurnard*, but entire, each terminating in a little point. *Colour* constantly red, with a conspicuous black spot on the upper part of the first dorsal, extending from the second to the fifth spine. From the *T. Pini*, it is easily distinguished, by the serratures of the lateral line, and the absence of the transverse striæ.

D. 8—19 ; A. 18 ; C. 11, besides short ones ; P. 11, and 3 ; V. 1/5.

Number of vertebræ thirty-seven.
Never attains the size of the *T. Gurnardus*, and is not so abundant. The above description is from specimens taken at Weymouth. It is doubtful whether the *Cuculus* of Willughby* be this species or the *T. Pini*.

GEN. 5. COTTUS, *Linn.*

16. C. *Gobio*, Linn. (*Bull-head.*) — Head nearly smooth : preopercle armed with a single spine.

C. Gobio, *Linn. Syst. Nat.* tom. i. p. 452. *Bloch, Ichth.* pl. 39. f. 2. *Don. Brit. Fish.* vol. iv. pl. 80.   *Flem. Brit. An.* p. 216.   Bull-head, or Miller's Thumb, *Will. Hist. Pisc.* p. 137. tab. H. 3. f. 3. River Bull-head, *Penn. Brit. Zool.* vol. iii. p. 216. pl. 39. *Id.* (Edit. 1812.) vol. iii. p. 291. pl. 43. *Bowd. Brit. fr. wat. Fish.* Draw. 24. *Yarr. Brit. Fish.* vol. i. p. 56.   Le Chabot de rivière, *Cuv. et Val. Poiss.* tom. iv. p. 106.   *Cuv. Reg. An.* tom. ii. p. 162.

Length. From three to four, rarely five, inches.
Descript. (*Form.*) Head very large, occupying one-third of the whole length ; as broad as long ; somewhat depressed above, rounded in front : body gradually tapering from behind the nape : greatest depth beneath the first dorsal, equalling one-fifth of the entire length : gape very wide ; jaws equal ; both armed, as well as the fore part of the vomer, with fine teeth like velvet : eyes small, placed on the upper part of the head, a little inclined ; somewhat nearer together in the *male* than in the *female :* head smooth, without spines, covered, as well as the whole body,

* *Hist. Pisc.* p. 281. §. 7.

with a soft naked skin: preopercle only, with a single curved spine at its
posterior angle, the point turning upwards; opercle terminating behind
in a flat blunt point: lateral line nearly straight, its course at one-third
of the depth; formed by a row of small lozenge-shaped elevations of the
skin, from thirty to thirty-five in number: first dorsal commencing a little
behind the base of the pectorals; all the rays somewhat soft and flexible,
but not articulated: second dorsal twice as high, and nearly three times
as long, as the first; the membrane continuous from one to the other;
most of the rays simple, but all flexible and articulated: anal commencing a little more backward than the second dorsal, and not extending so far towards the caudal: this last rounded: pectorals broad
and rounded, equalling one-fourth of the entire length; most of the rays
simple, but all articulated: ventrals immediately under them, the first or
spinous ray enveloped in a membrane, which causes it to appear longer
and thicker than it really is:

> D. 6—16; A. 13; C. 11, and some short ones; P. 14; V. 1/3.

Number of vertebræ thirty-two. (*Colours.*) Brownish gray, occasionally
marbled with yellow and dusky spots; belly silvery white: fins barred
and varied with brown.

Common in fresh waters, especially clear streams which have a gravelly bottom. Lurks under stones, and swims with great rapidity. Swells
out its head when in danger by raising the gill-covers, thus causing the
former to appear broader than it is. Food, principally aquatic insects.
Spawns in March and April, according to Bloch and some other authors;
but, according to Cuvier, during the months of May, June, and July.
Has no air-bladder.

17. C. *Scorpius*, Bl. (*Sea-Scorpion, or Father-lasher.*) —Two erect spines before the eyes: preopercle with three spines; the first barely one-fifth the length of the head.

> C. Scorpius, *Bloch, Ichth.* pl. 40. *Don. Brit. Fish.* vol. II. pl. 35.
> *Flem. Brit. An.* p. 216. Sea-Scorpion, *Yarr. Brit. Fish.* vol. I.
> p. 60. Le Chaboisseau de mer commun, *Cuv. et Val. Poiss.*
> tom. IV. p. 117.

LENGTH. Rarely exceeds eight or nine inches.
DESCRIPT. (*Form.*) Head less depressed than in the last species; its
breadth not much more than half its length: eyes larger in proportion,
situate a little nearer the nose than the nape; the space between them
concave, and not equal to their diameter: mouth large; jaws equal: in
front of the space between the eyes, two small, but sharp, erect spines:
at the posterior part of the upper margin of the orbit a small tubercle,
more or less pointed, whence proceeds a slightly elevated crest on each
side of the occiput towards the nape, terminating there in another
tubercle; the space included between these ridges is of an oblong form
approaching to square: preopercle terminating behind in a strong sharp
spine directed backwards and a little upwards, its length barely one-fifth
that of the head; beneath it one smaller; and at the anterior extremity
of the lower margin a third still smaller, directed downwards and somewhat forwards: opercle likewise ending in a sharp spine: subopercle
with two small spines, one directed backwards and the other downwards:
scapulars and clavicles also each with a single spine directed backwards:
first dorsal commencing nearly in a line with the base of the pectorals:
second immediately behind it: somewhat longer as well as higher: this

last with all the rays simple, but flexible and articulated: anal a little
more backward than the second dorsal, and generally terminating nearer
the caudal: pectorals very broad, rounded at the extremity; rays simple;
seventh and eighth longest: ventrals narrow; the spinous ray and first
soft one so intimately united as to appear but one:

D. 9 or 10—14 or 15; A. 11; C. 12, and some short; P. 17; V. 1/3.

Number of vertebræ thirty-four or thirty-five. (*Colours*.) Reddish gray,
marbled and spotted with dusky and brown; belly whitish: fins pale,
with spots and specks of brown forming irregular transverse bars.

Apparently not so abundant on the British coasts as the next species,
with which it was for a long time confounded. Inhabits rocky shores,
and is of solitary habits. Swells out its head when attacked, endeavour-
ing to wound with the spine on the preopercle. Is very voracious, and
feeds on crustacea. Colours variable. According to Bloch, spawns in
December and January.

18. C. *Bubalis*, Euph. (*Four-spined Father-lasher*.)
—Two erect spines before the eyes: preopercle with four
spines; the first one-third the length of the head.

C. Bubalis, *Cuv. et Val. Poiss.* tom. IV. p. 120. pl. 78. *Cuv. Reg.
An.* tom. II. p. 163. *Yarr. in Zool. Journ.* vol. IV. p. 470.
Father-lasher, *Will. Hist. Pisc.* p. 138. tab. H. 4. f. 3. *Penn.
Brit. Zool.* vol. III. p. 218. *Id.* (Edit. 1812.) vol. III. p. 294. *Yarr.
Brit. Fish.* vol. I. p. 63.

Length. From seven to nine inches.
Descript. (*Form*.) Very similar to the last species, but differs in the
following points. The head is rougher; the space between the eyes narrower, and more concave: the occipital ridges are closer together, more
prominent, and very finely denticulated; behind, they terminate each in
a sharp, strong, and well-defined point; the space included between the
ridges is twice as long as it is broad: the great spine on the preopercle is
nearly one-third the length of the head; beneath it are three, instead of
only two, smaller ones: spine on the opercle, as well as the tubercles
which form the lateral line, rough and granulated, in some cases finely
denticulated: second dorsal with only eleven or twelve, very rarely thirteen rays: anal with only nine; terminating before the second dorsal,
instead of after it, as in the last species.

D. 8—11 to 13; A. 9; C. 10; P. 16; V. 1/3.

(*Colours*.) Similar to those of the *C. Scorpio*, but the belly, lower part
of the sides, and membranes of the anal and pectoral fins, with a bright
red tinge, rarely observed in the other species.

First distinguished as British by Mr. Yarrell. Common on many parts
of the coast, and having the same habits as those of the last species. Is
evidently the one described by Willughby as well as Pennant, though the
figure of this last author on the whole more resembles the *C. Scorpio.*
Food, crustacea and the fry of other fish. Spawns in January.

19. C. *quadricornis*, Linn. (*Four-horned Father-lasher.*)
—Four tubercles on the occiput: preopercle with three spines.

C. quadricornis, *Linn. Syst. Nat.* tom. I. p. 451. *Bloch, Ichth.*
pl. 108. *Cuv. et Val. Poiss.* tom. IV. p. 123. *Cuv. Reg. An.*
tom. II. p. 163. Four-horned Cottus, *Yarr. Brit. Fish.* vol. I. p. 68.

LENGTH. From ten to twelve inches.

DESCRIPT. (*Form.*) Distinguished principally by four large, rough,
osseous tubercles, which take the place of the four occipital spines in the
C. Scorpius: head broader than in this last species; cranium broader
than long, and without the lateral ridges: first infraorbital much more
pitted, or hollowed out; on the second, often a small spine: preopercle
with three strong spines; the uppermost longer than the others, and
bending a little outwards: spine on the opercle, as well as that on the
supra-scapular, likewise a little curved; this last large: scales on the
lateral line, large, osseous, rectangular, with two concave impressions, one
above the other: above the lateral line a row of osseous tubercles, round,
a little raised in the middle, and finely granulated in streaks; a few
others scattered beneath: in most other respects the two species are
similar:

D. 7—14; A. 15; C. 11; P. 17; V. 1/3.   CUV.

Individuals of this species, recently observed in the London market
amongst Sprats from the mouth of the Thames, are now in the British
Museum. Has not hitherto occurred in any other instance in our seas.
Common in the Baltic, and said generally to keep near the mouths of
rivers where the salt and fresh waters mix. Habits resembling those of
the *C. Scorpius.*

GEN. 6. ASPIDOPHORUS, *Lacép.*

20. A. *cataphractus,* Nob. (*Pogge.*) — Dorsals contiguous : upper jaw longest : chin and branchiostegous
membrane furnished with numerous thread-like filaments.

A. armatus, *Lacép. Hist. Nat. des Poiss.* tom. III. p. 222. A. Euro-
pæus, *Cuv. et Val. Poiss.* tom. IV. p. 147. Cottus cataphractus,
Linn. Syst. Nat. tom. I. p. 451. *Bloch, Ichth.* pl. 39. f. 1. *Don.
Brit. Fish.* vol. I. pl. 16. Cataphractus Schoneveldii, *Flem. Brit.
An.* p. 216. Pogge, *Will. Hist. Pisc.* p. 211. tab. N. 6. f. 2, & 3.
Armed Bull-head, *Penn. Brit. Zool.* vol. III. p. 217. pl. 39. *Id.*
(Edit. 1812.) vol. III. p. 293. pl. 43. *Yarr. Brit. Fish.* vol. I. p. 70.

LENGTH. From four to six inches.

DESCRIPT. (*Form.*) Distinguished from the *Cotti* by the osseous
plates on the body, which form longitudinal sharp angular ridges, and
render it somewhat octagonal: head broad, and much depressed: body
tapering behind: depth at the nape about one-seventh the entire length:
breadth across the head one-fourth: space between the eyes concave:
snout slightly recurved, and armed at its extremity with four small acute
spines directed upwards: upper jaw projecting over the lower: both jaws
with fine sharp velvet-like teeth: pharyngeans the same: tongue and palate
smooth: infra-orbitals with three blunt tubercles on their lower margin;
beyond them a sharp spine directed backwards: preopercle with a similar
but larger spine; opercle small and unarmed: branchiostegous membrane,
as well as the chin, furnished with numerous small fleshy filaments in the
form of threads: body only octagonal from the vent to the termination of
the second dorsal and anal; at that point the two uppermost carinæ, and
the two lowermost, unite respectively to form one; and beyond, the body
is hexagonal: the lateral line is first parallel to the upper series, then
takes a bend opposite to the vent where the second series commences,
and passing between this and the third, proceeds straight to the caudal:
four uppermost carinæ rough and sharp: the four lower ones almost

smooth: first dorsal rounded; the rays flexible but not articulated: second immediately behind it; all the rays simple: anal answering exactly to the second dorsal: finless portion of the tail one-third of the entire length: caudal rounded: pectorals also rounded, about the length of the head; all the rays simple: ventrals immediately under them, narrow and pointed, the spinous ray closely attached to the first articulated one:

D. 5—6; A. 7; C. 11; P. 16; V. 1/2.

(*Colours.*) Brown above; beneath white; more or less marked with dusky spots.

Common on many parts of the coast, concealing itself in the sand, or amongst stones. Feeds on small crustacea and marine insects. Spawns, according to Bloch, in May.

GEN. 7.   SCORPÆNA, *Linn.*

(1. Sebastes, *Cuv.*)

21. S. *Norvegica*, Cuv. (*Northern Sebastes.*)—Dorsal fin with fifteen spinous and fifteen soft rays; the longest of the spinous rays scarcely more than one-fifth the depth of the body.

Sebastes Norvegicus, *Cuv. et Val. Poiss.* tom. iv. p. 240. pl. 87. *Cuv. Reg. An.* tom. ii. p. 166.   Serranus Norvegicus, *Flem. Brit. An.* p. 212.   Sea-Perch, *Penn. Brit. Zool.* vol. iii. p. 258. pl. 48. *Id.* (Edit. 1812.) vol. iii. p. 349. pl. 59.   Bergylt, & Norway Haddock, *Yarr. Brit. Fish.* vol. i. p. 73.

Length. Two feet and upwards. Cuv.

Descript. (*Form.*) Resembles the *Perch:* body oblong, a little compressed; dorsal and ventral lines slightly convex: mouth oblique; lower jaw longest: depth at the pectorals contained not quite three times and a half in the length; thickness not half the depth: snout a little convex: space between the eyes flat: infra-orbitals somewhat pitted, but not armed: one small spine on the edge of the orbit in front; behind it, on the cranium, three others also small: on each side of the occiput a slightly elevated crest, terminating likewise in a small spine: scapular and supra-scapular each with a single spine: two on the opercle; preopercle rounded, with five sharp, but rather short, spines; subopercle and interopercle each also with one small spine at the point where they meet: fine velvet-like teeth in both jaws, as well as on the vomer and palatines: dorsal commencing above the supra-scapular; spinous portion equalling nearly one-third of the whole length; rays strong but short: soft portion only half the length of the spinous, but twice as high : anal commencing in a line with the soft portion of the dorsal; first three rays spinous, the first only half the length of the two others; soft rays branched, twice as long as the spinous ones: caudal nearly even : pectorals equalling one-fifth of the whole length, rounded, as broad as long : the first ten rays branched, the rest simple: ventrals a little behind the pectorals, and not quite so long :

D. 15/15; A. 3/8; C. 14; P. 19; V. 1/5 :

scales covering the whole head and body; a few small ones on the soft portions of the dorsal and anal fins, as well as on the caudal: lateral line parallel to the back: its course at one-fourth of the depth. Number of vertebræ thirty-one. Cuv.

This species, the *Perca marina* of Linnæus, frequents high latitudes, and is little known as a native of our own seas, excepting along the northern coasts of the Island. Has been met with on the coasts of Aberdeenshire and Berwickshire; also in Zetland by Dr. Fleming. Food, according to Cuvier, crustacea and small fish. *Obs.* It is very doubtful whether the *Sea-Perch* of Willughby* be referable to this species.

GEN. 8. GASTEROSTEUS, *Linn.*

(1. Gasterosteus, *Cuv.*)

* *Sides more or less protected by transverse scaly plates.*

22. G. *aculeatus*, Linn. (*Three-spined Stickleback.*) —Three dorsal spines.

> G. aculeatus, *Linn. Syst. Nat.* tom. i. p. 489. *Bloch, Ichth.* pl. 53. f. 3. *Don. Brit. Fish.* vol. i. pl. 11. *Flem. Brit. An.* p. 219. G. trachurus, leiurus, & semiarmatus, *Cuv. et Val. Poiss.* tom. iv. p. 352. pl. 98, & p. 361. G. trachurus, & gymnurus, *Cuv. Reg. An.* tom. ii. p. 170. Stickleback, *Will. Hist. Pisc.* p. 341. tab. X. 14. f. 1. *Bowd. Brit. fr. wat. Fish.* Draw. 20. Three-spined Stickleback, *Penn. Brit. Zool.* vol. iii. p. 261. pl. 50. no. 129. *Id.* (Edit. 1812.) vol. iii. p. 353. pl. 61.

Length. From two to two and a half, rarely three inches.

Descript. (*Form.*) Oval; rather elongated; sides compressed; tail slender: dorsal and ventral lines equally convex: greatest depth about the middle, rather more than one-fifth of the entire length; head one-fourth; thickness a little more than half the depth: eyes large: cranium more or less striated, the striæ formed of minute granulations: mouth protractile: when closed, the lower jaw advancing a little beyond the upper: both jaws with fine small teeth, but none on the tongue, vomer, or palatines: opercle large and triangular, the posterior margin rounded: no true scales, but the sides protected by a series of oblong osseous plates, varying in number, disposed in transverse bands; a similar plate, ascending from the base of the ventrals, reposes on the third and fourth of the above series; there is also another placed longitudinally on each side of the breast, and a large triangular one on the belly, having its base in a line with the ventrals, and its apex directed towards the vent; all these scaly plates more or less granulated in lines: instead of a first dorsal three free strong spines, a little distant from each other, more or less serrated at their edges, varying in length, but the second always longest; first spine above the first or second of the lateral scaly plates; second above the fourth; the third, which is much smaller than the other two, in a line with the apex of the triangular plate on the belly: soft dorsal commencing immediately behind this last spine; all the rays, except the first, branched: anal about half the length of the soft dorsal, with one short, curved, free spine immediately before the first ray: caudal rounded: ventrals consisting of one strong serrated spine, united by a delicate membrane to one slender soft ray scarcely one-third of its own length:

D. 3—10 to 13; A. 1/8 to 10; C. 12; P. 10; V. 1/1:

sides of the tail sometimes furnished with a horizontal expansion of the skin forming a keel. Number of vertebræ thirty-three. (*Colours.*)

---

* *Hist Pisc.* p. 327.

Back and sides olivaceous, sometimes passing into yellowish brown or dusky blue: throat and breast, in some individuals, bright fiery red: belly and flanks silvery, with a pearly lustre.

Var. α. G. trachurus, *Cuv. Yarr. in Mag. of Nat. Hist.* vol. III. p. 522. fig. 127. *a.* Rough-tailed Stickleback, *Yarr. Brit. Fish.* vol. I. p. 76. The scaly plates extending the whole length of the sides; in number about thirty.

Var. β. G. semiarmatus, *Cuv. Yarr. in Mag. of Nat. Hist.* vol. III. p. 522. fig. 127. *b.* Half-armed Stickleback, *Yarr. Brit. Fish.* vol. I. p. 80. Lateral plates extending to a vertical line joining the vent and commencement of the soft dorsal; in number from twelve to fifteen.

Var. γ. G. leiurus, *Cuv. Yarr. in Mag. of Nat. Hist.* vol. III. p. 522. fig. 127. *c.* Smooth-tailed Stickleback, *Yarr. Brit. Fish.* vol. I. p. 81. Lateral plates from four to six; extending only as far as the pectoral fins, when these last are laid back.

Var. δ. G. brachycentrus, *Cuv.?* Short-spined Stickleback, *Yarr. Brit. Fish.* vol. I. p. 82. Lateral plates not extending beyond the pectorals: dorsal and ventral spines very short.

Common throughout the country in rivers and streams, as well as in stagnant waters. Is also found occasionally in the sea. Of active and lively habits. Is very voracious, and preys on worms and aquatic insects. Spawns, according to Bloch, in April and June; according to Cuvier, in July and August.

Obs. The above species is subject to great variation, not only in the number of the lateral plates, but in several other less obvious respects. The former may occasionally be found of every intermediate number between that which characterizes the *G. leiurus*, Cuv. and that which appears in the *G. trachurus* of the same author. This number, moreover, is sometimes found constant in specimens which differ remarkably in other respects; at other times, varying, when all other characters remain the same. From these circumstances combined, I feel satisfied that the above are mere varieties, notwithstanding the high authorities on which they stand recorded as distinct species. Perhaps it may be useful to state the result of a close comparison of a large number of individuals with each other from different localities.

(1.) *Specimens from the Thames, procured by W. Yarrell, Esq.* These agreed in having the eyes very large; the space between rugose, with granulations disposed in lines; the teeth rather prominent; the osseous disk between the opercle and pectoral large; the lateral plates varying in number, but well-defined and very regularly disposed; the ventral plate narrow, more than twice as long as broad at the base; the dorsal and ventral spines long, the latter equalling two-thirds of the depth of the body, nearly straight, and often with serrated edges; sides of the tail generally, but not always, carinated.

(2.) *Specimens from Wilbraham in Cambridgeshire.* Depth greater in proportion to the length than in no. 1; eyes much smaller; the rugose lines between as before; teeth similar; osseous disk behind the opercle much smaller; lateral plates few in number but well-defined; ventral plate very large, its breadth at the base contained only once and a half in its length; spines, the ventral especially, nearly as long as in the above, equally serrated, but not so straight, being slightly curved from their base; sides of the tail perfectly smooth.

(3.) *Specimens from the pond in the Botanic Garden, Cambridge.* Eyes intermediate in size between those of nos. 1 & 2, but varying slightly in different individuals; rugose lines on the vertex generally indistinct, sometimes wholly wanting with the exception of two, one above each eye, which are always present; teeth varying a little, but generally smaller than in either of the above; osseous plate behind the opercle generally larger than in the last, but seldom so large as in the Thames specimens; lateral plates varying in number, sometimes extending the whole length of the sides, but generally few, and irregularly disposed; ventral plate about twice as long as broad at the base; spines varying a little, but always much shorter (the dorsal especially) than in no. 1; ventrals equalling half the depth; sides of the tail, except in one or two instances, not carinated.

(4.) *Specimens from the North of Ireland, procured by W. Thompson, Esq.* Of very large size, measuring full three inches. Eyes large, but less than in the Thames specimens; the space between smooth, with the exception of two deeply impressed lines, one above each eye; teeth moderate; osseous disk between the opercle and pectoral rather large; lateral plates five in number, and regularly disposed; ventral plate twice as long as broad at the base, its apex very obtuse; dorsal and ventral spines strong, but much shorter in proportion than in any of the former specimens, a little curved, their margins finely serrated; sides of the tail smooth, without any trace of a keel.

From the above details it will be seen how each character varies in its turn, and at the same time how little connection there is between the variation of one part and that of the others*.

23. G. *spinulosus*, Yarr. and Jen. (*Four-spined Stickleback.*)—Four dorsal spines.

Four-spined Stickleback, *Edinb. New Phil. Journ.* Apr. 1831. p. 386. *Yarr. Brit. Fish.* vol. i. p. 83.

LENGTH.   One inch and a quarter.

DESCRIPT. Differs in no essential particulars from the last species, excepting in being smaller, and having an additional dorsal spine, situate half-way between the second and third of the ordinary ones; this spine is very small, and even shorter than that which precedes the soft fin: in my specimen, there are only two lateral plates, and these not very well developed; they occupy that portion of the side which lies beneath the first and second spines: the ascending plate from the base of the ventrals is longer and narrower than in young specimens of the *G. aculeatus* of the same size: none of the spines are serrated, and the sides of the tail shew no appearance of a keel:

D. 4—11; A. 1/10; C. 12; P. 10; V. 1/1.

Discovered in some plenty near Edinburgh, by Dr. James Stark (to whom I am indebted for a specimen), in September 1830. Possibly a mere variety of the last species, which is said to have been numerous in the same pond. In the number of the spines, it resembles the *G. tetracanthus* of Cuvier; but this last is represented as having the spines shorter than in the common species, and the ventral plate broader.

** *Sides naked.*

24. G. *Pungitius*, Linn. (*Ten-spined Stickleback.*)—Dorsal spines nine or ten.

---

* It is more than probable that some of the other foreign *Gasterostei* described by Cuvier are mere varieties of this species. They hardly differ more from those described above than these last do from each other.

G. Pungitius, *Linn. Syst. Nat.* tom. I. p. 491. *Bloch, Ichth.* pl. 53.
f. 4. *Don. Brit. Fish.* vol. II. pl. 32. *Flem. Brit. An.* p. 219.
Lesser Stickleback, *Will. Hist. Pisc.* p. 342. Ten-spined Stickleback, *Penn. Brit. Zool.* vol. III. p. 262. pl. 50. no. 130. *Id.* (Edit.
1812.) vol. III. p. 355. pl. 61. *Yarr. Brit. Fish.* vol. I. p. 85.
L'Epinochette, *Cuv. et Val. Poiss.* tom. IV. p. 370. *Cuv. Reg.
An.* tom. II. p. 170.

LENGTH. Rarely exceeding two inches.

DESCRIPT. (*Form.*) Rather more elongated than the *G. aculeatus;*
the depth not so great: differs essentially from that species in wanting
the lateral scaly plates, although it possesses the triangular one on the
belly, and the ascending branch from the base of the ventral spines:
dorsal spines generally nine, but sometimes ten in number, much shorter
than in the *G. aculeatus*, all equal, placed at equal distances from each
other, erect but somewhat inclining alternately to the right and to the
left, without denticulations at the edges, each furnished with a small
membrane at the base: ventral spines barely so long as the ventral plate,
likewise without serratures: ventral plate somewhat narrower and more
pointed than in the *G. aculeatus:* sides of the tail keeled; each keel
being furnished with several slender scales, which themselves appear
keeled under the microscope:

D. 9 or 10—10 or 9; A. 1/9; C. 12; P. 10; V. 1/1.

(*Colours.*) Back and portion of the sides yellowish green, marked occasionally with transverse dusky bands; abdomen silvery; the whole more
or less dotted with black specks: fins pale. A *variety* sometimes occurs
of a uniform dark bluish black, pervading the whole body above and
below.

Var. β. G. lævis, *Cuv. Reg. An.* vol. II. p. 170. Sides of the tail
smooth.

Equally abundant with the *G. aculeatus*, and as generally distributed.
Like that species, is occasionally found in salt water. Habits similar.
Obs. The *G. lævis* of Cuvier differs in no respect excepting in wanting
the carinated scales at the sides of the tail, and is evidently a mere
variety.

(2. SPINACHIA, *Flem.*)

25. G. *Spinachia*, Linn. (*Fifteen-spined Stickleback.*)

G. Spinachia, *Linn. Syst. Nat.* tom. I. p. 492. *Bloch, Ichth.* pl. 53.
f. 1. *Don. Brit. Fish.* vol. II. pl. 45. Spinachia vulgaris, *Flem.
Brit. An.* p. 219. Aculeatus marinus, *Will. Hist. Pisc.* p. 340.
tab. X. 13. f. 2. Fifteen-spined Stickleback, *Penn. Brit. Zool.*
vol. III. p. 263. pl. 50. no. 131. *Id.* (Edit. 1812.) vol. III. p. 356.
pl. 61. *Yarr. Brit. Fish.* vol. I. p. 87. Le Gastré, *Cuv. et Val.
Poiss.* tom. IV. p. 373. *Cuv. Reg. An.* tom. II. p. 170.

LENGTH. From five to six inches.

DESCRIPT. (*Form.*) Very much elongated; the entire length ten
times the depth and the breadth: lateral line marked throughout its
course by a series of carinated scales, which form ridges, and render the
posterior half of the body quadrangular; before the vent it is pentagonal:
head one-fourth of the whole length; snout very much produced; mouth
rather small; lower jaw longest: vertex flat, with a sulcus between the
eyes: dorsal spines commencing immediately above the pectorals, fifteen

in number, of equal length, the last excepted, which is longer and more
hooked than the others; space occupied by these spines equalling nearly
one-fourth of the whole length: soft fin immediately behind the last spine,
of a triangular form; its greatest height equalling the depth of the body;
all the rays branched: anal exactly answering to it in form, situation, and
number of rays; before the first ray a short hooked spine: portion of the
tail beyond these fins very much depressed, sharp at the sides, and equalling one-third of the whole length: caudal square at the end; all the rays
branched: pectorals rounded, one-ninth of the entire length; the rays
simple: ventral spines short and slightly curved, each accompanied by
two small soft rays:

D. 15—6 or 7; A. 1/6; C. 12; P. 11; V. 1/2:

about forty-two scales in the lateral line, all sharply keeled and slightly
granulated; those on the sides of the tail forming the sharpest edge: rest
of the skin smooth. Number of vertebræ forty-one. (*Colours.*) Greenish
brown; silvery beneath: dorsal and anal fins, each with a round black
spot.

    Found only in the sea, never ascending rivers. Not uncommon on
some parts of the coast. Feeds on worms and small crustacea, as well as
on the eggs and fry of other fish. Spawns in Spring. Is stated by Cuvier
to be the only known species belonging to this sub-genus.

———

GEN. 9.   SCIÆNA, Cuv.

(1. SCIÆNA, Cuv.)

26. S. Aquila, Cuv. (Maigre.)

    S. Aquila, *Cuv. et Val. Poiss.* tom. v. p. 21. pl. 100. *Neill in
    Edinb. New Phil. Journ. Apr.* 1826. p. 135. *Flem. Brit. An.* p. 213.
    S. Umbra, *Cuv. Reg. An.* tom. II. p. 172. The Maigre, *Yarr.
    Brit. Fish.* vol. I. p. 90.

    LENGTH. From three to five, sometimes six feet.

    DESCRIPT. (*Form.*) General appearance resembling that of the Basse
(*Perca Labrax*): head one-fourth of the entire length; greatest depth
rather more than one-fifth: profile descending obliquely, inclining to
convex at the nape, concave at the forehead: snout blunt and slightly
protuberant: each jaw with a row of sharp, somewhat hooked teeth,
separate from each other, with several smaller ones amongst them in
the lower, behind them in the upper jaw: none on the tongue, vomer, or
palatines: preopercle with the posterior margin denticulated when young,
but not afterwards: opercle terminating in two flat, but rather sharp
points, with an emargination between: first dorsal with the third spine
longest, equalling half the depth of the body: second dorsal more than
twice the length of the first, immediately behind it, the membrane of
the latter continuous with that of the former: pectorals and ventrals
nearly one-sixth of the entire length: anal very small in proportion to
the second dorsal, with only one slender spinous ray almost concealed in
the edge of the fin, and eight soft ones: caudal with seventeen branched
rays: number of rays altogether,

D. 9—1/27 or 28; A. 1/8; C. 17; P. 16; V. 1/5:

lateral line nearly parallel to the back: the whole head and body covered
with scales; those on the back and sides large, deeply imbricated, and

set obliquely to the axis of the body. Number of vertebræ twenty-four. (*Colours.*) Of a uniform silvery gray, inclining to brownish on the back, and to white on the belly: first dorsal, pectorals, and ventrals, red; the other fins reddish brown. Cuv.

Common in the Mediterranean, where it attains a large size. Has not occurred in the British seas in more than four or five instances. One specimen recorded by Mr. Neill as having occurred off the Shetland coasts in November 1819. A second taken in the seine, at Start Bay, on the south coast of Devon, in August 1825*. A third, taken on the coast of Northumberland, is in the possession of Mr. J. Hancock of Newcastle. A fourth is mentioned by Mr. Yarrell as having occurred on the Kentish coast in November 1834. Said to swim in shoals, and when taken, to make a low grumbling noise like the Gurnards. Air-bladder, according to Cuvier, very large, extending the whole length of the abdomen, and remarkable for its branched lateral appendages. *Obs.* This species has been much misunderstood, and confounded with others by many authors, especially by Willughby, Artedi, and Linnæus, whose descriptions in consequence are rendered of no value. It is the only Europæan species belonging to this sub-genus.

(*Umbrina,* Cuv.)

(3.) *S. cirrhosa,* Linn. Syst. Nat. tom. i. p. 481. Bloch, Ichth. pl. 300. *Umbrina vulgaris,* Cuv. et Val. Poiss. tom. v. p. 127. *Bearded Umbrina,* Yarr. Brit. Fish. vol. i. p. 95.

According to an extract from the Minute-Book of the Linnæan Society, (*Linn. Trans.* vol. xvi. p. 751.) dated Nov. 20, 1827, a specimen of this fish, weighing one hundred weight, has been taken in the river Exe. As, however, there has been much confusion with respect to the species of this family, it is possible that this British individual may not have been different from the *Sciæna Aquila* described above. According to Cuvier, the *Umbrina* is never found so large as this last species, though it often exceeds two feet in length. It is common on the coasts of France, Spain, and Italy, and is easily distinguished by a short barbule attached to the symphysis of the lower jaw.

GEN. 10.   SPARUS, *Cuv.*

(1. Chrysophrys, *Cuv.*)

27. S. *Aurata,* Linn. (*Gilt-head.*) — Molars in four or five rows above, and three below: a golden spot on the preopercle, and a dark one on the opercle.

S. Aurata, *Bloch, Ichth.* pl. 266. *Flem. Brit. An.* p. 211. Chrysophrys Aurata, *Cuv. et Val. Poiss.* tom. vi. p. 62. pl. 145. *Cuv. Reg. An.* tom. ii. p. 182. Gilt-head, *Will. Hist. Pisc.* p. 307. tab. V. 5. *Yarr. Brit. Fish.* vol. i. p. 97. Lunulated Gilt-head, *Penn. Brit. Zool.* vol. iii. p. 240. but not pl. 42. *Id.* (Edit. 1812.) vol. iii. p. 327. but not pl. 46.

---

* *Proceedings of Comm. of Zool. Soc.* 1831. p. 112.

Z

LENGTH.   Fifteen inches.

DESCRIPT.   (*Form.*) Body oval, larger before than behind: depth nearly one-third of the entire length; thickness two-fifths of the depth: snout obtuse: eyes moderate, situate on the upper part of the cheek, at the distance of twice the length of their diameter from the end of the snout: opercle narrow, almost three times as high as broad; the scales which cover it larger than those on the preopercle: lower jaw a little the shortest; in each jaw six strong, hooked, conical, rather blunt, incisors; tubercular teeth in five rows above, in three below; the anterior ones small and rounded; at the back of the mouth, always one, often two, oval ones larger than the others, their longest diameter measuring sometimes four lines: (in *young* individuals the teeth are smaller and in only four rows:) lateral line nearly straight; its course a little above one-third of the depth: number of scales in a longitudinal line nearly eighty, in the depth about twenty-four: dorsal commencing at a distance from the end of the snout equalling nearly one-third of the whole length; spinous, higher than the soft, portion: anal in a line with the third soft ray of the dorsal, and not extending beyond that fin; its spinous rays shorter than those of the dorsal; second stoutest: caudal moderately forked: pectorals long, reaching beyond the vent, equalling nearly one-fourth of the whole length; their point of attachment a little before the dorsal: ventrals rather behind the pectorals, moderately broad:

B. 6; D. 11/13; A. 3/11; C. 17; P. 16; V. 1/5.

Number of vertebræ twenty-four. (*Colours.*) Back silvery gray; belly like polished silver; sides with from eighteen to twenty longitudinal golden bands; on the forehead, between the eyes, a crescent-shaped band of golden yellow; a large spot on the shoulder; opercle also dusky or violet. CUV.

This species must be considered as very rare in the British seas, if it have really any claim at all to a place in the Fauna. The descriptions of Pennant and Fleming apply with tolerable correctness, but the *Gilt-head* of Donovan and other English authors is only the *Pagellus centrodontus* of Cuvier hereafter described. Found in the Mediterranean along with another nearly allied species, and said, by Cuvier, to feed on the conchiferous mollusca.

(2. PAGRUS, *Cuv.*)

28.   S. *Pagrus*, Linn. (*Braize, or Becker.*) — Silvery, tinged with red: no golden crescent between the eyes; no black spot on the shoulder.

> Pagrus vulgaris, *Cuv. et Val. Poiss.* tom. VI. p. 104. pl. 148. Le Pagre de la Mediterranée, *Cuv. Reg. An.* tom. II. p. 182. Becker, *Couch in Linn. Trans.* vol. XIV. p. 79. Braize, or Becker, *Yarr. Brit. Fish.* vol. I. p. 102.

LENGTH ?

DESCRIPT. (*Form.*) Snout obtuse, like that of the S. *Aurata*, but the nape less elevated, and the body more elongated than in that species: head one-fourth of the entire length; depth a little more: eyes large and round: opercle more than twice as high as broad: four strong pointed teeth at the extremity of each jaw, with a group of small card-like teeth behind them; beyond these a row of five teeth obtusely conical, and four or five round ones; within, and parallel to this

series, another row of five or six teeth all round: pharyngeans strong
and card-like: lateral line more strongly marked than in the *Gilt-head*:
scales on the upper part of the head and on the gill-covers smaller;
those on the body larger in proportion: number in a longitudinal line
nearly sixty; in the depth twenty: dorsal when laid back almost entirely
concealed in a deep groove; the spinous rays compressed, and somewhat
flexible; the longest not one-third the depth of the body: anal answering
to the soft portion of the dorsal; the three spinous rays sensibly stronger
than those of that fin; along its base a slight scaly projection partly
concealing it: pectorals pointed, contained three times and a half in the
entire length, and reaching when laid back to the third spinous ray in
the anal fin: ventrals only half the length of the pectorals; the spinous
ray one-sixth shorter than the first soft one:

B. 6; D. 12/10; A. 3/8; C. 17; P. 15; V. 1/5.

Number of vertebræ twenty-four. (*Colours.*) Silvery, tinged with red:
no semilunar mark between the eyes, as in the last species, and no dark
patch on the shoulder, as in the *S. centrodontus*. Cuv.

So much confusion and misunderstanding prevails with respect to the
species of this family, that it is not easy to attach to each correctly its
proper synonyms. The present one appears to be the *Becker* of Mr.
Couch, which is stated by that gentleman to approach the Cornish coasts
during the Summer and Autumn. There is very little respecting it, at
least on which any dependence can be placed, in other British writers.
The *Pagrus vulgaris* of Fleming (the *Red Gilt-head* of Pennant) is probably only the *Pagellus centrodontus* of Cuvier and of this work. According to Cuvier, the present species is found in the Mediterranean,
along with two others belonging to the same sub-genus.

(3. Pagellus, *Cuv.*)

29. S. *Erythrinus*, Linn. (*Spanish Sea Bream.*)—
Rose-colour, with silvery reflections: a golden crescent
between the eyes, but no lateral spot.

> S. Erythrinus, *Bloch, Ichth.* pl. 274. *Shaw, Nat. Misc.* vol. xx.
> pl. 834. Pagellus Erythrinus, *Cuv. et Val. Poiss.* tom. vi. p. 126.
> pl. 150. Erythrinus Rondeletii, *Will. Hist. Pisc.* p. 311. tab. V. 6.
> Le Pagel commun, *Cuv. Reg. An.* tom. ii. p. 183. Spanish
> Bream, *Couch in Loud. Mag. of Nat. Hist.* vol. v. p. 17. *Yarr.
> Brit. Fish.* vol. i. p. 104.

LENGTH?
DESCRIPT. (*Form.*) Body oval, elongated, moderately compressed, a
little contracted towards the tail: depth to the right of the pectorals one-
third of the length: nape elevated; profile descending in a straight line,
a little obliquely, towards the snout: eyes large and round, placed half-
way between the end of the snout and the shoulder: infra-orbitals large:
preopercle also large, covering nearly the whole cheek; the ascending
margin rectilineal and nearly vertical: mouth scarcely protractile: lower
jaw a little the longest: both jaws with fine card-like teeth at their extremities, the outer row rather the strongest; nine or ten on each side;
molars behind small, in two or more rows, more numerous in the *adult*
than in the *young:* pharyngeans strong and hooked: lateral line strongly
marked; its course straight to the end of the dorsal, then turning in and
terminating at the caudal, passing a little above the middle of the tail:

number of scales in a longitudinal line nearly sixty; in the depth twenty-
one: dorsal with the fourth spinous ray longest, equalling nearly one-
third of the depth beneath; the succeeding ones gradually decreasing;
soft rays a little longer than the last of the spinous: anal answering to
the soft portion of the dorsal; the spinous rays stronger than those in
that fin: caudal deeply forked; the two lobes equal, covered for half
their length with small scales: pectorals narrow, contained three times
and a half in the length of the body: ventrals triangular, rather large, a
little behind the pectorals:

D. 12/10 ; A. 3/8 ; C. 17 ; P. 15 ; V. 1/5.

Number of vertebræ twenty-four. (*Colours.*) Fine carmine-red, passing
into rose-red on the sides; belly with silvery reflections: fins rose-red;
the anal and ventrals palest. Cuv.

Observed by Mr. Couch in two or three instances off the coast of
Cornwall, and said to be known to the fishermen there by the name
of *Spanish Bream.* Not mentioned by any other writer on British
Zoology; though, according to Mr. Yarrell, it appears to have been
met with at Teignmouth by the late Mr. Walcott. Common in the
Mediterranean. Stated by Cuvier to keep in small shoals, and to
feed on fish and conchiferous mollusca. Is always smaller than the
next species.

30. S. *centrodontus*, Laroche. (*Common Sea Bream.*)
—Flesh-red, with golden-yellow reflections: a crescent-
shaped mark above the eyes ; a large black spot on the
shoulder.

> S. centrodontus, *Laroche, Ann. du Mus.* tom. XIII. p. 345. pl. 23. f. 11.
> S. Pagrus, *Bloch, Ichth.* pl. 267. S. Aurata, *Don. Brit. Fish.*
> vol. IV. pl. 89. Pagellus centrodontus, *Cuv. et Val. Poiss.* tom. VI.
> p. 133. Pagrus vulgaris, *Flem. Brit. An.* p. 211. Sea Bream,
> *Will. Hist. Pisc.* p. 312. tab. V. 1. f. 5. *Yarr. Brit. Fish.* vol. I.
> p. 107. Red Gilt-head, *Penn. Brit. Zool.* vol. III. p. 242. Lunu-
> lated Gilt-head, *Id.* pl. 42. no. 112. Le Rousseau, *Cuv. Reg. An.*
> tom. II. p. 183.

Length. From fifteen to twenty inches, or more.

Descript. (*Form.*) Distinguished from the last species by its larger
size, more obtuse snout, larger eye, and finer as well as more numerous
teeth: body a little thicker, and more regularly oval: head rather more
than one-fourth of the entire length; profile descending obliquely from
the forehead, and still more rapidly from the nostrils to the lips, giving
the snout a remarkably blunt and convex appearance: diameter of the
eye one-third the length of the head; the distance between the eyes
equal to this diameter: jaws nearly equal: anterior teeth finer than in
the *S. Erythrinus;* molars smaller in proportion, disposed in three or
more rows above, and in two or three below: pharyngean tubercles
larger: infra-orbitals and preopercle very narrow from the great size
of the eyes: lateral line strongly marked; its course parallel to the
curvature of the back at one-fourth of the depth: number of scales
in a longitudinal line nearly eighty; in the depth more than twenty:
dorsal rising from a shallow groove, and commencing just above the
insertion of the pectorals; its length half the entire length; fourth, fifth,
and sixth spinous rays longest, equalling nearly one-third of the depth;
soft portion of the fin half the length of the spinous, the rays much of a

height, being a little longer than the last of the spinous : anal answering
to this soft portion: caudal moderately forked ; lobes equal ; the base of
the fin scaly : pectorals long and pointed, about the length of the head :
ventrals immediately under them; the spinous ray strong, and shorter
than the others :

B. 6 ; D. 12 or 13/13 or 12 ; A. 3/12 ; C. 17 ; P. 16 or 17 ; V. 1/5.

Number of vertebræ twenty-three. (*Colours.*) Flesh-colour, with a bright
golden-yellow lustre; the red tint most conspicuous from the ridge of
the back to the lateral line; belly very pale reddish yellow: fins flesh-
red ; the ventrals palest: upper part of the head deep purplish flesh-red,
with a faint golden lunulated mark above the eyes : infra-orbitals, upper
and lateral portions of the snout, preopercle, and margin of the opercle,
bright silvery : at the commencement of the lateral line, above the pectorals, a black patch.

Common on many parts of the southern and western coasts: off Hastings and Weymouth in great abundance. Small specimens taken at the
former place in the month of September, of the length of eight inches,
were observed to be without the black spot on the shoulder, which is said
not to be acquired till during the second year. This species is probably
the *Bream* of Mr. Couch.* Its habits are similar to those of the last.

GEN. 11.   DENTEX, *Cuv.*

31. D. *vulgaris*, Cuv. (*Toothed Gilt-head.*)—Silvery ;
back tinged with sky-blue : dorsal bluish yellow ; pectorals
and caudal reddish.

D. vulgaris, *Cuv. et Val. Poiss.* tom. vi. p. 163. pl. 153.   *Flem.*
Brit. An. p. 212. Sparus Dentex, *Bloch, Ichth.* pl. 268. *Don.*
Brit. Fish. vol. iv. pl. 73. Dentex, *Will. Hist. Pisc.* p. 312.
tab. V. 3. Toothed Gilt-head, *Penn. Brit. Zool.* (Edit. 1812.)
vol. iii. p. 331. but not pl. 54. Four-toothed Sparus, *Yarr. Brit.*
Fish. vol. i. p. 111. Le Denté vulgaire, *Cuv. Reg. An.* tom. ii.
p. 184.

LENGTH.   From two to three feet.

DESCRIPT. (*Form.*) Body oval, elongated; dorsal line more convex
than the ventral: depth contained three times and three-quarters in
the length; thickness twice and a half in the depth: head large; its
length equalling the depth of the body: profile from the forehead convex; the snout, however, somewhat pointed: eyes moderate, high on the
cheeks, nearly at equal distances from the end of the snout and the point
of the opercle: infra-orbitals very large, occupying nearly half the cheeks;
preopercle occupying nearly the other half, pitted in front, and covered
with small smooth scales: scales on the opercle and subopercle rather
larger than those on the preopercle: jaws but little protractile; in each
four strong hooked canine teeth, behind which are others much smaller,
like velvet; beyond, on the edges of the jaws, a row of strong large
teeth, which are short and straight: palate and tongue smooth: lateral
line parallel to the curve of the back; its course at one-fourth of the
depth: about fifty scales in a longitudinal line, and twenty-four in the
depth: length of the dorsal rather more than one-third that of the body :
spinous rays moderate : vent nearly in the middle: anal short, com-

* *Linn. Trans.* vol. xiv. p. 79.

mencing a little behind it; first spinous ray in this fin shorter than
the second; second shorter than the third; this last equal to the soft
rays: caudal forked; upper lobe a little the longest: pectorals long and
narrow: ventrals triangular, placed a little behind them:

B. 6; D. 11/11; A. 3/7; C. 17; P. 14; V. 1/5.

Number of vertebræ twenty-four. Cuv.

An individual of this species, taken off the coast of Hastings in Sussex,
was obtained by Donovan in the Billingsgate market, April 9, 1805.
There is no other recorded instance of its having occurred in the British seas. Not uncommon in the Mediterranean, where it attains a
weight of twenty pounds and upwards.

GEN. 12.   CANTHARUS, *Cuv.*

32. C. *griseus*, Cuv. (*Black Sea Bream.*) — Silvery
gray, with bluish reflections; on each flank twenty-four
dark longitudinal lines.

C. griseus, *Cuv. et Val. Poiss.* tom. VI. p. 249.   Sparus lineatus,
    Mont. in Wern. Mem. vol. II. p. 451. pl. 23.   Pagrus lineatus,
    Flem. Brit. An. p. 211.   Black Bream, *Yarr. Brit. Fish.* vol. I.
    p. 114.

LENGTH.   Fifteen to eighteen, rarely twenty, inches.
DESCRIPT.   (*Form.*) Deeper in proportion to its length than the
Sparus centrodontus; the back more arched; the dorsal line falling more
abruptly: length of the head, and depth at the nape, equal, each contained about four times and a half in the total length: jaws equal: teeth
card-like, somewhat crowded in front, in several rows, the outer row longest; no molars, but on each side of the jaws above and below, one single
row of small card-like teeth: eyes moderate, their diameter contained
four times and a half in the length of the head; the space between them
a little convex: infraorbital broad, and deeply notched on that part of the
margin which answers to the extremity of the maxillary: scales on the
cheeks in six rows: lateral line broad and strongly marked; its course
parallel to the curvature of the back at one-third of the depth: dorsal
commencing in a line with the pectorals; the spinous portion of the fin
twice the length of the soft; fourth ray longest, exactly equalling the
depth to the lateral line; succeeding rays nearly of the same length:
caudal much forked; the upper lobe a little the longest: anal commencing nearly in a line with the soft portion of the dorsal, and terminating at the same distance from the caudal; spinous rays stronger
than those of the dorsal, shorter than the soft ones; these last branched,
the last two springing from one root: pectorals reaching to the vent,
one-fourth of the whole length, narrow and pointed; fifth ray longest:
ventrals a little behind the pectorals; in the axilla of each a long narrow
pointed scale, and on the belly between the two, another similar but
broader scale, of a triangular form, not present in the *Sparus centrodontus:*

B. 5; D. 12/11; A. 3/11 ; C. 17; P. 16; V. 1/5.

Number of vertebræ twenty-two.   (*Colours.*) Lead-gray, with a very
faint tinge of golden yellow; becoming paler on the belly: sides marked
with twenty-four or twenty-five longitudinal lines, darker than the
ground colour, but narrower and less conspicuous than the lateral line,

which last assumes the appearance of a broad brown band: fins dark
gray: a faint golden lunulated mark with blue reflections on each side
of the nape continuous with the gill-opening: beneath this line, immediately above the eyes, an irregular spot, presenting the same colours:
no lateral dark spot.

First noticed by Montagu, who states that "it is by no means an
uncommon fish on the south coast of Devon." Is also found occasionally
off Hastings, but is not distinguished by the fishermen from the *Sparus
centrodontus*, which is taken there in much greater plenty. According
to Cuvier, there are two or three other allied species met with in the
Mediterranean, which may not improbably also occur at times in the
British seas. Food, according to that same author, at least in part,
vegetable substances.

(4.) *Old Wife*, *(Sparus Vetula,)* Couch in Linn. Trans. vol. XIV. p. 79.
Loud. Mag. of Nat. Hist. vol. v. p. 743.

> As considerable doubt attaches to this species, I have thought it proper
> to place it at the end of the present family, to which it certainly belongs.
> The following is Mr. Couch's description. "Body deep, compressed, and
> bearing a considerable resemblance to the *S. Pagrus* (of Couch): lips
> fleshy; jaws furnished with a pavement of teeth, those in front the
> longest; gill-membrane with five rays; gill-covers and body covered
> with large scales; ten first rays of the dorsal fin spinous; the anal also has
> four spinous rays, after which it becomes more expanded; tail concave.
> This fish has a membranous septum across the palate, as in the Wrasse
> genus. When in high season, the colour behind the head is a fine green,
> towards the tail reddish orange; the belly has a lighter tinge of the same
> colour. When out of season, the whole is a dusky lead colour. Weight
> about three pounds." I should have had no hesitation in referring this
> fish to the species last described, with which Mr. Couch himself has since
> associated it, had it not been said to possess "a pavement of teeth." This
> character, which is common to nearly all the other British *Sparidæ*, is
> inapplicable to the *Canthari*, in which there are no rounded molars what-
> ever. I am more inclined to think from such a circumstance, that it
> will prove to be a species of the sub-genus *Sargus*, Cuv.

GEN. 13. BRAMA, *Schn.*

33. B. *Raii*, Cuv. *(Ray's Sea Bream.)*

B. Raii, *Cuv. et Val. Poiss.* tom. VII. p. 210. pl. 190. Sparus Raii,
Bloch, Ichth. pl. 273. *Don. Brit. Fish.* vol. II. pl. 37. S. niger,
Turt. Brit. Faun. p. 98. Brama marina caudâ forcipatâ, *Ray,
Syn. Pisc.* p. 115. *Will. Hist. Pisc. App.* p. 17. tab. V. 12.
B. marina, *Flem. Brit. An.* p. 210. La Castagnole, *Cuv. Reg.
An.* tom. II. p. 194. Toothed Gilt-head, *Penn. Brit. Zool.*
vol. III. p. 243. pl. 43. *Mont. in Linn. Trans.* vol. VII. p. 292.
Ray's Bream, *Yarr. Brit. Fish.* vol. I. p. 117.

LENGTH. From twenty-six to thirty inches. CUV.

DESCRIPT. *(Form.)* Body deep, compressed, elongated posteriorly:
snout rounded, very obtuse, the profile falling rapidly from the forehead: mouth oblique, approaching to vertical when the jaws are closed:
upper jaw with an outer row of sharp slender teeth, and a narrow band
of smaller ones behind; in the lower jaw two rows of similar teeth, with
smaller ones between; those in the inner row curving inwards and

stronger than the others; more particularly two or four in front of the lower jaw so much produced as to appear like true canines: palatines also with card-like teeth, but none on the vomer or tongue: eyes very large: cheeks and gill-covers scaly: lateral line indistinct; its course parallel to the back at one-fourth of the depth: dorsal commencing above the insertion of the pectorals; its length nearly half the entire length; three spinous rays gradually increasing; second and third soft rays longest, equalling nearly one-third of the depth; fourth to the ninth gradually decreasing; rest of the fin even, its height at this part only one-third that of the anterior portion: anal resembling the dorsal in form; commencing a little behind it, but terminating in the same vertical line: caudal crescent-shaped; the lobes long and pointed, equal: pectorals one-fourth of the whole length, pointed; sixth and seventh rays longest: ventrals very small, only one-quarter the length of the pectorals, placed immediately beneath them; at the base of their external margin a large triangular scaly plate; beneath, on the inner margin, another smaller one:

D. 3/33; A. 2/28; C. 26; P. 19; V. 1/6:

all the vertical fins with nearly their whole surface covered with small scales. (*Colours.*) Dull silvery, towards the back tinged with brown : vertical fins brownish ground, with silvery scales: pectorals and ventrals yellowish.

First described by Ray from a specimen found on the sands at low water near the mouth of the Tees, Sept. 18, 1681. Since then several other individuals have occurred at different times on various parts of the British coast. Common in the Mediterranean. Weight from ten to twelve pounds. According to Cuvier, spawns in Summer, and during that season is much tormented by intestinal worms. The only Europæan species belonging to this genus. *Obs.* Cuvier is of opinion that the *Chætodon* mentioned by Mr. Couch as taken at Looe in Cornwall, Aug. 1821*, was only an individual of this species.

<hr>

GEN. 14.   SCOMBER, Cuv.

(1. SCOMBER, Cuv.)

34. S. *Scomber*, Linn. (*Common Mackarel.*) — First dorsal with twelve rays : lower part of the sides and abdomen plain silvery.

> S. Scomber, *Linn. Syst. Nat.* tom. I. p. 492. *Bloch, Ichth.* pl. 54. *Don. Brit. Fish.* vol. V. pl. 120. S. vulgaris, *Flem. Brit. An.* p. 217. Mackrell, *Will. Hist. Pisc.* p. 181. tab. M. 3. Common Mackrell, *Penn. Brit. Zool.* vol. III. p. 264. pl. 51. Mackerel, *Yarr. Brit. Fish.* vol. I. p. 121. Le Maquereau commun, *Cuv. et Val. Poiss.* tom. VIII. p. 5.

LENGTH. From sixteen to eighteen inches.
DESCRIPT. (*Form.*) Body compressed, fusiform, tapering to a point before the caudal fin: depth behind the ventrals one-sixth of the length, caudal excluded; thickness half the depth: head a compressed cone, one-fourth of the whole length, measured to the commencement of the

caudal fork: profile slightly convex: jaws about equal: teeth small but numerous, in a single row on the edge of each jaw, as well as on the palatines; longer and more slender teeth on the pharyngeans: diameter of the eye one-fifth the length of the head: interopercle and subopercle much developed: first dorsal commencing at one-third of the whole length from the end of the snout; of a triangular form; rising from a deep groove, in which it is entirely concealed when laid back; second ray longest, equalling two-thirds of the depth beneath; last ray extremely short: space between the dorsals one-sixth of the whole length: second only half as high as the first; its length twice its height; first ray spinous; the next two articulated but simple; the rest articulated and branched: between it and the caudal five spurious finlets; each consisting of one branched ray; the last double: anal similar to the second dorsal, and followed by the same number of spurious finlets; before its base, immediately behind the vent, a small free spine: caudal forked nearly to the extremity of the tail itself; the central rays only one-fourth the length of the lateral ones: pectorals small, not half the length of the head; third ray longest; all except the first two branched: ventrals a little behind them, and somewhat shorter, triangular; first ray spinous but slender; the rest soft and branched:

B. 7; D. 11 or 12—1/11, and V false; A. 1—1/11, and V false; C. 17; P. 18 or 19; V. 1/5:

two cutaneous ridges on each side of the tail, forming a double keel. Number of vertebræ thirty-one. (*Colours.*) Back, and sides above the lateral line, rich green varied with blue, with dark transverse bands; belly and lower part of the sides silvery white: dorsal, caudal, and pectoral fins, dusky; ventrals and anal reddish.

Gregarious: approaches the shore in large shoals to spawn early in the Spring, and retires at the end of the Summer to deep water. Weight about two pounds; but, according to Pennant, has been known in one instance to exceed five. Has no swimming bladder.

35. S. *maculatus*, Couch. (*Spanish Mackarel.*) — First dorsal with seven rays: sides and belly thickly covered with small dusky spots.

S. maculatus, *Couch in Loud. Mag. of Nat. Hist.* vol. v. p. 22. f. 8.
Le Maquereau Colias, *Cuv. et Val. Poiss.* tom. viii. p. 29. pl. 209. ?
Spanish Mackarel, *Couch, l. c. Yarr. Brit. Fish.* vol. i. p. 131.

Descript.   " Figure round and plump, six inches and a half in compass near the pectoral fins (in a specimen fourteen inches and a half in length), the thickness of its figure being carried far towards the tail: mouth large; jaws of equal length; teeth small; tongue moveable and pointed: head large and long: eye large, one inch and one-eighth from the snout, and itself six eighths of an inch wide; from the snout to the pectoral three inches and a half: rays of the gill-membrane six, concealed: lateral line at first slightly descending, then straight: scales on the superior plate of the gill-covers as well as on the body: first dorsal in a chink, with seven rays, the first shorter, second and third of equal lengths: spurious fins six above and below, the anterior not high: tail divided, and at its origin doubly carinated: vent prominent. *Colour* dark blue on the back, striped like the *Mackarel*, but more obscurely and with fewer stripes; a row of large dark spots from the pectoral fin to the tail; sides and belly thickly covered with smaller dusky spots: tail, gill-covers, and sides, and behind the eye, bright yellow. From the *Mackarel*, which it resembles, this fish differs in the markings of the head, longer snout, larger eye and gape, longer head, and in having scales on the anterior gill-covers: the body is not nearly so much attenuated posteriorly; the ventral fins are sharp and slender, those of the *Mackarel* wider and more blunt: in the former, the pectorals lie close to the body; in the latter, they stand off; in the latter, also, is a large angular plate, the point directed backward, close above the pectoral fins, which does not exist in the Spanish Mackarel." Couch.

The above species will probably prove to be the *S. Colias* of Cuvier and Valenciennes, which is found in the Mediterranean, and is remarkably distinguished from the *S. Scomber* by having a swimming bladder. For the present, however, I have thought it proper to retain the name given to it by Mr. Couch, and to annex his description. This gentleman observes that it is scarce, but that some are taken every year off the coast of Cornwall. It attains the weight of four or five pounds, but is in no estimation as food. It is called by the fishermen *Spanish Mackarel*.

(5.) *S. Colias*, Turt. Brit. Faun. p. 100. sp. 76.

> Under this name Turton speaks of a species which is "found frequently in the Weirs about Swansea, and which very much resembles the *Common Mackarel*, except in size, which seldom exceeds six or seven inches in length. Its colours are much richer, and it does not appear to come in shoals." Whether this be any thing more than the young state of one of the foregoing species can only be determined by a closer examination of its characters.

(2. Thynnus, *Cuv.*)

36. S. *Thynnus*, Linn. (*Common Tunny.*) — Nine spurious finlets above and below: pectorals falcate; contained five times and a half in the entire length.

S. Thynnus, *Linn. Syst. Nat.* tom. I. p. 493. *Bloch, Ichth.* pl. 55. *Don. Brit. Fish.* vol. I. pl. 5. *Flem. Brit. An.* p. 218. Thynnus vulgaris, *Cuv. et Val. Poiss.* tom. VIII. p. 42. pl. 210. Tunny-Fish, *Will. Hist. Pisc.* p. 176. tab. M. 1. f. 3. Tunny, *Penn. Brit. Zool.* vol. III. p. 266. pl. 52. *Yarr. Brit. Fish.* vol. I. p. 134. Le Thon commun, *Cuv. Reg. An.* tom. II. p. 197.

LENGTH. From three to seven feet; sometimes more.
DESCRIPT. (*Form.*) General form resembling that of the *Mackarel*, but thicker in proportion to its length, and shorter in the snout: head a little less than one-fourth of the entire length; profile slightly convex; lower jaw a little longer than the upper; each with a single row of small sharp teeth, slightly curving inwards and backwards; about forty on each side above and below; a few fine teeth like velvet also on the palatines and fore part of the vomer: diameter of the eye one-seventh the length of the head: cheeks covered with long narrow pointed scales, which cause them to appear wrinkled: gill-covers, as well as all the rest of the head, naked: lateral line irregularly and slightly flexuous, marked throughout its length by large scales similar to those which form the corselet: pectorals sickle-shaped, contained five times and a half in the whole length: ventrals scarcely more than half as long: first dorsal commencing nearly in a line with the base of the pectorals, rising from a groove, and extending nearly to the second; first spine longest; the others decreasing rapidly to the sixth, afterwards more slowly: second

dorsal with one small concealed spine; the soft rays which follow elevated anteriorly and pointed; those behind rapidly decreasing: anal similar to the second dorsal, and nearly opposite to it, with two spines concealed in its anterior margin: nine or ten spurious finlets above, and nine below: caudal crescent-shaped:

B. 7; D. 14—1/13, and IX; A. 2/12, and VIII; C. 19, and 16 or 17; P. 31; V. 1/5:

sides of the tail keeled. Number of vertebræ thirty-nine. (*Colours.*) Upper part of the body bluish black; corselet inclining to whitish: sides of the head whitish: belly grayish, with silvery whitish spots: first dorsal, pectorals, and ventrals, dusky; caudal somewhat paler; second dorsal and anal inclining to flesh-colour, with silvery reflections; spurious finlets sulphur-yellow, edged with black. Cuv.

According to Pennant, "not uncommon in the Lochs on the western coast of Scotland; where they come in pursuit of Herrings." Rare southwards. Donovan mentions three which were captured near the mouth of the Thames in the Summer of 1801, and brought to Billingsgate market. Very abundant in the Mediterranean. Usually swim in large shoals. Feed on other fish. Weight of one examined by Pennant, measuring seven feet ten inches in length, four hundred and sixty pounds.

37. S. *Pelamys*, Linn. (*Bonito.*) — Eight spurious finlets above, seven below : sides of the abdomen with four longitudinal dusky bands.

S. Pelamys, *Linn. Syst. Nat.* tom. I. p. 492. Thynnus Pelamys, *Cuv. et Val. Hist. Nat. des Poiss.* tom. VIII. p. 82. pl. 214. Bonito, *Yarr. Brit. Fish.* vol. I. p. 140. Bonite des Tropiques, ou Thon à ventre rayé, *Cuv. Reg. An.* tom. II. p. 198.

LENGTH. Rarely exceeds thirty inches. YARR.

DESCRIPT. "Girth close behind the pectoral fins (in a specimen twenty-nine inches long) twenty inches; head conical, ending in a point at the nose; under jaw projecting; teeth few and small; tongue flat and thin; nostrils obscure, not in a depression; from the nose to the eye two inches and a half; gill-covers of two plates: body round to the vent, from thence tapering to the tail; near the tail depressed; lateral line at first descending and waved, becoming straight opposite the anal fin, from thence ascending and terminating in an elevated ridge, with another above and below the lateral line near the tail: eye elevated, round; iris silvery: from the nose to the pectoral fin eight inches and three-fourths, the fin pointed, four inches long, received into a depression: first dorsal fin seven inches long, four inches high, lodged in a groove; the first two rays stout, the others low: the body is most solid opposite the second dorsal, which fin and the anal are falcate: tail divided and slender: ventral fins in a depression. *Colour* a fine steel blue, darker on the back; sides dusky, whitish below: behind the pectoral fins is a bright triangular section of the surface, from which begin four dark lines, that extend along each side of the belly to the tail. Scales few, like the *Mackarel*." COUCH, as quoted by YARR. Number of fin-rays, according to Cuvier,

D. 15—1/12, and VIII ; A. 2/12, and VII ; C. 35; P. 27 ; V. 1/5.

Specimens of this fish, which is the *Bonito* of the Tropics so well known to navigators, are stated by Mr. Couch to have occurred occasionally on the Cornish coast. According to Stewart*, it has been also taken, though rarely, in the Frith of Forth; and, according to Dr. Scouler†, in the Frith of Clyde. In the two last instances, however, it is doubtful whether the present species be intended, or the *Pelamys Sarda* of Cuvier‡, to which also the name of *Bonito* has been applied. This last is found principally in the Mediterranean, and is characterized by a variable number of *obliquely-transverse* bands extending from the top of the back to a little below the lateral line. The species here described inhabits the Ocean, and is particularly distinguished by four *longitudinal* bands on each side of the abdomen: the teeth are also much weaker than in the *Pelamys Sarda*.

GEN. 15. XIPHIAS, *Linn.*

38. X. *Gladius*, Linn. (*Common Sword-Fish.*)

X. Gladius, *Linn. Syst. Nat.* tom. I. p. 432. *Bloch, Ichth.* pl. 76. *Flem. Brit. An.* p. 220. *Cuv. Reg. An.* tom. II. p. 201. *Cuv. et Val. Poiss.* tom. VIII. p. 187. pls. 225, & 226. X. Rondeletii, *Leach in Wern. Mem.* vol. II. p. 58. pl. 2. f. 1. *Leach, Zool. Misc.* vol. I. p. 62. pl. 27.   Sword-Fish, *Will. Hist. Pisc.* p. 161. tab. I. 27. f. 2. *Penn. Brit. Zool.* vol. III. p. 160. pl. 26. *Id.* (Edit. 1812.) vol. III. p. 216. pl. 30. *Knox in Edinb. Journ. of Nat. and Geog. Sci.* vol. II. p. 427. *Yarr. Brit. Fish.* vol. I. p. 143.

LENGTH. From ten to fifteen feet; sometimes more.

DESCRIPT. (*Form.*) Body elongated, nearly round posteriorly, a little compressed in front: depth increasing with the age from one-tenth to one-sixth of the entire length, reckoning this last from the end of the sword to the extremity of the lobes of the tail: sword three-tenths: upper part of the cranium flat or slightly convex; profile falling gently; sides of the head vertical: eye round; its diameter nearly two-thirds of the breadth of the cranium above it: sword terminating in a sharp point; the edges cutting, and finely denticulated: lower jaw likewise pointed, extending to where the upper surface of the sword becomes horizontal: no teeth in either of the jaws: pharyngeans only with fine teeth like shorn velvet: no true tongue: gill-opening large; the branchiostegous membrane with seven rays: pectorals inserted very low down, sickle-shaped, one-seventh of the entire length, this last being reckoned as before: ventrals none: dorsal commencing above the gill-opening, and extending in *young* subjects to within a short distance of the caudal; its anterior portion very much elevated and pointed; rays rapidly decreasing from the fifth to the eleventh, continuing low beyond that point to the thirty-ninth or fortieth; last three or four again elevated: all the intermediate or low portion of the fin extremely delicate, and with the rays more slender than those at the two extremities; in *adult* individuals often found very much torn, or even entirely destroyed, causing the two elevated ends which are left to appear like two distinct fins: anal somewhat similar in shape to the dorsal, but much shorter, only commencing in a line with its last third portion: caudal crescent-shaped:

B. 7; D. 3/40; A. 2/15; C. 17; P. 16:

---

* *Elem. of Nat. Hist.* vol. I. p. 363.
† *Loudon's Mag. of Nat. Hist.* vol. VI. p. 529.
‡ *Cuv. et Val. Hist. Nat. des Poiss.* tom. VIII. p. 108. pl. 217.

the whole head and body covered with a somewhat rough skin, the roughness arising from very minute scales; opercle smooth: lateral line scarcely visible: on each side of the tail a projecting horizontal keel. Number of vertebræ twenty-five. (*Colours.*) All the under parts fine silvery white: upper parts tinged with dusky blue. *Young individuals from twelve to eighteen inches in length,* have the whole body covered with little tubercles, disposed in longitudinal rows: these disappear first on the back, and afterwards on the belly: they are no longer visible in individuals of three feet. CUV.

Occasionally taken in the British seas, off various parts of the coast. Common in the Mediterranean, where it is much sought after as an article of food. Attacks other fish, on which it is said to prey; but, according to Bloch, feeds also on vegetable substances. The stomach of one examined by Fleming contained the remains of the *Loligo sagittata.* But little is known on the subject of its reproduction. When the intermediate part of the dorsal fin is worn away, it becomes the *X. Rondeletii* of Leach.

GEN. 16. CENTRONOTUS, *Lacép.*

(1. NAUCRATES, *Cuv.*)

39. C. *Ductor,* Nob. (*Common Pilot-Fish.*)

Naucrates Ductor, *Cuv. et Val. Poiss.* tom. VIII. p. 229. pl. 232. Gasterosteus Ductor, *Linn. Syst. Nat.* tom. I. p. 489. Scomber Ductor, *Bloch, Ichth.* pl. 338. Pilot-Fish, *Will. Hist. Pisc. App.* tab. 8. f. 2. *Yarr. Brit. Fish.* vol. I. p. 149.

LENGTH. One foot.

DESCRIPT. (*Form.*) General contour a little like that of the *Mackarel:* depth one-fifth of the whole length: length of the head contained in this last four times and a half: profile slightly convex; snout transversely obtuse; lower jaw projecting a little beyond the upper: each jaw with a narrow band of teeth like shorn velvet; the same on the palatines, vomer, and middle of the tongue: diameter of the eye one-fifth the length of the head: opercular pieces much as in the *Mackarel:* pectorals attached a little below the middle; oval, contained seven times and a half in the whole length: ventrals very close together, a little behind the insertion of the pectorals, of about the same length: first dorsal represented by three, rarely four, very small free spines, commencing nearly in a line with the extremity of the pectorals: second dorsal commencing about the middle of the body; anterior rays longest, equalling a little more than one third of the depth: anal of a similar form to this last fin, and commencing beneath the middle of its length; before it two small free spines, the first hardly perceptible: caudal forked to the middle; the lobes rather broad, and moderately pointed:

B. 7; D. 3 or 4—1/26 to 28; A. 2/16 or 17; C. 17, and 8; P. 18; V. 1/5:

cheeks, upper part of the opercle, and the whole body, excepting a triangle above the base of the pectoral, covered with small oval scales; forehead, snout, jaws, and greater portion of the opercular pieces, without scales: lateral line curved, marked by a narrow series of very small elevations: sides of the tail with a projecting horizontal keel. Number of vertebræ twenty-six. (*Colours.*) Silvery bluish gray; deeper on the back, paler on the belly: sides with five broad transverse bands of deep violet. CUV.

Mr. Couch states* that "two of this species a few years since accompanied a ship from the Mediterranean into Falmouth, and were taken in a net." It has been observed in the British seas, under similar circumstances, in a few other instances. Is well known for its habit of following vessels to a considerable distance, in order to get what falls from them.

(*LICHIA,* Cuv. ?)

(6.) *Albacore,* Couch in Linn. Trans. vol. XIV. p. 82.   *Lichia glaucus,* Cuv. et Val. Poiss. tom. VIII. p. 263. pl. 234.?

> Mr. Couch states that he believes the *Albacore* to be not uncommon in the Summer off Cornwall, though keeping at a distance from the shore, and but rarely taken. This name, however, having, like that of *Bonito,* been applied to more than one species, it does not appear with certainty to which he alludes. If he refer to the *Scomber glaucus* of Linnæus, this last is synonymous with the *Lichia glaucus* of Cuvier and Valenciennes. I have accordingly annexed a corresponding reference to their work for a description and figure.

GEN. 17.   CARANX, *Cuv.*

40.   C. *Trachurus*, Lacép. (*Scad.*)

C. Trachurus, *Cuv. et Val. Poiss.* tom. IX. p. 9. pl. 246.   Scomber Trachurus, *Bloch, Ichth.* pl. 56.   *Don. Brit. Fish.* vol. I. pl. 3. Trachurus vulgaris, *Flem. Brit. An.* p. 218.   Scad, *Will. Hist. Pisc.* p. 290. tab. S. 22.   *Penn. Brit. Zool.* vol. III. p. 269. pl. 51. Scad, or Horse-Mackerel, *Yarr. Brit. Fish.* vol. I. p. 154.

LENGTH.   From twelve to sixteen inches.

DESCRIPT. (*Form.*) General form resembling that of the *Mackarel:* tail slender; head a little pointed, rather less than one-fourth of the entire length: greatest depth a little more than one-fifth; thickness half the depth: lower jaw projecting beyond the upper, inclining upwards at an angle of forty-five degrees: in each jaw one very narrow row of extremely minute teeth, more sensible to the touch than to the eye; the same on the vomer and palatines: eyes large, above the middle of the cheek: cranium, cheeks, and all the body, covered with small scales; snout, jaws, and opercular pieces, without scales, the upper half of the opercle excepted: lateral line parallel to the back, at one-fourth of the depth, till opposite the commencement of the second dorsal; then bending obliquely downwards and backwards; when in a line with the ninth ray of that fin, passing off straight to the caudal at half the depth; protected throughout its course by a series of large scaly laminæ, seventy-two in number, three or four times as high as broad, closely compacted; the last forty of these laminæ with keels terminating backwards in sharp points, the keels more elevated and the points sharper as they get nearer the caudal: first dorsal commencing at about one-third of the length, triangular, its length and height about equal; third and fourth rays longest; before it a small, but sharp, horizontal spine, with the point directed forwards: second dorsal immediately behind the first; three times its length; its height at first nearly the same, but afterwards falling, and remaining low throughout the rest of its length: behind the vent, two stout, sharp spines, united at their base by a short membrane; then the true anal, exactly similar to the second dorsal excepting in being shorter, and com-

---

* *Linn. Trans.* vol. XIV. p. 82.

mencing nearer the caudal; its point of termination in the same vertical line: both dorsal and anal arising from deep grooves in the back and abdomen respectively: pectorals falcate, very much pointed, of the length of the head: ventrals a little behind them, scarcely more than half as long: caudal deeply forked:

B. 7; D. 8—1/30; A. 2—1/25; C. 17, and 10; P. 21; V. 1/5.

(*Colours.*) Lead-gray, variegated with blue and green; beneath silvery; a black spot on the upper part of the opercle; irides golden.

Common throughout the Summer, according to Mr. Couch, off the coast of Cornwall. Occurs also at Hastings, and off other parts of the English, as well as Scotch, coast. Preys on other fish. *Obs.* Cuvier and Valenciennes describe this fish as varying greatly in the number of scaly laminæ on the lateral line, as well as in the degree of curvature of this last, and seem to think that possibly two or more species may have been hitherto confounded. For this reason I have been the more particular in the above description, which is taken from specimens obtained at Hastings, Sept. 1833.

GEN. 18.   ZEUS, *Linn.*

(1. Zeus, *Cuv.*)

41.   Z. *Faber,* Linn. (*Dory.*)

Z. Faber, *Linn. Syst. Nat.* tom. I. p. 454. *Bloch, Ichth.* pl. 41. *Don. Brit. Fish.* vol. I. pl. 8. *Flem. Brit. An.* p. 218. Doree, *Will. Hist. Pisc.* p. 294. tab. S. 16. *Penn. Brit. Zool.* vol. III. p. 221. pl. 41. Dory or Doree, *Yarr. Brit. Fish.* vol. I. p. 162. Le Dorée, *Cuv. Reg. An.* tom. II. p. 211.

LENGTH. From twelve to eighteen inches.

DESCRIPT. (*Form.*) Oval, very much compressed; tail suddenly contracting immediately before the caudal: greatest depth half the entire length; thickness four times and a half in the depth: head very large, but greatly compressed, one-third of the entire length: profile falling regularly from the nape in nearly a straight line, and making a right angle with the lower jaw, when the mouth is closed: this last very protractile; gape large; upper lip reflexed: lower jaw a little longer than the upper, bifurcated behind, and terminating in two small sharp spines: both jaws with fine velvet-like teeth: eyes large, very high on the cheeks: opercle small, triangular, without spines: clavicular bone behind the opercle terminating in a sharp spine: two spines behind the eye directed backwards, and one on each side of the occiput: a row of spines on each side of the base of the dorsal and anal fins, at first simple, afterwards forked; between the ventrals and anal, a double row of large strongly serrated scales, the serratures directed backwards; pectoral ridge before the ventrals with three rows of the same serratures: scales on the cheeks and body, small, deeply impressed: lateral line continually descending from the supra-scapulars for two-thirds of its course, then suddenly passing off straight to the caudal: dorsal commencing in a line with the posterior angle of the opercle; the spinous and soft portions divided by a deep notch; third spine longest, equalling half the depth; all except the last attended by filamentous prolongations of the membrane nearly as long as themselves*; soft portion only half as high as the spinous; all the rays

---

* Judging from the descriptions of other authors, it would appear that these filaments vary very much in length, and that they are sometimes found twice or thrice the length of the spines themselves.

simple: anal with the first four rays strongly spinous; the soft portion
separate as in the dorsal, and answering to the same part in that fin:
caudal oblong, even at the end: pectorals small, rather less than one-
third the length of the head, of an oblong rounded form, the middle rays
a little the longest; all simple: ventrals a little before the pectorals,
more than twice their length; first ray strongly spinous; third and
fourth longest; all the soft rays except the last branched:

B. 6; D. 10/24; A. 4/23; C. 12, and 2 short; P. 13; V. 1/7.

(*Colours.*) Yellowish, varied with olive and lead-gray; in the middle of
each side an oval black spot: the whole tinged with a golden lustre.
    Not uncommon on some parts of the southern and western coasts.
Occasionally attains a considerable size. Pennant speaks of one which
weighed twelve pounds. According to Bloch, is very voracious, and
keeps near the shore in order to prey on the fish which come there
to spawn.

(2. Capros, *Lacép.*)

42. Z. *Aper*, Gmel. (*Boar-Fish.*)

Z. Aper, *Gmel. Linn.* tom. i. part iii. p. 1225. Capros Aper, *Lacép.*
Hist. Nat. des Poiss. tom. iv. p. 591. *Riss. Hist. Nat. de*
l'Eur. Mér. tom. iii. p. 380. *Cuv. Reg. An.* tom. ii. p. 211.
Proceed. of Comm. of Zool. Soc. 1833. p. 114. Aper Rondeletii,
Will. Hist. Pisc. p. 296. tab. I. 4. f. 4. Boar-Fish, *Yarr. Brit.*
Fish. vol. i. p. 169.

Length. The British specimens have not exceeded seven inches.
    Descript. (*Form.*) "Body a shorter oval than that of the *Dory:*
mouth protruding: a band of minute teeth considerably within each jaw:
eye very large, placed at the distance of its own diameter from the end
of the nose when the mouth is shut: nostrils large, just anterior to the
edge of the orbit: origin of the first dorsal, pectoral, and ventral, fins,
nearly in the same plane: the base of the first dorsal about as long as
its third spine, which is the longest: the base of the second dorsal equal
to that of the first, the rays very slender and flexible, the membrane
only extending up one-third of the length of the rays: pectoral fin as
long as the third ray of the first dorsal, slender and delicate in structure:
ventral with one strong spine, the other rays flexible and branched, the
membrane not extending the whole length of the rays: anal with all
the characters observable in the second dorsal, and ending at the same
distance from the tail: the caudal rays slender, and twice as long as the
fleshy portion of the tail: number of fin-rays,

D. 9/24; A. 3/24; C. 12; P. 14; V. 1/5.

No lateral line observable: body quite smooth when the finger is passed
from before backwards, but rough to the touch in the contrary direction,
from numerous small scales which are minutely ciliated." (*Colours.*)
"Upper part of the back and sides pale carmine, still lighter below, and
passing to silvery white on the belly: body divided by seven transverse
orange-coloured bands reaching three-fourths of the distance from the
back downwards: irides orange; the pupil bluish black: all the fin-rays
the same colour as the back; the membranes much lighter." Yarr.
Obs. In one of the British specimens there were no transverse bands.
    This species, which is a native of the Mediterranean, has twice occurred
in the British seas. The first individual is recorded by Dr. Henry Boase

as having been taken in Mount's Bay, in October 1825. The second is said to have been obtained in Bridgewater fish-market, on the 18th of April 1833. Dr. Boase's specimen is described in the " Proceedings of the Zoological Society," *l. c.* Of its habits little appears to be known, excepting that (according to Risso) it spawns in April.

GEN. 19.   LAMPRIS, *Retz.*

43.   L. *Luna,* Risso. (*Opah, or King-Fish.*)

L. Luna, *Riss. Hist. Nat. de l'Eur. Mérid.* tom. III. p. 341. *Flem. Brit. An.* p. 219. Zeus Luna, *Gmel. Linn.* tom. I. part iii. p. 1225. *Don. Brit. Fish.* vol. v. pl. 97. Z. imperialis, *Shaw, Nat. Misc.* vol. IV. pl. 140. Chrysotosa Luna, *Lacép. Hist. Nat. des Poiss.* tom. IV. p. 587. pl. IX. f. 3. Lampris guttatus, *Cuv. Reg. An.* tom. II. p. 211. Opah, *Penn. Brit. Zool.* vol. III. p. 223. pl. 42. *Id.* (Edit. 1812.) vol. III. p. 299. pl. 46. *Sow. Brit. Misc.* pl. 22. Opah, or King-Fish, *Yarr. Brit. Fish.* vol. I. p. 173.

LENGTH. From three to four feet and a half.

DESCRIPT. (*Form.*) Body oval, compressed, greatly diminishing at the tail, which is almost cylindrical: greatest depth (in a specimen three feet six inches long) nearly two feet: thickness not above six inches: mouth small: jaws without teeth: tongue thick, set with reflected prickles: eyes remarkably large: pectorals broad, about eight inches long: dorsal commencing a little behind their insertion, and extending nearly to the caudal; elevated anteriorly to the height of seven inches, but sloping away very suddenly, then continuing low till just at its termination, where it again becomes slightly elevated: ventrals very strong, placed near the middle of the body: anal narrow, running from the vent to the tail: caudal forked, expanding twelve inches. PENN.

D. 54; A. 26; C. 30; P. 28; V. 10:

skin smooth: scales scarcely perceptible: lateral line irregular, and somewhat curved at its commencement. DON. (*Colours.*) Back deep blue, inclining to regal purple, below which the purple is glossed in various directions of light with a reddish and golden hue, blending into green upon the sides, and the green fading to yellow as it approaches the silvery white of the abdomen: the whole body covered with numerous large distinct oval silvery spots: all the fins fine scarlet. DON.

Rare; but has been taken in the British seas in several instances; in some cases been found stranded on the coast after storms. Most of the individuals have occurred off Scotland and the Orkney Islands, but one or two on the western coast of England. Has been known to attain the weight of one hundred and forty pounds. Donovan, who has figured a specimen taken in the Frith of Forth, describes the pectoral fins as much longer than usually represented by British writers: he states that when placed erect, they reach even above the back. This species is not noticed either by Willughby or Ray.

GEN. 20.   CORYPHÆNA, *Linn.*

(1. CENTROLOPHUS, *Lacép.*)

44.   C. *Morio*, Cuv. (*Black-Fish.*)

Centrolophus Morio, *Cuv. et Val. Poiss.* tom. IX. p. 254.   C. niger,
Lacép. Hist. Nat. des Poiss. tom. IV. pp. 441, & 442. pl. 10. f. 2.
Perca nigra, *Gmel. Linn.* tom. I. part iii. p. 1321.   Black-Fish,
Borl. Nat. Hist. of Cornw. p. 271. pl. 26. f. 8.   *Yarr. Brit. Fish.*
vol. I. p. 158.   Black Ruffe, *Penn. Brit. Zool.* vol. III. p. 260.
Black Perch, *Id.* (Edit. 1812.) vol. III. p. 351.

DESCRIPT. " Smooth, with very small thin scales; fifteen inches long,
three-quarters of an inch (three or four inches?) broad besides the fin;
head and nose like a *Peal* or *Trout;* little mouth; very small teeth; a
full and bright eye; only one fin on the back, beginning from the nose
four inches and three-quarters, near six inches long; a forked tail; a
large double nostril." BORL.

" Fifteen inches long : (a second specimen measured two feet eight
inches in length, and weighed nearly fourteen pounds:) blunt and
rounded over the snout, flattened on the crown; mouth small; tongue
rather large; teeth in the jaws fine; nostrils double, that nearest the eye
large and open; eye prominent and bright; five gill-rays; though soft,
the membrane of the preopercle had a free edge, somewhat incised: body
compressed, about three inches deep; a thin elevated ridge, which makes
it appear deeper on the back, on which the dorsal fin is seated: this
fin begins at four and a half inches from the snout, and reaches to the
distance of twelve inches from it; the rays fleshy at the base, many of
them obsolete; vent six and a half inches from the lower jaw; pectoral
fins pointed; ventral fins bound down by a membrane; tail forked: lateral
line somewhat crooked at its commencement: body covered with minute
scales, which when dry appear curiously striated. *Colour* of the whole
black, the fins intensely so, very little lighter on the belly; somewhat
bronzed at the origin of the lateral line. While employed in drawing a
figure, the side on which it lay changed to a fine blue." COUCH, as quoted
by YARR.

We have as yet but an imperfect knowledge of this species, which was
originally described by Borlase from the papers of Mr. Jago, who obtained two specimens at Looe, May 26, 1721.   Cuvier seems to enter-
tain no doubt of Jago's fish being the same as the *Centrolophus niger*
of Lacépede, which last he thinks may prove to be the adult state of
his *C. Pompilus,* the *Pompilus* of Rondeletius.   This idea receives confirmation from a statement of Mr. Couch, who has lately rediscovered
this species in the Cornish seas, and, apparently without knowledge of
Cuvier's work, gives it as his opinion that it is the *Pompilus* of Gesner
and Ray*.   For the present, however, Cuvier considers these two species
as distinct, and if he be right in so doing, it is just possible that they
may both occur in our seas, and that Jago may have seen one, and
Mr. Couch the other.   For this reason I have annexed the descriptions
given by both these authors.   Mr. Couch's specimens were obtained
in 1830 and 1831.   His notice of them, in the work just referred to,
is accompanied by a remark, that there is "an error in Borlase's original description, of three-fourths of an inch, instead of three or four
inches," and that this "has chiefly led to the continued mistake respect-

* *Loud. Mag. of Nat. Hist.* vol. v. p. 315.

ing this fish." Some further particulars respecting this species, from
Mr. Couch, will be found in Yarrell's " British Fishes," 1. c., to which
the reader is referred.

GEN. 21.   LEPIDOPUS, *Gouan.*

45.   L. *argyreus*, Cuv. (*Scale-foot, or Scabbard-Fish.*)

L. argyreus, *Cuv. Reg. An.* tom. ii. p. 217.   *Cuv. et Val. Poiss.*
tom. viii. p. 163. pl. 223.   L. Lusitanicus, *Leach, Zool. Misc.*
vol. ii. p. 7. pl. 62.   L. tetradens, *Flem. Brit. An.* p. 205.   Vandellius Lusitanicus, *Shaw, Gen. Zool.* vol. iv. p. 199.   Zipotheca
tetradens, *Mont. in Wern. Mem.* vol. i. p. 82. pls. 2, & 3.   *Id.*
vol. ii. p. 432.   Scabbard-Fish, *Penn. Brit. Zool.* (Ed. 1812.)
vol. iii. p. 210.   *Yarr. Brit. Fish.* vol. i. p. 176.

LENGTH.   From four to six feet.
DESCRIPT.   (*Form.*) Ensiform, much compressed, and equally carinated above and below, except the head, which is flat on the top:
depth at the gills (in a specimen five feet six inches long) four inches
and a half, continuing nearly the same to the vent, from thence decreasing, at first gradually, but afterwards more suddenly ; portion of the tail
beyond the termination of the anal nearly round : head porrected, conic ;
lower jaw longest by half an inch, terminating in a callous fleshy projection : in each jaw an irregular row of extremely sharp-pointed teeth,
standing very conspicuous, even when the jaws are closed ; those below,
about twenty on each side ; above, not quite so numerous, but in this jaw
four large teeth in front, not found in the other ; two fore-teeth approximating ; and two larger canine, rather crooked and compressed, with a
slight process or barb on the inside near the point : tongue smooth : a
row of minute teeth on each palatine : eyes very large, lateral, independent, not covered with the common skin : pectorals five inches long ;
the lower rays twice the length of the upper ones : instead of ventrals,
two oblong silvery scales, half an inch in length, partly detached from
the body, and connected at the base ; their situation considerably behind
the pectorals : vent in the middle : anal commencing at about one-sixth
of the entire length from the posterior extremity, and running nearly
to the caudal : dorsal commencing at the nape and extending uninterruptedly till opposite the termination of the anal : caudal forked :

D. 105 ; A. 17 ; P. 12.

Lateral line slightly elevated : skin quite smooth, destitute of scales.
(*Colour.*) Like burnished silver, with a bluish tint. MONT.
First described as British by Montagu, from a specimen taken in Salcomb Harbour on the coast of South Devon, June 4th, 1808. Said to have
been swimming with great velocity, with its head above water. According to Fleming, a second individual, only ten inches in length, occurred
on the Devon coast in February 1810.   Mr. Yarrell mentions two others
which were also obtained from the southern shores of England.   *Obs.*
Cuvier, in his description of this species, observes that the number of
large hooked teeth in the upper jaw ought to be six, but that two or
three are generally found broken.   He also speaks of a triangular moveable scale a little behind the vent, not noticed by Montagu ; and states
further, that in his specimens, the anal rays amounted to twenty-five, but

that some of the anterior ones are so small and slender as easily to be overlooked. Number of vertebræ given as one hundred and eleven Cuvier would seem to be of an opinion, that there is no other well ascertained species belonging to this genus.

GEN. 22.   TRICHIURUS, *Linn.*

46.   T. *Lepturus*, Linn. ?   (*Hair-Tail.*)

T. Lepturus, *Hoy in Linn. Trans.* vol. XI. p. 210. *Flem. Brit. An.* p. 204. *Bloch, Ichth.* pl. 158.?   *Cuv. Reg. An.* tom. II. p. 218.?   *Cuv. et Val. Poiss.* tom. VIII. p. 173.?   Silvery Hair-Tail, *Yarr. Brit. Fish.* vol. I. p. 182.

LENGTH. Twelve feet and upwards. HOY.

DESCRIPT. " Length, from the gills to the extremity of the tail, twelve feet nine inches: breadth, eleven inches and a quarter, nearly equal for the first six feet in length from the gills, diminishing gradually from thence to the tail, which ended in a blunt point: greatest thickness two inches and a half: distance from the gills to the anus forty-six inches: dorsal fin extending from the head to the tail: no ventrals or anal; but the thin edge of the belly closely muricated with small hard points, scarcely visible through the skin, but plainly felt. Both sides of the fish white, with four longitudinal bars of a darker colour; the one imme-diately below the dorsal fin about two inches broad; each of the other three about three-fourths of an inch. Side-line straight along the middle." HOY.

The above fish, originally described by Mr. Hoy, *l.c.*, was found on the beach of the Moray-Frith, near the fishing village of Port Gordon in Scotland, November 12, 1812. Its head had been broken off, and was quite gone, and a small bit of the gills only remained about the upper part of the throat. A fish, supposed to be of a similar kind, had been cast upon the same shore two years previously, and Mr. Hoy commences his account with a description of this last individual. From the great difference, however, which appears in their relative proportions, as stated by this gentleman, I am inclined to Dr. Fleming's opinion, that the indi-vidual last alluded to was a distinct species, if not belonging to a different genus. There can be no doubt that the one described above was a true *Trichiurus*, and probably the *T. Lepturus* of Linnæus and other authors; but as the description is rather imperfect, and the species of this genus ill determined, it is impossible to speak with certainty on this last point. It is worth noting, however, that neither Cuvier nor Bloch describe this species as exceeding three feet. The *T. Lepturus* is found in the Atlantic Ocean, and, like the *Lepidopus argyreus*, appears to have a wide geogra-phical range. It is erroneously said by Bloch to inhabit fresh waters.

GEN. 23.   GYMNETRUS, *Bl.*

47.   G. *arcticus*, Cuv.   (*Deal-Fish.*)

Gymnogaster arcticus, *Cuv. Reg. An.* (1st Ed.) tom. II. p. 246. Deal-Fish of Orkney, *Flem. in Mag. of Nat. Hist.* vol. IV. p. 215. *with fig.* Vaagmaer, or Deal-Fish, *Yarr. Brit. Fish.* vol. I. p. 191.

Length. From four to six feet.

Descript. " Body excessively compressed, particularly towards the back, where it does not exceed a table-knife in thickness: breadth (in a specimen three feet long) nearly five inches, tapering to the tail: colour silvery, with minute scales; the dorsal fin of an orange-colour, occupying the whole ridge from the head to the tail, with the rays of unequal sizes: caudal fin forked, the rays of each fork about four inches long : pectorals very minute: no ventral or anal fins whatever: vent immediately under the pectoral fins, and close to the gill-openings: head about four inches and a half long, compressed like the body, with a groove in the top: gill-lids formed of transparent porous plates: eyes one inch and a quarter in diameter: both jaws armed with small teeth: lateral line rough ; and, towards the tail, armed with minute spines pointing forwards, and these are the only spines on the body." (*Another specimen.*) " Length four feet and a half: breadth eight inches: thickness one inch, thin at the edges, viz. back and belly: length of the head five inches, terminating gradually in a short snout: tail consisting of eight or nine fin-bones or rays, the third ray seven inches long, the rest four inches: dorsal fin reaching from the neck to the tail, rays four inches long: on each side of the fish, from head to tail, a row of prickles pointing forward, distance between each half an inch: under edge fortified by a thick ridge of blunt prickles: pectorals one inch long, lying upwards: skin rough, without scales (?): colour a leaden or silvery lustre; dorsal fin and tail blood-colour: the skin or covering of the head like that of a herring: several small teeth: gills red, consisting of four layers." Flem. *l. c.*

The above descriptions were communicated to Dr. Fleming by Dr. Alexander Duguid of Kirkwall, Orkney, in April and October 1829. They relate to a species of fish, which it would seem is not unfrequently cast on the shores of the Island of Sanday during bad weather, and which is called there the *Deal-Fish.* Dr. Fleming considers it as identical with the *Vaagmaer* of Olafsen, the *Gymnogaster arcticus* of Brunnich, and of Cuvier's first edition of the " Regne Animal," though afterwards referred by this last author to the genus *Gymnetrus,* Bl.*, under the belief that the ventrals, usually considered as wanting in the Vaagmaer, were only accidentally lost in the specimens hitherto observed†. The Vaagmaer is found off Iceland. Nothing is known of it as a British species beyond what Dr. Fleming has recorded in the work above referred to.

(7.) *G. Hawkenii*, Bloch, Ichth. pl. 423. *Blochian Gymnetrus*, Shaw, Gen. Zool. vol. iv. p. 197. pl. 29. *Ceil Conin*, Couch in Linn. Trans. vol. xiv. p. 77. *Hawken's Gymnetrus*, Yarr. Brit. Fish. vol. i. p. 188.

A doubtful native. Said to have been "drawn on shore in a net at Newlin in Cornwall, in Feb. 1791. The extremity of the tail was wanting ; the length of what remained was eight feet and a half, the depth ten inches and a half, thickness two inches and three quarters; weight forty pounds." *Couch.* The species itself is an obscure one, and not well ascertained. Bloch and Shaw have both figured the caudal fin from imagination, that part having been deficient in the specimens hitherto obtained.

---

* *Reg. An.* 2nd Edit. tom. ii. p. 219.

† This opinion, that the Vaagmaer possesses ventrals, when not mutilated by accident, has been confirmed by Professor Reinhardt, who has recently published a notice respecting a nearly perfect specimen of this fish, which had been cast ashore during the foregoing year, on the coast of Skagen. See *L'Institut*, 1834, p. 158.

GEN. 24.   CEPOLA, *Linn.*

48.   C. *rubescens*, Linn.   (*Red Band-Fish.*)

C. rubescens, *Mont. in Linn. Trans.* vol. VII. p. 291. pl. 17.   *Don. Brit. Fish.* vol. v. pl. 105.   *Flem. Brit. An.* p. 204.   *Cuv. Reg. An.* tom. II. p. 221.   C. Tænia, *Bloch, Ichth.* pl. 170.?   Serpens rubescens, *Will. Hist. Pisc.* p. 118. c. 13.   Red Band-Fish, *Penn. Brit. Zool.* (Edit. 1812.) vol. III. p. 285.   *Yarr. Brit. Fish.* vol. I. p. 195.   Red Snake-Fish, *Couch in Linn. Trans.* vol. XIV. p. 76.

LENGTH.   From ten to fifteen inches.

DESCRIPT.   (*Form.*) Long, slender, smooth, sub-pellucid, somewhat compressed, tapering gradually from the head to the tail: depth behind the head (in a specimen ten inches long) rather more than three-quarters of an inch; breadth half an inch: head not larger than the body, sloping from the eye to the end of the upper jaw: under jaw longest, sloping upwards: mouth large: both jaws with one row of distant, subulate, curved teeth at their very edge, the front ones projecting forwards: eyes large, placed high on the cheeks: pectorals small, rounded: ventrals small, oval; the first ray short and spinous, with a filament adjoining longer than the other rays, and detached from them; close together, and rather before than immediately under the pectorals: dorsal commencing just behind the head, immediately above the gill-opening, and continuing uninterruptedly to unite with the caudal: anal commencing just behind the vent, which is scarcely an inch from the ventral fins, and like the dorsal, continuing the whole length to unite with the caudal: this last lanceolate, the middle ray being much the longest, and gradually shortening on each side, till the distinction is lost in the dorsal and anal fins:

B. 4; D. 70; A. 61; C. 12; P. 16; V. 1/5:

lateral line a little curved near the head, but afterwards running quite straight to the tail: skin smooth, but when examined by a lens appearing finely punctured.   (*Colours.*) Pale carmine, darkest above and towards the tail; gill-covers, and undulated transverse lines along the sides, silvery: fins of the same colour as the body, except the ventrals, which are nearly white. MONT.

First noticed as a British species by Montagu, who obtained two specimens from Salcomb Bay, on the south coast of Devonshire.   Several others have since occurred off Cornwall, where it is represented by Mr. Couch as being not very uncommon.   In the Mediterranean it is well known.

––––––––––

GEN. 25.   MUGIL, *Linn.*

49.   M. *Capito*, Cuv.   (*Gray Mullet.*) — Maxillary visible when the mouth is closed: orifices of the nostril near together: the skin at the margin of the orbit not advancing upon the eye: scale above the pectoral short and obtuse.

M. Capito, *Cuv. Reg. An.* tom. ii. p. 232.   M. Cephalus, *Don. Brit. Fish.* vol. i. pl. 15.   *Flem. Brit. An.* p. 217.   Mullet, *Will. Hist. Pisc.* p. 274. tab. R. 3.?   *Penn. Brit. Zool.* vol. iii. p. 329. pl. 66. Gray Mullet, *Yarr. Brit. Fish.* vol. i. p. 200.

LENGTH.   From fifteen to twenty inches.

DESCRIPT.   (*Form.*) Back but little elevated: ventral line more convex than the dorsal: greatest depth beneath the first dorsal, about one-fourth of the whole length, excluding caudal: greatest thickness nearly two-thirds of the depth: head broad and depressed; snout short, transversely blunt and rounded, but vertically sharp: mouth very protractile, transverse, angular; teeth, in the jaws scarcely perceptible, on the tongue, vomer, and palatines, more developed: maxillary visible when the mouth is closed, and not retiring beneath the infra-orbital: upper lip rather thick and fleshy, margined with a number of close-set minute pectinations: eyes rather high up; the skin at the anterior and posterior margins of the orbit not advancing over any portion of the iride: nostrils double on each side; the two orifices placed near together, the anterior one round, the posterior one oblong: head smooth; all the upper part covered with large polygonal scales: scales on the body large, but smaller than the above, deciduous: first dorsal commencing about the middle; its height twice its length; spines strong; the first two equal and longest: second dorsal considerably behind the first; its height and length the same as in that fin; all the rays except the first branched: caudal forked· anal rather in advance of the second dorsal, somewhat longer than that fin, but of the same height: pectorals about three-fourths of the length of the head; second, third, and fourth rays longest; all the rays except the first branched: ventrals a little behind the pectorals, close together, somewhat shorter; first ray strongly spinous; second soft ray longest:

B. 6; D. 4—9; A. 3/9; C. 14, and some short; P. 17; V. 1/5.

(*Colours.*) Back dusky blue: sides and belly silvery; the former marked with several parallel longitudinal dark lines.

Several species of this genus are noticed by Cuvier in his " Regne Animal", confounded by previous authors under the general name of *M. Cephalus.* That which occurs most abundantly in our own seas, appears to be his *M. Capito,* to which species he himself refers the *Mullet* of Willughby and Pennant*. This is not uncommon on many parts of the coast, and is often found in estuaries. Spawns, according to Mr. Couch, about Midsummer.

50.   M. *Chelo*, Cuv.   (*Thick-lipped Gray Mullet.*)— Lips very large and fleshy, the margins ciliated; teeth penetrating into their substance like so many hairs: maxillary curved, showing itself behind the commissure. Cuv.

M. Chelo, *Cuv. Reg. An.* tom. ii. p. 232.   Thick-lipped Gray Mullet, *Yarr. Brit. Fish.* vol. i. p. 207.

LENGTH.   Ten inches.   COUCH.   Probably attains a larger size.

DESCRIPT.   (*Form.*) " Head wide, depressed: eyes (in a specimen ten inches long) one inch apart, and three-eighths of an inch from the angle

* Dr. Hancock appears to have been the first of our own naturalists to remark that the *Gray Mullet* of the British coasts was not the true *Mugil Cephalus.* He named it *M. Britannicus.* See *Lond. Quart. Journ. of Sci.* 1830. p. 129, &c.

of the mouth, not connected with any membrane : nostrils close together, and while the fish is alive, moveable on each contraction of the mouth : a prominent superior maxillary bone, minutely notched at its lower or posterior edge : upper lip protuberant and fleshy, with a thin margin minutely notched or ciliated ; the lip appears behind as projecting under the maxillary : carina of the under jaw prominent and square ; edge of the lower lip fine and simple : body solid, round over the back : pectoral fins high on the side, pointed, rounded below, the first rays short : the first dorsal fin five inches and three-eighths from the snout, the origin of the first three rays approximate, the first ray the longest : the first two rays of the anal fin short : tail broad, concave : scales large." (*Colours.*) "Head and back greenish ; all besides silvery, with six or seven parallel lines along the sides of the same colour as the back." COUCH, as quoted by YARR.

This species would seem, from Mr. Couch's MSS. communicated to Mr. Yarrell, to be not uncommon on the coast of Cornwall. Said to be "gregarious, frequenting harbours and the mouths of rivers in the winter months in large numbers." It does not appear, hitherto, to have been observed by any other of our own naturalists.

51. M. *curtus*, Yarr. (*Short Gray Mullet.*)

M. curtus, *Yarr. Brit. Fish.* vol. I. p. 210.

DESCRIPT. "Length of the head compared with that of the body and tail as one to three, the proportion in the *Common Gray Mullet* being as one to four : the body deeper in proportion than in *M. Capito*, being equal to the length of the head : head wider, the form of it more triangular, and also more pointed anteriorly : eye larger in proportion : fin-rays longer, particularly those of the tail : the ventral fins placed nearer the pectorals ; also a difference in the number of some of the fin-rays :

D. 4—1/8 ; A. 3/8 ; C. 14 ; P. 11 ; V. 1/5.

The *colours* of the two species are nearly alike ; and in other respects, except those named, they do not differ materially." YARR.

A new species described by Mr. Yarrell, of which only one specimen has hitherto been obtained. This, which is probably quite young, measuring but little more than two inches in length, was taken, in company with the fry of the Common Gray Mullet, between Brownsey Island and South Haven, at the mouth of Poole Harbour.

(8.) *M. Cephalus*, Cuv. Reg. An. tom. II. p. 231.

Whether the true *M. Cephalus* of Cuvier be found in the British seas, must be left doubtful, until naturalists shall have more closely examined and compared our native species. It may, perhaps, assist in determining this point, just to point out its distinguishing characters. These consist (according to Cuvier) in the eyes being partly covered by a fatty membrane adhering to the anterior and posterior margins of the orbit ; in the maxillary being entirely concealed beneath the infra-orbital, when the mouth is closed ; and in the base of the pectoral fin being surmounted by a long carinated scale* ; the orifices of the nostril are also separate from each other, and the teeth are tolerably well developed.

* See a representation of this scale in the vignette at the foot of page 201 of Yarrell's *British Fishes.*

GEN. 26.   ATHERINA, *Linn.*

52. A. *Presbyter*, Cuv. (*Atherine.*)—Anal with fifteen soft rays : fifty-one vertebræ.

A. Presbyter, *Cuv. Reg. An.* tom. II. p. 235. A. Hepsetus, *Don. Brit. Fish.* vol. IV. pl. 87. *Flem. Brit. An.* p. 217. Atherine, *Penn. Brit. Zool.* vol. III. p. 328. pl. 65. *Id.* (Edit. 1812.) vol. III. p. 434. pl. 76. Atherine, or Sand-Smelt, *Yarr. Brit. Fish.* vol. I. p. 214.

LENGTH. From four to six inches.

DESCRIPT. (*Form.*) Elongated; head and back in nearly the same horizontal line; abdomen rather more convex: greatest depth one-sixth of the entire length; thickness two-thirds of the depth: snout short; lower jaw projecting beyond the upper, and ascending to meet it at an angle of forty-five degrees with the axis of the body: mouth very protractile; both jaws, as well as the vomer and base of the tongue, with very fine velvet-like teeth; pharyngeans rather stronger: eyes large; their diameter contained two and a half times in the length of the head; distance from them to the end of the snout equalling scarcely more than half their diameter; space between, and upper part of the snout, with several longitudinal ridges and corresponding depressions: first dorsal commencing a little before the middle; its length rather less than its height; spines weak and slender; second and third longest: second dorsal remote, longer and more elevated than the first; first ray spinous; the rest soft; second longest: anal answering to second dorsal, but somewhat longer than that fin, commencing a little in advance of it: caudal deeply forked: pectorals a little shorter than the head: ventrals shorter than the pectorals, and about in a line with the tips of those fins when laid back :

B. 6; D. 7 to 9—1/12; A. 1/15; C. 17; P. 15; V. 1/5:

vent a little behind the middle. Number of vertebræ fifty-one. (*Colours.*) A longitudinal silver band on each side, running straight from behind the eye to the commencement of the caudal, bounded above by a narrow dusky or purplish line; breadth of the band about one-sixth of the depth: back, and portion of the sides above the band, pellucid grayish white, freckled with black; along the dorsal ridge an interrupted yellowish line : belly, and portion of the sides beneath the band, pellucid white, without spots: above the snout, and between the eyes, yellowish, spotted with black: fins pellucid, with minute black specks: irides silvery white.

According to Cuvier, the present genus, like the last, embraces several species hitherto confounded by naturalists. Our British specimens, at least those found on the southern coast, whence the individuals were obtained which furnished the above description, appear to belong to his *A. Presbyter.* Not uncommon at East Bourne and Brighton, where they are termed *Sand-Smelts.* Taken in most abundance during the spring months. Spawn in May and June. According to Pennant and Donovan, they are also found at Southampton and on the coast of Devonshire. *Obs.* The *Atherine* of Bloch (pl. 393. f. 3.) is probably distinct from our British species.

GEN. 27. BLENNIUS, *Linn.*

(1. BLENNIUS, *Cuv.*)

* *Head with two or more tentaculiform appendages.*

53. B. *ocellaris*, Bl. (*Ocellated Blenny.*)—Head with two principal appendages : dorsal bilobated ; the anterior lobe much elevated, marked with an ocellated spot.

> B. ocellaris, *Bloch, Ichth.* pl. 167. f. 1.   *Mont. in Wern. Mem.* vol. II. p. 443. pl. 22. f. 2.   *Flem. Brit. An.* p. 206.   Butterfly-Fish, *Will. Hist. Pisc.* p. 131. tab. H. 3. f. 2.   Ocellated Blenny, or Butterfly-Fish, *Yarr. Brit. Fish.* vol. I. p. 223.   Le Blennie papillon, *Cuv. Reg. An.* tom. II. p. 237.

LENGTH. From four to six inches.

DESCRIPT. (*Form.*) Sides much compressed : greatest depth contained three times and a half in the whole length, caudal excluded : thickness rather more than half the depth : head rounded anteriorly, very obtuse ; snout short ; profile nearly vertical : jaws equal : teeth numerous, closely compacted, the last in the series on each side above and below hooked, and longer than the others : eyes large, high on the cheeks ; the space between narrow and concave : above each eye a narrow tentaculiform appendage, slightly branched on its posterior margin, equalling in length one-third that of the head ; considerably behind the eyes, on each side of the occiput, a minute membranaceous flap : lateral line proceeding from the upper angle of the opercle at one-fourth of the depth, but bending suddenly down about the middle of the body, where it alters its course to half the depth : dorsal commencing at the occiput, and extending very nearly to the caudal, with which, however, it is not continuous, as in the next species ; the first eleven rays soft, but not articulated ; first much longer than any of the others, and more than equalling the whole depth of the body ; the succeeding ones gradually decreasing to the eleventh, which is the shortest in the whole fin ; beyond the eleventh the rays again lengthen, the twelfth being twice the length of the preceding one ; all the rays in this portion of the fin articulated, but not branched : anal commencing under the twelfth ray of the dorsal, and answering to the posterior lobe of that fin ; the two fins terminating exactly in the same line : caudal rounded ; rays branched ; the two outermost above and below excepted : pectorals the length of the head, slightly pointed ; all the rays simple : ventrals one-fourth shorter than the pectorals, narrow and pointed, of three simple rays, the middle one longer than the other two :

> D. 11/15 ; A. 17 ; C. 11, and 2 short ; P. 12 ; V. 3.

(*Colours.*) " Pale rufous brown, mixed with bluish gray, and slightly tinged with green in some parts ; the sides of the head, throat, and branchiostegous rays, spotted with rufous brown : the dorsal fin also a little spotted and barred with olive-brown and white ; between the sixth and eighth rays, a roundish purple-black spot, sometimes surrounded with white." MONT.

First noticed as a British species by Montagu, who obtained three specimens from an oyster-bed at Torcross, on the south coast of Devon, in 1814. A fourth, likewise British, from which the above description was taken, is in the collection of Mr. Yarrell. This last occurred among the rocks of the Island of Portland. In one of Montagu's examples the

ocellated spot was so ill-defined, that he was led to suspect it may sometimes be altogether wanting. He observed that those in which the ocellated spot was most perfect, had the first dorsal ray very long. Not an
uncommon species in the Mediterranean.

54. B. *Gattorugine*, Mont. (*Gattoruginous Blenny.*) —Head with two appendages: dorsal nearly even throughout, continuous with the caudal.

> B. Gattorugine, *Mont. in Wern. Mem.* vol. ii. p. 447. *Flem.*
> *Brit. An.* p. 206. Gattorugine, *Will. Hist. Pisc.* p. 132. c. xx.
> tab. H. 2. f. 2. *Penn. Brit. Zool.* vol. iii. p. 207. pl. 35. no. 91.
> Gattoruginous Blenny, *Yarr. Brit. Fish.* vol. i. p. 226.

LENGTH. From five to seven inches, sometimes more.

DESCRIPT. (*Form.*) Snout not so obtuse as in the last species, the
profile falling more gradually: teeth even throughout, the last in the
series not longer than the others: eyes very high on the cheeks, rising
above the level of the crown; the intervening space longitudinally impressed with a deep sulcus, conducting to another placed transversely
immediately behind the eyes; beyond this is a slight gibbosity in front of
the dorsal fin: over each eye a broad compressed tentaculiform appendage,
much palmated on both its margins, in length more than one-third that
of the head: lateral line as in the *B. ocellaris:* dorsal extending the
whole length of the body, and uniting with the base of the caudal; nearly
even throughout, having only a slight indentation about the middle;
posteriorly somewhat rounded; the first thirteen rays soft but not articulated, the first and thirteenth being the shortest; fourteenth one-third
longer than the preceding; this and all the succeeding ones articulated *,
but simple: anal commencing under the thirteenth ray of the dorsal, not
extending quite so far as that fin, and leaving a small space between
it and the caudal: this last as in the *B. ocellaris:* pectorals equalling
the head in length; all the rays simple; the two middle ones longer than
the others: ventrals of only two simple articulated rays, without even the
rudiment of a third; the inner ray longer and stouter than the outer one:

D. 13/20; A. 23; C. 11, and two short; P. 14; V. 2.

(*Colours.*) " Plain rufous brown, without any markings, paler on the
belly, as far as the vent: throat and fins orange-red, except the base of
the dorsal and pectorals: irides, and cirrhi over the eyes, orange." MONT.

The species of this genus, especially the British ones, have hitherto
been but ill-determined. There is reason to believe that two or more
have been confounded under the name of *B. Gattorugine.* The above
description, from a specimen taken at Weymouth, appears to agree with
the *Gattorugine* of Willughby and Pennant, which is probably quite
distinct from the species described by Linnæus under that name†. It
also accords with the *B. Gattorugine* of Montagu, and of Fleming, who
copies from him, but not with that of Donovan, as hereafter shown. Apparently not very common, at least on all parts of the coast. Pennant's
specimen was taken on the coast of Anglesea: Montagu's in a crab-pot
on the south coast of Devon. Others have since occurred in Cornwall to
Mr. Couch. Mr. Yarrell has also specimens from Poole Harbour, and
from other localities on the south coast.

---

* The articulations are not easily seen, except the membrane investing the rays be dissected
off, and the fin viewed against a strong light.

† On this point, see *Bull. des Sci. Nat.* 1828. tom. xv. no. 120.

(9.) *B. Gattorugine*, Don. Brit. Fish. vol. IV. pl. 86. *B. Gattoru-gine*, Linn. Syst. Nat. tom. I. p. 442.? Bloch, Ichth. pl. 167. f. 2.? Turt. Brit. Faun. p. 92.

"The anterior half of the lateral line double; its lower limb extending in a straight direction from the gills to the tail; midway between this and the back is an arched lateral line originating at the hind part of the head, and curving down to the former, with which it is united a little behind the tip of the pectoral fin: D. 32: P. 13: V. 2, of nearly equal length, with a very small lateral appendage: A. 20: C. 12: the rays of the tail branched: all the rest simple or undivided." DON.

The double lateral line, the small lateral appendage to the ventrals forming a kind of third ray, and the four palmated membranes on the head, clearly characterize this as a distinct species from either of the two last, supposing Donovan's figure and description to be correct. The four appendages on the head associate it with the *B. Gatt.* of Linnæus and Bloch. The double lateral line is probably an error. Bloch and Turton describe the lateral line as straight. In the *B. Gattorugine* of this work it is curved. Donovan does not state whence his specimen was obtained.

55. B. *palmicornis*, Cuv. (*Crested Blenny.*) — Head with four appendages: dorsal even throughout, continuous with the caudal: ventrals very small.

B. palmicornis, *Cuv. Reg. An.* tom. II. p. 237. B. Galerita, *Flem. Brit. An.* p. 207. B. Pennantii, *Jen. Cat. of Brit. Vert. An.* 24. sp. 54. Crested Blenny, *Penn. Brit. Zool.* vol. III. pl. 35. no. 90. but not p. 206. *Yarr. Brit. Fish.* vol. I. p. 233.

LENGTH. Four or five inches.

DESCRIPT. (*Form.*) Much more elongated than either of the pre-ceding species: depth, which is tolerably uniform throughout, equalling not more than one-seventh of the entire length: body considerably com-pressed: snout short and obtuse; the profile descending in a curve: "outline of the mouth, when viewed from above, forming a semicircle; viewed laterally, the angle of the mouth is depressed," the lower jaw ascending to meet the upper: gape rather wide; "lips capable of exten-sive motion:" teeth small and short; rather irregularly disposed, and not all exactly of the same length: eyes high on the cheeks, but not elevated above the crown: the intervening space flat, ornamented with four fimbriated tentaculiform appendages; the first pair of appendages are placed, one at the anterior margin of each eye, and are connected at the base by a low transverse membrane or fold of the skin; the second pair, which are twice the length of the first, and rather more fimbriated, are placed further back near the posterior margins of the eyes: "nape of the neck, and for some distance towards the commencement of the dorsal fin, the skin is smooth, with the exception of various small papillæ:" orbits surrounded by a circle of large, open, conspicuous pores: a row of similar pores at the upper part of the opercle, falling in with the commencement of the lateral line: dorsal commencing in a line with the upper angle of the opercle, and extending quite to the caudal, with which it is conti-nuous; its height, which is uniform throughout, equalling rather more than half the depth of the body; the first ray a little shorter than the second; all the rays simple, and apparently all spinous or inarticulated, their extreme tips projecting beyond the connecting membrane; the first three, however, accompanied by short filamentous prolongations of the membrane, which extend further than themselves: anal commencing in a line with the fourteenth ray of the dorsal, and extending, like that fin, quite to the caudal; the first ray only half the length of the second; the second a little shorter than the third; this last and the rest nearly equal, and of about the same length as the dorsal fin-rays, the ends, however, projecting further, the connecting membrane not being so deep: caudal rounded, with the principal rays branched: pectorals a little shorter than the head, also rounded: ventrals very small, consisting of only three rays, the longest being scarcely more than one-third the length of the head: number of fin-rays;

D. 51; A. 37; C. 14, and some short ones; P. 14; V. 3.

(*Colours.*) "General colour of the body and fins pale brown, mottled on the sides with darker brown; the head, the anterior part of the body, the ventral and pectoral fins, being darker than the other parts." YARR.

Of this species, I have only seen the specimen, taken at Berwick-upon-Tweed, which has been already described by Mr. Yarrell, and which was kindly lent me for examination. To that gentleman we are indebted for having cleared up some part of its history. It is probably the same as the *B. Galerita* of Dr. Fleming, which was found by him in Loch Broom. It is also clearly identical with Pennant's *figure* of the *Crested Blenny,* but the *description* of that species, in the "British Zoology," is in part borrowed from Willughby, and belongs to the *Alauda cristata* of that author, which last I am inclined to think is synonymous with the next species.

** *Head with one principal, transverse, crest-like, appendage.*

56. B. *Galerita,* Mont. (*Montagu's Blenny.*)

B. Galerita, *Mont. in Wern. Mem.* vol. i. p. 98. pl. 5. f. 2. Alauda cristata sive Galerita, *Will. Hist. Pisc.* p. 134. B. Montagui, *Flem. Brit. An.* p. 206. Diminutive Blenny, *Penn. Brit. Zool.* (Edit. 1812.) vol. iii. p. 277. Montagu's Blenny, *Yarr. Brit. Fish.* vol. i. p. 219.

LENGTH. From one and a half to two and a half inches. MONT.

DESCRIPT. "Body rather more slender than the *Smooth Blenny:* head much sloped; eyes high up, approximating, gilded; the upper lip furnished with a bony plate that projects at the angles of the mouth into a thin lamina that turns downwards, the ends of which are orange-coloured: on the top of the head, between the eyes, a transverse, fleshy, fimbriated membrane; the *fimbriæ* of a purplish brown colour, tipped with white: nostrils furnished with a minute bifid appendage: behind the crest several minute, erect, filiform *appendiculæ,* between that and the dorsal fin, placed longitudinally: lateral line considerably curved near the head: pectorals large and ovate, of twelve rays, reaching as far as the vent: ventrals, two unconnected rays: dorsal extending from the head to the tail, of thirty rays, and appearing like two distinct fins, by reason of the slope to the thirteenth ray, which is not above half the length of the anterior ones, and the sudden elongation of the fourteenth ray; this fin is very broad, and in one specimen there was an ovate black spot between the first and second ray, and another obscure one between the next rays, but this is not a constant character: anal fin equally broad, and extending from the vent to the tail, consisting of eighteen rays usually margined with black, and tipped with white: caudal slightly rounded, composed of fourteen rays.

D. 30; A. 18; C. 14; P. 12; V. 2.

The *colour* above generally olive-green spotted with pale blue shaded to white; the belly white, and the pectoral fins spotted with orange." MONT.
Montagu observes that this species is occasionally taken, with the *B. Gattorugine* and *B. Pholis*, among the rocks on the south coast of Devon, in the pools left by the receding tide. Several specimens seem to have been noticed by him. It has also occurred to Mr. Couch in Cornwall; but none of our other naturalists appear to have met with it. I see no reason for supposing it distinct from the *Alauda cristata* of Willughby, the *Galerita* of Rondeletius, although not the same as the *B. Galerita* of Linnæus, who (as Mr. Yarrell has pointed out) has confounded this species with that which has been since termed by Cuvier *B. palmicornis*. Should the contrary hereafter appear, it will then be proper to exchange the name of *Galerita* for that of *Montagui*, first adopted by Fleming.

*** *Head without appendages.*

57. B. *Pholis*, Linn. (*Smooth Blenny, or Shan.*)—
Dorsal notched in the middle; not continuous with the caudal.

> B. Pholis, *Linn. Syst. Nat.* tom. I. p. 443.   *Bloch, Ichth.* pl. 71. f. 2.   *Don. Brit. Fish.* vol. IV. pl. 79.   Alauda non cristata, *Will. Hist. Pisc.* p. 133. c. xxi. tab. H. 6. f. 2.   Pholis lævis, *Flem. Brit. An.* p. 207.   Smooth Blenny, *Penn. Brit. Zool.* vol. III. p. 208. pl. 36.   *Id.* (Edit. 1812.) vol. III. p. 280. pl. 40.   Shanny, or Smooth Shan, *Yarr. Brit. Fish.* vol. I. p. 230.   Baveuse commune, *Cuv. Reg. An.* tom. II. p. 238.

LENGTH. From four to five inches.
DESCRIPT. (*Form.*) Thicker anteriorly than the *B. Gattorugine;* the head less compressed; the body not quite so deep: depth contained a little more than four times in the whole length, caudal excluded: thickness two-thirds of the depth: snout short and obtuse; profile almost vertical: teeth crowded, with one or two longer than the others, and hooked, at the end of each series above and below: eyes smaller, and not so high on the cheeks as in *B. Gattorugine;* the space between wider, with only a very slight longitudinal depression: no appendages on the head of any kind, or transverse sulcus on the nape, which last part is rather convex: lateral line similar, taking a sweep over the pectorals: dorsal commencing at a greater distance from the end of the snout by one-fourth, and terminating a little before it reaches the caudal; first ray a little shorter than the second; fifth, sixth, and seventh, slightly the longest, equalling not quite half the depth; eighth and following ones decreasing to the twelfth, which is the shortest in the whole fin, and only half the length of the thirteenth, or first of the articulated rays; rest of the fin nearly even: anal answering to the posterior portion of the dorsal, but terminating a little sooner, leaving a larger space between it and the caudal: pectorals rather more rounded than in the *B. Gattorugine*, the middle rays being not so much elongated: ventrals rather shorter, of only two rays, the inner one longest:

> B. 6; D. 12/19; A. 20; C. 11, and 4 short; P. 13; V. 2.

(*Colours.*) Marbled and variegated with dusky and olive-brown, occasionally more or less spotted with white: rays of the anal always tipped with this last colour.

The most common species in the genus. Found on many parts of the coast, lurking beneath stones and sea-weed near low-water mark. Is tenacious of life, and will live for some time out of the water. Said to feed on small crustacea and marine worms. Spawns in Summer.

(2. GUNNELLUS, *Flem.*)

58. B. *Gunnellus*, Linn. (*Spotted Gunnel.*)—A row of dark ocellated spots along the base of the dorsal fin.

B. Gunnellus, *Linn. Syst. Nat.* tom. i. p. 443. *Bloch, Ichth.* pl. 71. f. 1. *Don. Brit. Fish.* vol. ii. pl. 27. Gunnellus Cornubiensium, *Will. Hist. Pisc.* p. 115. c. ix. tab. G. 8. f. 3. G. vulgaris, *Flem. Brit. An.* p. 207. Spotted Blenny, *Penn. Brit. Zool.* vol. iii. p. 210. pl. 35. *Id.* (Edit. 1812.) vol. iii. p. 282. pl. 39. Spotted Gunnel, or Butter-Fish, *Yarr. Brit. Fish.* vol. i. p. 239.

LENGTH. From six to eight, rarely ten, inches.

DESCRIPT. (*Form.*) Body elongated, and very much compressed throughout : greatest depth rather exceeding one-eighth of the entire length : thickness half the depth : head and back in one horizontal line ; the former small, not more than one-ninth of the whole length, excluding caudal : snout more pointed than in the true Blennies ; mouth small ; lower jaw sloping considerably upwards ; teeth minute : eyes placed rather high ; the space between forming an elevated ridge : nape, behind the eyes, a little depressed : dorsal fin commencing a little behind the nape, at a distance from the end of the snout equalling one-eighth of the entire length, and extending quite to the caudal, with which it is continuous ; all the rays simple and inarticulated, flexible, of the same height throughout, equalling scarcely more than one-sixth of the depth, projecting a little beyond the connecting membrane : anal commencing at about the middle of the whole length, likewise continuous with the caudal ; the first two rays spinous ; the rest articulated and branched : caudal rounded, with fifteen branched rays, and six simple ones shorter than the others, four above and two below : pectorals short, scarcely more than half the depth, rounded ; all the rays articulated, and, except the first and last, branched : ventrals extremely small, scarcely one-third the length of the pectorals, reduced to a single spine united to one small soft ray of about its own length :

D. 77 ; A. 2/40 ; C. 15, and 6 ; P. 12 ; V. 1/1 :

vent exactly in the middle. (*Colours.*) Deep olive, with a row of dark ocellated spots, varying in number, but generally from ten to twelve, along the line of the back, extending partly on to the dorsal fin : belly whitish : pectorals yellow.

Variety. Purple Blenny, *Low, Faun. Orc.* p. 203. " Reddish purple ; fins lightest. Likewise the spots on the back ; instead of eleven, has only a single one, and that placed near the beginning of the back fin." Low.

Not uncommon ; particularly off the coasts of Cornwall and Anglesea. Habits similar to those of the last species.

GEN. 28.   ZOARCES, *Cuv.*

59.   Z. *viviparus*, Cuv.   (*Viviparous Blenny.*)

Z. viviparus, *Cuv. Reg. An.* tom. ii. p. 240.   Blennius viviparus,
Linn. Syst. Nat. tom. i. p. 443.   *Bloch, Ichth.* pl. 72.   *Don. Brit.
Fish.* vol. ii. pl. 34.   Mustela vivipara, *Will. Hist. Pisc.* p. 122.
tab. H. 3.   f. 5.   Gunnellus viviparus, *Flem. Brit. An.* p. 207.
Viviparous Blenny, *Penn. Brit. Zool.* vol. iii. p. 211. pl. 37.   *Id.*
(Ed. 1812.) vol. iii. p. 283. pl. 41.   *Yarr. Brit. Fish.* vol. i. p. 243.

LENGTH.   From ten to twelve, rarely fifteen, inches.

DESCRIPT. (*Form.*) Slender, elongated : body sub-cylindric anteriorly,
compressed and tapering behind : skin smooth and naked : head small,
equalling about one-sixth of the entire length : snout blunt ; upper jaw
thick, projecting a little beyond the lower : teeth conical, sharp, very
minute : lateral line indistinct ; its course straight, at half the depth of
the body : dorsal commencing at the nape and extending the whole
length, nearly even till just before its union with the caudal, where it
becomes suddenly depressed and appears notched ; all the rays soft and
articulated, but simple : anal commencing a little before the middle, even
throughout, also uniting with the caudal : this last rounded : pectorals
large and rounded : ventrals jugular, very small and narrow, of three
rays :

B. 6 ; D., A. and C., about 150 ; P. 18 ; V. 3.

(*Colours.*) Back and sides yellowish brown, stained and spotted with
dusky : a series of dark spots more or less well-defined along the dorsal
fin : under parts, and anal, yellowish.

Found on many parts of the coast both in England and Scotland.
Stated by Pennant to be common in the mouth of the river Esk at
Whitby, Yorkshire.   Keeps at the bottom, lurking beneath stones.
Feeds on small crabs.   Is ovoviviparous.   The young, according to
Bloch, from two to three hundred in number, are excluded in the
month of June.

GEN. 29.   ANARRHICHAS, *Linn.*

60.   A. *Lupus*, Linn.   (*Wolf-Fish.*)

A. Lupus, *Linn. Syst. Nat.* tom. i. p. 430.   *Bloch, Ichth.* pl. 74.
Don. Brit. Fish. vol. i. pl. 24.   *Flem. Brit. An.* p. 208.   Wolf-
Fish, *Will. Hist. Pisc.* p. 130. c. xviii. tab. H. 3. f. 1.   Ravenous
Wolf-Fish, *Penn. Brit. Zool.* vol. iii. p. 151. pl. 24.   *Id.* (Edit.
1812.) vol. iii. p. 201. pl. 27.   Wolf-Fish, *Yarr. Brit. Fish.* vol. i.
p. 247.

LENGTH.   From four to six feet ; sometimes more.

DESCRIPT. (*Form.*) " Head a little flatted on the top : nose blunt ;
nostrils very small ; eyes small, and placed near the end of the nose : fore-
teeth strong, conical, diverging a little from each other, standing far out
of the jaws ; commonly six above, and the same below, though sometimes
only five in each jaw ; these are supported within side by a row of lesser
teeth, which makes the number in the upper jaw seventeen or eighteen,
in the lower eleven or twelve : grinding teeth of the under jaw higher
on the outer than the inner edges, which inclines their surfaces inward :
they join to the canine teeth in that jaw, but in the upper are separate
from them : in the centre are two rows of flat strong teeth, fixed on an

oblong basis upon the bones of the palate and nose: body long, a little
compressed: skin smooth and slippery: pectorals consisting of eighteen
rays: dorsal extending from the hind part of the head almost to the tail;
the rays in the fresh fish not visible: anal extending as far as the
dorsal: caudal rounded, of thirteen rays. (*Colours*.) Sides, back, and fins,
of a livid lead-colour; the first two marked downwards with irregular
obscure dusky lines: these in different fish have different appearances.
Young of a greenish cast." Penn.

A powerful and ferocious species, most abundant in the northern parts
of the globe, where it is said to attain to a larger size than in the British
seas. Not unfrequently met with off the coasts of Scotland and Yorkshire. Feeds on shell-fish and crustacea, which it readily crushes by
means of its strong molars. According to Pennant, is full of roe in
February, March, and April, and spawns in May and June.

GEN. 30.   GOBIUS, *Linn*.

61. G. *niger*, Linn. (*Black Goby*.) — Dorsals contiguous: lower jaw a very little the longest: distance
between the eyes not equal to their diameter.

G. niger, *Linn. Syst. Nat.* tom. I. p. 449.   Sea-Gudgeon, *Will.
Hist. Pisc.* p. 206. tab. N. 12. f. 1.   Black Goby, *Penn. Brit.
Zool.* vol. III. p. 213. pl. 38.   *Yarr. Brit. Fish.* vol. I. p. 251.
Le Boulereau noir, *Cuv. Reg. An.* tom. II. p. 243.

Length. From four to five, rarely six, inches.
Descript. (*Form.*) Elongated, the anterior extremity depressed, the
posterior compressed and tapering: depth one-sixth of the entire length;
thickness more than three-fourths of the depth: line of the back nearly
straight; abdominal line bellying a little behind the ventral fins: head
rather large, as broad as the body, somewhat more than one-fourth of the
whole length: snout blunt and rounded; gape wide; lower jaw a very
little the longest: fine card-like teeth, in several rows, the inner rows
much smaller than the outer: eyes large, placed on the upper part of the
head, approximating; the distance between barely three-fourths of their
diameter: gill-opening much contracted: head naked: marked on the
cheeks and before the eyes with several dotted lines, consisting of very
minute papillæ: from the occiput to the first dorsal a shallow groove:
body covered with large scales of a semicircular form, the free edges of
which are finely ciliated: lateral line straight along the middle, rather
indistinct: first dorsal commencing at one-third of the whole length,
excluding caudal; spines very slender and flexible, a little unequal in
height; fourth longest, equalling three-fourths of the depth; from the
last ray the membrane passing on, falls gradually till it terminates at the
base of the first ray in the second dorsal: this last with fourteen rays,
nearly of equal height, the middle ones somewhat exceeding the others,
equalling the longest of the spinous rays; all articulated, and, except the
first, branched: anal answering to second dorsal, but commencing a little
nearer the tail, and not extending quite so far; rays similar: caudal
rounded; rays branched: pectorals the length of the head, of an oval-
oblong form, with the middle rays longest; all the rays branched: ventrals forming by their union a funnel-shaped cavity; rays very unequal;
the central ones, which are longest, somewhat shorter than the pectorals:

B. 5; D. 6—14; A. 12; C. 13, and some short; P. 19; V. 10, when united:

vent exactly in the middle; immediately behind it a little conical papilla. (*Colours.*) Deep olive-brown, variegated with dusky spots and streaks: dorsals dusky brown, variegated with whitish.

Found on many parts of the coast, but not in any abundance. Sometimes called *Rock-Fish*, from the power which they are said to possess of affixing themselves to the rocks by means of their united ventrals, though, according to Fleming, these fins are not capable of acting as a sucker. It is probable that under the name of *Gobius niger* several species have been confounded. That figured by Bloch is evidently distinct from our British one, differing from it in having sixteen rays in the second dorsal, and the jaws of equal length. The *G. niger* of Donovan and Fleming refers to the next species.

62. G. *bipunctatus*, Yarr. (*Two-spotted Goby.*) — Dorsals nearly contiguous: lower jaw considerably the longest: distance between the eyes more than equal to their diameter.

> G. bipunctatus, *Yarr. Brit. Fish.* vol. I. p. 255. G. niger, *Don. Brit. Fish.* vol. v. pl. 104. *Flem. Brit. An.* p. 206.

LENGTH. From two to four inches.
DESCRIPT. (*Form.*) General form resembling that of the last species, but rather more elongated in proportion to the depth: eyes further asunder, and placed more laterally; the distance between them rather more than equal to their diameter: head moderately depressed: lower jaw considerably more projecting: dorsals not approximating quite so nearly; the first with a ray more; the second with three (in Donovan's two) less; posterior rays of this last rather the longest. In other respects the forms of the two species are similar.

> D. 7—11; A. 11; C. 12, and 2 short; P. 18; V. 12.

(*Colours.*) Testaceous, or yellowish white, all the scales on the back and upper part of the sides edged with brown; towards the top of the back this last colour prevails almost entirely: on the lateral line, beneath the commencement of the first dorsal, a conspicuous black spot; a similar one on each side of the base of the caudal fin: fins grayish white, with obsolete dusky bars.

Perhaps more common than the last species, with which it has, until lately, been confounded. The above description is taken from a specimen in the collection of Mr. Yarrell. Donovan's was from the coast of Devonshire.

63. G. *minutus*, Pall. (*Spotted Goby.*) — Dorsals remote; the second with the fourth and succeeding rays gradually decreasing: eyes closely approximating.

> G. minutus, *Gmel. Linn.* tom. I. part iii. p. 1199. *Don. Brit. Fish.* vol. II. pl. 38. *Flem. Brit. An.* p. 206. Spotted Goby, *Penn. Brit. Zool.* vol. III. p. 215. pl. 37. *Yarr. Brit. Fish.* vol. I. p. 258. Le Boulereau blanc, *Cuv. Reg. An.* tom. II. p. 243.

LENGTH. From two to three inches.
DESCRIPT. (*Form.*) More elongated and tapering than the *G. niger*, and not so much compressed: greatest depth, in the region of the pec-

torals, contained six times and a half in the entire length: thickness the
same as the depth: head depressed; snout short; lower jaw projecting
beyond the upper: eyes full and prominent, closely approximating on the
upper part of the head, the space between reduced to a shallow groove,
less than one-fourth of their diameter: opercle large, of an irregular
square form, with the lower angle rounded off; the ascending margin
nearly vertical; reaching nearly to the base of the rays of the pectorals:
scales small: first dorsal with the first four rays nearly equal; fifth and
sixth decreasing; all inclining backwards: space between the dorsals
equalling half the depth of the body: second dorsal with the first ray
a little shorter than the second; second, third, and fourth, equal and
longest; the succeeding rays gradually decreasing to the last, which is
scarcely more than half the length of the third and fourth: anal answering to the second dorsal, commencing and terminating nearly in the
same line; the rays, however, with the exception of the first, which is
much shorter than the others, more nearly of a height: caudal nearly
even.

B. 5; D. 6—11; A. 12; C. 13, and 2 short; P. 20; V. 12.

(*Colours.*) Yellowish white, and somewhat pellucid; the back and sides
obscurely spotted and mottled with ferruginous; three or four of these
spots, larger than the others, are placed at intervals on the lateral line;
that which is most distinct being just at the base of the caudal: rays
of the caudal and dorsal fins spotted with the same colour, giving the
appearance of transverse bars when the fins are close; anal and ventrals
plain: opercle with silvery reflections.

Common on many parts of the coast where it is sandy, and often taken
in the shrimp-nets. Is probably, however, frequently confounded with
the next species. Pennant considers it as the *Aphua Cobites* of Willughby, but as this last is represented as having seventeen rays in the
second dorsal, this opinion is probably incorrect.

64. G. *gracilis*, Jenyns. (*Slender Goby.*) — Dorsals
remote ; the second with the posterior rays longest : eyes
closely approximating.

G. gracilis, *Jen. Cat. of Brit. Vert. An.* 25. sp. 63. Slender Goby,
    Yarr. Brit. Fish. vol. i. p. 260.

LENGTH. Three inches two lines.
DESCRIPT. (*Form.*) Closely resembling the last species, but more
elongated and slender throughout: greatest depth barely one-seventh
of the whole length: snout rather longer: opercle approaching more
to triangular, the lower angle being more cut away, and the ascending
margin more oblique; a larger space between it and the pectorals: the
two dorsals further asunder: rays of the second dorsal longer; these
rays also gradually *increasing* in length, instead of *decreasing*, the posterior ones being the longest in the fin, and rather more than equalling
the whole depth: rays of the anal in like manner longer than in the
G. minutus :

D. 6—12; A. 12; C. 13, and 2 short; P. 21; V. 12:

in all other respects similar. (*Colours.*) Also resembling those of the
last, with the exception of the anal and ventral fins, which are dusky,
approaching to black in some places, instead of plain white, as in the
G. minutus.

Apparently a new species; though probably of not less frequent occurrence than the last, with which it may be easily confounded. My specimens were obtained from Colchester, and were supposed to have been taken somewhere off the Essex coast.

GEN. 31.   CALLIONYMUS, *Linn.*

65. C. *Lyra*, Linn. (*Gemmeous Dragonet.*)—Distance from the end of the snout to the posterior margin of the orbit, and thence to the first dorsal fin-ray, equal: first ray of the first dorsal greatly prolonged.

C. Lyra, *Linn. Syst. Nat.* tom. I. p. 433. *Bloch, Ichth.* pl. 161.
Don. Brit. Fish. vol. I. pl. 9. *Flem. Brit. An.* p. 208. Dracun-
culus, *Will. Hist. Pisc.* p. 136. tab. H. 6. f. 3. Gemmeous
Dragonet, *Penn. Brit. Zool.* vol. III. p. 164. pl. 27. *Id.* (Edit.
1812.) vol. III. p. 221. pl. 31. *Yarr. Brit. Fish.* vol. I. p. 261.
Le Savary ou Doucet, *Cuv. Reg. An.* tom. II. p. 247.

LENGTH. From nine to twelve inches.
DESCRIPT. (*Form.*) Head depressed, oblong-triangular, broader than the body, equalling one-fourth of the entire length: body elongated, gradually tapering from the nape to the caudal: eyes approximating, directed upwards, removed twice their diameter from the end of the snout; the distance from the end of the snout to the posterior margin of the orbit equalling the distance from this last point to the first dorsal fin-ray: gape wide: intermaxillary very protractile: upper jaw longest: both jaws with velvet-like teeth; none on the vomer or palatines: pre-opercle prolonged backwards, and terminating in three short but strong spines, the two innermost of which are directed upwards: opercle concealed beneath the investing skin, which is carried all round and nearly closes the branchial aperture, leaving only a small round hole on each side of the nape for the egress of the water: lateral line at first slightly descending, but afterwards straight: skin smooth and naked: first dorsal commencing at a little less than one-third of the whole length, caudal excluded; first ray prolonged into a slender filament, varying in length, but often reaching, when laid back, to the base of the caudal; the three succeeding rays much shorter, and rapidly decreasing, the last scarcely equalling the depth of the body; membrane of the fin extending beyond the last ray, and terminating at the base of the first ray in the second dorsal: this last fin three times as long as the first; all the rays articulated but simple; of moderate and nearly equal height, the last two only being a little the longest: both dorsals rise from a shallow groove which is continued on to the caudal: this last rounded; the uppermost ray and the two lowermost simple, the rest branched: anal similar to the second dorsal, but placed rather more backward, and with the rays not quite so long: pectorals somewhat pointed; the middle rays longest; all, except the first, branched: ventrals jugular, very far asunder, broader than the pectorals, to which they are partly united at the base by a membrane; first ray short and spinous; articulated rays very much branched:

D. 4—10; A. 9; C. 10, and 2 short; P. 20; V. 1/5:

vent rather before the middle; furnished with a conical papilla as in the last genus. (*Colours.*) " Predominant colour a fine pellucid brown, with marks and spots of pale blue, white, yellow, and black, disposed with peculiar elegance, especially about the head and dorsal fin: ventrals dark purple, finely contrasting with the pellucidity and whiteness of the pectorals: throat black." Don.

Found on many parts of the coast, but seldom in any plenty. Pennant states that it is not unfrequent off Scarborough, where it is taken by the hook in thirty or forty fathoms water. *Obs.* Both Willughby and Bloch represent this species with all the rays of the first dorsal nearly equally elongated. In our British specimens it is only the first ray which is so extraordinarily developed. This circumstance seems to suggest the possibility of their species being different from ours.

66.   C. *Dracunculus*, Linn.   (*Sordid Dragonet.*) — Distance from the end of the snout to the posterior margin of the orbit only half that from the eye to the first dorsal fin-ray: first ray of the first dorsal moderate.

C. Dracunculus, *Linn. Syst. Nat.* tom. i. p. 434. *Bloch, Ichth.* pl. 162. f. 2. *Don. Brit. Fish.* vol. iv. pl. 84. *Turt. Brit. Faun.* p. 89. Sordid Dragonet, *Penn. Brit. Zool.* vol. iii. p. 167. pl. 28. *Id.* (Edit. 1812.) vol. iii. p. 224. pl. 32. *Yarr. Brit. Fish.* vol. i. p. 266.

Length.   From six to eight and a half inches; rarely more.

Descript.   (*Form.*) Differs from the *C. Lyra,* which it closely resembles, in the following particulars: head shorter, and more decidedly triangular: eyes removed from the end of the snout by a space equalling not more than once their diameter; the distance from the end of the snout to the posterior margin of the orbit equalling only half the distance from this last point to the first dorsal fin-ray: gape much smaller: lateral line not so strongly marked: first dorsal with the first ray only one-third longer than the second, not prolonged into an extended filament. Number of fin-rays,

D. 4—10; A. 10; C. 10, and a short one; P. 21; V. 1/5.

(*Colours.*) Back and sides reddish brown, sometimes cinereous brown, mottled with darker spots; lower portion of the sides with a faint gloss of metallic gold: beneath white, with the posterior half pellucid: irides pale gold.

Considered by Neill* and Fleming† as only the female of the last species. This seems, however, hardly probable, from its being of much more frequent occurrence than the *C. Lyra,* invariably smaller, and with the colours very different.   Common on most parts of the coast, and, when small, often taken in the shrimp-nets.   Is sometimes called a *Fox.*

GEN. 32.   LOPHIUS, *Linn.*

67.   L. *piscatorius,* Linn.   (*Common Angler.*)

L. piscatorius, *Linn. Syst. Nat.* tom. i. p. 402. *Bloch, Ichth.* pl. 87. *Turt. Brit. Faun.* p. 115. *Don. Brit. Fish.* vol. v. pl. 101. *Flem. Brit. An.* p. 214. *Shaw, Nat. Misc.* vol. xi. pl. 422. Rana piscatrix, *Will. Hist. Pisc.* p. 85. tab. E. 1. Common Angler, *Penn.*

Brit. Zool. vol. III. p. 120. pl. 18.   *Id.* (Edit. 1812.) vol. III. p. 159. pl. 21. Fishing-Frog, *Yarr. Brit. Fish.* vol. I. p. 269. La Baudroye commune, *Cuv. Reg. An.* tom. II. p. 251.

LENGTH. From three to five feet.

DESCRIPT. (*Form.*) Head enormously large, occupying more than one-third of the entire length, broad and very much depressed: body tapering suddenly from behind the pectorals: snout obtuse and rounded; gape excessively wide; lower jaw considerably the longest, fringed along its edge with numerous short filaments: teeth conical, of various lengths and sizes, numerous and very sharp; two closely approximating rows in the lower jaw; the same above, but more widely separated; palatines, pharyngeans, and middle of the tongue, likewise bristling with teeth: eyes moderate, placed towards the upper part of the head, equally distant from the end of the snout and from each other: orbits above the eyes armed with a number of tooth-like processes, which forming two rows extend backwards to meet on the nape, but do not project through the skin: also two erect spines on each side of the end of the snout: gill-opening in the form of a wide, loose, purse-like cavity immediately beneath the pectorals; opercle small, not appearing externally: skin every-where soft and naked: above the nose, in front of the eyes, two long erect filamentous processes, one before the other, nearly half the length of the head; further down the mesial line, and about as far behind the eyes as the above are before them, another single filament about one-fourth shorter; after the same interval again two others about half the length of the first ones, and a third very short one; these three are sometimes connected at the base by a low membrane, forming a first dorsal; second dorsal commencing after a similar interval taken the third time, of a somewhat semicircular form, its length twice its height and half the length of the head; membrane enveloping the rays thick and fleshy, extending beyond the fin nearly to the caudal; this last even: pectorals in a line with the first of the three posterior dorsal filaments, of an oblong form, the rays of equal length, appearing truncated; their length one-third that of the head: anal similar to the second dorsal, but placed a little nearer the caudal: ventrals a little before the pectorals; the distance between them equalling their own length:

B. 6; D. 2—1—3—11; A. 9 or 10; C. 7 or 8; P. 24 to 26; V. 5.

(*Colours.*) All the upper parts brown, inclining to dusky: beneath white.

Taken occasionally on most parts of the coast. Keeps wholly at the bottom, and is very destructive to other fish. Has no swimming-bladder. *Obs.* Cuvier speaks of another species belonging to this genus, which may possibly also occur in the British seas. It is principally characterized by having the second dorsal less elevated, and only twenty-five vertebræ, the present species having thirty.

(10.) *L. Cornubicus,* Shaw, Gen. Zool. vol. v. p. 381. *Fishing-Frog of Mount's Bay,* Borl. Cornw. p. 266. pl. 27. f. 6. *Long Angler,* Penn. Brit. Zool. vol. III. p. 123.

In the opinion of Cuvier this supposed species is only an altered individual of the common one[*]. " Found on the shore of Mount's Bay, Aug. 9, 1757." BORL.

---

[*] The same may probably be said of the *Rana Piscatrix,* figured in Leigh's "Natural History of Lancashire," &c. (p. 186. pl. 6. f. 5.)

GEN. 33. LABRUS, *Linn.*

(1. Labrus, *Cuv.*)

* *Dorsal with twenty or twenty-one spinous rays.*

68. L. *maculatus*, Bloch. (*Ballan Wrasse.*)—Ascending margin of the preopercle oblique: soft portion of the dorsal more than twice the height of the spinous: dorsal and anal terminating nearly in the same line.

L. maculatus, *Bloch, Ichth.* pl. 294. L. Tinca, *Shaw, Nat. Misc.* vol. xi. pl. 426. *Id. Gen. Zool.* vol. iv. p. 499. pl. 72. *Don. Brit. Fish.* vol. iv. pl. 83. L. Balanus, *Flem. Brit. An.* p. 209. Ballan Wrasse, *Penn. Brit. Zool.* vol. iii. p. 246. pl. 44. *Id.* (Edit. 1812.) vol. iii. p. 334. pl. 55. *Yarr. Brit. Fish.* vol. i. p. 275. La Vieille tachetée, *Cuv. Reg. An.* tom. ii. p. 255.

LENGTH. From twelve to eighteen inches.

DESCRIPT. (*Form.*) Oblong-oval, narrowing at the tail beyond the termination of the dorsal and anal fins: body thick and bulky: depth one-fourth of the entire length: back not much elevated; dorsal line nearly straight from the commencement of the dorsal fin backwards, but in advance of that point falling gradually to the snout; no depression at the nape: head one-fourth of the whole length, caudal excluded: snout short and conical: mouth very protractile; lips double, the anterior pair thick and fleshy, and partially reflexed, shewing the teeth: jaws equal: teeth rather small, conical, the anterior ones longest, amounting to about eighteen in each jaw: distance from the eye to the end of the snout equalling twice the diameter of the eye; space between the eyes convex, without any depression or sulcus, equalling two diameters and a half: preopercle with the ascending margin inclined, this last forming with the basal margin an obtuse angle: lateral line bending a little downwards beneath the termination of the dorsal fin; its previous course nearly straight at one-third of the depth: dorsal commencing at a distance from the end of the snout equalling one-fourth of the entire length; space occupied by the fin nearly equalling half the entire length; spinous portion three-fourths of the whole, the height of this part one-fourth of the depth of the body; soft portion more than twice the height of the spinous: anal commencing in a line with the soft portion of the dorsal, and terminating also nearly in the same line with that fin; first three rays spinous, stronger than the dorsal spines, shorter than the soft rays which follow: caudal slightly rounded, its base scaly, beyond which are rows of scales between the rays for one-fourth of their length: pectorals rounded, two-thirds the length of the head: ventrals a little shorter: all the fins very stout; the membranes enveloping the rays thick and fleshy:

D. 20/11; A. 3/9; C. 13; P. 15; V. 1/5.

(*Colours.*) Back and sides bluish green, becoming paler on the belly; all the scales margined with orange-red: head and cheeks bluish green, reticulated with orange-red lines; lips flesh-colour; irides bluish green: all the fins greenish blue, with a few scattered red spots; the dorsal with spots along the base only; the blue on the caudal passing into dusky at the tip.

Not an uncommon species in the British seas. Pennant and Donovan obtained their specimens from Scarborough; where, according to the former author, "they appear during Summer in great shoals off Filey-Bridge; the largest weighing about five pounds." Donovan states that he has also received it from Cornwall; from the Skerry Islands, north of Anglesea, and from Scotland. Mr. Yarrell mentions various parts of the Irish coast, the eastern coast of England, and the shores of Dorsetshire and Devonshire, as other localities for this species. The description given above is that of a specimen in the collection of the Zoological Society, from the London market. Frequents rocky ground, and feeds principally on crustacea. Spawns, according to Mr. Couch, in April. *Obs.* The colours in this, and in all the other species of this family, are liable to much variation.

69. L. *lineatus*, Don. (*Streaked Wrasse.*)

L. lineatus, *Don. Brit. Fish.* vol. iv. pl. 74. *Turt. Brit. Faun.* p. 99. *Flem. Brit. An.* p. 209. L. Psittacus, *Riss. Hist. Nat. de l'Eur. Mérid.* tom. iii. p. 304.? Green-streaked Wrasse, *Yarr. Brit. Fish.* vol. i. p. 279.

LENGTH. Seven inches. DON.
DESCRIPT. Body green, with numerous longitudinal yellowish lines: fins greenish. Number of fin-rays,

D. 20/10; A. 3/8; C. 15; P. 14; V. 8.   DON.

Obtained by Donovan from the coast of Cornwall, where it is said to be provincially known by the name of *Green-Fish.* According to Mr. Yarrell, it appears also to have been met with on the Devonshire coast by Montagu. It is probably the *L. Psittacus* of Risso*, but Donovan's description is too imperfect to speak with certainty on this point.

70. L. *pusillus*, Jenyns. (*Corkling.*) — Ascending margin of the preopercle very oblique; a few obsolete denticulations about the lower angle: soft portion of the dorsal a little higher than the spinous: dorsal extending a little beyond the anal.

L. pusillus, *Jen. Cat. of Brit. Vert. An.* 25. sp. 69.

LENGTH. Four inches.
DESCRIPT. (*Form.*) Distinguished by its small size. Back but little elevated, sloping very gradually towards the snout; ventral line more convex than the dorsal; sides compressed: depth contained about three times and three-quarters in the entire length; thickness half the depth, or barely so much: head one-fourth of the entire length: snout rather sharp; jaws equal: teeth of moderate size, conical, regular, about sixteen or eighteen in each jaw: eyes rather high in the cheeks, situate half-way between the upper angle of the preopercle and the margin of the first upper lip; the space between about equal to their diameter, marked with a depression; a row of elevated pores above each orbit: preopercle with the ascending margin very oblique; the basal angle,

---

* By an error, the *L. Psittacus* was inserted in my Catalogue as British, independently of Donovan's species.

which falls a little anterior to a vertical line from the posterior part
of the orbit, very obtuse, and remarkably characterized by a few minute
denticulations, which further on become obsolete, and in some specimens
are scarcely anywhere obvious: lateral line a little below one-fourth of
the depth; nearly straight till opposite the end of the dorsal, then bending rather suddenly downwards, and again passing off straight to the
caudal: number of scales in the lateral line about forty-five: dorsal commencing at one-third of the length, excluding caudal; spinous portion
nearly three-fourths of the whole fin, the spines very slightly increasing
in length from the first to the last, which last is not quite one-third
of the depth of the body; soft portion a little higher than the spinous,
of a somewhat rounded form, the middle rays equalling nearly half the
depth: anal commencing a little anterior to the soft portion of the dorsal,
and terminating a little before it; the first three rays spinous, the third
being the longest, but the second the stoutest spine; soft rays resembling
those of the dorsal: caudal nearly even, with rows of scales between the
rays for nearly half their length: pectorals rounded, about two-thirds
the length of the head, immediately beneath the commencement of the
dorsal; all the rays soft and articulated, and, except the first, branched:
ventrals a little shorter; the first ray spinous, shorter than the second
and third, which are longest; all the soft rays branched; the last ray
united to the abdomen by a membrane for half its length:

B. 5; D. 20/10 or 11; A. 3/9; C. 13; P. 14; V. 1/5.

(*Colours of specimens in spirits.*) Yellowish brown, with irregular
transverse fuscous bands: dorsal irregularly spotted with fuscous; anal
light brown; the other fins pale.

This species, which is the smallest in the genus, is possibly the *Turdus
minor* or *Corkling* of Mr. Jago*. It is apparently quite distinct from
any of those described by other authors. Though belonging to the present section, which it is convenient to retain, it would seem to form the
transition to the *Crenilabri*, to which its near affinity is indicated by the
rudimentary denticulations on the margin of the preopercle. The only
specimens I have seen, amounting to four or five, were obtained at
Weymouth by Professor Henslow, and are now in the Museum of the
Cambridge Philosophical Society. One of these is very minute, and quite
young, but the two largest, measuring four inches, have all the appearance of being full-grown fish.

(11.) *Comber*, Penn. Brit. Zool. vol. III. p. 252. pl. 47. Id. (Edit.
1812.) vol. III. p. 342. pl. 58. *Comber Wrasse*, Yarr. Brit. Fish.
vol. I. p. 289. *Labrus Comber*, Gmel. Linn. tom. I. part iii.
p. 1297. Turt. Brit. Faun. p. 99. Flem. Brit. An. p. 209.

An obscure and doubtful species. Pennant's fish, which was obtained
from Cornwall, is thus characterized. "Of a slender form: dorsal fin
with twenty spinous, and eleven soft, rays: pectoral with fourteen: ventral
with five: anal with three spinous and seven soft: tail round. Colour of
the back, fins, and tail, red: belly yellow: beneath the lateral line ran
parallel a smooth even stripe from gills to tail, of a silvery colour."
Mr. Couch is recorded to have met with a single individual of this
species several years since, but his account of it, as given in the " British
Fishes" of Mr. Yarrell, is scarcely more explicit. He observes that
"compared with the *Common Wrasse*, the Comber is smaller, more slender,
and has its jaws more elongated: the two upper front teeth are very long:
a white line passes along the side from head to tail, unconnected with the
lateral line: it has distinct blunt teeth in the jaws and palate: the ventral
fins are somewhat shorter than in others of the genus."

* *Ray, Syn. Pisc.* p. 165.

It may be observed that Pennant supposed his fish to be the *Comber* of
Mr. Jago*. This, however, must be considered very doubtful, Ray men-
tioning nothing respecting Jago's fish, except that it was small, scaly, and
of a red colour. Cuvier† regarded it as a red variety of the *L. maculatus,*
with a series of white spots along the flank.

** *Dorsal with from sixteen to eighteen spinous rays: form elongated.*

71. L. *variegatus,* Gmel. (*Striped Wrasse.*)—Ascending margin of the preopercle nearly vertical: dorsal extending a little beyond the anal; the soft portion scarcely
higher than the spinous: branchiostegous membrane with
five rays.

L. variegatus, *Gmel. Linn.* tom. I. part iii. p. 1294. *Don. Brit. Fish.*
vol. I. pl. 21. *Turt. Brit. Faun.* p. 99. Turdus perbelle pictus,
Will. Hist. Pisc. p. 322. tab. X. 3. Sparus formosus, *Shaw, Nat.
Misc.* vol. I. pl. 31. Striped Wrasse, *Penn. Brit. Zool.* vol. III.
p. 240. pl. 45. *Id.* (Edit. 1812.) vol. III. p. 337. pl. 57. Blue-striped
Wrasse, *Yarr. Brit. Fish.* vol. I. p. 281. La Vieille rayée, *Cuv.
Reg. An.* tom. II. p. 255.

LENGTH. From twelve to fourteen inches.

DESCRIPT. (*Form.*) More elongated than any of the former species:
back not much elevated: greatest depth contained four times and one-
third in the entire length: thickness rather less than half the depth:
dorsal line continuous with the profile; no depression at the nape: head
more than one-fourth of the whole length: teeth numerous, conical,
sharp, the anterior ones longest, slightly curved; about twenty in the
upper, and thirty-five in the lower jaw; a few smaller ones behind:
distance from the eye to the end of the snout equalling twice and
a half the diameter of the eye; distance between the eyes equalling two
diameters; the intervening space very slightly concave: no elevated pores
above the orbits: ascending margin of the preopercle nearly vertical,
forming with the basal margin a slightly obtuse angle, which angle falls
behind the eye and not immediately under it, as in the next species:
lateral line high, its course at rather below one-fifth of the depth, bending
downwards opposite the termination of the dorsal, but much more gradually than in the species of the first section: number of scales in the
lateral line forty-six: dorsal commencing in a line with the pectorals and
posterior angle of the opercle; the soft portion scarcely higher than the
spinous: anal commencing in a line with the soft portion of the dorsal,
but terminating a little before that fin: caudal nearly even, with rows of
scales between the rays extending for half their length: pectorals not
half the length of the head: ventrals equal to them:

B. 5; D. 17/12 or 13; A. 3/11; C. 13; P. 15; V. 1/5.

(*Colours.*) Back and sides for two-thirds of their depth olivaceous brown,
with spots and interrupted longitudinal lines of bluish gray; remainder
of the sides orange: head and cheeks like the back; lower jaw, and all
beneath the head, bluish gray: dorsal orange, with a large oblong space

of bluish gray on its anterior half, occupying three-fourths of the height of the fin; on the posterior half, three round spots of the same colour: anal and ventrals orange-yellow, edged with bluish gray: caudal variegated with the same colours.

Not a very common species. Pennant and Donovan obtained specimens from the coast of Anglesea off the Skerry Islands. According to Mr. Thompson of Belfast, it is occasionally met with on the coast of Ireland. The individual described above is one of two, in the collection of Mr. Yarrell, from the London market. The colours are very variable, and at certain periods of the year extremely beautiful.

72. L. *Vetula*, Bloch? (*Sea-Wife*.)—Ascending margin of the preopercle oblique, forming with the basal an obtuse angle : dorsal nearly of equal height throughout : branchiostegous membrane with four rays.

L. Vetula, *Bloch, Ichth.* pl. 293.?   Sea-Wife, *Yarr. Brit. Fish.* vol. I. p. 284.

LENGTH.   Thirteen inches.

DESCRIPT. (*Form.*) Very similar to the last species, but rather more bulky in proportion to its length: depth about the same: thickness somewhat greater: teeth smaller, and more numerous, especially in the upper jaw: ascending margin of the preopercle more oblique, forming a more obtuse angle with the basal margin; this angle more immediately under the eye, a vertical from it forming a tangent to the posterior part of the orbit; (a line similarly drawn in *L. variegatus* is nearly coincident with the ascending margin, and falls behind the eye at a distance equalling the diameter of the eye:) branchiostegous membrane with only four rays*: lateral line rather lower, its course at one-fourth of the depth: scales somewhat larger: dorsal similar, nearly of equal height throughout: anal with the rays of the terminating fourth portion rather longer than the others; (in *L. var.* the rays are equal throughout:) caudal, pectorals, and ventrals, similar :

D. 16/13; A. 3/11; C. 13; P. 14; V. 1/5.

(*Colours of a specimen in spirits.*) Back and sides for three-fourths of their depth dark brown without spots; lower portion of the sides and belly pale orange-yellow: anterior half of the dorsal with a large oblong space at the base of dusky blue; remainder of the fin pale, with a row of dark spots, one at the base of nearly every ray: anal and ventrals orange-yellow, edged with dusky blue: caudal pale, with some of the exterior rays tipped with the same colour.

The individual described above, the only one which I have seen of this species, is in the collection of the Zoological Society. It was procured in the London market. It so nearly resembles the *L. variegatus*, that had I not seen the two together and compared them closely, I should have hesitated about admitting them as distinct. I do not feel certain that it is the *L. Vetula* of Bloch, but it approaches more nearly to that species than any other described one with which I am acquainted.

---

* The *Labrus Vetula* of Bloch is represented by that author as having six branchiostegous rays.

73. **L. *trimaculatus*, Gmel. (*Trimaculated Wrasse*.)**— Ascending margin of the preopercle oblique : dorsal with the posterior rays a little the longest : body red ; with three dark spots on each side, two at the base of the dorsal fin, and one between the dorsal and the caudal.

> L. trimaculatus, *Gmel. Linn.* tom. I. part III. p. 1294. *Shaw, Nat. Misc.* vol. XIX. pl. 786. *Don. Brit. Fish.* vol. III. pl. 49. *Turt. Brit. Faun.* p. 99. L. carneus, *Bloch, Ichth.* pl. 289. Trimaculated Wrasse, *Penn. Brit. Zool.* vol. III. p. 248. pl. 46. *Id.* (Edit. 1812.) vol. III. p. 336. pl. 56. Red Wrasse, *Yarr. Brit. Fish.* vol. I. p. 286. La Vieille couleur de chair, *Cuv. Reg. An.* tom. II. p. 256.

LENGTH. From eight to twelve inches.

DESCRIPT. (*Form.*) Oblong, elongated, and rather slender ; the back and profile nearly in a straight line : snout longer and more produced than in either of the two last species : greatest depth contained about four times and a half in the entire length : teeth numerous, conical, the anterior ones longest : ascending margin of the preopercle oblique, forming with the basal a much more obtuse angle than in the *L. variegatus* : course of the lateral line rather above one-fourth of the depth : dorsal and anal much as in *L. variegatus ;* the former with the posterior rays a little the longest : anal terminating a little before the dorsal : caudal even, or very slightly rounded, with rows of scales between the rays :

$$\text{D. } 18/13 \text{ ; A. } 3/11 \text{ ; C. } 13 \text{ ; P. } 15 \text{ ; V. } 1/5.$$

(*Colours.*) "Pervading colour a fine orange, varying to red upon the back, and becoming paler and whiter towards the belly : dorsal and tail a rich orange ; the former strongly marked with dark purplish black, and prettily edged with blue ; the rest of the fins of a paler hue : the three dark spots at the posterior extremity of the back of a rich blackish purple ; contiguous to these are four other spots of a delicate rose-colour ; two disposed in the space between the three dark ones, and the third and fourth placed one at each extremity of the outermost ones, so as to form together a series of seven spots, alternately of a pale rose-colour and a very deep purple." DON.

Apparently a rare species in the British seas. Pennant's specimen was taken on the coast of Anglesea ; Donovan's on the south coast of Devonshire near Exmouth. It has also occurred in Cornwall, and in the Frith of Forth. *Obs.* Fleming has erroneously considered this species and the Striped Wrasse as mere varieties of the *L. maculatus.*

(12.) *L. bimaculatus*, Linn. Syst. Nat. tom. I. p. 477. *Bimaculated Wrasse*, Penn. Brit. Zool. vol. III. p. 247. Id. (Edit. 1812.) vol. III. p. 335.

> This must be considered a very doubtful species, especially as British. Pennant does not appear to have seen it himself, but to have inserted it simply on the authority of Brunnich, who is said to have observed it at Penzance. No one has met with it since.

(13.) *Cook* (i. e. Coquus) *Cornubiensium*, Ray, Syn. Pisc. p. 163. f. 4. Penn. Brit. Zool. vol. III. p. 253. Id. (Edit. 1812.) vol. III. p. 340. *Lab. Coquus*, Turt. Brit. Faun. p. 99. Flem. Brit. An. p. 209.

> Ray's description of this species, which is one of those discovered by Mr. Jago on the coast of Cornwall, is so short and imperfect as hardly to

admit of its being identified with certainty. It is, however, in all probability the same as the *L. variegatus* already described. To the same species may be referred the *Cuckow-Fish* described by the editor of the last edition of the "British Zoology" (vol. III. p. 341.) Mr. Couch speaks of the *Cook** as a species with which he is familiar, but he has not added any description of the fish to which he alludes.

(LACHNOLAIMUS, Cuv. ?)

(14.) *L. Suillus*, Linn. Syst. Nat. tom. I. p. 476.   *Lachnolaimus suillus*, Cuv. Reg. An. tom. II. p. 257. note (1) ?   *Hog Wrasse*, Couch in Loud. Mag. of Nat. Hist. vol. v. p. 19.

Inserted by Mr. Couch in his "Fishes of Cornwall" on the authority of Osbeck, who mentions† "Rock-Fish (*Labrus Suillus*, Linn.)" amongst other species of fish which were brought on board his vessel by the people of the Scilly Islands. This bare statement, unaccompanied by any description of the fish alluded to, seems hardly sufficient ground for admitting the present species into the British Fauna.

(2. JULIS, *Cuv.*)

74. L. *Julis*, Linn. (*Rainbow Wrasse.*) — " Above fuscous and green; beneath white, with a fulvous dentated stripe on each side : two fore-teeth longest." Don.

L. Julis, *Linn. Syst. Nat.* tom. I. p. 476. *Bloch, Ichth.* pl. 287. f. 1.? *Don. Brit. Fish.* vol. IV. pl. 96. *Turt. Brit. Faun.* p. 99. Julis, *Will. Hist. Pisc.* p. 324. pl. X. 4. f. 1.? Julis vulgaris, *Flem. Brit. An.* p. 210. La Girelle, *Cuv. Reg. An.* tom. II. p. 257. Rainbow Wrasse, *Penn. Brit. Zool.* (Edit. 1812.) vol. III. p. 343. *Yarr. Brit. Fish.* vol. I. p. 291.

LENGTH. Rather exceeding seven inches. DON.
DESCRIPT. " Of a slender, or elongated form, and remarkable for the elegant distribution of its colours, which are changeable in various directions of light : a broad dentated stripe, extending along each side, from the head nearly to the tail, of a silvery and fulvous colour :

D. 9/13; A. 2/13; C. 13; P. 12; V. 1/5." DON.

Received by Donovan from the coast of Cornwall, in the year 1802. The only recorded instance in which it has hitherto occurred in the British seas. Inhabits the Mediterranean along with two other closely allied species.

(3. CRENILABRUS, *Cuv.*)

75. L. *Tinca*, Linn. (*Ancient Wrasse.*)—Dorsal line falling gradually to the snout : depth very nearly one-third of the length : denticulations of the preopercle moderate.

L. Tinca, *Linn. Syst. Nat.* tom. I. p. 477. *Turt. Brit. Faun.* p. 98. Crenilabrus Tinca, *Flem. Brit. An.* p. 208. Turdus vulgatissimus, *Will. Hist. Pisc.* p. 319. *Ray, Syn. Pisc.* p. 136. Ancient Wrasse, *Penn. Brit. Zool.* vol. III. p. 244. pl. 47. *Id.* (Edit. 1812.) vol. III. p. 332. pl. 58. Gilt-Head, *Yarr. Brit. Fish.* vol. I. p. 293.

---

* *Linn. Trans.* vol. XIV. p. 80.   † *Voyage to China*, vol. II. p. 122.

LENGTH.   From eight to ten inches.

DESCRIPT.   (*Form.*) General form resembling that of the species in
the first section of the first sub-genus: greatest depth contained a very
little more than three times in the entire length: thickness twice
and a half in the depth: dorsal line falling very regularly, continuous
with the profile; no depression at the nape: head contained three times
and a half in the whole length: jaws equal; teeth prominent, of moderate size, the middle anterior ones longest, about thirteen above and
fifteen below, with a secondary but imperfect row of smaller ones behind
in the upper jaw: eyes moderate, rather high up; their distance from
the end of the snout equalling twice their diameter; the space between
them a little concave, equal to the same: ascending margin of the preopercle sharply denticulated, but the denticulations not so much developed as in the next species; nearly vertical, and making a right angle
with the basal margin: opercle large; the margin entire, rounded below,
emarginated above: lateral line following the curvature of the back at
one-fourth of the depth, bending suddenly downwards opposite the termination of the dorsal fin: scales very large; number in the lateral line
thirty-six: dorsal commencing in a line with the pectorals, and posterior
angle of the opercle; soft portion rounded, higher than the spinous:
anal commencing a little anterior to the soft portion of the dorsal, but
terminating in a line with that fin: caudal rounded, scaly at the base,
but with no rows of scales between the rays: pectorals and ventrals
much as in the other species of this genus:

B. 5; D. 16/9; A. 3/10; C. 13, and 2 short; P. 14; V. 1/5.

(*Colours.*) Back, and upper part of the sides above the lateral line,
marked with alternate longitudinal lines of dull red and dusky blue;
sides beneath the lateral line bluish green, spotted with dull red;
abdomen the same, but paler: upper part of the head deep brownish
red, with undulating lines of bright azure-blue; cheeks and gill-covers
bluish green with longitudinal lines of red; throat and beneath the
pectorals paler, lined with red: irides bluish green, with an inner circle
of red: dorsal, caudal, anal and ventral fins, bluish green, spotted and
lined with red: pectorals pale without spots.

Found on many parts of the coast, and perhaps the most common of all
the British species belonging to this family. Chiefly frequents deep
water where the bottom is rocky, and is often taken in the prawn-pots.
Feeds principally on crustacea.   Spawns in April.   It is the *Common
Wrasse* of Couch*, and the *Old Wife* of some English authors.   Pennant
calls it *Ancient Wrasse*, but it must not be confounded with the Ancient
Wrasse of Donovan, which is clearly the *L. maculatus* of this work.

76. L. *Cornubicus*, Gmel. (*Goldsinny.*)—Depth considerably less than one-third of the length : denticulations
of the preopercle very much developed : a conspicuous black
spot on each side of the tail.

L. Cornubius, *Gmel. Linn.* tom. I. part iii. p. 1297.   *Don. Brit.
Fish.* vol. III. pl. 72.   *Turt. Brit. Faun.* p. 99.   Goldsinny, *Ray,
Syn. Pisc.* p. 163. fig. 3.   *Penn. Brit. Zool.* vol. III. p. 251. pl. 47.

* *Linn. Trans.* vol. XIV. p. 80.

Id. (Edit. 1812.) vol. III. p. 339. pl. 58.   Corkwing, *Couch in Loud. Mag. of Nat. Hist.* vol. v. p. 17. f. 4.   Goldfinny, *Yarr. Brit. Fish.* vol. I. p. 296.

LENGTH.   From four to four inches and a half.

DESCRIPT.   (*Form.*) Very similar to the last species, but much smaller, and slightly more elongated: depth contained nearly three times and a half in the entire length: thickness twice and a half in the depth: head one-fourth of the whole length: jaws and teeth similar, but the latter not so numerous in the upper jaw, only eight or ten, with no secondary row behind; those below in about the same number: denticulations of the preopercle longer and more conspicuous: all the other characters, including lateral line, form and relative position of the fins, number of fin-rays, &c. exactly the same in the two species.

B. 5; D. 16/9; A. 3/10; C. 13, and 2 short; P. 14; V. 1/5.

(*Colours.*) Somewhat similar to those of the last species, but in general much paler: a conspicuous dusky spot on each side of the tail, near the commencement of the caudal, and immediately below the lateral line: dorsal fin variegated with fuscous bands.

First observed by Mr. Jago on the coast of Cornwall, and communicated by him to Ray. Obtained since from the same locality by Donovan and Mr. Couch. Has been also found in Devonshire by Montagu. The specimens which furnished the above description were procured at Weymouth. *Obs.* This species is erroneously considered by Fleming as a mere variety of the last.

77. L. *gibbus*, Gmel. (*Gibbous Wrasse.*) — Depth considerably more than one-third of the length; dorsal line falling suddenly to the snout.

L. gibbus, *Gmel. Linn.* tom. I. part iii. p. 1295.   *Turt. Brit. Faun.* p. 98.   Crenilabrus gibbus, *Flem. Brit. An.* p. 209.   Gibbous Wrasse, *Penn. Brit. Zool.* vol. III. p. 250. pl. 46.   *Id.* (Edit. 1812.) vol. III. p. 338. pl. 56.   *Yarr. Brit. Fish.* vol. I. p. 298.

LENGTH.   Eight inches. PENN.

DESCRIPT.   (*Form.*) "Very deep and elevated, the back vastly arched, and very sharp or ridged: greatest depth three-eighths of the length: from the beginning of the head to the nose a steep declivity: teeth like those of the others: eyes of a middling size: the nearest cover of the gills finely serrated:

D. 16/9; A. 3/11; P. 13; V. 1/5:

caudal large, rounded at the end; the rays branched; the ends of the rays extending beyond the webs: lateral line incurvated towards the tail: gill-covers and body covered with large scales. (*Colours.*) Gill-covers most elegantly spotted, and striped with blue and orange, and the sides spotted in the same manner; but nearest the back the orange disposed in stripes: dorsal and anal sea-green, spotted with black; ventrals and tail a fine pea-green; pectorals yellow, marked at their base with transverse stripes of red." PENN.

This species appears to be known only from the description of Pennant, who obtained a specimen taken off Anglesea. Its great depth clearly distinguishes it from the *L. Cornubicus*, of which Mr. Couch seems inclined to think it a mere variety*.

78.   L. *luscus*, Linn. ? (*Scale-rayed Wrasse.*) — Very much elongated : between the rays of the dorsal, anal, and caudal fins, processes of imbricated scales.

> L. luscus, *Couch in Loud. Mag. of Nat. Hist.* vol. v. p. 18, & p. 742. fig. 121.   Scale-rayed Wrasse, *Yarr. Brit. Fish.* vol. I. p. 300.

DESCRIPT. "Length twenty-two inches; greatest depth, exclusive of the fins, two inches and a quarter: body plump and rounded: head elongated; lips membranous; teeth numerous, in several rows, those in front larger and more prominent, rather incurved: eyes moderately large: anterior gill-plate serrate; six gill-rays: body and gill-covers with large scales: lateral line nearer the back, descending with a sweep opposite the termination of the dorsal fin, thence backward straight: dorsal with twenty-one firm, and eight soft, rays: the fin connected with the latter expanded, reaching to the base of the tail: pectorals round, with fourteen rays: ventrals with six rays, the outermost simple, stout, firm, tipped; between these fins a large scale: anal with six firm, and eight soft, rays, the latter a soft portion expanded: caudal round, with fifteen rays: between each ray of the dorsal, anal, and caudal fins, a process formed of firm elongated imbricated scales.

D. 21/8 ; A. 6/8 ; C. 15 ; P. 14 ; V. 1/5.

Colour a uniform light brown, lighter on the belly; upper eye-lid black; at the edge of the base of the caudal fin a dark brown spot: pectorals yellow; all the other fins bordered with yellow." COUCH.

A single individual of this species is recorded by Mr. Couch to have been taken off Cornwall, in February 1830, at the conclusion of a very cold season. It appears to be particularly characterized by having rows of scales between the rays of the dorsal and anal fins, as well as the caudal; this last fin exhibiting the above character in many other species of the present family. Its identity, however, with the *L. Luscus* of Linnæus appears very questionable. Cuvier thinks† that the Linnæan *L. Luscus* is only a variety of *L. Turdus* of Gmelin.

GEN. 34.   CENTRISCUS, *Linn.*

79. C. *Scolopax*, Linn. (*Trumpet-Fish.*)

> C. Scolopax, *Linn. Syst. Nat.* tom. I. p. 415.   *Bloch, Ichth.* pl. 123. f. 1.   *Don. Brit. Fish.* vol. III. pl. 63.   *Turt. Brit. Faun.* p. 117. *Shaw, Nat. Misc.* vol. XIV. pl. 584.   *Flem. Brit. An.* p. 220. *Cuv. Reg. An.* tom. II. p. 268.   Trumpet-Fish, *Will. Hist. Pisc.* p. 160. c. xi. tab. I. 25. f. 2.   *Couch in Linn. Trans.* vol. XIV. p. 89.   *Yarr. Brit. Fish.* vol. I. p. 302.   Snipe-nosed Trumpet-Fish, *Penn. Brit. Zool.* (Edit. 1812.) vol. III. p. 190.

---

* *Mag. of Nat. Hist.* vol. v. p. 18.   † *Reg. An.* tom. II. p. 256. note (2).

LENGTH. From four to five inches.

DESCRIPT. (*Form.*) "Body oval, compressed: snout elongated, the jaw-bones forming a tube extending an inch and a half before the eyes; mouth placed at the extremity, small, without teeth: eyes large: back elevated, forming a slight ridge, and ending in a short spine just in advance of the long and strong denticulated spine of the first dorsal fin: scales on the body hard, rough, minutely ciliated at the free edge, the surface granulated: first dorsal with but three spinous rays (generally said to be four); the first three times as long as, and also much stronger than, the others, pointed, moveable, and toothed like a saw on the under part, constituting a formidable weapon of defence; the other spines short, but their points projecting beyond the membrane by which they are united: the rays of the second dorsal soft: anal elongated; the rays short: pectorals small: ventrals also small, with a depression behind in which they can be lodged.

D. 4—12; A. 18; C. 16; P. 17; V. 4.

(*Colours.*) Back red, the sides rather lighter; sides of the head and belly silvery, tinged with gold-colour: irides silvery, streaked with red; pupils black: all the fins grayish white." YARR. •

An individual of this species is recorded to have been thrown on shore at Menabilly near Fowey, Cornwall, early in the year 1804*. Donovan appears to have been acquainted with one or two other instances in which it had occurred on the western coasts of England. Common in the Mediterranean.

ORDER II.   MALACOPTERYGII.

(I. ABDOMINALES.)

GEN. 35. CYPRINUS, *Linn.*

(* 1. CYPRINUS, *Cuv.*)

* *With barbules.*

* 80. C. *Carpio*, Linn. (*Common Carp.*)—Mouth with two barbules on each side: caudal forked.

C. Carpio, *Linn. Syst. Nat.* tom. i. p. 525. *Bloch, Ichth.* pl. 16. *Turt. Brit. Faun.* p. 107. *Don. Brit. Fish.* vol. v. pl. 110. *Flem. Brit. An.* p. 185. Carp, *Will. Hist. Pisc.* p. 245. tab. Q. 1. f. 2. *Penn. Brit. Zool.* vol. iii. p. 353. pl. 70. *Id.* (Edit. 1812.) vol. iii. p. 467. pl. 81. *Bowd. Brit. fr. wat. Fish.* Draw. no. 2. *Yarr. Brit. Fish.* vol. i. p. 305. La Carpe vulgaire, *Cuv. Reg. An.* tom. ii. p. 271.

* *Linn. Trans.* vol. VIII. p. 358.

LENGTH. From one to one and a half, or even two, feet.

DESCRIPT. (*Form.*) Oval; body thick anteriorly; back moderately elevated; dorsal line more convex than the ventral, falling with the profile in one continuous curve, without any depression at the nape: greatest depth beneath the commencement of the dorsal fin, measuring rather more than one-third of the entire length: greatest thickness in the region of the gills, equalling half the depth: head large: jaws equal; lips thick, furnished with two barbules at the corners of the mouth, and two shorter ones above nearer the nose: mouth small: no teeth in the jaws; pharyngeans with flat teeth striated on the crown: eyes small, and rather high on the cheeks: opercle marked with radiating striæ: lateral line nearly straight; its course a very little below the middle: scales large; number, in the lateral line, thirty-eight; in the depth twelve, six and a half being above, and five and a half below, the lateral line * : dorsal commencing in a line with the end of the pectorals, and occupying a space equal to nearly one-third of the entire length; first two rays bony, and partaking of the nature of spines; the first not half the length of the second; this last very strong and serrated posteriorly; third (or first of the soft rays) longest, equalling rather more than one-third of the depth; succeeding ones gradually decreasing to the seventh or eighth, beyond which they remain even to the end; all the soft rays branched; the last two from one root† : anal short, opposite the last quarter of the dorsal, and terminating in a line with that fin; first two rays bony; the second strongly serrated; third ray longest, nearly equalling the third in the dorsal; fourth and succeeding ones decreasing; all the soft rays branched; the last two from one root: caudal forked for half its length; all the principal rays except the outer ones branched: pectorals attached low down, in a line with the posterior margin of the opercle; their length about three-fourths that of the head: all the rays soft, and, except the first, branched: ventrals similar to the pectorals but rather shorter, situate in a line with the first three soft rays of the dorsal:

B. 3; D. 2/22; A. 2/6; C. 19, and some short; P. 17; V. 9.

(*Colours.*) General colour olive-brown, tinged with gold; darkest on the head; belly yellowish white: fins, dorsal and caudal, dusky; ventrals and anal tinged with red.

Originally from the middle of Europe. Said to have been introduced into England about the year 1514, but was certainly known before that time. Common in lakes and ponds, as well as in some rivers. Attains to the weight of nearly twenty pounds, but arrives at a still larger size on some parts of the Continent. Spawns in May and June, and is very prolific. Food, insects, worms, and aquatic plants.

** *Without barbules.*

* 81. C. *Gibelio*, Gmel. (*Gibel.*)—Depth one-third of the entire length: lateral line bending slightly downwards: caudal crescent-shaped.

C. Gibelio, *Gmel. Linn.* tom. i. part iii. p. 1417. *Bloch, Ichth.* pl. 12. *Flem. Brit. An.* p. 185. Carassius, *Will. Hist. Pisc.* p. 249. Crucian,

---

* The number of scales in the lateral line, and the number of rows of scales in the depth, are characters of some importance in distinguishing the different species of *Cyprinidæ*. They are here adopted from Jurine, and I shall state, nearly in the words of that author, the exact method in which they are computed. In estimating the number of scales in the lateral line, the reckoning is confined to those scales which are marked with the tube-like projection, the small irregular scales at the insertion of the caudal being neglected. The number of rows in the depth is taken at the deepest part of the body, or in a line from the first rays of the dorsal fin to the base of the ventral. Such a line, however, being interrupted near the middle by the lateral line, it is divided into two parts, the dorsal portion containing the number of scales *above* the lateral line, the ventral portion the number *below* it. Moreover, one of the scales themselves being always divided by the lateral line, and this line serving as the boundary of the two portions, it follows that half a scale is given each way to be added to the number of entire scales that appear in these portions respectively. It may be further stated that the curved tile-like scale, which appears on the ridge of the back in most of the fish belonging to this family, being common to both sides, is not taken into the account; neither are the small incomplete irregular scales which may be often observed at the base of the rays of the dorsal fin. In like manner, the numerous small scales which appear at the bottom of the abdomen, and which could not be counted with precision, are omitted; the reckoning at this point commencing with the first entire scale above the long scale which is placed at the base of the ventral fin. See Jurine's memoir on the Fish of the Lake of Geneva, contained in the *Mém. de la Soc. de Phys. et d'Hist. Nat. de Genève*, tom. III. part i. pp. 143, 144.

† In all the *Cyprinidæ*, the last two rays in the dorsal and anal fins will be found to spring from one root. In computing the fin-ray formula they may be reckoned either as one or two. I have considered them as two.

Penn. Brit. Zool. vol. III. p. 364. pl. 72. no. 171.   *Bowd. Brit. fr. wat. Fish.* Draw. no. 23.   *Yarr. Brit. Fish.* vol. I. p. 311.   Gibele, *Penn. Brit. Zool.* (Edit. 1812.) vol. III. p. 480. pl. 83.   La Gibèle, *Cuv. Reg. An.* tom. II. p. 271.

LENGTH. From ten to twelve inches, or more.

DESCRIPT. (*Form.*) Back moderately elevated, the dorsal line more convex than the ventral : greatest depth one-third of the entire length : head about one-fifth : profile falling very regularly, and forming one continuous curve with the line of the back : snout short, and rather obtuse : jaws nearly equal, the lower one a little the longest when the mouth is open ; gape rather small : eyes small : opercle marked with radiating striæ : lateral line descending in a gentle curve a little below the middle : scales large ; number in the lateral line thirty-four ; in the depth twelve, six and a half above, and five and a half below, the lateral line : dorsal much as in the last species ; the first two rays bony ; the first very short ; the second strong and serrated, but the serratures very fine compared with those of the same ray in the *Carp :* anal short, also with the first two rays bony ; the first extremely short ; this fin terminating a little beyond the termination of the dorsal : caudal forked ; the depth of the fork about one-third of its length : pectorals and ventrals much as in the *Carp ;* the latter nearly in a line with the second bony ray of the dorsal.

D. 2/18 ; A. 2/7 ; C. 19, and some short ; P. 14 ; V. 9.

Number of vertebræ (according to Mr. Yarrell) thirty. (*Colours.*) Back, and sides above the lateral line, olive-brown ; lower part of the sides yellow, becoming paler on the belly ; the whole tinged with a bright golden lustre : irides golden : cheeks and gill-covers bright golden yellow : dorsal fin olivaceous ; caudal the same, tinged with orange-yellow ; anal, pectorals, and ventrals, bright orange-red.

Supposed to be a naturalized species in this country, but not exactly known when it was introduced. Found in some of the ponds about London, as well as in other parts of England. Usual weight about half a pound : has been known, however, to weigh upwards of two pounds. Said to spawn in April or May. Food, aquatic plants and worms. Is very tenacious of life.

(15.) *C. Carassius,* Linn. Syst. Nat. tom. I. p. 526. Bloch, Ichth. pl. 11. Turt. Brit. Faun. p. 108. *Le Carreau* ou *Carrassin,* Cuv. Reg. An. tom. II. p. 271.

Mr. Yarrell has reason to believe that he has more than once received this species from the Thames. Its claims, however, to a place in the British Fauna are not fully established. By Turton, it was probably confounded with the *C. Gibelio,* from which it may be distinguished by the greater depth of the body, straight lateral line, and nearly even caudal. It has also more rays in the dorsal and anal fins.

* 82.   C. *auratus,* Linn. (*Golden Carp.*) — Caudal deeply forked ; sometimes three or four lobed.

C. auratus, *Linn. Syst. Nat.* tom. I. p. 527. *Bloch, Ichth.* pls. 93, & 94. *Turt. Brit. Faun.* p. 108. *Flem. Brit. An.* p. 185. Gold-Fish, *Penn. Brit. Zool.* vol. III. p. 374. *Id.* (Edit. 1812.) vol. III. p. 490. Gold Carp, *Yarr. Brit. Fish.* vol. I. p. 315. La Dorade de la Chine, *Cuv. Reg. An.* tom. II. p. 272.

LENGTH. Seldom exceeds eight or ten inches.

DESCRIPT. (*Form.*) General form resembling that of the *Carp :* head short ; jaws equal ; eyes large ; nostrils tubular, placed near the eyes : body covered with large scales : lateral line straight, near the back : fins extremely variable in form and size, as well as in the number of the rays : dorsal often very small ; sometimes entirely wanting, or represented by a simple elevation on the ridge of the back : anal often double : caudal large, sometimes enormously developed ; deeply forked, or divided into three or more lobes. (*Colours.*) Black during the first year ; afterwards mottled with silver ; this last colour continually spreading till it occupies the entire fish: after a few years the red tint is assumed, which becomes more brilliant with age: sometimes red from birth, or before acquiring the silvery hue: fins scarlet: irides golden.

A native of China. According to Pennant, first introduced into England about the year 1691, but not generally known till 1728. Is now completely naturalized, and breeds freely in ponds in many parts of the country. Spawns in May.

(2. BARBUS, *Cuv.*)

83. C. *Barbus*, Linn. (*Barbel.*)

C. Barbus, *Linn. Syst. Nat.* tom. I. p. 525. *Bloch, Ichth.* pl. 18. *Don. Brit. Fish.* vol. II. pl. 29. *Turt. Brit. Faun.* p. 107. Barbus vulgaris, *Flem. Brit. An.* p. 185. Barbel, *Will. Hist. Pisc.* p. 259. tab. Q. 2. f. 1. *Penn. Brit. Zool.* vol. III. p. 357. pl. 71. *Id.* (Edit. 1812.) vol. III. p. 472. pl. 82. *Bowd. Brit. fr. wat. Fish.* Draw. no. 9. *Yarr. Brit. Fish.* vol. I. p. 321. Barbeau commun, *Cuv. Reg. An.* tom. II. p. 272.

LENGTH. From two to three feet.

DESCRIPT. (*Form.*) Rather elongated; the back but little elevated: dorsal line continuous with the profile, and falling in one gradual slope to quite the end of the snout: greatest depth beneath the commencement of the dorsal, equalling between one-fifth and one-sixth of the entire length: head one-fifth of the same; of a somewhat oblong form: snout rather pointed, and advancing considerably beyond the lower jaw: upper lip fleshy, furnished with four barbules; two at the corners of the mouth, and two shorter ones in front of the nose: eyes small; nostrils placed near them: lateral line nearly straight; its course along the middle: head smooth; scales on the body rather small, firmly attached to the skin, finely striated, with their free edges slightly scolloped: dorsal short, commencing at about the middle point between the end of the snout and the base of the caudal; first ray very short, second half the length of the third; this last strong and bony, with sharp serratures at the edges; succeeding rays all soft and branched, and gradually decreasing to the last, which is only half the length of the third: anal also short, commencing in a line with the tip of the dorsal when laid back; third ray longest; all the rays soft, and, except the first two, branched: caudal forked for more than half its length: pectorals shorter than the head: ventrals attached beneath the middle of the dorsal, a little shorter than the pectorals; in the axilla of each a long narrow pointed scale:

B. 3; D. 3/9; A. 8; C. 20; P. 16; V. 9.

(*Colours.*) Back and sides olivaceous brown, with more or less of a golden-yellow lustre; belly white: irides golden-yellow: dorsal brown, tinged with red; anal and ventrals reddish yellow; caudal deep purplish red; pectorals pale brown.

Common in rapid streams and rivers, especially those with a hard gravelly bottom. Lives in society. Conceals itself during the day in hollows and amongst large stones: roves about at night in quest of food. At the approach of Winter retires down the river to deep water. Food aquatic mollusca, worms, and small fish. Spawns in May and June; but, according to Bloch, is not capable of breeding till towards the fourth or fifth year. Grows quickly, and attains a large size: has been known to weigh as much as eighteen pounds.

(3. Gobio, *Cuv.*)

84.   C. *Gobio*, Linn. (*Gudgeon.*)

C. Gobio, *Linn. Syst. Nat.* tom. I. p. 526.   *Bloch, Ichth.* pl. 8.
f. 2.   *Don. Brit. Fish.* vol. III. pl. 71.   *Turt. Brit. Faun.* p. 107.
Gobio fluviatilis, *Flem. Brit. An.* p. 186.   Gudgeon, *Will. Hist.
Pisc.* p. 264. tab. Q. 8. f. 4.   *Penn. Brit. Zool.* vol. III. p. 361.
Id. (Edit. 1812.) vol. III. p. 476.   *Bowd. Brit. fr. wat. Fish.*
Draw. no. 15.   *Yarr. Brit. Fish.* vol. I. p. 325.

LENGTH.   From six to eight inches.
DESCRIPT.   (*Form.*)   Of an elongated form, resembling that of the
Barbel: greatest depth beneath the commencement of the dorsal, equalling one-fifth of the entire length; thickness half the depth: head large,
approaching to conical, a little depressed, with a transverse groove across
the nose, beyond which, at the extremity of the snout, is a small elevation;
its length about equal to the depth of the body: mouth wide: upper jaw
very protractile, projecting beyond the lower when the mouth is closed,
and furnished with a short barbule at each angle: nostrils a little in
advance of the eyes: these last moderately large: head smooth and
naked: scales on the body large, thin, firmly attached to the cuticle,
semicircular, the free portion radiated, and crenated at the margin:
lateral line at first very slightly descending, but afterwards straight,
along the middle of the side: number of scales in the lateral line forty;
in the depth nine; five and a half above, and three and a half below, the
lateral line: dorsal commencing exactly in the middle of the whole
length, caudal excluded; its length half, and its greatest height three-
fourths, of the depth of the body; first and second rays simple, the
others branched; second and third longest: anal similar to the dorsal,
but smaller; commencing nearly in a line with the extremity of that fin
when laid back: caudal forked for about half its length: pectorals about
three-fourths the length of the head; second and third rays longest; all
the rays except the first branched: ventrals in a line with the third
dorsal ray, a little shorter than the pectorals, but of a similar form;
rays similar: vent about midway between the ventrals and the anal:

B. 3; D. 10; A. 9; C. 19, and some short ones; P. 16; V. 8.

(*Colours.*) Back, and upper part of the sides, olivaceous brown, spotted
with black; gill-covers greenish white; lower part of the sides silvery;
belly white: dorsal and caudal spotted; the other fins plain.
Common in rivers and gentle streams, preferring those with a sandy
or gravelly bottom.   Frequents shallows during the warm months, but
retires to deep water at the approach of Winter.   Generally keep in
shoals.   Pennant mentions one taken near Uxbridge which weighed
half a pound: usually much smaller.   Food, worms, mollusca, and
aquatic plants.   Spawns in April or May.

(4. TINCA, *Cuv.*)

85.   C. *Tinca*, Linn. (*Tench.*)

C. Tinca, *Linn. Syst. Nat.* tom. I. p. 526.   *Bloch, Ichth.* pl. 14.
Turt. Brit. Faun. p. 108.   *Don. Brit. Fish.* vol. V. pl. 113.   Tinca
vulgaris, *Flem. Brit. An.* p. 186.   Tench, *Will. Hist. Pisc.* p. 251.
c. vi. tab. Q. 5. f. 1.   *Penn. Brit. Zool.* vol. III. p. 359.   *Id.* (Edit.

1812.) vol. III. p. 474. *Bowd. Brit. fr. wat. Fish.* Draw. no. 13. *Yarr. Brit. Fish.* vol. I. p. 328. Tanche vulgaire, *Cuv. Reg. An.* tom. II. p. 273.

LENGTH.   From twelve to eighteen inches; sometimes more.

DESCRIPT. (*Form.*) Thick and bulky in proportion to its length: back moderately elevated: dorsal line continuous with the profile, falling in one regular curve to the end of the snout: greatest depth a little before the dorsal, contained about three times and a half in the entire length: thickness exceeding half the depth: head about one-fourth of the whole length, excluding caudal: snout rather broad and rounded when viewed from above: eyes small and somewhat sunk in the head, directed downwards: jaws equal: a minute barbule at each corner of the mouth: lateral line descending in a gentle curve from the upper part of the opercle to the middle of the body, then passing off straight to the base of the caudal: scales very small, invested with a slimy mucus: dorsal commencing a little beyond the middle; its greatest height rather more than half the depth of the body; its length a little less than its height; first ray scarcely more than half the length of the second; this last and the next three nearly equal; the succeeding ones slightly decreasing; all except the first two branched: anal similar to the dorsal, but smaller; commencing beyond the termination of that fin: caudal broad, rather thick and fleshy, the end nearly even: pectorals large and rounded, about two-thirds the length of the head; the fifth, sixth, and seventh rays longest: ventrals exactly half-way between the pectorals and the anal; in shape and length similar to the former:

B. 3; D. 11; A. 10; C. 19, &c,; P. 18; V. 10.

(*Colours.*) Head, back, and sides, deep olive-green, tinged with golden-yellow; abdomen sordid yellow: irides orange-red: all the fins deep purplish brown, inclining to dusky.

Inhabits lakes, ponds, and other still waters. Keeps near the bottom, and remains in a tranquil state buried in the mud during the winter months. Usually from four to six pounds in weight, but has been known to exceed eleven. Spawns in June. Very tenacious of life.

(5. ABRAMIS, *Cuv.*)

86.   C. *Brama*, Linn. (*Yellow Bream.*)——Depth one-third of the whole length : number of scales in the lateral line fifty-seven: anal with twenty-eight or twenty-nine rays.

C. Brama, *Linn. Syst. Nat.* tom. I. p. 531.   *Bloch, Ichth.* pl. 13. *Don. Brit. Fish.* vol. IV. pl. 93.   *Turt. Brit. Faun.* p. 108. Abramis Brama, *Flem. Brit. An.* p. 187. Bream, *Will. Hist. Pisc.* p. 248. tab. Q. 10. f. 4.   *Penn. Brit. Zool.* vol. III. p. 362. pl. 70. no. 169.   *Id.* (Edit. 1812.) vol. III. p. 478. pl. 81.   *Bowd. Brit. fr. wat. Fish.* Draw. no. 18.   *Yarr. Brit. Fish.* vol. I. p. 335. Brême commune, *Cuv. Reg. An.* tom. II. p. 274.

LENGTH.   From one to two feet, or upwards.

DESCRIPT. (*Form.*) Body very deep in proportion to its length; the depth increasing suddenly at the shoulder; greatest above the ventrals, where it equals one-third of the entire length: sides much compressed; the greatest thickness contained three times and one-third in the depth: back sharp: dorsal line forming a salient angle at the commencement

of the dorsal fin, thence falling very obliquely to the nape, from which
point the profile falls less obliquely, causing a depression at the nape:
ventral line less convex than the dorsal: head small, about one-fifth of
the entire length: mouth remarkably small in proportion: jaws nearly
equal: distance from the eye to the end of the snout rather greater than
the diameter of the eye; distance between the eyes nearly equal to twice
their diameter: scales smaller than in the next species, of a broad oblong
form, the basal portion with the margin somewhat sinuous, without radii,
the free portion with ten or twelve diverging radii: lateral line sloping
downwards from the upper part of the opercle, and curved throughout;
midway, its course is at two-thirds of the entire depth: number of scales
in the lateral line about fifty-seven; in the depth eighteen, twelve and a
half being above, and five and a half below, the lateral line: the whole of
the dorsal behind the middle, as well as behind the ventrals; first ray
only half the length of the second; both these simple; the rest branched;
last two from one root: anal twice the length of the dorsal; first ray very
small and easily overlooked; second half the length of the third; third
and fourth longest; the succeeding ones decreasing to the twelfth, beyond which they remain even: caudal crescent-shaped; the lower lobe
longer than the upper: pectorals reaching to the ventrals: these last
extending to the vent:

B. 3; D. 12 or 13; A. 28 or 29; C. 19, &c.; P. 17; V. 9.

(*Colours.*) Back dusky, passing into bluish green; sides yellowish white,
with a slight golden lustre; belly almost plain white: irides yellowish
white: all the fins dusky, the pectorals alone faintly tinged with red.

Found in large lakes and slow rivers; generally in shoals. Keeps near
the bottom. Food, worms and aquatic vegetables. Attains to a large
size: weight sometimes exceeding twelve pounds. Spawns in May.

87. C. *Blicca*, Bl.? (*White Bream.*) — Depth three
times and a half in the entire length: number of scales in
the lateral line not exceeding fifty-one: anal with from
twenty-two to twenty-four rays.

C. Blicca, *Bloch, Ichth.* pl. 10.? C. latus, *Gmel. Linn.* tom. I. part iii.
p. 1438.? *Jen. Cat. of Brit. Vert. An.* 26. sp. 86. La Bordelière,
Cuv. Reg. An. tom. II. p. 274.? White Bream, or Bream-Flat,
Yarr. Brit. Fish. vol. I. p. 340.

Length. Rarely exceeding ten or twelve inches.
Descript. (*Form.*) Not so deep as the last species; the back much
less elevated: depth, at the commencement of the dorsal, contained three
times and a half in the entire length: greatest thickness very little more
than three times in the depth: dorsal line falling less obliquely, and
continued in one regular slope to the end of the snout, without any
depression at the nape: eyes relatively larger; the distance from them
to the end of the snout not nearly equal to their diameter; the distance between them not equal to one and a half times their diameter: scales larger;
the number in the lateral line about fifty or fifty-one, scarcely exceeding
this last number: number in the depth fifteen; nine and a half being
above, and five and a half below, the lateral line: anal shorter, with five
or six fewer rays; dorsal and pectorals also with one or two rays less
in number: in all other respects similar to the last.

D. 10 or 11; A. 22 to 24; C. 19, &c.; P. 15; V. 9.

(*Colours.*) Back dusky, tinged with bluish green; sides of a silvery bluish white, with scarcely any of the golden yellow lustre observable in the last species: irides silvery: all the fins dusky, but sometimes very pale; pectorals and ventrals occasionally tinged with reddish.

This species, very distinct from the last, though closely resembling it, agrees in all respects with the *C. Blicca* of Bloch, excepting that I never saw the pectorals and ventrals of so deep a red as represented by that author. It is without doubt the same as the *White Bream* alluded to by Sheppard in the Linnæan Transactions*. It is of very common occurrence in the Cam, and is found in some parts of that river in which the *C. Brama* is not met with. It is known to the fishermen about Ely by the name of *Bream-Flat*. It never attains to the size of the last species, rarely exceeding a pound in weight.

(6. LEUCISCUS, *Klein.*)

* *Dorsal immediately above the ventrals.*

88.   C. *Rutilus*, Linn.   (*Roach.*)—Body deep: jaws equal: dorsal with twelve rays: irides, and all the fins, red.

> C. Rutilus, *Linn. Syst. Nat.* tom. I. p. 529. *Bloch. Ichth.* pl. 2, *Don. Brit. Fish.* vol. III. pl. 67. *Turt. Brit. Faun.* p. 108. Leuciscus Rutilus, *Flem. Brit. An.* p. 188. Roach, *Will. Hist. Pisc.* p. 262. tab. Q. 10. f. 5. *Penn. Brit. Zool.* vol. III. p. 365. *Id.* (Édit. 1812.) vol. III. p. 482. *Bowd. Brit. fr. wat. Fish.* Draw. no. 3. *Yarr. Brit. Fish.* vol. I. p. 348. La Rosse, *Cuv. Reg. An.* tom. II. p. 275.

LENGTH. From twelve to fifteen inches.

DESCRIPT. (*Form.*) Oval; the back much elevated, and sharply ridged: greatest depth at the commencement of the dorsal fin, about one-third of the length, excluding caudal: greatest thickness not twice and a half in the depth: dorsal line very convex, falling gradually to the nape, whence the profile falls less obliquely and in nearly a straight line, causing a slight depression at the part just mentioned: head contained about four times and three-quarters in the whole length, caudal excluded: mouth small: jaws equal: eyes moderate; the distance between equal to twice and a half their diameter: lateral line commencing at the upper part of the opercle, and taking a descending course below the middle, but not quite so low as two-thirds of the depth: head and gill-covers smooth and naked; scales on the body broad, marked with numerous very fine circular concentric striæ, and with a few deeper and more distinct lines radiating anteriorly and posteriorly; number in the lateral line forty-three; above it seven and a half; beneath three and a half: dorsal commencing a very little behind the middle point between the end of the snout and the base of the caudal; its greatest height equalling half the depth; its length nearly the same; first ray only one-third the length of the second, which is longest; third and succeeding rays gradually decreasing; all except the first two branched; the last two from one root: anal commencing a little beyond the termination of the dorsal; of a similar form; second ray longest; all the rays except the first branched; last two from one root: caudal deeply forked: pectorals rather more than three-fourths of the length of the head; first ray

longest; all the rays except the first branched: ventrals in a line with
the commencement of the dorsal, about equal to the pectorals, rounded;
first ray simple; the others branched; in their axilla a triangular pointed
scale:

B. 3; D. 12; A. 13; C. 19, and 4 or 6 short ones; P. 16; V. 9.

(*Colours.*) Upper part of the head and back dusky green, with blue
reflections; sides and belly silvery: cheeks and gill-covers silvery white:
dorsal and caudal dusky, tinged with red; anal, pectorals, and ventrals,
bright red: irides reddish yellow.

Common in lakes and still deep rivers throughout the country. Keeps
in large shoals. Usual weight from a pound to a pound and a half:
sometimes, however, exceeding two, or even three, pounds. Spawns in
May or June, at which season the scales are rough to the touch: is very
prolific. Food, worms and aquatic vegetables.

89. C. *Dobula*, Linn. (*Dobule.*) — Elongated: head broad; snout blunt and rounded: upper jaw longest: anal, pectorals, and ventrals, red.

C. Dobula, *Linn. Syst. Nat.* tom. I. p. 528.   *Bloch, Ichth.* pl. 5.
   Yarrell in Linn. Trans. vol. XVII. p. 9.   Le Meunier, *Cuv. Reg.
   An.* tom. II. p. 275.   Dobule Roach, *Yarr. Brit. Fish.* vol. I.
   p. 346.

LENGTH. That of the specimen described below was six inches and
a half: gets to a larger size.

DESCRIPT. (*Form.*) "Body slender in proportion to its length: the
head, compared with the length of the head and body alone, without
the caudal rays, as two to nine: depth of the body equal to the length
of the head: diameter of the eye compared with the length of the head
as two to seven: nose rather rounded: upper jaw longest: the ascending
line of the nape and back more convex than any other portion of the
dorsal or abdominal line: the first ray of the dorsal fin arising half-way
between the anterior edge of the orbit of the eye, and the edge of the
fleshy portion of the tail; the first ray half as long as the second, which
is the longest, and is as long again as the last ray of this fin, the length
of the last ray being equal to the length of the base of the fin: the pectoral fin rather long and narrow: ventrals arising just in advance of the
line of the origin of the first ray of the dorsal fin; the distance from the
origin of the ventrals to the origin of the anal fin, and from the origin of
the last ray of the anal fin to the end of the fleshy portion of the tail,
equal; the first ray of the anal fin nearly as long again as the last: tail
considerably forked, the external rays being as long again as those in the
centre: scales of the body moderate in size, fifty forming the lateral line,
with an oblique row of seven scales above it under the dorsal fin, and four
below it; the lateral line itself concave to the dorsal line throughout its
whole length.

D. 9; A. 10; C. 19; P. 16; V. 9.

(*Colours.*) Top of the head, nape, and back, dusky blue, becoming
brighter on the sides, and passing into silvery white on the belly: dorsal
and caudal fins dusky brown; pectoral, ventral, and anal fins, pale orange-
red: irides orange: cheeks and opercle silvery white." YARR.

A single individual of this species was obtained by Mr. Yarrell in
August 1831, whilst fishing in the Thames below Woolwich. No other

has hitherto occurred in this country. According to Bloch, it prefers
clear rivers and large lakes, in which it deposits its spawn in the months
of March and April. Food, worms and aquatic mollusca. In general
appearance it somewhat resembles the last species, but is much less deep
for its length, and darker in colour. Said rarely to exceed half a pound
in weight.

90. C. *Leuciscus*, Linn. (*Dace*.)—Elongated; depth
rather more than one-fifth of the length: upper jaw longest:
dorsal with ten rays: anal, pectorals, and ventrals, pale:
irides yellowish.

> C. Leuciscus, *Linn. Syst. Nat.* tom. I. p. 528. *Bloch, Ichth.* pl. 97.
> *Don. Brit. Fish.* vol. IV. pl. 77. *Turt. Brit. Faun.* p. 109.
> Leuciscus vulgaris, *Flem. Brit. An.* p. 187. Dace, *Will. Hist.
> Pisc.* p. 260. tab. Q. 10. f. 3. *Penn. Brit. Zool.* vol. III. p. 366.
> *Id.* (Edit. 1812.) vol. III. p. 483. *Bowd. Brit. fr. wat. Fish.*
> Draw. no. 11. *Yarr. Brit. Fish.* vol. I. p. 353. La Vandoise,
> *Cuv. Reg. An.* tom. II. p. 275.

LENGTH. From eight to ten inches; sometimes more.
DESCRIPT. (*Form.*) More elongated than the *Roach;* the back but
slightly elevated: greatest depth one-fourth of the entire length, excluding caudal; thickness half the depth: dorsal line continuous with
the profile, and deviating but little from a straight line: head small;
one-fifth of the entire length, measured quite to the extremity of the
longest caudal rays: snout rather acute, viewed laterally, but somewhat
rounded when viewed from above; upper jaw projecting beyond the
lower: eyes moderate; distant from the end of the snout a little more
than the length of their diameter; the distance from one to the other
scarcely more than one diameter and a half: lateral line slightly descending; its course, beneath the commencement of the dorsal, at just
two-thirds of the depth: scales smaller than in the *Roach*, with the
radiating striæ posteriorly finer and more numerous; number in the
lateral line fifty-one; above it eight and a half; beneath four and a
half: dorsal commencing a little behind the middle point between the
extremity of the snout and the base of the caudal; second ray longest,
equalling rather more than two-thirds of the depth: anal similar to the
dorsal, commencing in a line with the tip of that fin when folded back:
caudal deeply forked: pectorals and ventrals as in the *Roach*, the latter
a very little in advance of the first ray of the dorsal:

D. 10; A. 11; C. 19, &c.; P. 16; V. 9.

(*Colours.*) Upper part of the head and back dusky, with a bluish cast;
this last tint terminating at about one-third of the depth by a tolerably
well-defined line: sides beneath, and belly, silvery: dorsal and caudal
fins dusky; pectorals, ventrals, and anal, very pale red: irides yellowish.
Common in deep rivers and other clear waters, but not so plentiful as
the Roach. Is gregarious. According to Bloch, spawns in June, but
according to other authors in February and March. Is very prolific, and
multiplies fast. Seldom attains to the weight of a pound, though Pennant mentions one which weighed a pound and a half. Is sometimes
called a *Dare*.

91. C. *Lancastriensis*, Shaw. (*Graining*.)—Elongated;
depth one-fifth of the length: back and upper part of the
sides pale drab: fins yellowish white.

C. Lancastriensis, *Shaw, Gen. Zool.* vol. v. p. 234.   Leuciscus
Lancastriensis, *Yarr. in Linn. Trans.* vol. xvii. p. 7. pl. 2. f. 1.
Graining, *Penn. Brit. Zool.* vol. iii. p. 367. *Id.* (Edit. 1812.)
vol. iii. p. 484.   *Yarr. Brit. Fish.* vol. i. p. 355.

Length.   From seven to nine inches.

Descript.   (*Form.*) "Length of the head, compared with the whole
length of the head, body, and tail, as one to six: depth of the body, compared with the whole length, as one to five: nose more rounded than in
the *Dace;* the upper line of the head straighter: eye rather larger: the
inferior edge of the preopercle less angular: the dorsal line less convex:
dorsal fin commencing exactly half-way between the point of the nose
and the end of the fleshy portion of the tail; the first ray short; the
second longest: pectorals longer in proportion than in the *Dace:* ventrals placed, on a vertical line, but little in advance of the first ray of the
dorsal fin: anal commencing, on a vertical line, under the termination of
the dorsal fin-rays when that fin is depressed; the first ray short; the
second longest; the last double: the fleshy portion of the tail long and
slender; the caudal rays also long and deeply forked: all the fins a little
longer than those of the *Dace:* scales of a moderate size, rather larger
than those of the *Dace,* the diameter across the line of the tube greater,
and the radiating lines less numerous; the number in the series forming
the lateral line forty-eight; those in an oblique line up to the base of the
dorsal fin eight, and downwards to the origin of the ventral fins four:
lateral line descending from the upper edge of the opercle by a gentle
curve to the middle of the body, and thence to the centre of the tail
in a straight line: number of fin-rays,

D. 9; A. 11; C. 19; P. 17; V. 10.

(*Colours.*) Top of the head, back, and upper part of the sides, of a pale
drab-colour, tinged with bluish red, separated from the lighter coloured
inferior parts by a well-defined boundary line: irides yellowish white:
cheeks and gill-covers shining silvery white, tinged with yellow: all the
fins pale yellowish white." Yarr.

Originally observed by Pennant in the Mersey near Warrington.
Mr. Yarrell has since obtained it from the same locality, and pointed
out its claims to rank as a distinct species.   According to this last gentleman, it is met with in considerable abundance in several streams
connected with the above river; but is not known to exist in ponds.
In its habits and food it is said to resemble the Trout.   Weight not
commonly exceeding half a pound.

** *Dorsal above the space intervening between the ventrals
and anal.*

92. C. *Cephalus*, Linn. (*Chub.*) — Elongated: body
thick; snout broad and rounded; upper jaw longest:
dorsal and anal with ten rays: pectorals, ventrals, and
anal, pale red.

C. Cephalus, *Linn. Syst. Nat.* tom. I. p. 527.   C. Jeses, *Don. Brit. Fish.* vol. v. pl. 115.   Leuciscus Cephalus, *Flem. Brit. An.* p. 187. Chub or Chevin, *Will. Hist. Pisc.* p. 255. tab. Q. 4. f. 2.   Chub, *Penn. Brit. Zool.* vol. III. p. 368. pl. 73.   *Id.* (Edit. 1812.) vol. III. p. 485. pl. 84.   *Bowd. Brit. fr. wat. Fish.* Draw. no. 6.   *Yarr. Brit. Fish.* vol. I. p. 358.

LENGTH. From sixteen to eighteen inches.

DESCRIPT. (*Form.*) Oblong-oval, elongated, and subcylindrical: greatest depth contained four times and a half in the entire length ; thickness two-thirds of the depth: dorsal line continuous with the profile and nearly straight ; ventral rather more convex: head one-fifth of the entire length: snout broad and rounded: gape large ; upper jaw projecting beyond the lower: eyes rather small ; the space between them flat, equalling three times their diameter: nostrils large: lateral line descending, following the curve of the ventral line at about two-thirds of the depth: scales large ; the free portion finely striated across, with six or eight diverging radii from the centre ; the basal with finer and more numerous diverging radii, the margin lobed ; number in the lateral line forty-five ; above it seven and a half ; beneath three and a half: dorsal commencing about the middle of the back ; second ray longest, equalling two-thirds of the depth: anal similar to the dorsal, commencing in a line with the tip of that fin when folded down: caudal forked for nearly half its length: pectorals and ventrals much as in the *Roach;* the latter a very little in advance of the first ray of the dorsal, and having a narrow elongated pointed scale in their axilla:

B. 3 ; D. 10 ; A. 10; C. 19, &c. ; P. 19 ; V. 9.

(*Colours.*) Back dusky green ; the sides and belly silvery: lateral scales with the free portion dotted with black: cheeks and gill-covers with gold reflections: irides pale yellow, almost white: dorsal and caudal fins dusky ; pectorals pale ; anal and ventrals tinged with red, with the exception of two or three of the last rays.

Found principally in rivers. Lurks in holes and near the roots of trees. Food, insects, worms, and the young of other fish.   Spawns in April and May.   Attains to a weight of four or five pounds, sometimes more.   *Obs.* The *C. Jeses* of Linnæus and Bloch, and which this last author supposes to be the *Chub* of Pennant, is evidently distinct from this species, and has not hitherto been identified as a native of Britain.

93. C. *Erythrophthalmus*, Linn. (*Rudd, or Red-Eye.*) —Body deep : lower jaw longest: anal with fourteen rays : sides and abdomen gilded ; caudal, ventrals, and anal, bright vermilion.

C. Erythrophthalmus, *Linn. Syst. Nat.* tom. I. p. 530.   *Bloch, Ichth.* pl. 1.   *Don. Brit. Fish.* vol. II. pl. 40.   *Turt. Brit. Faun.* p. 108.   Leuciscus Erythroph. *Flem. Brit. An.* p. 188.   Rudd or Finscale, *Will. Hist. Pisc.* p. 252. tab. Q. 3. f. 1.   Rud, *Penn. Brit. Zool.* vol. III. p. 363. pl. 72.   *Id.* (Edit. 1812.) vol. III. p. 479. pl. 83.   *Bowd. Brit. fr. wat. Fish.* Draw. no. 21.   Red-Eye, *Yarr. Brit. Fish.* vol. I. p. 361.   Le Rotengle, *Cuv. Reg. An.* tom. II. p. 276.

LENGTH. From twelve to fourteen inches.

DESCRIPT. (*Form.*) General appearance resembling that of the *Roach*, but the body deeper and thicker; the back more arched, and forming a slightly salient angle at the commencement of the dorsal fin; ventral line very convex anteriorly, but behind, along the base of the anal, nearly straight; tail suddenly contracting before the caudal: head small: snout short; lower jaw projecting beyond the upper: lateral line bending downwards; its course, beneath the commencement of the dorsal, at about two-thirds of the depth: scales large; number in the lateral line forty-one; above it seven and a half: beneath three and a half: dorsal fin entirely behind the middle, as well as the ventrals; first ray only half the length of the second; all the rays except the first two branched: anal commencing a little beyond the termination of the dorsal: caudal deeply forked: pectorals about the length of the head: ventrals a little shorter, situate exactly half-way between the pectorals and the vent:

$$D. 11; A. 14; C. 19, \&c.; P. 14; V. 9.$$

(*Colours.*) Back olivaceous; sides and belly golden-orange, the metallic lustre very brilliant in the living fish, but fading soon after death: irides orange: dorsal and pectorals dusky, tinged with red; ventrals, anal, and caudal, bright vermilion, the two former pale at the base.

Found in rivers and other deep waters, not uncommonly. Recorded by Willughby as inhabiting the lakes of Yorkshire and Lincolnshire, and the river Cherwell in Oxfordshire. Is also met with in the Thames, Stour, and Cam; very abundantly in some parts of the river last mentioned, where it is called a *Shallow*. Feeds on worms, mollusca, and vegetable substances. Spawns in April or May. Weight from one to one and a half, rarely two pounds.

94. C. *cœruleus*, Nob. (*Azurine.*)—Depth moderate: lower jaw longest: anal with fourteen rays: sides and abdomen silvery; all the fins plain white.

Leuciscus cœruleus, *Yarr, in Linn. Trans.* vol. XVII. p. 8. pl. 2. f. 2. Azurine, *Id. Brit. Fish.* vol. I. p. 365.

DESCRIPT. (*Form.*) General form resembling that of the last species: greatest depth rather more than one-fourth of the entire length; head one-fifth; thickness not half the depth: back arched; the dorsal line descending in one regular curve to the end of the snout, without any depression at the nape; ventral line much less convex: snout blunt; mouth small; lower jaw a little the longest: eyes rather large: nostrils midway between them and the upper lip: opercle marked with radiating lines: lateral line descending; its lowest point at rather more than two-thirds of the depth: scales large, oval, marked anteriorly and posteriorly with a variable number of radiating striæ; number in the lateral line about forty-one; above it seven and a half; beneath three and a half: dorsal entirely behind the middle, commencing half-way between the posterior edge of the orbit of the eye and the base of the caudal fin; its form and rays as in the last species: anal commencing nearly in a line with the last ray of the dorsal; first ray extremely short and easily overlooked; second half the length of the third; third and fourth equal and longest; all except the first three branched: caudal forked for half its length: pectorals and ventrals as in the *Red-Eye*, the latter

altogether in advance of the dorsal, which last is directly over the intervening space between them and the anal:

D. 12; A. 14; C. 20; P. 16; V. 9.

(*Colours.*) Upper part of the head, back, and sides, slate-blue, passing
into silvery white beneath, and both shining with metallic lustre: irides
white, tinged with pale straw-yellow: all the fins plain white; the dorsal
and caudal inclining to dusky.

A new species described by Mr. Yarrell from specimens received along
with the Graining from Knowsley in Lancashire. Not much at present
known of its habits. Said to be hardy, tenacious of life, and to spawn
in May. Weight of the largest individual hitherto obtained about a
pound.

95. C. *Alburnus*, Linn. (*Bleak*.)—Elongated: lower jaw longest, ascending: anal with about nineteen rays: bright silvery; fins pellucid white.

C. Alburnus, *Linn. Syst. Nat.* tom. I. p. 531.   *Bloch, Ichth.* pl. 8.
f. 4.   *Don. Brit. Fish.* vol. I. pl. 18.   *Turt. Brit. Faun.* p. 109.
Leuciscus Alburnus, *Flem. Brit. An.* p. 188.   Bleak, *Will. Hist.
Pisc.* p. 263. tab. Q. 10. f. 7.   *Penn. Brit. Zool.* vol. III. p. 370.
pl. 73.   *Id.* (Edit. 1812.) vol. III. p. 487. pl. 84.   *Bowd. Brit. fr.
wat. Fish.* Draw. no. 4.   *Yarr. Brit. Fish.* vol. I. p. 368.   L'Ablette,
Cuv. Reg. An. tom. II. p. 276.

LENGTH. From six to seven, rarely eight, inches.

DESCRIPT. (*Form.*) General form resembling that of the *Dace*, but
more elongated: greatest depth exactly one-fifth of the entire length;
greatest thickness about half the depth: ventral line more convex than
the dorsal, rising rather abruptly posteriorly along the base of the anal
fin: head contained five times and a half in the entire length: forehead
flat: eyes large; their diameter very nearly one-third the length of the
head: snout short; lower jaw projecting, ascending to meet the upper:
lateral line descending in a sweep from the upper angle of the
opercle till it reaches the middle of its course, thence passing off nearly
straight to the caudal; above the ventrals its course is at just two-thirds
of the depth: scales of moderate size, thin, finely striated, easily detached; number in the lateral line about forty-eight; above it seven and
a half; beneath three and a half: dorsal entirely behind the middle; its
greatest height about two-thirds of the depth; its length scarcely more
than half its height; first ray only half the length of the second; second
and third rays longest; the succeeding ones decreasing; the first two
simple, the rest branched: anal commencing in a line with, or rather
in advance of, the last ray of the dorsal; longer than in any of the preceding species of this sub-genus, and occupying half the space between
the vent and the origin of the caudal fin; the first two rays very short;
third and fourth longest, about equalling the longest rays in the dorsal;
the first three simple, the others branched: caudal forked for half its
length; the lower lobe of the fin a very little longer than the upper:
pectorals shorter than the head, not reaching to the ventrals when laid
back: ventrals shorter than the pectorals, considerably before the dorsal,
and not reaching to the vent.

D. 10 or 11; A. 19 to 21; C. 19, &c.; P. 16; V. 9.

(*Colours.*) Back olivaceous green; sides and belly bright silvery; the two colours separated by a well-defined line: cheeks and gill-covers silvery white: all the fins pale; the anal and ventrals nearly pure white: irides pale yellow.

Common in rivers, swimming in large shoals near the surface. Spawns in May and June. Food, principally insects.

96. C. *Phoxinus*, Linn. (*Minnow.*) — Body slender, rounded: jaws equal: scales very minute: anal with ten or eleven rays: fins pale.

C. Phoxinus, *Linn. Syst. Nat.* tom. i. p. 528. *Bloch, Ichth.* pl. 8. f. 5. *Don. Brit. Fish.* vol. iii. pl. 60. *Turt. Brit. Faun.* p. 109. Leuciscus Phoxinus, *Flem. Brit. An.* p. 188. Pink or Minim, *Will. Hist. Pisc.* p. 268. tab. Q. 8. f. 7. Minnow, *Penn. Brit. Zool.* vol. iii. p. 373. *Id.* (Edit. 1812.) vol. iii. p. 489. *Bowd. Brit. fr. wat. Fish.* Draw. no. 8. *Yarr. Brit. Fish.* vol. i. p. 372. Le Véron, *Cuv. Reg. An.* tom. ii. p. 276.

LENGTH. From three to four inches.

DESCRIPT. (*Form.*) Body elongated and rounded, tapering posteriorly: dorsal line but slightly curved; ventral more convex: greatest depth a little before the ventrals, equalling about one-fifth of the entire length: head rather less than one-fifth: thickness exceeding half the depth: snout short; jaws equal: eyes small: nostrils wide, approximating: lateral line very slightly descending; its course a little below the middle: scales very minute: dorsal entirely behind the middle, as well as the ventrals; first ray half the length of the second; second and third longest; the first two and the last in the fin simple; the rest branched: anal commencing in a vertical line with the last ray of the dorsal; first ray very short and easily overlooked; second not half the length of the third; third and fourth longest; the first three and the last ray of all simple; the others branched: caudal forked for nearly half its length: pectorals about three-fourths the length of the head: ventrals a little shorter:

D. 10; A. 10 or 11; C. 19, &c.; P. 16; V. 8.

(*Colours.*) Back, and upper half of the sides, deep olive-brown, sometimes spotted with black; lateral line often of a golden hue; lower portion of the sides and belly yellowish white, but (in the *males ?*) during the spawning season of a rich crimson: dorsal and caudal fins pale brown; generally a large dusky spot at the base of the caudal; anal, pectorals, and ventrals, lighter.

Common in rivers, more especially those with a gravelly bottom. Keeps in shoals. Spawns the end of May, or beginning of June, at which season the head is covered with small tubercles. Food, worms, insects, and aquatic plants.

(16.) C. *Idus*, Linn. Syst. Nat. tom. i. p. 529.   Bloch, Ichth. pl. 36. *The Ide*, Yarr. Brit. Fish. vol. i. p. 344.

Said to have been found by the late Dr. Walker in the mouth of the Nith*. Its claims, however, to a place in the British Fauna do not appear to have been confirmed by any subsequent observer.

---

* *Stew. El. of Nat. Hist.* vol. i. p. 382.

(17.) *C. Orfus*, Linn. Syst. Nat. tom. i. p. 530. Bloch, Ichth. pl. 96.

This species having been confounded by Willughby and Ray with the Rud, (*C. Erythrophthalmus*), Linnæus was led, apparently on their authority, to consider it as inhabiting the English rivers. This error has been reproduced in the "British Animals," (p. 186), of Dr. Fleming, who attaches the name of *Barbus Orfus* to the *Rud* or *Finscale* of Willughby, a species undoubtedly the same as the *Red-Eye* of Donovan. It is almost certain that the true *C. Orfus*, which is a native of Germany, has no claim whatever to a place in the British Fauna.

GEN. 36.   COBITIS, *Linn.*

97. C. *barbatula*, Linn. (*Bearded Loach.*) — Body rounded anteriorly, compressed behind : sides of the head unarmed.

C. barbatula, *Linn. Syst. Nat.* tom. i. p. 499. *Bloch, Ichth.* pl. 31. f. 3. *Don. Brit. Fish.* vol. i. pl. 22. *Turt. Brit. Faun.* p. 103. *Flem. Brit. An.* p. 189. Loche, *Will. Hist. Pisc.* p. 265. tab. Q. 8. f. 1. *Penn. Brit. Zool.* vol. iii. p. 282. pl. 58. *Id.* (Edit. 1812.) vol. iii. p. 379. pl. 69. *Bowd. Brit. fr. wat. Fish.* Draw. no. 12. *Yarr. Brit. Fish.* vol. i. p. 376. La Loche franche, *Cuv. Reg. An.* tom. ii. p. 278.

LENGTH.   From four to four and a half, rarely five, inches.

DESCRIPT. (*Form.*) Elongated; subcylindric anteriorly, compressed towards the tail: greatest depth one-seventh of the entire length; thickness, in the region of the pectorals, about two-thirds of the depth: head small, a little depressed, the profile gently sloping: snout blunt; upper jaw projecting over the lower; mouth small, placed beneath, furnished with six short barbules, two at the corners, and four in front of the upper lip; those at the corners longest, equalling rather more than one-third the length of the head: eyes small; the intervening space flat: body covered with very small scales, and invested with a mucous secretion: lateral line straight: dorsal commencing midway between the end of the snout and base of the caudal; its height nearly equalling the depth of the body; first ray very short and easily overlooked; second not half the length of the third; fourth and fifth longest; the first three simple; the rest branched : anal commencing beyond the tip of the dorsal when laid back, somewhat smaller than that fin, but in other respects similar; first ray very small; fourth and fifth longest: caudal slightly rounded: pectorals attached low down, about the length of the head, rounded; second and third rays longest, and, as well as the fourth, much stouter than the others: ventrals in a line with the commencement of the dorsal, somewhat shorter than the pectorals; third ray longest:

B. 3 ; D. 10 ; A. 9 ; C. 17 ; P. 12 ; V. 8 :

vent in a line with the tip of the dorsal when laid back. (*Colours.*) Back and sides yellowish brown, mottled and spotted with dusky; abdomen and lateral line whitish : dorsal, caudal, and pectoral fins, spotted ; anal and ventrals nearly plain.

Not uncommon in rivers and streams with a gravelly bottom. Feeds on aquatic insects. Spawns in March and April.

98. C. *Tænia*, Linn. *(Groundling.)* — Body much
compressed throughout : beneath each eye a forked spine.

C. Tænia, *Linn. Syst. Nat.* tom. I. p. 499.   *Bloch, Ichth.* pl. 31.
  f. 2.   *Berken. Syn.* vol. I. p. 79.   *Turt. Brit. Faun.* p. 103.
  Flem. Brit. An. p. 189.   C. barbatula aculeata, *Will. Hist. Pisc.*
  p. 265. tab. Q. 8. f. 3.   Spinous Loche, *Penn. Brit. Zool.* (Edit.
  1812.) vol. III. p. 381.   Spined Loche, *Yarr. Brit. Fish.* vol. I.
  p. 381.   La Loche de rivière, *Cuv. Reg. An.* tom. II. p. 278.

LENGTH. From three to four inches.
DESCRIPT. *(Form.)* Much more compressed than the last species,
especially about the head, which is also smaller: thickness of the body
only half the depth: profile more convex, the snout appearing somewhat ·
truncated: barbules shorter and less conspicuous: eyes smaller, placed
very high on the cheeks; the intervening space contracted into a narrow
elevated ridge: beneath each eye, but a little in advance, a sharp moveable forked spine directed backwards: dorsal and anal fins similar, and
similarly situated, but the former with a ray more: pectorals relatively
shorter and less developed, not equal to the length of the head; the
second, third, and fourth rays not stouter than the others: ventrals likewise smaller:

D. 11; A. 9; C. 15; P. 8; V. 7.

(Colours.) Yellowish, tinged with orange; the back and upper half of
the sides spotted and mottled with brown; more particularly a longitudinal series of large round spots on the lateral line, a second on the
dorsal ridge, and a third intermediate between these two; those on the
lower part of the back, between the dorsal and caudal fins, sometimes
assume the appearance of short transverse bars: dorsal and caudal spotted; the other fins plain.
Much less frequent than the last species. Found in the Trent in
Nottinghamshire, and, according to Turton, in the clear streams of
Wiltshire. I have also met with it in some plenty in the Cam, as
well as in fish-ponds at Ely. Keeps near the bottom, and appears to
reside more in the mud than the *C. barbatula.* Spawns, according to
Bloch, in April and May. Is very tenacious of life.

GEN. 37. ESOX, Cuv.

99. E. *Lucius*, Linn. *(Pike.)*

E. Lucius, *Linn. Syst. Nat.* tom. I. p. 516.   *Bloch, Ichth.* pl. 32.
  Don. Brit. Fish. vol. v. pl. 109.   *Turt. Brit. Faun.* p. 105.
  Flem. Brit. An. p. 184.   *Cuv. Reg. An.* tom. II. p. 282.   Pike,
  Will. Hist. Pisc. p. 236. tab. P. 5. f. 2.   *Penn. Brit. Zool.*
  vol. III. p. 320. pl. 63.   *Id.* (Edit. 1812.) vol. III. p. 424. pl. 74.
  Bowd. Brit. fr. wat. Fish. Draw. no. 17.   *Yarr. Brit. Fish.*
  vol. I. p. 383.

LENGTH. From two to three feet; sometimes more.
DESCRIPT. *(Form.)* Oblong, rather elongated, suddenly narrowing
behind the dorsal and anal fins; sides compressed: depth nearly uniform throughout, about one-sixth of the entire length: head large, rather
more than one-fourth: cranium flat, a little concave between the eyes;

D D

snout broad and depressed, rounded at the extremity; lower jaw projecting beyond the upper: intermaxillaries, vomer, palatines, tongue,
pharyngeans, and branchial arches, armed with sharp card-like teeth
of unequal lengths; also a series of long sharp teeth on the sides of
the lower jaw: eyes moderate, situate half-way between the end of the
snout and posterior edge of the opercle: nostrils a little in advance: above
and below each orbit, beneath the lower jaw on each side, and along the
margin of the preopercle, a row of pores: gill-opening very large: cheeks
and upper part of the opercle covered with small scales; scales on the
body moderate, oblong-oval, with the basal margin three-lobed: lateral
line at first slightly descending, but afterwards straight: dorsal placed
very far back, commencing at about two-thirds of the entire length; first
six rays simple, gradually increasing in length, the first being very short;
seventh longest; this and all the succeeding ones branched: anal similar
to the dorsal, and answering to it: caudal forked: pectorals attached low
down, not half the length of the head, rounded; fourth ray longest:
ventrals equal to the pectorals, placed at about the middle of the entire
length; third ray longest:

> B. 14 or 15 *; D. 21; A. 18 or 19; C. 19, &c.; P. 15; V. 11.

(*Colours.*) Head, back, and sides, bright olive-green spotted with yellow,
or, when out of season, greenish gray with pale spots; more or less of
a metallic gloss; belly white: fins dusky, spotted and variegated with
red.

Probably indigenous, though usually supposed to have been introduced into England during the reign of Henry VIII. Found in rivers,
lakes, and most stagnant waters. Very voracious, preying on other fish,
including its own species, as well as water-rats and young water-fowl.
Is very long-lived. Grows rapidly, and occasionally attains a weight of
thirty, forty, or even sixty pounds. Spawns in March and April.

GEN. 38.   BELONE, *Nob.*

(1. BELONE, *Cuv.*)

100.   B. *vulgaris*, Flem. (*Common Gar-Fish.*)

> B. vulgaris, *Flem. Brit. An.* p. 184.   Esox Belone, *Linn. Syst.
> Nat.* tom. I. p. 517.   *Bloch, Ichth.* pl. 33.   *Don. Brit. Fish.*
> vol. III. pl. 64.   *Turt. Brit. Faun.* p. 105.   Horn-Fish or Gar-
> Fish, *Will. Hist. Pisc.* p. 231. tab. P. 2. f. 4.   Gar-Pike, *Penn.
> Brit. Zool.* vol. III. p. 324. pl. 63.   *Id.* (Edit. 1812.) vol. III. p. 429.
> pl. 74.   Gar-Fish, *Yarr. Brit. Fish.* vol. I. p. 391.

LENGTH.   From eighteen inches to two feet; rarely more.

DESCRIPT.   (*Form.*) Subcylindrical, slender, and very much elongated: depth nearly uniform till past the commencement of the dorsal
and anal fins, contained seventeen times and a half in the entire length:
thickness rather more than two-thirds of the depth: abdomen flat,
bounded on each side by a longitudinal series of large scales, forming
a sort of lateral keel which runs the whole length of the body: head,
snout included, contained a little more than three times and a half in the
entire length; cranium flat and horizontal; cheeks vertical; snout produced into a long, slender, sharp-pointed beak; the lower jaw considerably the longest: both jaws armed at their edges with a single row of

---

* One individual was found to have fourteen on one side and fifteen on the other.

fine sharp card-like teeth; none on any other part of the mouth: eyes large, placed high, a little behind the corners of the mouth: nostrils wide, immediately in advance of them: lateral line nearly straight, not very distinct: head and opercle without scales; those on the body, with the exception of the longitudinal row on each side of the abdomen, thinly scattered and not very conspicuous: dorsal very far behind, commencing at three-fourths of the entire length; first ray only half the length of the second, which is longest; third and three following ones decreasing; beyond the sixth, the rays remain low and nearly even to the termination of the fin; all except the first branched: anal similar to the dorsal, and answering to it: caudal forked: pectorals small, in length scarcely exceeding the depth of the body, attached about half-way down, a little behind the gill-opening; second ray longest: ventrals still smaller; their point of attachment exactly half-way between the posterior part of the opercle and the end of the fleshy portion of the tail:

D. 18; A. 21; C. 19, &c.; P. 13; V. 7.

(*Colours.*) Head, back, and upper part of the sides, fine rich bluish green: gill-covers, and all below the lateral line of the body, bright silvery.

Common on many parts of the coast, appearing in shoals about April, and remaining till late in Autumn. At the approach of Winter, retires to deep water. From its usually preceding the Mackerel, is sometimes called the *Mackerel-Guide.* Said to deposit its spawn close to the shore, among rocks and sea-weed. The bones are well known for acquiring a green colour when boiled.

(18.) *Little Gar* (*Esox Brasiliensis*), Couch in Linn. Trans. vol. XIV. p. 85.

> A doubtful species, taken by Mr. Couch " in the harbour at Polperro, in July 1818, as it was swimming with agility near the surface of the water. About an inch in length: head somewhat flattened at the top; the upper jaw short and pointed; the inferior much protruded, being at least as long as from the extremity of the upper jaw to the back part of the gill-covers: the mouth opened obliquely downwards; but that part of the under jaw which protruded beyond the extremity of the upper, passed straight forward in a right line with the top of the head: body compressed, lengthened, and resembling that of the Gar-Pike (*E. Belone*): one dorsal and one anal placed far behind, and opposite to each other: tail straight. Colour of the back bluish green, with a few spots; the belly silvery." COUCH.
>
> Mr. Couch conceived that this species might be the *Esox Brasiliensis* of Linnæus. It seems, however, more likely to have been the young of some species of *Hemiramphus,* Cuv.

(2. SCOMBERESOX, Lac.)

101. B. *Saurus*, Nob. (*Saury, or Skipper.*)

Esox Saurus, *Rackett in Linn. Trans.* vol. VII. p. 60. tab. 5. *Turt. Brit. Faun.* p. 105. *Neill in Wern. Mem.* vol. I. p. 541. *Don. Brit. Fish.* vol. v. pl. 116. Lacertus vel Saurus, *Will. Hist. Pisc.* p. 232. Scomberesox Saurus, *Flem. Brit. An.* p. 184. Skipper (Cornubiensium), *Ray, Syn. Pisc.* p. 165. Saury, *Penn. Brit. Zool.* vol. III. p. 325. pl. 64. *Id.* (Edit. 1812.) vol. III. p. 424. pl. 75. Saury Pike, *Yarr. Brit. Fish.* vol. I. p. 394.

LENGTH. From fifteen to eighteen inches. NEILL.

DESCRIPT. (*Form.*) " Body long and slender, agreeing precisely with that of the common *Gar-Fish:* snout subulate, fine, toothless, and slightly curving upwards: jaws of unequal length, the lower longest, and bending upwards at the tip: body smooth; the scales with which it is covered being thin and glabrous: the lower part of the body from the gills to the tail marked with a longitudinal carina or keel, which terminates at the latter part in a somewhat protuberant manner: all the fins small: the dorsal placed far down the back, and containing eleven rays: between this and the tail five distinct pinnules or spurious fins: pectorals somewhat falcated, containing eleven rays: ventrals with six rays: anal opposite to the dorsal, of eleven rays: between this and the tail seven distinct pinnules: caudal of twenty-two rays.

D. 11, and V false; A. 11, and VII false; C. 22; P. 11; V. 6.

(*Colours.*) Back of a most lovely azure blue, changing to green, and glossed with purple and yellow; the lower parts silvery." Don.

Rare on the southern coast, but, according to Mr. Neill, not uncommon in the North of Scotland, entering the Frith of Forth almost every Autumn in considerable shoals. Mr. Rackett's specimen was taken near the Isle of Portland in Dorsetshire. This species derives its English name of *Skipper* from its habit of leaping out of the water, and passing over a considerable space (Mr. Couch says thirty or forty feet) before returning to that element. It is not noticed either by Linnæus, Gmelin, or Bloch.

GEN. 39.　EXOCŒTUS, *Linn.*

102.　E. *volitans*, Linn.? (*Flying-Fish.*) — Ventrals small, placed before the middle.

E. volitans, *Linn. Syst. Nat.* tom. I. p. 520. *Don. Brit. Fish.* vol. II. pl. 31. *Turt. Brit. Faun.* p. 106. E. evolans, *Bloch, Ichth.* pl. 398. Winged Flying-Fish, *Penn. Brit. Zool.* vol. III. p. 333. pl. 67. *Id.* (Edit. 1812.) vol. III. p. 441. pl. 78. Flying-Fish, *Yarr. Brit. Fish.* vol. I. p. 398.

According to Pennant, a fish of this genus was caught in June 1765, at a small distance below Caermarthen, in the river Towy. A second individual is said to have occurred in July 1823, in the Bristol channel, ten miles from Bridgewater*. Others are recorded to have been seen off Portland Island, in August 1825, by a vessel going down channel†. Although referred, by Pennant, in the first instance, to the *E. volitans* of Linnæus, in none of these cases does the species appear to have been determined with certainty. Pennant seems to suppose his to be the one so common in the Mediterranean: but, according to Cuvier, this last is the *E. exiliens* of Bloch ‡, which is distinguished from the *E. volitans* by its much longer ventrals, placed beyond the middle of the body. For this reason I have not annexed any detailed description.

* *Ann. of Phil.* vol. XXII. p. 152.
† *Lond. Quart. Journ. of Sci.* vol. XX. p. 412.
‡ *Ichth.* pl. 397.

(2.) SILURUS, Arted.

(19.) <i>S. Glanis</i>, Linn. Syst. Nat. tom. i. p. 501.   Bloch, Ichth.
pl. 34.   Cuv. Reg. An. tom. ii. p. 291.   <i>Sly Silurus</i>, Yarr. Brit.
Fish. vol. i. p. 403.

> This species, or at least a fish bearing the same name, is included by
> Sibbald in his list of Scottish fishes*, but it is not said on what authority.
> Possibly some other species (perhaps the <i>Burbot</i>) may be intended.   At
> any rate it is not likely that it should exist in that country at the present
> day, as from its great size, it would in that case hardly have been for so
> long a time overlooked by our naturalists.   It is the largest fresh-water
> fish found in Europe, attaining the weight of from one to two hundred
> pounds and upwards.   It is met with in the rivers of Germany, Hungary,
> and other parts of the Continent.

GEN. 40.   SALMO, Cuv.

103.   S. <i>Salar</i>, Linn. (<i>Common Salmon.</i>) — Posterior
margin of the gill-cover forming a semicircle: vomerine
teeth confined to the anterior extremity: caudal forked:
ventrals dusky on their inner surface.

S. Salar, <i>Linn. Syst. Nat.</i> tom. i. p. 509.   <i>Bloch, Ichth.</i> pls. 20,
& 98.   <i>Turt. Brit. Faun.</i> p. 103.   <i>Flem. Brit. An.</i> p. 179.   <i>Jard.
in Edinb. New Phil. Journ.</i> vol. xviii. p. 46.   Salmo, <i>Will. Hist.
Pisc.</i> p. 189. tab. N. 2. f. 1.   Salmon, <i>Penn. Brit. Zool.</i> vol. iii.
p. 284. pl. 58. no. 143.   <i>Id.</i> (Edit. 1812.) vol. iii. p. 382. pl. 69.
Le Saumon, <i>Cuv. Reg. An.</i> tom. ii. p. 302.

Length.   From two to three feet, sometimes three feet and a half.
Descript.   (<i>Form.</i>) Oval; moderately elongated; with the head and
back in nearly the same line: greatest depth a little before the dorsal,
contained about five times and a half in the entire length, increasing,
however, with age: thickness half the depth: head small, about one-
sixth of the entire length: snout rather sharp: jaws, in <i>young fish</i>,
nearly equal; but in <i>old males</i>, the lower one longest, and curving
upwards in a hook: a row of sharp teeth along both sides of each jaw,
as well as on the palatines; but those on the vomer confined to its anterior extremity, and in some specimens rather obsolete; two rows of teeth
on the tongue: eyes directly above the posterior extremity of the maxillary, and nearer the end of the snout than the furthest point of the gill-
cover by one-third: gill-cover with the posterior margin more curved
than in the next species, and forming a semicircle; opercle oblong, the
basal margin slightly ascending posteriorly; subopercle about one-third
the size of the opercle; a line drawn from the extremity of the upper jaw
to the furthest point of the gill-cover passes through the eyes: lateral
line perfectly straight, dividing the body into two nearly equal parts:
scales small: dorsal occupying a middle position between the end of the
snout and the end of the fleshy part of the tail; rather longer than high,
its greatest elevation not equalling half the depth of the body; first ray
very short; fourth longest; first two rays simple, the rest branched; last
two from one root: adipose small; much nearer the caudal than the anal:

* Scot. Illust. part ii. vol. ii. p. 25.

anal similar to the dorsal, more in advance than the adipose, terminating in a line with this last fin; third ray longest; first two simple; the rest branched; the last two from one root: tail, between the adipose and the caudal fin, more slender than in the next species; the end of the fleshy portion cut square, appearing truncated: caudal very much forked when young, gradually becoming less so as age advances, but never (except perhaps in *very old* fish) quite even: pectorals more than half the length of the head; their inferior margin rather concave; second and third rays longest: ventrals beneath the middle of the dorsal; rather shorter than the pectorals; the axillary scale half their own length: number of rays,

B. 12; D. 14; A. 12; C. 19, and some short ones; P. 15; V. 9.

Number of vertebræ sixty. (*Colours.*) Bluish gray or lead-colour; abdomen silvery; here and there, principally above the lateral line, a few dusky spots: dorsal, caudal, and pectorals, dark gray; ventrals deeply stained, especially on their inner surface, with the same colour; anal less so, nearly white. *In the fry, till about five or six inches long,* the sides shew more or less indication of dark transverse bands. *The adult male, during and after the spawning season,* acquires a reddish tinge.

Found both in the sea and in rivers. Principal fisheries carried on in Scotland, the North of England, and Ireland, where the species is very abundant. Begins to ascend rivers in April; at the approach of Autumn, pushes up towards their sources in order to spawn, springing up cataracts, and surmounting any other obstacles which oppose its progress. Spawning season principally from October to February, but varying much in different rivers. Male and female pair for the occasion, and excavate a furrow in the gravelly or sandy beds of shallows, in which the spawn and milt are deposited simultaneously. After spawning, both sexes return to the sea in a very reduced state; the males going down sooner than the females: at this season, the former are called *Kippers,* the latter *Kelts.* Young fry, termed *Smolts* or *Samlets,* appear about March, and keep going down to sea from the end of that month to the middle of May: after remaining in the sea some weeks, they return to the rivers, having attained to the weight of from a pound and a half to four or five pounds: fish of the former weight, and up to two pounds, termed *Peal;* of the latter weight *Grilse,* which last name they retain till they have spawned once, when they are called *Salmon.* From the time of their first return to the rivers, they increase rapidly in size. Greatest weight which the species attains to forty or fifty pounds, sometimes more: Pennant mentions one which weighed seventy-four pounds. Food at sea, according to Fleming, principally the Sand-eel. *Obs.* According to M. Agassiz*, the *S. hamatus* of Cuvier is only an old fish of this species: the *S. Gœdenii* of Bloch (*Ichth.* pl. 102.†) the young.

104.   S. *Eriox,* Linn. (*Bull Trout, or Grey.*)—Posterior margin of the gill-cover very little curved: vomerine teeth confined to the anterior extremity: caudal even: back and sides spotted with purplish gray; ventrals plain white.

S. Eriox, *Linn. Syst. Nat.* tom. i. p. 509.   *Turt. Brit. Faun.* p. 103.   *Flem. Brit. An.* p. 180.   S. Cambricus, *Don. Brit. Fish.*

---

* *Edinb. New Phil. Journ.* vol. xvii. p. 385.
† Mr. Yarrell thinks that the fish figured by Bloch on his 102nd plate is only S. *Fario.*

vol. IV. pl. 91. The Grey, *Will. Hist. Pisc.* p. 193. *Penn. Brit.*
Zool. vol. III. p. 295. *Id.* (Edit. 1812.) vol. III. p. 394.? Sea
Trout, *Penn. Brit. Zool.* vol. III. p. 296. *Id.* (Edit. 1812.) vol. III.
p. 397.

LENGTH. Two feet eight inches.

DESCRIPT. (*Form.*) Closely resembling the *Salmon*, but of a more
clumsy make: head and nape somewhat thicker: curvature of the posterior margin of the gill-cover much less considerable; margin of the
preopercle more sinuous; subopercle larger with respect to the opercle,
the basal margins of both nearly parallel to the axis of the body; a line
drawn from the extremity of the upper jaw to the furthest point of the
gill-cover passes beneath the eyes: vomer with only two or three teeth
at its anterior extremity: tail, beyond the adipose, more bulky and muscular than in the *Salmon:* caudal even; in *old fish* rather convex: the
other fins similar: number of fin rays,

D. 12 to 14; A. 11; C. 19, &c.; P. 14; V. 10.

Number of vertebræ fifty-nine. The *female* is characterized by having
shorter jaws than the *male*, with the teeth less developed. (*Colours.*)
For the most part similar to those of the *Salmon*, but the back and sides,
above the lateral line, more spotted; the spots being most abundant in
the *female*. In the spawning season, the *male* acquires a red tinge: the
female remains gray.

Migratory like the Salmon. A common species in the Tweed, where
the young are called *Whitlings.* Found also in the rivers of Wales,
Dorsetshire, and Cornwall. Apparently the same as the *Sewen* of Donovan, (*S. Cambricus*,) said by that author to be found in such great plenty
on the coasts of Glamorganshire and Caermarthenshire; this last (which,
according to Donovan, rarely exceeds twelve or fifteen inches in length,
and from one to two pounds in weight) only a younger fish. It is also
probable that the *S. Hucho* of Fleming, and other British authors, is not
distinct from the present species. Flesh inferior to that of the Salmon;
cutting yellow.

105. S. *Trutta*, Linn. (*Sea Trout.*) — Gill-cover
slightly produced behind; the margin rounded: vomerine
teeth extending the whole way: caudal forked: back and
sides with X-shaped dusky spots: ventrals plain white.

S. Trutta, *Linn. Syst. Nat.* tom. I. p. 509. *Bloch, Ichth.* pl. 21.
Turt. Brit. Faun. p. 103. *Flem. Brit. An.* p. 180. *Jard. in*
Edinb. New Phil. Journ. vol. XVIII. p. 49. Trutta lacustris,
Will. Hist. Pisc. p. 198. tab. N. 1. fig. 5. Truite saumonée,
Cuv. Reg. An. tom. II. p. 304.

LENGTH. From one to two feet, or rather more.

DESCRIPT. (*Form.*) Not so slender as a *Salmon* of the same size:
jaws nearly equal: teeth rather larger; those on the vomer extending
all along the ridge of the palate, and forming by pressure a groove in the
tongue between the two rows of lingual teeth: eyes rather nearer the
extremity of the snout: gill-cover more produced behind than in either
of the preceding species, the margin more curved; basal margins of the
opercle and subopercle sloping obliquely upwards to form a considerable
angle with the axis of the body: position of the fins much as in the
Salmon; but the adipose rather larger; the caudal, with the outer rays

shorter, and not so much forked; in *very old fish* nearly even: the fleshy portion of the tail rounded at its extremity: pectorals with their inferior margin straight: number of fin-rays,

D. 13; A. 11; C. 19, &c.; P. 14; V. 10.

Number of vertebræ fifty-eight.   (*Colours.*) Darker in the body, and lighter in the fins, than the *Salmon:* back and sides, above the lateral line, more thickly spotted; the spots assuming the form of the letter X; those above sometimes surrounded by a pale circle; gill-covers and cheeks spotted, as well as the dorsal and adipose fins: ventrals always plain white.

A common species, inhabiting the sea and rivers.   Enters these last about the end of May or beginning of June.   Is the *Salmon-Trout* of the London markets.   Flesh red, and highly esteemed.   Food, according to Sir W. Jardine, principally the *Talitrus Locusta*, or common Sandhopper.   *Obs.* According to Agassiz, the *S. Lemanus* of Cuvier is the same as the present species.   The *Sea-Trout* of Pennant appears to be identical with the species last described, which is called by the above name in some rivers.

(20.) *S. albus*, Flem. Brit. An. p. 180.   Jard. in Edinb. New Phil. Journ. vol. XVIII. p. 50.   *White*, Penn. Brit. Zool. vol. III. p. 302.   Id. (Edit. 1812.) vol. III. p. 396.   *Herling*, Jard. in Proceed. of Berwicksh. Nat. Club, p. 50.

> This is held to be a distinct species by Sir W. Jardine, and some other of our naturalists.   I must confess, however, that I have been unable to discern any appreciable difference between it and the last, of which, in the opinion of Mr. Yarrell and myself, it is only the young of the first year. Found in the Solway, the Tweed, the Esk, and a few other rivers in the North.   Is sometimes called a *Whiting* or *Phinoc*.   Pennant says it never exceeds a foot in length.   According to Sir W. Jardine, the fish in the Solway average from a pound to a pound and a half in weight, very seldom reaching two pounds.   It is added by this last gentleman, that "one of the most marked appearances of this fish, is the great proportional breadth of the back, and the peculiar grayish green colour of the upper parts."

106. S. *Fario*, Linn. (*Common Trout*.) — Gill-cover produced behind into a rounded angle: vomerine teeth extending the whole way: maxillaries reaching to a vertical line from the posterior part of the orbit: caudal slightly forked: back and sides with numerous red spots.

S. Fario, *Linn. Syst. Nat.* tom. I. p. 509.   *Bloch, Ichth.* pls. 22, & 23.   *Turt. Brit. Faun.* p. 103.   *Don. Brit. Fish.* vol. IV. pl. 85. *Flem. Brit. An.* p. 181.   *Jard. in Edinb. New Phil. Journ.* vol. XVIII. p. 51.   Trutta fluviatilis, *Will. Hist. Pisc.* p. 199. Trout, *Penn. Brit. Zool.* vol. III. p. 297. pl. 59. no. 146.   River Trout, *Id.* (Edit. 1812.) vol. III. p. 399. pl. 70.   *Bowd. Brit. fr. wat. Fish.* Draw. no. 14.   Truite commune, *Cuv. Reg. An.* tom. II. p. 304.

LENGTH.   From one to two feet; sometimes more.

DESCRIPT.   (*Form.*) General proportions resembling those of the *S. Trutta:* differs from that species in the form of the gill-cover, which is much more produced behind, forming at its distal extremity a rounded angle; basal margins of the opercle and subopercle rising more obliquely:

snout short; but the jaws, which are nearly equal, becoming more lengthened in the spawning season: maxillary reaching to a vertical line forming a tangent to the posterior part of the orbit, by which character it
is distinguished from the *S. Salmulus*: teeth on the whole length of the
vomer: dorsal and adipose fins placed as in the *Salmon*; the former with
the first three rays, the first especially, very small and inconspicuous, but
gradually increasing in length; sixth and seventh longest; first four
simple, the rest branched: anal entirely in advance of a vertical line
from the adipose: caudal not so much forked as in the *S. Trutta*, or
so square as in *S. Eriox*: number of fin-rays,

$$\text{D. 14; A. 11; C. 19, \&c.; P. 13; V. 9.}$$

(*Colours.*) Back dusky: sides and belly, the former more especially,
yellow, tinged with gold, and also with green: a row of red spots along
the lateral line: dorsal fin, and above the lateral line, spotted with dusky.
In *young fish*, more or less indication of transverse dusky bands on the
sides.

Var. β. Gillaroo Trout, *Sow. Brit. Misc.* pl. 61.

A common species in lakes and rivers, attaining in some localities to a
large size. Has been known to weigh from sixteen to twenty pounds,
though usually much smaller. In many places seldom exceeds a pound
or a pound and a half. Spawns in September and October; ascending
to the sources of rivers for this purpose. Is very voracious. Feeds on
worms, small fish, and insects, especially *Ephemeræ* and *Phryganeæ*.
The variety, called the *Gillaroo Trout*, is distinguished by its strong
muscular stomach, resembling the gizzard of birds, resulting from feeding
principally on shells. It is found in some of the lakes in Ireland.

Obs. The above species exhibits very great variation in colours*, and
in some measure in form also, according to the locality in which it is
found. Possibly two or more species may have been hitherto confounded,
but in the present state of the science it is almost impossible to decide
this point. Sir W. Jardine, who has paid great attention to the whole
family, and from whom we may expect much light upon the subject, has
particularized some remarkable varieties found in Sutherlandshire, in
the "Edinburgh New Philosophical Journal," l. c., to which I refer the
reader. According to Agassiz, the *S. punctatus* of Cuvier, the *S. marmoratus* of the same author, and the *S. alpinus* of Bloch, all belong to
this species.

107. S. *ferox*, Jard. and Selb. (*Great Lake Trout.*)

Great Lake Trout (S. ferox), *Jard. in Encycl. Brit.* (7th Edit.)
Art. Angling, p. 142. *Id. in Edinb. New Phil. Journ.* vol. xviii.
p. 55.

Descript. (*Form.*) "Principally distinguished by its large size,
square tail in all its stages of growth, the form of the gill-covers and
teeth, the relative position of the fins, the form of the scales, particularly those composing the lateral line, and in the generally delicate skin
which is spread over the outside of the body being extremely strong and
tough, and from under which the perfectly transparent scales can be

---

* Sir H. Davy was of opinion, that when Trout feed much on hard substances, such as larvæ
and their cases, and the ova of other fish, they have more red spots, and redder fins : and that
when they feed most on small fish, as minnows, and on flies, they have more tendency to become
spotted with small black spots, and are generally more silvery. See *Salmonia*, (2nd Edit.) p. 41.

extracted. The fins may be stated nearly thus, though a greater variation may occur;

D. 13 to 15 ; A. 12; P. 14 ; V. 11 ; gill-covers, 12:

the greatest variation occurs in the dorsal fin." JARD. (*Colours.*)
" Deep purplish brown on the upper parts, changing into reddish gray,
and thence into fine orange-yellow, on the breast and belly: the whole
body, when the fish is newly caught, appearing as if glazed over with
a thin tint of rich lake-colour, which fades rapidly away as the fish dies:
gill-covers marked with large dark spots: the whole body covered with
markings of different sizes, and varying in amount in different individuals; the markings, in some, few, scattered, and of a large size; in others,
thickly set, and of smaller dimensions: each spot surrounded by a paler
ring, which sometimes assumes a reddish hue: the spots more distant
from each other as they descend beneath the lateral line: lower parts
of the fish spotless." JARD.

A new species first identified as distinct from the Common Trout by
Sir W. Jardine and Mr. Selby. The former of these gentlemen states
that it is generally distributed in all the larger and deeper lochs of Scotland, but that it seldom ascends or descends the rivers running into or
out of them, and never migrates to the sea. Very voracious, feeding
nearly entirely upon small fish. Average weight from ten to twenty
pounds: has been known, however, to reach twenty-eight pounds.
Spawns in Autumn. *Obs.* It is probably the same as the *S. lacustris*
of Berkenhout*, though (in the opinion of M. Agassiz) not of continental authors.

108. S. *Salmulus*, Turt. (*Samlet*.) —Vomerine teeth extending the whole way: maxillaries reaching to beneath the centre of the orbit: caudal forked for half its length: sides marked with long, narrow, transverse, bluish bands.

S. Salmulus, *Turt. Brit. Faun.* p. 104. *Jard. in Edinb. New Phil.
Journ.* vol. XVIII. p. 56. Salmulus, *Will. Hist. Pisc.* p. 192.
tab. N. 2. fig. 2. *Ray, Syn. Pisc.* p. 63. Samlet and Parr, *Penn.
Brit. Zool.* vol. III. p. 303. pl. 59. no. 148. & pl. 66. no. 78. *Id.*
(Edit. 1812.) vol. III. p. 404. pls. 70, & 77.

LENGTH. From six to eight inches.
DESCRIPT. (*Form.*) Closely resembling the *young Trout*, but differing
in the following particulars. Body somewhat deeper in proportion to its
length: snout blunter: teeth weaker and less developed; those on the
tongue not very conspicuous: maxillary shorter, not reaching beyond a
vertical line from the centre of the orbit; also broader at its posterior
extremity: gill-cover not so much produced into an angle, the hinder
margin being more regularly rounded, as in the *Salmon:* "scales, taken
from the lateral line below the dorsal fin, altogether larger, the length
greater by nearly one-third, the furrowing more delicate, and the form of
the canal not so apparent or so strongly marked towards the basal end of
the scale †:" caudal more deeply forked, the fork extending about half
its length: pectorals larger. (*Colours.*) "The row of blue marks on
the sides, which are also found in the *young Trout*, and in the *young*

---

* *Syn.* vol. I. p. 79.
† This character, which I have not had an opportunity of verifying myself, is taken from
Sir W. Jardine.

of several of the *Salmonidæ*, in this species are narrower and more lengthened. The general spotting seldom extends below the lateral line, and two dark spots on the gill-cover are a very constant mark." JARD. According to Pennant, "the adipose fin is never tipped with red; nor is the edge of the anal white."

This fish, which is common in many of the rivers of Wales and Scotland, as well as in some of those in England, has been regarded by different observers as the young, either of the Salmon, the Sea Trout, or the common Trout. It is, however, now pretty well ascertained to be a distinct species, always remaining of a small size. Is called in some places a *Parr*, in others a *Skirling* or *Brandling*. Said, by Sir W. Jardine, "to frequent the clearest streams, delighting in the shallower fords or heads having a fine gravelly bottom, and hanging there in shoals, in constant activity, apparently day and night." According to Dr. Heysham*, the adult fish go down to the sea after spawning, which takes place, as in the other migratory species of this genus, in the depth of Winter.

109. S. *Umbla*, Linn. (*Charr.*)—Vomerine teeth confined to the anterior extremity : dorsal midway between the end of the snout and the base of the caudal : anal commencing beyond the tip of the reclined dorsal : axillary scale nearly half the length of the ventrals.

S. Umbla, *Linn. Syst. Nat.* tom. I. p. 511. *Bloch, Ichth.* pl. 101. S. alpinus, *Don. Brit. Fish.* vol. III. pl. 61. *Turt. Brit. Faun.* p. 104. *Flem. Brit. An.* p. 180. Charr, *Penn. Brit. Zool.* vol. III. p. 305. pl. 60. *Id.* (Edit. 1812.) vol. III. p. 407. pl. 71.

LENGTH. From twelve to fourteen inches.

DESCRIPT. (*Form.*) Elongated; the line of the back nearly straight; profile sloping gently downwards from the nape: greatest depth about one-fifth of the entire length : head contained five times and a half in the same: snout short and somewhat obtuse: jaws nearly equal, except in the spawning season, when the lower one becomes longest: teeth small and sharp; those on the vomer confined to the anterior extremity: eyes moderate; their diameter rather less than one-fourth the length of the head; the distance between them equalling twice their diameter: gill-cover produced behind into a rounded lobe; the basal margin sloping very obliquely upwards: lateral line arising at the upper angle of the opercle, at first slightly descending, but afterwards nearly straight, its course being a little above the middle : scales small : dorsal a little before the middle of the entire length; the distance from the first ray to the end of the snout, when measured behind, not reaching beyond the base of the caudal; of a somewhat triangular form, the posterior rays being not more than half the length of the anterior ones; fifth ray longest, equalling a little more than half the depth of the body: adipose so placed, that two-thirds of the distance between the dorsal and caudal lie before it, one-third behind it : anal commencing considerably beyond a vertical line from the tip of the reclined dorsal: pectorals just three-fourths the length of the head : ventrals beneath the middle of the dorsal; in their axillæ a long narrow pointed scale, nearly half their own length:

B. 10 or 11; D. 14; A. 13; C. 19, and some short ones; P. 11; V. 9.

* *Catalogue of the Animals of Cumberland*, p. 31.

(*Colours.*) Back, and upper part of the sides, bluish gray, tinged with olivaceous; flanks and belly flesh-colour: above the lateral line spotted with white; beneath the same, spots more obscure: dorsal, anal, and caudal, dusky, the latter darkest; pectorals and ventrals dark red. *In the spawning season*, the flanks and abdomen are bright crimson-red; the whole of the sides, above and below the lateral line, spotted with deeper red; the anal, pectorals, and ventrals, are also deep red, the first rays of the anal and ventrals excepted, which are bluish white.

Found in the lakes of Cumberland and Westmoreland, especially in Winander Mere, in the latter county; also in Crummock and Coniston Waters in Lancashire, and, according to Sir W. Jardine, in many of the northern lochs of Scotland. Frequents clear and deep waters, keeping near the bottom. Feeds on insects. Varies much in its colours at different seasons, a circumstance which has obtained for it several different names. In its ordinary state, it is the *Case Charr* of Pennant and other authors: when exhibiting the bright crimson belly which it assumes before spawning, it is called *Red Charr*: when out of season, the spawn having been shed, it is distinguished by the name of *Gilt Charr*. *Obs.* According to Agassiz, the *S. Umbla*, the *S. Salvelinus*, the *S. alpinus*, and the *S. Salmarinus*, of Linnæus, are all referable to this species in its different states.

110. S. *Salvelinus*, Don. (*Torgoch*.)—Vomerine teeth confined to the anterior extremity: dorsal exactly in the middle of the entire length: anal commencing in a line with the tip of the reclined dorsal: axillary scale not one-third the length of the ventrals.

S. Salvelinus, *Don. Brit. Fish.* vol. v. pl. 112.

LENGTH. Six inches.

DESCRIPT. (*Form.*) Differs from the last species as follows: not so much elongated in proportion to its depth: head larger; contained not more than four times and a half in the entire length: teeth, those on the tongue especially, stronger and more developed: eyes larger; their diameter rather more than one-fourth the length of the head; the distance between them not equalling above one diameter and a half: posterior lobe of the opercle not so much produced: all the fins relatively larger: dorsal exactly in the middle of the entire length; the distance in front, when measured behind the fin, reaching to the end of the caudal; fifth and sixth rays longest, equalling at least three-fourths of the depth of the body: posterior portion of the fin very little less elevated than the anterior: adipose nearer the middle point between the dorsal and the caudal: anal commencing exactly in a line with the tip of the reclined dorsal: pectorals longer: scale in the axillæ of the ventrals much shorter, not one-third the length of the fin:

B. 9; D. 14 or 15; A. 13; C. 19, &c.; P. 13; V. 8.

(*Colours.*) Probably as variable as in the last species. The following were those of the specimens examined: head, back, and upper part of the sides, dark olivaceous-green; lower part of the sides yellowish, passing into bright orange-red on the abdomen: above the lateral line spotted with yellowish white; yellow of the sides, beneath the lateral line, spotted with red: dorsal, caudal, and pectorals, dark olivaceous; first and last rays of the pectorals, and the whole of the anal and ventrals, bright red.

Whether this species be found on the Continent, or be the same as any of those described by foreign authors, it is not easy to determine, owing to the great confusion which prevails in this genus. I have, however, little hesitation in considering it as the *S. Salvelinus* of Donovan, though not of Turton and Fleming, who appear to have confounded it with the *Red Charr* of Pennant, which is only a variety of the last species. The same may be said of Willughby, who has comprised them both under the title of *Umbla minor*, Gesn.* That it is distinct from the *S. Umbla* of this work (the *S. alpinus* of most English authors) no one, I conceive, can doubt, who has had an opportunity of comparing the two. My examination was made from specimens in the possession of Mr. Yarrell, who obtained them from Corsygiddel Lake near Barmouth. According to Donovan, it is found in the Waters of Llyn Quellyn, one of the alpine lakes on the west side of Snowdon; he adds, that formerly it was also met with in the Llanberris Lake, on the opposite side of the mountain, but that of late years it has disappeared in the locality last mentioned. The species appears to be confined to Wales, in which country it is said to be called *Torgoch*, a Welch term signifying *Red-belly*.

GEN. 41.   OSMERUS, *Art.*

111.   O. *Eperlanus*, Flem. (*Smelt.*)

O. Eperlanus, *Flem. Brit. An.* p. 181. Eperlanus, *Will. Hist. Pisc.* p. 202. tab. N. 6. f. 4. Salmo Eperlanus, *Linn. Syst. Nat.* tom. i. p. 511. *Bloch, Ichth.* pl. 28. figs. 1, 2. *Don. Brit. Fish.* vol. ii. pl. 48. *Turt. Brit. Faun.* p. 104. Smelt, *Penn. Brit. Zool.* vol. iii. p. 313. pl. 61. no. 151. *Id.* (Ed. 1812.) vol. iii. p. 416. pl. 72.

LENGTH. From eight to ten, rarely twelve, inches. Pennant mentions one which measured thirteen inches.

DESCRIPT. (*Form.*) Elongated; the back straight, and in the same line with the profile: greatest depth one-seventh of the entire length; thickness rather more than half the depth: head small, one-fifth of the entire length, somewhat conical: lower jaw longest, curving upwards when the mouth is closed: gape wide, extending to beneath the eyes: maxillary teeth sharp, but very fine; those in the lower jaw curved, and much longer; two rows of teeth on each palatine; also some very strong long curved teeth on the tongue and front of the vomer: eyes large: gill-cover produced posteriorly into an obtuse lobe: lateral line at first slightly descending, but afterwards straight: scales small, deciduous: dorsal commencing exactly half-way between the extremity of the upper jaw and the end of the fleshy part of the tail; its height nearly twice its length, and about equal to, or rather less than, the depth of the body; third ray longest; first two rays simple, the others branched; the last two from one root: adipose small, a little nearer the caudal than the dorsal: anal commencing a little beyond the tip of the reclined dorsal, much longer than that fin, and extending beyond a vertical line from the adipose; first ray very short; fourth longest; first three simple, the rest branched: the last two from one root: caudal deeply forked: pectorals

* *Hist. Pisc.* p. 196.

attached low down, and just below the produced lobe of the gill-cover: ventrals beneath the commencement of the dorsal: number of fin-rays,

D. 11; A. 17; C. 19; P. 11; V. 8.

(*Colours*.) Back whitish, tinged with green; upper part of the sides varied with blue; lower part of the sides, and belly, bright silvery: irides silvery; pupil black: fins pale.

A common species on the British coasts, ascending rivers in December, January, and February, for the purpose of spawning, which takes place in March and April. Food, according to Bloch, worms, and small shells. Varies greatly in size; a circumstance which has induced the author just mentioned to form two species of it. Derives its English name of *Smelt* from a peculiar scent which it emits, and which has been compared by some to cucumbers, by others to violets. Is sometimes called a *Sparling**.

GEN. 42. THYMALLUS, *Cuv*.

112. T. *vulgaris*, Nilss. (*Grayling*.)

T. vulgaris, *Nilss. Prod. Ichth. Scand.* p. 13. Thymallus, *Will. Hist. Pisc.* p. 187. tab. N. 8. Salmo Thymallus, *Linn. Syst. Nat.* tom. I. p. 512. *Bloch, Ichth.* pl. 24. *Don. Brit. Fish.* vol. IV. pl. 88. *Turt. Brit. Faun.* p. 104. Coregonus Thymallus, *Flem. Brit. An.* p. 181. Grayling, *Penn. Brit. Zool.* vol. III. p. 311. pl. 61. no. 150. *Id.* (Edit. 1812.) vol. III. p. 414. pl. 72. Ombre commune, *Cuv. Reg. An.* tom. II. p. 306.

LENGTH. From ten to fifteen, rarely eighteen, inches.

DESCRIPT. (*Form*.) Back slightly elevated at the commencement of the dorsal fin, from which point it falls gradually to the snout: greatest depth one-fifth of the entire length; thickness not quite half the depth: head contained five times and a half in the entire length: snout rather short; obtuse, and rounded: gape small: upper jaw a little the longest: maxillary, and all the other teeth, small and fine: lateral line at first slightly descending, afterwards straight: scales large, disposed in longitudinal rows; seven and a half above the lateral line, the same number below it: dorsal commencing at one-third, and occupying about one-fourth, of the entire length; being twice as long as high; its greatest elevation three-fourths of the depth of the body; anterior rays gradually increasing from the first, which is very short, to the eighth and ninth, which are longest; tenth and succeeding rays slightly decreasing; first eight simple, the rest branched: adipose situate at nearly two-thirds of the distance from the dorsal to the base of the caudal: anal commencing a little beyond the tip of the reclined dorsal; shaped like that fin, but much smaller; first five rays simple, the rest branched: caudal deeply forked: pectorals three-fourths the length of the head: ventrals about the same; attached beneath the middle of the dorsal; with a long narrow scale in their axillæ:

B. 10; D. 22; A. 15, the last double; C. 19, and some short ones; P. 15; V. 11.

(*Colours*.) Upper part of the head dusky; back and sides silvery gray, marked with longitudinal dusky streaks: dorsal spotted; the spots arranged in longitudinal lines: other fins plain.

---

* I may state in this place that the *Mallotus villosus*, or *Capelin*, was inserted by error in my Catalogue as a doubtful inhabitant of the British seas. There is no recorded authority for such insertion.

An inhabitant of streams and rivers, in which it remains stationary all the year, though asserted by Donovan to be migratory*. Partial to clear and rapid waters. Found in Derbyshire, in some of the rivers in the North, and in a few other parts, of England. Food, insects, testaceous mollusca, small fish, &c. Spawns in April and May. Has been known to attain the weight of five pounds†, but is usually found much smaller.

GEN. 43. COREGONUS, *Cuv.*

113. C. *Lavaretus*, Flem.? (*Gwiniad.*)—Jaws equal; snout scarcely advancing beyond them.

C. Lavaretus, *Flem. Brit. An.* p. 182.? *Nilss. Prod. Ichth. Scand.* p. 15.? Salmo Wartmanni, *Bloch, Ichth.* pl. 105.? Gwiniad, *Will. Hist. Pisc.* p. 183. *Penn. Brit. Zool.* vol. iii. p. 316. pl. 62. *Id.* (Ed. 1812.) vol. iii. p. 419. pl. 73. Le Lavaret, *Cuv. Reg. An.* tom. ii. p. 307.?

LENGTH. From ten to twelve inches.

DESCRIPT. (*Form.*) Extremely similar in form to the *Common Herring.* Back slightly arched: greatest depth about one-fifth of the entire length: head triangular, also about one-fifth: snout moderate, scarcely advancing beyond the jaws; these last equal, and without teeth; a few very fine velvet-like teeth on the tongue: eyes round, and large; their diameter contained three times and a half in the length of the head; the distance between them about equal to their diameter: gill-opening very large: opercle of a somewhat triangular form, the basal margin ascending very obliquely; subopercle approaching to oblong, rounded beneath: lateral line straight, dividing the sides into two nearly equal parts: scales large; of an oval or roundish form, marked with close concentric circles, but without radiating lines: dorsal occupying about the middle of the entire length; the distance from the end of the snout to the first ray, when measured behind the fin, reaching a little beyond the end of the fleshy part of the tail; anterior part of the fin elevated, the fourth ray, which is longest, equalling three-fourths of the depth of the body; fifth and succeeding rays rather rapidly decreasing; length of the fin about two-thirds of its greatest height; first three rays simple, the rest branched: space between the dorsal and adipose three times that between this last and the caudal: anal commencing considerably beyond the tip of the reclined dorsal, terminating in a line with the adipose; similar to the dorsal in form, but longer and less elevated: caudal deeply forked: pectorals inserted low down, a little shorter than the head: ventrals attached beneath the middle of the dorsal; axillary scale nearly one-third their own length:

B. 10; D. 13; A. 16; C. 19, and some short ones; P. 17; V. 11.

(*Colours.*) "Head dusky; pupil deep blue: gill-covers silvery, powdered with black: back, as far as the lateral line, glossed with deep blue and purple, but towards the line assuming a silvery cast, tinged with gold, beneath which those colours entirely prevail: lateral line marked by a series of distinct dusky spots: ventrals, in some, of a fine sky-blue, in others, as if powdered with blue specks; the ends of the other lower fins tinged with the same colour." PENN.

---

* According to Sir H. Davy, "the Grayling will not bear even a brackish water, without dying." *Salmonia*, (2d Edit.) p. 207.
† Daniel's *Rural Sports*, vol. ii. p. 280.

This species is found in Bala Lake, Merionethshire, as well as in the North of England and Scotland.   I do not feel certain that it is identical with the *C. Lavaretus* of continental authors (synonymous with the *Salmo Wartmanni* of Bloch), there being several other allied species, the characters of which have not as yet been determined with precision. The above description is from specimens in the collection of W. Yarrell, Esq.   By Turton and some other English authors, it appears to have been confounded with the *Salmo Lavaretus* of Bloch (*S. Oxyrhinchus*, Linn.), a very distinct species, in which the snout is furnished with a soft conical projection at its extremity extending beyond the jaws, and which is not, that I am aware, a native of this country.   According to Pennant, the Gwiniad is a gregarious fish, and spawns in December.

114.   C. *Pollan*, Thomps.   (*Pollan.*)

C. Pollan, *Thomps. in Proceed. of Zool. Soc.* June 9, 1835.

LENGTH.   From ten to twelve inches. THOMPS.

DESCRIPT.   (*Form.*)   " Differs from the *Gwiniad* in the snout not being produced; in the scales of the lateral line; in having fewer rays in the anal fin, and in its position being rather more distant from the tail; in the dorsal, anal, and caudal fins, being of less dimensions; and in the third ray of the pectoral fin being longest; (the first being of the greatest length in the *Gwiniad*.)   Relative length of the head to that of the body as one to about three and a half: depth of the body equal to the length of the head: jaws equal; both occasionally furnished with a few delicate teeth; the tongue with many teeth : lateral line sloping downwards for a short way from the opercle, and thence passing straight to the tail: nine rows of scales from the dorsal fin to the lateral line, and the same number thence to the ventral fin; the row of scales on the back and that of the lateral line not reckoned: the third ray of the pectoral fin longest:

B. 9; D. 14; A. 13; C. 19; P. 16; V. 12.

Number of vertebræ fifty-nine.   (*Colours.*)   Colour to the lateral line dark blue; thence to the belly silvery : dorsal, anal, and caudal fins, towards the extremity tinged with black; pectoral and ventral fins of crystalline transparency, excepting at their extremities, which are faintly dotted with black: irides silvery; pupil black." THOMPS.

The above description is that of a species of *Coregonus*, lately brought under the notice of the Zoological Society by Mr. W. Thompson of Belfast, who considers it distinct from those hitherto published by authors.   It is found in Lough Neagh in Ireland, in which district it is said to be known by the name of *Pollan*.   Not having given it myself a close examination, I forbear offering any opinion about it.   Judging, however, from the description, it certainly appears different from the last species, with which it was probably confounded by Fleming, who gives Lough Neagh as a locality for the *C. Lavaretus*.

115.   C. *Marænula*, Jard.   (*Vendace.*) — Lower jaw longest, obliquely ascending.

C. Marænula, *Jard. in Edinb. Journ. of Nat. and Geog. Sci.* vol. III. p. 4. pl. 1.   Salmo Marænula, *Bloch, Ichth.* pl. 28. f. 3.?   *Gmel. Linn.* tom. I. part iii. p. 1381.?   S. albula, *Stew. El. of Nat. Hist.* vol. I. p. 373.   La Vemme, *Cuv. Reg. An.* tom. II. p. 307.?

LENGTH. From four to ten inches. JARD.

DESCRIPT. (*Form.*) Differs essentially from the *C. Lavaretus* in having the lower jaw longest, and ascending at an angle of forty-five degrees to meet the upper, which receives it as in a groove: general outline similar: greatest depth exactly one-fourth of the entire length, caudal excluded: head small; "the crown heart-shaped, and so transparent that the form of the skull and brain may be seen through the integuments*:" maxillaries and lower jaw without teeth: tongue, which is small and triangular, and placed far back, rough to the touch, with a few, almost invisible, velvet-like teeth: eyes large and brilliant; their diameter contained three times and a half in the length of the head; the intervening space scarcely equal to their diameter: gill-opening very large: lateral line straight: "scales of considerable size, oval, and nearly smooth on the outer surface:" dorsal commencing at the middle of the entire length; very much elevated and pointed anteriorly, its greatest height being nearly twice its length; first ray very short; fourth longest; fifth and succeeding rays rapidly decreasing; the last not half the length of the fourth; first three simple, the rest branched: space between the dorsal and adipose more than double that between the adipose and caudal: anal commencing a little beyond the tip of the reclined dorsal, and terminating in a line with the adipose; first ray very minute; fourth and fifth longest; first four simple, the rest branched: caudal very much forked: pectorals attached low down: ventrals opposed to the anterior half of the dorsal; the axillary scale scarcely more than one-fourth of their length:

B. 9; D. 12; A. 14; C. 19, &c.; P. 15; V. 11.

"Number of vertebræ fifty to fifty-two." (*Colours.*) "Upper parts of a delicate greenish brown, shading gradually into a clear silver lustre: irides and cheeks silvery: dorsal fin greenish brown; the lower fins all bluish white." JARD.

First distinguished as a British species by Sir W. Jardine. By previous authors in this country it appears to have been confounded with the *C. Lavaretus*. The only locality known for it "is the lochs in the neighbourhood of Lochmaben, in Dumfries-shire;" into which (according to tradition) it was introduced by Mary Queen of Scots. "General habits resembling those of the Gwiniad. Swims in large shoals, retiring to the depths of the lakes in warm and clear weather. Spawns about the commencement of November."

(3.) *SCOPELUS*, Cuv.

(21.) *S. Humboldti*, Cuv. Reg. An. tom. II. p. 315. *S. borealis*, Nilss. Prod. Ichth. Scand. p. 20. *Sheppy Argentine*, Penn. Brit. Zool. vol. III. p. 327. pl. 65. no. 156. Id. (Edit. 1812.) vol. III. p. 432. pl. 76.

> Cuvier considers the *Sheppy Argentine* of Pennant, an obscure species of which little is known, to be the same as the *Serpes Humboldti* of Risso, this last being the type of his genus *Scopelus*. The following is Pennant's description of his fish, which he obtained from the sea near Downing, in 1769. "Length two inches and one-fourth. Eyes large; irides silvery: lower jaw sloped much: teeth small: body compressed, and of an equal depth almost to the anal fin: tail forked. Back of a dusky green: the sides and covers of the gills as if plated with silver. Lateral line in the middle and quite straight. On each side of the belly a row of circular punctures: above them another, ceasing near the vent."
>
> Whether the *Argentine* of Low† be the same as Pennant's fish, can scarcely, from his imperfect description, be determined.

GEN. 44.   CLUPEA, *Linn.*

(1. Clupea, *Linn.*)

116.   C. *Harengus*, Linn.   (*Common Herring.*) —
Minute teeth in both jaws: infra-orbitals and gill-covers
veined: subopercle rounded at bottom: dorsal behind
the centre of gravity: ventrals beneath the middle of the
dorsal.

> C. Harengus, *Linn. Syst. Nat.* tom. i. p. 522. *Bloch, Ichth.* pl. 29.
> f. 1.   *Turt. Brit. Faun.* p. 106.   *Flem. Brit. An.* p. 182.
> Harengus, *Will. Hist. Pisc.* p. 219. tab. P. 1. f. 2.   British
> Herring, *Penn. Brit. Zool.* vol. iii. p. 335. pl. 68. no. 160. Com-
> mon Herring, *Id.* (Edit. 1812.) vol. iii. p. 444. pl. 79.   Hareng
> commun, *Cuv. Reg. An.* tom. ii. p. 317.

Length.   Ten to twelve inches; sometimes more.
Descript.   (*Form.*) Oval; rather elongated: dorsal and ventral lines
equally convex: greatest depth one-fifth of the entire length, excluding
caudal: thickness half the depth: sides compressed: belly sharply carinated, but without any sensible serratures: head triangular, very much
compressed; one-fifth of the entire length, this last being measured to
the base of the caudal fork: lower jaw longer than the upper, with a few
minute teeth confined to its extremity; upper jaw with the lower half
of the maxillaries finely serrated: a few minute teeth on the tongue, as
well as on the vomer: eyes large; their diameter contained about four
times and a half in the length of the head: infra-orbitals, preopercle, and
upper part of the opercle, marked with fine vein-like striæ: subopercle
rounded beneath: gill-opening extremely large: lateral line not very
distinct; its course nearly straight, and rather above the middle: scales
large, very deciduous: dorsal fin behind the centre of gravity, commencing exactly half-way between the end of the snout and base of
the caudal rays; rays rapidly increasing from the first, which is very
short, to the fifth, which is longest; then gradually decreasing; the
first four simple, the succeeding ones branched: anal commencing beyond the tip of the dorsal, this last being laid back; of about the same
length as that fin, but not so high; fourth and some of the succeeding
rays longest: caudal deeply forked: pectorals rather narrow, more than
half the length of the head: ventrals attached beneath the middle of
the dorsal, a vertical line from the first dorsal ray falling considerably
in advance of them:

> B. 8; D. 19; A. 17; C. 19, and 5 or 6 short ones; P. 17; V. 9.

Number of vertebræ fifty-six.   (*Colours.*) Back and upper portion of
the sides deep sky-blue, tinged with sea-green: belly and flanks bright
silvery: irides, cheeks, and gill-covers, tinged with gold.
A common and well-known species visiting our coasts in large shoals
towards the end of Summer. Deposits its roe in October and November,
after which it retires again into deep water. Food, according to Pennant,
small crustacea; sometimes the fry of its own species.

117.   C. *Leachii*, Yarr.   (*Leach's Herring.*)

> C. Leachii, *Yarr. in Proceed. of Zool. Soc.* (1831.) p. 34.   *Id. in
> Zool. Journ.* vol. v. p. 278. pl. 12.

Descript. " Much deeper in proportion than the *Common Herring*, an adult fish eight inches long, being one inch and seven-eighths deep, while a *Common Herring* of the same depth measures ten inches and a half in length: dorsal and abdominal lines much more convex; the latter keeled, but without serration: under jaw with three or four prominent teeth placed just within the angle formed by the symphysis; the upper maxillæ with their edges slightly crenated: eye large: scales smaller than in the other species: no distinct lateral line. Back and sides deep blue, with green reflections, passing into silvery white beneath. Dorsal fin behind the centre of gravity; but not so far behind it as in the *Common Herring:* number of fin-rays,

D. 18; A. 16; C. 20; P. 17; V. 9.

Number of vertebræ fifty-four." Yarr.

A new species, obtained by Mr. Yarrell in 1831, from fishermen engaged in taking Sprats at the mouths of the Thames and Medway. Found heavy with roe on the 31st of January: probably does not spawn till the middle of February. Flesh said to be much milder than that of the Common Herring.

Obs. From the statements made by Mr. Yarrell in the " Zoological Journal*", it seems probable that there may be yet another species of Herring, larger than either of those described above, occasionally met with in the British seas. Pennant also speaks of one, seen by Mr. Travis, which measured twenty-one inches and a half in length.

118. C. *Sprattus*, Bloch. (*Sprat.*)—Teeth in the lower jaw obsolete: infra-orbitals and gill-covers not veined: dorsal further back than in the *Herring;* the ventrals beneath its anterior margin : keel of the abdomen serrated : anal with eighteen rays.

C. Sprattus, *Bloch, Ichth.* pl. 29. f. 2. *Turt. Brit. Faun.* p. 107. Sprattus, *Will. Hist. Pisc.* p. 221. Sprat, *Penn. Brit. Zool.* vol. iii. p. 346. *Id.* (Edit. 1812.) vol. iii. p. 457. Melet, Esprot ou Harenguet, *Cuv. Reg. An.* tom. ii. p. 318.

Length. Five inches.

Descript. Proportions nearly the same as those of the *adult Herring*, but the depth (equalling one-fifth of the entire length, caudal included) considerably greater than in a *young Herring* of the same length: keel of the abdomen more sharply serrated than in that species: teeth in the lower jaw more obsolete, scarcely sensible to the touch: subopercle of nearly the same form; but the veins on the infra-orbitals and preopercle not so distinct: scales larger: dorsal placed a little further back, commencing at the middle point between the end of the snout and the base of the caudal fork: ventrals, in consequence, *relatively* more forward, being slightly in advance of a vertical line from the first dorsal ray: number of fin-rays,

D. 17; A. 18; C. 19, &c.; P. 16; V. 7.

Number of vertebræ forty-eight.

* Vol. v. pp. 279, and 382.

This species has by many authors been confounded with the young of the Herring. Pennant was the first to point out its true distinguishing characters. It is very abundant in the Thames during the Winter, entering the river (according to Pennant) in the beginning of November, and leaving it in March. It is also found on other parts of the coast, but not every-where in plenty. Mr. Couch states *, that he never saw above one specimen of the true Sprat in Cornwall; though the Cornish fishermen apply this name to the young of both the Herring and the Pilchard.

119.   C. *alba*, Yarr.   (*White-Bait*.) ——Minute teeth in both jaws : dorsal further back than in the *Herring :* ventrals beneath the middle of the dorsal : keel of the abdomen serrated : anal with sixteen rays.

C. alba, *Yarr. in Zool. Journ.* vol. iv. pp. 137, and 465. pl. 5. f. 2. C. latulus, *Cuv. Reg. An.* tom. ii. p. 318.   White-Bait, *Penn. Brit. Zool.* vol. iii. p. 371. pl. 69. no. 176.   *Id.* (Edit. 1812.) vol. iii. p. 465. pl. 80.

LENGTH.   Three to four inches, rarely four inches nine lines.

DESCRIPT.   Body more compressed than in the *Herring*, the thickness being less than half the depth : abdominal serratures much sharper than in either the *Herring* or *Sprat*, but not so sharp as in the *Shad*, in which last species they are also of a different form : head one-fourth of the entire length : lower jaw longest : teeth very minute; those in the lower jaw confined to the extremity; upper jaw with the lower half of the maxillaries finely serrated : eyes large; their diameter nearly one-third the length of the head : lateral line distinctly marked, and straight : dorsal a very little further back than in the *Herring* †; ventrals immediately beneath it : number of fin-rays,

D. 19; A. 16; C. 19, &c.; P. 16; V. 9.

Number of vertebræ fifty-six.

Supposed formerly to have been the young of the Shad ‡, but clearly proved by Mr. Yarrell to be a distinct species.   Found only in the Thames, which river it ascends in April, sometimes as early as the end of March.   Abundant throughout the Summer about Greenwich and Blackwall, but never found higher up the river than the locality last mentioned.   Supposed to deposit its spawn during Winter.   Swims near the surface.   Food minute shrimps.

120.   C. *Pilchardus*, Bloch.   (*Pilchard*.) ——Teeth obsolete : infra-orbitals and opercular pieces strongly veined : subopercle square at bottom : dorsal exactly in the centre of gravity : ventrals beneath the posterior half of the dorsal.

---

* Loudon's *Mag. of Nat. Hist.* vol. v. p. 315.
† Cuvier says *plus avancée*, but I have not found it so in our English specimens, at least in those which I have examined.
‡ The *White-Bait* represented in Donovan's *British Fishes* (vol. v. pl. 98.) are really young Shads, and not the above species.

C. Pilchardus, *Bloch, Ichth.* pl. 406.   *Don. Brit. Fish.* vol. III.
pl. 69.   *Turt. Brit. Faun.* p. 106.   *Flem. Brit. An.* p. 183.
Harengus minor, sive Pilchardus, *Will. Hist. Pisc.* p. 223.
tab. P. 1. f. 1.   Pilchard, *Penn. Brit. Zool.* vol. III. p. 343.
pl. 68. no. 161. *Id.* (Edit. 1812.) vol. III. p. 453.   Le Pilchard,
ou le Célan, *Cuv. Reg. An.* tom. II. p. 319:

LENGTH.   Nine to eleven, rarely twelve, inches.

DESCRIPT.   General form resembling that of the *Herring:* the body,
however, somewhat thicker and rounder; the depth greater, the dorsal
line being more curved: belly not so sharp as in that species, although
the abdominal serratures, more especially those in front of the ventrals,
are rather more produced: head shorter: lower jaw not so long with
respect to the upper: scarcely any perceptible teeth; the maxillaries
simply with a few very fine denticulations quite at their lower extremity: diameter of the eye about one-fourth the length of the head:
subopercle cut square at bottom, and forming with the preopercle an
oblong (not a semicircle as in the *Herring*); both opercle and preopercle, but the last especially, with strongly-marked radiating striæ:
scales larger than in the *Herring:* dorsal more forward, and placed
exactly in the centre of gravity; the distance from the end of the snout
to the first ray, equalling the distance from the last ray to the base of the
caudal: caudal deeply forked: pectorals two-thirds the length of the
head, attached low down, beneath the subopercle: ventrals rather behind a vertical line from the middle of the dorsal:

B. 6; D. 18; A. 18; C. 19, &c.; P. 16; V. 8.

Principally taken off the coast of Cornwall, where they appear in large
shoals towards the end of Summer. The fishery for them commences
(according to Mr. Couch) towards the end of July, and terminates about
the time of the autumnal equinox. Food undetermined, but thought by
Mr. Couch to be the seeds of *fuci*.

(2. ALOSA, *Cuv.*)

121. C. *Finta*, Cuv. (*Shad.*)—Distinct teeth in both
jaws: a row of dusky spots along each side of the body.

C. Finta, *Cuv. Reg. An.* tom. II. p. 320.   C. Alosa, *Bloch, Ichth.*
pl. 30. f. 1.   *Don. Brit. Fish.* vol. III. pl. 57.   *Turt. Brit. Faun.*
p. 106.   *Flem. Brit. An.* p. 183. ? Shad, *Will. Hist. Pisc.* p. 227.
tab. P. 3. f. 1.   *Penn. Brit. Zool.* vol. III. p. 348. pl. 69. no. 164.
Id. (Edit. 1812.) vol. III. p. 460. pl. 80.   *Yarr. in Zool. Journ.*
vol. IV. pp. 137, and 465. pl. 5. f. 1. (Young.) Thames Shad,
Bowd. Brit. fr. wat. Fish. Draw. no. 19.

LENGTH.   From ten to sixteen inches; occasionally rather more.

DESCRIPT.   (*Form.*) Much larger in all its dimensions than either the
Herring or the *Pilchard:* body thicker; also somewhat deeper in proportion to its length: ventral line more convex than the dorsal: abdomen
sharply carinated; the serratures much sharper and stronger than in
any of the true *Clupeæ*, most developed between the ventrals and the
anal: head somewhat triangular; measuring rather more than one-fifth
of the entire length: snout short; under jaw relatively longer than in
the *Pilchard*, but not so long as in the *Herring:* intermaxillary deeply
notched: maxillaries sharply serrated with fine teeth along their whole
margin; lower jaw likewise with three or four teeth, much stronger than

the others, on each side near the extremity: tongue smooth, of a triangular form, free, and terminating in a blunt point: eyes placed high
on the cheeks; much smaller than in the *Pilchard*, their diameter being
scarcely more than one-fifth the length of the head; the distance from
them to the edge of the maxillary just equal to their diameter: subopercle as in the *Herring*, but rounded off at bottom more obliquely;
preopercle more resembling that of the *Pilchard*, and marked with radiating striæ as in that species, though not quite so distinctly: lateral line
scarcely perceptible: scales of moderate size: dorsal placed further back
than in the *Pilchard*, but more advanced than in the *Herring*, the distance from the snout to its commencement, when brought behind the fin,
reaching to nearly one-third of the caudal; fifth ray longest; the preceding ones gradually increasing from the first, which is very short; first
three simple, the rest branched; the last two from one root: anal longer
than in the *Pilchard*, and not approaching quite so near the caudal; the
intervening space one-seventh of the entire length of the body, caudal
excluded: caudal deeply forked: pectorals more than half the length of
the head: ventrals beneath the middle of the dorsal.

B. 8; D. 20; A. 21; C. 19, &c.; P. 15; V. 9.

Number of vertebræ fifty-five. (*Colours.*) Back, and upper part of the
sides, dusky blue: lower part of the sides, and belly, silvery white, or
yellowish, glossed with golden hues: a row of dusky spots, generally five
or six in number, but varying in different individuals, along each flank.
Obs. The *young* of this species are distinguished from *White-Bait* by
their greater depth in proportion to their length, smaller eye, bifid snout,
the presence of teeth along the *whole* margin of the maxillary, more forward dorsal fin, much sharper, as well as differently formed, abdominal
serratures, and by the row of spots on the sides, the first of which, immediately behind the opercle, is never wanting.
   A migratory species, entering rivers in May for the purpose of spawning, and returning to the sea about the end of July. Very abundant
in the Thames and Severn. In the former river is found as high up
as Putney and Hammersmith, where the White-Bait is unknown. Feeds,
according to Bloch, on worms, insects, and small fish. Spawns about the
first week in July. Flesh coarse and insipid. In the Severn is called
a *Twaite*, the name of *Shad* being reserved for the next species.

122.   C. *Alosa*, Cuv.   (*Allis.*)—Jaws without distinct
teeth: a single black spot behind the gills.

C. Alosa, *Cuv. Reg. An.* tom. II. p. 319.   Shad, *Bowd. Brit. fr.
   wat. Fish.* Draw. no. 27.

LENGTH. From two to three, sometimes four, feet.
   DESCRIPT. (*Form.*) Depth greater than in the last species, equalling
rather more than one-fourth of the entire length: maxillaries rough at
the edges, but without any distinct teeth: anal a little longer: in all
other respects nearly similar: number of fin-rays,

D. 19; A. 26; C. 19, &c.; P. 15; V. 9.

(*Colours.*) Resembling those of the *C. Finta*, but with rarely more than
a single dusky spot behind the gills, which is always present.
   This species abounds in the Severn, and is also occasionally, though
rarely, taken in the Thames, in which last river it is called *Allis*. It

is more esteemed for the table than the *C. Finta:* it also attains to a larger size, weighing from four to five, sometimes even as much as eight, pounds. *Obs.* Either this or the last species is the *Chad* of Jesse*.

GEN. 45.   ENGRAULIS, *Cuv.*

123.   E. *Encrasicholus,* Flem. (*Anchovy.*)

E. Encrasicholus, *Flem. Brit. An.* p. 183.   Clupea Encrasicholus, *Linn. Syst. Nat.* tom. i. p. 523.   *Bloch, Ichth.* pl. 30. f. 2. *Don. Brit. Fish.* vol. iii. pl. 50.   *Turt. Brit. Faun.* p. 107. Encrasicholus, *Will. Hist. Pisc.* p. 225. tab. P. 2. f. 2.   Anchovy, *Penn. Brit. Zool.* vol. iii. p. 347. pl. 67.   *Id.* (Edit. 1812.) vol. iii. p. 441. pl. 78.   L'Anchois vulgaire, *Cuv. Reg. An.* tom. ii. p. 322.

Length.   Six inches and a half. Penn.
Descript.   (*Form.*) Body slender, but thicker in proportion than the *Herring:* eyes large : under jaw much shorter than the upper : teeth small ; a row in each jaw, and another on the middle of the tongue : the tongue doubly ciliated on both sides : dorsal consisting of twelve rays, transparent, and placed nearer the nose than the tail : scales large and deciduous : edge of the belly smooth : tail forked. (*Colours.*) Back green, and semipellucid : sides and belly silvery, and opaque : irides white, with a cast of yellow. Penn.   According to Donovan, the number of the fin-rays is as follows :

D. 15 ; A. 14 ; C. 24 ; P. 15 ; V. 7.

Apparently a rare species in the British seas.   First obtained by Ray from the estuary of the Dee.   Pennant mentions a few which were taken near his house at Downing, in Flintshire, in 1769.   Donovan procured a specimen from the coast of Hampshire.   More recently single individuals have occurred on the coasts of Norfolk and Durham.   Common in the Mediterranean, where there is also (according to Cuvier) a second and smaller species, distinguished by the profile being less convex.   Both this last and the British one belong to that section of the genus, in which the belly is smooth without a sharp edge, and the dorsal opposite the ventrals.

(4.) *LEPISOSTEUS,* Lacép.

(22.)   L. *Gavialis,* Lacép. Hist. Nat. des Poiss. tom. v. p. 333. *Esox osseus,* Linn. Syst. Nat. tom. i. p. 516.   Bloch, Ichth. pl. 390.   Berkenh. Syn. vol. i. p. 81.   Don. Brit. Fish. vol. v. pl. 100.

> Berkenhout was the first to include this species in the British Fauna. He gives us to understand that it had occurred on the Sussex coast.   The only other author who has mentioned any locality for it is Stewart, who states† that it has been taken in the Frith of Forth.   It is probable, how-ever, that in both these instances there is some error, as the species is a native of America, where it is said to inhabit lakes and large rivers.

---

* *Gleanings in Nat. Hist.* Second Series, p. 129.
† *Elements of Nat. Hist.* vol. i. p. 374.

(II. SUBBRACHIALES.)

GEN. 46. GADUS, *Linn.*

124. G. *Morrhua*, Linn. (*Common Cod.*)—Back and sides spotted with yellow and brown; lateral line white: jaws nearly equal.

G. Morrhua, *Linn. Syst. Nat.* tom. I. p. 436. *Bloch, Ichth.* pl. 64. *Turt. Brit. Faun.* p. 89. *Don. Brit. Fish.* vol. v. pl. 106. Asellus major vulgaris, *Will. Hist. Pisc.* p. 165. tab. L. m. 1. n. 1. f. 4. Morhua vulgaris, *Flem. Brit. An.* p. 191. Common Cod-Fish, *Penn. Brit. Zool.* vol. III. p. 172. *Id.* (Edit. 1812.) vol. III. p. 231. La Morue, *Cuv. Reg. An.* tom. II. p. 331.

LENGTH. Two to four feet. Has been known (according to Pennant) to reach five feet eight inches.

DESCRIPT. (*Form.*) Oval; elongated; thickest behind the pectorals; somewhat tapering posteriorly: greatest depth about one-fifth of the entire length: dorsal line nearly straight beyond the commencement of the first fin, in front of which it slopes gently downwards to the snout; ventral line more bellying: head large; rather more than one-fourth of the entire length: snout rounded: jaws nearly equal; but sometimes the upper a little the longest: both jaws, as well as the fore part of the vomer, armed with small, sharp, card-like teeth in several rows, of unequal lengths: beneath the symphysis of the lower jaw a single barbule about one inch and a half in length: eyes moderate: head smooth and naked: body covered with small soft scales: a longitudinal groove on the nape extending from behind the eyes to the commencement of the first dorsal: lateral line arising from the upper part of the opercle, curving gently downwards till beneath the twelfth ray of the second dorsal, then passing off straight to the caudal; beneath the first dorsal, its course is about one-fifth of the depth: three dorsals; the first commencing at nearly one-third of the length; of a somewhat triangular form; its length rather greater than its height, which last equals about one-third of the depth of the body; first ray only half the length of the second; third, fourth, and fifth, rays longest; succeeding ones gradually diminishing; the last ray very small.: second dorsal almost immediately behind the first, of the same height, but its length half as much again; third, fourth, and fifth, rays longest: third dorsal resembling the first, but rather longer; fourth, fifth, and sixth, rays longest; the first ray very short: two anals; the first nearly corresponding to the second dorsal, beginning a little backwarder, but terminating in the same line; first ray very small, and easily overlooked; seventh and eighth longest: second anal answering exactly to the third dorsal: caudal nearly even at the extremity; the rays proceeding principally from the sides of the tail, which is prolonged into the middle of the fin: pectorals rounded, rather less than half the length of the head; fifth ray longest; all the rays, except the first two, branched: ventrals a little shorter than the pectorals, placed before them, narrow, and pointed; third ray longest: number of fin-rays,

D. 12—20—19; A. 19—17; C. 34, and several short ones; P. 19; V. 6.

(*Colours.*) Back, head, and upper half of the sides, cinereous brown, obscurely spotted with yellow; lower half of the sides, and abdomen, white: lateral line forming a narrow white band, very conspicuous on the dusky ground: fins dusky; ventrals pale, approaching to white.

A common species on most parts of the coast, but said to increase in numbers towards the North. According to Dr. Fleming, the most extensive fisheries in our seas are off the Western Isles and the coast of Zetland. Spawns in the early part of the Spring. Food, worms, crustacea, shell-fish, &c. Has been known to attain the weight of seventy-eight pounds.

(23.) *G. Callarias*, Linn. Syst. Nat. tom. I. p. 436. Bloch, Ichth. pl. 63. Berkenh. Syn. vol. I. p. 67. Turt. Brit. Faun. p. 89. Nilss. Prod. Ichth. Scand. p. 40. *Asellus varius vel striatus*, Will. Hist. Pisc. p. 172. tab. L. m. l. n. l. f. 1. *Variable Cod-Fish*, Penn. Brit. Zool. (Ed. 1812.) vol. III. p. 239. *Le Dorsch*, Cuv. Reg. An. tom. II. p. 332.

> This species, which is common in the Northern seas, especially in the Baltic, has been included in the British Fauna by Berkenhout, Turton, and the Editor of the last edition of Pennant's Zoology; its claims to insertion, however, must be considered as rather doubtful. It is probable that by some observers it has been confounded with a variety of the last species, in which the upper jaw projects a little beyond the lower, though never so much as in the *G. Callarias*, in which this character forms a striking feature. According to Cuvier, the true *G. Callarias* is usually of much smaller size than the *G. Morrhua*. Nilsson states its length to be from one to two feet. The same observer has annexed a distinguishing character between the two species, which it may be well to repeat here for the guidance of our own naturalists, in the event of the *G. Callarias* being really an inhabitant of the British seas. He remarks, that in the *G. Morrhua*, the length of the lower jaw equals half that of the head, also equals the distance from the snout to the posterior margin of the orbit: in the *G. Callarias*, it is shorter than half the length of the head, and equals the distance from the snout to the middle of the eye. The colours of this last species, upon which some authors appear to have relied, are said to be extremely variable.
>
> The *G. Callarias* has been sometimes distinguished by the English name of *Dorse*. Its flesh (according to Cuvier) is reckoned superior to that of the Common Cod.

125. G. *Æglefinus*, Linn. (*Haddock.*)—Lateral line, and a large spot behind the pectorals, black: upper jaw longest.

G. Æglefinus, *Linn. Syst. Nat.* tom. I. p. 435. *Bloch, Ichth.* pl. 62. *Don. Brit. Fish.* vol. III. pl. 59. *Turt. Brit. Faun.* p. 89. Onos, *Will. Hist. Pisc.* p. 170. tab. L. m. l. n. 2. Morhua Æglefinus, *Flem. Brit. An.* p. 191. Hadock, *Penn. Brit. Zool.* vol. III. p. 179. *Id.* (Edit. 1812.) vol. III. p. 241. L'Egrefin, *Cuv. Reg. An.* tom. II. p. 331.

LENGTH. From eighteen inches to two feet; rarely more.

DESCRIPT. (*Form.*) Rather more elongated, in proportion to its depth, than the *Common Cod:* barbule on the chin shorter: nape with an elevated ridge instead of a groove: upper jaw considerably the longest: lateral line hardly so much curved: first dorsal more decidedly triangular; the second and third rays longest, and more elevated above the others: third dorsal of the same length as the first, but not so

high, the rays being more nearly equal: first and second anals answering to the second and third dorsals respectively: insertion of the
pectorals in a line with the first ray of the first dorsal: ventrals narrow
and pointed; the second ray longest: caudal forked: number of fin-rays,

D. 16—21—19; A. 24—20; C. about 40, besides short ones; P. 21; V. 6:

scales small; firmly attached to the skin: vent in a line with the commencement of the second dorsal. (*Colours.*) Dusky brown: belly, and
lower part of the sides, silvery: lateral line black: a large black spot
on each side of the body, behind the pectorals and beneath the first
dorsal.

An abundant species on all parts of the coast, particularly during
Winter. Migrates northwards in Spring. Keeps in large shoals.
Spawns in February.

> (24.) *G. punctatus*, Turt. Brit. Faun. p. 90. *Morhua punctatus*,
> Flem. Brit. An. p. 192.
>
> "Body eighteen inches long, slightly arched on the back, a little pro-
> minent on the belly, covered above with numerous gold-yellow roundish
> spots, beneath with dusky specks which are stellate under a glass: head
> large, gradually sloping: teeth small, in several rows in the upper jaw,
> in the lower a single row: nostrils double: iris reddish, pupil black:
> chin with a single beard: nape with a deep longitudinal groove: lateral
> line nearer the back, curved as far as the middle of the second dorsal fin,
> growing broader and whiter towards the end: upper fins and tail brown,
> with obscure yellowish spots, and darker towards the ends; lower ones
> tinged with green: vent near the middle of the body: scales small; under
> a glass minutely speckled with brown: gill-covers of two pieces: lower
> jaw with five obscure punctures on each side:
>
> D. 14—20—18; A. 19—16; P. 18; V. 6,
>
> the first ray shorter than the second, and divided a little way down; C. 36,
> even at the extremity." Turt.
> This supposed species, which I am not acquainted with, is stated by
> Dr. Turton as being frequently taken in the Weirs at Swansea. No other
> author appears to have noticed it. I would venture to suggest that it is
> only a variety of the *G. Morhua.*

126. G. *luscus*, Linn. (*Bib*, or *Pout.*)—Depth one-fourth of the length: first anal commencing nearly in a line with the first dorsal.

> G. luscus & barbatus, *Linn. Syst. Nat.* tom. I. p. 437. *Turt.
> Brit. Faun.* p. 90. G. barbatus, *Bloch, Ichth.* pl. 166. G. luscus,
> *Don. Brit. Fish.* vol. I. pl. 19.? Morhua lusca & barbata,
> *Flem. Brit. An.* p. 191. Bib and Blinds (Cornubiensibus), Asellus
> luscus, *Will. Hist. Pisc.* p. 169. Asellus mollis latus, *Id. App.*
> p. 22. tab. L. m. 1. n. 4. Pout & Bib, *Penn. Brit. Zool.* vol. III.
> pp. 183, & 184. pl. 30. no. 76. *Id.* (Edit. 1812.) vol. III. pp. 246,
> & 247. pl. 34.

Length. From ten to twelve inches, seldom more.
Descript. (*Form.*) Remarkable for the great depth of the body,
equalling, at least, one-fourth of the entire length: sides compressed:
back slightly arched, and somewhat carinated; nape in particular offering
a sharp ridge, which commences in a line with the eyes, and extends
nearly to the dorsal: head about one-fourth of the entire length, ex-

cluding caudal: snout obtuse and rounded: upper jaw a little the longest: a single row of sharp moderately long teeth in the lower jaw; the
same in the upper with a band of smaller teeth behind: barbule at the
chin about one-fourth the length of the head: eyes large; their diameter
one-third the length of the head; invested with a loose membranous skin
capable of inflation; the distance between the eyes less than their diameter: scales not particularly large: lateral line curved, the flexure
taking place beneath the commencement of the second dorsal; anterior
to which its course is at rather more than one-fourth of the depth: beneath the lower jaw, on each side, a row of seven or eight open pores:
fins thick, fleshy at the base, invested with a loose skin: first dorsal
commencing at about one-third of the entire length, excluding caudal;
second and third rays longest; fourth and succeeding ones gradually
decreasing; the last very short; greatest height of this fin about two-
thirds the depth of the body: second dorsal commencing at a very short
interval after the first; more than twice its length; third ray longest:
third dorsal closely following the second; in length, a little exceeding
the first; third and fourth rays longest: first anal commencing in a line
with the second ray of the first dorsal, and terminating in a line with
the last ray of the second dorsal; the rays gradually increasing to the
eleventh, which is longest, the first being very short: second anal immediately following the first; answering to the third dorsal; fourth ray
longest: caudal nearly even: pectorals about three-fourths the length
of the head; third and fourth rays longest: ventrals long and narrow;
the first two rays very much produced beyond the others, terminating
in slender filaments; the second, which is the longer, rather more than
equalling the length of the pectorals: number of fin-rays,

D. 12—23—19; A. 35—21; C. 31, and some short ones; P. 18; V. 6:

vent directly beneath the commencement of the first dorsal. (*Colours.*)
Whitish, inclining to dusky olivaceous on the back; sides tinged with
yellow: fins dusky, becoming paler at the base; a dusky spot at the root
of the pectorals.

Common all along the southern coast, where it is taken in considerable
quantities for the table. Found also in other places. It is the *Whiting
Pout* of the London market. *Obs.* I have ventured to bring together (as
Bloch has already done before me) the *G. luscus* and *G. barbatus* of
authors, under a strong belief that they form but one species*. Should

---

* This opinion has not been adopted hastily. I have in vain sought for any author who has
described both the supposed species from his own observation, and after a due comparison of
their respective characters. The error of considering them as distinct appears to have originated
with Ray, the Editor of Willughby's Ichthyology. It would seem that Willughby was the first
to describe a fish (called in Cornwall *Bib* or *Blinds*) under the name of *Asellus luscus*, a species
evidently the same as the *Pout* of the Southern coast, to which Willughby's description, as far as
it goes, applies exactly. After that the body of his work was printed, Ray, his Editor, appears to
have received from Martin Lister, along with other novelties, a short account of the *Whiting
Pout* of the London market, to which he gave a separate place in the Appendix, never suspecting
that it might be the same as what had been already described by Willughby under the name of
Bib. Hence the two nominal species, which were afterwards perpetuated by Ray in his "Synopsis
Piscium;" and either to that work or Willughby's, the descriptions of all succeeding authors,
so far as regards one of the species, when they have noticed both, may ultimately be traced.
This is the case with Artedi, in the instance of the *G. luscus.* He simply refers to Ray and Willughby, annexing a short character, apparently taken from the description by the author last
mentioned. This character is repeated by Linnæus in his "Systema Naturæ," accompanied by
a reference to Artedi. Pennant's account of the two species is partly copied, and partly original:
his description of the *Pout* is perhaps his own; but that of the *Bib* is in a great measure taken
from Willughby, and although he has made one or two additional remarks, as well as annexed
a figure, I question whether these were derived from any fish *specifically* distinct from his
Whiting Pout. Gmelin, who, with respect to the *G. luscus*, only compiles from Willughby and
Pennant, appears to have suspected that the two fish were not really different. Berkenhout
states nothing beyond what is mentioned either by Willughby, Pennant, or Gmelin. Turton's
descriptions of the two species are evidently compiled from Pennant and Gmelin, excepting as
regards the number of fin-rays in the *G. luscus*, in which there is manifestly some error. Lastly.

I be wrong in holding this opinion, the minute description which I have given above of the *Pout*, as it occurs at Hastings, where my specimens were obtained, will not be without its use in enabling future observers to point out more precisely than has been hitherto done, the essential differences between it and the true *G. luscus.*

(25.) *Lord-Fish.*

Mr. Yarrell possesses the drawing of a fish (itself, unfortunately, not preserved) which was brought to him some years since, under the above name, by the Thames fishermen, and which was said to have been taken at the mouth of that river. In general form, it approaches the *G. luscus*, but it differs remarkably from that species, in having the first anal much shorter, and more rounded, commencing at a further distance from the head, and leaving a considerable space between itself and the second anal ; the vent also, which in *G. luscus* is in a line with the commencement of the first dorsal, is here in a line with the commencement of the second dorsal, or hardly so far advanced, being nearer the tail than the head. The number of fin-rays is as follows :

D. 14—19—18; A. 17—11; C. 24; P. 14; V. 6.

It is impossible to do more than thus briefly indicate the existence of a fish, which, if not a case of accidental deformity*, may hereafter turn out to be an undescribed species.

127. G. *minutus,* Linn. (*Poor.*)——Depth one-fifth of the length : first dorsal entirely before the first anal.

G. minutus, *Linn. Syst. Nat.* tom. I. p. 438. *Bloch, Ichth.* pl. 67.
f. 1. *Turt. Brit. Faun.* p. 90. Asellus mollis minor, *Will. Hist.
Pisc.* p. 171. tab. L. m. l. n. l. f. 2. Morhua minuta, *Flem. Brit.
An.* p. 191. Poor or Power, *Jago in Ray's Syn. Pisc.* p. 163. fig. 6.
Penn. Brit. Zool. vol. III. p. 185. pl. 30. no. 77. *Id.* (Edit. 1812.)
vol. III. p. 249. pl. 34.

LENGTH. From six to eight inches.

DESCRIPT. The smallest species in the genus, but more elongated in proportion than the *G. luscus :* greatest depth one-fifth of the entire length : head contained nearly five times in the same : lateral line nearly straight : a row of very distinct open pores, six or seven in number, commencing near the corner of the mouth, on each side of the head, and extending along the margin of the preopercle : distance from the end of the snout to the commencement of the first dorsal considerably less than one-third of the entire length : vent in a line with the tenth ray of the fin just mentioned : first dorsal entirely before the first anal, this last commencing nearly in the same line as that in which the former terminates : number of fin-rays,

D. 13—24—20 ; A. 28—24; P. 18; V. 6.

In all other respects, the *form* of this species is similar to that of the *G. luscus.* The *colour,* according to Pennant, is light brown on the back, and dirty white on the belly.

Dr. Fleming compiles from Willughby and Pennant. I would beg to ask, after this statement, what is the value of our authority for considering these species as distinct? In further confirmation of their identity, I may add that Mr. Yarrell has received from Mr. Couch, of Cornwall, a drawing of the fish which is called *Bib* on that coast, and that it proves in every respect to be the same as the *Whiting Pout* of the London market.

* This has been suggested by Mr. Yarrell, who hints that it may possibly be only a *monstrous* variety of the *G. Morrhua,* and that the name of *Lord-Fish,* given it by the fishermen, may be due to this circumstance.

First noticed as a British species by Jago, who obtained it on the Cornish coast, where it has been since observed by Mr. Couch. The specimen described above was caught at Weymouth, and measured eight inches in length, considerably exceeding the size usually assigned by authors to this species. Said to be very abundant in the Mediterranean, and to go in large shoals. According to Willughby and Bloch, it is peculiarly characterized internally by the *peritonæum* being black.

GEN. 47.   MERLANGUS, *Cuv.*

128.   M. *vulgaris*, Flem.   (*Whiting.*) — Upper jaw longest : lateral line nearly straight.

M. vulgaris, *Flem. Brit. An.* p. 195.   Gadus Merlangus, *Linn. Syst. Nat.* tom. I. p. 438.   *Bloch, Ichth.* pl. 65.   *Turt. Brit. Faun.* p. 91.   *Don. Brit. Fish.* vol. II. pl. 36.   Asellus mollis major seu albus, *Will. Hist. Pisc.* p. 170. tab. L. m. 1. n. 5. Whiting, *Penn. Brit. Zool.* vol. III. p. 190.   *Id.* (Edit. 1812.) vol. III. p. 255.   Merlan commun, *Cuv. Reg. An.* tom. II. p. 332.

LENGTH.   From twelve to sixteen, rarely twenty, inches.

DESCRIPT.   (*Form.*) More slender and elongated than the *Common Cod :* greatest depth one-sixth of the entire length : head about one-fourth : snout a little pointed ; upper jaw very sensibly the longest : teeth above in several rows ; the outer row longer than the others, and appearing beyond those in the lower jaw, when the mouth is closed ; these last forming but a single row : eyes round, large ; their diameter about one-fifth the length of the head : no longitudinal groove on the nape : lateral line nearly straight, showing only a slight flexure beneath the commencement of the second dorsal : scales small : first dorsal commencing at about one-third of the entire length ; of a triangular form, its length and greatest elevation about the same, equalling two-thirds of the depth of the body ; third, fourth, and fifth, rays longest : second dorsal commencing after a very short interval, much longer than the first, but in other respects similar : third dorsal resembling the second, and commencing after about the same interval ; fourth and fifth rays longest : vent in a line with the fourth ray of the first dorsal ; first anal commencing immediately behind it, and terminating a little beyond a vertical line from the end of the second dorsal ; first seven rays gradually increasing in length from the first, which is extremely short ; eighth and some of the succeeding rays longest, and nearly even ; last five or six gradually decreasing : second anal answering to the third dorsal : caudal nearly even : pectorals a little in 'advance of the first dorsal ; rather more than half the length of the head ; third and fourth rays longest : ventrals narrow and tapering, rather shorter than the pectorals ; second ray much longer than the others : number of fin-rays,

D. 15—19—20 ; A. 32—21 ; C. 31, and some short ones ; P. 19 ; V. 6.

(*Colours.*) Back, and upper part of the sides, pale brown, or reddish gray, generally without spots : belly silvery : lateral line whitish : a dusky spot at the roots of the pectoral fins.

A common species, taken in large quantities for the table during the spring and summer months.   Said to keep in large shoals at the distance of two or three miles from the shore.

129. M. *Pollachius*, Flem. (*Pollack.*) — Lower jaw
considerably the longest : lateral line curved, and dark
coloured : caudal slightly forked.

> M. Pollachius, *Flem. Brit. An.* p. 195. Gadus Pollachius, *Linn.
> Syst. Nat.* tom. I. p. 439. *Bloch, Ichth.* pl. 68. *Don. Brit. Fish.*
> vol. I. pl. 7. *Turt. Brit. Faun.* p. 91. Asellus Huitingo-Polla-
> chius, *Will. Hist. Pisc.* p. 167. Pollack, *Penn. Brit. Zool.* vol. III.
> p. 188. *Id.* (Edit. 1812.) vol. III. p. 254. Lieu, ou Merlan jaune,
> *Cuv. Reg. An.* tom. II. p. 333.

LENGTH. From two feet to two feet nine inches.

DESCRIPT. (*Form.*) Not so much elongated as the *Whiting :* depth
greater, equalling (beneath the first dorsal) one-fourth of the whole
length, excluding caudal: ventral line more convex than the dorsal:
head long, contained three times and a half in the entire length: snout
a little depressed : lower jaw considerably the longest: teeth smaller and
finer than those of the *Whiting :* lateral line with a strongly marked
flexure beneath the termination of the first dorsal; its course, before the
bend, running at one-fourth of the depth, after it, at nearly one-half:
first dorsal resembling that of the *Whiting :* second twice the length of
the first : third rather more than half the length of the second : vent in a
line with the third ray of the first dorsal: anals much as in the *Whiting :*
caudal slightly forked: length of the pectorals about half that of the
head: ventrals much smaller than in the *Whiting;* only one-third the
length of the pectorals; second ray longest: number of fin-rays,

> D. 14—21—19 ; A. 28—21 ; C. 31, and several short ones ; P. 19 ; V. 6.

(*Colours.*) Upper part of the head, back, and a portion of the sides, grayish or dusky brown, sometimes inclining to green; the rest of the sides,
and lower part of the body, whitish; these two colours separated by a
well-defined line, coinciding with the lateral line along the first half of its
course, but leaving it at the flexure: lips and fins dusky, with a tinge of
dull red.

Not uncommon off Weymouth and Scarborough, and other rocky parts
of the British coast. The specimen described above was caught at Hastings, and measured thirty-three inches in length, being above the usual
size of this species. Is sometimes called a *Whiting Pollack.*

130. M. *Carbonarius*, Flem. (*Coal-Fish.*) — Lower
jaw longest : lateral line straight, and white : caudal deeply
forked.

> M. Carbonarius, *Flem. Brit. An.* p. 195. Gadus Carbonarius,
> *Linn. Syst. Nat.* tom. I. p. 438. *Bloch, Ichth.* pl. 66. *Don. Brit.
> Fish.* vol. I. pl. 13. *Turt. Brit. Faun.* p. 91. Asellus niger,
> *Will. Hist. Pisc.* p. 168. tab. L. m. 1. n. 3. Coal-Fish, *Penn.
> Brit. Zool.* vol. III. p. 186. pl. 31. no. 78. *Id.* (Edit. 1812.) vol. III.
> p. 250. pl. 35. Merlan noir, *Cuv. Reg. An.* tom. II. p. 332.

LENGTH. Two to three feet.

DESCRIPT. (*Form.*) Resembling the last species, but more elongated;
greatest depth about one-fifth of the entire length: head a little shorter :
profile rather more convex: lower jaw not projecting so far beyond the

upper: lateral line perfectly straight throughout its whole course: dorsal and anal fins much as in the *Pollack*: ventrals rather longer than in that species: caudal more deeply forked: number of fin-rays,

D. 14—19—22; A. 26—24; C. 31, &c.; P. 21; V. 6.

(*Colours*.) Head, back, upper part of the sides, and dorsal fins, brown, dusky, or deep black; varying in different specimens: lateral line, belly, ventral and anal fins, whitish. According to Pennant, the dark colour of the back and sides deepens with age.

Equally common with the last species, but taken in most abundance on the northern coasts of the Island. Said by Pennant to swarm about the Orkneys, where the young are much used by the poor as an article of food. Is called in Cornwall a *Rauning Pollack*.

131. M. *virens*, Flem. (*Green Cod.*)—" Jaws equal: lateral line straight."

M. virens, *Flem. Brit. An.* p. 195. Gadus virens, *Linn. Syst. Nat.* tom. I. p. 438. *Neill in Wern. Mem.* vol. I. p. 532. Green Cod, *Penn. Brit. Zool.* (Edit. 1812.) vol. III. p. 253.

LENGTH. Less than a foot. FLEM.
DESCRIPT. "Smooth; dusky green on the back, silvery in every other part: jaws of equal length; side-line straight; tail forked." PENN. According to Fleming, the number of fin-rays is,

D. 15—24—19; A. 27—22; P. 22; V. 6.

This species, which I have not seen, is said by Mr. Neill to resemble the young Coal-Fish. Pennant first included it in the British Fauna, on the authority of Sir John Cullum. Dr. Fleming states, that it is frequently taken in the Frith of Forth, during Summer.

GEN. 48.   MERLUCCIUS, *Cuv.*

132. M. *vulgaris*, Flem. (*Common Hake.*)—" Whitish, grayish on the back: lower jaw longest."

M. vulgaris, *Flem. Brit. An.* p. 195. Gadus Merluccius, *Linn. Syst. Nat.* tom. I. p. 439. *Bloch, Ichth.* pl. 164. *Don. Brit. Fish.* vol. II. pl. 28. *Turt. Brit. Faun.* p. 91. Asellus primus, *Will. Hist. Pisc.* p. 174. tab. L. m. 2. n. 1. Hake, *Penn. Brit. Zool.* vol. III. p. 191. *Id.* (Edit. 1812.) vol. III. p. 257. Merlus ordinaire, *Cuv. Reg. An.* tom. II. p. 333.

LENGTH. From eighteen inches to nearly three feet. PENN.
DESCRIPT. Of a slender elongated form: head large, broad, and flattish: mouth very wide: lower jaw longest: teeth very long and sharp, particularly those of the lower jaw: near the eyes four small perforations: lateral line straight, nearer the back, beginning with several small tubercles near the head: vent nearer the head: first dorsal small, and pointed: the second reaching from the base of the first almost to the tail; the last rays highest: pectorals and ventrals pointed: caudal nearly even: number of fin-rays,

D. 9—38 to 40; A. 36 to 39; C. 18; P. 12 to 15; V. 7 or 8.   PENN. & TURT.

Said to be found in vast abundance on many of our coasts, particularly those of Ireland. Rare, according to Fleming, in Scotland. A coarse fish, and seldom admitted to table.

GEN. 49.   LOTA, *Cuv.*

133. L. *Molva*, Nob. (*Ling.*)—Above gray, inclining to olive ; beneath silvery : upper jaw longest.

> Gadus Molva, *Linn. Syst. Nat.* tom. I. p. 439.   *Bloch, Ichth.* pl. 69.
> *Turt. Brit. Faun.* p. 91.   *Don. Brit. Fish.* vol. v. pl. 102.   Asellus
> longus, *Will. Hist. Pisc.* p. 175. tab. L. m. 2. n. 2.   Molva vul-
> garis, *Flem. Brit. An.* p. 192.   Ling, *Penn. Brit. Zool.* vol. III.
> p. 197.   *Id.* (Edit. 1812.) vol. III.   p. 262.   Lingue, *Cuv. Reg. An.*
> tom. II. p. 333.

LENGTH. (*Average.*) From three to four feet. Pennant mentions having heard of one which reached seven feet.

DESCRIPT. (*Form.*) Body slender, more elongated than that of the *Hake*, roundish : head flat : gape large : lower jaw shorter than the upper, with a single barbule at its extremity : teeth in the upper jaw small, and very numerous ; those in the lower longer and larger, forming but a single row : lateral line straight : scales small, firmly adhering to the skin : two dorsals ; of equal height : first short, commencing near the head, not pointed as in the *Hake*, but with most of the rays even : second long, immediately behind the first, reaching nearly to the caudal ; the posterior portion the most elevated : vent in a line with the eighth or ninth ray of the second dorsal : anal immediately behind it, long, resem-bling the second dorsal, and terminating in the same line with that fin posteriorly : caudal rounded at the extremity : number of fin-rays,

> D. 15—65 ; A. 67 ; C. 40 ; P. 15 ; V. 6*.

(*Colours.*) Back and sides gray, inclining to olive ; sometimes cinereous, without the olivaceous tinge ; belly silvery : ventrals white ; dorsal and anal edged with white ; caudal marked near the end with a transverse black bar, the extreme tip white.

Not an uncommon species on many parts of the coast. Said by Pen-nant to abound about the Scilly Isles, on the coasts of Scarborough, and those of Scotland and Ireland. Approaches the land in January and February, according to Mr. Couch, in order to deposit its spawn. Very prolific. Feeds on other fish.

134. L. *vulgaris*, Nob. (*Burbot.*) — Yellowish or olivaceous brown, with darker blotches : jaws equal.

> Gadus Lota, *Linn. Syst. Nat.* tom. I. p. 440.   *Bloch, Ichth.* pl. 70.
> *Turt. Brit. Faun.* p. 91.   *Don. Brit. Fish.* vol. IV. pl. 92.   Mustela
> fluviatilis, *Will. Hist. Pisc.* p. 125. tab. H. 3. f. 4.   Molva Lota,
> *Flem. Brit. An.* p. 192.   Burbot, *Penn. Brit. Zool.* vol. III. p. 199.
> *Id.* (Edit. 1812.) vol. III. p. 265.   Barbolt, *Bowd. Brit. fr. wat. Fish.*
> Draw. no. 30.   Lotte commune, *Cuv. Reg. An.* tom. II. p. 334.

LENGTH. From one to two feet ; sometimes more.

DESCRIPT. (*Form.*) Body elongated, thick and roundish anteriorly, but much compressed behind : dorsal line nearly straight, but the ventral

---

* The above fin-ray formula is from Turton.

rather convex: greatest depth between one-fifth and one-sixth of the entire length: head broad and depressed: snout short and rounded: jaws equal; each with a band of rasp-like teeth: beneath the chin a single barbule, not one-third the length of the head: gape large: eyes round, moderate: gill-opening large; the membrane uniting with that on the opposite side under the throat: head naked: scales on the body minute, deeply imbedded, and invested with a slimy mucus: lateral line straight, not very distinct: dorsals of equal height; the first short, and slightly rounded, commencing at one-third of the entire length; the second long, closely following the first, and carried on quite to the caudal, to the base of which it is united; height of the second dorsal uniform throughout, only the first and last rays shorter than the others: vent a little before the middle of the entire length, excluding caudal; anal immediately behind it, carried on likewise very nearly to the caudal, but not ex-tending quite so far as the second dorsal: caudal rounded: pectorals rounded, shorter than the head: ventrals of about the same length, narrow and pointed; the second ray much longer than the others: number of fin-rays,

D. 13—71; A. 68; C. 48, including short ones; P. 20; V. 7.

(*Colours.*) Yellowish brown, blotched and stained with dark olivaceous brown; sometimes of a uniform dark olivaceous brown: head approaching to dusky: belly yellowish white.

The only species of this family inhabiting fresh water. Not uncommon in Cambridgeshire, where it is called an *Eel-Pout*. Found also (accord-ing to Pennant) in the Trent, in the river Witham, and in the great East Fen in Lincolnshire; but not generally distributed over the country. Frequents lakes and rivers. In England, seldom attains a greater weight than three pounds, but on the Continent is said sometimes to reach ten or twelve. Spawns (according to Bloch) in the months of December and January. Feeds on other fish, worms, and aquatic insects. Very tenacious of life: will live a long time out of water. Flesh excellent eating.

GEN. 50.   MOTELLA, *Cuv.*

135. M. *tricirrata*, Nilss. (*Three-bearded Rock-Ling.*) —Reddish yellow, spotted with black: two barbules on the snout; and one at the symphysis of the lower jaw.

M. tricirrata, *Nilss. Prod. Ichth. Scand.* p. 48. Gadus tricirratus, *Bloch, Ichth.* pl. 165. *Don. Brit. Fish.* vol. I. pl. 2. *Turt. Brit. Faun.* p. 92. *Flem. Brit. An.* p. 193. Rock-Ling, *Jago in Ray's Syn. Pisc.* p. 164. fig. 9. Three-bearded Cod, *Penn. Brit. Zool.* vol. III. p. 201. pl. 33. no. 87. *Id.* (Edit. 1812.) vol. III. p. 267. pl. 36. Mustèle commun, *Cuv. Reg. An.* tom. II. p. 334.

LENGTH. (*Average.*) From twelve to fifteen inches. According to Pennant, sometimes reaches nineteen inches.

DESCRIPT. (*Form.*) Body elongated; approaching cylindric ante-riorly, compressed behind: depth tolerably uniform throughout, equal-ling, behind the pectorals, one-seventh of the entire length: thickness, at the same part, more than three-fourths of the depth: head depressed;

rather more than one-fifth of the entire length: snout short, broad, and rounded: gape wide: upper jaw a little the longest: a broad band of velvet-like teeth in each jaw, with a single row of longer conical ones *behind* them in the lower, *before* them in the upper: sharp card-like teeth on the front of the vomer: two barbules on the upper part of the snout, in advance of the nostrils; a third at the symphysis of the lower jaw; these three barbules of equal length, each measuring one-fourth that of the head: gill-opening large; the membranes uniting under the throat as in the *Burbot:* scales very small; the skin every-where soft, and covered with a mucosity: lateral line bending downwards beneath the commencement of the second dorsal, and gradually altering its course from one-fifth to one-half the depth: first dorsal commencing in a line with the gill-opening, situate in a deep groove, about half the length of the head; all the rays detached, fine and hair-like, scarcely showing themselves above the groove, numerous; the first ray stouter and longer than the others: second or true dorsal immediately behind the first; long, running nearly to, but not connected with, the caudal; its height, except just at its commencement, uniform, being rather more than one-third the depth of the body: vent exactly in the middle of the entire length, caudal excluded; anal immediately behind it, resembling the second dorsal, and terminating in the same line with that fin: caudal, and also the pectorals, rounded: ventrals narrow; the first two rays longer than the others, with the intervening membrane deeply divided: number of fin-rays,

2nd D. 56; A. 48; C. 24, and some short ones; P. 19; V. 7.

(*Colours.*) "Head and body reddish yellow, marked above the lateral line with large black spots: dorsal fin and caudal darker; anal of a brighter red, but all spotted." PENN.

Frequents rocky shores, but is far more rare in the British seas than the next species. The specimen which furnished the above description, was taken at Weymouth. Is sometimes called the *Whistle-Fish.*

136. M. *Mustela*, Nilss. (*Five-bearded Rock-Ling.*) —Olive-brown: four barbules on the snout; and one at the symphysis of the lower jaw.

M. Mustela, *Nilss. Prod. Ichth. Scand.* p. 49. Gadus Mustela, *Linn. Syst. Nat.* tom. I. p. 440. *Don. Brit. Fish.* vol. I. pl. 14. *Turt. Brit. Faun.* p. 92. *Flem. Brit. An.* p. 193. Five-bearded Cod, *Penn. Brit. Zool.* vol. III. p. 202. pl. 33. no. 88. *Id.* (Edit. 1812.) vol. III. p. 268. pl. 36.

LENGTH. About the same as that of the last species.

DESCRIPT. (*Form.*) Differs from the last species, which it closely resembles, in having two additional barbules, rather shorter than the other ones, at the extremity of the upper lip: head shorter: upper jaw more projecting: teeth not quite so strongly developed: eyes smaller: all the fins similar, but the first ray of the first dorsal much longer and stouter with relation to the other rays in that fin: number of fin-rays,

2nd D. 51; A. 43; C. 24, &c.; P. 16; V. 7;

the dorsal and anal always containing fewer than in the *M. tricirratus,* by about five rays. (*Colours.*) Back and sides deep olive-brown, sometimes inclining to green; generally without spots: belly whitish, tinged with silvery.

Much more abundant than the *M. tricirrata*, and met with on most parts of the British coast. By Willughby, the two species were considered simply as varieties of one, which he describes under the general name of *Mustela vulgaris* *. Some modern authors, amongst whom may be reckoned Mr. Couch †, are inclined to the same opinion.

137. M. *glauca*, Nob. (*Mackerel Midge.*) — " Back bluish green ; all besides silvery : five barbules."

Midge (Ciliata glauca), *Couch in Loud. Mag. of Nat. Hist.* vol. v. p. 15. fig. 2. and p. 741.

LENGTH. One inch three lines. COUCH.

DESCRIPT. " Body moderately elongated, the proportions much resembling those of the *Whiting :* head obtuse : upper jaw longest, having four barbs, the under jaw one ; teeth in both jaws : gill-membrane with seven rays : eyes large and bright : pectoral and ventral fins rather large for the size of the fish : a ciliated membrane placed in a chink behind the head : the dorsal and anal fins reaching almost to the tail, which last is large and straight : scales deciduous. Colour of the back bluish green : belly and fins silvery." COUCH.

This fish, which has been noticed only by Mr. Couch, will probably prove eventually to be the fry of some other species. This gentleman states that it is found in multitudes on the Cornish coast, swimming near the surface ; and that it is migratory, making its first appearance about the middle of May. When Winter approaches, they disappear ; he is disposed, however, to think that they do not go to a great distance.

(26.) *Gadus argenteolus*, Mont. in Wern. Mem. vol. II. p. 449. Flem. Brit. An. p. 193.

" Head obtuse : cirri three ; two before the nostrils, and one on the chin : upper jaw longest : eyes lateral ; irides silvery : all the fins of a pale colour, and the whole fish of a silvery resplendence, except the back, which is blue, changeable to dark green : pectorals rounded, with sixteen or eighteen rays : ventrals with six or seven, the middle ray considerably the longest, and placed much before the pectorals : first dorsal commencing above the gills ; the rays very minute and obscure, the first excepted, which is much the longest, but more than thirty have been counted : second dorsal commencing close to the other, in a line with the end of the pectorals, and terminating close to the caudal ; the rays innumerable : anal beginning immediately behind the vent, and terminating even with the dorsal : caudal nearly even at the end. Length about two inches." MONT.

This fish is supposed by Montagu to constitute a new species. He mentions having noticed many of them thrown upon the shore in the South of Devonshire, in the Summer of 1808, and adds, that he had taken two or three since. The fishermen, he observes, called it *White-Bait*. It has, however, so much the character of the fry of some larger species, that it cannot be viewed without doubt. Had it not been said to possess but three cirri, I should have thought it the same as Mr. Couch's *Mackerel Midge*. Montagu appears to be quite certain that it is not the young of the *Motella tricirrata*.

---

* *Hist. Pisc.* p. 121.   † See *Linn. Trans.* vol. XIV. p. 73.

GEN. 51. BROSMUS, *Flem.*

138. B. *vulgaris*, Flem. (*Torsk.*)

B. vulgaris, *Flem. Brit. An.* p. 194. Gadus Brosme, *Gmel. Linn.* tom. I. part iii. p. 1175. *Don. Brit. Fish.* vol. III. pl. 70. *Turt. Brit. Faun.* p. 92. *Nilss. Prod. Ichth. Scand.* p. 47. Torsk, *Penn. Brit. Zool.* vol. III. p. 203. pl. 34. *Id.* (Edit. 1812.) vol. III. p. 269. pl. 37. *Low, Faun. Orc.* p. 200.

LENGTH. From eighteen inches to two feet; rarely three feet. NILSS. Largest specimen observed by Low, three feet and a half.

DESCRIPT. (*Form.*) Greatest depth (in a specimen twenty inches and a half in length) four inches and a half: head small: upper jaw a little the longest: both jaws with numerous small teeth: on the chin a small single beard: belly, from the throat, growing suddenly very prominent, continuing so to the vent, where it grows smaller to the tail; body, beyond the vent, pretty much compressed: from the head to the dorsal fin a broad furrow: lateral line scarcely discernible, but running nearer the back than the belly, till about the middle of the fish, where it bends a little downward, and then runs straight to the tail: dorsal running the whole length of the back, within about an inch of the tail: anal beginning at the vent, and ending at the tail, but not joined with it: the rays of the dorsal and anal fins numerous, but from their softness, and from the thickness of the skin, not easily counted with exactness: caudal rounded: pectorals broad, and rounded: ventrals small, thick, fleshy, ending in four points, or cirri. Low. The following is the number of rays in the several fins, according to Donovan:

D. 49; A. 37; C. 35; P. 21; V. 5.

(*Colours.*) "Head dusky; back and sides yellow, the yellow becoming lighter by degrees, and losing itself in the white of the belly: edges of the dorsal, anal, and caudal fins white; the other parts dusky: pectorals brown." Low.

A native of the northern seas. Represented by Low as being extremely common on the coast of Shetland, where it forms a considerable article of commerce. According to Pennant, it has not been discovered lower than the Orkneys. Is sometimes called a *Tusk.*

GEN. 52. PHYCIS, *Arted.*

139. P. *furcatus*, Flem. (*Common Fork-Beard.*) — First dorsal more elevated than the second; the first ray very much elongated: ventrals twice as long as the head. Cuv.

P. furcatus, *Flem. Brit. An.* p. 193. P. blennoides, *Nilss. Prod. Ichth. Scand.* p. 49. Asellus Callarias, *Will. Hist. Pisc.* p. 205. *Ray, Syn. Pisc.* p. 75. Barbus major, *Jago in Ray's Syn. Pisc.* p. 163. fig. 7. Blennius Phycis, *Turt. Brit. Faun.* p. 93. Forked Hake, *Penn. Brit. Zool.* vol. III. p. 193. pl. 31. no. 82. *Id.* (Edit. 1812.) vol. III. p. 259. pl. 35. Greater forked Beard, *Couch in Linn. Trans.* vol. XIV. p. 75. Merlus barbu, *Cuv. Reg. An.* tom. II. p. 335.

Length. Eleven inches and a half. Penn. Eighteen inches and a half. Borlase.

Descript. Greatest depth (in a specimen eleven inches and a half long) three inches: head sloping down to the nose as in the rest of the *Gadidæ:* mouth large: besides the teeth in the jaws, a triangular congeries of small teeth in the roof of the mouth: at the end of the lower jaw a small beard: first dorsal triangular; the first ray* extending far beyond the rest, and very slender: second dorsal commencing just behind the first, and extending almost to the tail: ventrals three inches long; consisting of only two rays, joined at the bottom, and separated or bifurcated towards the end: vent in the middle of the body: anal extending from thence† just to the tail: lateral line incurvated: tail rounded. *Colour* cinereous brown. Penn. Number of fin-rays,

D. 10—62; A. 56; P. 12.   Flem.

First obtained by Mr. Jago from the coast of Cornwall, where it has been since observed by Mr. Couch. According to this last gentleman, it keeps in deep water, and is not common: is called by the Cornish fishermen a *Hake's Dame.* Pennant's specimen was taken on the shores of Flintshire. It has also occurred near St. Andrew's in Scotland‡. *Obs.* The specific character of this fish given above from Cuvier, is requisite in order to distinguish it from a nearly allied species found in the Mediterranean, (*P. Mediterraneus,* Laroche,) in which the first dorsal is round, and not elevated above the second, and the ventrals nearly of the same length with the head. According to Cuvier, it is this last species, which is the *Blennius Phycis* of Linnæus, and not the one described and figured by Pennant, as supposed by many of our English authors.

GEN. 53.   RANICEPS, *Cuv.*

140.   R. *trifurcatus,* Flem. (*Tadpole-Fish.*)

R. trifurcatus, *Flem. Brit. An.* p. 194. Blennius trifurcatus, *Turt. Brit. Faun.* p. 93. Trifurcated Hake, *Penn. Brit. Zool.* vol. iii. p. 196. pl. 32. Trifurcated Tadpole-Fish, *Id.* (Edit. 1812.) vol. iii. p. 272. pl. 38.

Length. From eight to twelve inches. Davies.

Descript. (*Form.*) Head depressed and very broad: eyes large: mouth very wide, with irregular rows of incurvated teeth ; in the roof of the mouth likewise a congeries of teeth: no tongue, a broad abrupt rudiment only supplying the defect: body compressed, but remarkably so as it approaches the tail: above the pectoral fins, on each side, a row of tubercles, nine or ten in number, from the last of which commences the lateral line, which descends in a curved direction at the middle, and from thence continues straight to the tail: first dorsal placed in a furrow, rudimentary, consisting of three slender feeble rays easily overlooked: second dorsal reaching almost to the tail, with sixty-two rays: anal corresponding, with fifty-nine: caudal rounded, with thirty-six: pectorals also rounded, with twenty-three: ventrals with six rays, the last three of which are very slender and short, and the whole connected by a very delicate membrane.

D. 3—62; A. 59; C. 36; P. 23; V. 6.

* According to Nilsson, it is the *third* ray which is so much elongated beyond the others.
† According to Mr. Couch, "a few spines are placed before the anal fin."
‡ *Wern. Mem.* vol. vi. p. 569.

(*Colour.*) Deep brown, the folding of the lips excepted, which are snow-white: irides yellowish. DAVIES.

Pennant's description of this species was taken from a specimen sent him from Beaumaris by Mr. Hugh Davies, which gentleman has given some additional particulars respecting it in the last edition of the "British Zoology." Within these last three or four years, it has been obtained from Berwick Bay by Dr. Johnston *, a circumstance conclusive as to the existence of the species †, though it is still but little known to many of our naturalists.

> (27.) *R. Jago,* Flem. Brit. An. p. 194.   *Barbus minor,* (*The Lesser Forked-Beard*), Jago in Ray's Syn. Pisc. p. 164. fig. 8.   Couch in Linn. Trans. vol. XIV. p. 75.   *Lest Hake,* Penn. Brit. Zool. vol. III. p. 195.   Id. (Edit. 1812.) vol. III. p. 261.

> There is great reason for believing that this supposed species, obtained by Mr. Jago from the coast of Cornwall, where it has been since found by Mr. Couch, is identical with the *R. trifurcatus* last described. Jago says but little of his fish by which it can be recognized. He has, however, annexed a figure, which, allowing for the rude style in which drawings were executed in those days, might easily be intended for the species just mentioned. The following is Mr. Couch's description of his own specimen. " Length ten inches : head wide and flat: eyes forward and prominent: under jaw shortest: teeth in the jaws and palate, sharp and incurved, and some in the throat: a small barb at the under jaw: body compressed, smooth: first dorsal fin triangular and extremely small; second dorsal fin and the anal fin long, ending in a point: tail round : ventral fins with several rays, of which the two outermost are much elongated, the longest measuring two inches: the fins all covered with the common skin: a furrow passing above the eyes to the back.   Stomach firm, with longitudinal folds: no appendix to the intestines: air-bladder large, and of unusual form.   In the intestines were the remains of an *Echinus.*"

GEN. 54.   PLATESSA, *Cuv.*

141. P. *vulgaris,* Flem. (*Common Plaice.*) — Rhomboidal : a row of osseous tubercles on the eye-side of the head : lateral line curved above the pectoral : body smooth : teeth blunt, contiguous.

> P. vulgaris, *Flem. Brit. An.* p. 198.   Pleuronectes Platessa, *Linn. Syst. Nat.* tom. I. p. 456.   *Bloch, Ichth.* pl. 42.   *Don. Brit. Fish.* vol. I. pl. 6.   *Turt. Brit. Faun.* p. 96.   Passer Bellonii, *Will. Hist. Pisc.* p. 96. tab. F. 3.   Plaise, *Penn. Brit. Zool.* vol. III. p. 228.   *Id.* (Edit. 1812.) vol. III. p. 304.   Plie franche, ou Carrelet, *Cuv. Reg. An.* tom. II. p. 338.

LENGTH. From twelve to eighteen inches.

DESCRIPT. (*Form.*) Subrhomboidal; the tail very much contracted before the caudal : greatest breadth just half the length, fins excluded :

---

head a little less than one-fourth of the entire length: dorsal curve not
carried on continuously to the mouth, but very much depressed behind
the eyes: snout a little sharp; mouth small, ascending; the lower jaw
longest: teeth small, closely set, cutting, even, and rather obtuse: eyes on
the right side; full and prominent; both equally advanced towards the
end of the snout; the intervening space narrow, with an osseous ridge in
the middle, which, behind the eyes, becomes interrupted, giving rise to a
flexuous row of tubercles five or six in number: lateral line commencing
at the upper part of the opercle where the tubercles terminate, slightly
arched above the pectoral, but afterwards continued straight along the
middle of the body: both sides of the body smooth: scales minute, and
deeply impressed in the cuticle, causing the skin, except on the lateral
line, to appear pitted: dorsal commencing behind the eye, and extending
the whole length of the back, leaving, however, a small space between it
and the caudal; greatest elevation about the middle, equalling rather
more than one-fourth of the breadth of the body; all the rays simple,
and projecting a little beyond the webs: pectorals immediately behind
the posterior angle of the opercle, rounded, small; that on the right side
of the body not half the length of the head; fifth and sixth rays longest;
the first two and the last simple, the rest branched; pectoral on the left
side a little shorter and smaller than the other: anal commencing a little
beyond a vertical line from the pectorals, similar to the dorsal, and terminating in the same line with that fin; before it a short stiff spine
directed forwards: caudal oblong, even or slightly rounded at the extremity; its length equalling nearly half the breadth of the body; the
three outermost rays above and below simple, the rest branched: ventrals
a little shorter than the pectorals, and rather in advance of those fins;
third and fourth rays longest; all the rays simple:

B. 6; D. 72; A. 53; C. 20; P. 11; V. 6.

(*Colours.*) Upper part of the body and fins olivaceous brown, marked
with large bright orange spots; also, occasionally, a few oblong dusky
blotches, or stains of a darker brown than the ground colour: beneath
white.

Very abundant on most parts of the British coast. The largest said to
be found off Rye, on the coast of Sussex. According to Pennant, has
been known to weigh fifteen pounds. Feeds on small fish and testaceous
mollusca. Spawns in February and March. *Obs.* Cuvier notices a second
species of Plaice* (*Platessa lata*, Cuv.), which is sometimes taken, though
rarely, upon the French coast, closely resembling the common sort, and
possessing the same row of tubercles on the head, but differing in the
greater breadth of its body, which is not contained more than once and a
half in the entire length. Possibly this species may occur in our own
seas; though I am not aware that it has ever been observed hitherto.

142. P. *Flesus*, Flem. (*Flounder.*) — Rhomboidal: a
row of tubercular asperities along the base of the dorsal
and anal fins; lateral line slightly curved, and rough with
denticulated scales; rest of the body smooth: teeth blunt.

P. Flesus, *Flem. Brit. An.* p. 198. Pleuronectes Flesus, *Linn. Syst.
Nat.* tom. i. p. 457. *Bloch, Ichth.* pl. 44. *Don. Brit. Fish.* vol. iv.
pl. 94. *Turt. Brit. Faun.* p. 96. Passer fluviatilis, *Will. Hist.*

* *Reg. An.* tom. ii. p. 339.

Pisc. p. 98. tab. F. 5. Flounder, *Penn. Brit. Zool.* vol. III. p. 229.
Id. (Edit. 1812.) vol. III. p. 307. *Bowd. Brit.fr. wat. Fish.* Draw.
no. 25. Flet ou Picaud, *Cuv. Reg. An.* tom. II. p. 339.

LENGTH. Twelve inches and upwards.

DESCRIPT. (*Form.*) Resembling the *Plaice*, but rather more elongated; greatest breadth contained more than twice in the length, fins
excluded; body, in the adult fish, thicker. Dorsal and ventral lines
equally curved: profile depressed above the eyes: snout rather sharp;
mouth small; lower jaw longest, ascending obliquely at an angle of forty-
five degrees: teeth small and cutting, the summits obtuse: eyes large,
approximating, nearly equally in advance, the lower one a little the most
so: immediately behind the eyes, an elevated ridge of minute tubercular
asperities passing off to the upper part of the opercle, there to unite with
the lateral line, which last takes a slight bend over the pectoral before
passing off straight to the extremity of the caudal: greater part of the
head rough from the scales being denticulated; region of the lateral line
also rough from a band of similar scales extending along its whole length
immediately above and below it; there is also a row of tuberculated asperities along the basal margins of the dorsal and anal fins: rest of the body
smooth; the scales small, and very adherent: dorsal commencing above
the eye, and extending nearly the whole length of the back, as in the
Plaice; greatest elevation a little behind the middle, equalling one-
third of the depth of the body: anal as in the *Plaice;* immediately
before it a strong sharp spine directed forwards: caudal oblong, slightly
rounded at the extremity: pectoral on the right side rather more than
half the length of the head; that on the left smaller: ventrals much
smaller than the pectorals: number of fin-rays,

D. 61; A. 43; C. 18; P. 10; V. 6.

(*Colours.*) Extremely variable: upper surface generally olivaceous brown,
more or less deep; sometimes entirely dusky; occasionally flesh-coloured
or yellowish, or with brown spots upon a ground of one of these colours;
or with one-half of the body deep brown, the rest pale; more rarely
entirely flesh-colour, with scattered spots of a deep rose-red: under side
of the body generally whitish, but sometimes nearly as dark as above.

Var. β. Pleuronectes Passer, *Bloch, Ichth.* pl. 50. Eyes and lateral
line on the left side.

Equally common with the last species, and often found in rivers. Very
abundant in the Thames, where they are taken in considerable quantities
during the spring months. Such generally held in more estimation for
the table than those met with in the sea. Has been known to weigh
(according to Pennant) six pounds. Spawns in April and May. *Obs.*
The sinistral variety is not very uncommon. The *Pleuronectes roseus* of
Shaw *, and the *Platessa carnaria* of Brown †, are mere varieties of this
species, distinguished by a peculiarity of colouring; the former being of
a uniform delicate rose-colour; the latter flesh-red, with irregular, deep,
rose-coloured, distant spots.

143. P. *Limanda*, Flem. (*Dab.*) — Subrhomboidal :
lateral line strongly curved above the pectoral : body rough
throughout; the scales with ciliated margins : teeth sharp,
a little distant from each other.

* *Nat. Misc.* vol. VII. pl. 238.
† *Edinb. Journ. of Nat. and Geog. Sci.* vol. II. p. 99. pl. 2.

P. Limanda, *Flem. Brit. An.* p. 198. Pleuronectes Limanda, *Linn. Syst. Nat.* tom. I. p. 457. *Bloch, Ichth.* pl. 46. *Don. Brit. Fish.* vol. II. pl. 44. *Turt. Brit. Faun.* p. 96. Passer asper sive squamosus, *Will. Hist. Pisc.* p. 97. tab. F. 4. Dab, *Penn. Brit. Zool.* vol. III. p. 230. *Id.* (Edit. 1812.) vol. III. p. 308. Limande, *Cuv. Reg. An.* tom. II. p. 339.

LENGTH. From six to nine, rarely twelve, inches.

DESCRIPT. (*Form.*) General form similar to that of the *Flounder:* greatest breadth contained about twice and a half in the length, including caudal: head contained five times in the same: dorsal line nearly continuous with the profile, suffering very little depression above the eyes: teeth small; sharper and narrower than in either of the foregoing species, and not set quite so closely together: eyes large, but rather less prominent than in the *Plaice;* both equally advanced towards the mouth; between them a slightly projecting ridge, passing backwards in an ascending direction, but not accompanied by any osseous tubercles: lateral line at first strongly curved, but after passing the pectoral, straight to the end of its course: both sides of the body rough, but the upper one much the most so; the scales having their free edges ciliated: dorsal, anal, and caudal fins as in the *Plaice:* before the anal, a small, sharp, reclined spine, directed forwards: pectorals more than half the length of the head; first ray only half the length of the second; third longest; the first two and the last simple; the others branched: ventrals small, nearly in a line with the pectorals: number of fin-rays,

D. 72; A. 57; C. 18; P. 10; V. 6.

(*Colours.*) Upper side of a uniform pale brown; sometimes clouded with shades of a darker tint, or with a few ill-defined spots: beneath white.

Rather less abundant than either of the foregoing species, and never attaining to so great a size. Found, nevertheless, on most parts of the British coast. Feeds on marine worms and small crustacea. Spawns in May and June.

144. P. *microcephala*, Flem. (*Lemon Dab.*)—Oblong-oval: lateral line slightly curved above the pectoral: body smooth: head and mouth very small: jaws equal: teeth obtuse.

P. microcephala, *Flem. Brit. An.* p. 198. Pleuronectes microcephalus, *Don. Brit. Fish.* vol. II. pl. 42. *Turt. Brit. Faun.* p. 96. P. lævis, *Id. l. c.* P. microstomus, *Nilss. Prod. Ichth. Scand.* p. 53.? Rhombus lævis Cornubiensis maculis nigris, (A Kitt,) *Jago in Ray's Syn. Pisc.* p. 162. fig. 1. Smear-Dab, *Penn. Brit. Zool.* vol. III. p. 230. but not pl. 41. no. 106. *Id.* (Edit. 1812.) vol. III. p. 309. pl. 47. New species of Sole, *Edinb. New Phil. Journ.* no. 37. July, 1835. p. 209.

LENGTH. From twelve to eighteen inches.

DESCRIPT. (*Form.*) Oblong-oval; more elongated than any of the preceding species: greatest breadth, dorsal and anal fins excluded, contained twice and three-quarters in the entire length: head very small, not more than one-seventh of the entire length: dorsal curve continuous with the profile, falling regularly to the extremity of the snout: mouth extremely small; lips a little projecting: jaws equal: teeth cutting, set closely

together, their summits nearly even, and rather obtuse: eyes moderately
large, approximating, situate close behind the mouth, and both equally
advanced towards it; between them an osseous ridge, which, however,
is not produced behind as in the last species: lateral line commencing
higher up than the gill-opening, curved above the pectoral, but afterwards straight; the degree of curvature less than in the *Dab,* but greater
than in the *Plaice:* both sides of the body smooth: scales small, their
free edges scarcely ciliated: dorsal commencing above the eyes, and
extending the whole length of the back, and very nearly to the caudal;
greatest elevation one-fourth of the breadth of the body: anal commencing in a line with the pectorals, and answering to the dorsal; the
spine before it scarcely perceptible: caudal much as in the *Plaice:* the
two pectorals of equal size, and more than half the length of the head:
ventrals very small, a little in advance of the pectorals, and about three-
quarters of their length; second ray longest: number of fin-rays,

D. 92; A. 69; C. 19; P. 9; V. 5.

(*Colours.*) Above light brown, sometimes mottled with yellow and
dusky: beneath white. Pennant says, "belly white, marked with five
large dusky spots;" but, according to the editor of the last edition of the
"British Zoology," this spotting is not a constant character.

Met with occasionally on the southern and western coasts, but much
less plentiful than any of the preceding species. The specimen from
which the above description was taken, was obtained at Hastings. Said
to be frequent on the coast of Cornwall. According to Hanmer*, it is
known at Bath by the name of the *Lemon Sole;* at Plymouth, by that of
the *Merry Sole;* at Looe, by that of the *Kitt;* and at Penzance, by that
of the *Queen,* or *Queen-Fish. Obs.* This species is probably the *Pleuronectes microstomus* of Faber and Nilsson, but this last author has
noticed another, the *P. Cynoglossus* of Linnæus, which also approaches
very nearly to it. Possibly both these species may occur in our own seas.
Donovan appears to have considered it as the *Vraie Limandelle* of Duha-
mel, but, according to Cuvier, this last is synonymous with the *Platessa
Pola* next described. I may add that I can see no difference between
the present species and the supposed *New Species of Sole* lately characterized by Mr. Parnell †, of which I have seen a specimen in the possession of Mr. Yarrell.

145. P. *Pola*, Cuv. (*Pole.*)—Oblong-oval: lower eye
more advanced than the upper one: lateral line straight
throughout its course: body everywhere smooth: lower
jaw longest: teeth cutting.

P. Pola, *Cuv. Reg. An.* tom. II. p. 339. New species of Platessa,
Edinb. New Phil. Journ. no. 37. July, 1835. p. 210.

LENGTH. Seventeen to nineteen inches.
DESCRIPT. (*Form.*) Oblong-oval, approaching the form of the *Sole:*
greatest breadth, dorsal and anal fins excluded, rather exceeding one-
third of the entire length; body narrowing both ways from that point,
but more towards the tail than the head: length of the head half the

* See Hanmer's observations on the genus *Pleuronectes,* in the Appendix (No. 5.) to the third
volume of the last edition of Pennant's "British Zoology."
† *Edinb. New Phil. Journ.* l. c.

breadth of the body: mouth very small; lower jaw longest; commissure
of the lips, when the mouth is closed, nearly vertical: teeth cutting, set
closely together, with even summits, extending the whole length of the
jaws: eyes on the right side, large, placed obliquely, the lower one being
more advanced than the upper, close together, with an osseous ridge
between; diameter of the orbit equalling one-third the length of the
head: lateral line almost perfectly straight throughout its whole course,
but not exactly parallel to the axis of the body, inclining slightly upwards
anteriorly; half-way, its course is found to be a very little above the
mesial line: skin smooth above and below: scales large: dorsal fin
commencing above the eye, at a distance from the end of the snout
equalling nearly half the length of the head; rays short at first, but
doubling their length beyond the line of the pectorals; from that point
nearly even throughout; greatest elevation of the fin contained five times
and a half in the breadth of the body: anal commencing just opposite the
point at which the dorsal rays begin to lengthen, answering to that fin,
and terminating in the same line, a little before the caudal: caudal rounded
at the extremity; its length equalling half the breadth of the body: pectorals attached just behind the posterior angle of the opercle, their length
about half that of the head: ventrals immediately beneath them, of the
same length.

D. 109; A. 93*; C. 19; P. 12; V. 6.

(*Colour of a specimen in spirits.*) Yellowish brown.

The above description of this species, which is a recently acquired addition to the British Fauna, was taken from a specimen in the Museum
of the Zoological Society, procured in the London market, in May, 1833.
Mr. Yarrell has another from the Frith of Forth, sent him by Mr. R. H.
Parnell, by whom it appears to have been considered as an undescribed
species †. This last gentleman states that it is known to the fishermen
in that neighbourhood under the appellation of *Craig Fluke.* I have
ventured to suggest the English name of *Pole,* as being in unison with
the Latin name which it has received from Cuvier.

146. P. *Limandoides,* Nob. (*Sandnecker.*) — Oblong-oval: both eyes equally advanced towards the mouth: lateral line straight: body rough; the scales with ciliated margins: teeth conical, and sharp-pointed.

> Pleuronectes Limandoides, *Bloch, Ichth.* pl. 186. *Gmel. Linn.* tom. I.
> part iii. p. 1232. *Nilss. Prod. Ichth. Scand.* p. 57. P. liman-
> danus, *Edinb. New Phil. Journ.* no. 37. July, 1835. p. 210.

LENGTH. From ten to twelve inches.

DESCRIPT. (*Form.*) Oblong-oval; the body more elongated than in the
last species: greatest breadth, dorsal and anal fins excluded, about one-third
of the entire length; head rather more than half the breadth: mouth considerably larger than in the *P. Pola;* lower jaw longest, ascending obliquely
to meet the upper; teeth conical, sharp-pointed, a little distant from each
other: eyes on the right side, and both equally advanced towards the
mouth; between them an osseous ridge, produced behind, and falling in
with the commencement of the lateral line; diameter of the orbit one-
fourth the length of the head: lateral line straight throughout its course:

---

* The numbers of rays in the dorsal and anal fins are taken from Mr. Parnell.
† See *Edinb. New Phil. Journ.* l. c.

scales large, with their free edges ciliated, communicating a marked
roughness to both sides of the body: dorsal commencing above the
upper eye, and extending nearly to the caudal; highest part of the fin
a little beyond the middle: caudal rounded: anal and other fins, much
as in the *P. Pola:* number of fin-rays,

D. 82; A. 64; C. 18; P. 10; V. 6.

(*Colour.*) Of a uniform pale brown, or yellowish brown, above; white
beneath.

This species, which, like the last, has been only recently added to our
Fauna, has been obtained from Berwick Bay by Dr. Johnston, and from
the Frith of Forth by Mr. Parnell. In the last-mentioned locality, particularly on the Fifeshire coast, it is represented as not very uncommon,
and as known to most of the fishermen by the name of *Sandnecker*, or
Long Fluke. It appears to be a northern species, inhabiting, according
to Bloch, sandy bottoms, and preying upon young crabs and small lob-
sters. Flesh stated by the same author to be white, and of good eating.
Obs. In its general form this species resembles the *Holibut*, with which,
perhaps, it ought properly to be associated.

GEN. 55. HIPPOGLOSSUS, *Cuv.*

147. H. *vulgaris*, Flem. (*Holibut.*) — Eyes on the
right side: lateral line arched above the pectorals: body
oblong; smooth.

> H. vulgaris, *Flem. Brit. An.* p. 199. Pleuronectes Hippoglossus,
>   *Linn. Syst. Nat.* tom. I. p. 456. *Bloch, Ichth.* pl. 47. *Don. Brit.*
>   *Fish.* vol. IV. pl. 75. *Turt. Brit. Faun.* p. 95. *Nilss. Prod.*
>   *Ichth. Scand.* p. 57. Hippoglossus, *Will. Hist. Pisc.* p. 99.
>   tab. F. 6. Holibut, *Penn. Brit. Zool.* vol. III. p. 226. *Id.* (Edit.
>   1812.) vol. III. p. 302. Le grand Flétan ou Helbut, *Cuv. Reg. An.*
>   tom. II. p. 340.

LENGTH. From three to six feet, and upwards.

DESCRIPT.* (*Form.*) Body oblong; of a more elongated form than
in the last sub-genus, tapering much towards the tail: greatest breadth,
dorsal and anal fins excluded, rather more than one-third of the entire
length: head small, a little more than one-sixth of the same: mouth
large; both jaws armed with several long, sharp, curved, distant, teeth:
eyes large, approximating, situate on the right, very rarely on the left,
side of the head: gill-cover of three pieces; the gill-opening large, with
the membrane exposed: lateral line arched above the pectoral, but afterwards running straight to the caudal fin: body smooth: both sides
covered with small, soft, oblong, scales, strongly adhering, and invested
with a slimy mucus: dorsal commencing above the eyes, and reaching
very nearly to the caudal: vent further removed from the head, than
in the other species belonging to this family: before the anal a long
spine: pectorals oblong: caudal crescent-shaped:

B. 7; D. 107; A. 82; C. 16; P. 15; V. 7.

(*Colours.*) "Dusky brown, most commonly inclining to a liver-colour,
and free from spots; the tint variable, and said to be blackest, or more
dusky, in fish of poor condition: lower surface uniformly white." DON.

---

* Not having any original description of this species, the above has been compiled from Bloch,
Gmelin, Donovan, and Nilsson. The fin-ray formula is from Bloch.

Not uncommon on some parts of the coast, and occasionally exposed for sale in the London markets. Attains to a very large size. One taken off the Isle of Man in April 1828, is said to have measured seven feet and a half in length, and to have weighed three hundred and twenty pounds*. Said to be very voracious, preying upon other fish, and on crustacea. Spawns, according to Bloch, in the Spring. Flesh poor, and not much esteemed. In the northern parts of Britain, is called a *Turbot*.

GEN. 56. PLEURONECTES, *Flem.*

148.   P. *maximus*, Linn.   (*Turbot.*) — Body rhomboidal, and nearly as broad as long: the eye-side beset with small, subacute, osseous, tubercles.

P. maximus, *Linn. Syst. Nat.* tom. I. p. 459.   *Bloch, Ichth.* pl. 49. *Don. Brit. Fish.* vol. II. pl. 46.   *Flem. Brit. An.* p. 196.   P. tuber‑ culatus, *Turt. Brit. Faun.* p. 97.   Rhombus maximus asper non squamosus, *Will. Hist. Pisc.* p. 94. tab. F. 2.   Turbot, *Penn. Brit. Zool.* vol. III. p. 233.   *Id.* (Edit. 1812.) vol. III. p. 315. pl. 49.   Le Turbot, *Cuv. Reg. An.* tom. II. p. 341.

LENGTH. From eighteen inches to two feet; sometimes more.

DESCRIPT. (*Form.*) Body rhomboidal, approaching to round: greatest breadth, dorsal and anal fins included, almost equalling the entire length without the caudal: head broad: dorsal curve carried on continuously to the mouth, without any depression before or behind the eyes; forming with the ventral curve, at the extremity of the snout, a right angle: lower jaw longest, ascending obliquely to meet the upper: both jaws armed with small card-like teeth: eyes on the left side of the head; both equally advanced towards the mouth; a little remote from each other, the intervening space nearly flat: basal and ascending margins of the preopercle meeting at a right angle; gill-opening large: lateral line commencing behind the orbit of the upper eye, forming a considerable arch above the pectoral, but afterwards straight, dividing the body into two equal parts: both sides of the body smooth, but studded with small, subacute, osseous, tubercles; the tubercles on the upper or eye-side larger and more numerous than those on the lower: scales small: dorsal commencing in front of the eye, immediately above the upper jaw, and extending very nearly to the caudal; greatest elevation of the fin about the middle, attained gradually: anal commencing nearly in a line with the posterior lobe of the opercle, and answering to the dorsal: ventrals appearing like a continuation of the anal; a small space intervening, in which the vent is situate: caudal rounded: number of fin-rays,

D. 67 ; A. 45 ; C. 17 ; P. 12 ; V. 6.

(*Colours.*) Upper side yellowish brown, mottled and spotted with darker brown: under side white.

Found on many parts of the British coast, in some places, in considerable abundance. Attains to a larger size than any other species in this family, the Holibut excepted. Weight from fifteen to twenty pounds, sometimes as much as thirty, or even more. Flesh firm, and highly esteemed for the table. Food, according to Bloch, insects and worms.

---

* Loudon's *Mag. of Nat. Hist.* vol. I. p. 84.

**149. P. *Rhombus*, Linn. (*Brill.*) — Body broadly
oval ; smooth, without tubercles : first rays of the dorsal
half free, and branched at their extremities.**

P. Rhombus, *Linn. Syst. Nat.* tom. I. p. 458.   *Bloch, Ichth.* pl. 43.
  Don. Brit. Fish. vol. IV. pl. 95.   *Sow. Brit. Misc.* pl. 50.   *Turt.
  Brit. Faun.* p. 97.   *Flem. Brit. An.* p. 196.   Rhombus non
  aculeatus squamosus, *Will. Hist. Pisc.* p. 95. tab. F. 1.   Pearl,
  Penn. Brit. Zool. vol. III. p. 238.   *Id.* (Edit. 1812.) vol. III. p. 321.
  pl. 50.   La Barbue, *Cuv. Reg. An.* tom. II. p. 341.

LENGTH.   From twelve to eighteen inches; sometimes more.
DESCRIPT.   (*Form.*) Very similar to the *Turbot*, but of a more oval
form : breadth not so great, contained about once and a half in the
entire length : upper surface perfectly smooth, without any osseous
tubercles : lateral line arched above the pectorals, but the curvature not
so great as in that species : the first four or five rays of the dorsal fin
half free, and divided at their extremities : in most other respects the
two species are similar :

D. 71 ; A. 57 ; C. 16 ; P. 12 ; V. 6.*

(*Colours.*) Rather darker than the *Turbot :* upper surface deep brown,
with numerous dusky and white spots ; sometimes intermixed with
yellowish : beneath white : fins spotted.
    Met with in the same localities as the last species, and more abund-
antly.   Does not attain to so great a size.   Flesh less esteemed.   Is
sometimes called a *Pearl.*

**150. P. *punctatus*, Bloch. (*Bloch's Top-Knot.*) —
Roundish oval : both sides of the body rough ; the edges
of the scales denticulated : the first ray of the dorsal fin
elongated : ventrals and anal separate.**

P. punctatus, *Bloch, Ichth.* pl. 189.   *Flem. in Wern. Mem.* vol. II.
  p. 241.   *Id. Phil. Zool.* pl. III. f. 2.   *Id. Brit. An.* p. 197.

LENGTH.   Five inches and a half.
DESCRIPT.   (*Form.*) Roundish oval, the dorsal and ventral lines
equally convex : greatest breadth, fins excluded, just half the length :
head a little less than one-third of the same : profile notched immediately
before the eyes : mouth of moderate size, very protractile ; jaws nearly
equal ; the lower one a very little the longest, and ascending obliquely
at an angle of rather more than forty-five degrees : teeth so fine as to
be scarcely visible : eyes large, remarkably full and prominent, their
diameter about one-fourth the length of the head ; placed on the left
side ; approximating ; the lower one rather more advanced than the
upper ; between them a projecting ridge : basal and posterior margins
of the preopercle meeting at a very obtuse angle, the former rising
obliquely to meet the latter : lateral line commencing at the upper part
of the opercle, at first very much arched, but afterwards straight : both
sides of the body, but more especially the upper, extremely rough ;
scales minute ; those on the upper side having their free margins set
with from four to six longish denticles ; those beneath having the

---

* The above fin-ray formula is from Bloch.

denticles finer and more numerous : dorsal commencing immediately in
advance of the upper eye, and extending very nearly to the caudal, at
the same time passing underneath the tail, where the rays become very
delicate ; greatest elevation of the fin near its retral extremity ; first ray
very much produced, nearly three times the length of those which
follow ; most of the rays divided at their tips ; some of the last in the
fin branched from the bottom : anal commencing in a line with the
posterior angle of the preopercle, answering to the dorsal, and terminating in the same manner beneath the tail ; greatest elevation corresponding : caudal oblong, the extremity rounded : pectorals inserted behind
the posterior lobe of the opercle, a little below the middle ; the first ray
very short ; the next three or four longest ; the succeeding ones nearly
as long ; pectoral on the eye-side rather larger than that on the side
opposite : ventrals immediately before the anal, and appearing like a
continuation of that fin, but not connected with it, as in the next species ;
vent situate between the two last pairs of rays : the rays of all the fins
covered with rough scales nearly to their tips :

D. 87 ; A. 68 ; C. 16 ; P. (Left) 12, (Right) 11 ; V. 6.

(*Colours.*) Above brown, or reddish brown, mottled and spotted with
black ; a large round spot, more conspicuous than the others, in the
middle of the side towards the posterior part of the body ; fins spotted :
beneath, plain white.

This species, which I believe to be the same as the *P. punctatus* of
Bloch, was confounded by that author with the *P. Megastoma.* More
recently, it has been confounded by several naturalists, including Cuvier,
Nilsson, Hanmer, and Fleming, with that next described. The elongated
first dorsal ray, and the ventrals, disjoined from the anal, will, however,
always serve to distinguish it. It is evidently to the present species that
Fleming's fish, procured in Zetland, belongs. The only other British
specimen I know of, is in the Museum of the Cambridge Philosophical
Society. This last, from which the above description was taken, was
obtained by Professor Henslow at Weymouth. The Top-Knot of Hanmer
belongs to the next species.

151. P. *hirtus*, Mull. (*Muller's Top-Knot.*)—Round-
ish oval : eye-side of the body rough ; the edges of the
scales denticulated : jaws equal : the first dorsal ray not
longer than the succeeding ones : ventrals and anal
united.

P. hirtus, *Mull. Zool. Dan.* vol. III. p. 36. pl. 103. Smear-Dab,
Penn. Brit. Zool. vol. III. pl. 41. no. 106. (*No description annexed.*) Top-Knot, *Hanmer in Penn. Brit. Zool.* (Edit. 1812.)
vol. III. p. 322. pl. 51. Whiff, *Couch in Linn. Trans.* vol. XIV.
p. 78.

LENGTH. Seven inches nine lines.

DESCRIPT. (*Form.*) In general appearance very similar to the last
species, but differing in the following particulars : profile without the
notch before the commencement of the dorsal fin : mouth rather smaller,
and more oblique ; when closed, the maxillaries assuming nearly a
vertical position : jaws more nearly equal : eyes not so prominent, nor
so close together ; the lower one rather more in advance with respect
to the upper, a tangent to the posterior part of the orbit of the former

nearly bisecting the latter into two equal parts: the space between more
flattened, or with very little of a projecting ridge: basal and posterior
margins of the preopercle meeting at a less angle, the former being more
nearly parallel to the axis of the body: upper side of the body less
rough; the lower one perfectly smooth: scales on the upper side
smaller, with more numerous and shorter denticles; the two middle
denticles, however, longer than the others; the scales on the lower side
without any denticles: dorsal fin almost in close contact with the mouth;
the first ray not longer than the succeeding ones: ventrals united, at
their posterior margins, to the anal, from which, at first sight, they are
scarcely to be distinguished; the vent placed between them: fleshy
portion of the tail not so long, or not so much projecting from the oval
of the body; the dorsal and anal fins approaching one another more
closely on its under surface:

D. 96; A. 73; C. 16; P. 12; V. 6.

(*Colours.*) For the most part similar to those of the *P. punctatus:* the
dark spots and markings are however better defined; more particularly
a black, slightly angulated, band, passing across the head through the
eyes, and a large spot beyond the extremity of the pectoral, upon the
lateral line.

Muller is the only author, so far as I am aware, who has distinguished
this from the last species. It appears to have been more often met with
in our seas than the *P. punctatus.* Pennant has evidently figured it
under the name of *Smear-Dab,* though the corresponding description
belongs to the *Platessa microcephala* of this work. A better representation of it is given in the last edition of the "British Zoology," from a
specimen obtained by Mr. Hanmer from the coast near Plymouth. More
recently it has been noticed on the Cornish coast by Mr. Couch, and on
the coast of Berwickshire by Dr. Johnston. It has also occurred near
the mouth of the Medway. Mr. Couch observes that it keeps in rocky
ground, and rarely, if ever, takes a bait.

152. P. *Megastoma*, Don. (*Whiff.*) — Body oblong;
the eye-side rough, with the scales finely ciliated: gape
large; lower jaw longest: first rays in the dorsal fin
free, but simple.

P. Megastoma, *Don. Brit. Fish.* vol. III. pl. 51. *Turt. Brit. Faun.*
p. 97. *Flem. Brit. An.* p. 196. P. Cardina, *Cuv. Reg. An.*
tom. II. p. 341. Passer Cornubiensis asper, magno oris hiatu,
(A Whiff,) *Jago in Ray's Syn. Pisc.* p. 163. fig. 2. Whiff, *Penn.*
Brit. Zool. vol. III. p. 238. *Id.* (Edit. 1812.) vol. III. p. 324. pl. 52.
Carter, or Lantern-fish, *Couch in Linn. Trans.* vol. XIV. p. 78.

LENGTH. From twelve to eighteen inches.
DESCRIPT. (*Form.*) Body oblong, the tail suddenly contracting before
the caudal; thin, and rather pellucid: greatest breadth, dorsal and anal
fins excluded, not quite one-third of the entire length: head large, nearly
one-fourth of the same: dorsal curve falling regularly to the end of the
snout; the profile slightly concave before the eyes: gape extremely
large; lower jaw longest, ascending obliquely, furnished with a blunt
tubercle beneath the symphysis: both jaws with very fine velvet-like
teeth: eyes very large; their diameter at least one-fourth the length
of the head; placed on the left side; approximating; the lower one most

in advance; between them an osseous ridge, passing upwards behind to unite with the lateral line: gill-opening large; opercle small, of a triangular form; subopercle and interopercle much developed: lateral line very much arched above the pectoral, afterwards straight, and carried on to quite the end of the caudal: scales large; those on the eye-side of the body with their free edges finely ciliated, communicating a roughness to the touch; those on the opposite side smooth, with their margins entire; scales on the lateral line with a slightly elevated oblong tubercle: dorsal commencing about half-way between the extremity of the snout and the upper eye, and carried on very nearly to the caudal; greatest elevation of the fin a little beyond the middle, where it equals one-third of the depth of the body; most of the rays simple, some of the longest only divided at their tips; the first four or five nearly free, the connecting membrane being very low: anal commencing in a line with the posterior lobe of the opercle, and answering to the dorsal; before it a blunt point: caudal oblong, rounded at the extremity, its length rather more than half the depth of the body; all the principal rays, except the two outermost, branched: pectorals inserted a little below the middle of the depth, and in a line with the commencement of the anal; very unequal; that on the eye-side rather more than half the length of the head; the opposite one more than one-third shorter; first ray very short; third and fourth rays longest: ventrals entirely in advance of the pectorals, and appearing like a portion of the anal, only double, from which they are separated by the vent:

B. 7; D. 85; A. 71; C. 15, and 4 short; P. (Left) 12, (Right) 10; V. 6.

(*Colours.*) Upper side light reddish brown, here and there mottled and spotted with dusky and darker brown: under side white.

First observed by Mr. Jago on the coast of Cornwall, where it has been since represented, by Mr. Hanmer and Mr. Couch, as very common. Occasionally met with on other parts of the southern, as well as on the western, coast. The specimen from which the above description was taken occurred at Hastings. In Cornwall called a *Lantern-Fish*.

153. P. *Arnoglossus*, Schn. (*Scald-Fish.*) — Body oblong-oval : scales large, deciduous, finely ciliated : jaws equal : lower eye most in advance : before the anal a strong sharp spine.

P. Arnoglossus, *Flem. Brit. An.* p. 197. *Bonap. Faun. Ital.*
Fasc. IV. Arnoglossus, vel Solea lævis, *Will. Hist. Pisc.* p. 102.
tab. F. 8. f. 7.? Smooth Sole, *Penn. Brit. Zool.* vol. III. p. 232.
Scald-Fish, *Id.* (Edit. 1812.) vol. III. p. 325. pl. 53.

LENGTH. Five to six inches.
DESCRIPT. (*Form.*) Oblong-oval; the body narrowing behind more gradually than in the last species, and not so suddenly contracted before the caudal; thin and somewhat pellucid: greatest breadth, dorsal and anal fins excluded, one-third of the entire length: head one-fourth of the same, excluding caudal: profile slightly emarginated before the eyes: gape moderate; jaws nearly equal; lower one obliquely ascending; both with fine velvet-like teeth: eyes placed as in the *Whiff*, but not so large in proportion: lateral line arched above the pectoral, afterwards straight: scales large, thin, very deciduous; their free edges finely ciliated, and emarginated; those on the lateral line with an oblong tubercle as in the

G G

Whiff, but not so much elevated: dorsal and anal as in that species; before the anal a strong, sharp, triangular, spine or lamina, directed downwards and backwards: caudal rounded: pectorals unequal; that on the upper side about three-fourths the length of the head: ventrals consisting of a double row of rays; that on the upper side more advanced than the other: the rays of all the fins slender and bristly; the connecting membranes very delicate, and easily broken:

D. 85; A. 66; C. 17; P. 10; V. 6.

(*Colour.*) " Upper side pale brown, or dirty white." Hanmer.

Apparently not common in the British seas. Hitherto noticed only by Mr. Hanmer, who states that it occurs at Plymouth, though very rarely. The Museum of the Cambridge Philosophical Society possesses specimens from Weymouth, where it is called *Megrim**. The name of *Scald-Fish* has arisen from the peculiarly smooth naked appearance of the sides, when divested of the scales, which adhere so slightly as to yield to the slightest friction. Inhabits the Mediterranean, along with one or two other closely allied species.

(28.) *P. Cyclops*, Don. Brit. Fish. vol. iv. pl. 90. *Platessa Cyclops*, Flem. Brit. An. p. 199.

Eyes on the left side: left eye subvertical, and visible on both sides. Body very broad, smooth; marked with dusky spots, surrounded by a whitish ring: head elongated: lateral line much curved above the pectoral fin: scales inconspicuous: dorsal commencing behind the eye: middle rays of both dorsal and anal longest: caudal rounded:

D. 66; A. 52; C. 16; P. 11; V. 7.

Length, one inch and three-eighths. Don.

An obscure and doubtful species. Sent to Donovan by Captain Merrick, of Aberfraw, in Anglesea, North Wales, who obtained it on that coast. Probably the fry of some other species. The backward commencement of the dorsal fin associates it with the last genus; but the sinistral position of the eyes with this.

GEN. 57.   SOLEA, *Cuv.*

(1. Solea, *Cuv.*)

154. S. *vulgaris*, Flem. (*Common Sole.*) — Greatest breadth not half the length: upper side of the body dark brown; the pectoral tipped with black.

S. vulgaris, *Flem. Brit. An.* p. 197. Pleuronectes Solea, *Linn. Syst. Nat.* tom. i. p. 457. *Bloch, Ichth.* pl. 45. *Don. Brit. Fish.* vol. iii. pl. 62. *Turt. Brit. Faun.* p. 96. Buglossus, seu Solea, *Will. Hist. Pisc.* p. 100. tab. F. 7. Sole, *Penn. Brit. Zool.* vol. iii. p. 231. *Id.* (Edit. 1812.) p. 311.

Length. From twelve to eighteen inches; sometimes two feet, or more.

Descript. (*Form.*) Oblong-oval; very much rounded anteriorly; body narrowing behind: dorsal line carried on in one continuous curve to the

---

* According to Mr. Hanmer, the name of *Megrim* is sometimes given to the last species. Several other instances might be pointed out, in which the same English name is applied on different parts of the coast, to two or more totally distinct species.

mouth: greatest breadth, dorsal and anal fins excluded, rather more than one-third of the entire length : length of the head just half the breadth of the body: snout obtuse and rounded, projecting beyond the mouth; this last appearing distorted on the side opposed to the eyes, and furnished on that side only with fine velvet-like teeth; upper jaw the longest: eyes small; distant from each other about twice their diameter; the lower one immediately above the corner of the mouth, the upper one further advanced towards the end of the snout; the space between them flat: nostrils tubular, placed a little above the lip, one on the upper and the other on the under side of the head: side of the head opposed to the eyes bearded with numerous white fleshy cirri: lateral line arising above the upper eye, and, after making a great curve, descending to the upper part of the opercle; thence running straight to the caudal along the middle of the side: scales small, of an oblong form; their free edges ciliated, the denticles about ten in number: dorsal commencing a little above the mouth, and extending along the whole ridge of the back quite to the caudal; its greatest elevation less than one-seventh of the breadth of the body; all the rays simple, of a compressed conical form, and scaly for the greater part of their length: pectorals one-third the length of the head, both of equal size, placed just behind the upper part of the gill-opening; narrow and rounded, with the middle rays longest; first and last rays simple, the others branched: anal commencing a little in advance of the insertion of the pectorals; answering to the dorsal: caudal oblong, slightly rounded at the extremity: ventrals very small, about two-thirds the length of the pectorals; situate just in advance of the anal; third ray longest:

B. 6; D. 84; A. 67; C. 18; P. 8; V. 5.

"Number of vertebræ forty-seven [*]." (*Colours.*) All the upper side of the body dark brown, the scales edged with a deeper tint, causing a reticulated appearance; the pectoral on that side tipped with black: under side of the body white: irides golden yellow.

Common on all parts of the coast, particularly in the West and South of England, where it attains a large size. Weight, according to Pennant, sometimes so much as six or seven pounds; usually, however, very much less. Keeps almost entirely at the bottom, and feeds on the eggs and fry of other fish.

155. S. *Pegusa*, Yarr. (*Lemon Sole.*) — Greatest breadth, dorsal and anal fins included, half the length: upper side of the body light orange-brown, freckled with dark brown spots; pectoral tipped with black.

S. Pegusa, *Yarr. in Zool. Journ.* vol. iv. p. 467. pl. 16.

Length.　Eight to ten and a half inches.

Descript.　(*Form.*) Wider in proportion to its whole length than the *Common Sole*, and also somewhat thicker: greatest breadth (in a specimen eight inches long), not including the dorsal and anal fins, three inches, including both fins, four inches: head obtuse, shorter and wider: mouth arched: opercle formed externally of a single piece, circular in shape, and less deep: under surface of the head almost smooth, without any of the papillary eminences so remarkable in the *Common Sole;* the nostril on that side pierced in a prominent tubular projection, wanting in

---

[*] Yarrell. (*Zool. Journ.* vol. iv. p. 468.)

the other species: scales differing both in character and general arrangement; the appearance of them more strongly marked upon the under
than upon the upper surface: lateral line straight, but not very strongly
marked: tail narrower than in the *Common Sole*, though composed of
the same number of rays:

D. 81; A. 69; C. 17; P. 8; V. 5.

Number of vertebræ forty-three. (*Colours.*) Upper surface a mixture of
orange and light brown, freckled over with small circular spots of very
dark brown, presenting a mottled appearance; tip of the pectoral black:
under surface white. YARR.

First obtained by Mr. Yarrell at Brighton, where it is said to be
"occasionally taken with the Common Sole by trawling over a clear
bottom of soft sand, about sixteen miles from the shore." Is known
there by the name of *Lemon Sole*. Has since been met with, in a
few instances, in the London market. *Obs.* This species is not the
Pleuronectes Pegusa of Risso, as was at first supposed by Mr. Yarrell.
It appears to be undescribed by any of the continental authors.

(2. MONOCHIRUS, Cuv.)

**156. S. *Lingula*, Nob. (*Red-backed Sole.*)—Eye-side
of the body light reddish brown; dorsal, anal, and caudal
fins with dusky spots.**

Pleuronectes Lingula, *Hanmer in Penn. Brit. Zool.* (Edit. 1812.)
vol. III. p. 313. pl. 48.   P. variegatus, *Don. Brit. Fish.* vol. v.
pl. 117.   Solea variegata, *Flem. Brit. An.* p. 197.   S. Mangilii,
Bonap. Faun. Ital. Fasc. v.   Solea parva, sive Lingula, *Rondel.
Pisc.* p. 324.   *Will. Hist. Pisc.* p. 102. tab. F. 8. fig. 1.

LENGTH. From six to nine inches.
DESCRIPT. (*Form.*) Very much resembling the *Common Sole*, but
remarkably distinguished by the small size of the pectorals, that on the
eye-side being less than one-eighth the length of the head, that on the
side opposite scarcely perceptible: body rather thicker in proportion than
in that species; the breadth hardly so great, equalling just one-third of
the entire length, excluding caudal: eyes rather nearer together; the
upper one a little in advance: scales of a different form; oblong, but
always contracted about the middle; their free edges set with more
numerous denticles, varying from eighteen to twenty-one in number:
dorsal and anal fins with fewer rays, and not approaching quite so
near the caudal:

B. 6; D. 77; A. 62; C. 19; P. (Right) 4; V. 5.

In other respects the two species are similar. (*Colours.*) "Upper side a
very light brown, tinged with red; the scales shewing a pattern, something like that of the *Common Sole*, though in proportion coarser; the
dorsal, anal, and caudal fins, marked with brown or blackish spots, extending some lines to the body of the fish." HANMER.

A local species obtained by Mr. Hanmer from the coast near Plymouth,
where it is said to be common in the Spring. It is probably the same as
the *Pleuronectes variegatus* of Donovan, which was procured by that
naturalist in Billingsgate market, and which is said to have been since
found at Rothsay, in Scotland*. The specimen from which the above

<hr />

* Loudon's Mag. of Nat. Hist. vol. VI. p. 530.

description was taken, was caught at Weymouth, and is now in the
Museum of the Cambridge Philosophical Society. It appears to differ
from Mr. Hanmer's fish in its colours, but as these were not observed in
the recent state, and may possibly have been altered by the preserving
liquor, I have suppressed any notice of them, and substituted a part of
Mr. Hanmer's description. It also differs in the larger number of dorsal
fin-rays, which amount, in Mr. Hanmer's fish, to about sixty-eight. Further observation is necessary in order to decide whether, in this instance,
I have confounded two nearly allied species.

GEN. 58. LEPADOGASTER, *Gouan*.

157. L. *Cornubiensis*, Flem. (*Cornish Sucker*.) —
A double cirrus in front of each eye: dorsal and anal
fins connected by a membrane with the caudal.

> L. Cornubiensis, *Flem. Brit. An.* p. 189. Cyclopterus Cornubicus,
> *Shaw, Gen. Zool.* vol. v. p. 397. C. ocellatus, *Don. Brit. Fish.*
> vol. IV. pl. 76. *Turt. Brit. Faun.* p. 116. Small Suck-Fish,
> *Borl. Cornw.* p. 269. pl. 25. f. 28, & 29. Jura Sucker, *Penn. Brit.*
> *Zool.* vol. III. p. 137. pl. 22. no. 59. *Id.* (Edit. 1812.) vol. III.
> p. 181. pl. 25. *Couch in Linn. Trans.* vol. XIV. p. 87.

LENGTH. Four inches.

DESCRIPT. (*Form.*) Head and anterior part of the body broad and
depressed; towards the caudal compressed and tapering: snout very
much produced, spatula-shaped, narrower and more flattened than the
head; gape wide; jaws nearly equal, the lower one a little the shortest;
both furnished with minute sharp teeth; lips a little reflected: length of
the head rather more than one-third of the entire length: eyes lateral;
the space between them equalling about twice their diameter; immediately in advance of the anterior angle of each a membranous cirrus
with a second minute filament branching out from its base; behind the
cirrus a small fleshy tubercle: gill-opening small: skin smooth and
naked: pectorals large, placed immediately behind the gill-opening, and
extending downwards to the lower surface of the body, where the rays
become suddenly stronger, and the membrane, doubling forwards, passes
on to unite with that of the opposite fin under the throat; the membranes
of the pectorals thus united enclose a disk, and form an hemispherical
cavity; behind this cavity is a second, larger, circular, concave disk,
formed by the united ventrals: dorsal commencing beyond the middle of
the entire length, and reaching very nearly to the caudal, with which its
membrane is connected: anal shorter, commencing further back, united
in like manner to the caudal: rays of both fins articulated but simple:
caudal rounded: number of fin-rays,

D. 19; A. 11; C. 14; P. 18, and 4 stouter ones.

(*Colours.*) Dusky, or purplish brown, (according to Mr. Couch, sometimes crimson,) with minute inconspicuous spots; flesh-coloured beneath:
on the nape, behind the eyes, two ocellated spots; "each consisting of a
large obovate spot of deep purple, enclosed within a broad pale brownish

ring, and embellished in the centre with a brilliant blue dot, or pupil*: "
dorsal, anal, and caudal fins, bright purplish red.

First observed by Borlase on the coast of Cornwall, where it has been
since noticed by Mr. Couch. Found by Pennant in the Sound of Jura;
by Montagu †, in some plenty, at Milton, on the coast of Devonshire,
adhering to the rocks at low water. *Obs.* All our English authors represent this species as having only eleven rays in the dorsal fin; and this is
made by Fleming a ground of distinction between it and the *L. Gouani*
of Risso, which is said to have a larger number. In the only two British
specimens, however, which I have had an opportunity of examining,
they amounted to no less than nineteen. Possibly we may have two species in our seas, which have been hitherto confounded ‡. I may add, that
in the above specimens, although there were two filaments before each
eye, the second was extremely minute compared with the first, and much
smaller than represented and described by Donovan.

158. L. *bimaculatus*, Flem. (*Bimaculated Sucker*.)
—No cirri before the eyes: dorsal and anal fins short;
not connected with the caudal: behind the pectoral, on
each side, a purple spot.

> L. bimaculatus, *Flem. Brit. An.* p. 190.   Cyclopterus bimaculatus,
> *Turt. Linn.* vol. I. p. 907. *Don. Brit. Fish.* vol. IV. pl. 78. *Turt.*
> *Brit. Faun.* p. 115. Bimaculated Sucker, *Penn. Brit. Zool.* vol. III.
> App. p. 397. pl. 22. *Id.* (Edit. 1812.) vol. III. p. 181. pl. 25. *Mont.*
> *in Linn. Trans.* vol. VII. p. 293.

LENGTH. An inch and a half; rarely more.

DESCRIPT. (*Form.*) General form resembling that of the last species,
but the head and anterior part of the body more depressed: snout conical,
with the sides not so much hollowed out: jaws equal; teeth more developed, those in the lower jaw sharp and curved: eyes further asunder,
and placed more laterally; no cirri in front of them: pectorals, and the
two disks which form the organs of adhesion, similar: dorsal short, and
placed far behind: anal answering to it: both fins terminating at a small
distance from the caudal, with which they are not in any way connected:
caudal narrow, the end nearly even:

D. 6; A. 6; C. 12; P. about 20, and 4.

(*Colours.*) Back and sides pink or rose-colour, with spots and interrupted
fasciæ of white: behind the pectoral fin, on each side, a purple spot, surrounded by a ring of white: irides pink, surrounded by a dark purplish
ring: fins variegated with pink and white: under surface of the body
whitish. According to Montagu §, "the fry are of a green colour, minutely speckled with blue, and without the smallest trace of the pectoral
spots."

First obtained at Weymouth by the late Dowager Dutchess of Portland.
Has been since taken at the same place by Professor Henslow. Not very
uncommon, according to Montagu, at Torcross in Devonshire, adhering
to stones and old shells; procured by deep dredging. By the same means

---

* Donovan.   † *Linn. Trans.* vol. VII. p. 294.
‡ Several others, allied to our British one, are noticed by Risso.
§ *Wern. Mem.* vol. I. p. 92.

Mr. W. Thompson has procured several specimens in Belfast Bay. Has also occurred on the coasts of Kent and Cornwall. Apparently unknown except in the British seas.

GEN. 59. CYCLOPTERUS, *Linn.*

(1. Cyclopterus, *Cuv.*)

159.   C. *Lumpus*, Linn.   (*Common Lump-Fish.*) — Three longitudinal rows of osseous tubercles on each side: a tuberculated ridge on the back, representing a first dorsal fin.

C. Lumpus, *Linn. Syst. Nat.* tom. I. p. 414. *Bloch, Ichth.* pl. 90. *Don. Brit. Fish.* vol. I. pl. 10. *Turt. Brit. Faun.* p. 115. *Flem. Brit. An.* p. 190. Lumpus Anglorum, *Will. Hist. Pisc.* p. 208. tab. N. 11. Lump-Sucker, *Penn. Brit. Zool.* vol. III. p. 133. pl. 21. no. 57. *Id.* (Edit. 1812.) vol. III. p. 176. pl. 24. Le Lump, *Cuv. Reg. An.* tom. II. p. 346.

Length. From eighteen inches to two feet.

Descript. (*Form.*) Body deep, and at the same time remarkably thick and fleshy: back sharp and elevated, with a salient ridge of osseous tubercles, occupying the place of, and representing, a first dorsal fin; the tubercles ten in number, of a somewhat conical form, striated, and sharp-pointed: three longitudinal rows of similar tubercles on each side of the body; the first commencing a little above the eye, and extending nearly to the caudal; the second commencing behind the gills, and reaching to the same distance; the third, a short row of five tubercles, placed at the side of the abdomen, and terminating near the commencement of the anal fin: there are also two very short rows of tubercles, placed one on each side of the space intervening between the dorsal ridge and the dorsal fin: belly, included between the two rows of abdominal tubercles, flat: head short; forehead broad, rising very obliquely: mouth wide; lips thick and fleshy; jaws furnished with numerous small sharp teeth, besides which are some small rough tubercles on the pharyngean bones, and near the root of the tongue: nostrils single, tubular, about half-way between the mouth and the eyes: skin without scales, but every-where rough with small sharp points: second or true dorsal placed far behind; its length a little exceeding its height; extending to near the caudal, but leaving a small intervening space: anal answering to the dorsal: ventrals united, forming together a circular disk, with a funnel-shaped cavity in the middle: pectorals very large, passing downwards and forwards beneath the throat, and surrounding the disk of the ventrals:

B. 6; D. 11; A. 10; C. 12; P. 21.

(*Colours.*) Back and sides dusky olive, here and there tinged with red-dish; belly crimson: caudal and anal fins purplish red, spotted with dusky: pectorals bright orange.

Var. β. C. pavoninus, *Shaw, Nat. Misc.* vol. IX. pl. 310. " Back of a fine azure, deepening towards the ridge: the sides tinged with crimson: mouth, sides of the head, and all the under parts to the tail, of a delicate sea-green, with a silvery tinge on the cheeks, the pectoral fins, and the part of the body next the tail: irides likewise silvery; pupil black: fins and tail terminating in a fine pale yellow." Davies.

Not an uncommon species on many parts of the British coast, but taken in most abundance northwards. Spawns, according to Bloch, in March.

Power of adhesion, by means of the ventral disk, very great. *Var. β* was taken near Bangor in Caernarvonshire, in 1797, and sent to Shaw by Mr. Hugh Davies of that place. It measured only six inches in length. *Obs.* The *Lumpus gibbosus* of Willughby* (*Cyclopt. pyramidatus,* Shaw†), characterized by a pyramidal hump on the back, and said to be found in the Scotch seas, owes its origin, in the opinion of Cuvier, to a badly-stuffed specimen of the present species.

(2. LIPARIS, *Arted.*)

160. C. *Liparis,* Linn. (*Common Sea-Snail.*)—Dorsal and anal fins united to the caudal.

> C. Liparis, *Linn. Syst. Nat.* tom. I. p. 414.   *Bloch, Ichth.* pl. 123. f. 3.   *Don. Brit. Fish.* vol. II. pl. 47.   *Turt. Brit. Faun.* p. 115. Liparis nostras, *Will. Hist. Pisc. App.* p. 17. tab. H. 6. fig. 1. Liparis vulgaris, *Flem. Brit. An.* p. 190.   Unctuous Sucker, *Penn. Brit. Zool.* vol. III. p. 135. pl. 21. no. 58.   *Id.* (Edit. 1812.) vol. III. p. 179. pl. 24.

LENGTH. From three to five inches.

DESCRIPT. (*Form.*) Body elongated, thick and rounded anteriorly, but much compressed behind: belly very protuberant: head large, broad, a little depressed in front, and somewhat inflated about the gills; its length contained about four times and a half in the entire length: snout blunt and rounded: mouth moderately large; upper lip with two short cirri: in each jaw a band of rasp-like teeth: tongue thick and fleshy: eyes small, and rather high on the cheeks: nostrils double: gill-opening very small; the opercle produced behind into a cartilaginous spine: head and body every-where covered with a smooth, soft, naked, unctuous, semi-transparent, skin: dorsal fin commencing a little behind the nape, and extending to the base of the caudal, with which it is just united; rays slender and simple, the anterior ones rather shorter than those which follow, but on the whole the rays nearly of a length: anal commencing at about half the length of the body, and also uniting to the caudal, but at a point beyond that at which the dorsal terminates: caudal slightly rounded: pectorals large, extending downwards and forwards to unite under the throat; two or three rays, just at the turn of the fin beneath the body, very much elongated, and considerably produced beyond those on each side of them: ventral disk concave, and nearly circular; placed on the throat, and partly encircled by the pectorals; the circumference set with twelve or thirteen flattened tubercles, the central portion impressed with four or five curved lines branching out on each side of a longitudinal diameter:

D. 36; A. 26; C. 12; P. 32‡.

(*Colours.*) "Pale brown, sometimes finely streaked with darker brown." PENN. In a *variety*, met with by Donovan, "the head and body were strongly marked with longitudinal streaks and waves of white, edged with blue, and disposed on a ground of testaceous or rather chestnut-colour." It is observed by this last author, that this species "differs very considerably in colour at different seasons of the year, as well as in its various stages of growth: small specimens have occurred in which the sides and belly were white; in some pale yellow, and in others rosy; the sides of the head usually partaking of the same tints as those of the body."

---

* *Hist. Pisc.* p. 209. tab. N. 10. fig. 2.
† *Gen. Zool.* vol. v. part ii. p. 390. pl. 167.
‡ The above fin-ray formula is from Donovan.

Common on many parts of the coast, and generally found near the mouths of rivers. When taken out of the water, said rapidly to dissolve and melt away. Food, according to Bloch, aquatic insects, young shells, and small fish. Spawns early in the year: found by Pennant heavy with roe in January. Arrives at a much larger size in the northern seas than in our own.

161. C. *Montagui*, Don. (*Montagu's Sea-Snail.*)— Dorsal and anal fins unconnected with the caudal: upper lip marked with several indentations.

C. Montagui, *Don. Brit. Fish.* vol. III. pl. 68. (Young.) *Mont. in Wern. Mem.* vol. I. p. 91. pl. 5. f. 1. C. Montacuti, *Turt. Brit. Faun.* p. 115. Liparis Montagui, *Flem. Brit. An.* p. 190. Montagu's Sucker, *Penn. Brit. Zool.* (Edit. 1812.) vol. III. p. 183.

Length. From two to three inches.

Descript. (*Form.*) General form similar to that of the *C. Liparis:* body very much rounded as far as the vent, beyond which it becomes suddenly compressed: head more depressed than in that species, and much inflated at the gills: snout, jaws, and teeth, similar: eyes small, placed high: front of the head, above the upper lip, scalloped with about six indentations: rest of the head, and body, very smooth: dorsal fin commencing a little behind the nape, and extending to the base of the caudal, with which, however, it is not in any way connected; rays at first very short and inconspicuous, but gradually increasing in length to just before the caudal, where the fin is broadest, and presents a rounded appearance: anal similar, and likewise separate from the caudal: pectorals and ventral disk much as in the *C. Liparis:* vent about half-way between the posterior margin of the disk and the commencement of the anal fin:

D. about 26; A. about 24; C. 12; P. about 29 *.

(*Colours.*) "Purplish brown in appearance to the naked eye; but by the assistance of a lens, the ground-colour is dull orange, covered with minute confluent spots of the former: the under parts are paler, and about the throat and sucker white: irides golden; pupil dark blue." Mont.

Discovered by Montagu, at Milton, on the south coast of Devon, where a few specimens were obtained at extraordinary low tides, among the rocks. Has been since found on the coast of Ireland by Mr. W. Thompson of Belfast: also on the coast of Berwickshire by Dr. Johnston. Apparently a rarer species than the foregoing.

GEN. 60. ECHENEIS, *Linn.*

162. E. *Remora*, Linn. (*Common Remora.*)—Shield on the head with about eighteen transverse bars: caudal crescent-shaped.

E. Remora, *Linn. Syst. Nat.* tom. I. p. 446. *Bloch, Ichth.* pl. 172. *Turt. Brit. Faun.* p. 94. *Cuv. Reg. An.* tom. II. p. 347. Remora, *Will. Hist. Pisc.* p. 119. App. tab. 9. f. 2. Mediterranean Remora, *Penn. Brit. Zool.* (Edit. 1812.) vol. III. App. p. 524.

* The above fin-ray formula is from Montagu.

LENGTH. From twelve to eighteen inches. BLOCH.

DESCRIPT. (*Form.**) Body moderately elongated; covered with small scales: head perfectly flat above; the shield consisting of from seventeen to nineteen transverse elevated bars divided into two series; the margin of the shield cartilaginous: eyes lateral: mouth wide and rounded: lower jaw advancing beyond the upper; furnished, as well as the intermaxillaries, with small card-like teeth; a very regular row of small teeth, resembling *cilia*, along the edge of the maxillaries, which form the outer margin of the upper jaw; the anterior margin of the vomer furnished with a band of card-like teeth, and its whole surface, as well as that of the tongue, rough: four orifices near the upper lip; the anterior pair cylindrical, the posterior oval: gill-opening very large: lateral line, which is scarcely visible, taking a curve towards the end of the pectoral fin: dorsal single, commencing a little beyond the middle of the length: anal opposite: vent nearer the caudal than the head: caudal crescent-shaped: all the fin-rays soft, much branched, and invested with a thick membrane.

B. 9†; D. 21; C. 20; P. 22; V. 4.

(*Colour.*) "Dusky brown." TURT.

This species, which is well known for its power of adhering, by means of the shield on the head, to other fish, and to the bottoms of vessels, is found in the Mediterranean, as well as in various parts of the ocean. In a single instance it has occurred in the British seas, Dr. Turton having taken a specimen at Swansea, from the back of a Cod-Fish, in the summer of 1806.

(III. A P O D E S.)

GEN. 61.   ANGUILLA, *Cuv.*

(1. ANGUILLA, *Cuv.*)

163.   A. *acutirostris*, Yarr. (*Sharp-nosed Eel.*) — Snout sharp, compressed at the sides; gape extending to beneath the middle of the eye: about one-third of the entire length before the dorsal, and between one-eighth and one-ninth before the pectorals.

A. acutirostris, *Yarr. in Proceed. of Zool. Soc.* 1831. p. 133. *Riss. Hist. Nat. de l'Eur. Mérid.* tom. III. p. 198.?   A. vulgaris, *Turt. Brit. Faun.* p. 87.   *Flem. Brit. An.* p. 199.   Muræna Anguilla, *Bloch, Ichth.* pl. 73. Common Eel, *Penn. Brit. Zool.*

----

* Compiled from Cuvier and Bloch.

† The above formula is from Bloch: according to Cuvier, there are but eight rays in the branchiostegous membrane.

vol. iii. p. 142.   *Id.* (Edit. 1812.) vol. iii. p. 191.   *Bowd. Brit.
fr. wat. Fish.* Draw. no. 7.   Sharp-headed Eel, *Yarr. in Zool.
Journ.* vol. 4. p. 469.   L'Anguille long-bec, *Cuv. Reg. An.* tom. ii.
p. 349.

LENGTH.   Usual length from two to three, sometimes four, feet: has
been known to attain to six feet three inches.

DESCRIPT.   (*Form.*) Very much elongated; body thick, approaching
to cylindrical; the depth and thickness nearly uniform for three-fourths
of the entire length; the last quarter compressed and slightly tapering:
depth, taken at the commencement of the dorsal fin, equalling about one-
sixteenth of the entire length: head, measured from the end of the snout
to the branchial orifice, contained nearly eight times and three-quarters
in the same; convex, and slightly elevated, at the nape, from which point
the profile slopes forward, becoming much depressed above the eyes:
snout sharp and attenuated, compared with that of the two next species;
the sides rather compressed: jaws gradually narrowing towards their
extremities, which are slightly rounded; the lower one a little the longest; both furnished with a broad band of velvet-like teeth, the band above
dilating on to the fore part of the vomer: gape small; the commissure of
the lips not extending to a vertical line drawn as a tangent to the posterior part of the orbit: eyes small; the distance from them to the end
of the snout not equalling twice their diameter; the space between them
rather less than the above distance: nostrils double; the anterior orifice
tubular, situate on the edge of the upper lip, the posterior one a simple
pore immediately in advance of the eye: a row of pores above the upper
lip on each side, and another forming the commencement of the lateral
line; which last arises a little above the pectorals, and passes off straight
to the extremity of the tail: gill-opening reduced to a small round aperture, immediately before, and a little below, the pectoral fin: scales very
minute, scarcely visible, deeply imbedded in a thick, soft, slimy skin:
dorsal commencing at about (sometimes a little before) one-third of the
entire length; low, preserving throughout the same elevation, which
equals scarcely more than one-fourth of the depth: vent before the
middle of the entire length by a space equalling the depth of the body;
anal commencing immediately behind it, similar to the dorsal: both
dorsal and anal are carried quite to the extremity of the tail, forming by
their union a pointed caudal: pectorals small and rounded, not half the
length of the head; the distance from the line of their insertion to the
end of the snout contained eight times and a half in the entire length,
and about twice and three-quarters in the portion anterior to the commencement of the dorsal fin: ventrals wanting.   Number of vertebræ
one hundred and thirteen*.   (*Colours.*) Upper part of the head, back,
and a large portion of the sides, dark olivaceous green, tinged with
brown; lower part of the sides paler: throat, belly, and a portion of the
anal fin, yellowish white.

Common in rivers, lakes, and other fresh-waters, throughout the country.   Attains to a larger size than either of the two following species, with
which it was formerly confounded.   Two taken some years since in a fen-
dyke near Wisbeach, in Cambridgeshire, weighed together fifty pounds;
the heaviest twenty-eight, the other twenty-two pounds.   Usually, however, much smaller.   Generally considered as viviparous, but, from the
observations of Mr. Yarrell, it is probable that this is not the case†.   In
the Autumn, migrates down the rivers, in order, it is said, to pass the

---

* The number of vertebræ rests on the authority of Mr. Yarrell.
† See on this subject *Proceed. of Zool. Soc.* 1831. p. 133; also Jesse's *Glean. in Nat. Hist.*
(Second Series), p. 57, &c.

Winter in the brackish water, and to deposit its spawn; the young fry
migrating up the river in the Spring. Many, however, certainly remain
in ponds all the year, and breed there. Roves about, and feeds, principally in the night. Said to quit its native element occasionally, and
to cross meadows, in search of other waters, as well as for the purpose of
feeding on worms and snails*. Very tenacious of life. *Obs.* This species
varies a good deal in colour, according to the nature of the water in which
it is found. Those in which the belly is of a clear white are called sometimes *Silver Eels.*

164. A. *latirostris*, Yarr. (*Broad-nosed Eel.*)—Snout
broad and rounded; gape extending to a vertical line from
the posterior part of the orbit: more than one-third of the
entire length before the dorsal, and about one-seventh
before the pectorals.

> A. latirostris, *Yarr. in Proceed. of Zool. Soc.* 1831. p. 133. *Riss.
> Hist. Nat. de l'Eur. Mérid.* tom. III. p. 199.? Blunt-headed Eel,
> *Yarr. in Zool. Journ.* vol. IV. p. 469. Glut Eel, *Bowd. Brit. fr.
> wat. Fish.* Draw. no. 22. L'Anguille pimperneaux, *Cuv. Reg.
> An.* tom. II. p. 349.

LENGTH. From one to two, perhaps sometimes three, feet.
DESCRIPT. (*Form.*) Body much larger and thicker anteriorly than in
the last species, but more compressed behind; thickness not uniform
beyond the commencement of the dorsal, from which point the compres-
sion of the sides rapidly increases: depth greatest at the nape: head
large, appearing, when viewed from above, broader than the body: snout
blunt and rounded, flattened before the eyes: jaws broad; the lower one
wider and longer than the upper: gape large; the commissure reaching
to, or almost beyond, a tangent to the posterior part of the orbit: lips
thick and fleshy at the sides of the mouth, and partially reflexed: eyes
larger than in the *A. acutirostris;* the distance from them to the end
of the snout equals at least twice their diameter; the distance between
them rather less: dorsal commencing at a point beyond one-third of the
entire length; both that and the anal thicker in substance and more
elevated than the same fins in the *A. acutirostris,* their height equalling
nearly half the depth: vent before the middle by a space equalling about
three-fourths of the depth: tail broader, and more rounded at its extremity: pectorals somewhat larger, and placed, as well as the branchial
orifices, further behind; the distance from the line of their insertion to
the end of the snout is contained not more than seven times in the entire
length, and not so much as twice and a half in the portion anterior to the
commencement of the dorsal fin. Number of vertebræ one hundred and
fifteen†. (*Colours.*) Back and sides of a darker colour than in the
A. acutirostris, and having more of a bluish than a greenish tinge; the
lateral line, however, forms a pale green stripe down each side: underneath, including a portion of the anal, white, without any yellow tinge.
The colours, however, are variable, as in the last species.
This species, which is probably the *Grig‡* or *Glut Eel* of Pennant, is
nearly as common as the last. It has not been known, however, to exceed

* See an instance mentioned by Dr. Hastings in his *Nat. Hist. of Worcestersh.* p. 134.

† The number of vertebræ rests on the authority of Mr. Yarrell.

‡ I am informed by Mr. Yarrell, that the term *Grig* is applied in many places generally to all
small-sized Eels. Too much reliance, therefore, must not be placed upon the mere name.

a weight of five pounds. Independently of the above external differences, Mr. Yarrell has observed others " in the size and character of the bones of the head and *vertebræ ;* those of the present species being nearly as large again as the same parts of the *A. acutirostris* in examples of the same length * "

165. A. *mediorostris*, Yarr. (*Snig Eel.*)—Snout rather long, and moderately broad ; gape extending not quite to a vertical line from the posterior part of the orbit: rather less than one-third of the entire length before the dorsal, and between one-seventh and one-eighth before the pectorals.

A. mediorostris, *Yarrell's Mss.* Snig Eel, *Yarr. in Jesse's Glean. of Nat. Hist.* (2nd Series) pp. 75, & 76.

LENGTH. The length of my specimen is nineteen inches.

DESCRIPT. (*Form.†*) More slender and elongated in proportion to the depth and thickness than either of the preceding species: depth at the commencement of the dorsal fin not exceeding one-nineteenth of the entire length: nape but little elevated, and nearly in the same horizontal line with the profile: snout and jaws somewhat resembling those of the *A. acutirostris*, but longer and broader than in that species, though not so broad as in the *A. latirostris :* both jaws rounded at their extremities ; the lower one longest: teeth longer and more developed than in the *A. acutirostris :* gape more capacious, owing to the greater length of the jaws ; commissure nearly, but not quite, extending to a tangent to the posterior part of the orbit: the distance from the eye to the end of the snout equalling full twice the diameter of the former: dorsal commencing rather before one-third of the entire length ; its height about one-third of the depth of the body: vent nearer the middle than in either of the two former species: caudal moderately pointed at its extremity: pectorals small ; the distance from the line of their insertion to the end of the snout contained seven times and a half in the entire length. (*Colours.*) Upper parts dark greenish brown, passing by a lighter olive-green to yellowish white below.

This species was first distinguished by Mr. Yarrell, who received it from the river Avon in Hampshire. Said to be known there by the name of *Snig.* Does not attain to a large size, seldom exceeding half a pound in weight. Said to differ from the other eels in its habit of roving and feeding during the day. Presents also some osteological peculiarities, " the first five cervical vertebræ being smooth and round, and entirely destitute of superior or lateral spinous processes, both of which are possessed by the two other species ‡."

(29.) *Grig Eel*, Bowd. Brit. fr. wat. Fish. Draw. no. 28. *L'Anguille Plat-Bec*, Cuv. Reg. An. tom. II. p. 349.

Being unacquainted with this species, I am unable to point out its distinguishing characters. According to Mrs. Bowdich, it is the smallest of

* *Proceed. of Zool. Soc.* 1831. p. 133.

† The above description having been drawn up with reference to a single specimen, the only one I have had an opportunity of examining, possibly some of the characters may not be found constant in all cases.

‡ Yarr. *l. c.*

the Eel tribe, and is caught plentifully in the Thames, but more especially in Berkshire and Oxfordshire. She thinks that Pennant has confounded it with the *Glut Eel*. Mr. Yarrell informs me, he considers it as distinct from the last species.

(2. CONGER, *Cuv.*)

166. A. *Conger*, Shaw. (*Conger Eel*.) — Dorsal and anal fins margined with black : lateral line spotted with white.

> A. Conger, *Shaw, Gen. Zool.* vol. IV. part i. p. 20. pl. 1. *Turt.
> Brit. Faun.* p. 87. *Flem. Brit. An.* p. 200. Muræna Conger,
> *Linn. Syst. Nat.* tom. I. p. 426. *Bloch, Ichth.* pl. 155. *Don.
> Brit. Fish.* vol. V. pl. 119. Conger, *Will. Hist. Pisc.* p. 111,
> tab. G. 6. *Penn. Brit. Zool.* vol. III. p. 147. *Id.* (Edit. 1812.)
> vol. III. p. 196. *Yarr. in Proceed. of Zool. Soc.* (1831.) p. 158.

LENGTH. From five to six feet : said to reach occasionally as much as ten feet, or upwards.

DESCRIPT. (*Form.*) General form resembling that of the *Common Eel* : body thick, and nearly cylindrical anteriorly, compressed and tapering behind : head larger than in that species, being a little less than one-seventh of the entire length : crown flat ; snout a little depressed, narrowing towards the extremity, and rather pointed : upper jaw a little the longest : both jaws with a band of sharpish card-like teeth, "forming three rows, of which those in the middle line are much the largest ; numerous smaller teeth, more uniform in size, occupy the line of the vomer, but do not extend far backwards :"* lips fleshy : gape wide ; not extending quite so far as a tangent to the posterior part of the orbit : eyes much larger than in the *Common Eel* : nostrils double ; the first orifice placed a little before the eye ; the second, which is tubular, at the extremity of the snout : a row of mucous pores along the upper lip ; several pores also between the corner of the gape and the gill-opening : dorsal commencing a little behind the pectorals, or at about one-fifth of the entire length : vent (in a specimen measuring thirty-one inches and a half in length) about three inches before the middle : anal commencing immediately behind the vent, and extending quite to the extremity of the tail, where it unites with the dorsal (prolonged in a similar manner) to form a pointed caudal :

"B. 10 ; D. A. & C. 306 ; P. 19"†.

(*Colours.*) Of a uniform pale brownish gray above, passing into a dirty white beneath : dorsal and anal fins whitish, margined with deep bluish black : lateral line spotted with white.

A common inhabitant of the British seas, and found on most parts of the coast in considerable abundance. Attains to a very large size : has been known to weigh upwards of one hundred pounds. Frequents rocky ground. Is very voracious, preying on other fish and on crustacea.

> (30.) *A. Myrus*, Shaw, Gen. Zool. vol. IV. part i. p. 24. Turt. Brit.
> Faun. p. 87. Flem. Brit. An. p. 200. *Muræna Myrus*, Linn.
> Syst. Nat. tom. I. p. 426. Berkenh. Syn. vol. I. p. 64. *Le Myre*,
> Cuv. Reg. An. tom. II. p. 350.

* Yarrell.   † Bloch.

This species, which is found in the Mediterranean, has been included in the British Fauna by Berkenhout and Turton, but it is not said on what authority. It is distinguished from the last by its smaller size, and by some spots on the snout, a transverse band on the occiput, and two rows of dots on the nape, of a whitish colour *.

(5.) *OPHISURUS*, Lacép.

(31.) *O. Ophis*, Lacép. Hist. Nat. des Poiss. tom. ii. p. 196. Flem. Brit. An. p. 200. *Murǽna Ophis*, Linn. Syst. Nat. tom. i. p. 425. Bloch, Ichth. pl. 154. Berkenh. Syn. vol. i. p. 64. *Ophis maculata*, Turt. Brit. Faun. p. 87.

Like the last, a very doubtful native. Given as British by Berkenhout, but without any remarks. Of a whitish or silvery colour, with several longitudinal rows of dark oval spots. Length from three to four feet. Inhabits the European seas.

(32.) *O. Serpens*, Lacép. Hist. Nat. des Poiss. tom. ii. p. 198. *Murǽna Serpens*, Linn. Syst. Nat. tom. i. p. 425. *Serpens marinus*, Merr. Pinax, p. 185. Sibb. Scot. Illust. part ii. tom. ii. p. 23.

Whether this, or the last, be the species alluded to by Merrett and Sibbald under the name of *Serpens marinus*, is very doubtful. Neither is it known on what authority either of these naturalists has inserted it in the British Fauna. The *O. Serpens* of Lacépede is distinguished from the *O. Ophis*, by its being without spots. It also grows to a larger size, attaining the length of five or six feet. A native of the Mediterranean.

GEN. 62. MURÆNA, *Thunb.*

167. M. *Helena*, Linn. (*Common Murœna.*) —. Olivaceous brown, marbled with yellow.

M. Helena, *Linn. Syst. Nat.* tom. i. p. 425. *Bloch, Ichth.* pl. 153. Muræna, *Will. Hist. Pisc.* p. 103. tab. G. 1. Roman Muræna, *Shaw, Gen. Zool.* vol. iv. part i. p. 26. pl. 2. Murène commune, *Cuv. Reg. An.* tom. ii. p. 252.

LENGTH. Three feet and upwards. Cuv.

DESCRIPT. (*Form.*) Body, in *old* fish, compressed at the sides, in *young*, round : head small : mouth large : jaws armed with sharp pointed teeth, a little distant from each other : palate also armed with teeth : two tubular orifices near the eyes, and two at the extremity of the snout : gill-opening large : dorsal, anal, and caudal, united; forming together a low fleshy fin, invested by the common skin, commencing on the back at a pretty considerable distance from the head, passing round the tail, and terminating underneath the body at the vent: no pectorals or ventrals. BLOCH. (*Colours.*) "Of a dusky greenish brown, pretty thickly variegated on all parts with dull yellow subangular marks or patches, disposed in a somewhat different manner in different individuals, and generally scattered over with smaller specklings of brown; the whole forming a kind of obscurely reticular pattern." SHAW.

An individual of this species, measuring four feet four inches in length, was caught by a fisherman at Polperro, in Cornwall, in October 1834†. I am not aware that it had been ever taken previously in our

---

* Cuvier, *l. c.*

† This circumstance was com unicated by Mr. Couch to Mr. Yarrell, to which latter gentleman I am indebted for the knowledge of it.

seas*. Common in the Mediterranean, and well-known as the *Muræna* of the Romans. Said to live with equal facility in fresh and salt water, though principally found at sea. Is very voracious.

GEN. 63. LEPTOCEPHALUS, *Gronov.*

168. L. *Morrisii*, Gmel. (*Anglesea Morris.*)

L. Morrisii, *Gmel. Linn.* tom. I. part iii. p. 1150. *Turt. Brit. Faun.* p. 88. *Mont. in Wern. Mem.* vol. II. p. 436. pl. 22. f. 1. *Leach, Zool. Misc.* vol. III. p. 10. pl. 126. *Flem. Brit. An.* p. 200. *Cuv. Reg. An.* tom. II. p. 358. *Deere in Loud. Mag. of Nat. Hist.* vol. VI. pp. 530, & 531. Ophidium pellucidum, *Couch in Loud. Mag.* vol. V. pp. 313, & 742. Anglesea Morris, *Penn. Brit. Zool.* vol. III. p. 158. pl. 25. no. 67. *Id.* (Edit. 1812.) vol. III. p. 212. pl. 28.

LENGTH.   From five inches to six inches and a quarter.

DESCRIPT. (*Form.*) Body ribband-shaped, extremely thin and compressed, semipellucid: greatest depth, which is tolerably uniform throughout, diminishing only near the head and tail, one-twelfth of the entire length: thickness (according to Montagu) not exceeding the sixteenth part of an inch: head small; the profile sloping a little downwards from the line of the back, which is nearly straight: snout short: jaws nearly equal: teeth (according to Montagu) numerous, and all inclining forwards: eyes large: gill-opening, a small transverse aperture before the pectorals: lateral line straight, and nearly in the middle: sides of the body marked with a double series of oblique lines which meet in the lateral line at an acute angle; these lines are parallel to each other in the same series, and the angles formed by their union with those of the other series are directed forwards: dorsal commencing a little beyond one-third of the entire length, low and rather obscure, the rays extremely delicate, and not easily counted: vent about the middle†: anal commencing immediately behind it; in form, similar to the dorsal: both dorsal and anal are carried on to the extremity of the tail, where they unite to form a caudal: pectorals extremely small, scarcely a line in length, but sufficiently obvious, if carefully sought for: ventrals wanting. (*Colours.*) Pale colourless white, with a row of minute black dots along the margins of the back and abdomen: a few similar dots, arranged in a longitudinal series, down the mesial line of each side.

First discovered, in the sea near Holyhead, by the late Mr. William Morris, who communicated it to Pennant. Has been since met with in several instances in our seas. Four specimens taken near Beaumaris, by the Rev. Hugh Davies; one by Mr. Lewis Morris, at Penrhyn Dyfi; two by Mr. Anstice in the river Pervet, near Bridgewater; one by Mr. Deere, at Slapton, near Dartmouth; and four by Mr. Couch, on the coast of Cornwall. Mr. Thompson has also recently recorded the occurrence, at different times, of six specimens on the coast of Ireland. The pectorals are so small, as to have been thought wanting by Pennant, a circum-

---

* A species of *Muræna*, three feet in length, is figured in Nash's "Collections for the History of Worcestershire" (vol. I. p. lxxxvi.), along with some other fish from the Severn, but nothing positive is stated respecting its capture, or the circumstances which have led to its being introduced into that work. It appears not very dissimilar to the *M. Helena* described above.

† So it appeared to be in the specimen examined by me. Montagu says, "situated a trifle nearer the head than the tail."

stance which has led to some little confusion amongst naturalists, in their attempts to identify his fish. There can be little doubt, however, that in all the above instances, the same species has been observed. At the same time it may be added, that several others have been detected in warmer latitudes, though I am not aware that their essential and distinguishing characters have been hitherto established.

GEN. 64.   OPHIDIUM, *Linn.*

(33.) *O. barbatum*, Linn. Syst. Nat. tom. i. p. 431.   Bloch, Ichth. pl. 159. f. 1.   Berkenh. Syn. vol. i. p. 66.   Turt. Brit. Faun. p. 88.   *Donzelle commune*, Cuv. Reg. An. tom. ii. p. 359.

Two pair of small barbules attached to the extremity of the hyoid bone, the anterior pair shorter than the other. Flesh-colour; the dorsal and anal fins edged with black. Length from eight to ten inches*. Cuv.
Introduced into the British Fauna by Berkenhout, but without the mention of any authority for its insertion. Must, in consequence, be considered as a very doubtful native. Found in the Mediterranean, along with another closely allied species.

169.   O. *imberbe*, Mont.   (*Beardless Ophidium.*) — Lower jaw without barbules.

O. imberbe, *Mont. in Wern. Mem.* vol. i. p. 95. pl. 4. f. 2.   *Flem. Brit. An.* p. 201.   Beardless Ophidium, *Penn. Brit. Zool.* vol. iii. App. p. 398. and vol. iv. pl. 93.?   *Id.* (Edit. 1812.) vol. iii. p. 208. pl. 29.

Length. About three inches. Mont.
Descript. (*Form.*) Body ensiform, considerably compressed towards the tail, and in shape not unlike *Cepola rubescens:* depth about one-twelfth of the entire length: head very obtuse, rounded in front: mouth, when closed, inclining obliquely upwards; lips marginated: eyes large, placed forward, and lateral: gill-membranes inflated beneath: lateral line nearly in the middle, arising at the angle of the gill-cover, but rather obscure: vent nearly in the middle: pectorals rounded: dorsal commencing immediately above the base of the pectoral, at first not so broad, and usually not so erect, as the other part: anal commencing at the vent, and, together with the dorsal, uniting with the caudal fin, which is cuneiform, but obtusely pointed:

D. about 77 ; A. 44; C. 18 or 20; P. 11.

(*Colours.*) Purplish brown, disposed in minute speckles; along the base of the anal fin about ten small bluish white spots, regularly placed, but scarcely discernible without a lens, and possibly peculiar to young specimens: all the fins of the same colour as the body, except the pectoral and caudal; the first of which is pale, the last yellowish: irides dark, with a circle of silver round the pupil. Mont.
The above fish was obtained on the south coast of Devon by Montagu, who considered it as the *Ophidium imberbe* of Linnæus. Cuvier, however, appears to have entertained some doubts as to its identity with that species †. Whether it be the same as the *Beardless Ophidium* of Pennant, which was sent to that naturalist from Weymouth by the late

---

* Cuvier says "eight or ten inches at the most," but Bloch, "from twelve to fourteen inches."
† See *Reg. An.* tom. ii. p. 359. note (2).

H h

Dowager Dutchess of Portland, and which also was referred to the
Linnæan species, is likewise uncertain; Pennant having published no
description of his fish, and his figure being very unlike that given by
Montagu. Montagu's specimen was taken among rocks at low water.

GEN. 65.   AMMODYTES, *Linn.*

170.   A. *Tobianus,* Bloch. (*Wide-mouthed Launce.*)
— Gape large; maxillaries long; the pedicels of the
intermaxillaries very short : dorsal commencing in a line
with the extremities of the pectorals.

> A. Tobianus, *Bloch, Ichth.* pl. 75. f. 2.   *Turt. Brit. Faun.* p. 87.
> *Cuv. Reg. An.* tom. ii. p. 360.   A. lanceolatus, *Lesauv. in Bull.
> des Sci. Nat.* (1825.) tom. iv. p. 262.   A. Anglorum verus, *Jago
> in Ray's Syn. Pisc.* p. 165. pl. 2. f. 12.   Sand Launce, *Penn. Brit.
> Zool.* vol. iii. p. 156. but not pl. 25. no. 66.   *Id.* (Edit. 1812.)
> vol. iii. p. 206. but not pl. 28.

LENGTH.   From ten to fifteen inches and a half.

DESCRIPT.   (*Form.*) Slender, and very much elongated : body square,
but with the angles somewhat rounded, approaching cylindrical, and of
nearly equal thickness throughout : greatest depth contained about sixteen times in the entire length : head an elongated cone, forming one-
fifth of the same : lower jaw projecting far beyond the upper, and
terminating in a point ; the upper one slightly rounded at its extremity :
scarcely any perceptible teeth, excepting two long sharp teeth on the
front of the vomer directed downwards : gape very wide on account of the
great length of the maxillaries ; intermaxillaries (compared with those of
the next species) with the pedicels very short : when the mouth is fully
opened, the upper jaw turns up at its extremity, and the maxillaries
become vertical, drawing after them the sides of the lower jaw, which,
ascending from behind, become vertical also, and parallel to the former :
gill-opening very large : pieces of the opercle all considerably developed,
but especially the subopercle, which is produced beyond the true opercle
in the form of a projecting lobe, having its descending margin sinuated,
and its surface elegantly marked with several diverging striæ; true
opercle forming an equilateral triangle : head naked ; body covered with
minute scales : lateral line arising on each side of the nape, and running
parallel with the dorsal fin a very little below it ; marked by a series
of oblong slightly elevated tubercles : along the middle of each side a
second impressed line formed by the division of the muscles : dorsal commencing at about, or a little beyond, one-fourth of the entire length,
exactly in a line with the extremities of the pectorals, and terminating
a little before the caudal ; height tolerably uniform throughout, equalling
not quite half the depth of the body ; rays very slender ; all simple, but
articulated : vent some little way beyond the middle of the entire length ;
anal commencing immediately behind it, similar to the dorsal, and terminating in the same line with that fin : caudal forked for nearly half its
length ; the rays much branched, with the exception of the outermost
above and below : pectorals inserted just below the produced lobe of the
subopercle, and equalling one-third the entire length of the head ; fourth
and fifth rays longest ; the middle ones branched ; two or three of the
lateral ones above and below simple.

B. 7 ; D. 58 ; A. 31 ; C. 15, and a few short ones : P. 15.

(Colours.) Back, and upper part of the sides, brown, a little varied with blue and green: one or two dusky lines running parallel with the dorsal fin: lower part of the sides, and belly, silvery.

Not so common on the British coast as the next, with which it was confounded previously to M. Lesauvage, who first pointed out (l. c.) the distinguishing characters of the two species. Generally keeps near the shore, burying itself in the sand, at the ebb of the tide, to the depth of one or two feet. Food, marine worms, and, according to Bloch, the young of its own species. Is much used as a bait for other fish. Said to spawn in May.

171. A. *Lancea,* Cuv. (*Small-mouthed Launce.*)— Gape not so large; maxillaries short; the pedicels of the intermaxillaries very long: dorsal commencing before the extremities of the pectorals.

A. Lancea, *Cuv. Reg. An.* tom. ii. p. 360. A. Tobianus, *Don. Brit. Fish.* vol. ii. pl. 33. *Swains. Zool. Illust.* vol. i. pl. 63. upper fig. *Lesauv. in Bull. des Sci. Nat.* (1825.) tom. iv. p. 262. *Flem. Brit. An.* p. 201. Ammodytes, *Will. Hist. Pisc.* p. 113. Launce, *Penn. Brit. Zool.* vol. iii. pl. 25. no. 66. but not p. 156. Sand-Launce, *Id.* (Edit. 1812.) vol. iii. pl. 28. but not p. 206.

LENGTH. From five to eight inches; rarely more.

DESCRIPT. (*Form.*) Much thicker in proportion than the *A. Tobianus;* in a fish measuring one-fourth less in length, the depth and thickness remain the same as in that species: head a perfect cone, contained five times and a half in the entire length: lower jaw not produced so far beyond the upper, and less pointed: the two teeth on the vomer much less developed: gape smaller, the maxillaries being much shorter; the pedicels of the intermaxillaries, on the contrary, are considerably longer, very much increasing the protractility of the upper jaw, which, when the mouth is opened, instead of turning back as in the last species, protrudes itself forwards and downwards, the maxillaries never becoming vertical: the pieces of the opercle not so much developed, nor produced so far backwards, but preserving the same form: dorsal commencing a little nearer the head, in a line with the commencement of the last quarter of the pectorals: both dorsal and anal contain fewer rays: pectorals exactly half the length of the head: in other respects the forms of the two species are similar.

D. 53 or 54; A. 28; C. 15, &c.; P. 13.

(*Colours.*) Similar to those of the *A. Tobianus,* only paler.

Common on all our sandy shores, in which it may be found buried at the ebb of the tide. Habits resembling those of the last species. *Obs.* Willughby has erroneously figured this species (tab. G. 8. f. 1.) with two dorsal fins: his description, however, is correct.

ORDER III. OSTEODERMI.

GEN. 66. SYNGNATHUS, *Cuv.*

* *Anal, caudal, and pectoral fins, all present.*

172. S. *Acus,* Linn. (*Great Pipe-Fish.*)—Body heptangular anteriorly : crown with an elevated longitudinal
ridge ; profile descending in a sinuous curve : snout much
narrower, vertically, than the head.

S. Acus, *Linn. Syst. Nat.* tom. I. p. 416. *Bloch, Ichth.* pl. 91.
f. 2. *Turt. Brit. Faun.* p. 116. *Flem. Brit. An.* p. 175. Pipe-
Fish, *Penn. Brit. Zool.* vol. III. pl. 23. no. 60. lower fig. but not
p. 138. Shorter Pipe-Fish, *Id.* (Edit. 1812.) vol. III. pl. 26. no. 60.
lower fig. *Low, Faun. Orc.* p. 181.

LENGTH. From twelve to sixteen inches : according to Bloch, from
two to three feet.

DESCRIPT. (*Form.*) Very much elongated, slender, tapering behind :
greatest depth and thickness about equal ; each contained thirty-seven
times in the entire length : body, from the head to the vent, heptangular ;
thence to the termination of the dorsal fin, hexangular ; thence to the
caudal, quadrangular : the heptangular portion presents two longitudinal
ridges on each side, one on each side of the middle of the back, and one
down the middle of the belly ; this last terminates at the vent ; the
dorsal ridges terminate at the end of the dorsal fin, and the upper pair
of lateral ridges rise to take their place ; beyond the vent, the under surface of the tail is very flat, with the margins rather dilated, and, in the
male, contains a long purse-like cavity, for the reception of the *ova,*
opening by a longitudinal slit : body protected by transverse, striated,
shields or plates, sixty-three in number ; nineteen occupying that portion
of the trunk between the gills and the vent, forty-four the remainder of
the length : head compressed, contained (snout included) about seven
times and a half in the entire length : occiput rising into a longitudinal
elevated ridge, continued over the crown ; the profile falling thence in
a sinuous curve to the base of the snout : eyes large, protected above
by a sharp osseous ridge ; the intervening space concave ; in front of each
a sharp spinous process : snout elongated, nearly twice the length of
the rest of the head, compressed, much narrower than the head in a
vertical direction ; mouth very small, situate quite at the extremity ;
lower jaw longest, ascending : no teeth : opercle large, marked with
diverging striæ, closed on all sides by a continuous membrane, the gill-
opening being reduced to a small hole on each side of the nape : dorsal
so placed as to terminate exactly at the middle point of the entire length ;
length of the fin about equal to that of the head ; its height equalling

the depth of the body, and nearly uniform throughout, the anterior rays being slightly shorter than the succeeding ones; all the rays simple: vent in a line with the seventh dorsal ray; anal immediately behind it, very small and inconspicuous, consisting of only three short simple rays: caudal moderate, rounded; the rays simple and articulated: pectorals a little behind the gills, not very large, of a rounded form; all the rays simple.

D. 42; A. 3; C. 10; P. 12.

(*Colours.*) Pale yellowish brown, with transverse bands of darker brown: belly whitish.

Not uncommon on many parts of the coast, frequenting chiefly the shallower places. I am not aware, however, that in the British seas it ever attains to the length which Bloch assigns to it. This and several other species in the present genus are remarkable for the *males* carrying the ova, until hatched, and even the young themselves for a short time after they have been hatched, in a peculiar longitudinal pouch beneath the tail, into which the former are received, at the time of their exclusion by the *female**. The present species breeds in Summer, and at a very early age, sometimes when not exceeding four inches in length. *Obs.* This and the next were considered by Pennant as mere varieties of one species, to which he applied the name of *Shorter Pipe-Fish.* The same opinion appears to have been entertained by Montagu†.

173. S. *Typhle*, Linn. (*Lesser Pipe-Fish.*) — Body hexangular anteriorly: crown flat; profile nearly in the same line: snout almost as broad, vertically, as the head.

S. Typhle, *Linn. Syst. Nat.* tom. I. p. 416. *Bloch, Ichth.* pl. 91. f. 1.? *Don. Brit. Fish.* vol. III. pl. 56. *Turt. Brit. Faun.* p. 116. *Flem. Brit. An.* p. 175. Acus Aristotelis, *Will. Hist. Pisc.* p. 158. tab. I. 25. f. 1. Pipe-Fish, *Penn. Brit. Zool.* vol. III. pl. 23. no. 60. upper fig. but not p. 138. Shorter Pipe-Fish, *Id.* (Ed. 1812.) vol. III. pl. 26. no. 60. upper fig.

LENGTH. One foot: rarely more.

DESCRIPT. (*Form.*) Thicker in proportion to its length than the last species; the ventral carina not so prominent, causing the anterior part of the body to appear more hexangular than heptangular: number of transverse shields between the gills and the vent the same, but from the vent to the caudal only thirty-six or thirty-seven: head larger; the crown nearly flat, without any elevated ridge; the profile passing off almost in a straight line to the mouth, with very little sinuosity: snout every-way larger; longer, and, measured vertically, nearly as broad as the head; very much compressed: spinous process before the eyes smoother, and less projecting: head, including the snout, rather more than one-sixth of the entire length: opercle much larger: dorsal placed further back, being exactly in the middle of the entire length: anal very minute: caudal and pectorals similar.

D. 39; A. 3; C. 10; P. 15.

(*Colours.*) "Varying from greenish olive, to olivaceous yellow, and brown, variegated sometimes with dark or bluish lines." DON.

---

* See on this subject *Procced. of Zool. Soc.* (1834.) p. 118.
† *Wern. Mem.* vol. I. p. 86.

Found in the same situations as the last species, and equally, if not more, common. *Obs.* I feel some hesitation in considering the *S. Typhle* of Bloch to be the same as that of English authors. His figure, as Donovan has observed, resembles more nearly the *S. Acus* in a young state.

(34.) *S. pelagicus*, Don. Brit. Fish. vol. III. pl. 58. Turt. Brit. Faun. p. 117. Flem. Brit. An. p. 176,

> I very much doubt whether this supposed species be any thing more than the young of *S. Acus*. During a stay at East Bourne, in Sept. 1833, I obtained three specimens, taken in the shrimp-nets at that place, which appeared exactly to coincide with Donovan's figure, but which, I am tolerably satisfied, are only what I have stated above. Two of these were *females*, and possessed an extremely minute anal fin; but the third, which was a *male*, exhibited no vestige of it whatever, even when examined carefully with a lens. In this last individual, though measuring only three inches and a half in length, the caudal pouch was full of newly-hatched young. What the *S. pelagicus* of Linnæus may be, I do not pretend to say.

** *Anal and pectoral fins wanting; caudal obsolete.*

174. S. *æquoreus*, Linn. (*Æquoreal Pipe-Fish.*) — Body octangular anteriorly: snout short; much narrower, vertically, than the head: dorsal and vent nearly in the middle of the entire length.

S. æquoreus, *Linn. Syst. Nat.* tom. I. p. 417.   *Mont. in Wern. Mem.* vol. I. p. 85. pl. 4. f. 1.   *Flem. Brit. An.* p. 176.   Acus nostras caudâ serpentinâ, *Sibb. Scot. Illust.* part ii. tom. II. p. 24. tab. 19.   Æquoreal Pipe-Fish, *Penn. Brit. Zool.* (Edit. 1812.) vol. III. p. 188.

Length. From twenty to twenty-four inches.

Descript. (*Form.*) Readily distinguished from both the foregoing species by the want of the pectoral and anal fins. Form slender, and very much elongated: body compressed, with an acute dorsal and abdominal ridge; also with three slight ridges on each side; hence the trunk from the gills to the vent is octangular; the tail is obsoletely quadrangular, becoming almost round towards the tip, which is extremely tapering: transverse shields or plates, between the gills and the vent, twenty-eight in number; from the vent to the extremity of the tail, sixty or more (Montagu says about sixty-six), but, from the extreme minuteness of the last few, not admitting of being counted with exactness : head not more than one-twelfth of the entire length; without any elevated ridge on the occiput: snout narrower than the head, similar in shape to that of *S. Acus*, but much shorter in relation to the entire length of the fish: dorsal occupying nearly a middle position in the entire length; the distance from the last ray to the end of the tail at the same time a little exceeding that from the end of the snout to the commencement of the fin: vent a very little before the middle, being nearly in a vertical line with the commencement of the last quarter of the dorsal fin: tail compressed at the extremity, showing a very small rudimentary caudal fin; the rays however so obsolete, and so much enveloped in the common skin, as to be scarcely distinguishable.

D. about 40; A. 0; C. 0?; P. 0.

(*Colours.*) "Yellowish, with transverse pale lines, with dark margins, one in each joint, and another down the middle of each plate, giving it the appearance of possessing double the number of joints it really has: these markings, however, cease just beyond the vent. Mont.

This species, which had been previously observed by Sibbald in the Frith of Forth, was obtained by Montagu at Salcomb, in 1807. A second specimen, he mentions, was afterwards picked up on the same coast. That from which the above description was taken was procured in Berwick Bay by Dr. Johnston: it is now in the collection of W. Yarrell, Esq.

175. S. *Ophidion*, Bloch. (*Snake Pipe-Fish.*)—Body round, very obsoletely octangular anteriorly: snout short; much narrower, vertically, than the head : dorsal and vent considerably before the middle of the entire length.

S. Ophidion, *Bloch, Ichth.* pl. 91. f. 3. *Gmel. Linn.* tom. i. part iii. p. 1456.? S. anguineus, *Jen. Cat. of Brit. Vert. An.* 30. sp. 176. Snake Pipe-Fish, *Shaw, Gen. Zool.* vol. v. part ii. p. 453. pl. 179. (Copied from Bloch.) Longer Pipe-Fish, *Low, Faun. Orc.* p. 179.

Length. From twelve to fourteen inches. According to Bloch, from one to two feet.

Descript. (*Form.*) Very similar to the last species, but more slender and tapering in proportion to its length: anterior part of the body scarcely thicker than a goose-quill, presenting the same angles as the *S. æquoreus*, but with these angles so ill-defined and obsolete, as to appear on the whole nearly round; beyond the vent the body is obsoletely quadrangular, becoming quite round near the extremity of the tail, the tip of which is compressed into a very minute rudimentary caudal fin: transverse shields on the trunk not separated by any well-marked lines, so as scarcely to admit of being counted: head one-twelfth of the entire length, similar, as is also the snout, to that of the *S. æquoreus*: dorsal much forwarder than in that species, entirely before the middle; the distance from the last ray to the end of the tail more than half as long again as that from the end of the snout to the commencement of the fin: vent considerably before the middle of the entire length, but in relation to the dorsal situate as in the *S. æquoreus*, three-fourths of that fin lying in advance of it: rays of the caudal too minute and obsolete to be distinguished.

D. 38; A. 0; C. 0?; P. 0.

(*Colour of a specimen in spirits.*) Of a uniform yellowish brown, paler beneath: no indication of the transverse bands which appear in the last species.

This species, which has evidently been confounded with the next by many authors *, I have, since the publication of my Catalogue, ascertained to be the true *S. Ophidion* of Bloch, and probably of Gmelin, but whether of Linnæus also is doubtful, as his very concise description applies nearly equally well to both. It is closely allied to the *S. æquoreus*, from which it scarcely differs, excepting in its slenderer and rounder form, and much forwarder position of the dorsal and vent. It is indeed with this last species, that it has probably been confounded by Low, under

---

* Montagu was the first to point out the great discrepancies which appear in the different figures and descriptions given by different authors of S. Ophidion, and it is entirely owing to his observations on this subject, that I was led to detect the existence of another species. See *Wern Mem.* vol. i. p. 89.

the name of *Longer Pipe-Fish*, part of whose description, borrowed from Sibbald and Pennant, does not belong to it: the proportions, however, which Low assigns to his own specimens, convince me that this is the species which had occurred to himself, and, according to his own account, in great plenty. The only specimens which I have seen, amounting, however, to several, are in the collection of W. Yarrell, Esq. I am ignorant as to the locality whence they were obtained.

(35.) *Longer Pipe-Fish*, Penn. Brit. Zool. vol. iii. p. 138. pl. 23. no. 61. Id. (Edit. 1812.) vol. iii. p. 184. pl. 26. no. 61. *Syngnathus barbarus*, Turt. Brit. Faun. p. 117. Flem. Brit. An. p. 176.

I consider this as a very doubtful species. The figure, in which there is neither anal nor pectoral fins, approaches so closely the *S. æquoreus*, that I feel confident it was taken either from that species, or the *S. Ophidion* last described. In the description, however, mention is made both of anal and pectorals, but of no caudal, a combination of characters which is not only at variance with every British species I am acquainted with, but which will not accord with any of those given by Linnæus, (not even with the *S. barbarus*, in which, according to that author, there is no anal,) and which, moreover, will not fall under any of the sections into which Cuvier has divided the genus. There is strong ground for believing that, by some unaccountable accident, Pennant has mixed up under one name the descriptions of two totally distinct species.

*** *Anal, pectoral, and caudal fins, all wanting.*

176. S. *lumbriciformis*, Nob. (*Worm Pipe-Fish*.) — Body round, slightly compressed anteriorly : snout very short ; of nearly equal breadth with the head : dorsal and vent at about the middle of the entire length.

S. Ophidion, *Flem. Brit. An.* p. 176.   Acus lumbriciformis, *Will. Hist. Pisc.* p. 160.   Little Pipe-Fish, *Penn. Brit. Zool.* vol. iii. p. 141. pl. 33. no. 62.   Id. (Edit. 1812.) vol. iii. p. 187. no. 62.

LENGTH. From six to nine inches.

DESCRIPT. (*Form.*) Extremely slender; the trunk or anterior half slightly compressed, but without angles; the tail round, and tapering to a very fine point, without even the rudiment of a caudal fin: transverse shields smooth, and of a membranous nature, somewhat resembling the segments of the common earth-worm; between the gills and the vent twenty-eight; thence to the end of the tail upwards of sixty, perhaps near seventy, but towards the tip so minute as scarcely to admit of being counted with exactness: head small, scarcely one-seventeenth of the entire length; crown flat, without any elevated ridge; snout very short, blunt at the tip, compressed, with a sharp keel above and below; its breadth, vertically, not much less than that of the head: dorsal at about the middle of the entire length, but rather more of the fin behind than before it, the distance from the end of the snout to the first ray being a little greater than that from the last ray to the extremity of the tail: vent also almost exactly in the middle, if any thing a little behind it; with respect to the dorsal, it is forwarder than in the two last species, only one-third of that fin lying in advance of it.

D. 39; A. 0; C. 0; P. 0.

(*Colour.*) Of a uniform olive, sometimes yellowish, brown; in my specimen, a longitudinal dark fascia extends from behind the gill-cover over
the first six segments of the trunk.

As this species would seem not to be the *S. Ophidion* of continental
authors, I have restored to it the name originally given it by Willughby,
in consideration of its worm-like appearance. I believe it is not uncommon
on many parts of the coast, and is said to be called in Cornwall a *Sea-
Adder*. The *ova* are not carried by the male in a caudal pouch, but "in
hemispheric depressions on the external surface of the abdomen, anterior
to the vent." *Obs.* The *S. Ophidion* of Berkenhout and Turton may be
intended either for this species or the last.

GEN. 67.   HIPPOCAMPUS, *Cuv.*

177.   H. *brevirostris*, Cuv. ?   (*Sea-Horse.*) — Snout
short.

> H. brevirostris, *Cuv. Reg. An.* tom. ii. p. 363.   H. Rondeletii,
> *Will. Hist. Pisc.* tab. I. 25. fig. 3.

Pennant states his having been informed that the *Syngnathus Hippocampus* of Linnæus had been found on the southern shores of this kingdom*. More recently, Messrs. C. and J. Paget have recorded † that it is
occasionally met with at Yarmouth. As, however, several species have
been confounded by authors under the above name, and the British one
has not hitherto been correctly ascertained, it is impossible to annex any
description. At the same time there is ground for believing that it will
prove to be the *Hippocampus brevirostris* of Cuvier.

ORDER IV.   GYMNODONTES.

GEN. 68.   TETRODON, *Linn.*

178.   T. *stellatus*, Don.   (*Stellated Globe-Fish.*) —
Above blue; beneath silvery: abdomen spinous; each
spine arising from a stellated root, of four rays. Don.

> T. stellatus, *Don. Brit. Fish.* vol. iii. pl. 66.   *Turt. Brit. Faun.*
> p. 116.   *Flem. Brit. An.* p. 174.   Globe Diodon, *Penn. Brit.
> Zool.* vol. iii. p. 132.. pl. 20.   Globe Tetrodon, *Id.* (Edit. 1812.)
> vol. iii. p. 174. pl. 23.

---

* *Brit. Zool.* vol. iii. p. 141.   † *Nat. Hist. of Yarm.* p. 18.

DIMENS. Entire length one foot seven inches; length of the belly, when distended, one foot; the whole circumference in that situation two feet six inches. PENN.

DESCRIPT. (*Form.*) Body usually oblong, but when alarmed the fish has the power of inflating the belly to a globular shape of great size: the whole surface of the abdomen, down to the vent, armed with small sharp spines, each arising from a distinct stellated root of four processes: mouth small: back from head to tail almost straight, or at least very slightly elevated: dorsal placed low on the back; anal opposite: caudal almost even, divided by an angular projection in the middle: pectorals present; ventrals wanting.

D. 11; A. 10; C. 6; P. 14.

(*Colours.*) Back of a rich deep blue: belly and sides silvery white; the spines of a rich carmine-colour: tail and fins brown: irides white, tinged with red. PENN. and DON.

An individual of this species is recorded by Pennant as having been taken at Penzance in Cornwall. A second specimen, also captured on the Cornish coast, is figured by Donovan. Bloch appears to have considered Pennant's fish as his *T. lagocephalus**, but by Donovan it was thought, and apparently with some reason, to be distinct. This and all the other species in this genus have the power of inflating the abdomen to a large size, at the same time that they erect the spines with which it is armed, by which means they defend themselves against the attacks of their enemies.

GEN. 69.   ORTHAGORISCUS, *Schn.*

179.   O. *Mola,* Schneid.   (*Short Sun-Fish.*) — Depth about two-thirds of the length: skin rough.

O. Mola, *Flem. Brit. An.* p. 175.   Mola Salviani, *Will. Hist. Pisc.* p. 151.   tab. I. 26.   Tetrodon Mola, *Linn. Syst. Nat.* tom. I. p. 412.   *Don. Brit. Fish.* vol. II. pl. 25.   Diodon Mola, *Bloch, Ichth.* pl. 128.   Cephalus brevis, *Shaw, Gen. Zool.* vol. v. part ii. p. 437.   pl. 175.   *Turt. Brit. Faun.* p. 116.   Sun-Fish, *Borl. Cornw.* p. 267.   pl. 26.   f. 6.   Short Diodon, *Penn. Brit. Zool.* vol. III.   p. 131.   pl. 19.   Short Tetrodon, *Id.* (Edit. 1812.) vol. III. p. 172.   pl. 22.

LENGTH.   From three to four feet.

DESCRIPT. (*Form.*) Oblong, approaching orbicular, truncated behind: sides very much compressed; the dorsal and ventral lines presenting a sharp edge: depth behind the pectorals about two-thirds of the entire length; thickness rather more than one-third of the depth: head not distinguishable from the trunk; mouth small; jaws exposed; the lamellated substance undivided: eyes moderate, about equidistant from the corner of the mouth and the branchial aperture, which last is of an oval form, and situate immediately before the pectoral fin: skin destitute of scales, but every-where very rough with minute granulations: no lateral line: dorsal placed at the further extremity of the body, short but very much elevated, its height equalling two-thirds or more of the depth of the body, terminating upwards in a point; rays very much branched: anal opposite and exactly similar to the dorsal: caudal with the posterior

* *Ichth.* pl. 140.

margin slightly rounded, very short, but its depth (or breadth, measured vertically) nearly equalling that of the body, extending from the dorsal to the anal, with both of which fins it is connected: pectorals small, rounded, attached horizontally: ventrals wanting.

D. 17; A. 16; C. 14; P. 13*.

(*Colours.*) Back dusky gray; belly and sides silvery.

Rare in the British seas, but has been captured, in different instances, upon various parts of the coast. Attains to an enormous size; sometimes weighing as much as four hundred or five hundred pounds. Generally distributed over the European seas. According to Bloch, occasionally reposes on one side, in which situation, when surprised, it is easily taken. Flesh bad, but yielding a great deal of oil.

180. O. *oblongus*, Schneid. (*Oblong Sun-Fish.*) — Length more than twice the depth of the body: skin smooth.

O. truncatus, *Flem. Brit. An.* p. 175.   *Jen. Cat. of Brit. Vert. An.* 31. sp. 181.   Tetrodon truncatus, *Gmel. Linn.* tom. I. part iii. p. 1448.   *Don. Brit. Fish.* vol. II. pl. 41.   Cephalus oblongus, *Shaw, Gen. Zool.* vol. v. part ii. p. 439. pl. 176.   *Turt. Brit. Faun.* p. 116.   Sun-Fish from Mount's Bay, *Borl. Cornw.* p. 268. pl. 26. f. 7.   Oblong Diodon, *Penn. Brit. Zool.* vol. III. p. 129. pl. 19. Oblong Tetrodon, *Id.* (Edit. 1812.) vol. III. p. 170. pl. 22.

DESCRIPT. (*Form.*) Represented by authors as closely resembling the last species, excepting in its more oblong and elongated shape, the entire length being more than twice (according to Turton, nearly three times) the depth of the body: skin smooth: branchial aperture lunulate.   The number of fin-rays, according to Donovan, stands thus:

D. 12; A. 15; C. 17; P. 14.

(*Colours.*) "Back dusky, with some variegations; abdomen silvery; between the eyes and the pectoral fins a few dusky streaks pointing downwards." SHAW.

Apparently more rare in the British seas than the last species. First noticed by Borlase, who has figured a specimen from Mount's Bay in Cornwall.   The same author speaks of another taken at Plymouth in 1734, which weighed five hundred pounds. Since then other individuals have occasionally been met with.   In the stomach of one, obtained by Donovan from the Bristol Channel, there were found fragments of testaceous and crustaceous animals.   *Obs.* Both this and the *O. Mola* have a bright glistening appearance when taken fresh out of the water, to which circumstance is to be attributed their English name of *Sun-Fish.*   At night they are said to be phosphorescent.

* The above fin-ray formula is from Bloch.

ORDER V. SCLERODERMI.

GEN. 70. BALISTES, Cuv.

181. B. *Capriscus*, Gmel. (*Mediterranean File-Fish.*)

B. Capriscus, *Gmel. Linn.* tom. I. part iii. p. 1471. *Cuv. Reg. An.*
tom. II. p. 372. Capriscus Rondeletii, *Will. Hist. Pisc.* p. 152.
tab. I. 19. Le Baliste Caprisque, *Lacép. Hist. Nat. des Poiss.*
tom. I. p. 372. Mediterranean File-Fish, *Shaw, Gen. Zool.* vol. v.
part ii. p. 411. pl. 168 *.

LENGTH. From one to two feet. SHAW.

DESCRIPT. (*Form.*) Head very much compressed: mouth small: each
jaw armed with eight large, broad, strong teeth, forming a continuous
series: eyes round, placed high: first dorsal nearly in the middle of the
back, consisting of three strong spines connected by a membrane, the
first three times as large as either of the other two; when at rest, concealed in an osseous furrow on the ridge of the back; these spines are so
articulated as only to admit of being elevated or depressed all together:
second dorsal long, reaching nearly to the caudal: anterior portion of the
abdomen armed with a sharp strong recurved bone, the extremity of which
projects out of the skin in a backward direction; between this and the
vent are several other much smaller, but moderately strong, serrated
spines: anal similar to the dorsal, commencing behind the vent: tail
becoming suddenly narrow, and terminating in a broad fin: pectorals
small and round: ventrals wanting. WILL. (*Colours.*) "General colour
violaceous gray, sometimes variegated both on the body and fins with blue
or red spots: irides yellow; pupils blue." SHAW.

A single individual of this species, which is a native of the Mediterranean, as well as of the American seas, is recorded by Mr. Children
to have been captured on the Sussex coast in August, 1827 *. Not previously known as a British fish. Most of the species belonging to this
genus inhabit the Tropics.

* See his *Address to the Zoological Club of the Linnæan Society.* p. 6.

(II. CARTILAGINEI.)

ORDER VI. ELEUTHEROPOMI.

GEN. 71. ACIPENSER, *Linn.*

182. A. *Sturio*, Linn. (*Common Sturgeon.*)—Osseous
tubercles in five longitudinal rows; strong and spinous.

A. Sturio, *Linn. Syst. Nat.* tom. I. p. 403. *Bloch, Ichth.* pl. 88.
Turt. Brit. Faun. p. 114. *Don. Brit. Fish.* vol. III. pl. 65. *Flem.
Brit. An.* p. 173. Sturio, *Will. Hist. Pisc.* p. 239. tab. P. 7. fig. 3.
Sturgeon, *Penn. Brit. Zool.* vol. III. p. 124. pl. 19. no. 53. Com-
mon Sturgeon, *Id.* (Edit. 1812.) vol. III. p. 164. pl. 22. L'Esturgeon
ordinaire, *Cuv. Reg. An.* tom. II. p. 379.

LENGTH. From six to eight feet, sometimes more.

DESCRIPT. (*Form.*) Body elongated; somewhat pentagonal; with five
longitudinal rows of osseous tubercles, one on the back, two at the sides,
and two at the edges of the abdomen; tubercles marked with radiating
striæ, broad at the base, terminating above in a sharp strong spine
directed backwards; those on the back more developed than the others:
the whole skin rough with minute points and tubercles independently
of the above larger ones: abdomen flat: head long, covered above
with lozenge-shaped plates; a longitudinal sulcus on the forehead:
snout depressed, somewhat conical, rather slender and sharp-pointed;
mouth placed beneath, cylindrical, small, without teeth, bordered by a
protractile cartilage instead of lips: four pendent barbules on the under
surface of the snout, nearer its extremity than the mouth: eyes small:
nostrils near the eyes, double; the anterior orifice round, the posterior
one elongated: gill-opening large; the gill-cover marked with numerous
striæ, radiating in all directions: a single dorsal, of a somewhat triangular form, small, placed very far back near the tail: anal also small,
and nearly opposite: caudal forked; the upper lobe long and pointed,
produced very much beyond the lower: pectorals oval: vent beneath the
commencement of the dorsal; the ventral fins a little in advance of it.

D. 35; A. 23; C. 125; P. 28; V. 24*.

(*Colours.*) Upper parts gray, variegated with dusky, sometimes inclining
to olivaceous; the central part of the tubercles white: beneath silvery
white.

A migratory species, residing in the sea during the winter months,
but entering rivers at the approach of Spring to spawn. Very abundant
in many parts of the Continent, but seldom in any plenty in this country.
Attains to the weight of from one hundred to three hundred pounds.

* The above fin-ray formula is from Donovan.

Pennant mentions one, taken in the Esk, which weighed four hundred and sixty pounds. Arrives at a still larger size abroad, according to Bloch and other authors. Feeds principally upon the smaller fish. Flesh much esteemed. The well known article of food termed *Caviar* is prepared from the roe.

ORDER VII. ACANTHORRHINI.

GEN. 72. CHIMÆRA, *Linn.*

183. C. *monstrosa*, Linn. (*Sea-Monster.*)

C. monstrosa, *Linn. Syst. Nat.* tom. I. p. 401. *Bloch, Ichth.* pl. 124. *Turt. Brit. Faun.* p. 114. *Don. Brit. Fish.* vol. v. pl. 111. *Flem. Brit. An.* p. 172. Galeus Acanthias Clusii, *Will. Hist. Pisc.* p. 57. tab. B. 9. f. 6. Northern Chimæra, *Penn. Brit. Zool.* (Edit. 1812.) vol. III. p. 159*.   La Chimère arctique, *Cuv. Reg. An.* tom. II. p. 382.

LENGTH. From two to three feet. CUV.

DESCRIPT. (*Form.*) " Body compressed: head blunt; the snout subascending, blunt: a narrow crenulated grinder on each side in the lower jaw, and a broad tubercular one corresponding above: nostrils immediately above the upper lip, contiguous, each with a cartilaginous complicated valve: branchial openings in front of the pectorals: eyes large, lateral: lateral line connected with numerous waved anastomosing grooves on the cheeks and face: on the crown, in front of the eyes, a thin osseous plate, bent forwards, with a spinous disk at the extremity on the lower side: the first dorsal fin above the pectorals, narrow, with a strong spine along the anteal edge: second dorsal arising immediately behind the first, narrow, and continued to the caudal, where it terminates suddenly: pectorals large, subtriangular: ventrals rounded; in front of each a broad recurved osseous plate, with recurved spines on the ventral edge: claspers pedunculated, divided into three linear segments, the anteal one simple, the retral ones having the opposite edges covered with numerous small reflected spines: a small anal fin opposite the extremity of the second dorsal: caudal fin above and below, broadest near the origin, gradually decreasing to a linear produced thread." FLEM. (*Colours.*) " The whole body dark brown above, varied with yellowish brown and silvery; the lower parts of a bright silver colour: eyes green, with silvery irides, and very brilliant, or shining with phosphoric splendour." DON.

Found principally in the northern seas of Europe, and but rarely met with in those of Great Britain. The above description was taken by Dr. Fleming from a specimen captured in the Zetland seas, where it is said to be known by the name of *Rabbit-Fish.* Food, according to Bloch, medusæ and crabs.

ORDER VIII. PLAGIOSTOMI.

GEN. 73.   SQUALUS, *Linn.*

(1. S<small>CYLLIUM</small>, *Cuv.*)

184.   S. *Canicula*, Linn. (*Spotted Dog-Fish.*)—Spots small and numerous : ventrals cut obliquely at their posterior margin : valves of the nostrils united, partly covering the mouth.

S. Canicula, *Linn. Syst. Nat.* tom. I. p. 399.   S. Catulus, *Bloch, Ichth.* pl. 114. *Don. Brit. Fish.* vol. III. pl. 55.   *Blainv. Faun. Franç.* p. 69. pl. 17. f. 1.   Scyllium Catulus, *Flem. Brit. An.* p. 165. (Male.)   S. stellare, *Id.* (Fem.)   S. Canicula, *Bon. Faun. Ital.* Fasc. VII.   Catulus major vulgaris, *Will. Hist. Pisc.* p. 62. tab. B. 4. f. 1.   Catulus minor, *Id.* p. 64.?   Lesser Spotted Dog-Fish, (Male), and Spotted Dog-Fish, (Fem.) *Penn. Brit. Zool.* vol. III. pp. 115, and 113. pl. 15.   Lesser Spotted Shark, (Male,) and Spotted Shark, (Fem.) *Id.* (Ed. 1812.) vol. III. pp. 150, and 148. pl. 19.   La grande Roussette, *Cuv. Reg. An.* tom. II. p. 386.

L<small>ENGTH</small>.   From two to three feet ; sometimes three feet and a half, or even more.

D<small>ESCRIPT</small>. (*Form.*) Body elongated, tapering from behind the pectorals, where the thickness is greatest : head blunt, depressed : snout short and rounded : nostrils on the under surface of the snout, near the mouth, large, prolonged in a channel to the margin of the lips, and almost entirely closed by a fleshy valve or lobe of the skin ; each valve unites with its fellow on the opposite side, the two together forming a large flap, emarginated in front, which extends over the upper lip and entirely conceals it : mouth beneath, behind the nostrils, of a semicircular form : both jaws with several rows of small, but sharp, teeth, inclining backwards ; each tooth furnished with a long point in the middle, and smaller denticulations at the sides : eyes large, oblong-oval, at equal distances from each other and the end of the snout ; behind each a small temporal orifice, or spiracle, communicating with the mouth : branchial openings at the sides of the neck, five in number, parallel to, and equally distant from, each other, arranged in a longitudinal series, the first as far behind the eyes, as these last are distant from the end of the snout, the last immediately above the pectoral fin ; first four openings nearly of equal size, the fifth smaller : skin somewhat glistening, very rough when the hand is passed from tail to head, but only slightly so in the opposite direction, the roughness proceeding from very minute denticulated scales : no distinct lateral line : two dorsals ; both placed very much behind : the first commencing about the middle of the entire length, of a trapezoidal form, cut square behind, its greatest height about equal to the depth of the body, the space which it occupies about two-thirds of the same :

second dorsal rather before the middle point between the first dorsal and
the end of the caudal, shaped like the first, but rather smaller: anal
answering to the space between the two dorsals, commencing a little
beyond the termination of the first, and terminating nearly in a line with
the commencement of the second; somewhat triangular, with the posterior
portion produced backwards in the form of a lanceolate process: tail very
long, equalling a little more than half the entire length; terminating in
a caudal fin; upper lobe of the caudal commencing a little beyond the
termination of the second dorsal, low at first, but gradually widening
towards its extremity, which is truncated; lower lobe a little distant from
the upper, and of a triangular form: pectorals large, of about the length
of the head, broadest at their posterior margin, which is cut square: ventrals a little in advance of the first dorsal, attached horizontally, much
smaller than the pectorals, obliquely truncated behind, their posterior
margins meeting at an acute angle; together they form a kind of lozenge,
in the middle of which is the vent.   The *male* is characterized by having
the ventrals larger than in the other sex, and united throughout their
length by an intermediate membrane; they are also furnished on their
inner margins with fusiform appendages, not extending beyond the fin in
young subjects, but lengthening in adults: in the *female*, the ventrals
have the last third portions of their inner margins separate. (*Colours.*)
Back, upper portion of the head, and the whole of the sides, reddish gray,
or dirty flesh-red, with very numerous small dark brown spots; the spots
on the posterior portion of the body more scattered: fins coloured like the
back, but the spots larger and less numerous; anal almost without spots:
under portion of the head and body whitish, free from spots.   *Obs.* The
spots are generally less numerous, and rather larger, in the *female* than
in the *male*.

A common species on all parts of the coast. Does not attain to any
great size. Very voracious, preying on almost any animal substance.
Oviparous: produces, according to Pennant, about nineteen young at
a time. Very tenacious of life.   *Obs.* I have ventured to bring together the *Spotted Dog-Fish* and the *Lesser Spotted Dog-Fish* of Pennant,
under a strong suspicion that they are simply the two sexes of the present species. The female has probably been confounded by some authors
with the following.

185.   S. *stellaris*, Linn. (*Rock Dog-Fish.*) — Spots
large and scattered: ventrals cut square at their posterior
margin: valves of the nostrils separate, not reaching to the
mouth.

S. stellaris, *Linn. Syst. Nat.* tom. i. p. 399. *Blainv. Faun. Franç.*
p. 71. pl. 17. f. 2.   Scyllium stellare, *Bon. Faun. Ital.* Fasc. vii.
Catulus maximus, *Will. Hist. Pisc.* p. 63.?   Spotted Dog-Fish,
Var. *Penn. Brit. Zool.* vol. iii. p. 114.   *Id.* (Edit. 1812.) vol. iii.
p. 150.   Rock Shark, *Shaw, Gen. Zool.* vol. v. part ii. p. 336.
La petite Roussette ou Rochier, *Cuv. Reg. An.* tom. ii. p. 386.

LENGTH.   From two to three feet.   According to Blainville, it attains
to a larger size than the last species.

DESCRIPT.   (*Form.*) Closely resembling the *S. Canicula*, but differing
essentially in the structure of the lobes of the nostrils, and in the form of
the ventrals: the former are not united as in that species, and of a smaller
size, leaving the whole of the mouth and the upper lip visible: the vent-

rals, instead of being cut obliquely, are cut nearly square, their posterior margins meeting at a very obtuse angle; they are united or separate, according to the sex, in a similar manner: the snout is rather more elongated; and, according to some authors, the tail rather shorter, giving the dorsal a more backward position, but this last character I have not noticed myself. (*Colours.*) Upper parts brownish gray, with very little of the red tinge observable in the last species: back, flanks, and tail, sparingly marked with large brown spots of a deep brown or black colour: under parts whitish.

Whether this species be common on our coasts or not, I am ignorant. In many instances it appears to have been confounded with the last. The only specimens I have seen were obtained at Weymouth. According to Blainville, resides amongst rocks. Skin very rough, and said to be of more value in the arts than that of the *S. Canicula.* In some of its characters it appears to resemble the *Catulus maximus* of Willughby, to which I have annexed a reference above.

(36.) *Scyllium melanostomum,* Bon. Faun. Ital. Fasc. viii. *Squalus melastomus,* Blainv. Faun. Franç. p. 75.

A drawing of a third British species of this sub-genus, supposed to be the *S. melanostomum* of the Prince of Musignano, has been recently forwarded to Mr. Yarrell by Mr. Couch of Cornwall. It was taken upon that coast. Not possessing any description of it, and not being able to speak with certainty as to its identity with the above species, I have confined myself to giving this notice of its existence in the British seas. The following are the characters given to the *S. melanostomum* in the " Fauna Italica."—" Sc. *rufocinereus; maculis magnis obscurioribus oblongis; pinnis ventralibus obliquè truncatis; ore intus nigro-cæruleo.*"

(2. Carcharias, *Cuv.*)

186. S. *Carcharias,* Linn. (*White Shark.*) — Teeth triangular, the sides straight and denticulated; those in the lower jaw narrower than those above.

S. Carcharias, *Linn. Syst. Nat.* tom. i. p. 400. *Turt. Brit. Faun.* p. 113. *Blainv. Faun. Franç.* p. 89. Canis Carcharias, *Will. Hist. Pisc.* p. 47.? Carcharias vulgaris, *Flem. Brit. An.* p. 167. White Shark, *Penn. Brit. Zool.* vol. iii. p. 106. *Id.* (Edit. 1812.) vol. iii. p. 139. *Low, Faun. Orc.* p. 174. Le Requin, *Cuv. Reg. An.* tom. ii. p. 387.

Length. Twenty-five feet. Cuv.

Descript. (*Form.*) " Body very much elongated: skin very hard and granulated: the first dorsal placed before the middle of the back, elevated, rounded above; the second small, nearly in the middle of the tail: anal nearly opposite to this last: head flattened; snout rounded, pierced with a large number of pores: tail moderate, terminating in a falciform caudal of two lobes, the upper lobe double the lower one: mouth very large, semicircular, entirely beneath: vent nearly in the middle: form of the nostrils unknown: eyes lateral, small, round: jaws large, bent: teeth in five or six rows in both jaws, above and below; of a triangular form, compressed, finely denticulated at the edges, which are perfectly straight; the lower ones a little narrower than the upper ones: no temporal orifices: branchial openings five in number, but their form and proportion unknown: pectorals very large, in the form of an

I i

isosceles triangle, extending beyond the base of the first dorsal: ventrals small, a little nearer the second dorsal than the first. (*Colours.*) Cinereous brown above, whitish beneath, with two rows of black dots on the sides." Blainv.

This species appears to be very rare in the British seas; nor am I aware of any description of a native specimen on record. Grew has incidentally thrown out a remark* that it is sometimes found upon the Cornish coast. Low states, that according to information given to him, it is met with in the neighbourhood of the Orkneys. None of our other English authors, that I am aware, have specified any localities in which it has occurred. The description given above is taken from the *Faune Français.* It may be of use in enabling future observers to identify this species: it must be remembered, however, that two or more appear to have been confounded under the name of *Carcharias;* and possibly it may not belong to the one which has been met with in the British seas. According to Cuvier, the *S. Carcharias* of Bloch† is very distinct. The present species is widely distributed, and attains to a very large size. It is very voracious, and much dreaded by navigators.

187. S. *Vulpes,* Gmel. (*Sea-Fox,* or *Thresher.*)— Teeth triangular, pointed; the edges not denticulated: caudal with the upper lobe nearly as long as the body.

S. Vulpes, *Gmel. Linn.* tom. i. part iii. p. 1496. *Turt. Brit. Faun.* p. 112. *Blainv. Faun. Franç.* p. 94. pl. 14. f. 1. Vulpes marina, *Will. Hist. Pisc.* p. 54. tab. B. 6. f. 2. Carcharias Vulpes, *Flem. Brit. An.* p. 167. Long-tailed Shark, *Penn. Brit. Zool.* vol. iii. p. 110. pl. 14. *Id.* (Edit. 1812.) vol. iii. p. 145. pl. 17. La Faux, ou Renard, *Cuv. Reg. An.* tom. ii. p. 388.

Length. Thirteen feet; the tail alone measuring more than six. Penn.

Descript. (*Form.*) "Body fusiform, appearing very much elongated in consequence of the relatively great size of the tail: skin very finely and equally shagreened above and below: first dorsal moderately large, triangular, elevated, adhering by almost the whole length of its base, in the middle of the back: the second exceedingly small, triangular, inclined, terminating behind in a very sharp point, and falling in a vertical line with the base of the anal, which is similar to itself: head small, rounded: snout short, conical: tail exceedingly long, in consequence of the great development of the caudal fin, which is in the form of a long sithe; upper lobe enveloping the extremity of the vertebral column, and separated by a notch from the lower lobe, which is moderately broad at its origin: mouth moderate, of a horse-shoe form, entirely beneath: vent at nearly the anterior third of the entire length: nostrils beneath, at the posterior third of the length of the snout, transversely oval: eyes lateral, large, occupying three-fifths of the length of the upper jaw: teeth similar in both jaws, triangular, very sharp, not denticulated, broad at the base, without any accessory points or tubercles: no temporal orifices: branchial openings nearly equal, entirely lateral, the last two somewhat smaller, nearer together, and reaching beyond the anterior margin of the pectoral fin: pectorals narrow, very much elongated, triangular; the adhering side much smaller, and attached for nearly its whole length: ventrals small, triangular, horizontal, adhering by two-thirds of their inner edge, the remaining portion free. (*Colours.*) Bluish gray above, white be-neath; the pectoral fins attached to the white portion." BLAINV.

Met with occasionally on the British coast, but not very plentiful. Derives its English name of *Thresher* from its supposed habit of attack-ing and striking the Grampus with its long fox-like tail*.

188. S. *glaucus*, Linn. (*Blue Shark*.) — Teeth in the upper jaw triangular, and curved; · in the lower jaw longer, and more straight; all denticulated: body slender, slate-blue above.

S. glaucus, *Linn. Syst. Nat.* tom. I. p. 401. *Watson in Phil. Trans.*
    (1778.) vol. LXVIII. p. 789. pl. 12. *Bloch, Ichth.* pl. 86. S. cæruleus,
    Blainv. Faun. Franç. p. 90. Galeus glaucus, *Will. Hist. Pisc.*
    p. 49. Carcharias glaucus, *Flem. Brit. An.* p. 167. Blue Shark,
    Penn. Brit. Zool. vol. III. p. 109. *Id.* (Edit. 1812.) vol. III. p. 143.
    Le Bleu, *Cuv. Reg. An.* tom. II. p. 388.

LENGTH. Six or seven feet. PENN. (Edit. 1812.)

DESCRIPT. (*Form.*) Elongated: the skin less rough than in the others of this genus: snout long, sharp, depressed, not pellucid at the extremity, punctured above and below with numerous pores: (mouth large, widely cleft; teeth numerous, in four or five rows; the upper ones broadest, curving a little backwards, and denticulated at the edges; the lower ones narrower, straight, in the form of a scalene triangle, and finely denticulated:) nostrils long, transverse, (equally distant from the edge of the jaw, and the extremity of the snout:) eyes elliptic, but the irides exactly circular, the pupils lenticular and transverse: (commissure of the lips extending far beyond them:) no temporal orifices: five branchial openings, (moderate, elevated, lateral; the first three largest and furthest asunder; the last two, especially the fifth, smaller, and closer together:) two dorsals; the first at about the middle of the length, excluding caudal; (triangular, of moderate size, the base longer than the two other sides:) second dorsal not far from the setting on of the caudal, (much smaller, equally triangular, and much more inclined:) anal answering to this last fin: tail (scarcely equalling half the body,) with a triangular excavation at the upper part of the base of the caudal fin; this last with two lobes, the upper extending very far beyond the lower, and terminating in an acute angle: vent at the distance of more than one-third of the length from the setting on of the caudal: pectorals large, (falciform,) very long, terminating in an acute angle: ventrals small, (cut square behind.) (*Colours.*) Back of a fine, moderately deep, blue: belly silvery. WILL. and BLAINV.

Said to be not uncommon on some parts of the coast, particularly that of Cornwall during the pilchard season. The specimen described by Dr. Watson in the "Philosophical Transactions" was taken on the coast of Devonshire. It is possible, however, that in the case of this species, as in that of the *S. Carcharias*, two or more have been confounded under one name. The above description is from Willughby, and is that of a specimen observed by him at Penzance. It appears to be the same as the *S. cæruleus* of Blainville, from whom I have borrowed some addi-tional characters†. The *S. glaucus* of this last author is a closely allied species, in which the teeth are not denticulated at the sides.

---

* *Borl. Cornw.* p. 265.

† The parts borrowed from Blainville are included in parenthesis.

(3. LAMNA, *Cuv.*)

189.   S. *Cornubicus*, Gmel.   (*Porbeagle Shark.*)

S. Cornubicus, *Gmel. Linn.* tom. I. part iii. p. 1497. *Turt. Brit.
Faun.* p. 113. *Don. Brit. Fish.* vol. v. pl. 108. *Neill in Wern.
Mem.* vol. I. p. 549. *Blainv. Faun. Franç.* p. 96. pl. 14. f. 2.?
S. Selanonus, *Leach in Wern. Mem.* vol. II, p. 64. pl. 2. f. 2.?
Lamna Cornubica, *Flem. Brit. An.* p. 168.   Porbeagle, *Borl.
Cornw.* p. 265. pl. 26. f. 4.   *Penn. Brit. Zool.* vol. III. p. 117. *Id.*
(Edit. 1812.) vol. III. p. 152.   *Goodenough in Linn. Trans.* vol. III.
p. 80. pl. 15.   Le Squale Nez, *Cuv. Reg. An.* tom. II. p. 389.?

LENGTH.   Said to attain the length of from five to nine feet.

DESCRIPT.   (*Form.*) Girth (in the thickest part of a specimen three
feet nine inches in length) two feet one inch : body very thick and deep,
but extremely slender and flattened just at the setting on of the tail ; the
sides near that part distended and sloping, thinning off to a sharp angle
or elevated line : snout very long, slender towards the extremity, sharp-
pointed, and punctured beneath : nostrils near the mouth, at about one-
fourth of the distance between it and the end of the snout ; of a lunar
form, the extremities pointing backwards : mouth semicircular : teeth
very sharp, smooth, two-edged, with a little acute process at the base on
either side ; (the process in some concealed within the gums ;) arranged
(according to Goodenough) in the upper jaw, in two rows, except in the
front, where the two middle ones stand single ; in the under jaw, in two
rows also, except in the front, where the two middle teeth have a triple
row ; the inner row bent inwards, the others all turned outwards ; (according to Pennant) in three rows in the upper jaw ; the same on the
sides of the lower, but only two rows in the front of the latter : eyes about
four inches * from the extremity of the snout, and upon an exact level
with the surface of the body : branchial openings five in number, placed
in a regular series ; the apertures perpendicular, and about three inches
long : skin, when stroked backwards, a little roughish, with an obsolete
line of minute tubercles running from the head down the sides, and at
length ending in the thick elevated ridge, which takes place at the depression of the body near the tail : first dorsal placed nearly in the middle,
erect, its height not quite equal to its length : second dorsal pretty near
the tail, much smaller : anal nearly opposite to this last, of the same
length and size : above and below the tail, near the base of the caudal,
a semicircular or lunar impression, the points directed backwards : caudal
of a lunar form, vertical, the upper lobe nearly one-third longer (Pennant
says, a little longer †) than the lower : pectorals immediately behind the
branchial openings, and equalling rather more than one-sixth of the
entire length ; of a semilunar form behind : ventrals small, also of a
semilunar form behind. GOODEN. and PENN.   (*Colours.*) " Colour of
the whole upper part, the sides, fins, and tail, dusky, tinged obscurely
with green and blue ; beneath, from the tip of the nose, and also part
of the sides, entirely white." PENN.

This species was first noticed by Jago, from whose drawing of it Borlase's
figure was engraved.   Since his time many other individuals have occurred on different parts of the coast.   Dr. Goodenough's specimen was
obtained at Hastings ; Pennant's at Brighton.   Mr. Neill states that it
is occasionally met with in the Frith of Forth : Mr. Couch remarks that

---

* This is with reference to Dr. Goodenough's specimen : as, however, there was only an inch
difference in length between his and Pennant's, this difference would not much affect the relative
proportions.

† According to Cuvier and Blainville, the lobes are nearly equal.

on the coast of Cornwall it is not uncommon. It appears to be the same
as the *S. Selanonus* of Leach. Is ovoviviparous. Hunts its prey (according to Mr. Couch) in companies, from which circumstance it has received
its common name. *Obs.* Authors do not agree in all the characters which
they assign to this species, but it is probable that some of these, especially
the number of rows of teeth, may vary with age.

190.   S. *Monensis*, Shaw.   (*Beaumaris Shark.*)

> S. Monensis, *Shaw, Gen. Zool.* vol. v. part ii. p. 350.   S. Cornu-
>   bicus, β, *Gmel. Linn.* tom. i. part iii. p. 1497.   *Turt. Brit. Faun.*
>   p. 113.   Beaumaris Shark, *Penn. Brit. Zool.* vol. iii. p. 118.
>   pl. 17.   *Id.* (Edit. 1812.) vol. iii. p. 154. pl. 20.   Le Beaumaris,
>   *Cuv. Reg. An.* tom. ii. p. 389. note (2).

Length. Seven to nine and a half feet. Dav.
Descript. (*Form.*) " Snout and body of a cylindrical form : greatest
circumference (in a specimen seven feet long) four feet eight inches :
nose blunt : nostrils small : mouth armed with three rows of slender
teeth, flatted on each side, very sharp, and furnished at the base with two
sharp processes ; the teeth are fixed to the jaws by certain muscles, and
are liable to be raised or depressed at pleasure : first dorsal two feet eight
inches distant from the snout, of a triangular form : second dorsal very
small, and placed near the tail : pectorals strong and large : ventrals
and anal small : the space between the second dorsal and the tail much
depressed, the sides forming an acute angle ; above and below a transverse fossule or dent : tail crescent-shaped, but the horns of unequal
lengths ; the upper, one foot ten inches ; the lower, one foot one inch :
skin comparatively smooth, being far less rough than that of the lesser
species of this genus. (*Colour.*) The whole fish lead-colour." Dav.
No one appears to have met with this species excepting Mr. Davies,
who communicated to Pennant, by whom it was first published, a drawing
and description of one taken near Beaumaris. In the last edition of the
" British Zoology," there are some further particulars by Mr. Davies,
including an account of a second individual stranded near Bangor Ferry,
on the Anglesea side of the Menai, in June 1811. This last differed from
the former specimen in having the nose smaller (though itself a larger
fish), and more abruptly tapering. It was a female, and contained in its
belly four young ones, each about twenty-eight or thirty inches long.
Obs. By Gmelin and Turton, this species is made a variety of the last,
from which it scarcely seems to differ, excepting in its blunt snout.
Donovan and Fleming regard them as the same fish. Further observation alone can determine whether either of these opinions is
correct.

(4. Galeus, *Cuv.*)

191.   S. *Galeus*, Linn.   (*Common Tope.*)

> S. Galeus, *Linn. Syst. Nat.* tom. i. p. 399.   *Bloch, Ichth.* pl. 118.
>   *Turt. Brit. Faun.* p. 112.   *Blainv. Faun. Franç.* p. 85.   Canis
>   Galeus, *Will. Hist. Pisc.* p. 51. tab. B. 6. f. 1.   Galeus vulgaris,
>   *Flem. Brit. An.* p. 165.   Tope, *Penn. Brit. Zool.* vol. iii. p. 111.
>   *Id.* (Edit. 1812.) vol. iii. p. 146. but not pl. 18.   Le Milandre,
>   *Cuv. Reg. An.* tom. ii. p. 389.

Length. From three to five feet.
Descript. (*Form.*) Body fusiform, elongated : head moderately large,
depressed, behind the eyes broad, but narrowing anteriorly ; snout pro-

duced, of a somewhat triangular form, very much flattened, and in front
of the nostrils somewhat pellucid: nostrils beneath, nearly midway between the mouth and the extremity of the snout, partly covered by a
small membranous flap: mouth wide: jaws moderately bent: teeth small,
sharp-pointed, of a triangular form, with some smaller denticulations on
the outer edge only; in several rows, and nearly similar above and below:
eyes about half-way between the end of the snout and the first branchial
opening; behind each a small temporal orifice: branchial openings five
in number, rather small, near together; the first four of nearly equal
size; the fifth smaller, and placed immediately above the base of the
pectoral: skin moderately rough from tail to head, but smooth in the
opposite direction: two dorsals: the first not very large, commencing at
exactly one-third of the entire length; its height and length about equal;
of a triangular form, but with a projecting point at its posterior extremity
directed towards the tail: second dorsal just half-way between the first
and the extremity of the tail; of a similar form, but smaller: anal present, resembling the second dorsal; nearly opposite to that fin, but placed
a little backwarder: caudal with a large projecting lobe on its lower
margin; the upper lobe terminal, and obliquely truncated at its extremity: pectorals moderate, approaching triangular, the distance from
their insertion to the end of the snout considerably more than equal
to their length: ventrals exactly in the middle of the entire length, and
answering to the middle of the space between the two dorsals; only half
the size of the pectorals, and obliquely truncated at their extremities.
(*Colours.*) Of a uniform deep slate-gray above; yellowish white beneath.

Common in the Mediterranean, but apparently of not very frequent
occurrence in the British seas. Willughby speaks of its being met with
on the Cornish coast, where it has been since observed by Mr. Couch.
Pennant's specimen was taken on the coast of Flintshire, and weighed
twenty-seven pounds, its length being five feet. Dr. Johnston has procured it on the coast of Berwickshire [*]. The individual described above
was obtained, with others, at Weymouth, by Professor Henslow. According to Bloch, it sometimes attains to the weight of one hundred pounds.
It is stated by this same author, that it usually lives in society, and in
deep water.

(5. MUSTELUS, *Cuv.*)

192.   S. *Mustelus*, Linn.   (*Smooth Hound.*)

S. Mustelus, *Linn. Syst. Nat.* tom. I. p. 400.   *Turt. Brit. Faun.*
p. 112.   *Blainv. Faun. Franç.* p. 81. pl. 20. f. 1.   Mustelus
lævis, *Will. Hist. Pisc.* p. 60. tab. B. 5. f. 2.   *Flem. Brit. An.*
p. 166.   Galeus Mustelus, *Leach in Wern. Mem.* vol. II. p. 63.
pl. 2. f. 3.   Smooth Hound, *Penn. Brit. Zool.* vol. III. p. 116.
pl. 16.   Smooth Shark, *Id.* (Edit. 1812.) vol. III. p. 151.   Tope
Shark, *Id.* pl. 18.   L'Emissole commune, *Cuv. Reg. An.* tom. II.
p. 390. note (1).

LENGTH.   From three to four feet; sometimes more.
DESCRIPT.   (*Form.*) General form very similar to that of the *S.
Galeus;* the snout, however, not quite so much produced: nostrils
midway between the mouth and the end of the snout, partly covered
by a small cutaneous membrane: teeth small and numerous, obtuse,
forming a closely compacted pavement, disposed in a quincuncial order:
eyes large, oval: behind each a temporal orifice of moderate size:

branchial openings large, set nearer together as they are more behind; the middle three larger than the others; the last smallest, and placed above the base of the pectoral: skin less rough, when rubbed from tail to head, than in the *S. Galeus;* in the opposite direction perfectly smooth: all the fins in shape and situation exactly the same as in that species; only the second dorsal relatively somewhat larger, while the projecting lobe on the lower portion of the caudal is smaller. (*Colours.*) Upper parts of a uniform pearl-gray; paler, or almost white, beneath.

Tolerably common on most parts of the coast. According to Mr. Couch, "keeps near the bottom, and feeds chiefly on crustaceous animals, which its blunt teeth are well calculated to crush."

(37.) *S. Hinnulus,* Blainv. Faun. Franç. p. 83. pl. 20. f. 2. *Mustelus stellatus,* Riss. Hist. Nat. de l'Eur. Mérid. tom. III. p. 126. *L'Emissole tachetée de blanc, ou Lentillat,* Cuv. Reg. An. tom. II. p. 390. note (1).

Length three feet. *Form* almost the same as that of the last species; only the lateral line more distinctly marked. *Colour* brownish ash, with a row of small whitish spots from the eye towards the first of the branchial openings; lateral line indistinctly spotted with white; also a moderate number of small scattered white spots between the lateral line and the dorsal ridge.

The above notes, made at Weymouth in Aug. 1832, relate to a species of shark, not unfrequently captured on that coast, which appears to be identical with the *S. Hinnulus* of Blainville. I have since seen a drawing of a similar fish in the possession of Mr. Yarrell, to whom it was sent by Mr. Couch of Cornwall. Not being aware at the time of the existence of a second species of *Mustelus,* and having had no opportunity of comparing a recent specimen with Blainville's description, I restrict myself to this notice of the circumstance, without positively asserting the *S. Hinnulus* to be British. It is, however, a great question, whether this last be any thing more than a variety of the *S. Mustelus.* As such it is considered by the Prince of Musignano in his *Fauna Italica.*

(6. SELACHE, *Cuv.*)

193. S. *maximus,* Linn. ? (*Basking Shark.*)

S. maximus, *Linn. Syst. Nat.* tom. I. p. 400. ? *Home in Phil. Trans.* (1809.) p. 206. pl. 6. f. 1. *Flem. Brit. An.* p. 164. Basking Shark, *Penn. Brit. Zool.* vol. III. p. 101. pl. 13. *Id.* (Edit. 1812.) vol. III. p. 134. pl. 16. *Low, Faun. Orc.* p. 171.

LENGTH. Thirty feet and upwards.

DESCRIPT. (*Form.*) Form rather slender: snout short, blunt, and pierced full of small holes: mouth large: teeth small and numerous; (according to Low, in five or six rows;) those before much bent, those more remote in the jaws conic and sharp-pointed: eyes small: branchial apertures five in number, large, reaching from the neck to the throat: first dorsal very large, not directly in the middle, but rather nearer the head: the second small, situate near the tail: pectorals (in a specimen twenty-three feet long) nearly four feet: ventrals smaller, placed just beneath the hind fin of the back: a small anal: tail very large; (according to Pennant,) the upper lobe remarkably longer than the lower; (according to Low,) the lobes equal in length, only the upper one somewhat broader and blunter than the lower: skin rough, like shagreen, but less so on the belly than the back. PENN. and LOW. The following characters are added from Sir E. Home. " Nostrils opening on the edge

of the upper lip: eyes very small; the pupils perfectly round: half-way
between the eye and the gills, on each side, the orifice of a canal, leading
into the mouth: pectorals situate a little behind the posterior gills: dorsal
situate nearly opposite to the middle space between the pectoral and anal
(ventral?) fins: posterior dorsal small, and situate half-way between the
anal (ventral?) fins, and the setting on of the tail: the two anal (ventral?)
fins attached on their upper edge for about half their extent each to the
lower side of a long projecting body peculiar to the *male*: all the fins
have a thick round edge anteriorly, and become gradually thinner towards the posterior part, which is partially serrated: a deep sulcus at the
setting on of the tail, and, on each side of the fish, a scabrous ridge
extending from this sulcus as far forwards as the posterior dorsal fin."
(Colours.) "Upper part of the body deep lead-colour: belly white."
PENN.

A large species of Shark, referred by authors to the *Squalus maximus*
of Linnæus, has been repeatedly noticed in the British seas. Pennant
observes that such a fish has been long known to the inhabitants of the
South and West of Ireland and Scotland, and those of Caernarvonshire
and Anglesea; that they are seen in the Welsh seas in most summers,
sometimes in vast shoals; that they also appear in the Frith of Clyde,
and among the Hebrides, in the month of June, in small droves, but
oftener in pairs. Mr. Neill states[*] that they are common in the Scottish
seas, occasionally, though seldom, entering the Frith of Forth. Low
speaks of their being also common in the Orkneys. Dr. Shaw notices
one which was taken at Abbotsbury in Dorsetshire[†]. Sir E. Home has
described another captured at Hastings. Mr. Couch mentions another
taken on the coast of Cornwall[‡]. Whether, however, in these and other
instances the same species has been observed, from the want of more
accurate descriptions, it is impossible to determine. Blainville is of
opinion that no less than four distinct species have been confounded
by naturalists under the name of *Squalus maximus* [§]. Cuvier, on the
other hand, thinks that the differences observable in the figures and
descriptions which authors have given of this fish, may have arisen from
incorrect observation, and from the difficulty which attends a close exa-
mination of such large animals [||]. These points can only be cleared up
by further investigation into the real characters of such individuals as

---

* *Wern. Mem.* vol. I. p. 550.   † *Gen. Zool.* vol. v. part ii. p. 330.
‡ *Linn. Trans.* vol. XIV. p. 91.
§ See two memoirs on this subject, one contained in the *Journal de Physique* for Sept. 1810,
(vol. LXXI. p. 248); the other in the *Annales du Muséum* for 1811. (vol. XVIII. p. 88.) In the
first of these, Blainville has briefly characterized what he considers as three species, under the
following names: (1.) *Squalus Gunnerianus;* distinguished principally by the want(?) of temporal orifices, and the presence of an anal fin: this, which is intended for the original fish observed by Gunner, (with reference to which Linnæus established his species,) he regards the same
as Pennant's, supposing this last to have been without temporal orifices, on which point Pennant
is silent. (2.) *S. peregrinus;* in which there are neither temporal orifices nor anal: the type of
this species is in the Museum at Paris. (3.) *S. Homianus;* distinguished by the presence of
temporal orifices, and the absence of an anal: this species is founded upon the specimen described
by Sir E. Home, and named after him. In his *second* memoir, Blainville has given a most
elaborate description of both the external and internal characters of a large shark brought to
Paris in 1810, which he regards as a fourth species, characterized by the presence of both temporal
orifices and anal, the former, however, being extremely small. With this last he associates the
" Basking Shark, *male*," figured by Shaw in his " General Zoology," (vol. v. part ii. pl. 149.), the
drawing of which was probably taken from the specimen mentioned by that author as having
been captured on the coast of Dorsetshire, but to which no description is annexed.
    It is much to be desired that any of our own naturalists who may have an opportunity of
observing any individuals of the species usually termed *Basking Shark*, would take as accurate
and detailed a description as possible of the several parts, and compare it afterwards with that
given by Blainville in the second of the above memoirs, which should serve as a standard of comparison in all future cases. It is particularly important that they note the presence or absence
of temporal orifices and an anal fin, which are so small (compared with the entire bulk of the
animal) as to be easily overlooked. They should also attend to the form of the teeth, the nature
of the skin, and the size, as well as form and position, of the branchial openings.
    || See *Reg. An.* tom. II. p. 391. note (1).

may be hereafter met with. The Basking Shark of Pennant is represented as a tame and inoffensive species, deriving its English name from
its habit of "lying as if to sun itself on the surface of the water." Its
food is supposed to consist entirely of marine plants, no remains of fish
having ever been discovered in the stomach. Pennant thinks that it
is migratory.

(7. Spinax, *Cuv.*)

194.   S. *Acanthias*, Linn. (*Picked Dog-Fish.*)

S. Acanthias, *Linn. Syst. Nat.* tom. i. p. 397. *Bloch, Ichth.* pl. 85.
  Don. Brit. Fish. vol. iv., pl. 82. *Turt. Brit. Faun.* p. 114.
  Blainv. Faun. Franç. p. 57. *Cuv. Reg. An.* tom. ii. p. 392.
  Galeus Acanthias sive Spinax, *Will. Hist. Pisc.* p. 56. tab. B. 5.
  f. 1. Spinax Acanthias, *Flem. Brit. An.* p. 166. Picked Dog-
  Fish, *Penn. Brit. Zool.* vol. iii. p. 100. Picked Shark, *Id.* (Edit.
  1812.) vol. iii. p. 133.

Length.   From three to three and a half, sometimes four, feet.
Descript.   (*Form.*) General form much resembling that of the *S.
Mustelus:* body moderately elongated: head depressed; snout long,
conical, obtuse at the extremity: nostrils beneath, more remote from the
mouth than in the *S. Mustelus*, partly covered by a minute cutaneous
flap: jaws bent: teeth in two rows, small, sharp, the edges cutting and
not denticulated, bending from the middle each way towards the corners
of the mouth, the points short and inclining backwards: eyes large,
oblong: temporal orifices large, round, placed higher than in the *S. Mustelus:* branchial openings five in number, small, a little decreasing in
size from the first to the last; placed in a line with the base of the pectorals, the last opening being immediately in advance of those fins: skin
very rough when rubbed from tail to head, but nearly smooth in the
opposite direction: lateral line tolerably well-defined, straight: two dorsals; in form and situation much as in the *S. Mustelus*, but before each
a sharp strong spine; the spine of the second stronger and longer than
that of the first: caudal unequally forked, the upper lobe projecting far
beyond the lower: no anal: pectorals broad, triangular, cut square behind, reaching when laid back to a vertical line from the spine of the first
dorsal: ventrals a little behind the middle of the entire length, much
smaller than the pectorals, obliquely truncated. (*Colours.*) Of a uniform
reddish gray, or grayish brown above; whitish beneath. The *young*,
according to Bloch and Cuvier, are spotted with white.
  A common species on all parts of the British coast. Is very voracious,
preying on other fish. Ovoviviparous.

(38.) *S. Spinax*, Linn. Syst. Nat. tom. i. p. 398.   Stew. El. of
  Nat. Hist. vol. i. p. 319.   Blainv. Faun. Franç. p. 60.   Nilss.
  Prod. Ichth. Scand. p. 118. *Acanthias Spinax*, Riss. Hist. Nat.
  de l'Eur. Mérid. tom. iii. p. 132.

>   This species is marked as British by Stewart, in both Editions of his
> "Elements of Natural History," but on what authority he does not men-
> tion. It is not noticed, that I am aware, by any other of our English
> authors. Said to be distinguished from the last, which it closely resembles,
> principally by the abdomen being nearly black, and the nostrils at the ex-
> tremity of the snout. According to Nilsson, it is the smallest species in
> the genus, not exceeding a length of sixteen inches. It is found in the
> Northern seas, as well as in the Mediterranean.

(8. SCYMNUS, *Cuv.*)

195. S. *borealis*, Scoresb. (*Greenland Shark.*)

S. borealis, *Scoresb. Arct. Reg.* vol. I. p. 538. and vol. II. pl. 15.
f. 3, & 4.   S. Norvegianus, *Blainv. Faun. Franç.* p. 61.   S. glacialis, *Nilss. Prod. Ichth. Scand.* p. 116.   Scymnus borealis,
Flem. Brit. An. p. 166.

LENGTH. From twelve to fourteen feet, sometimes more. SCORESB.
DESCRIPT. "Circumference (in a specimen fourteen feet long) about
eight feet. Colour gray: eye blue; pupil emerald-green. Mouth wide.
Teeth in the upper jaw, broad at the base, suddenly becoming narrow
and lanceolate with the cutting edges rough; in the lower jaw pyramidal,
compressed, the cutting edges crenulated, a little convex on the fore-edge,
and subangularly concave on the hind-edge. Tongue broad and short.
First dorsal fin larger than the second; more advanced than the ventrals: pectorals large: ventrals elongated; the two sides nearly parallel."
FLEM. According to Blainville, the general form of the body is exactly
similar to that of the *S. Acanthias*, differing principally in the want
of spines before the dorsal fins, and in the peculiar character of the teeth,
which are arranged in two rows in the lower jaw, and in three in the
upper.
This species, which is the *S. Carcharias* of Gunner and Fabricius, and
perhaps of Bloch, is a native of the Northern seas. Dr. Fleming, however, mentions two instances in which it has occurred in those of our own
Islands. One of the specimens was caught in the Pentland Frith in
1803: the other was found dead at Burra Firth, Unst, in July 1824. Said
to be very voracious. Food, according to Scoresby, dead whales, as well
as small fishes and crabs. Nilsson states that it resides principally in
the deepest parts of the sea, rarely coming to the surface.

Obs. It is uncertain to which of the foregoing sub-genera the following
two doubtful species belong.

(39.) S. *Selanoneus*, Flem. Brit. An. p. 168.

An obscure species, found (according to Fleming) by the late Dr.
Walker, in Lochfyne, in Argyleshire, where it is said to appear during
the herring season. The following description is quoted by Dr. Fleming,
from the original Mss. of that naturalist.
  "*Caput*, maxilla subæqualis, superiore prominente, rostrata. Maxilla
superior crassissima, apice truncata marginata, angulo superiori obtuso
suberecto. Maxilla inferior angusta. Dentes numerosi, acuti. Oculi
super cantham oris positi sunt. *Corpus*, octo-pedale, oblongum, tere-
tiusculum, cute aspera. Spiracula quinque, antico breviore, erecta,
lineari-lunata: margine postico curvato. Tria spiracula postica super
pinnam pectoralem positi sunt, duo altera ante pinnam pectoralem versus
oculum. *Pinnæ:* dorsum suberectum, muticum, bipinne. Pinna dorsalis
antica erecta, subpedalis, circa medium corporis. Pinna dorsalis postica,
multo minor, medium inter pinnam anticam et caudam occupat. Pinnæ
pectorales pedem longitudine superant, et ante pinnam anticam dorsalem
positæ sunt. Pinnæ ventrales spatium ante pinnam dorsalem posticam
occupant. *Cauda*, perpendicularis, furcata, segmentis subæqualibus
subacutis; superiori longiori. Sore prolato, maxillis subæqualibus; superiore truncata emarginata."
  From the circumstance of Dr. Walker's taking no notice either of the
anal fin or temporal orifices, Dr. Fleming infers the absence of both. He
thinks that in consequence this species may claim to rank as a new genus,
occupying a place between *Carcharias* and *Lamna*.

(40.) *Rashleigh Shark*, Couch in Linn. Trans. vol. xiv. p. 91.

" Twenty-nine feet four inches long; twenty-four feet round; the fork of the tail seven feet: weight four tons. Eye in front, under a snout that projects and is turned upward: mouth two feet and a half wide. Head deep : the first dorsal fin much elevated." Couch.

The above notes relate to a species of Shark, a drawing and memorandum of which are said to be in the possession of W. Rashleigh, Esq. of Mena-billy. Mr. Couch observes that it seems to resemble the *Basking Shark*, but differs from it in the form of the head and situation of the eye.

GEN. 74. ZYGÆNA, *Cuv.*

196. Z. *Malleus*, Val.? (*Hammer-head.*)

Z. Malleus, *Valenc. in Mém. du Mus.* (1822.) tom. ix. p. 223. pl. 11. f. 1.? *Cuv. Reg. An.* tom. ii. p. 393. Squalus Zygæna, *Linn. Syst. Nat.* tom. i. p. 399.

A fish of this genus is recorded by Messrs. C. and J. Paget* as having occurred at Yarmouth, October 1829. The head is said to be preserved in the Norwich Museum. As, however, the species have been much confounded, and the exact one in this instance does not appear to have been determined, it is impossible to annex any description. The *Squalus Zygæna* of Linnæus is the *Z. Malleus* of M. Valenciennes, who has published detailed descriptions of four species of this genus, accompanied in each case by a representation of the head, which appears to offer the best characteristic marks for distinguishing them. In the *Z. Malleus*, which is the most common species, the head is in the form of a very long rectangle; the anterior margin straight, and deeply notched near the external angle; the nostrils immediately beneath the notch. The *Squalus Zygæna* of Bloch is a distinct species, in which the nostrils are removed much further from the eyes. M. Valenciennes' memoir, which is referred to above, should be consulted by those who may have an opportunity of seeing any other British specimen of this genus.

GEN. 75. SQUATINA, *Dumér.*

197. S. *Angelus*, Cuv. (*Angel-Fish.*)

S. Angelus, *Cuv. Reg. An.* tom. ii. p. 394. *Blainv. Faun. Franç.* p. 53. S. vulgaris, *Flem. Brit. An.* p. 169. Squatina, *Will. Hist. Pisc.* p. 79. tab. D. 3. Squalus Squatina, *Linn. Syst. Nat.* tom. i. p. 398. *Bloch, Ichth.* pl. 116. *Don. Brit. Fish.* vol. i. pl. 17. *Turt. Brit. Faun.* p. 114. *Shaw, Nat. Misc.* vol. xxi. pl. 906. Angel-Fish, *Penn. Brit. Zool.* vol. iii. p. 98. pl. 12*. Angel Shark, *Id.* (Edit. 1812.) vol. iii. p. 130. pl. 15.

LENGTH. From five to seven, sometimes eight, feet.

DESCRIPT. (*Form.*) Broad and depressed anteriorly, elongated and tapering behind: upper part of the body convex; lower part flat: great-est breadth in the region of the pectorals, equalling, those fins being excluded, not quite one-fourth of the entire length; being included, more than half of the same: head nearly round, broader than the body, from which it is separated by a deepish notch on each side: mouth large, terminal, transverse: jaws but little bent, and nearly of equal length:

* *Nat. Hist. of Yarm.* p. 17.

teeth numerous, in five rows above and below, broad at the base, each
terminating upwards in a sharp slender point: nostrils almost at the
margin of the upper lip, covered by a membrane terminating in two filaments; between the nostrils the snout is slightly notched: eyes on the
upper part of the head, very small, not half the size of the large temporal
orifices, which last are of a lunulate form, the horns of the crescent being
directed backwards: branchial openings rather small, situate on each
side of the neck, between the head and the trunk: skin very rough,
covered with numerous small prickly tubercles; some larger tubercles
of a similar nature above the eyes, and along the mesial line of the back:
two dorsals, placed very much behind on the upper part of the tail; both
small, and nearly of the same size: caudal large, obliquely bifurcated,
the upper lobe being a little the longest: no anal: pectorals very large,
attached horizontally, broadest at their posterior margin, projecting forwards on each side of the neck in the form of an acute shoulder: ventrals of a somewhat similar form, but smaller. *Obs.* Cuvier and Fleming
describe this species as having the pectorals armed with short curved
spines near their margins. In the few specimens which I have examined
they were not present. Probably, as in the family of the *Rays*, they
are merely a sexual character. (*Colours.*) Upper parts more or less deep
gray: lower parts dirty white.

A common and very voracious species, preying upon other fish. Keeps
near the bottom. Attains to a large size. Pennant mentions having seen
them of near an hundred weight. According to Bloch, produces in the
Spring and Autumn from seven to eight young. On some parts of the
coast is called a *Monk-Fish;* on others, a *Kingstone.*

(41.) *Lewis,* Couch in Linn. Trans. vol. xiv. p. 90.

Under the above name, Mr. Couch notices a fish, which he states is not
unfrequently taken with a line on the coast of Cornwall, and which bears
some resemblance to the *Squatina Angelus,* but which he seems disposed
to consider as a distinct species. The following is his description:
"Somewhat smaller than the *Monk.* Head large, flat, the jaws of equal
length, forming a wide mouth; the upper jaw falls in somewhat at the
middle, so that at this part the lower jaw seems a little the longest; both
are armed with several rows of sharp teeth; the tongue is small. The
head is joined to the body by something which resembles a neck; the
body is flat so far back as the ventral fins, beyond these it is round: the
pectoral and ventral fins are very large; the former are flat, and both have
near their extremities a number of spines. The two dorsal fins are placed
far behind: the lobes of the tail are equal and lunated. There are five
spiracula: the eyes are very small, and the nictitating membrane, which
is of the colour of the common skin, contracts over the eye, leaving a linear
pupil. The body is slightly rough, of a sandy brown colour: the under
parts white. It is about five feet long, and keeps near the bottom."
Judging from the above description, I must confess I hardly see in what
respects it differs from the last species.

(6.) *PRISTIS,* Lath.

(42.) *P. antiquorum,* Lath. in Linn. Trans. vol. ii. p. 277. pl. 26.
f. 1. (Rostrum.) Cuv. Reg. An. tom. ii. p. 395. *Squalus Pristis,*
Linn. Syst. Nat. tom. i. p. 401. *Pristibatis antiquorum,* Blainv.
Faun. Franç. p. 50.

According to the late Dr. Walker*, this species has been found sometimes in Loch Long. It does not appear, however, to have been noticed

---

* On the authority of Dr. Fleming. See *Brit. An* p. 164.

in our seas by any other naturalist. It is distinguished from some other allied species by the rostral spines not exceeding in number from eighteen to twenty-four on each side. It is found in various parts of the Ocean, as well as in the Mediterranean. Attains to the length of from twelve to fifteen feet.

GEN. 76. RAIA, *Linn.*

(1. TORPEDO, *Dumér.*)

198. R. *Torpedo*, Linn. (*Electric Ray.*)

R. Torpedo, *Linn. Syst. Nat.* tom. i. p. 395. *Bloch, Ichth.* pl. 122.? *Don. Brit. Fish.* vol. iii. pl. 53. *Turt. Brit. Faun.* p. 110. *Blainv. Faun. Franç.* p. 44. Torpedo vulgaris, *Flem. Brit. An.* p. 169. Torpedo, *Walsh in Phil. Trans.* (1774.) p. 464. Electric Ray, *Penn. Brit. Zool.* vol. iii. p. 89. pl. 10. *Id.* (Edit. 1812.) vol. iii. p. 118. pl. 12.

LENGTH. From two to four feet.

DESCRIPT. (*Form.*) "Head and body indistinct, and nearly round: greatest breadth two-thirds of the entire length: thickness, in the middle, about one-sixth of the breadth, attenuating to extreme thinness on the edges: mouth small; teeth minute, spicular: eyes small, placed near each other: behind each a round spiracle, with six small cutaneous rags on their inner circumference: branchial openings five in number: skin every-where smooth: two dorsal fins on the trunk of the tail: tail one-third of the entire length, pretty thick and round; the caudal fin broad and abrupt: ventrals below the body, forming on each side a quarter of a circle. (*Colours.*) Cinereous brown above; white beneath." PENN.

First ascertained to be a native of the British seas by Mr. J. Walsh, who obtained specimens from Torbay. According to Pennant, it is not unfrequently taken on that coast; has been also caught off Pembroke, and sometimes near Waterford in Ireland. Donovan mentions the coast of Cornwall; where it has been since noticed by Mr. Couch, though, according to this last gentleman, it is extremely rare. I may add that it occurs also occasionally off Weymouth, where it is called the *Numb-Fish.* It must be stated, however, that, in the opinion of Risso and Cuvier, several species have been confounded under the name of *Raia Torpedo,* and the exact one met with in our seas, or whether more than one has occurred, are points not hitherto ascertained[*]. Fleming thinks that the

---

[*] It may assist future observers in determining the British species, to state the leading characters of four established by Risso in his "Histoire Naturelle de l'Europe Méridionale."

(1.) *Torpedo Narke,* Riss. tom. iii. p. 142. *T.* Corpore supra rubro luteo, maculis quinque ocellatis, in pentagoni figura dispositis.

    La Torpille à taches œillées, *Cuv. Reg. An.* tom. ii. p. 397. *Blainv. Faun. Franç.* pl. 10. f. 2.

(2.) *T. unimaculata,* Riss. tom. iii. p. 143. pl. 4. f. 8. *T.* Corpore fulvo, albido punctulato; ocello unico, oblongo, in medio dorso; cauda elongata, gracili.

    La Torpille à une tache, *Blainv. Faun. Franç.* pl. 10. f. 1.

Risso states that in this species the spiracles are large, and without the tooth-like processes : the electrical apparatus scarcely visible, and giving but very slight shocks.

British species belongs to the *Torpedo marmorata* of Risso. According to Blainville, who regards Risso's species as mere varieties, the *T. Galvani* of that author is the one most commonly met with on the shores of the Mediterranean. This fish, at least the British species, attains to a large size: according to Pennant, it has been known to weigh above eighty pounds. The exact use of the electrical apparatus is not well ascertained. It is generally supposed to serve as a means of defence, or to assist the fish in securing its prey, which is said to consist of other fish. Mr. Couch imagines that it is connected with the functions of digestion *.

(2. RAIA, *Cuv.*)

* *Snout sharp ; more or less elongated.*

199. R. *Batis*, Linn. (*Skate.*)—Snout sharp, conic, the lateral margins not parallel: skin granulated above : one or three rows of spines on the tail ; the points of the lateral rows, when present, directed forwards : colour beneath gray, with black specks.

R. Batis, *Linn. Syst. Nat.* tom. I. p. 395. *Bloch, Ichth.* pl. 79. *Turt. Brit. Faun.* p. 110. *Flem. Brit. An.* p. 171. *Blainv. Faun. Franç.* p. 13.? R. lævis undulata, seu cinerea, *Will. Hist. Pisc.* p. 69. tab. C. 5. *Ray, Syn. Pisc.* p. 25. Skate, *Penn. Brit. Zool.* vol. III. p. 82. pl. 9. *Id.* (Edit. 1812.) vol. III. p. 111, and Sharp-nosed Ray, pl. 11. La Raie blanche ou cendrée, *Cuv. Reg. An.* tom. II. p. 398.

LENGTH. From two to four feet ; sometimes more.

DESCRIPT. (*Form.*) Form rhomboidal ; the transverse diameter greater by one-third than the length, this last being measured from the extremity of the snout to the vent : body thin, in proportion to its bulk : snout considerably elongated, sharp, conical, the lateral margins never becoming parallel, but approaching gradually to form an acute angle : teeth numerous, in several rows, rather closely compacted, oval and broad at the base, each terminating above in a sharp conical point, hooked, the hooks inclining backwards, and most developed in the inner rows, and on the central teeth in those rows : nostrils in a line with the angles of the mouth, with which they are connected by means of a prolonged channel, and placed at less than one-third of the distance from the mouth to the margin of the pectorals : eyes of moderate size : spiracles large : skin finely granulated above, communicating a slight roughness to the touch ;

(3.) T. *marmorata*, Riss. tom. III. p. 143. pl. 4. f. 9. *T.* Corpore carneo, maculis fuscis, fasciisque sinuosis, marmorato ; cauda crassa, summitate rotundata.

La Torpille marbrée, *Blainv. Faun. Franç.* pl. 9.

Spiracles surrounded by seven tooth-like processes : branchial openings crescent-shaped : electrical apparatus very distinct.

(4.) T. *Galvani*, Riss. tom. III. p. 144. *T.* Corpore fulvo, immaculato, nigro-marginato.

La Torpille Galvanienne, *Cuv. Reg. An.* tom. II. p. 397.

Differs from the three preceding species in its much larger dimensions, and in the upper part of the body being constantly of a uniform red colour, without any spots or markings whatever.

* See *Linn. Trans.* vol. XIV. pp. 89, 90.

under surface mostly smooth, but a little rough in places, more especially
beneath the snout: a row of strong spines along the mesial line of the
tail, with the points directed backwards; a lateral row on each side of
the same, with the points standing out or directed forwards; sometimes
the lateral rows are wanting, or simply indicated (especially in *young*
specimens) by small osseous tubercles: generally no spines above or
behind the eyes, or on any part of the back: tail as long as the body,
depressed, not very stout; furnished with two moderately-sized finlets
near the extremity, a little remote from each other; scarcely the rudiment of a caudal: pectorals broad, rounded at their lateral extremities,
the anterior margins nearly straight, the posterior rather convex: ventrals moderate, divided into two lobes; the upper lobe polliciform; the
appendages (of the *male*) very small in young specimens, and not extending so low as the ventrals themselves, but in the adults much longer and
more developed. *Obs.* The *Males* in this, and in all the other species of
this family, besides possessing the ventral appendages, are characterized
by several parallel rows of sharp hooked spines on the anterior lobe, and
at the angle, of each of the pectorals. These spines are always very
much reclined, and partly concealed, with the points directed inwards.
They are quite independent of the other, generally larger and more erect,
spines, which are more or less characteristic of the particular species.
The number of rows, and the number in each row, depend upon age,
being greatest in the oldest individuals: sometimes, in *very young males*,
these sexual spines (as they may be termed) hardly show themselves at
all. It may be added that the teeth also often differ in the two sexes;
the *males* generally having them sharper and more pointed than the
other sex; in the *young*, however, they are sometimes similar in both
sexes. (*Colours.*) Upper surface of a uniform dusky brown, tinged with
cinereous: under surface dusky gray, sometimes grayish white, studded
with black specks, having a white centre, most abundant beneath the
snout. The colours of both sides become paler with age.
    Not uncommon on many parts of the coast, though less plentiful than
some other species. Attains to a very large size, weighing sometimes
nearly two hundred pounds. According to Pennant, the ova, or *purses*,
are cast by the females from May to September. The young are sometimes called *Maids*. *Obs.* By some authors the skin of this species is
represented as smooth; and it is not quite certain whether two have not
been in some instances confounded under the name of *R. Batis*.

200.   R. *Oxyrhynchus*, Linn.? (*Sharp-nosed Ray.*)—
Snout sharp, slender, and very much elongated, the lateral
margins parallel near the tip: skin smooth: one or three
rows of spines on the tail: colour beneath plain white,
without spots.

> R. Oxyrhinchus, *Mont. in Wern. Mem.* vol. ii. p. 423.   Sharp-
> nosed Ray, *Penn. Brit. Zool.* vol. iii. p. 83.   *Id.* (Edit. 1812.)
> vol. iii. p. 113. but not pl. 11.

Length.   Six feet and upwards.
    Descript.   (*Form.*) Differs from the last species, which it very much
resembles, in having the snout more slender, the margins, in a moderately sized fish, running nearly parallel to each other for three or four
inches at the extremity: teeth longer, and not so closely compacted:
skin perfectly smooth: three rows of spines on the tail, when arrived at

maturity. (*Colours.*) Upper parts of a plain brown colour, without spots or lines : under parts white, also without spots. MONT.

This species, which I have not seen, is represented by authors as not very uncommon in the British seas. Like the last it attains to a great size. One obtained by Pennant in the Menai measured nearly seven feet in length, and five feet two inches in breadth. Montagu states that specimens have been taken of which the computed weight was above five hundred pounds. As, however, in the case of *R. Batis*, the name of *Oxyrhinchus* has been applied at different times to two or more perfectly distinct species, and it is impossible to state whether we have not in our seas more than one to which that name has been given. For this reason I have not annexed any references except to Pennant and Montagu, whose descriptions alone (of all our English authors) appear original, and can with any certainty be referred to the same species. Whether the present one be synonymous with the *Oxyrhinchus* of Willughby *, I consider very doubtful. The *R. Oxyrhinchus* of Bloch †, as well as that of Blainville ‡, appear quite distinct.

201. R. *marginata*, Lacép. (*Bordered Ray.*)—Snout sharp, slender, moderately elongated, the lateral margins nearly parallel at the tip: skin smooth: three rows of spines on the tail : colour beneath white, with a broad dusky border.

R. marginata, *Lacép. Hist. Nat. des Poiss.* tom. v. p. 663. pl. 20. f. 2.   *Blainv. Faun. Franç.* p. 19: pl. 3. f. 2.   *Flem. Brit. An.* p. 172.

DIMENS. The following are those of an English specimen. Total length fifteen inches six lines : length of the head (measured from the end of the snout to the spiracles behind the eyes) three inches six lines; of the tail (from the vent to its extremity) seven inches nine lines : greatest breadth (across the pectorals) eleven inches three lines. *Obs.* The total length of Blainville's specimen was two feet.

DESCRIPT. (*Form.*) Rhomboidal; the transverse diameter rather more than one-third greater than the length from the end of the snout to the vent: snout elongated, projecting considerably from between the pectorals, terminating in a sharp point, with the lateral margins nearly parallel for the last quarter of their length: mouth moderately wide; jaws transverse; teeth numerous, closely set, in several rows, roundish, or somewhat quadrilateral, at the base, each terminating in a sharp point: nostrils in a line with the corners of the mouth, and rather more than half-way between them and the upper margins of the pectorals; a channel from the nostrils to the mouth, covered by a membranous flap: eyes and spiracles both large: skin perfectly smooth above; also beneath, excepting along the anterior margins of the pectorals and the surface of the snout, which are set with very minute spines and denticles: one large sharp spine above each eye, inclining backwards, and another smaller one behind each eye: no spines on any part of the back, but three rows on the tail, one occupying the middle ridge, the two others the sides; the spines on these rows strong and sharp, and mostly inclining backwards: tail scarcely longer than the body, depressed, rather stout, with two moderately-sized finlets, of equal size and form, nearly contiguous; scarcely the rudiment of a caudal: pectorals broad, with the anterior margin hollowed out, and not prolonged beyond the basal half of the snout: ventrals moderate, deeply notched or bilobated. (*Colours.*) General colour of the upper parts reddish brown, somewhat paler on the pectorals, with a faint indication of round whitish spots: beneath white, with a broad border all round, especially beneath the angles of the pectorals, of dark reddish brown, approaching to dusky: tail entirely black.

First described by Lacépede from specimens sent him by M. Noel from Dieppe, Liverpool and Brighton. The individual described above was obtained at Weymouth by Professor Henslow, and is now in the Museum of the Cambridge Philosophical Society. In the same collection is a young one, extracted from the purse, which is very large, compared with those of other species. Blainville states that he has seen several from the Channel, the Ocean, and the Mediterranean. He thinks that it never attains to a very large size.

202. R. *chagrinea*, Mont. (*Shagreen Ray.*) — Snout long and sharp: skin rough above: only two principal rows of spines on the tail, the ridge being without spines: colour beneath white.

R. chagrinea, *Mont. in Wern. Mem.* vol. ii. p. 420. pl. 21.   R. aspera nostras, *Will. Hist. Pisc.* p. 78.   R. aspera, *Flem. Brit. An.* p. 172. *Blainv. Faun. Franç.* p. 22.?   Shagreen Ray, *Penn. Brit. Zool.* vol. iii. p. 87.   *Id.* (Édit. 1812.) vol. iii. p. 117.

DIMENS.   The following were those of Montagu's specimen. Entire length three feet; length of the tail seventeen inches: breadth twenty-four inches. According to Pennant, it attains to the size of the Skate.

DESCRIPT.   (*Form.*) Form narrower than that of the common kinds: greatest breadth two-thirds of the entire length: snout long and sharp, much resembling that of the *R. Oxyrhinchus*: teeth slender and very sharp: the whole upper surface rough, covered closely with minute shagreen-like tubercles, resembling the skin of the *Dog-Fish*; under surface smooth, except the head, breast and tail: nine or ten spines above the eye, but in the middle of the brow a vacancy; on the snout several tubercular spines, but scarcely definable, in two rows: behind the head, seven or eight spines on the dorsal ridge, extending so far back as to be in a line with the branchiæ: two rows of strong spines on the tail, one on each side of the ridge, projecting outwards, the points much hooked backwards, and extremely sharp; some smaller spines on each side of the tail, intermixed with innumerable little spicula.   In Montagu's specimen, which was a *male*, there were the usual four series of hooked spines, very sharp-pointed, each series consisting of two rows: the ventral appendages were nearly half as long as the tail. (*Colours.*) Upper surface of a uniform cinereous brown; in one instance, with a few black spots: under surface white. PENN and MONT.

This species, which appears very distinct and well characterized, I have not seen. Judging from the descriptions given of it by Pennant and Montagu, I am inclined to consider it the same as the *R. aspera* of Willughby, who expressly mentions the double row of spines on the tail. It is also the *R. aspera* of Fleming, and perhaps of Blainville, but it would be hazardous to annex any other synonyms.   Pennant met with

K k

it at Scarborough, where, he observes, it is called the *French Ray*. He
says that it is fond of *Launces*, or Sand-eels, which it takes greedily as
a bait. Montagu speaks of having seen several of both sexes on the
coast of Devon, but none larger than that which he has described. He
adds that it is known to some of the west country fishermen by the name
of *Dun-Cow*.

** *Snout short, and rather obtuse.*

203. R. *maculata*, Mont. (*Spotted Ray*.) — Teeth,
in the *adult*, sharp-pointed : skin smooth : generally three
rows of spines on the tail, the middle row continued along
the back : colour above brown, with distinct roundish
dusky spots.

> R. maculata, *Mont. in Wern. Mem.* vol. II. p. 426.   R. Rubus,
> *Don. Brit. Fish.* vol. I. pl. 20.   Fuller Ray, *Penn. Brit. Zool.*
> vol. III. p. 86.   (Synonyms excluded.)   *Id.* (Edit. 1812.) vol. III.
> p. 116. (Syn. excl.)

LENGTH. From two to three feet.

DESCRIPT. (*Form.*) Rhomboidal; the transverse diameter more than
one-third greater than the length from the end of the snout to the vent :
snout short and obtuse, projecting very little beyond the pectorals, the
anterior margins of which meet in front at more than a right angle :
jaws transverse, moderate : teeth small, numerous, very closely compacted; in several longitudinal, somewhat oblique, rows; roundish at
the base, each terminating above in a minute fine sharp point, the
points inclining inwards, and much more developed on the inner than
on the outer rows, on which last they are sometimes entirely wanting;
in *young* fish all the teeth are obtuse, the points not shewing themselves
till afterwards : nostrils much nearer to the mouth than to the anterior
angles of the pectorals : eyes moderate : spiracles large : skin perfectly
smooth above and below, excepting along the anterior margins of the
pectorals and the upper ridge of the snout, which are rough with very
minute spines : two strong spines at the corners of each eye; an interrupted series of spines down the line of the back, with one isolated spine
on each side of the series, about mid-way between the eyes and the
posterior margin of the pectorals; the dorsal series of spines is continued down the middle of the tail, at the sides of which are more or less
indication of two lateral rows; sometimes, in small specimens, these last
are wholly wanting; all the above spines incline a little backwards :
tail about the length of the body, rather stout, depressed, with two
moderate finlets, of similar size and form, nearly but not quite contiguous;
merely the rudiment of a caudal : pectorals broad, the lateral angles
rather obtuse, the posterior margin rounded, the anterior margin straight
or nearly so : ventrals moderate. The *male*, in addition to the spines
mentioned above, has the usual series of curved spines on the pectorals,
which, however, do not shew themselves till a certain age. (*Colours.*)
Upper parts brown, sometimes reddish brown, distinctly marked all over
with roundish dusky spots : under parts plain white. A *variety* is not
uncommon, in which the usual spots are nearly obsolete, but there is
more or less trace of one ocellated spot in the middle of each pectoral :
Montagu has noticed two kinds of this last variety; one, with a large

dark spot surrounded with a white circle; the other with a black spot within a white circle, the whole surrounded by five equidistant dark spots. *Another variety* is in the Museum of the Cambridge Philosophical Society, in which the upper parts are pale orange-yellow, with light rufous brown spots.

This species, although very common and well characterized, has been so misunderstood and confused with others, that it is extremely difficult to attach to it its proper synonyms. For this reason I have adopted the name given it by Montagu, who was the first in this country to point out its true distinguishing characters. It is undoubtedly the *R. Rubus* of Donovan, and most probably the *Fuller Ray* of Pennant, who describing from an adult male, appears to have considered the sexual spines as characteristic of the species. It is impossible to identify it with certainty in the works of Turton and Fleming*, both of whom appear to have mixed up the description of this with that of other species. It is known on some parts of the coast by the name of *Hommelin*, on others by that of *Sand-Ray*.

204. R. *microcellata*, Mont. (*Small-eyed Ray*.) — Teeth obtusely cuneiform : skin rough with minute spines : one row of small hooked spines on the tail, continued along the dorsal ridge to the head : eyes remarkably small.

R. microcellata, *Mont. in Wern. Mem.* vol. ii. p. 430. *Flem. Brit. An.* p. 171.

Dimens. Total length twenty inches: length of the tail nine inches: breadth fourteen inches. Mont.

Descript. (*Form.*) Resembling in shape the *R. maculata*, but rather more obtuse in front, and particularly distinguished by the comparative smallness of the eyes†: teeth obtusely cuneiform, with a broad edge, that feels rough to the finger as it is withdrawn from the mouth; in one jaw fifty-three, in the other fifty-six, longitudinal rows, closely connected: skin on the upper side rough with minute spines; the under side smooth: in one specimen there was observed a single large spine, with a broad base, before one of the eyes; (possibly in older fish that part may be more spinous;) above the eyes, the spinulæ were rather larger than those which cover the whole upper surface: one row of small hooked spines on the tail, continued along the dorsal ridge to the head. (*Colours.*) Upper parts plain brown, with the exception of a few scattered pale spots and lines on the margins of the wings: under parts white. Mont.

This species appears to have been observed hitherto only by Montagu, who obtained two females, the largest not exceeding the dimensions above given. He states that it appears to be confounded with the *R. chagrinea*, both being indiscriminately called *Dun-Cow* by the fishermen in the West of England. Whether it be the same as any of those described by continental authors is uncertain.

---

* The *R. Fullonica* of Turton is partly applicable to the above species and partly to the *R. chagrinea* last described. The *R. Rubus* of the same author may be the same as the *R. maculata* of Montagu, but the description is not quite correct, nor sufficiently precise to enable one to speak with certainty on this point. The *R. Rubus* of Fleming agrees with the *R. maculata* in some of its characters, but not in others.

† Montagu says, "The eyes of the specimen described did not exceed half an inch in diameter from the opposite angles of the eye-lids; whereas the *R. maculata*, and most others of similar size, have eyes nearly double that diameter."

205. R. *clavata*, Linn. (*Thorn-Back*.)—Teeth sharp
in the *male?*, blunt in the *female:* skin rough; studded
with large osseous tubercles terminating in strong spines;
one, three, or five, rows of such tubercles on the tail.

R. clavata, *Linn. Syst. Nat.* tom. I. p. 397.   R. Rubus, *Bloch, Ichth.*
pl. 84. (Male.)   R. clavata, *Id.* pl. 83. (Female.)   R. clavata,
Don. Brit. Fish. vol. II. pl. 26.   *Turt. Brit. Faun.* p. 111.   *Mont.
in Wern. Mem.* vol. II. p. 416.   *Flem. Brit. An.* p. 170.   *Blainv.
Faun. Franç.* p. 33.   Thorn-Back, *Penn. Brit. Zool.* vol. III. p. 93.
pls. 11, and 12.   *Id.* (Edit. 1812.) vol. III. p. 122. pls. 13, and 14.
La Raie bouclée, *Cuv. Reg. An.* tom. II. p. 398.

LENGTH.   From two to three feet; sometimes more.
DESCRIPT. (*Form.*) General form resembling that of the *R. maculata:* snout short, and rather obtuse: mouth wide, transverse: teeth
larger than in the above species, and not so closely compacted; set in
oblique rows; each with a broad round head, terminating, in the *male?*,
in a strong curved point; in the *female*, all blunt, with scarcely any trace
of a point or cutting edge: nostrils, eyes, and spiracles, much as in the
R. maculata: body rather thick, convex above; the whole of the upper
surface extremely rough with minute hooked spines and asperities, besides which are a greater or less number of large osseous tubercles, each
terminating upwards in a strong hooked spine, or tooth-like process, very
sharp at the extremity; these spinous tubercles, which are of an oval
form, and very broad at the base, are scattered about in rather an irregular manner, and very variable in number; almost always one or two
above the eyes, and a row down the middle of the back, continued along
the ridge of the tail; also one or two on each side of the dorsal series
about the middle; sometimes three complete rows on the back, and three
or five on the tail; occasionally, especially in large fish, the under surface
of the body is studded with tubercles as well as the upper; more rarely
the tubercles are almost wanting altogether: tail a little longer than the
body, depressed, rather stout, and very rough with minute asperities independently of the spinous tubercles; two finlets near its extremity, much
as in *R. maculata*, besides the rudiment of a caudal: pectorals and ventrals the same as in that species. (*Colours.*) Variable: generally bluish
gray above, tinged with reddish brown; the whole sparingly sprinkled
with large, but ill-defined, whitish spots: beneath white. A *variety*
sometimes occurs, shewing more or less appearance of an ocellated spot
on the middle of each pectoral.

Common as this species is on all parts of the coast, its true characters, at least those which distinguish the sexes, are involved in a little
obscurity. Montagu was led to regard the *R. Rubus* of authors, in which
the teeth are sharp, as the male of *R. clavata*, in which they are blunt,
from the circumstance of his not being able to discover a female of the
former, nor a male of the latter, species. As far as my own observation goes, which, however, has been but limited, it confirms Montagu's
opinion. I have never seen a *male Thorn-Back* with *blunt* teeth, but
I have seen, in the collection of Mr. Yarrell, two fish perfectly similar in
every respect, excepting that in one, a *male* with long ventral appendages, the teeth were sharp, in the other, a *female*, the teeth were blunt.
These I was led to regard as the sexes of the common *Thorn-Back*. Yet
both Risso and Blainville speak of the sexes of the *R. clavata*, without
any allusion to the teeth being otherwise than blunt, in the *male*, as well

as in the *female*. Moreover, they both give the *R. Rubus* as a distinct species. Further observation is necessary in order to clear up this difficulty. According to Pennant, the Thorn-Back preys on all sorts of flat fish, as well as on Herrings and Sand-eels, of which it is said to be particularly fond; also on crabs. Produces its young in July and August, which, until a certain age, are called, in common with the young of the *R. Batis*, by the name of *Maids*. It is taken in large quantities for the table.

206. **R. *radiata*, Don. (*Starry Ray*.)** — Teeth sharp in both sexes: skin smooth; but thickly studded with strong conical spines, intermixed with more numerous smaller ones, radiating at the base; one or three rows on the tail.

R. radiata, *Don. Brit. Fish.* vol. v. pl. 114. *Flem. Brit. An.* p. 170.

DIMENS. The following were those of the specimen described below. Entire length eighteen inches nine lines: length of the head (measured from the end of the snout to the spiracles) three inches two lines; of the tail (measured from the vent) nine inches three lines: breadth, across the pectorals, thirteen inches three lines.

DESCRIPT. (*Form.*) General form similar to that of the last species: snout short and obtuse, projecting very little beyond the pectorals: teeth much larger than in the *R. maculata*, not so closely compacted, and terminating above in a sharper and longer point; from those of the *R. clavata*, they differ in being rather smaller at the base, more widely separate, and strongly pointed in *both* sexes: ground of the back smooth, but thickly studded with strong sharp hooked spines, arising from a conical furrowed base, intermixed with smaller ones, which spread out at bottom in a radiating or stellate manner; of the larger spines a row occupies the mesial ridge of the tail, and is continued along the back to behind the eyes; there are also two on each side of the centre of the back, one before the eyes, and two at the posterior angles of the same; the smaller radiating spines form a parallel and more numerous series on each side of the central row of larger ones, commencing at the middle of the back, and extending nearly to the extremity of the tail; (in Donovan's specimen these lateral rows appear to have been wanting;) they are also irregularly but thickly scattered over the wings of the pectorals, becoming smaller and more numerous towards the margins: the other characters resemble those of the last species. (*Colours.*) Upper surface brown, with a slight reddish tinge; beneath, white.

Since publishing my Catalogue of British *Vertebrata*, I have seen a pair of this species, male and female, in the collection of Mr. Yarrell, who received them from the Frith of Forth. The same gentleman possesses a third specimen sent him by Dr. Johnston of Berwick. I am inclined, now, to regard it as a well-marked species, quite distinct from any of the foregoing ones, but perhaps not specifically different from the *R. Rubus* of Blainville*, of which it is considered as a variety by that author. Both sexes are equally thorny on their upper surface, the under surface being, in both, smooth. The male specimen above alluded to had the ventral appendages half the length of the tail. Donovan's example of this species was caught on the north coast of Britain.

* *Faun. Franç.* p. 21.

(43.) *R. Miraletus*, Don. Brit. Fish. vol. v. pl. 103.   Turt. Brit.
Faun. p. 111.   *R. oculata*, Flem. Brit. An. p. 172.

This is probably nothing more than the ocellated variety of the *R. maculata* already alluded to.   As such it was regarded by Montagu *.   Blainville, however, makes it the same as his *R. Speculum†*.   Procured by
Donovan in the London market, and supposed to have come from the
coast of Sussex.

(44.) *Rough Ray*, Penn. Brit. Zool. vol. iii. p. 85.   Id. (Edit. 1812.)
vol. iii. p. 115.

"Length from the nose to the tip of the tail two feet nine inches : the
tail almost of the same length with the body.   Nose very short: before
each eye a large hooked spine, and behind each another, beset with lesser.
Upper part of the body of a cinereous brown colour, mixed with white,
and spotted with black ; and entirely covered with small spines.   On the
tail three rows of great spines ; all the rest of the tail irregularly beset
with lesser.   The fins, and under side of the body, equally rough with
the upper.   Teeth flat and rhomboidal."   Penn.
This species was taken by Pennant in Loch Broom, in the shire of Ross.
It is doubtful whether it be distinct from all those already described.
Blainville appears to consider it as his *R. Rubus*, but the "flat rhomboidal
teeth" seem rather at variance with the characters which he ascribes to
those of that species.

(45.) *R. Cuvieri*, Lacép. Hist. Nat. des Poiss. tom. i. p. 141. pl. 7.
f. 1.   Neill in Wern. Mem. vol. i. p. 554.   Flem. Brit. An. p. 172.
Cuvier Ray, Penn. Brit. Zool. (Edit. 1812.) vol. iii. p. 124.

This supposed species, which was first noticed by Lacépede, was obtained by Mr. Neill, in a single instance, on the Scottish coast in 1808.
Its distinguishing character consists in the first dorsal fin being on the
middle of the back.   Cuvier‡, however, regards it as nothing more than
an accidental variety, or rather monstrosity, observed by him in more than
one species.   Blainville§ speaks with confidence as to Lacépede's fish
being nothing more than a variety of the *R. clavata*.   As tending to con-
firm this opinion, it is worth noticing that Mr. Neill's specimen is said to
have been obtained from among a large cargo of *Thorn-Backs*.

(3. Trygon, *Adans.*)

207.   R.   *Pastinaca*,   Linn.   (*Sting-Ray.*) — Back
smooth.

R. Pastinaca, *Linn. Syst. Nat.* tom. i. p. 396.   *Bloch, Ichth.* pl. 82.
Don. Brit. Fish. vol. v. pl. 99.   *Turt. Brit. Faun.* p. 112.
Blainv. Faun. Franç. p. 35. pl. 6.   Pastinaca marina, *Will. Hist.
Pisc.* p. 67. tab. C. 3.   Trygon Pastinaca, *Flem. Brit. An.* p. 170.
Sting-Ray, *Penn. Brit. Zool.* vol. iii. p. 95.   *Id.* (Edit. 1812.)
vol. iii. p. 125.   Pasténague commune, *Cuv. Reg. An.* tom. ii.
p. 399.

Length.   From two to three feet; rarely more.
Descript.   (*Form.*) Disk of the body more approaching to orbicular
than in the last sub-genus; very thick and convex in the middle, but
growing thin towards the edges : the transverse diameter scarcely more
than one-fourth greater than the length measured from the end of the
snout to the vent: snout sharp, but very short, scarcely projecting beyond

---

* *Wern. Mem.* vol. ii. p. 429.   † *Faun. Franç.* p. 29. pl. 4. f. 1.
‡ *Reg. An.* tom. ii. p. 399.   § *Faun. Franç.* p. 35.

the pectorals, the anterior margins of which meet at an obtuse angle: mouth small; teeth small, arranged in oblique rows, appearing granulated on the surface: eyes moderate: skin entirely smooth above and below, "excepting a few small tubercles along the mesial line of the back and tail, as well as on the upper and posterior part of the pectoral fins*:" tail varying in length, less than, equal to, or very much exceeding, half the entire length; slender, tapering at the extremity to a fine point, without any trace of fins, but armed, at about the first third of its length, with a very strong, sharp, serrated spine, the serratures directed backwards: pectorals large, rounded posteriorly and at the lateral angles: ventrals small, entire. *Obs.* Occasionally the tail is found armed with two spines, owing, it is said, to the circumstance of the spine being annually renewed, and the new one sometimes appearing before the old one drops off. (*Colours.*) "Upper part of the body dirty yellow, the middle of an obscure blue; lower part white; the tail and spine dusky." Penn. According to Donovan, small specimens are more or less spotted.

Met with principally on the southern coasts, and rather less frequently than some of the other species. It occurs at Weymouth, as well as on the coast of Cornwall. The spine is capable of inflicting a severe wound, but is not poisonous. Flesh said to be rank and disagreeable. The liver is large, and yields a great deal of oil.

(Myliobatis, Dumér.)

(46.) *Whip Ray,* Penn. Brit. Zool. vol. iii. p. 88. Id. (Edit. 1812.) vol. iii. p. 128. *R. Aquila,* Linn. Syst. Nat. tom. i. p. 396.? Blainv. Faun. Franç. p. 38. pl. 7.? *L'Aigle de Mer,* Cuv. Reg. An. tom. ii. p. 401.?

> Pennant states that in 1769, Mr. Travis, of Scarborough, had brought to him by a fisherman of that town the tail of a Ray (the body having been flung away) which was above three feet long, extremely slender and taper, and destitute of a fin at the end. It is conjectured by the Editor of the last edition of the "British Zoology," that this fish must have been the *R. Aquila* of Linnæus, a species which is found in the Mediterranean, and which attains to a large size. It is, however, equally probable that it may have belonged to the next sub-genus. The *R. Aquila* cannot, therefore, be considered otherwise than as a doubtful native.

(4. Cephaloptera, Dumér.)

208.   R. *Giorna,* Lacép.? (*Giorna Ray.*) — "Body smooth: the margins of the fins straight; horns of one colour: spine very long, situate at the base of the tail." Riss.

Cephaloptera Giorna, *Riss. Hist. Nat. de l'Eur. Mérid.* tom. iii. p. 163. pl. 5. Cephaloptera, *Thompson in Proceed. of Zool. Soc.* June 9, 1835.

A fish of this sub-genus is stated by Mr. Thompson, in a recent communication to the Zoological Society, to have been taken about five years ago on the southern coast of Ireland, and thence sent to the Royal Society of Dublin, in the Museum of which public body it is at present

---

* Blainville.

preserved. In breadth it is about forty-five inches. The specimen being imperfect, and the characters of some of the species ill-defined, Mr. Thompson hesitates applying to it a specific name. He states, however, that it somewhat resembles the *C. Giorna* figured by Risso. I have accordingly annexed a reference to that author.

ORDER IX.   CYCLOSTOMI.

GEN. 77. PETROMYZON, *Linn.*

209.   P. *marinus*, Linn. (*Sea Lamprey.*) — Greenish or yellowish brown, marbled with dusky : dorsals separate; the posterior one rounded, just reaching to the caudal.

P. marinus, *Linn. Syst. Nat.* tom. I. p. 394. *Bloch, Ichth.* pl. 77. *Don. Brit. Fish.* vol. IV. pl. 81. *Turt. Brit. Faun.* p. 109. *Flem. Brit. An.* p. 163. *Blainv. Faun. Franç.* p. 5. pl. 1. f. 1, & 2. Lampetra Rondeletii, *Will. Hist. Pisc.* p. 105. tab. G. 2. f. 2. Sea Lamprey, *Penn. Brit. Zool.* vol. III. p. 76. pl. 8. no. 27. *Id.* (Edit. 1812.) vol. III. p. 102. pl. 10. *Bowd. Brit. fr. wat. Fish.* Draw. no. 26.   La Grande Lamproye, *Cuv. Reg. An.* tom. II. p. 404.

LENGTH.   From two to two and a half feet; sometimes more.

DESCRIPT. (*Form.*) Anguilliform: body thick and cylindric anteriorly, compressed and somewhat tapering beyond the commencement of the dorsal fin: head indistinct, obtuse and obliquely truncated in front, rather depressed above the eyes: mouth very large, circular, bordered by a fleshy lip, studded on the inside with corneous conical tooth-like papillæ disposed in concentric rows, and gradually increasing in size as they advance inwards; beyond these one large tooth below, in the middle, with six or eight points, the extreme points being the most developed; answering to it above a similar tooth, also in the middle, but with only two points; tongue with two pairs of crenated teeth: eyes large, lateral, a little in advance of the first branchial opening: a single nostril on the top of the head, in the middle, a little in advance of the eyes, moderately large: line of the branchial apertures a little below the level of the eyes, and rather inclining downwards posteriorly: skin every-where smooth and naked: two distinct dorsals: the first commencing beyond the middle of the entire length, short and low, of a somewhat semicircular form: second commencing a little behind the first, more elevated, and attaining its greatest height rather suddenly, afterwards sloping gradually off, and finally terminating immediately before the caudal: vent very much behind, beneath the anterior portion of the second dorsal, and at nearly, but not quite, three-fourths of the entire length: caudal rounded at the extremity, giving a truncated appearance to the tail, the fleshy portion of which, however, is pointed; underneath, the caudal is continued for a little way towards the vent, sinking gradually into a low ridge representing the anal. (*Colours.*) Above, greenish or yellowish brown, marbled with dark brown and dusky: beneath, white, tinged with reddish.

A migratory species, entering rivers from the sea early in the Spring to spawn, and returning after the expiration of a few months. Common in many parts of Great Britain, but said to be more abundant in the Severn than in most other rivers. Attains to the weight of between four and five pounds. Has the power of adhering very firmly to stones with its circular mouth, by means of suction. Flesh much esteemed. *Obs.* It was formerly supposed that in this and the next species the two sexes were united in the same individual; this has, however, been since proved to be erroneous *.

210. P. *fluviatilis*, Linn. (*River Lamprey.*)—Dusky blue above, silvery beneath: dorsals widely separate; the posterior one angular, uniting with the caudal.

> P. fluviatilis, *Linn. Syst. Nat.* tom. I. p. 394. *Bloch, Ichth.* pl. 78. f. 1. *Don. Brit. Fish.* vol. III. pl. 54. *Turt. Brit. Faun.* p. 110. *Flem. Brit. An.* p. 163. *Blainv. Faun. Franç.* p. 6. pl. 2. f. 1. Lampetræ medium genus, *Will. Hist. Pisc.* p. 106. tab. G. 3. f. 2. Lesser Lamprey, *Penn. Brit. Zool.* vol. III. p. 79. pl. 8. no. 28. *Id.* (Edit. 1812.) vol. III. p. 106. pl. 10. *Bowd. Brit. fr. wat. Fish.* Draw. no. 16. Lamproye de rivière, *Cuv. Reg. An.* tom. II. p. 404.

Length. From twelve to fifteen inches.

Descript. (*Form.*) General form resembling that of the *P. marinus*, but more elongated in proportion: anterior half of the body thick and semicylindric; posterior portion much compressed: mouth similar; teeth less numerous; one large tooth above, in the middle, with two remote points; opposed to it below, a larger one, forming the arc of a circle, with seven or eight points, and having a crenated appearance; a few other smaller teeth at the corners of the mouth: eyes large: nostril single, in the middle of the upper part of the head, a little in advance of the eyes: line of the branchial apertures commencing nearly on a level with the eyes, but inclining a little downwards posteriorly: skin every-where smooth: a considerable space between the two dorsals: the first commencing at about, or a very little beyond, the middle of the entire length, low, and nearly of equal height throughout: the second commencing beyond two-thirds of the entire length, low at first, but elevated about the middle into a sharp projecting angle, then again sloping off to meet the caudal with which it unites: vent at exactly three-fourths of the entire length: anal narrow, extending to, and also uniting with, the caudal. (*Colours.*) Dusky blue above; silvery white beneath: fins whitish. Said, however, by Donovan, to be very variable.

Common in many of our rivers, especially in the Thames, about Mortlake, where large quantities are said to be caught annually, and sold

to the Dutch to be used as bait in the Cod-fisheries. Food, according
to Bloch, insects, worms, small fish, and the flesh of dead fish. Spawns
towards the end of April or beginning of May. Sometimes called a
Lampern.

211. P. *Planeri*, Cuv. (*Planer's Lamprey.*)—Dusky
blue above, silvery beneath: dorsals contiguous.

P. Planeri, *Cuv. Reg. An.* tom. II. p. 404. *Nilss. Prod. Ichth.
    Scand.* p. 122. Lampetra parva et fluviatilis, *Will. Hist. Pisc.*
    p. 104. tab. G. 2. f. 1.?

LENGTH. From eight to ten inches.
DESCRIPT. Differs from the *P. fluviatilis*, principally in having the
two dorsals contiguous, or with only a very small space between: the
first commences at about, or a little before, the middle of the entire
length; the second at exactly two-thirds of the same: the vent is,
relatively, a little further from the extremity of the tail; the body is
also somewhat thicker in proportion to its length. In all other respects,
including colours, armature of the mouth, &c., the two species are
identical.
    This species is evidently the *P. Planeri* of Cuvier and Nilsson, but
not of Bloch and Blainville. That of Bloch, Cuvier thinks is only the
young of *P. fluviatilis*. It is probably also the species described by
Willughby under the name of *Lampetra parva*, in which he expressly
speaks of the two dorsals being contiguous. Willughby, however, erroneously considered it as the *Pride* of Plot, a circumstance which has led
to some little confusion in the works of later authors with respect to the
synonyms of this last fish. Whether the *P. Planeri* be common in this
country I am not aware. My specimens were given to me by Mr. Yarrell,
who obtained them from a brook in Surrey. The same gentleman has
since received it from the Tweed.

(47.) *P. Juræ*, Mac Cull. West. Isl. vol. II. pp. 186, 187. pl. 29.
    f. 1.

    Under the above name, Dr. Mac Culloch has described a species of
    Petromyzon, which he considers distinct from those hitherto noticed by
    naturalists. He observes that "in size it approaches to the *P. fluviatilis*,
    which it also resembles in the proportion and disposition of the fins; but
    that it differs materially in the absence of the *annuli*, in the greater number
    of the teeth, and in the number and forms of the bony bodies which sur-
    round the opening of the throat." This fish was found adhering to the
    back of a gray gurnard on the Eastern shore of Jura: the specimen was
    not preserved. Dr. Fleming does not seem to allow that it is specifically
    different from the *P. fluviatilis**, an opinion in which I feel inclined to
    join.

GEN. 78. AMMOCŒTES, *Dumér.*

212. A. *branchialis*, Flem. (*Pride.*)

A. branchialis, *Flem. Brit. An.* p. 164. *Blainv. Faun. Franç.*
    p. 3. pl. 2. f. 3, & 4. Petromyzon branchialis, *Linn. Syst. Nat.*
    tom. I. p. 394. *Bloch, Ichth.* pl. 78. f. 2. *Turt. Brit. Faun.*
    p. 110. P. cæcus, *Couch in Loud. Mag. of Nat. Hist.* vol. v.

* *Brit. An.* p. 164.

p. 23. fig. 10.   Lampetra cæca, *Will. Hist. Pisc.* p. 107. tab. G. 3.
f. 1.   *Ray, Syn. Pisc.* p. 36.   Pride, *Plot, Oxfordsh.* p. 187.
pl. 10.   *Penn. Brit. Zool.* vol. III. p. 80. pl. 8. no. 29.   *Id.* (Edit.
1812.) vol. III. p. 107. pl. 10.   *Bowd. Brit. fr. wat. Fish.* Draw.
no. 32.   Lamprillon, *Cuv. Reg. An.* tom. II. p. 406.

LENGTH.   From six to eight inches.

DESCRIPT.   (*Form.*) Body more slender than in the last genus, vermiform, scarcely larger than a goose-quill, marked with numerous transverse
lines, subcylindric anteriorly, somewhat compressed and tapering beyond
the vent: jaws and lips soft and membranaceous; the upper lip semicircular, fleshy, prominent: inside of the mouth papillose, with a lingual
and palatine plate somewhat harder than the other portions, but no true
teeth: eyes obscure: a single nostril on the upper part of the head:
branchial orifices seven in number, situate in a kind of lateral groove;
the body at this part somewhat dilated: skin naked, and covered with an
abundant mucosity: two dorsals; the first small and low; the second
closely following, longer, and rather more elevated anteriorly, but sloping
off to a narrow edge before uniting to the caudal: fleshy portion of the
tail sharp at the extremity, but the fin rounded: vent rather anterior to
the commencement of the last quarter of the entire length; anal commencing a little beyond it, uniting with the caudal.   *Obs.* Two distinct
forms of this fish are not unfrequent: in one, the eyes are larger, the
mouth smaller, the snout more elongated, and the orifice on the crown
further removed from the extremity: in the other, the eyes are smaller
and very obsolete, the mouth larger, and the snout shorter: whether
these are merely sexes, or two different species, is not certain.   (*Colours.*)
" Bluish or reddish gray above, whitish beneath; fins of the same colour
and almost transparent*."   The following are those which Mr. Couch
assigns to his *Petromyzon cæcus:* "Colour dusky yellow, dark on the
back, light below; fins light."

Said to be frequent in the rivers near Oxford, particularly the Isis,
where it was first observed by Plot.   Found also in other parts of England, as well as in Ireland.   Buries itself in the soft mud.   Has not
the power of adhering by the mouth, like the Lamprey; although the
lips, according to Mr. Couch, are capable of extensive and complicated
motions.   Spawns at the end of April or the beginning of May.   Probably the *Stone Grig* of Merrett†.

GEN. 79.   MYXINE, *Linn.*

213.   M. *glutinosa*, Linn.   (*Glutinous Hag.*)

M. glutinosa, *Linn. Syst. Nat.* tom. I. p. 1080.   *Flem. Brit. An.*
p. 164.   *Nilss. Prod. Ichth. Scand.* p. 123.   Gastrobranchus
cæcus, *Bloch, Ichth.* pl. 413.   *Shaw, Nat. Misc.* vol. X. pl. 362.
Turt. Brit. Faun. p. 110.   Myxine cæca, *Blainv. Faun. Franç.*
p. 2.   Glutinous Hag, *Penn. Brit. Zool.* vol. IV. p. 39. pl. 20.
f. 15.   *Id.* (Edit. 1812.) vol. III. p. 109.   Le Gastrobranche, *Cuv.*
Reg. An. tom. II. p. 406.

LENGTH.   From ten to fifteen inches.

DESCRIPT.   (*Form.*) Body elongated, vermiform, thick and cylindric
anteriorly, compressed and slightly tapering behind: head scarcely
distinguishable, obliquely truncated in front: mouth large, circular,

---

* Blainville.   † *Pinax,* p. 188.

obliquely terminal, surrounded by eight barbules; in the middle of the
upper margin a single nostril or spiracle of a roundish form; a pair of
barbules are placed on each side of the spiracle; the remaining pairs
at the sides of the mouth: maxillary ring soft and membranaceous,
with a single curved tooth on the upper part; two rows of strong pectinated teeth on each side of the tongue: eyes wanting: branchiæ opening
externally by two small apertures, placed beneath, near the mesial line,
at a little beyond one-fourth of the entire length: a row of pores along
each side of the abdomen: skin naked, invested with an abundant
mucosity: a low and rather obscure fin commences beyond the middle
of the length, turns round the tail, and is continued along the under
surface of the body as far as the vent: this last placed at a great
distance from the head, scarcely one-twelfth of the entire length intervening between it and the posterior extremity. (*Colours.*) " Blue
above; whitish beneath." BLAINV.

This species was placed by Linnæus in his class *Vermes.* Its affinity,
however, to the other Cyclostomous Fishes is obvious. Inhabits the
northern seas, but is met with on some parts of the English and Scottish
coasts. Said by Pennant to be often taken at Scarborough, where it is
in the habit of "entering the mouths of other fish when on the hooks
attached to the lines which remain a tide under water, and totally devouring the whole except the skin and bones." The fishermen there
call it the *Hag.* According to Dr. Johnston, it occurs on the coast of
Berwickshire*.

* *Proceed. of Berwicksh. Nat. Club.* p. 7.

APPENDIX.

Since this Work went to press, I have been made acquainted with two recent additions to the British Fauna, too late for inserting them in their proper places. One of these belongs to the Class MAMMALIA, the other to the Class AVES.

GEN. LUTRA. Page 13.

9*. L. *Roensis*, Ogilby. (*Irish Otter.*)

L. Roensis, *Ogilby in Proceed. of Zool. Soc.* (1834.) p. 111.

By the above name, Mr. Ogilby has designated, provisionally, a species of Otter found in Ireland, chiefly along the coast of the county of Antrim, which he is disposed to regard as distinct from the *Common Otter* (*L. vulgaris*) of England. The difference is said to consist in the intensity of its colouring, which approaches nearly to black both on the upper and under surface; in the less extent of the pale colour beneath the throat; in the relative size of the ears, and in the proportions of other parts. Mr. Ogilby adds that it is further distinguished by the peculiarity of its habitation and manners. " It is, in fact, to a considerable extent a marine animal*, living in hollows and caverns formed by the scattered masses of the basaltic columns on the coast of Antrim, and constantly betaking itself to the sea when alarmed or hunted. It feeds chiefly on the Salmon." No detailed description of it has been yet published.

----

* Possibly this species may be the *Sea Otter*, which, according to Pennant, was noticed by Sir Robert Sibbald. See *Brit. Zool.* vol. I. p. 95.

GEN. NOCTUA.   Page 93.

(SURNIA. p. 93.)

26*.   N. *funerea*, Nob. (*Canada Owl.*)—Upper parts
spotted with brown and white; beneath white, with transverse brown bars: quills spotted with white; tail marked
with distant, transverse, narrow, white bars.

> Strix funerea, *Lath. Ind. Orn.* vol. i. p. 62.   *Temm. Man. d'Orn.*
> tom. i. p. 86.   *Faun. Bor. Amer.* part ii. p. 92.   Little Hawk
> Owl, *Edw. Nat. Hist.* pl. 62.   Hawk Owl, *Lath. Syn.* vol. i.
> p. 143.

DIMENS. Entire length fourteen inches two or three lines: length of
the tail six inches six lines. TEMM.

DESCRIPT. Forehead dotted with white and brown; a black band
arises behind the eyes, surrounds the orifice of the ears, and terminates
on the sides of the neck: upper parts marked with brown and white
spots of various forms; edge of the wing with similar white spots
upon a brown ground: throat whitish; the rest of the under parts white,
with transverse streaks of cinereous brown: a large spot of dusky brown
at the insertion of the wings: tail-feathers cinereous brown, with distant
zigzag streaks forming narrow transverse bands: bill yellow, varied
with black spots according to age: irides pale yellow: feet feathered to
the claws. The *female* differs only in being somewhat larger, and in
having the colours less pure. TEMM.

An individual of this species, which inhabits the Arctic Regions, is
recorded by Mr. Thompson, in a recent communication to the Zoological
Society*, to have been taken on board a collier, a few miles off the coast
of Cornwall, in March, 1830, being at the time in so exhausted a state
as to allow itself to be captured by the hand. According to Temminck,
the species appears occasionally as a bird of passage in Germany, and
more rarely in France, but never in the southern provinces. Said to
feed principally on mice and insects. Builds in trees, and lays two white
eggs.

* *Proceed. of Zool. Soc.* June 9, 1835.

ALPHABETICAL LATIN INDEX.

THE names of the Genera are printed in LARGE CAPITALS: those of the Sub-genera in SMALL CAPITALS: those of the Species in Roman Characters: the Synonyms, as well as the names of a few Species incidentally mentioned, and those of the principal Varieties, in *Italics*.

The first of the two numbers attached to the names of the Genera and Sub-genera, refers to the page containing the generic and sub-generic characters.

A.

ABRAMIS, *Cuv.* 316, 406
   Brama, Flem. 406
Acanthias Spinax, Riss. 505
ACCENTOR, *Bechst.* 53, 102
   alpinus, *Bechst.* 102
   modularis, *Cuv.* 103
ACCIPITER, *Vig.* 50, 85
ACCIPITER, *Will.* 50, 85
   fringillarius, *Will.* 85
   palumbarius, *Will.* 85
ACERINA, *Cuv.* 307, 334
   vulgaris, Cuv. 334
ACIPENSER, *Linn.* 326, 493
   Sturio, *Linn.* 493
Aculeatus marinus, Will. 351
Acus Aristotelis, Will. 485
   lumbriciformis, Will. 488
   nostras caudâ serpentinâ,
   Sibb. 486
Alauda cristata sive Galerita,Will. 381
   non cristata, Will. 382
ALAUDA, *Linn.* 56, 126
   alpestris, *Linn.* 126
   arborea, *Linn.* 127
   arvensis, *Linn.* 127
   petrosa, Mont. 119
   rubra, *Gmel.* 126
ALCA, *Linn.* 76, 260
   impennis, *Linn.* 261
   Pica, Gmel. 261
   Torda, *Linn.* 260
ALCEDO, *Linn.* 61, 157

ALCEDO, Ispida, *Linn.* 157
ALOSA, *Cuv.* 318, 437
AMMOCŒTES, *Dum.* ... 329, 522
   branchialis, *Flem.* 522
AMMODYTES, *Linn.* ... 323, 482
   Anglorum verus, Jag. 482
   Lancea, *Cuv.* 483
   lanceolatus, Les. 482
   Tobianus, *Bl.* 482
   Tobianus, Don. 483
Ammodytes, Will. 483
ANARRHICHAS, *Linn.*... 314, 384
   Lupus, *Linn.* 384
ANAS, *Linn.* 72, 230
   acuta, *Linn.* 232
   adunca, *Linn.* 234
   albeola, Wils. 246
   albifrons, Temm................ 223
   Anser ferus, Temm. 222
   Bernicla, Temm................ 224
   bicolor, Don.................... 230
   Boschas, *Linn.*................. 233
   Clangula, Temm.............. 245
   Clypeata, *Linn.* 230
   Crecca, *Linn.* 235
   Cygnus, Temm. 227
   dispar, Gmel. 243
   ferina, Temm. 241
   frœnata, Sparm. 244
   Fuligula, Temm................ 244
   fusca, Temm. 239
   glacialis, Temm. 247

ANAS, *Glaucion*, Linn............. 246
   glocitans, *Pall*................... 232
   histrionica, Temm. 246
   hyemalis, Linn. 248
   leucocephala, Temm. 240
   leucophthalmos, Temm. 242
   leucopsis, Temm. 224
   Marila, Temm. 244
   minuta, Linn. 246
   mollissima, Temm. 237
   moschata, Linn. 230
   nigra, Temm. 239
   Olor, Temm................... 228
   Penelope, Temm............... 236
   perspicillata, Temm. 240
   Querquedula, *Linn*............. 234
   ruficollis, Temm. 225
   rufina, Temm. 240
   rutila, Temm. 229
   Segetum, Temm............... 222
   spectabilis, Temm. 238
   Strepera, *Linn*................. 231
   Tadorna, Temm............... 229
ANAS, *Swains*................... 73, 230
ANGUILLA, *Cuv.* 322, 474
   acutirostris, *Yarr.* 474
   Conger, *Shaw*, ... 478
   latirostris, *Yarr.* 476
   mediorostris, *Yarr.* 477
   Myrus, *Shaw*, ... 478
   vulgaris, Turt................... 474
ANGUILLA, *Cuv.* 322, 474
ANGUIS, *Cuv.*................ 288, 295
   Eryx, *Linn*...................... 296
   fragilis, *Linn.* 295
ANOUS, *Steph*............... 77, 270
ANSER, *Briss.* 72, 222
   Ægyptiacus, *Briss*. 225
   albifrons, *Steph*. 223
   Canadensis, Bon............... 227
   ferus, *Steph*.................... 222
   Gambensis, *Briss*. 226
   Guineensis, Briss. 226
   Leucopsis, *Bechst*. 224
   ruficollis, *Pall*................. 225
   Segetum, *Steph*. 222
   torquatus, *Frisch*, ... 224
ANSER, *Steph*................... 72, 222

ANTHUS, *Bechst.* 55, 117
   aquaticus, Temm. 119
   arboreus, *Bechst*. 118
   petrosus, *Flem.* 118
   pratensis, *Bechst*.............. 117
   Richardi, *Vieill.* 117
Aper Rondeletii, Will............. 368
AQUILA, *Briss.* 49, 80
   Albicilla, *Briss.* 80
   Chrysaëtos, *Vig.* 80
   Haliæetus, *Mey.* 81
AQUILA, *Cuv.* 50, 80
ARDEA, *Linn.* 66, 186
   æquinoctialis, Mont............ 188
   alba, *Linn.* 187
   Caspica, Gmel................ 187
   Cayennensis, Linn. 192
   cinerea, *Lath.* 186
   comata, Pall. 189
   Gardeni, Gmel................ 192
   Garzetta, *Linn*................ 187
   lentiginosa, *Mont.* 191
   minuta, *Linn.* 189
   Nycticorax, *Linn.* 191
   purpurea, *Linn.* 186
   Ralloides, *Scop.* 189
   russata, *Wag.* 188
   stellaris, *Linn.* 190
ARDEA, *Steph*................... 67, 186
Arnoglossus, vel Solea lævis, Will. 465
ARVICOLA, *Lacép.* 6, 33
   agrestis, *Flem*. 33
   amphibia, *Desm.* 33
   aquatica, Flem. 33
   ater, Macg. 33
   riparia, *Yarr.* 34
   vulgaris, Desm. 33
Asellus Callarias, Will. 452
   Huitingo-Pollachius, Will.... 446
   longus, Will. 448
   luscus, Will................... 442
   major vulgaris, Will. 440
   mollis latus, Will. 442
   mollis major seu albus, Will.. 445
   mollis minor, Will. 444
   niger, Will. 446
   primus, Will. 447
   varius vel striatus, Will....... 441

ASPIDOPHORUS, *Lac.* . 308, 346
   armatus, Lac. 346
   Cataphractus, *Nob.* 346
   Europæus, Cuv. 346
ASTUR, *Vig.* 50, 85
ATHERINA, *Linn.* 313, 377
   Hepsetus, Don. 377
   Presbyter, *Cuv.* 377

B.

BALÆNA, *Lac.* 9, 46
BALÆNA, *Linn.* 9, 46
   Boops, *Linn.* 47
   Musculus, *Linn.* 47
   Mysticetus, *Linn.* 46
   Physalus, *Linn.* 47
   rostrata, Fab. 48
BALÆNOPTERA, *Lac.* 9, 47
   Boops, Flem. 47
   Gibbar, Scoresb. 47
   Jubartes, Scoresb. 47
   Musculus, Flem. 47
   Rorqual, Scoresb............. 47
BALISTES, *Cuv.* 325, 492
   Capriscus, *Gmel.* 492
BARBASTELLUS, *Gray,* 6, 28
BARBUS, *Cuv.* 316, 404
   Orfus, Flem. 416
   vulgaris, Flem. 404
Barbus major, Jag. 452
   minor, Jag. 454
BELONE, *Cuv.* 317, 418
BELONE, *Nob.* 317, 418
   Saurus, *Nob.* 419
   vulgaris, *Flem.* 418
BERNICLA, *Steph.* 72, 224
BLENNIUS, *Cuv.* 313, 378
BLENNIUS, *Linn.* 313, 378
   Galerita, Flem. 380
   Galerita, *Mont.* 381
   Gattorugine, *Don.* 380
   Gattorugine, Linn. 380
   Gattorugine, *Mont.* 379
   Gunnellus, *Linn.*............. 383
   Montagui, Flem............... 381
   ocellaris, *Bl.*................. 378
   palmicornis, *Cuv.*............ 380
   Pennantii, Jen. 380

BLENNIUS, Pholis, *Linn.* 382
   Phycis, Linn. 453
   Phycis, Turt. 452
   trifurcatus, Turt.............. 453
   viviparus, Linn. 384
BOMBYCILLA, *Briss.* ... 56, 125
   garrula, *Bon.* 125
Bombycivora garrula, Temm. ... 125
BOS, *Linn.* 7, 36
   Taurus, *Linn.* 36
BOSCHAS, *Swains.* 73, 232
BOTAURUS, *Steph.* 67, 190
BRAMA, *Schn.* 310, 359
   marina caudâ forcipatâ, Ray, 359
   marina, Flem. 359
   Raii, *Cuv.* 359
BROSMUS, *Flem.* 320, 452
   vulgaris, *Flem.* 452
BUBO, *Cuv.* 51, 90
BUBO, *Geoff.*................... 51, 90
   maximus, *Flem.* 90
   Scops, *Nob.* 91
BUFO, *Laur.*.................. 299, 301
   Calamita, *Laur.* 302
   Rubeta, Flem. 302
   terrestris, *Rœs.*.............. 301
   terrestris fœtidus, Rœs. 302
   vulgaris, *Flem.*............... 301
Bufo, Ray, ... 301
Buglossus, seu Solea, Will......... 466
BUTEO, *Bechst.*................ 50, 87
BUTEO, *Nob.* 50, 87
   apivorus, *Ray,* 88
   cineraceus, *Flem.*............ 90
   cyaneus, *Nob.* 89
   Lagopus, *Vig.* 87
   rufus, *Nob.* 88
   vulgaris, *Will.*............... 87

C

Cæcilia, Ray, ... 295
CAIRINA, *Flem.* 72, 230
   moschata, *Flem.* 230
CALAMOPHILUS, *Leach,* 55, 125
   biarmicus, *Leach,* 125
CALIDRIS, *Ill.* 66, 183
   arenaria, *Ill.* 183
CALLIONYMUS, *Linn.*... 314, 388

CALLIONYMUS, Dracunculus,
  Linn.............................. 389
  Lyra, Linn. .... 388.
Canis Carcharias, Will. 497
  Galeus, Will.................. 501
CANIS, Flem. 4, 13
CANIS, Linn....................... 4, 13
  familiaris, Linn. 13
  Lupus, Linn. 14
  Vulpes, Linn. 14
CANTHARUS, Cuv. 310, 358
  griseus, Cuv. 358
CAPRA, Linn. 7, 37
  Hircus, Linn. 37
CAPRIMULGUS, Linn. ... 61, 160
  Europæus, Linn................ 160
Capriscus Rondeletii, Will. 492
CAPROS, Lac................. 311, 368
  Aper, Lac. 368
CARANX, Cuv. 311, 366
  Trachurus, Lac. 366
Carassius, Will. 402
Carbo Cormoranus, Temm. 262
  cristatus, Temm................ 262
  Graculus, Temm. 263
CARCHARIAS, Cuv. 327, 497
  glaucus, Flem. 499
  vulgaris, Flem. 497
  Vulpes, Flem. 498
CARDUELIS, Briss............ 57, 137
CASTOR, Linn. 7, 34
  Fiber, Linn...................... 34
Cataphractus Schoneveldii, Flem. 346
Cataractes parasiticus, Flem...... 282
  vulgaris, Flem. 280
Cathartes Percnopterus, Temm.... 79
CATODON, Lac. 9, 44
  macrocephalus, Lac............ 44
  Sibbaldi, Flem................. 45
  Trumpo, Lac. 44
Catulus major vulgaris, Will...... 495
  maximus, Will. 496
  minor, Will. 495
CAVIA, Gmel. 7, 36
  Cobaya, Gmel.................. 36
CENTRISCUS, Linn. 315, 400
  Scolopax, Linn. 400
CENTROLOPHUS, Lac...... 312, 370

CENTROLOPHUS, Morio, Cuv.... 370
  niger, Lac. 370
  Pompilus, Cuv. 370
CENTRONOTUS, Lac. ...311, 365
  Ductor, Nob. 365
CEPHALOPTERA, Dum. ... 329, 519
  Giorna, Riss. 519
Cephalus brevis, Shaw, ... 490
  oblongus, Shaw, ... 491
CEPOLA, Linn.................312, 374
  rubescens, Linn. 374
  Tænia, Bl. 374
Cernua fluviatilis, Flem............ 334
CERTHIA, Linn. 59, 152
  familiaris, Linn. 152
CERVUS, Linn.................... 8, 37
  Capreolus, Linn. 38
  Dama, Linn. 38
  Elaphus, Linn................. 37
Chætodon, Couch, ... 360
CHARADRIUS, Linn.........65, 177
  Cantianus, Lath. 180
  Hiaticula, Linn. 179
  minor, Mey. 179
  Morinellus, Linn.............. 178
  pluvialis, Linn................ 177
CHAULIODUS, Swains.........73, 231
Chelona imbricata, Flem........... 290
CHELONIA, Brongn. 287, 290
  imbricata, Gray, ... 290
Chenalopex Ægyptiaca, Steph..... 225
CHIMÆRA, Linn. 326, 494
  monstrosa, Linn............... 494
CHRYSOPHRYS, Cuv.........310, 353
  Aurata, Cuv. 353
Chrysotosa Luna, Lac. 369
CICONIA, Briss. 67, 192
  alba, Ray,................... 192
  nigra, Ray,.................. 193
Ciliata glauca, Couch,............. 451
CINCLUS, Bechst.................53, 98
  aquaticus, Bechst.............. 98
CIRCUS, Bechst.................. 51, 88
CLANGULA, Flem............74, 245
  albeola, Steph. 246
  chrysophthalmos, Steph....... 245
  histrionica, Steph.............. 246
CLUPEA, Cuv..................318, 434

CLUPEA, *Linn.*..............318, 434
   alba, *Yarr.* 436
   Alosa, Bl...................... 437
   Alosa, *Cuv.* 438
   Encrasicholus, Linn. 439
   Finta, *Cuv.* 437
   Harengus, *Linn.* 434
   latulus, Cuv.................... 436
   Leachii, *Yarr.*................. 434
   Pilchardus, *Bl.*................. 436
   Sprattus, *Bl.*.................. 435
COBITIS, *Linn.* 316, 416
   barbatula, *Linn.* 416
   barbatula aculeata, Will...... 417
   Tænia, *Linn*..................... 417
COCCOTHRAUSTES, *Briss.* ... 57, 136
COCCYZUS, *Vieill.*............60, 155
   Americanus, *Bon*.............. 155
Coluber Berus, Linn. 298
   Berus, Turt. 297
   Cæruleus, Shepp. 298
   Chersea, Linn................. 298
   Dumfrisiensis, Sow. 297
   Natrix, Linn................. 296
   Prester, Linn. 298
COLUMBA, *Linn.*............ 62, 161
   Livia, *Briss*..................... 162
   migratoria, *Linn*.............. 163
   Oenas, *Linn*................... 161
   Palumbus, *Linn*.............. 161
   Turtur, *Linn.* 162
COLUMBA, *Swains.* 62, 161
COLYMBUS, *Lath.* 75, 255
   arcticus, *Linn*................. 256
   glacialis, *Linn*................. 255
   Immer, Linn. 256
   septentrionalis, *Linn.* 257
   stellatus, Gmel................ 257
   Urinator, Linn. 252
CONGER, *Cuv*..................322, 478
Conger, Will. 478
Coquus Cornubiensium, Ray,...... 396
CORACIAS, *Linn.*............ 60, 156
   garrula, *Linn.* 156
COREGONUS, *Cuv*......... 318, 431
   Lavaretus, *Flem*.............. 431
   Marænula, *Jard.*.............. 432
   Pollan, *Thomps.* 432

COREGONUS, *Thymallus,* Flem. 430
Coriudo coriacea, Flem. 290
CORVUS, *Cuv.* 58, 145
CORVUS, *Linn.*............. 58, 145
   Corax, *Linn*..................... 145
   Cornix, *Linn.* 146
   Corone, *Linn.* 145
   frugilegus, *Linn.* 146
   glandarius, Temm. 148
   Monedula, *Linn.* 147
   Pica, *Linn.* 147
CORYPHÆNA, *Linn.*....... 311, 370
   Morio, *Cuv.* 370
COTTUS, *Linn.* 308, 343
   Bubalis, *Euph*.................. 345
   Cataphractus, Linn........... 346
   Gobio, *Linn*.................... 343
   quadricornis, *Linn.* 345
   Scorpius, *Bl*.................... 344
COTURNIX, *Briss*..............64, 174
   dactylisonans, Temm. 174
CRENILABRUS, *Cuv.*........315, 397
   gibbus, Flem. 399
   Tinca, Flem.................... 397
CREX, *Bechst*.................. 71, 217
   Baillonii, *Selb.* 219
   Porzana, *Selb.*................. 218
   pratensis, *Bechst*.............. 217
   pusilla, *Selb*................... 219
Cuculus lineatus, Ray, ... 339
CUCULUS, *Linn.*..............60, 154
   Americanus, Linn. 155
   canorus, *Linn.* 154
   Carolinensis, Wils. 155
   cinerosus, Temm................ 155
CURRUCA, *Bechst*..............54, 108
CURSORIUS, *Lath.* 64, 176
   isabellinus, *Mey.*.............. 176
CYCLOPTERUS, *Cuv.* 322, 471
CYCLOPTERUS, *Linn....* 321, 471
   bimaculatus, Turt. 470
   Cornubicus, Shaw, ... 469
   Liparis, *Linn.* 472
   Lumpus, *Linn.* 471
   Montacuti, Turt................ 473
   Montagui, *Don.* 473
   occllatus, Don. 469
   pavoninus, Shaw, ... 471

CYCLOPTERUS, *pyramidatus*,
Shaw, ... 472
CYGNUS, *Mey.* 72, 226
Bewickii, *Yarr.* 226
Canadensis, *Steph.* 227
ferus, *Ray,* 227
Guineensis, *Nob.* 226
Olor, *Steph.* 228
CYPRINUS, *Cuv* 316, 401
CYPRINUS, *Linn.* 316, 401
Alburnus, *Linn.* 414
auratus, *Linn.* 403
Barbus, *Linn.* 404
Blicca, *Bl.* 407
Brama, *Linn.* 406
cæruleus, *Nob.* 413
Carassius, *Linn.* 403
Carpio, *Linn.* 401
Cephalus, *Linn.* 411
Dobula, *Linn.* 409
Erythrophthalmus, *Linn.* ... 412
Gibelio, *Gmel.* 402
Gobio, *Linn.* 405
Idus, *Linn.* 415
Jeses, Don. 412
Lancastriensis, *Shaw,* 411
latus, Gmel. 407
Leuciscus, *Linn.* 410
Orfus, *Linn.* 416
Phoxinus, *Linn.* 415
Rutilus, *Linn.* 408
Tinca, *Linn.* 405
CYPSELUS, *Ill.* 61, 159
alpinus, *Temm.* 159
Apus, *Flem.* 159
murarius, Temm 159

D.

DAFILA, *Leach,* 73, 232
DELPHINAPTERA, *Lac.* 9, 43
albicans, Flem................. 43
DELPHINUS, *Cuv.* 9, 40
DELPHINUS, *Linn.* 9, 40
albicans, *Fab.* 43
Chemnitzianus, Desm......... 44
Deductor, Scoresb. 42
Delphis, *Linn.* 40
edentulus, Desm. 44

DELPHINUS, *Gladiator,* Lac... 42
globiceps, Cuv. 42
Grampus, Desm. 42
Hunteri, *Desm.* 44
Hyperoodon, Desm............ 44
Leucas, Desm. 43
melas, *Traill,* 42
Orca, *Fab* 42
Phocœna, *Linn.* 41
Sowerbyi, Desm. 44
truncatus, Mont. 41
Tursio, *Fab.* 41
ventricosus, Lac. 42
DENDRONESSA, *Swains...* 73, 237
Sponsa, *Swains.* 237
DENTEX, *Cuv* 310, 357
vulgaris, *Cuv.* 357
Dentex, Will. 357
Diodon Mola, Bl. 490
Dracunculus, Will. 388

E.

ECHENEIS, *Linn.* 322, 473
Remora, *Linn.* 473
ECTOPISTES, *Swains.* 62, 163
ELANUS, *Sav.* 50, 86
EMBERIZA, *Linn.* 56, 128
calcarata, Temm............... 128
Ciris, *Linn.* 133
Cirlus, *Linn.* 131
Citrinella, *Linn.* 131
Hortulana, *Linn.* 132
Lapponica, *Nilss.* 128
Miliaria, *Linn.* 130
montana, Gmel. 130
mustelina, Gmel. 130
nivalis, *Linn.* 129
Schœniclus, *Linn.* 130
Schœniculus, Temm. 130
EMBERIZA, *Mcy.* 56, 130
Encrasicholus, Will. 439
ENGRAULIS, *Cuv.* 318, 439
Encrasicholus, *Flem.* 439
Eperlanus, Will. 429
EQUUS, *Linn.* 8, 39
Asinus, *Linn.* 39
Caballus, *Linn.* 39
ERINACEUS, *Linn.* 5, 19

ERINACEUS, Europæus, *Linn.* 19
ERITHACA, *Swains*............ 54, 103
Erythrinus Rondeletii, Will....... 355
ESOX, *Cuv.* 316, 417
   Belone, Linn. 418
   Brasiliensis, Couch,........... 419
   Lucius, Linn. 417
   osseus, Linn..................... 439
   Saurus, Rack. 419
   Sphyræna, Linn. 337
EXOCŒTUS, *Linn.*......... 317, 420
   evolans, Bl. 420
   exiliens, Bl. 420
   volitans, *Linn.*.................. 420

F.

FALCO, *Linn.* 50, 81
   æruginosus, Linn.............. 89
   Æsalon, Gmel.................. 83
   Albicilla, Temm.............. 80
   Buteo, Temm. 87
   cineraceus, Mont. 90
   cyaneus, Temm. 89
   fulvus, Temm.................. 80
   furcatus, Linn.............. 86
   gentilis, Gmel.................. 85
   Haliæetus, Temm. 81
   Islandicus, Lath. 81
   Lagopus, Temm...... 87
   Lanarius, Temm.............. 82
   Milvus, Temm. 86
   Nisus, Temm.................. 85
   ossifragus, Linn. 81
   palumbarius, Temm.......... 85
   peregrinus, *Gmel*.............. 82
   ɼufipes, *Bechst*................ 83
   rufus, Temm. 89
   Subbuteo, *Linn.* 82
   Tinnunculus, *Linn*............ 84
FELIS, *Linn.* 4, 14
   Catus, *Linn*..................... 14
   maniculata, *Rüpp.* 15
FRATERCULA, *Briss*...... 75, 260
   arctica, *Steph.* 260
FREGILUS, *Cuv.* 58, 144
   Graculus, *Selb*.................. 144
FRINGILLA, *Cuv.* 57, 133
FRINGILLA, *Linn.* 56, 133

FRINGILLA, cannabina, *Linn.* 139
   Carduelis, *Linn.* 137
   Chloris, *Temm*.................. 136
   Coccothraustes, *Temm.* 136
   Cœlebs, *Linn.* 133
   domestica, *Linn.* 134
   Linaria, *Linn.*................... 138
   Linota, Gmel. 140
   montana, *Linn.* 135
   Montifringilla, *Linn.* 134
   Montium, *Gmel.* 140
   Spinus, *Linn.* 137
FULICA, *Linn.* 71, 221
   atra, *Linn*...................... 221
   aterrima, Linn................. 221
FULIGULA, *Ray,*........... 74, 240
   cristata, *Steph*... 244
   dispar, *Steph.* 243
   ferina, *Steph*.................... 241
   Marila, *Steph.* 243
   Nyroca, *Steph*.................. 242
   rufina, *Steph*.................... 240

G.

GADUS, *Linn.* 319, 440
   Æglefinus, *Linn.*.............. 441
   argenteolus, *Mont.* 451
   barbatus, Linn. 442
   Brosme, Gmel................. 452
   Callarias, *Linn.* 441
   Carbonarius, Linn. 446
   Lota, Linn. 448
   luscus, *Linn*.................... 442
   Merlangus, Linn.............. 445
   Merluccius, Linn.............. 447
   minutus, *Linn*.................. 444
   Molva, Linn. 448
   Morrhua, *Linn.* 440
   Mustela, Linn.................. 450
   Pollachius, Linn.............. 446
   punctatus, *Turt.* 442
   tricirratus, Bl.................. 449
   virens, Linn.................... 447
GALEUS, *Cuv.* 327, 501
   Acanthias Clusii, Will. 494
   Acanthias, sive Spinax, Will. 505
   glaucus, Will. 499
   Mustelus, Leach,.............. 502

GALEUS, *vulgaris*, Flem. 501
GALLINULA, *Lath.* 71, 220
   Baillonii, Temm. 219
   chloropus, *Lath.* 220
   Crex, Temm. 217
   Porzana, Temm. 218
   pusilla, Temm. 219
GALLUS, *Briss.* 63, 165
   Bankiva, Temm. 165
   crispus, *Briss.* 166
   cristatus, Temm. 165
   domesticus, *Briss.* 165
   ecaudatus, *Temm.* 166
   lanatus, *Temm.* 166
   Morio, *Temm.* 166
   pentadactylus, Temm. 165
   Pumilio, Temm. 165
   pusillus, Temm. ....:. 165
GARRULUS, *Briss.* 58, 148
   glandarius, *Flem.* 148
   Picus, Temm. 148
GASTEROSTEUS, *Cuv.* 309, 348
GASTEROSTEUS, *Linn*...308, 348
   aculeatus, *Linn.* 348
   brachycentrus, Cuv. 349
   Ductor, Linn. 365
   gymnurus, Cuv. 348
   lævis, Cuv. 351
   leiurus, Cuv. 349
   Pungitius, *Linn.* 350
   semiarmatus, Cuv. 349
   Spinachia, *Linn.* 351
   spinulosus, *Yarr.* 350
   trachurus, Cuv. 349
Gastrobranchus cæcus, Bl. 523
Gattorugine, Will. 379
GLAREOLA, *Briss.* 70, 216
   Pratincola, *Leach*, ... 216
   torquata, Temm. 216
GOBIO, *Cuv.* 316, 405
   fluviatilis, Flem. 405
GOBIUS, *Linn.* 314, 385
   bipunctatus, *Yarr.* 386
   gracilis, *Jen.* 387
   minutus, *Pall.* 386
   niger, Don. 386
   niger, *Linn.* 385
GRUS, *Pall.* 66, 185

GRUS, cinerea, *Bechst.* 185
GUNNELLUS, *Flem.* 313, 383
   Cornubiensium, Will. 383
   viviparus, Flem. 383
   vulgaris, Flem. 384
GYMNETRUS, *Bl.* 312, 372
   arcticus, *Cuv.* 372
   Hawkenii, *Bl.* 373
Gymnogaster arcticus, Cuv......... 372

H.

HÆMATOPUS, *Linn.* 66, 184
   ostralegus, *Linn.* 184
HALIÆETUS, *Sav.* 50, 80
HARELDA, *Leach*, ... 74, 247
   glacialis, *Steph.* 247
Harengus, Will. 434
   minor, sive Pilchardus, Will. 437
HIMANTOPUS, *Briss.* 68, 201
   melanopterus, *Temm.* 201
HIPPOCAMPUS, *Cuv.* ... 324, 489
   brevirostris, *Cuv.* 489
   Rondeletii, Will. 489
HIPPOGLOSSUS, *Cuv.* ... 321, 460
   vulgaris, *Flem.* 460
Hippoglossus, Will. 460
HIRUNDO, *Linn.* 61, 157
   riparia, *Linn.* 158
   rustica, *Linn.* 157
   urbica, *Linn.* 158
HYPEROODON, *Lac.* 9, 44
   bidens, *Flem.* 44
   Butskopf, Lac............... 44

I.

IBIS, *Lac.* 67, 194
   Falcinellus, *Temm.* 194

J.

JULIS, *Cuv.* 315, 397
   vulgaris, Flem. 397
Julis, Will. 397

L.

LABRAX, *Cuv.* 307, 331
   Lupus, Cuv...................... 331
LABRUS, *Cuv.* 315, 391
LABRUS, *Linn.* 315, 391

LABRUS, *Balanus*, Flem......... 391
 bimaculatus, *Linn.* 396
 carneus, Bl. 396
 Comber, *Gmel.*.................. 393
 Coquus, Turt. 396
 Cornubicus, *Gmel.* 398
 gibbus, *Gmel.* 399
 Julis, *Linn.* 397
 lineatus, *Don.* 392
 luscus, *Linn.* 400
 maculatus, *Bl.*.................. 391
 Psittacus, Riss. 392
 pusillus, *Jen.* 392
 Suillus, *Linn.* 397
 Tinca, *Linn.* ....,...... 397
 Tinca, Shaw, ... 391
 trimaculatus, *Gmel.* 396
 variegatus, *Gmel.* 394
 Vetula, *Bl.* 395
LACERTA, *Cuv.*........... 288, 291
 agilis, *Berk.*................... 292
 anguiformis, *Shepp.*........... 294
 aquatica, Linn.............. ... 304
 arenicola, Daud. 292
 maculata, Shepp. 304
 muralis, Latr. 295
 ocellata, Daud. 294
 œdura, *Shepp.* 294
 palustris, Linn. 303
 Stirpium, *Daud.* 291
 viridis, *Daud.* 292
 vulgaris, Shepp. 304
Lacertus viridis, Ray,.............. 292
Lacertus, Will...................... 419
LACHNOLAIMUS, *Cuv*...... 315, 397
 Suillus, Cuv.................... 397
LAGOPUS, *Vieill.*.............. 63, 170
LAMNA, *Cuv.*................. 327, 500
 Cornubica, Flem.............. 500
Lampetra cœca, Will.............. 523
 parva et fluviatilis,Will...... 522
 Rondeletii, Will.............. 520
Lampetræ medium genus, Will.... 521
LAMPRIS, *Retz*............ 311, 369
 guttatus, Cuv. 369
 Luna, *Riss.* 369
LANIUS, *Linn.*................. 52, 95
 Collurio, *Linn.*.................. 96

LANIUS, Excubitor, *Linn.* 95
 rufus, *Briss.*..................... 96
LARUS, *Linn.*.................. 77, 270
 arcticus, *Macg.*................. 279
 argentatus, *Brunn.* 276
 argentatus, Sab. 279
 Atracilla, *Linn.* 273
 canus, *Linn.* 275
 capistratus, *Temm.* 272
 eburneus, *Gmel.* 276
 fuscus, *Linn.*................... 277
 glaucus, *Brunn.* 279
 hybernus, Gmel. 276
 Islandicus, *Edmondst.*......... 279
 marinus, *Linn.*.................. 278
 minutus, *Pall.* 271
 parasiticus, Edmondst. 282
 ridibundus, *Linn.*.............. 272
 ridibundus, Wils............... 273
 Sabini, *Sab.* 270
 tridactylus, *Lath.*.............. 274
LARUS, *Steph.* 77, 271
LEPADOGASTER, *Gouan*, 321, 469
 bimaculatus, *Flem.* 470
 Cornubiensis, *Flem.*........... 469
 Gouani, Riss. 470
LEPIDOPUS, *Gouan*, ... 312, 371
 argyreus, *Cuv.*.................. 371
 Lusitanicus, Leach,........... 371
 tetradens, Flem. 371
LEPISOSTEUS, *Lac*....... 319, 439
 Gavialis, *Lac.* 439
LEPTOCEPHALUS, *Gron.* 323, 480
 Morrisii, *Gmel.*................. 480
LEPUS, *Linn.*..................... 7, 34
 albus, *Briss*..................... 35
 Cuniculus, *Linn.* 35
 timidus, *Linn.*.................. 34
 variabilis, Flem. 35
LESTRIS, *Illig.*............... 77, 280
 Cataractes, *Temm.* 280
 parasiticus, *Temm.* 283
 pomarinus, *Temm.* 281
 Richardsonii, *Swains.* 282
LEUCISCUS, *Klein*, ... 316, 408
 Alburnus, Flem. 414
 Cœruleus, Yarr. 413
 Cephalus, Flem. ....,.... 412

LEUCISCUS, *Erythrophthalmus,*
Flem........................... 412
Lancastriensis, Yarr. 411
Phoxinus, Flem. 415
Rutilus, Flem.................. 408
vulgaris, Flem................. 410
LICHIA, *Cuv.*.................. 311, 366
glaucus, Cuv. 366
LIMOSA, *Briss.*.............. 69, 202
melanura, *Leisl.* 203
rufa, *Briss.* 202
LINARIA, *Steph.*............. 57, 138
borealis, Selb. 139
canescens, Gould,............. 139
Lingula, Rond. 468
LIPARIS, *Art.* 322, 472
Montagui, Flem............... 473
nostras, Will. 472
vulgaris, Flem................ 472
LOBIPES, *Cuv.*.............. 69, 214
hyperboreus, *Steph.*........... 214
LOPHIUS, *Linn.*............. 314, 389
Cornubicus, *Shaw,* 390
piscatorius, *Linn.*........... 389
LOTA, *Cuv.*................. 319, 448
Molva, *Nob.*................... 448
vulgaris, *Nob.*................ 448
LOXIA, *Briss.* 57, 141
curvirostra, *Linn.*........... 141
falcirostra, Lath............ 143
leucoptera, *Gmel.*............ 143
Pytiopsittacus, *Bechst.*....... 142
Lumpus Anglorum, Will. 471
gibbosus, Will................ 472
LUTRA, *Cuv.*.................. 4, 13
Roensis, *Ogilb.*.............. 525
vulgaris, *Desm.* 13

M.

MACHETES, *Cuv.* 69, 207
MACRORAMPHUS, *Leach,* ... 69, 207
MANATUS, *Cuv.* 8, 40
borealis, Flem............... 40
MARECA, *Steph.* 73, 236
Penelope, *Selb.*.............. 236
Martes Abietum, Flem. 11
Fagorum, Flem. 11
MECISTURA, *Leach,* 55, 124

MELEAGRIS, *Linn.*......... 62, 164
Gallopavo, *Linn.*............. 164
MELES, *Cuv.*................... 3, 10
Taxus, *Flem.* 10
vulgaris, Desm. 10
MELIZOPHILUS, *Leach,* . 54, 112
provincialis, *Leach,*......... 112
MERGULUS, *Ray,* 75, 259
Alle, *Selb.* 259
MERGUS, *Linn.*............. 74, 248
albellus, *Linn.*.............. 250
Castor, Linn. 249
cucullatus, *Linn.*............ 249
Merganser, *Linn.*............ 248
minutus, Linn................ 251
Serrator, *Linn.*.............. 249
MERLANGUS, *Cuv.* 319, 445
Carbonarius, *Flem.* 446
Pollachius, *Flem.* 446
virens, *Flem.*................ 447
vulgaris, *Flem.*.............. 445
MERLUCCIUS, *Cuv.* 319, 447
vulgaris, *Flem.*.............. 447
MEROPS, *Linn.*............ 61, 156
Apiaster, *Linn.*.............. 156
MILVUS, *Bechst.*............. 50, 86
furcatus, *Nob.* 86
Ictinus, *Sav.*................ 86
MILVUS, *Vig.* 50, 86
Mola Salviani, Will. 490
Molva Lota, Flem. 448
vulgaris, Flem. 448
MONOCHIRUS, *Cuv.*......... 321, 468
MONODON, *Linn.*............. 9, 43
Monoceros, *Linn.*............. 43
Morhua Æglefinus, Flem. 441
barbata, Flem. 442
lusca, Flem. 442
minuta, Flem. 444
punctatus, Flem. 442
vulgaris, Flem. 440
Mormon Fratercula, Temm. 260
MOTACILLA, *Linn.*......... 55, 114
alba, *Linn.* 114
Boarula, *Linn.*............... 115
flava, *Ray,* 115
flava, Temm.................. 116
neglecta, *Gould,* 116

MOTELLA, *Cuv*............ 320, 449
   glauca, *Nob*.................... 451
   Mustela, *Nilss*............... 450
   tricirrata, *Nilss*. 449
MUGIL, *Linn*. 313, 374
   Britannicus, Hanc. 375
   Capito, *Cuv*...................:. 374
   Cephalus, *Cuv*.................. 376
   Cephalus, Don................ 375
   Chelo, *Cuv*. 375
   curtus, *Yarr*.................... 376
MULLUS, *Linn*. 307, 337
   barbatus, *Linn*. 338
   Surmuletus, *Linn*. 337
Mullus, Will. 338
   imberbis, Will................... 339
MURÆNA, *Thunb*.......... 323, 479
   Anguilla, Bl. 474
   Conger, Linn. 478
   Helena, *Linn*. 479
   Myrus, Linn. 478
   Ophis, Linn................... 479
   Serpens, Linn................... 475
Muræna, Will. 479
MUS, *Linn*. 6, 30
   decumanus, *Pall*............... 32
   messorius, *Shaw*, ... 31
   Musculus, *Linn*............... 31
   Rattus, *Linn*. 32
   sylvaticus, *Linn*............... 30
MUSCICAPA, *Linn*. 52, 97
   albicollis, *Temm*............... 98
   grisola, *Linn*. 97
   luctuosa, *Temm*. 97
MUSTELA, *Cuv*. 3, 11
Mustela fluviatilis, Will............ 448
   vivipara, Will.................. 384
   vulgaris, Will.................. 451
MUSTELA, *Linn*. 3, 11
   Erminea, *Linn*.................. 13
   Foina, *Linn*.................... 11
   Furo, *Linn*. 12
   Martes, *Linn*. 11
   Putorius, *Linn*. 11
   vulgaris, *Gmel*.................. 12
MUSTELUS, *Cuv*............. 327, 502
   lævis, Will. 502
   stellatus, Riss. 503

MYLIOBATIS, Dum.......... 329, 519
MYOXUS, *Gmel*.................. 6, 30
   avellanarius, *Desm*............ 30
MYXINE, *Linn*. 329, 523
   cæca, Blainv. 523
   glutinosa, *Linn*. 523

N.

NATRIX, *Flem*. 289, 296
   Dumfrisiensis, *Flem*. 297
   torquata, *Ray*, ... 296
Nauclerus furcatus, Vig. 86
NAUCRATES, *Cuv*........... 311, 365
   Ductor, Cuv...................... 365
NEOPHRON, *Sav*. 49, 79
   Percnopterus, *Sav*. 79
NOCTUA, *Sav*................ 51, 93
   funerea, *Nob*. 526
   nyctea, *Nob*...................... 93
   passerina, *Selb*.................. 94
   Tengmalmi, *Selb*............... 94
NOCTUA, *Selb*. 52, 94
NUCIFRAGA, *Briss*.......... 58, 149
   Caryocatactes, *Temm*......... 149
NUMENIUS, *Briss*. 68, 195
   arquata, *Lath*. 195
   Phæopus, *Lath*. 195
NUMIDA, *Linn*.............. 63, 168
   Meleagris, *Linn*. 168
NYCTICORAX, *Steph*.......... 67, 191

O.

ŒDICNEMUS, *Temm*....... 65, 177
   crepitans, *Temm*.............. 177
OIDEMIA, *Flem*. 73, 239
   fusca, *Flem*...................... 239
   leucocephala, *Steph*............ 240
   nigra, *Flem*...................... 239
   perspicillata, *Steph*............ 240
Onos, Will. 441
OPHIDIUM, *Linn*. 323, 481
   barbatum, *Linn*. 481
   imberbe, *Mont*.................. 481
   pellucidum, Couch, ... 480
Ophis maculata, Turt............... 479
OPHISURUS, *Lac*. 322, 479
   Ophis, *Lac*. 479
   Serpens, *Lac*. 479

ORIOLUS, *Linn.* 53, 102
   Galbula, *Linn.*.................. 102
ORTHAGORISCUS, *Schn.* 325, 490
   Mola, *Schn.* 490
   oblongus, *Schn.* 491
   truncatus, Flem. 491
ORTYX, *Steph*.................. 64, 173
OSMERUS, *Art.* 318, 429
   Eperlanus, *Flem.*............... 429
OTIS, *Linn*.................... 64, 174
   Tarda, *Linn*...................... 174
   Tetrax, *Linn.* 175
OTUS, *Cuv.* 51, 91
   Brachyotos, *Flem.* 92
   vulgaris, *Flem*.................. 91
OVIS, *Linn*.......................... 7, 37
   Aries, *Linn*....................... 37

P.

PAGELLUS, *Cuv.* 310, 355
   centrodontus, Cuv. 356
   Erythrinus, Cuv................. 355
PAGRUS, *Cuv.* 310, 354
   lineatus, Flem.................... 358
   vulgaris, Cuv. 354
   vulgaris, Flem.................. 356
PANDION, *Sav.*.................. 50, 81
PARUS, *Leach*, ... 55, 121
PARUS, *Linn*.................... 55, 121
   ater, *Linn*...................... 123
   biarmicus, Temm............... 125
   cæruleus, *Linn*.................. 122
   caudatus, *Linn*.................. 124
   cristatus, *Linn*.................. 122
   major, *Linn*...................... 121
   palustris, *Linn*.................. 123
Passer asper sive squamosus, Will. 457
   Bellonii, Will. 454
   Cornubiensis asper magno oris
      hiatu, Jag...................... 464
   fluviatilis, Will. 455
Pastinaca marina, Will. 518
PASTOR, *Temm.* 58, 144
   roseus, *Temm*.................... 144
PAVO, *Linn.* 63, 164
   cristatus, *Linn*............. 164
Pelecanus Onocrotalus, *Linn*....... 264
PERCA, *Cuv*.................. 307, 330

PERCA, *Linn.* 307, 330
   Cabrilla, *Linn*.................. 332
   Cernua, *Linn.* 334
   Channus, Couch, ... 332
   fluviatilis, *Linn.* 330
   Gigas, *Gmel*...................... 333
   Labrax, *Linn.* 331
   marina, Linn. 348
   nigra, Gmel.................. 370
   robusta, Couch, ... 333
PERDIX, *Briss*............... 63, 172
   borealis, Temm. 173
   cinerea, *Briss.* 172
   Coturnix, *Lath.* 174
   rubra, *Briss*.................. 172
   Virginiana, *Lath*............... 173
PERDIX, *Steph*.................64, 172
PERNIS, *Cuv*..................... 51, 88
PETROMYZON, *Linn.* ... 329, 520
   branchialis, Linn.............. 522
   cæcus, Couch, ... 522
   fluviatilis, *Linn.* 521
   Juræ, *Macc*.................. 522
   marinus, *Linn*.................. 520
   Planeri, Bl....................... 522
   Planeri, *Cuv*.................. 522
PHALACROCORAX, *Briss.*76, 262
   Carbo, *Steph.* 262
   cristatus, *Steph.* 262
   Graculus, *Steph.* 263
PHALAROPUS, *Briss*...... 70, 215
   hyperboreus, Temm............ 214
   lobatus, *Flem.* 215
   platyrhinchus, Temm.......... 215
   Williamsii, Simm. 214
PHASIANUS, *Linn*.......... 63, 166
   Colchicus, *Linn.* 166
   torquatus, *Temm*............... 167
PHILOMELA, *Swains*.......... 54, 107
PHOCA, Linn. 4, 15
   barbata, *Mull.* 16
   vitulina, *Linn*................... 15
PHOCÆNA, *Cuv*............... 9, 41
PHŒNICURA, *Swains*.......... 54, 104
   Tithys, Jard. 105
Pholis lævis, Flem.................. 382
PHYCIS, *Art.* 320, 452
   blennoides, Nilss. 452

PHYCIS, furcatus, *Flem*.......... 452
   Mediterraneus, Lar. 453
Physalis vulgaris, Flem............. 47
PHYSETER, *Lac.* 9, 45
PHYSETER, *Linn.* 9, 44
   Catodon, *Linn*................. 45
   macrocephalus, *Shaw*,......... 44
   microps, *Linn*................. 46
   Mular, Lac.................. 45
   Tursio, *Linn.* 45
PICA, *Cuv.* 58, 147
PICUS, *Linn.* 59, 149
   major, *Linn*..................... 150
   martius, *Linn*.................. 151
   minor, *Linn*.................... 151
   tridactylus, *Linn*.............. 151
   villosus, *Linn*................. 151
   viridis, *Linn.* 149
PLATALEA, *Linn.* 67, 193
   Leucorodia, *Linn.* 193
PLATESSA, *Cuv.* 320, 454
   carnaria, Brown,.............. 456
   Cyclops, Flem................. 466
   Flesus, *Flem.* 455
   lata, Cuv...................... 455
   Limanda, *Flem.* 456
   Limandoides, *Nob.* 459
   microcephala, *Flem.* 457
   Pola, *Cuv*..................... 458
   vulgaris, *Flem*................ 454
PLECOTUS, *Geoff.* 6, 27
PLECTROPHANES, *Mey.* ... 56, 128
   Lapponica, Selb. 128
PLECTROPTERUS, *Leach*, .. 72, 226
   Gambensis, Steph. 226
PLEURONECTES, *Flem*...321, 461
   Arnoglossus, *Schn.* 465
   Cardina, Cuv. 464
   Cyclops, *Don.* 466
   Cynoglossus, Linn. 458
   Flesus, Linn. 455
   Hippoglossus, Linn............ 460
   hirtus, *Mull*................... 463
   lævis, Turt. 457
   Limanda, Linn. 457
   limandanus, Parn.............. 459
   limandoides, Bl. 459
   Lingula, Han................... 468

PLEURONECTES, maximus,
   Linn. 461
   Megastoma, *Don*.............. 464
   microcephalus, Don........... 457
   microstomus, Nilss............ 457
   Passer, Bl. 456
   Pegusa, Riss. 468
   Platessa, Linn.................. 454
   punctatus, *Bl.* 462
   Rhombus, *Linn.* 462
   roseus, Shaw, ... 456
   Solea, Linn..................... 466
   tuberculatus, Turt. 461
   variegatus, Don. 468
PODICEPS, *Lath*............. 74, 251
   auritus, *Lath.* 253
   cornutus, *Lath*................. 252
   cristatus, *Lath*................. 251
   Hebridicus, Sow............... 254
   minor, *Lath*.................... 254
   obscurus, Lath................. 253
   rubricollis, *Lath*............... 252
Pompilus, Rond. 370
Pristibatis antiquorum, Blainv.... 508
PRISTIS, *Lath.*..............328, 508
   antiquorum, *Lath.* 508
PROCELLARIA, *Linn.* ... 77, 284
   Anglorum, *Temm.* 285
   fuliginosa, *Strickl.* 285
   glacialis, *Linn*................. 284
   Leachii, *Temm.* 286
   pelagica, *Linn*................. 285
   pelagica, Wils................. 286
   Puffinus, *Linn*................. 284
   Wilsoni, *Bon.* 286
PROCELLARIA, *Vig.* 78, 284
Psophia crepitans, *Linn*........... 185
PUFFINUS, *Ray*, ... 78, 284
   fuliginosus, Strickl............. 285
PUTORIUS, *Cuv*................... 3, 11
PYRGITA, *Cuv.* 57, 134
Pyrrhocorax Graculus, Temm. ... 144
PYRRHULA, *Briss*.......... 57, 140
   Enucleator, *Temm*............. 141
   vulgaris, *Temm.* 140

Q.

Querquedula glocitans, Vig......... 233

R.

RAIA, *Cuv*..................... 328, 510
RAIA, *Linn*.................. 328, 509
  Aquila, *Linn*. 519
  aspera, Flem. 513
  aspera nostras, Will. 513
  Batis, *Linn*....................... 510
  chagrinea, *Mont*. 513
  clavata, *Linn*. 516
  Cuvieri, *Lac*..................... 518
  Fullonica, Turt. 515
  Giorna, *Lac*...................... 519
  lævis undulata seu cinerea,
    Will......................... 510
  maculata, *Mont*. 514
  marginata, *Lac*. 512
  microcellata, *Mont*. 515
  Miraletus, *Don*. 518
  oculata, Flem. 518
  Oxyrhynchus, *Linn*........... 511
  Pastinaca, *Linn*. 518
  radiata, *Don*. 517
  Rubus, Bl. 516
  Rubus, Blainv. 517, 518
  Rubus, Don...................... 514
  Rubus, Flem..................... 515
  Rubus, Turt. 515
  Speculum, Blainv. 518
  Torpedo, *Linn*................... 509
RALLUS, *Linn*. 70, 217
  aquaticus, *Linn*. 217
RANA, *Laurent*. 299, 300
  aquatica, Ray, ... 300
  Bufo, Linn. 301
  Bufo, β, Gmel................. 302
  esculenta, *Linn*. 301
  fusca, Rœs. 300
  Rubetra, Turt. 302
  temporaria, *Linn*............... 300
  viridis, Rœs..................... 301
Rana piscatrix, Will. 389
RANICEPS, *Cuv*........... 320, 453
  Jago, *Flem*. 454
  trifurcatus, *Flem*.............. 453
Ranunculus viridis, Merr. 303
RECURVIROSTRA, *Linn*. 68, 201
  Avocetta, *Linn*. 201
REGULUS, *Cuv*. 54, 113

REGULUS, aurocapillus, *Selb*.... 113
  ignicapillus, *Nob*.............. 113
Remora, Will...................... 473
RHINOLOPHUS, *Geoff*. 5, 19
  bihastatus, Desm............... 20
  Ferrum-equinum, *Gmel*....... 19
  Hipposideros, *Bechst*. 20
  unihastatus, Desm. 19
*Rhombus lævis Cornubiensis macu-
  lis nigris*, Jag.................... 457
  *maximus asper non squa-
    mosus*, Will................... 461
  non aculeatus squamosus, Will. 462
RUSTICOLA, *Vieill*............ 69, 204

S.

Salamandra aquatica, Ray, ... 303
  cristata, Latr. 303
  punctata, Latr.................. 304
SALICARIA, *Selb*............ 54, 106
SALMO, *Cuv*. 317, 421
  albula, Stew.................... 432
  albus, *Flem*..................... 424
  alpinus, Bl. 425
  alpinus, Don.................... 427
  alpinus, Linn................... 428
  Cambricus, Don. 422
  Eperlanus, Linn............... 429
  Eriox, *Linn*..................... 422
  Fario, *Linn*..................... 424
  ferox, *Jard*..................... 425
  Gœdenii, Bl.................... 422
  hamatus, Cuv. 422
  Hucho, Flem. 423
  lacustris, Berk................... 426
  Lemanus, Cuv.................. 424
  Marænula, Bl.................. 432
  marmoratus, Cuv.............. 425
  punctatus, Cuv................. 425
  Salar, *Linn*..................... 421
  Salmarinus, Linn. 428
  Salmulus, Turt. 426
  Salvelinus, *Don*. 428
  Salvelinus, Linn............... 428
  Thymallus, Linn............... 430
  Trutta, *Linn*. 423
  Umbla, *Linn*................... 427
  Wartmanni, Bl................. 431

Salmo, Will............................ 421
Salmulus, Will. 426
Saurus, Will. 419
SAXICOLA, *Bechst.*......... 55, 119
   Œnanthe, *Bechst*............... 119
   Rubetra, *Bechst.* 120
   Rubicola, *Bechst.*............... 121
SCIÆNA, *Cuv*............... 309, 352
   Aquila, *Cuv*....................... 352
   cirrhosa, *Linn*................... 353
   Labrax, Bl. 331
   Umbra, Cuv. 352
Sciæna, *Cuv*.................... 309, 352
SCIURUS, *Linn*................. 6, 29
   vulgaris, *Linn*................... 29
SCOLOPAX, *Linn.* 69, 204
   Gallinago, *Linn.* 205
   Gallinula, *Linn.* 206
   grisea, *Gmel*..................... 207
   major, *Gmel.* 205
   Noveboracensis, Wils......... 207
   Rusticola, *Linn.* 204
   Sabini, *Vig.*...................... 204
Scolopax, *Vieill.* 69, 204
SCOMBER, *Cuv*............. 310, 360
   Colias, Cuv....................... 362
   Colias, *Turt*...................... 362
   Ductor, Bl. 365
   maculatus, *Couch*, ... 361
   Pelamys, *Linn*.................... 363
   Scomber, *Linn*................... 360
   Thynnus, *Linn*. 362
   trachurus, Bl. 366
   vulgaris, Flem.................. 360
Scomber, *Cuv*............... 310, 360
Scomberesox, *Lac.* 317, 419
   Saurus, Flem. 419
SCOPELUS, *Cuv*............ 318, 433
   borealis, Nilss.................. 433
   Humboldti, *Cuv*............... 433
Scops, *Sav*...................... 51, 91
SCORPÆNA, *Linn*......... 308, 347
   Norvegica, *Cuv.* 347
Scyllium, *Cuv.* 327, 495
   Canicula, Bon................... 495
   Catulus, Flem................... 495
   melanostomum, Bon............ 497
   stellare, Bon..................... 496

Scyllium, *stellare*, Flem......... 495
Scymnus, *Cuv*.............. 328, 506
   borealis, Flem................... 506
Sebastes, *Cuv*.............. 308, 347
   Norvegicus, Cuv............... 347
Selache, *Cuv*.............. 328, 503
Serpens rubescens, Will............. 374
   marinus, Merr. 479
Serpes Humboldti, Riss............ 433
Serranus, *Cuv*............... 307, 332
   Cabrilla, Cuv. 332
   Couchii, *Yarr*................... 334
   Gigas, Cuv....................... 333
   Norvegicus, Flem. 347
   Scriba, Cuv...................... 333
SILURUS, *Art.* 317, 421
   Glanis, *Linn.* 421
SITTA, *Linn*................... 60, 154
   Europæa, *Linn*. 154
SOLEA, *Cuv*............... 321, 466
   Lingula, *Nob.* 468
   Mangilii, Bon................... 468
   parva, sive Lingula, Rond.... 468
   Pegusa, *Yarr*.................... 467
   variegata, Flem. 468
   vulgaris, *Flem*.................. 466
Solea, *Cuv*. 321, 466
Solea, Will. 466
SOMATERIA, *Leach*, ... 73, 237
   mollissima, *Leach*, ... 237
   spectabilis, *Leach*, ... 238
SOREX, *Linn*.................... 5, 17
   Araneus, *Linn*................... 17
   ciliatus, Sow.................... 18
   Daubentonii, Desm............ 18
   fodiens, *Gmel.* 18
   remifer, *Geoff.* 18
SPARUS, *Cuv*.............. 309, 353
   Aurata, Don. 356
   Aurata, *Linn*. 353
   centrodontus, *Lar.* 356
   Dentex, Bl. 357
   Erythrinus, *Linn*............... 355
   formosus, Shaw, ... 394
   lineatus, Mont.................. 358
   niger, Turt...................... 359
   Pagrus, Bl. 356
   Pagrus, *Linn.* 354

SPARUS, *Raii*, Bl. 359
    Vetula, Couch, ... 359
SPHARGIS, *Merr.* 287, 290
    coriacea, *Gray*, ... 290
SPHYRÆNA, *Schn.* 307, 337
    Spet, Lac. 337
    vulgaris, *Cuv.* 337
SPINACHIA, *Flem.* 309, 351
    vulgaris, Flem. 351
SPINAX, *Cuv.* 328, 505
    Acanthias, Flem. 505
Sprattus, Will. 435
SQUALUS, *Linn.* 327, 495
    Acanthias, *Linn.* 505
    borealis, *Scoresb.* 506
    cæruleus, Blainv. 499
    Canicula, *Linn.* 495
    Carcharias, Bl. 498, 506
    Carcharias, Gunn. 506
    Carcharias, *Linn.* 497
    Catulus, Bl. 495
    Cornubicus, *Gmcl.* 500
    Cornubicus, β, Gmel. 501
    Galeus, *Linn.* 501
    glacialis, Nilss. 506
    glaucus, Blainv. 499
    glaucus, *Linn.* 499
    Gunnerianus, Blainv. 504
    Hinnulus, *Blainv.* 503
    Homianus, Blainv. 504
    maximus, *Linn.* 503
    melastomus, *Blainv.* 497
    Monensis, *Shaw*, ... 501
    Mustelus, *Linn.* 502
    Norvegianus, Blainv. 506
    peregrinus, *Blainv.* 504
    Pristis, Linn. 508
    Selanoneus, *Flem.* 506
    Selanonus, Leach, ... 500
    Spinax, *Linn.* 505
    Squatina, Linn. 507
    stellaris, *Linn.* 496
    Vulpes, *Gmel.* 498
    Zygœna, Bl. 507
    Zygœna, Linn. 507
SQUATAROLA, *Cuv.* 65, 181
SQUATINA, *Dum.* 328, 507
    Angelus, *Cuv.* 507

SQUATINA, *vulgaris*, Flem. ... 507
Squatina, Will. 507
STELLERUS, *Cuv.* 8, 40
STERNA, *Linn.* 77, 264
    Anglica, *Mont.* 269
    arctica, *Temm.* 267
    Cantiaca, *Gmel.* 265
    Caspia, *Pall.* 264
    Dougallii, *Mont.* 265
    Hirundo, *Linn.* 266
    minuta, *Linn.* 267
    nævia, Gmel. 269
    nigra, *Linn.* 268
    stolida, *Linn.* 270
STERNA, *Steph.* 77, 264
STREPSILAS, *Ill.* 65, 182
    collaris, Temm. 182
    Interpres, *Leach*, ... 182
STRIX, *Linn.* 51, 92
    Aluco, Temm. 93
    Brachyotos, Temm. 92
    Bubo, Temm. 90
    flammea, *Linn.* 92
    funerea, Lath. 526
    nyctea, Temm. 94
    Otus, Temm. 91
    passerina, Temm. 94
    pulchella, Don. 91
    Scops, Temm. 91
    Tengmalmi, Temm. 94
Sturio, Will. 493
STURNUS, *Linn.* 57, 143
    vulgaris, *Linn.* 143
SULA, *Briss.* 76, 263
    alba, Temm. 263
    Bassana, *Briss.* 263
SURNIA, *Selb.* 52, 93
SUS, *Linn.* 8, 39
    Scrofa, *Linn.* 39
    domestica, ... 40
SYLVIA, *Lath.* 53, 103
    arundinacea, *Lath.* 107
    Atricapilla, *Lath.* 108
    Cetti, *Temm.* 107
    cinerea, *Lath.* 109
    Curruca, *Lath.* 109
    Dartfordiensis, Mont. 112
    Hippolais, *Lath.* 111

SYLVIA, hortensis, *Lath*......... 108
  ignicapilla, Temm............. 114
  Locustella, *Lath*............... 106
  Luscinia, *Lath*................. 107
  Phœnicurus, *Lath*. 104
  Phragmitis, *Bechst*............ 106
  provincialis, Temm........... 112
  Regulus, Temm............... 113
  Rubecula, *Lath*. 103
  rufa, Temm.................. 112
  sibilatrix, *Bechst*............. 110
  Suecica, *Lath*. 104
  Tithys, *Scop*................. 105
  Trochilus, *Lath*. 111
  Troglodytes, Temm........... 153
SYLVIA, *Selb*. 54, 110
SYNGNATHUS, *Cuv*. ... 324, 484
  Acus, *Linn*.................... 484
  æquoreus, *Linn*. 486
  anguineus, Jen. 487
  barbarus, Turt................ 488
  Hippocampus, Linn........... 489
  lumbriciformis, *Nob*. 488
  Ophidion, *Bl*. 487
  pelagicus, *Don*............... 486
  Typhle, Bl. 486
  Typhle, *Linn*. 485
SYRNIUM, *Sav*. 51, 93
  Aluco, *Nob*.................... 93

T.

TADORNA, *Leach*, ... 72, 229
  Bellonii, *Steph*................ 229
  rutila, *Steph*.................. 229
TALPA, *Linn*. 4, 17
  Europæa, *Linn*. 17
Testudo coriacea, Linn. 290
imbricata, Linn. 290
TETRAO, *Linn*.............. 63, 168
  Lagopus, *Sab*. 170
  Lagopus, Temm............... 171
  medius, *Mey*. 169
  rupestris, *Sab*. 171
  Scoticus, *Temm*. 170
  Tetrix, *Linn*. 169
  Virginianus, Wils. 173
  Urogallus, *Linn*............... 168

TETRAO, *Steph*. 63, 168
TETRODON, *Linn*. 325, 489
  lagocephalus, Bl. 490
  Mola, Linn...................... 490
  stellatus, *Don*................. 489
  truncatus, Gmel. 491
THALASSIDROMA, *Vig*...... 78, 285
THYMALLUS, *Cuv*....... 318, 430
  vulgaris, *Nilss*............... 430
Thymallus, Will.................. 430
THYNNUS, *Cuv*. 310, 362
  Pelamys, Cuv................... 363
  vulgaris, Cuv. 362
TINCA, *Cuv*. 316, 405
  vulgaris, Flem.................. 405
TORPEDO, *Dum*. 328, 509
  Galvani, Riss. 510
  marmorata, Riss............... 510
  Narke, Riss................... 509
  unimaculata, Riss. 509
  vulgaris, Flem................. 509
TOTANUS, *Bechst*. 68, 196
  Calidris, *Bechst*. 196
  fuscus, *Leisl*................... 196
  Glareola, Markw.............. 197
  Glareola, *Temm*............... 198
  Glottis, *Bechst*................. 200
  Hypoleucos, *Temm*........... 199
  Macularia, *Temm*. 199
  Ochropus, *Temm*. 197
TRACHINUS, *Linn*. 307, 335
  Draco, Bl...................... 336
  Draco, *Linn*................... 335
  major, Don. 335
  Vipera, Cuv.................... 336
Trachurus vulgaris, Flem......... 366
TRICHECHUS, *Linn*. 4, 16
  Rosmarus, *Linn*. 16
TRICHIURUS, *Linn*...... 312, 372
  Lepturus, *Linn*. 372
TRIGLA, *Cuv*............. 308, 338
  Adriatica, Flem. 339
  Blochii, Yarr. 343
  Cuculus, *Bl*................... 343
  Gurnardus, *Linn*.............. 342
  Hirundo, *Bl*.................. 340
  lævis, Mont.................... 340

TRIGLA, lineata, *Gmel.* 339
 lineata, Mont. 339
 Lyra, *Linn.* 341
 Pini, *Bl.* 338
TRINGA, *Briss.* 69, 207
 Canutus, *Linn.* 213
 cinerea, Temm................. 213
 maritima, *Brunn.*............... 211
 minuta, *Leisl.* 212
 Morinella, Linn. 183
 nigricans, Mont. 211
 pectoralis, *Bon.*............... 210
 pugnax, *Linn.*.................. 207
 rufescens, *Vieill.* 214
 subarquata, *Temm.*........... 208
 Temminckii, *Leisl.* 211
 variabilis, *Mey.* 209
TRINGA, *Selb.* 69, 208
TRITON, *Laur.* 299, 303
 aquaticus, Flem. 304
 palustris, *Flem.*............... 303
 punctatus, *Bon.* 304
 vittatus, *Gray,*................ 305
 vulgaris, Flem.................. 305
TROGLODYTES, *Cuv.* ... 59, 153
 Europæus, *Selb.* 153
 vulgaris, Temm. 153
Trutta fluviatilis, Will. 424
 lacustris, Will................... 423
TRYGON, *Adans.* 329, 518
 Pastinaca, Flem............... 518
TURDUS, *Linn.* 53, 98
 iliacus, *Linn.* 100
 Merula, *Linn.* 101
 musicus, *Linn.*................. 100
 pilaris, *Linn.*.................... 99
 torquatus, *Linn.* 101
 varius, *Horsf.* 101
 viscivorus, *Linn.* 98
Turdus minor, Jag................. 393
 perbelle pictus, Will. 394
 vulgatissimus, Will........... 397

U.

Umbla minor, Will................. 429
UMBRINA, *Cuv.*.............. 309, 353
 vulgaris, Cuv. 353

UPUPA, *Linn.*.................. 60, 153
 Epops, *Linn.* 153
URIA, *Briss.* 75, 258
 Alle, Temm..................... 259
 Brunnichii, *Sab.* 258
 Grylle, *Lath.* 258
 Troile, *Lath.*................... 258
URSUS, *Linn.*.................... 3, 10
 Arctos, *Linn.* 10

V.

Vandellius Lusitanicus, Shaw, ... 371
VANELLUS, *Briss.* 65, 181
 cristatus, *Mey.*................. 182
 griseus, *Briss.* 181
 melanogaster, Temm. 181
VANELLUS, *Cuv.*.............. 65, 182
VESPERTILIO, *Geoff.*............. 5, 20
VESPERTILIO, *Linn.*.......... 5, 20
 auritus, *Linn.* 27
 Barbastellus, *Gmel.*........... 28
 Bechsteinii, *Leisl.*............. 21
 brevimanus, *Jen.*............... 28
 discolor, *Natt.* 24
 emarginatus, *Geoff.* 26
 Leisleri, *Kuhl,* 23
 minutus, Mont................. 20
 murinus, *Desm.* 20
 mystacinus, *Leisl.*............. 26
 Nattereri, *Kuhl,* 21
 Noctula, *Gmel.*................ 23
 Pipistrellus, *Gmel.* 24
 pygmæus, *Leach,*............. 25
 Serotinus, *Gmel.* 22
VIPERA, *Daud.* 289, 297
 communis, *Leach,* 297
Vipera, Ray, ... 297
VULPES, *Flem.* 4, 14
 vulgaris, Flem.................. 14
Vulpes marina, Will. 498

X.

XEMA, *Leach,* 77, 270
 Sabini, Leach, ... 270
XIPHIAS, *Linn.* 311, 364
 Gladius, *Linn.*................. 364
 Rondeletii, Leach, ... 364

Y.

YUNX, *Linn.* 59, 152
    Torquilla, *Linn.* 152

Z.

ZEUS, *Cuv* 311, 367
ZEUS, *Linn.* 311, 367
    Aper, *Gmel.* 368

ZEUS, Faber, *Linn.* 367
    imperialis, Shaw,.............. 369
    Luna, Gmel...................... 369
Zipotheca tetradens, Mont. 371
ZOARCES, *Cuv.* 313, 384
    viviparus, *Cuv*.................. 384
ZYGÆNA, *Cuv.* 328, 507
    Malleus, *Val.* 507

ALPHABETICAL ENGLISH INDEX.

The names adopted for the Species in this Work are printed in Roman characters: the Synonyms, and names of the Varieties, in *Italics*.

A.

ACCENTOR, alpine,	102
Hedge,	103
Adder, Sea,	489
Albacore,	366
Allis,	438
Anchovy,	439
Angel-fish,	507
Angler, common,	389
long,	390
Argentine, Sheppy,	433
Ass,	39
Atherine,	377
Auk, *little*,	259
black-billed,	261
great,	261
Razor-bill,	260
Avocet, scooping,	201
Azurine,	413

B.

Badger,	10
Band-fish, red,	374
Barbastelle,	28
Barbel,	404
Barbolt,	448
Basse, common,	331
Stone,	334
Bat, Barbastelle,	28
common,	24
great,	23
greater Horse-shoe,	19

Bat, greater long-eared,	27
Horse-shoe,	19
lesser Horse-shoe,	20
lesser long-eared,	28
long-eared,	27
Noctule,	23
Pipistrelle,	24
pygmy,	25
Serotine,	22
Bear, brown,	10
common,	10
Beaver,	34
common,	34
Becker,	354
Bee-eater, common,	156
Beluga,	43
Bergylt,	347
Bernicle, Brent,	224
common,	224
red-breasted,	225
Bib,	442
Bittern, American,	191
common,	190
little,	189
Blackbird,	101
Black-cap,	108
Black-fish,	370
Bleak,	414
Blenny, crested,	380
diminutive,	381
gattoruginous,	379
Montagu's,	381

Blenny, ocellated,...... 378
    purple, ... 383
    smooth,...... 382
    spotted, ... 383
    viviparous, ... 384
Blinds, ... 442
Blind-worm, ... 295
Boar, ... 39
    wild, ... 39
Boar-fish, ... 368
Bonito, ... 363
Bottle-head, ... 44
Braize, ... 354
Brambling, ... 134
Brandling, ... 427
Bream, *black*, ... 358
    black Sea,...... 358
    common Sea, ... 356
    Ray's, ... 359
    Ray's Sea, ... 359
    Spanish, ... 355
    Spanish Sea,...... 355
    white, ... 407
    yellow, ... 406
Bream-flat, ... 407
Brill, ... 462
Buck, ... 38
Bullfinch, common,...... 140
    Pine,...... 141
Bull-head, ... 343
    armed, ... 346
    River, ... 343
Bunting, Cirl, ... 131
    common, ... 130
    green-headed, ... 132
    Lapland, ... 128
    Ortolan, ... 132
    painted,...... 133
    Reed, ... 130
    Snow, ... 129
    yellow, ... 131
Burbot,...... 448
Bustard, great,...... 174
    little,...... 175
    thick-kneed, ... 177
Butter-fish, ... 383
Butterfly-fish, ... 378
Buzzard, common, ... 87

Buzzard, Honey, ... 88
    Moor, ... 89
    rough-legged, ... 87

C.

Cachalot, blunt-headed, ... 44
    high-finned, ... 45
    two-toothed, ... 44
Campagnol, Bank, ... 34
    Field, ... 33
    Water, ... 33
Cane,...... 12
Carp, common,...... 401
    golden, ... 403
Carter, ... 464
Cat, domestic, ... 15
    wild, ... 14
Cavy, variegated,...... 36
Ceil-conin, ... 373
Chad, ... 439
Chaffinch,...... 133
Charr, ... 427
    case, ... 428
    gilt, ... 428
    red, ... 428
Chatterer,...... 125
    Bohemian, ... 125
Chevin, ... 412
Chiff-chaff, ... 111
Chimæra, northern, ... 494
Chough, Cornish,...... 144
Chub, ... 411
Coal-fish, ... 446
Cock, *Bantam*,...... 165
    crested, ... 165
    domestic, ... 165
    Dorking, ... 165
    dwarf, ... 165
    frizzled,...... 166
    Negro, ... 166
    rumpless, ... 166
    Silk, ... 166
Cod, common, ... 440
    five-bearded, ... 450
    green, ... 447
    three-bearded, ... 449
Cod-fish, common, ... 440
    variable, ... 441

Colin, northern,	173
Comber,	393
Cook,	396
Coot, common,	221
greater,	221
Corkling,	392
Corkwing,	399
Cormorant, black,	263
common,	262
crested,	262
green,	263
Corn-crake,	217
Cormorant,	262
Cottus, four-horned,	345
Courser, cream-coloured,	176
Craig-fluke,	459
Crake, Baillon's,	219
Corn,	217
little,	219
Meadow,	217
spotted,	218
Crane, common,	185
Creeper, common,	152
Cross-bill, common,	141
Parrot,	142
white-winged,	143
Crow, Carrion,	145
hooded,	146
red-legged,	144
Crucian,	402
Cuckoo, Carolina,	155
common,	154
Cuckoo-fish,	397
Curlew, *Brazilian,*	194
common,	195
pigmy,	208
Whimbrel,	195

D.

Dab,	456
Lemon,	457
Smear,	457, 463
Dace,	410
Dare,	410
Deal-fish,	372
Deer, Fallow,	38
red,	37
Diodon, Globe,	489

Diodon, oblong,	491
short,	490
Dipper, European,	98
Diver, black-throated,	256
great northern,	255
northern,	255
red-throated,	257
second speckled,	257
speckled,	257
Dobule,	409
Dog,	13
Dog-fish, *lesser,*	495
picked,	505
Rock,	496
spotted,	495
spotted,	496
Dolphin, blunt-toothed,	41
common,	40
Doree,	367
Dormouse,	30
Dorse,	441
Dory,	367
Dotterel,	178
little Ring,	179
Dove, Ring,	161
Rock,	162
Stock,	161
Turtle,	162
Dragonet, gemmeous,	388
sordid,	389
Duck, bimaculated,	232
castaneous,	242
collared,	243
common,	234
common wild,	234
domestic,	234
Eider,	237
ferruginous,	229, 242
Harlequin,	246
hook-billed,	234
King,	238
long-tailed,	247
Muscovy,	230
Musk,	230
olive-tufted,	242
Pintail,	232
red crested,	240
Scaup,	244

Duck, Summer, ... 237
    tufted, ... 244
    Velvet, ... 239
    western,........................ 243
    white-throated, ... 240
Dun-cow, ... 514, 515
Dun-diver, ... 248
Dunlin,............................ 209

E.

Eagle, cinereous, ... 80
    golden, ... 80
Eel, *blunt-headed*, ... 476
    broad-nosed 476
    common, ... 474
    Conger,......................... 478
    Glut, ... 476
    Grig,............................ 477
    Grig, ... 476
    sharp-headed, ... 475
    sharp-nosed, ... 474
    silver, ... 476
    Snig,... 477
Eel-pout, ... 449
Eft, common, ... 304
    striped,......................... 305
    warty, ... 303
Egret, ... 187
    little,.......................... 187
Eider, common, ... 237
    King, ... 238
Elanus, swallow-tailed, ... 86
Ermine, ... 13

F.

Falcon, *ash-coloured*, ... 90
    gentil, ... 85
    Jer, ... 81
    peregrine, ... 82
    red-footed,..................... 84
    red-legged, ... 83
    rough-legged, ... 87
    spotted,........................ 82
    Stone, ... 83
    swallow-tailed, ... 86
Father-lasher, ... 344

Father-lasher, four-horned, ... 345
    four-spined, ... 345
Ferret, ... 12
Fieldfare, ... 99
File-fish, Mediterranean,........... 492
Finch, Chaf,....................... 133
    Gold, ... 137
    Haw, ... 136
    Lapland, ... 128
    Mountain, ... 134
    painted,........................ 133
Fin-fish, ... 47
Finscale, ... 412
Fishing-frog, ... 390
    of Mount's Bay, ... 390
Fitchet, ... 11
Flounder, ... 455
Fluke, Craig, ... 459
    long, ... 460
Flycatcher, pied, ... 97
    spotted, ... 97
    white-collared, ... 98
Flying-fish, ... 420
    winged,........................ 420
Fork-beard, common, ... 452
    greater,........................ 452
    lesser, ... 454
Fox, ... 14
Fox, ... 389
Frog, common,..................... 300
    edible, ... 301
    great, ... 302
    green, ... 301
    Tree,............................ 303
Fulmar, northern, ... 284

G.

Gadwall, ... 231
    common, ... 231
Gallinule, common, ... 220
    Crake, ... 217
    little,.......................... 219
    olivaceous, ... 219
    spotted 218
Gambet, ... 197
Gannet, common, ... 263
    Solan, ... 263

Gar-fish, common, ... 418
    little, ... 419
Garganey, ... 234
Garrot, buffel-headed, ... 246
    common Golden-eye, ... 245
    Golden-eye, ... 245
    Harlequin, ... 246
Gibel, ... 402
Gilt-head, ... 397
Gilt-head, ... 353
    lunulated, ... 353, 356
    red, ... 356
    toothed, ... 357
    toothed, ... 359
Globe-fish, stellated, ... 489
Goat, ... 37
    domestic, ... 37
Goatsucker, European, ... 160
Goby, black, ... 385
    slender, ... 387
    spotted, ... 386
    two-spotted, ... 386
Godwit, bar-tailed, ... 202
    black-tailed, ... 203
    Cambridge, ... 196
    cinereous, ... 201
    common, ... 203
    red, ... 202
Golden-eye, ... 245
Goldfinch, ... 137
Goldfinny, ... 399
Gold-fish, ... 403
Goldsinny, ... 398
Goosander, ... 248
Goose, Bean, ... 222
    Bernicle, ... 224
    Brent, ... 224
    Canada, ... 227
    Chinese, ... 226
    Cravat, ... 227
    domestic, ... 222
    Egyptian, ... 225
    Gray-lag, ... 222
    red-breasted, ... 225
    ruddy, ... 229
    spur-winged, ... 226
    Swan, ... 226
    white-fronted, ... 223

Goose, wild, ... 222
Goshawk, ... 85
Graining, ... 411
Grampus, ... 42
Grayling, ... 430
Grebe, *black-chin,* 254
    crested, ... 251
    dusky, ... 253
    eared, ... 253
    great-crested, ... 251
    horned, ... 253
    little, ... 254
    red-necked, ... 252
    Sclavonian, ... 252
    Tippet, ... 252
Green-fish, ... 392
Greenshank, ... 200
Grey, ... 422
Grig, ... 476
    Stone, ... 523
Grilse, ... 422
Grosbeak, common, ... 136
    green, ... 136
    Haw, ... 136
    Pine, ... 141
Groundling, ... 417
Grous, black, ... 169
    hybrid, ... 169
    red, ... 170
    spurious, ... 169
    Wood, ... 168
Gudgeon, ... 405
    Sea, ... 385
Guillemot, black, ... 258
    foolish, ... 258
    Franks, ... 258
    lesser, ... 258
    spotted, ... 259
Guinea-pig, ... 36
Gull, *arctic,* 282, 283
    black-backed, ... 278
    black-headed, ... 272
    black-toed, ... 282
    brown-headed, ... 272
    brown-headed, ... 273
    common, ... 275
    glaucous, ... 279
    great black-backed, ... 278

Gull, Herring, ... 276
  Iceland,........................... 279
  Iceland,........................... 280
  Ivory, ... 276
  Kittiwake, ... 274
  laughing, ... 273
  lesser black-backed, ... 277
  little,............................. 271
  masked, ... 272
  parasitic, ... 283
  red-legged, ... 272
  Sabine's, ... 270
  Skua, ... 280
  Tarrock, ... 275
  Wagel, ... 277
  Winter, ... 276
Gunnel, spotted, ... 383
Gurnard, *French,* 340
  gray, ... 342
  pine-leaved, ... 338
  Piper, ... 341
  red, ... 343
  red, ... 339
  sapphirine, ... 340
  streaked, ... 339
Gwiniad, ... 431
Gymnetrus, Blochian,.............. 373
  Hawken's, ... 373

H.

Haddock, ... 441
  Norway, ... 347
Hag, glutinous, ... 523
Hair-tail, ... 372
  silvery, ... 372
Hake, common, ... 447
  forked, ... 452
  lest, ... 454
  trifurcated, ... 453
Hake's-Dame, ... 453
Hammer-head,....................... 507
Hare, alpine, ... 35
  common, ... 34
  Irish, ... 35
  varying, ... 35
Hareld, long-tailed,.................. 247
Harrier, ash-coloured, ... 90

Harrier, Hen 89
  Marsh, ... 88
Hawk, Gos, ... 85
  Sparrow, ... 85
Hedgehog, ... 19
Herling, ... 424
Heron, buff-backed,.................. 188
  common, ... 186
  common Night, ... 191
  crested purple, ... 186
  freckled, ... 191
  great white, ... 187
  little,............................. 189
  little Egret, ... 187
  little white, ... 188
  purple, ... 186
  Squacco, ... 189
  white, ... 187
Herring, *British,*..................... 434
  common, ... 434
  Leach's........................... 434
Hind, ... 37
Hinny,................................. 39
Hobby,................................. 82
  orange-legged, ...,........... 84
Hog, common, ... 39
  domestic, ... 40
Holibut, ... 460
Hommelin, ... 515
Hoopoe, ... 153
Horn-fish, ... 418
Horse, ... 39
  common, ... 39
Hound, smooth, ... 502

I.

Ibis, *bay,* 194
  glossy, ... 194
  green, ... 194
Ide, ... 415
Imber, lesser, ... 256

J.

Jackdaw, ... 147
Jay, ... 148
Judcock, ... 206

K.

Kelt, ... 422
Kestril, ... 84
King-fish, ... 369
King-fisher, common, ... 157
Kingstone, ... 508
Kipper, ... 422
Kite, ... 86
Kitt, ... 457
Kittiwake, ... 274
Knobber, ... 37
Knot, ... 213

L.

Lampern, ... 522
Lamprey, lesser, ... 521
Planer's, ... 522
River, ... 521
Sea, ... 520
Lantern-fish, ... 464
Lanner, ... 82
Lapwing, crested, ... 182
green, ... 182
Lark, Field, ... 118, 119
Pipit, ... 118
red, ... 126
Rock, ... 119
Shore, ... 126
Sky, ... 127
Tit, ... 117
Tree, ... 118
Wood, ... 127
Lark-bunting, Lapland, ... 128
Launce, Sand, ... 482, 483
small-mouthed, ... 483
wide-mouthed, ... 482
Lestris, Richardson's, ... 282
Lewis, ... 508
Ling, ... 448
Linnet, brown, ... 139
common, ... 139
Mountain, ... 140
Lizard, brown, ... 305
common, ... 292
Sand, ... 291
scaly, ... 293
warty, ... 303

Loach, bearded, ... 416
Lobe-foot, red, ... 214
Loche, ... 416
spined, ... 417
spinous, ... 417
Long-beak, brown, ... 207
Longshanks, black-winged, ... 201
Lord-fish, ... 444
Lump-fish, common, ... 471
Lump-sucker, ... 471

M.

Mackerel, common, ... 360
Horse, ... 366
Spanish, ... 361
Mackerel-guide, ... 419
Mackerel-midge, ... 451
Magpie, ... 147
Maid, ... 511, 517
Maigre, ... 352
Mallard, ... 233
Martin, Bank, ... 158
common, ... 11
greatest, ... 159
House, ... 158
Pine, ... 11
Sand, ... 158
Megrim, ... 466
Merganser, hooded, ... 249
minute, ... 251
red-breasted, ... 249
Merlin, ... 83
Mermaid of the Shetland Seas, ... 40
Midge, ... 451
Miller's-thumb, ... 343
Minim, ... 415
Minnow, ... 415
Mole, ... 17
Monk-fish, ... 508
Morris, Anglesea, ... 480
Mouse, common, ... 31
Field, ... 30
Harvest, ... 31
House, ... 31
Meadow, ... 33
short-tailed, ... 33
Wood, ... 30

Mule, ... 39
Mule-bird, ... 167
Mullet, gray, ... 374
    plain red, ... 338
    short gray, ... 376
    striped red, ... 337
    thick-lipped gray, ... 375
Muræna, common, ... 479
    Roman, ... 479
Mysticete, fin-backed, ... 47
    pike-headed, ... 47

N.

Narwhal, ... 43
    small-headed, ... 43
Natter-jack, ... 302
Neophron Egyptian, ... 79
Newt, common, ... 305
    common Water, ... 304
    great Water, ... 303
    smaller Water, ... 304
    warted, ... 303
Night-heron, ... 191
Nightingale, ... 107
Night-jar, ... 160
Noctule, ... 23
Noddy, black, ... 270
Numb-fish, ... 509
Nutcracker, ... 149
Nuthatch, ... 154

O.

Old-wife, ... 359
Old-wife, ... 398
Opah, ... 369
Ophidium, beardless, ... 481
Oriole, golden, ... 102
Osprey, ... 81
Otter, ... 13
    Irish, ... 525
    Sea, ... 525
Otterpike, ... 337
Ouzel, Ring, ... 101
    rose-coloured, ... 144
    Water, ... 98
Owl, Canada, ... 526
    Eagle, ... 90
    great-eared, ... 90

Owl, little, ... 94
    little-horned, ... 91
    little Night, ... 94
    long-eared, ... 91
    scops-eared, ... 91
    short-eared, ... 92
    snowy, ... 93
    tawny, ... 93
    Tengmalm's, ... 94
    Tengmalm's Night, ... 94
    white, ... 92
Ox, common, ... 36
Oyster-catcher, common, ... 184
    pied, ... 184

P.

Parr, ... 426
Partridge, common, ... 172
    Guernsey, ... 172
    Maryland, ... 173
    red-legged, ... 172
    Virginian, ... 173
Pastor, rose-coloured, ... 144
Peacock, crested, ... 164
Peal, ... 422
Pearl, ... 462
Pelican, ... 264
Perch, black, ... 370
    common, ... 330
    dusky, ... 333
    Sea, ... 347
    smooth, ... 332
Pero, ... 167
Petrel, fork-tailed, ... 286
    Fulmar, ... 284
    Leach's, ... 286
    stormy, ... 285
Pettychaps, greater, ... 108
    lesser, ... 112
Phalarope, gray, ... 215
    red, ... 214, 215
    red-necked, ... 214
Pheasant, Bohemian, ... 167
    common, ... 166
    hybrid, ... 167
    Ring, ... 166, 167
    ring-necked, ... 167
Phinoc, ... 424

Pigeon, *domestic*, ... 162
  Passenger, ... 163
  wild, ... 162
Pike, ... 417
  Gar, ... 418
  Saury, ... 419
  Sea, ... 337
Pilchard, ... 436
Pilot-fish, common, ... 365
Pink, ... 415
Pintado, Guinea, ... 168
Pintail, ... 232
  common, ... 232
Pipe-fish, æquoreal, ... 486
  great, ... 484
  lesser, ... 485
  little, ... 488
  longer, ... 488
  longer, ... 487
  shorter, ... 484, 485
  Snake, ... 487
  Worm, ... 488
Piper, ... 341
Pipistrelle, ... 24
Pipit, Meadow, ... 117
  Richard's, ... 117
  Rock, ... 118
  Tree, ... 118
Plaice, common, ... 454
Plover, *cream-coloured*, ... 176
  Dotterel, ... 178
  golden, ... 177
  gray, ... 181
  great, ... 177
  Kentish, ... 180
  little ringed, ... 179
  long-legged, ... 201
  ringed, ... 179
Pochard, common, ... 241
  Nyroca, ... 242
  red-crested, ... 240
  red-headed, ... 241
  Scaup, ... 243
  tufted, ... 244
  western, ... 243
Pogge, ... 346
Pole, ... 458
Polecat, ... 11

Pollack, ... 446
  rauning, ... 447
  Whiting, ... 446
Pollan, ... 432
Poor, ... 444
Pope, ... 335
Porbeagle, ... 500
Porpesse, ... 41
Pout, ... 442
  Whiting, ... 443
Power, ... 444
Pratincole, *Austrian*, ... 216
  collared, ... 216
Pride, ... 522
Ptarmigan, common, ... 170
  red, ... 170
  Rock, ... 171
Puffin, ... 260
Purre, ... 209

Q.

Quail, common, ... 174
Queen-fish, ... 458

R.

Rabbet, ... 35
Rabbit-fish, ... 494
Rail, *Land*, ..., ... 217
  Water, ... 217
Rat, black, ... 32
  brown, ... 32
  Norway, ... 32
  Water, ... 33
Raven, ... 145
Ray, bordered, ... 512
  Cuvier's, ... 518
  electric, ... 509
  French, ... 514
  Fuller, ... 514
  rough, ... 518
  Sand, ... 515
  Shagreen, ... 513
  sharp-nosed, ... 511
  sharp-nosed, ... 510
  small-eyed, ... 515
  spotted, ... 514
  starry, ... 517

Ray, Sting, ... 518
    Whip, ... 519
Razor-bill, ... 260
Redbreast, ... 103
    blue-throated, or Swedish, ... 104
Red-eye, ... 412
Redpole, *greater,* 139
    lesser, ... 138
    mealy, ... 139
    mealy-backed, ... 139
    Stone, ... 139
Redshank, ... 197
    spotted, ... 196
Redstart, ... 104
    blue-throated, ... 104
    Tithys, ... 105
Redtail, black, ... 105
Redwing, ... 100
Regulus, fire-crested, ... 113
    gold-crested, ... 113
Remora, common, ... 473
    Mediterranean, ... 474
Ringtail, ... 89
Roach, ... 408
    Dobule, ... 409
Rock-fish, ... 386, 397
Rockling, five-bearded, ... 450
    three-bearded, ... 449
Roe, ... 38
Roe-buck, ... 38
Roller, garrulous, ... 156
Rook, ... 146
Rotche, common, ... 259
Rudd, ... 412
Ruff, ... 207
Ruffe, *black,* 370
    common, ... 334

S.

Salmon, common, ... 421
Salmon-trout, ... 424
Samlet, ... 426
Samlet, ... 422
Sanderling, ... 183
    common, ... 183
Sand-necker, ... 459
Sandpiper, *ash-coloured,* 213

Sandpiper, buff-breasted, ... 214
    common, ... 199
    dusky, ... 196
    equestrian, ... 208
    gray, ... 181
    green, ... 197
    Greenwich, ... 208
    little, ... 211
    long-legged, ... 198
    pectoral, ... 210
    purple, ... 211
    red, ... 209
    Redshank, ... 196
    Shore, ... 208
    spotted, ... 199
    Wood, ... 198
Saury, ... 419
Scabbard-fish, ... 371
Scad, ... 366
Scald-fish, ... 465
Scale-foot, ... 371
Scoter, black, ... 239
    Surf, ... 240
    Velvet, ... 239
    white-headed, ... 240
Sea-adder, ... 489
Sea-fox, ... 498
Sea-horse, ... 489
Seal, common, ... 15
    great, ... 16
    pied, ... 16
Sea-monster, ... 494
Sea-scorpion, ... 344
Sea-snail, common, ... 472
    Montagu's, ... 473
Sea-wife, ... 395
Sebastes northern, ... 347
Serotine, ... 22
Serranus, dusky, ... 333
    smooth, ... 332
Sewen, ... 423
Shad, ... 437
Shad, ... 438
    Thames, ... 437
Shag, crested, ... 262
Shag, ... 262
Shallow, ... 413
Shan, ... 382

Shan, *smooth*, ... 382
Shanny, ... 382
Shark, *Angel*, ... 507
  basking, ... 503
  Beaumaris, ... 501
  blue, ... 499
  Greenland, ... 506
  lesser spotted, ... 495
  long-tailed, ... 498
  picked, ... 505
  Porbeagle, ... 500
  Rashleigh, ... 507
  Rock, ... 496
  smooth, ... 502
  spotted, ... 495
  Tope, ... 502
  white, ... 497
Shearwater, cinereous, ... 284
  Manks, ... 285
Sheep, ... 37
Shieldrake, *Casarka*, ... 229
  common, ... 229
  ruddy, ... 229
Shoveller, common, ... 230
  red-breasted, ... 231
Shrew, common, ... 17
  fetid, ... 17
  oared, ... 18
  Water, ... 18
Shrike, cinereous, ... 95
  red-backed, ... 96
  Wood, ... 96
Silurus, sly, ... 421
Siskin, ... 137
Skate, ... 510
Skipper, ... 419
Skirling, ... 427
Skua, arctic, ... 283
  arctic, ... 282
  common, ... 280
  pomarine, ... 281
  Richardson's, ... 282
Slow-worm, common, ... 295
Smear-dab, ... 457, 463
Smelt, ... 429
  Sand, ... 377
Smew, ... 250
Smolt, ... 422

Snake, Aberdeen, ... 296
  ringed, ... 296
Snake-fish, red, ... 374
Snig, ... 477
Snipe, brown, ... 207
  common, ... 205
  great, ... 205
  Jack, ... 206
  Jadreka, ... 203
  Sabine's, ... 204
  spotted, ... 196
Sole, common, ... 466
  Lemon, ... 467
  Lemon, ... 458
  merry, ... 458
  red-backed, ... 468
  smooth, ... 465
Sparling, ... 430
Sparrow, House, ... 134
  Mountain, ... 135
  Tree, ... 135
Sparus, four-toothed, ... 357
Spoonbill, white, ... 193
Sprat, ... 435
Squirrel, common, ... 29
Stag, ... 37
Starling, common, ... 143
  rose-coloured, ... 144
Stickleback, fifteen-spined, ... 351
  four-spined, ... 350
  half-armed, ... 349
  lesser, ... 351
  rough-tailed, ... 349
  short-spined, ... 349
  smooth-tailed, ... 349
  ten-spined, ... 350
  three-spined, ... 348
Stilt, black-winged, ... 201
Stint, little, ... 212
  Temminck's, ... 211
Stoat, ... 13
Stone-basse, ... 334
Stone-chat, ... 121
Stone-grig, ... 523
Stork, black, ... 193
  white, ... 192
Sturgeon, common, ... 493
Sucker, bimaculated, ... 470

Sucker, Cornish, ... 469
    Jura, ... 469
    Lump, ... 471
    Montagu's, ... 473
    unctuous, ... 472
Suck-fish, small, ... 469
Sun-fish, *Mount's Bay,* 491
    oblong, ... 491
    short, ... 490
Surmullet, striped, ... 337
    red, ... 338
Swallow, Chimney, ... 157
Swan, Bewick's 228
    Canada, ... 227
    Guinea, ... 226
    mute, ... 228
    whistling, ... 227
    wild, ... 227
Swift, alpine, ... 159
    common, ... 159
    white-bellied, ... 159
Swift-foot, cream-coloured, ... 176
Sword-fish, common, ... 364

T.

Tadpole-fish, ... 453
    trifurcated, ... 453
Teal, ... 235
    bimaculated, ... 233
    common, ... 235
    Garganey, ... 234
    Summer, ... 236
Tench, ... 405
Tern, arctic, ... 267
    black, ... 268
    brown, ... 267
    Caspian, ... 264
    common, ... 266
    gull-billed, ... 269
    lesser, ... 267
    roseate, ... 265
    Sandwich, ... 265
Tetrodon, Globe, ... 489
    oblong, ... 491
    short, ... 490
Thick-knee, common, ... 177
Thorn-back, ... 516

Thresher, ... 498
Throstle, ... 100
Thrush, Missel, ... 98
    Redwing, ... 100
    solitary, ... 143
    Song, ... 100
Titlark, ... 117
Titmouse, bearded, ... 125
    blue, ... 122
    cole, ... 123
    crested, ... 122
    great, ... 121
    long-tailed, ... 124
    Marsh, ... 123
Toad, common, ... 301
    mephitic, ... 302
Tope, common, ... 501
Top-knot, Bloch's, ... 462
    Muller's, ... 463
Torgoch, ... 428
Torpedo, ... 509
Torsk, ... 452
Tortoise, coriaceous, ... 290
Tringa, buff-breasted, ... 214
    Curlew, ... 209
    minute, ... 212
    purple, or Rock, ... 211
    spotted, ... 199
    Temminck's, ... 211
Trout, Bull, ... 422
    common, ... 424
    Gillaroo, ... 425
    great Lake, ... 425
    River, ... 424
    Salmon, ... 424
    Sea, ... 423
Trumpeter, gold-breasted, ... 185
Trumpet-fish, ... 400
    snipe-nosed, ... 400
Tub-fish, ... 341
Tunny, common, ... 362
Tunny-fish, ... 362
Turbot, ... 461
Turbot, ... 461
Turkey, ... 164
Turnstone, common, ... 182
Turtle, *Canada,* 163
    coriaceous, ... 290

Turtle, imbricated, ... 290
  spotted-necked, ... 163
Tusk, ... 452
Twaite, ... 438
Twite, ... 140

U.

Umbrina, bearded, ... 353
Urchin, common, ... 19

V.

Vaagmaer, ... 372
Vendace, ... 432
Viper, black, ... 298
  blue-bellied, ... 298
  common, ... 297
  red, ... 298

W.

Wagtail, blue-headed, ... 116
  gray, ... 115
  pied, ... 114
  white, ... 114
  yellow, ... 115
Walrus, ... 16
  arctic, ... 16
Warbler, Black-cap, ... 108
  blue-throated, ... 104
  Dartford, ... 112
  Garden, ... 109
  Grasshopper, ... 106
  Hedge, ... 103
  passerine, ... 108
  Reed, ... 106
  Sedge, ... 106
Wax-wing, Bohemian, ... 125
Weasel, ... 12
  common, ... 12
Weever, common, ... 336
  great, ... 335
  lesser, ... 336
  little, ... 336
Whale, beaked, ... 44
  Bottle-nose, ... 41
  Bottle-nose, with two teeth, ... 44
  Ca'ing, ... 42

Whale, common, ... 46
  Fin, ... 47
  round-lipped, ... 47
  sharp-lipped, ... 47
  Spermaceti, ... 44
Wheatear, ... 119
Whiff, ... 464
Whiff, ... 463
Whimbrel, ... 195
Whin-chat, ... 120
Whistle-fish, ... 450
White, ... 424
White-bait, ... 436
White-bait, ... 451
Whitethroat, ... 109
  lesser, ... 109
Whiting, ... 445
Whiting, ... 424
Whiting-pollack, ... 446
Whiting-pout, ... 443
Whitling, ... 423
Wife, old, ... 359, 398
  old, ... 398
  Sea, ... 395
Wigeon, ... 236
  common, ... 236
Wolf, ... 14
Wolf-fish, ... 384
  ravenous, ... 384
Wood-chat, ... 96
Woodcock, ... 204
Woodpecker, barred, ... 151
  great black, ... 151
  great spotted, ... 150
  green, ... 149
  hairy, ... 151
  lesser spotted, ... 151
  middle spotted, ... 150
  pied, ... 150
  three-toed, ... 151
Wrasse, ancient, ... 397
  Ballan, ... 391
  bimaculated, ... 396
  blue-striped, ... 394
  Comber, ... 393
  common, ... 398
  Cook, ... 396
  gibbous, ... 399

Wrasse, *green-streaked*, ... 392
  Hog, ... 397
  Rainbow, ... 397
  red, ... 396
  scale-rayed, ... 400
  streaked, ... 392
  striped, ... 394
  trimaculated, ... 396
Wren, common, ... 153
  fire-crested, ... 114

Wren, *golden-crested*, ... 113
  Reed, ... 107
  Willow, ... 111
  Wood, ... 110
  yellow, ... 110, 111
Wryneck, ... 152

X.

Xeme, Sabine's, ... 270

Printed in the United States
By Bookmasters